NEUROBIOLOGY of
SENSATION and REWARD

FRONTIERS IN NEUROSCIENCE

Series Editors
Sidney A. Simon, Ph.D.
Miguel A.L. Nicolelis, M.D., Ph.D.

Published Titles

Apoptosis in Neurobiology
Yusuf A. Hannun, M.D., Professor of Biomedical Research and Chairman, Department
 of Biochemistry and Molecular Biology, Medical University of South Carolina, Charleston,
 South Carolina
Rose-Mary Boustany, M.D., tenured Associate Professor of Pediatrics and Neurobiology, Duke
 University Medical Center, Durham, North Carolina

Neural Prostheses for Restoration of Sensory and Motor Function
John K. Chapin, Ph.D., Professor of Physiology and Pharmacology, State University
 of New York Health Science Center, Brooklyn, New York
Karen A. Moxon, Ph.D., Assistant Professor, School of Biomedical Engineering, Science,
 and Health Systems, Drexel University, Philadelphia, Pennsylvania

Computational Neuroscience: Realistic Modeling for Experimentalists
Eric DeSchutter, M.D., Ph.D., Professor, Department of Medicine, University of Antwerp,
 Antwerp, Belgium

Methods in Pain Research
Lawrence Kruger, Ph.D., Professor of Neurobiology (Emeritus), UCLA School of Medicine and
 Brain Research Institute, Los Angeles, California

Motor Neurobiology of the Spinal Cord
Timothy C. Cope, Ph.D., Professor of Physiology, Wright State University, Dayton, Ohio

Nicotinic Receptors in the Nervous System
Edward D. Levin, Ph.D., Associate Professor, Department of Psychiatry and Pharmacology and
 Molecular Cancer Biology and Department of Psychiatry and Behavioral Sciences,
 Duke University School of Medicine, Durham, North Carolina

Methods in Genomic Neuroscience
Helmin R. Chin, Ph.D., Genetics Research Branch, NIMH, NIH, Bethesda, Maryland
Steven O. Moldin, Ph.D., University of Southern California, Washington, D.C.

Methods in Chemosensory Research
Sidney A. Simon, Ph.D., Professor of Neurobiology, Biomedical Engineering,
 and Anesthesiology, Duke University, Durham, North Carolina
Miguel A.L. Nicolelis, M.D., Ph.D., Professor of Neurobiology and Biomedical Engineering,
 Duke University, Durham, North Carolina

The Somatosensory System: Deciphering the Brain's Own Body Image
Randall J. Nelson, Ph.D., Professor of Anatomy and Neurobiology,
 University of Tennessee Health Sciences Center, Memphis, Tennessee

The Superior Colliculus: New Approaches for Studying Sensorimotor Integration
William C. Hall, Ph.D., Department of Neuroscience, Duke University, Durham, North Carolina
Adonis Moschovakis, Ph.D., Department of Basic Sciences, University of Crete, Heraklion, Greece

New Concepts in Cerebral Ischemia
Rick C. S. Lin, Ph.D., Professor of Anatomy, University of Mississippi Medical Center,
    Jackson, Mississippi

DNA Arrays: Technologies and Experimental Strategies
Elena Grigorenko, Ph.D., Technology Development Group, Millennium Pharmaceuticals,
    Cambridge, Massachusetts

Methods for Alcohol-Related Neuroscience Research
Yuan Liu, Ph.D., National Institute of Neurological Disorders and Stroke,
    National Institutes of Health, Bethesda, Maryland
David M. Lovinger, Ph.D., Laboratory of Integrative Neuroscience, NIAAA,
    Nashville, Tennessee

Primate Audition: Behavior and Neurobiology
Asif A. Ghazanfar, Ph.D., Princeton University, Princeton, New Jersey

Methods in Drug Abuse Research: Cellular and Circuit Level Analyses
Barry D. Waterhouse, Ph.D., MCP-Hahnemann University, Philadelphia, Pennsylvania

Functional and Neural Mechanisms of Interval Timing
Warren H. Meck, Ph.D., Professor of Psychology, Duke University, Durham, North Carolina

Biomedical Imaging in Experimental Neuroscience
Nick Van Bruggen, Ph.D., Department of Neuroscience Genentech, Inc.
Timothy P.L. Roberts, Ph.D., Associate Professor, University of Toronto, Canada

The Primate Visual System
John H. Kaas, Department of Psychology, Vanderbilt University
Christine Collins, Department of Psychology, Vanderbilt University, Nashville, Tennessee

Neurosteroid Effects in the Central Nervous System
Sheryl S. Smith, Ph.D., Department of Physiology, SUNY Health Science Center,
    Brooklyn, New York

Modern Neurosurgery: Clinical Translation of Neuroscience Advances
Dennis A. Turner, Department of Surgery, Division of Neurosurgery,
    Duke University Medical Center, Durham, North Carolina

Sleep: Circuits and Functions
Pierre-Hervé Luppi, Université Claude Bernard Lyon, France

Methods in Insect Sensory Neuroscience
Thomas A. Christensen, Arizona Research Laboratories, Division of Neurobiology,
    University of Arizona, Tuscon, Arizona

Motor Cortex in Voluntary Movements
Alexa Riehle, INCM-CNRS, Marseille, France
Eilon Vaadia, The Hebrew University, Jerusalem, Israel
Neural Plasticity in Adult Somatic Sensory-Motor Systems
Ford F. Ebner, Vanderbilt University, Nashville, Tennessee

Advances in Vagal Afferent Neurobiology
Bradley J. Undem, Johns Hopkins Asthma Center, Baltimore, Maryland
Daniel Weinreich, University of Maryland, Baltimore, Maryland

The Dynamic Synapse: Molecular Methods in Ionotropic Receptor Biology
Josef T. Kittler, University College, London, England
Stephen J. Moss, University College, London, England

Animal Models of Cognitive Impairment
Edward D. Levin, Duke University Medical Center, Durham, North Carolina
Jerry J. Buccafusco, Medical College of Georgia, Augusta, Georgia

The Role of the Nucleus of the Solitary Tract in Gustatory Processing
Robert M. Bradley, University of Michigan, Ann Arbor, Michigan

Brain Aging: Models, Methods, and Mechanisms
David R. Riddle, Wake Forest University, Winston-Salem, North Carolina

Neural Plasticity and Memory: From Genes to Brain Imaging
Frederico Bermudez-Rattoni, National University of Mexico, Mexico City, Mexico

Serotonin Receptors in Neurobiology
Amitabha Chattopadhyay, Center for Cellular and Molecular Biology, Hyderabad, India

TRP Ion Channel Function in Sensory Transduction and Cellular Signaling Cascades
Wolfgang B. Liedtke, M.D., Ph.D., Duke University Medical Center, Durham, North Carolina
Stefan Heller, Ph.D., Stanford University School of Medicine, Stanford, California

Methods for Neural Ensemble Recordings, Second Edition
Miguel A.L. Nicolelis, M.D., Ph.D., Professor of Neurobiology and Biomedical Engineering,
    Duke University Medical Center, Durham, North Carolina

Biology of the NMDA Receptor
Antonius M. VanDongen, Duke University Medical Center, Durham, North Carolina

Methods of Behavioral Analysis in Neuroscience
Jerry J. Buccafusco, Ph.D., Alzheimer's Research Center, Professor of Pharmacology and Toxicology,
    Professor of Psychiatry and Health Behavior, Medical College of Georgia,
    Augusta, Georgia

***In Vivo* Optical Imaging of Brain Function, Second Edition**
Ron Frostig, Ph.D., Professor, Department of Neurobiology, University of California,
    Irvine, California

Fat Detection: Taste, Texture, and Post Ingestive Effects
Jean-Pierre Montmayeur, Ph.D., Centre National de la Recherche Scientifique, Dijon, France
Johannes le Coutre, Ph.D., Nestlé Research Center, Lausanne, Switzerland

The Neurobiology of Olfaction
Anna Menini, Ph.D., Neurobiology Sector International School for Advanced Studies,(S.I.S.S.A.),
    Trieste, Italy

Neuroproteomics
Oscar Alzate, Ph.D., Department of Cell and Developmental Biology,
    University of North Carolina, Chapel Hill, North Carolina

Translational Pain Research: From Mouse to Man
Lawrence Kruger, Ph.D., Department of Neurobiology, UCLA School of Medicine, Los Angeles,
    California
Alan R. Light, Ph.D., Department of Anesthesiology, University of Utah, Salt Lake City, Utah

Advances in the Neuroscience of Addiction
Cynthia M. Kuhn, Duke University Medical Center, Durham, North Carolina
George F. Koob, The Scripps Research Institute, La Jolla, California

NEUROBIOLOGY of SENSATION and REWARD

Edited by

Jay A. Gottfried

Department of Neurology
Northwestern University, Feinberg School of Medicine
Chicago, Illinois

CRC Press
Taylor & Francis Group
Boca Raton London New York

CRC Press is an imprint of the
Taylor & Francis Group, an **informa** business

CRC Press
Taylor & Francis Group
6000 Broken Sound Parkway NW, Suite 300
Boca Raton, FL 33487-2742

First issued in paperback 2019

© 2011 by Taylor & Francis Group, LLC
CRC Press is an imprint of Taylor & Francis Group, an Informa business

No claim to original U.S. Government works

ISBN-13: 978-1-4200-6726-2 (hbk)
ISBN-13: 978-0-367-38293-3 (pbk)

Library of Congress Cataloging-in-Publication Data

Neurobiology of sensation and reward / [edited by] Jay A. Gottfried.
    p. ; cm. -- (Frontiers in neuroscience)
  Includes bibliographical references and index.
  ISBN 978-1-4200-6726-2 (hardcover : alk. paper)
  1. Sense organs--Physiology. 2. Neurobiology. I. Gottfried, Jay A. II. Series: Frontiers in neuroscience (Boca Raton, Fla.)
  [DNLM: 1. Sensation--physiology. 2. Brain--physiology. 3. Motivation. 4. Perception--physiology. 5. Reward. WL 702]

QP431.N48 2011
612.8--dc23   2011017680

Visit the Taylor & Francis Web site at
http://www.taylorandfrancis.com

and the CRC Press Web site at
http://www.crcpress.com

Contents

PART I   *First Principles*

PART II   *A Systems Organization of the Senses*

PART III   From Sensation to Reward

PART IV   Civilized Sensory Rewards (Distinctly Human Rewards)

Series Preface

Our goal in creating the Frontiers in Neuroscience Series is to present the insights of experts on emerging fields and theoretical concepts that are, or will be, in the vanguard of neuroscience. Books in the series cover genetics, ion channels, apoptosis, electrodes, neural ensemble recordings in behaving animals, and even robotics. The series also covers new and exciting multidisciplinary areas of brain research, such as computational neuroscience and neuroengineering, and describes breakthroughs in classical fields like behavioral neuroscience. We hope every neuroscientist will use these books in order to get acquainted with new ideas and frontiers in brain research. These books can be given to graduate students and postdoctoral fellows when they are looking for guidance to start a new line of research.

Each book is edited by an expert and consists of chapters written by the leaders in a particular field. Books are richly illustrated and contain comprehensive bibliographies. Chapters provide substantial background material relevant to the particular subject. We hope that as the volumes become available, the effort put in by us, the publisher, the book editors, and individual authors will contribute to the further development of brain research. The extent to which we achieve this goal will be determined by the utility of these books.

Sidney A. Simon, PhD
Miguel A.L. Nicolelis, MD, PhD
Series Editors

Preface

A TOAST: TO MAKING SENSATIONS MORE REWARDING AND REWARDS MORE SENSATIONAL

The Neurobiology of Sensation and Reward provides a comprehensive systems overview of sensory and reward processing in the brain. Although the topics of sensation and reward are critically intertwined at the most fundamental levels of animal behavior, they rarely receive double billing. In the field of contemporary neuroscience, the researchers who study sensation rarely consider the rewards these sensations evoke, and the researchers who study reward tend to ignore the sensations evoking these rewards. Over the last 70 years, beginning perhaps with the rise of Skinnerian behaviorism, the scientific treatment of sensation and reward has increasingly diverged. Particularly by the 1950s, when intracranial electrical and drug self-stimulation techniques came into vogue, the behavioral and neural bases of reward could be studied entirely in the absence of environmental (sensory) input. The role of the sensory stimulus became relegated to second-best.

In 2011, this rather arbitrary and counterproductive dualism continues to dominate the neuroscience landscape. For example, despite the wealth of exciting new information regarding reward learning in the brain, surprisingly little is said about how the sensory specificity of reward processing is even achieved. Surely the functional integrity of a reward system is only as secure as its driving sensory inputs. Thus, a synthesis is in order, and this book is the first attempt to reunite these drifting neuro-tectonic plates by highlighting the important and intimate links between sensation and reward. In examining the neurobiological interface between sensation and reward, the book will reveal how the intrinsic properties of sensory systems effectively define and constrain the ambitions of reward-related processing in human and non-human brains. What can reward processing teach us about the senses, and what can sensory processing teach us about reward?

Once upon a time, say 500 million years ago, sensations and rewards clearly occupied the same ecological niche. All detectable sensations promised or exemplified reward; in turn, all rewards were necessarily embodiments of sensory stimuli. (This is not to minimize the importance of *unrewarding* sensations, but the topic of aversive stimuli is reserved for another book.) Indeed, insofar as rewards are those things that promote (reinforce) behavior, any sensory detector unable to provide biologically adaptive (i.e., rewarding) information for an organism would have soon been traded out of the gene pool, a victim of natural selection. During the phlegmatic era of Precambrian coelenterates, the only utility of sensory detection was to increase the chances of survival, and if a jellyfish or hydroid happened to sprout a novel detector apparatus that had no bearing on its fitness or longevity, this feature would have disappeared under evolutionary pressures.

Yet a curious implication arises out of these considerations: basic reward processing does not actually require a brain, at least not in the cortical sense. If one is willing to accept an operational definition of reward as reinforcement, then this suggests that all manner of species are under the spell of rewarding sensory stimuli. Brain or no, any organism that modifies its behavior to maximize access to a reward should qualify. Humans, monkeys, dogs, rats, obviously; but also coelenterates, cellular slime molds, ciliated protozoa. Each of these species will orient itself, or migrate, toward a food source diffusing through an aqueous medium. Sensory rewards in the general adaptive sense can reinforce behavior of bacteria, viruses, even plants. After all, sunlight will induce phototaxis in a green plant, such that it will extend its shoots and fronds up into the sky. Is the sunlight not thus a kind of sensory "reward" for the plant, operating at an extremely slow timescale?

The preceding discussion raises intriguing questions that will form an important organizational theme of the book. Specifically: what distinct advantages does a central nervous system confer upon reward processing in an organism endowed with a brain? In other words, how does the processing of a reward actually benefit from the presence of a brain? Some possibilities include the ability to integrate information from multiple sensory modalities in the service of reward; the ability to predict and anticipate reward; the ability to defer the consumption of a reward for another time; the ability to select among several rewarding alternatives; the ability to overcome natural (innate) response tendencies elicited by a reward; the ability to assign meaning to a reward. These questions are masterfully considered in the chapter by Murray, Wise, and Rhodes in this volume.

The book is divided into four parts. Each chapter can be read on its own of course, though the full scope of the book is best appreciated as a passage through all chapters in sequence. Part I, "First Principles," sets the tone for the remaining chapters of the volume, and introduces historical, ecological, and evolutionary themes that will recur throughout the book. The opening chapter, by Norman Weinberger and Kasia Bieszczad, presents a forceful case for treating the neurobiology of sensation and reward as a continuum rather than as separate frontiers, and serves as a proximal "bookend" for the volume. In the following chapter, Larry Marks traces a brief, but not too brief, scientific history of sensation and reward, beginning with Greek philosophers and Egyptian pha-raohs, in an attempt to chart how scientific thinking about sensory perception and hedonics has con-verged and diverged throughout human history. Norman White then provides a scholarly unpacking of the concept of "reward," using his chapter to define rewards and reinforcers, to illustrate why not all rewards are reinforcing and why not all reinforcers are rewarding, and to outline how behavioral paradigms can be used to dissociate among different and competing forms of reward. Finally, as mentioned above, Elisabeth Murray, Steven Wise, and Sarah Rhodes delve deep into the question of how the evolutionary development of the brain confers progressively greater complexity of reward processing, comprehensively spanning the phylogenetic scale from invertebrates to humans.

Part II of the book, "A Systems Organization of the Senses," offers an overview of the five senses. All of the contributing authors in this section—Jay Gottfried and Don Wilson on "Smell," Don Katz and Brian Sadacca on "Taste," Steven Hsiao and Manuel Gomez-Ramirez on "Touch," Christina Konen and Sabine Kastner on "Sight," and Corrie Camalier and Jon Kaas on "Sound"—bring their sensory systems expertise to bear on questions of sensory coding and the implications for reward processing. Because the natural (real-world) form of sensory objects necessarily constrains the type of information that a central nervous system can extract from the environment, these chapters include sensory ecological and evolutionary perspectives as a bridge for understanding how brain systems generate and maintain perceptual codes of meaningful objects that ultimately subserve reward processing. The chapter by Hsiao and Gomez-Ramirez ("Touch") additionally contains a section on multisensory integration, where the important point is made that brain structures tradi-tionally thought to be unisensory are influenced by sensory information from other modalities. The final chapter in this section, "Sensory Agnosias," by H. Branch Coslett, focuses on neurobehavioral syndromes associated with higher-order sensory perceptual deficits, and highlights how patient lesion models can be used to gain a better mechanistic understanding of sensory perception and awareness in the human brain. In order to allay any false hopes that might reasonably arise from a book with the word "Reward" in the title, it should be noted that the sensory chapters do not include dedicated discussions of sensory-specific hedonics, valence coding, or neural hot spots of pleasant-ness, first, because there are many other recent fine review articles and books that do just this, and second, because a determination of hedonics in non-human animals relies on indirect measures and can be difficult to relate to human models.

Part III of the book, "From Sensation to Reward," serves as a counterpoint to Part II, with chapters that directly consider the interplay between reward systems and the sensory cues that drive them. In the first chapter of this section, Suzanne Haber takes a slightly unique approach in laying out the neuroanatomy of reward. Her starting point is the basal ganglia, and particularly the ventral striatum, a region that is perfectly poised to straddle sensations and rewards, being densely interconnected both with cortical sensory systems and with reward-based systems underlying learning and behavior. Ivan de Araujo follows with a challenge to the doctrine of "sweet taste = primary reward," his point being that there are numerous instantiations of a reward, both proximal and distal, pre-ingestive and post-ingestive, affording multiple opportunities for relating external (sensory) and internal (endocrine, autonomic) cues to neural representations of reward that can variously guide adaptive behavior. The subsequent chapter by Bernard Balleine describes the behavioral neuroscience of incentive learning, an associative learning mechanism that perfectly epitomizes the integration of sensations, rewards, and motivational states to control goal-directed actions. John O'Doherty presents an up-to-date overview of reward prediction. He specifically focuses on the different ways that a given (sensory) stimulus can enter into conditioned associations, including Pavlovian, instrumental, and habitual learning, and describes imaging and computational approaches to reveal how different types of learning lead to different representations of predictive reward signals in the brain. Fresh insights into the role of the orbitofrontal cortex in mediating adaptive behavior is the focus of Geoff Schoenbaum, Matthew Roesch, Tom Stalnaker, and Yuji Takahashi, whose chapter contends that this brain region signals predictive information about what the future holds, responding not only to reward, but to reward anticipation and to reward-predicting sensory cues. The last chapter of this section is by Lesley Fellows and complements Coslett's chapter in the prior section. This chapter illustrates the neuropsychological consequences of damage to brain reward systems and puts forward a bold new hypothesis that the decision-making deficits observed in patients with prefrontal injury underscore a neurology of value.

Part IV, "Civilized Sensory Rewards (Distinctly Human Rewards)," is the coda to the book. The chapter authors in this section have been encouraged to weave empirical data with unbridled speculation, to consider the neurobiology of these higher-order rewarding sensory experiences, and how they might fit into broader schemas of sensation and reward. Are the "civilized" sensory rewards of perfume, art, and music truly distinct to humans, and if so, why? What rewarding functions do they serve (if any), and by what mechanisms do they accomplish these functions? One could speculate that the sensory information transmitted by attar of roses, the Venus de Milo, or the Moonlight Sonata is hardwired to human reward systems. Or alternatively, perhaps humans are hardwired to create certain types of art objects that specifically engage our reward systems. In her chapter on "Perfume," Rachel Herz outlines the history and sociology of perfume and considers the putative roles of fine fragrances as human pheromone-like compounds, aphrodisiacs, and reproductive lures. The emotional potency of smells in general is considered in the context of associative learning processes and empirical neuroimaging data, linking back to previous chapters in the book. The chapter by Anjan Chatterjee on "Visual Art" discusses the interface between visual art and the brain, highlighting different levels of visual complexity as represented both in the brain and in art, and he provides a new framework for neuroaesthetics research that may usefully guide hypothesis-driven work on this topic in future. To conclude, Zald and Zatorre ponder in their chapter on "Music" why humans take pleasure from sequences of tones, and they articulate some of the possible mechanisms by which music exerts this effect. They end with the idea that the rewarding aspects of music lie in its ability to anticipate, defer, and confirm delivery of a known auditory outcome, in keeping with neurobiological models of reward prediction.

In bringing together contributions from leading investigators in the fields of sensation and reward, all under a single volume, this book is meant to inject the "sensory" back into the study of reward processing, and the "rewarding" back into the study of sensory processing. It is hoped that the combined perspectives of neurobiology, ecology, and evolution will create interest and enthusiasm for cross-disciplinary scientific collaborations that bridge the interface between sensation and reward.

Jay A. Gottfried, MD, PhD
Departments of Neurology and Psychology
Northwestern University Feinberg School of Medicine
Chicago, Illinois, USA

Acknowledgments

This idea of this book first took shape in 2006, when Sid Simon and Miguel Nicolelis, editors of the book series, Frontiers in Neuroscience, approached me to edit a volume on *The Neurobiology of Sensation and Reward*. They were kind enough to give me complete freedom in developing the themes and content of the book, and I am indebted to them for providing me with this opportunity—except during a few private moments of book chapter turmoil when I swore sweet vengeance on them for providing me with this opportunity.

The book slowly materialized, in fits and starts, as chapters began trickling in. I would like to thank Barbara Norwitz, Executive Editor at Taylor & Francis Group/CRC Press, for her gentle patience as the days turned into weeks, the weeks into months, and so on. Her generous flexibility with the submission date was part blessing and part amnesia, which too was a blessing. Pat Roberson and Kathryn Younce at Taylor & Francis Group were also invaluable in helping bring this book to completion.

The great heroes of the book are the contributing authors. There are 31 authors distributed across 19 chapters, and each of them deserves congratulations. They tolerated my tedious appeals to "try and integrate sensation and reward," they suffered my entreaties to provide outlines of their chapters, and they agreed to an extensive round of chapter revisions, long after the point that they ever wished to think about their chapters again. It is my good luck that all of the authors were willing to stick with the project, and in my opinion their combined efforts have amounted to much more than the sum of the parts.

Lastly the book would be nowhere without the support and understanding of my family: Hilary, Xander, Julian, and Pippin. Hilary in particular advised me that "just one hour a day" would be enough to see this project through to completion. This plan sounded so reasonable at the time, though my 2006 forecast did not predict a 2011 submission date. That's a lot of one-hours.

Jay A. Gottfried
Northwestern University

Contributors

Bernard W. Balleine, PhD
Brain & Mind Research Institute
University of Sydney
Camperdown, NSW, Australia

Kasia M. Bieszczad, PhD
Department of Neurobiology and Behavior
University of California
Irvine, California

Corrie R. Camalier
Psychology Department
Vanderbilt University
Nashville, Tennessee

Anjan Chatterjee, MD
Department of Neurology and Center for
    Cognitive Neuroscience
University of Pennsylvania
Philadelphia, Pennsylvania

H. Branch Coslett, MD
School of Medicine
University of Pennsylvania
Philadelphia, Pennsylvania

Ivan E. de Araujo, PhD
The John B. Pierce Laboratory
New Haven, Connecticut

Lesley K. Fellows, MD, CM, DPhil
Department of Neurology and Neurosurgery
McGill University
Montreal, Québec, Canada

Manuel Gomez-Ramirez, PhD
Krieger Mind/Brain Institute and the Solomon
    S. Snyder Department of Neuroscience
The Johns Hopkins University
Baltimore, Maryland

Jay A. Gottfried, MD, PhD
Department of Neurology
Northwestern University Feinberg School of
    Medicine
Chicago, Illinois

Suzanne N. Haber, PhD
Department of Pharmacology and Physiology
University of Rochester School of Medicine
Rochester, New York

Rachel S. Herz, PhD
Department of Psychology
Brown University
Providence, Rhode Island

Steven Hsiao, PhD
Krieger Mind/Brain Institute and the
    Solomon S. Snyder Department of
    Neuroscience
The Johns Hopkins University
Baltimore, Maryland

Jon H. Kaas, PhD
Psychology Department
Vanderbilt University
Nashville, Tennessee

Sabine Kastner, PhD
Department of Psychology
Princeton University
Princeton, New Jersey

Donald B. Katz, PhD
Department of Psychology and Volen Center
    for Complex Systems
Brandeis University
Waltham, Massachusetts

Christina S. Konen, PhD
Department of Psychology
Princeton University
Princeton, New Jersey

Lawrence E. Marks, PhD
John B. Pierce Laboratory
New Haven, Connecticut

Elisabeth A. Murray, PhD
Laboratory of Neuropsychology
National Institute of Mental Health
Bethesda, Maryland

John P. O'Doherty, PhD
Trinity College Institute of Neuroscience and
    School of Psychology
University of Dublin
Dublin, Ireland

Sarah E.V. Rhodes, PhD
Laboratory of Neuropsychology
National Institute of Mental Health
Bethesda, Maryland

Matthew R. Roesch, PhD
Department of Psychology
University of Maryland
College Park, Maryland

Brian F. Sadacca
Department of Psychology and Volen Center
    for Complex Systems
Brandeis University
Waltham, Massachusetts

Geoffrey Schoenbaum, MD, PhD
Department of Anatomy and Neurobiology
University of Maryland School of Medicine
Baltimore, Maryland

Tom A. Stalnaker, PhD
Department of Anatomy and Neurobiology
University of Maryland School of Medicine
Baltimore, Maryland

Yuji K. Takahashi, PhD
Department of Anatomy and Neurobiology
University of Maryland School of Medicine
Baltimore, Maryland

Norman M. Weinberger, PhD
Center for the Neurobiology of Learning and
    Memory
University of California
Irvine, California

Norman M. White, PhD
Department of Psychology
McGill University
Montreal, Québec, Canada

Donald A. Wilson, PhD
Emotional Brain Institute and Department of
    Child and Adolescent Psychiatry
New York University School of Medicine
    and Nathan Kline Institute for Psychiatric
    Research
Orangeburg, New York

Steven P. Wise, PhD
Laboratory of Systems Neuroscience
National Institute of Mental Health
Bethesda, Maryland

David H. Zald, PhD
Department of Psychology
Vanderbilt University
Nashville, Tennessee

Robert J. Zatorre, PhD
Department of Neuropsychology
McGill University
Montreal, Québec, Canada

Part I

First Principles

1 Introduction: From Traditional Fixed Cortical Sensationism to Contemporary Plasticity of Primary Sensory Cortical Representations

Norman M. Weinberger and Kasia M. Bieszczad

CONTENTS

1.1 SCOPE AND GOALS

This monograph constitutes a singular and bold initiative toward a synthesis of sensation and reward. Although these topics have developed in parallel, this may be largely a historical "accident." On the other hand, the complexity of each domain may be so great that an extensive separate research on sensation and reward was necessary before an attempted synthesis could be initiated (see chapter by Marks in this volume). In any event, the current state of affairs does not yet permit an overarching specification of lines of integration between these topics. Thus, while introductory chapters are often concerned with such matters, the Preface by Jay Gottfried adequately sets the rationale for and context of this volume. Accordingly, this introductory chapter has another goal, which is to deal with both very old and very new aspects of one of the issues specified in the Preface: "the ability to predict and anticipate reward." The very old issue concerns the long-established, but rarely discussed, exclusion of learning, memory, and related cognitive processes from primary sensory cortical fields in audition, somesthesis, and vision. The very new topic concerns recent surprising findings about the effects of reward level on associative plasticity in the primary auditory cortex. So, in one sense, this chapter serves as a set of proximal "bookends" for the contents of this monograph.

1.2 CONCERNING THE TRADITIONAL EXCLUSION OF LEARNING, MEMORY, AND COGNITION FROM PRIMARY SENSORY CORTICES

A foundation for traditional theories of cortical organization has been that primary sensory cortices provide only an analysis of the external world, i.e., sensations. In one scenario, the neural substrates of novel "neutral" sensations are conveyed from primary sensory cortical regions to cortical "association areas," where they are given meaning by uniting with sensations from rewards or punishments. This conception was initiated by Pavlov. Another, more dominant scenario is that neutral sensations are analyzed in primary sensory cortical zones, then passed on to "higher" sensory zones where processes of interpretation and comprehension take place. Finally, this information is combined elsewhere with sensations from reinforcers, at which point associations are formed. In short, the end-point of these processes is that sensations acquire the ability to predict and anticipate rewards with other reinforcers. However, the discovery, and its aftermath during the past 20 years or so, that primary sensory fields participate in all these processes, i.e., they are not simply analyzers of sensations, should render both conceptualizations moot. While awareness of the cognitive aspects of primary sensory cortices is increasing, this is a continuing, and in some circles contentious, enterprise.

The dominant traditional view, so deeply woven into the fabric of neuroscience that it is rarely actually discussed, is that primary sensory cortices *analyze* sensory input, while "higher order" sensory and association cortices *interpret and comprehend* the sensory analyses that they receive from primary cortical fields. This view is a heritage of the nineteenth century. Some of the first structural–functional relationships discovered concerned the spinal cord; Bell and Magendie are credited with finding that the dorsal roots are sensory and the ventral roots are motor (Fearing 1930). Much of the research program for the rest of the century concerned the extent to which the entire neuraxis was organized on sensory-motor principles (Young 1970). Subsequently, Fritsch and Hitzig discovered the motor cortex, and lesion studies provided the approximate locations of sensory cortices. By the end of the nineteenth century, the cerebral cortex was conceived of only in sensory-motor terms:

> For the cerebral hemispheres consist only of centres related respectively to the sensory and motor tracts... Ideas are revived associations of sensations and movements... thought is internal speech... intellectual attention is ideal vision. (Ferrier 1886)

As studies of the brain burgeoned, later workers concluded that sensory and motor areas did not comprise the entire neocortex, which also had "association" areas. Fleshsig (1901) reported differential myelination in the cortex: sensory and motor cortices exhibited myelination at birth, other areas thereafter. He concluded, erroneously as it turned out, that only the sensory and motor cortices received subcortical projections; the association areas were thought to receive inputs only from other cortical regions (Fleshsig 1901). In short, Fleshsig's schema was that the cortex consisted of (a) sensory-motor zones that were connected to the thalamus and brain stem and were functional at birth; and (b) the association cortex that was connected only to other cortical regions and was not functional until well after birth. The late Irving Diamond pointed out that as association cortical areas myelinate later, this sequence of myelination "is just what would be expected if an infant sees sensory qualities such as color and brightness before these impressions are associated with another to form the perception of objects" (Diamond and Weinberger 1986). Thus, Fleshsig had provided an anatomical basis for the distinction between "lower" (i.e., sensory-motor) and higher psychological functions.

The final step was to parse sensory cortices into finer-grain functions. This was supplied by the impressive cytoarchitectonic studies of A.W. Campbell's (1905) *Histological Studies on the Localization of Cerebral Function*. His influence has been profound. Campbell labeled the region now identified as primary visual cortex (V1) "visual sensory," and called regions nearby (e.g., areas

18 and 19) "visual psychic." Similarly, the region now known as A1 was termed "auditory sensory," while adjacent areas (in modern parlance, auditory "belt" areas) (Kaas and Hackett 2000) were "auditory psychic" (cf. chapter by Camalier and Kaas in this volume). In this, Campbell intended to make a clear distinction between cortical regions he considered to subserve *only sensations* from those he believed to concern the *comprehension* of sensations. Accordingly, this distinction, made purely on cytoarchitectonic grounds, removed learning, memory, and other cognitive functions from primary sensory fields (Diamond 1979). Campbell's "ghost" still walks the halls of neuroscience.

1.3 ASSOCIATIVE REPRESENTATIONAL PLASTICITY

There has been a sea change in the concept of the sensory cortex. It had long been known that primary sensory cortices develop increased responses to stimuli that gain behavioral importance via learning (e.g., classical or instrumental conditioning) (reviewed in Weinberger and Diamond 1987). However, such plasticity could not be understood within the context of sensation or sensory processing because a single stimulus (or two stimuli in the case of discrimination learning) constitutes too limited a stimulus set to reveal whether associative processes had modified receptive fields or other fundamental descriptors of sensory representation. New experimental paradigms, which combined the study of learning with standard analyses in sensory physiology, could do so. Initiated in the period 1985–1990, this "unified" approach revealed that associative learning was accompanied by *systematic changes in the representation of sound* in the primary auditory cortex (A1). More specifically, Pavlovian conditioned stimuli (CS) gained increased representation, as indexed both by tuning shifts of frequency receptive fields to the CS and gain in area within the tonotopic map (Diamond and Weinberger 1986; Gonzalez-Lima and Scheich 1986; Gonzalez-Lima 1989; Bakin and Weinberger 1990). For convenience, we refer to learning-dependent, systematic changes in cortical representation of a stimulus parameter as high-order (cortical) associative representational plasticity (HARP).

The involvement of A1 in learning and memory is extensive. HARP possesses all the major attributes of associative memory. In addition to being associative, it is highly specific to the signal frequency (within fractions of an octave), rapidly acquired (in as few as five trials), discriminative (increased responses to a reinforced [CS +] tone and decreased responses to an unreinforced [CS –] tone), consolidates (becomes stronger over hours and days in the absence of further training), and exhibits long-term retention (tracked for up to 8 weeks post-training) (reviewed in Weinberger 2007; Weinberger 2010 in press). Therefore, HARP is an excellent candidate for serving as a substrate of *specific auditory memory*, despite the fact that it develops in a primary sensory zone.

Subsequently, associative processes in both Pavlovian and instrumental conditioning were found to be responsible for HARP in A1 in a wide variety of tasks, for both reward and punishment, and for all other acoustic parameters investigated, e.g., stimulus level (Polley et al. 2004), rate of tone pulses (Bao et al. 2004), envelope frequency of frequency-modulated tones (Beitel et al. 2003), tone sequence (Kilgard and Merzenich 2002), and auditory localization (Kacelnik et al. 2006). Furthermore, HARP develops during associative learning in all investigated taxa, e.g., big brown bat (Gao and Suga 1998), cat (Diamond and Weinberger 1986), guinea pig (Bakin and Weinberger 1990), owl monkey (Recanzone, Schreiner, and Merzenich 1993), rat (Kisley and Gerstein 2001; Hui et al. 2009), and human (Molchan et al. 1994; Morris, Friston, and Dolan 1998; Schreurs et al. 1997). Similar findings have been reported in other sensory systems, as reported elsewhere in the sensory chapters in this volume. The contrast between the traditional "sensation" and contemporary "cognitive" conceptions of primary sensory cortical fields (Scheich et al. 2007) cannot be overemphasized.

1.4 AUDITORY CORTICAL PLASTICITY AND LEARNING STRATEGY

Neurophysiological correlates of learning have been found in virtually all brain structures studied over the past 60 years (e.g., reviewed in John 1961; Thompson, Patterson, and Teyler 1972; Farley and Alkon 1985; Nakahara et al. 1998; Suzuki and Eichenbaum 2000; Schouenborg 2004). An implicit assumption has been that the magnitude of learning-related plasticity is linked to the magnitude of learning, as indexed by appropriate behavioral performance measures. Recent findings in A1 have challenged this assumption. For example, whereas two-tone discrimination learning for reward in the owl monkey was found to be correlated with the amount of expanded representation of the training frequency band (Recanzone et al. 1993), similar discrimination learning for reward in the cat apparently yielded *no plasticity* in A1 (Brown, Irvine, and Park 2004). This "failure to replicate" has been interpreted as casting grave doubts about the importance of HARP in the auditory cortex, and has cast suspicion on associative plasticity in other sensory modalities. One explanation is that the failure to replicate is due to a "species difference" between the monkey and cat (Brown, Irvine, and Park 2004). If so, then while the findings in the monkey could be accepted, at best such cortical plasticity would not be a general process; at worst, the findings might be idiosyncratic to the species used.

However, unexpected findings from studies of "simple" auditory learning in the rat suggest another type of explanation, and indeed add another dimension to the search for understanding the role of sensory systems in reward learning. Animals were trained to bar press in the presence of a 10 sec tone to receive a water reward. (A maximum of two rewards could be obtained on each trial.) They were "punished" for bar pressing in silence (e.g., after tone offset) by a time-out period signaled by a flashing light (Figure 1.1a). While apparently a simple task, rats could solve this problem in two ways. One strategy would be to start bar pressing at tone onset and stop bar pressing at tone offset. For convenience, this is called the tone-duration strategy (T-Dur) (Figure 1.1b-1). Alternatively, they could use the strategy of beginning to bar press at tone onset, but continue to respond until a bar press after the rewarded period elicited the flashing light error signal, i.e., effectively ignoring tone offset; hereafter called the tone-onset-to-error strategy (TOTE) (Figure 1.1b-2). The standard protocol cannot easily distinguish between these strategies because the error signal can practically co-occur with tone offset (Figure 1.1a-1).

The two learning strategies can be disambiguated by inserting a "grace period" immediately following tone offset (Figure 1.1a-2). Bar presses during this period would not be rewarded and would not be "punished," although all subsequent responses during silence on that trial would produce a flashing light and time-out. This pattern of behavior would indicate the use of the TOTE strategy (Figure 1.1b-2).

In this initial study, different groups were trained with different protocols, either the standard protocol, or with the grace period protocol. Both groups learned to achieve about the same level of high performance (Figure 1.1c). Representational areas in A1 were determined by obtaining complete maps of the tonotopic organization of frequency in a terminal mapping experiment. Despite a high level of performance, the STD group exhibited *no detectable plasticity* (Figure 1.2).

This "negative finding" apparently supports the views of Brown, Irvine, and Park (2004) that HARP in A1 is not important. However, the group trained in the grace protocol exhibited HARP in the form of signal-specific *decreases in absolute threshold and bandwidth*, i.e., learning was accompanied by highly specific *increases in neural sensitivity and selectivity*, respectively (Figure 1.2).

The "grace group (GRC)" could have solved the problem of obtaining water by responding only during tones by using either the T-Dur or TOTE strategies. In fact, they tended to use the latter strategy, i.e., they continued to bar press after tone offset until they received the error signal. Insofar as both groups solved the problem to the same approximate level, the differences in cortical plasticity cannot be due to differences in performance, but rather appear to reflect the use of different learning strategies (Berlau and Weinberger 2008). (We will consider why the different strategies could lead to differential plasticity in A1 later.)

If the use of the TOTE strategy is responsible for plasticity, then a greater use of this strategy should produce greater plasticity. In the first study, the GRC group used the TOTE strategy about 20% of the time, and they developed specific decreases in cortical threshold and bandwidth, but

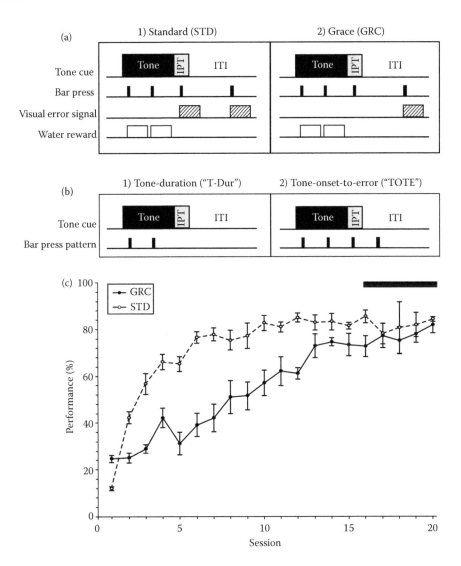

FIGURE 1.1 Learning to respond to tones for rewards can be solved using different behavioral strategies. (a) Two protocols were used to train animals to respond to tone cues for rewards. The grace protocol differed from the standard protocol only in the inclusion of an immediate post-tone (IPT) "null" period during which bar presses elicited neither rewards nor error signals (shaded boxes). (b) Behavioral response patterns can be used to identify two different learning strategies: (1) tone-duration, in which tone onset and tone offset are used to limit responses to the rewarded period; and (2) tone-onset-to-error, in which the only acoustic cue used to guide behavior is the tone onset. (c) STD and GRC groups trained in the "standard" and "grace" protocols, respectively, learn to solve the problem to the same asymptotic level of performance (black bar). (Reprinted from *Neurobiology of Learning and Memory*, 89, Berlau, K.M., and Weinberger, N.M., Learning strategy determines auditory cortical plasticity, 153–66, 2008, with permission from **Elsevier**.)

not in enlargement of representational area. We therefore hypothesized that HARP has several forms. The most modest would be local decreases in threshold and bandwidth. A stronger form would be local tuning shifts to the signal frequency. Finally, the strongest form would be an actual increase in the area of representation, probably reflecting global shifts in tuning to the signal frequency. To test this hypothesis, we used a protocol that greatly increased the likelihood that subjects would use the TOTE strategy. Indeed, the use of TOTE increased from about 20% to about 80% of trials. This was accompanied by a significant increase in the area of representation

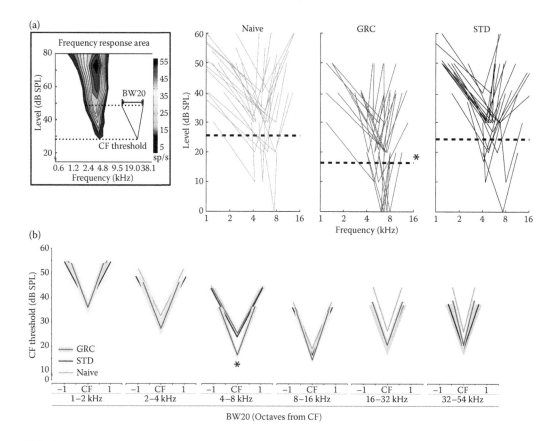

FIGURE 1.2 (See Color Insert) Tone-onset-to-error learning strategy results in reduced threshold and bandwidth in A1. Frequency response areas (FRA) were constructed for each recording site in A1 to determine threshold and bandwidth across A1 tonotopy (inset). (a) Individual threshold and bandwidth 20 dB SPL above threshold (BW20) at the characteristic frequency (CF) of the cortical site is represented by a "V" shape. Dashed lines show group mean CF threshold of the A1 population tuned within ± 0.5 octaves of the 5.0 kHz signal frequency. Only the GRC group using the TOTE strategy develops frequency-specific increases in neural sensitivity and selectivity. (b) Representational plasticity in threshold and BW20 is evident in the GRC group as the changes are specific only to A1 sites tuned near the signal frequency. Solid lines show group means with shaded areas showing ± standard error. Asterisks mark significant differences from a naive and STD group means. (Reprinted from *Neurobiology of Learning and Memory*, 89, Berlau, K.M., and Weinberger, N.M., Learning strategy determines auditory cortical plasticity, 153–66, 2008, with permission from **Elsevier**.)

of the signal frequency in A1, supporting the hypothesis of different forms of HARP (Bieszczad and Weinberger 2010c).

Overall, the findings reveal that a hitherto ignored factor can be critical for the formation of plasticity in associative reward learning. It is not sufficient to demonstrate learning, or even to obtain a high level of learning performance in order to obtain associative plasticity in the primary auditory cortex. Rather, it appears necessary to also determine *how* subjects learn to solve problems.

In addition to the general implications for understanding the neural bases of sensation and reward effects, *learning strategy* may explain the failure to replicate in the cat the development of HARP in the owl monkey (see above). It may indeed be the case that "species differences" are the cause, but only because cats and owl monkeys may use *different learning strategies*. Owl monkeys probably rely on onset transients in their con-specific vocalizations, and therefore would be expected to use the tone-onset-dependent TOTE strategy, which confers HARP in A1.

1.5 LEARNING STRATEGY TRUMPS LEVEL OF REWARD IN DETERMINING CORTICAL PLASTICITY

In the initial study, animals had been trained at a moderate level of motivation, i.e., with moderate water restriction. We expected that animals trained with a greater magnitude of reward, i.e., those with stricter water deprivation, would develop HARP of even greater magnitude. Therefore, we trained two groups of rats to bar press to tones for water rewards in an identical grace period protocol (Figure 1.1a-2) at different levels of water deprivation. One group was moderately motivated (ModMot). This group was expected to replicate our prior findings for learning-induced plasticity. Indeed, it did replicate the finding of signal-specific threshold and bandwidth decreases in A1, i.e., increases in neural sensitivity and selectivity, without area gain. The second group was highly motivated (HiMot), i.e., with higher water restriction. Because these animals were thirstier, water would be expected to have a higher subjective reward value. Indeed, the HiMot animals responded to tones faster and learned to a higher level than the ModMot group (Figure 1.3a). Remarkably, however, the HiMot animals *did not develop detectable plasticity*—either in the form of specific decreases in cortical threshold and bandwidth or in enlargements of representational area (Figure 1.3b). What might be the basis for this counter-intuitive result?

The underlying reason for the lack of HARP in these HiMot animals can be explained by returning to a consideration of learning strategy. An analysis of the patterns of behavior during learning revealed that the ModMot group used the TOTE strategy (Figure 1.1b-2), while the HiMot subjects used the T-Dur strategy (Figure 1.3c). In effect, the TOTE type of learning strategy "trumped" the level of subjective reward for the development of specific plasticity in A1. Thus, *how* animals learn to associate sensation and rewards has a dominant role for the formation of cortical plasticity, even over the potent behavioral influence of a reward's perceived magnitude (Bieszczad and Weinberger 2010b).

1.6 PLASTICITY IN SENSATION MAY DEPEND ON "MATCHING" OF CRITICAL CUES AND NEURONAL RESPONSE PROCLIVITIES

How is it possible that learning can take place apparently without plasticity in a structure thought to be critical for a signal's sensation? For example, animals that did not use the TOTE strategy still learned the association between tones and rewards but exhibited no HARP in A1. We suggest that the key to understanding sensory plasticity in the prediction of reward (and other reinforcers) is a "concordance" between the cue features that are the basis of a learning strategy and the particular response "proclivities" of sensory neurons to respond to those features. Thus, the TOTE strategy is based on responding to the onset of a tone, while ignoring its offset. As it happens, neurons in A1 appear to be particularly sensitive to onset transients (Masterton 1993; Heil and Irvine 1998; Phillips and Heining 2002). In this way, cells in A1 might develop specific plasticity during TOTE learning because of their proclivity to respond to acoustic onset transients. By contrast, subjects that employed other strategies apparently used additional behavioral guidance by tone offset, or exclusive guidance by offset, without the use of tone onset. We hypothesize that in both groups, learning was supported by the development of HARP in neurons whose response proclivities matched the cues that guided the learning strategies, i.e., in "on-off" cells for animals using an on-off strategy (i.e., T-Dur) and "off" cells when only tone offsets are used. Such off-sensitive cells could not be detected during mapping of A1 with animals necessarily under barbiturate anesthesia, which precludes detection of "off" responses. Also, other auditory cortical regions might contain such cells. Thus, one cannot assume that representations of relevant sensations will develop plasticity by virtue of serving as signals for reward. Rather, it may be necessary to determine *how* a problem is solved, and then match the critical cue with cells that are particularly "tuned" to that cue. The intriguing implication is that the expression of reward plasticity in a given sensory neuron (or set of neurons) is directly tied to its intrinsic physiological response profile evoked by a particular sensory input. This idea underscores the intimate, perhaps inextricable, interplay between processing of sensation and processing of reward, a recurrent theme appearing throughout this book.

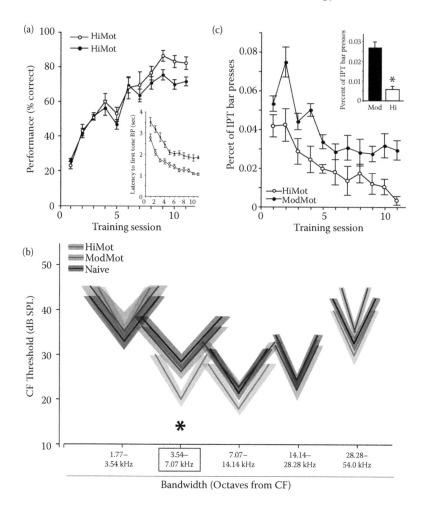

FIGURE 1.3 Learning strategy trumps reward level in determining HARP. (a) As expected, highly motivated animals (HiMot) attained a higher performance level at asymptote and responded faster throughout learning to tone signals that predict reward (inset) than animals trained while moderately motivated (ModMot). (b) Counter-intuitively, only the moderately motivated group developed signal-specific representational plasticity in decreased neural threshold and bandwidth (asterisk), while the highly motivated group with a higher subjective reward level had no detectable representational plasticity in threshold or bandwidth. (c) The basis for the lack of plasticity is in the learning strategies each group employed. The ModMot group exhibited a pattern of behavior that included bar presses throughout the immediate post-tone (IPT) null "grace" period, and thus ignored tone offsets. Thus, ModMot animals only used tone onsets to guide behavior, i.e., the TOTE learning strategy. By contrast, the HiMot group showed a reduced number of responses immediately after the tone (inset, asterisk). In effect, HiMot animals made use of tone offsets as well as tone onsets to limit responses to the rewarded tone period, i.e., they used a T-Dur strategy. Therefore, plasticity is dictated not by the subjective value of reward, but primarily by *how* sensory-reward problems are solved. In A1, a tone-onset-dependent strategy like TOTE appears critical for the development of HARP. (Reprinted from *Neurobiology of Learning and Memory*, 93, Bieszczad, K.M., and Weinberger, N.M., Learning strategy trumps motivational level in determining learning-induced auditory cortical plasticity, 229–39, 2010, with permission from **Elsevier**.)

1.7   CONCLUSIONS

We hope that this chapter has provided some insight into the reasons why sensory mechanisms for "the ability to predict and anticipate reward" have traditionally been excluded from primary sensory cortices. We further trust that we have adequately conveyed the importance of learning strategy for understanding how initially neutral sensations can produce specific reorganizations of their representations. We are well aware that the findings summarized here complicate the search for understanding the relationships between neural mechanisms of sensation and reward. Nonetheless, the finding that the effects of reward magnitude depend on *how* problems are solved, provides a promising entry point into this domain. Achievement of an adequate synthesis between sensation and reward may well require a generation. In this quest, we will need to become fully cognizant of the influences of many factors, some of which have received scant attention in the past. Learning strategy appears to be such a factor.

1.8   CODA

The leitmotif of this volume is the unification of sensation and reward, with the goal beyond its pages of a conceptual and neural synthesis of these two historically separate fields of inquiry. For navigation in such a quest, we propose sensory-specific HARP as an instrument for unison.

HARP posits that sensation and reward transcend cooperation, e.g., that the neural representation of a sensation's "sensory" characteristics and "behavioral" meaning are one and the same. Cortical topology along various modalities and dimensions of the sensory world identifies a sensation's specific features, but at the same time (and perhaps even more importantly), can attribute *feature-specific* subjective value (meaning) by the relative amount of cortical representation along a pertinent stimulus dimension.

Is a function of HARP to track the value of specific sensations? One might predict that the subjective value of sensation will be greatest for sensory events that predict the largest rewards, and that HARP will be greatest for signals of high reward magnitude. Furthermore, these salient sensations are likely to be most strongly remembered due to their behavioral importance. If so, the magnitude of a sensation's representational gain in the cortex should also index its memory strength. Remarkably, learning-induced gains in primary sensory cortical area have been found at the junction of sensation, reward value, and memory.

Recent work from our laboratory (Rutkowski and Weinberger 2005) showed that the more water deprived the subjects were during learning to respond to tones for water rewards, the larger the cortical expansion of tone-specific representational area in A1. Therefore, as the salience of the sensory tone signal increased (with the level of deprivation), the more representational area it gained in the cortex. Subsequently, Bieszczad and Weinberger (2010a) determined a tone signal's strength in memory by measuring its resistance to extinction. They found that animals with the greatest gains in tone-specific A1 representational area were the most resistant to response extinction. Therefore, gains in sensory cortical area conferred the strength of tone-signal memory. Overall, the findings indicate that HARP indexes a synthesis among sensation, reward, and strength of memory.

Thus, the representation of sensation-specific value might be intrinsic to sensory cortices by experience-dependent allocation of representational areas "casting bets" on the sensations with the highest probability of positive return. Sensory cortical over-representation of specific sensations and their enhanced presence in memory could provide the necessary neural impetus to elicit processes for behavioral reactions as complex as decision making, subjective preference, or aesthetic pleasure. Undoubtedly, there is an orchestration of competing sensory inputs in the reward-circuitry and executive-control areas of the brain. However, these control circuits also influence the inherent activity of their component sensory inputs. For example, acetylcholine is a potent modulator of auditory cortical plasticity (Ashe and Weinberger 1991; Miasnikov, Chen, and Weinberger 2008), as is dopamine (Bao, Chan, and Merzenich 2001; Schicknick et al. 2008; Hui et al. 2009), and the affective amygdala drives the nucleus basalis to release acetylcholine in the cerebral cortex (discussed in Chavez, Mcgaugh, and Weinberger 2009).

The linkage between sensation and reward via learning-induced representational plasticity and memory eases navigation toward their unification in concept and in the brain. That learning produces various forms of HARP in sensory systems and depends on factors such as how learning connects rewards to sensations, and under what motivational conditions, alludes to the possibility that one *function* of representational plasticity is, in fact, to unite sensation identity with its reward value.

ACKNOWLEDGMENTS

We thank Gabriel K. Hui and Jacquie Weinberger for assistance. This research was supported by research grants from the National Institutes of Health/National Institute on Deafness and Other Communication Disorders (NIDCD), DC-02938, DC-05592, and DC-010013.

REFERENCES

Ashe, J.H., and Weinberger, N.M. 1991. Acetylcholine modulation of cellular excitability via muscarinic receptors: Functional plasticity in auditory cortex. In *Activation to Acquisition: Functional Aspects of the Basal Forebrain Cholinergic System*, ed. R.T. Richardson, 189–246. Boston: Birkhauser.

Bakin, J.S., and Weinberger, N.M. 1990. Classical conditioning induces CS-specific receptive field plasticity in the auditory cortex of the guinea pig. *Brain Res* 536:271–86.

Bao, S., Chan, V.T., and Merzenich, M.M. 2001. Cortical remodelling induced by activity of ventral tegmental dopamine neurons. *Nature* 412:79–83.

Bao, S., Chang, E.F., Woods, J., and Merzenich, M.M. 2004. Temporal plasticity in the primary auditory cortex induced by operant perceptual learning. *Nat Neurosci* 7:974–81.

Beitel, R.E., Schreiner, C.E., Cheung, S.W., Wang, X., and Merzenich, M.M. 2003. Reward-dependent plasticity in the primary auditory cortex of adult monkeys trained to discriminate temporally modulated signals. *Proc Natl Acad Sci USA* 100:11070–75.

Berlau, K.M., and Weinberger, N.M. 2008. Learning strategy determines auditory cortical plasticity. *Neurobiol Learn Mem* 89:153–66.

Bieszczad, K.M., and Weinberger, N.M. 2010a. Representational gain in cortical area underlies increase of memory strength. *Proc Natl Acad Sci USA* 107:3793–98.

———. 2010b. Learning strategy trumps motivational level in determining learning-induced auditory cortical plasticity. *Neurobiol Learn Mem* 93:229–39.

———. 2010c. Remodeling the cortex in memory: Increased use of a learning strategy increases the representational area of relevant acoustic cues. *Neurobiol Learn Mem* 94(2):124–44.

Brown, M., Irvine, D.R.F., and Park, V.N. 2004. Perceptual learning on an auditory frequency discrimination task by cats: Association with changes in primary auditory cortex. *Cereb Cortex* 14:952–65.

Campbell, A.W. 1905. *Histological Studies on the Localization of Cerebral Function*. Cambridge: Cambridge University Press.

Chavez, C.M., Mcgaugh, J.L., and Weinberger, N.M. 2009. The basolateral amygdala modulates specific sensory memory representations in the cerebral cortex. *Neurobiol Learn Mem* 91:382–92.

Diamond, D.M., and Weinberger, N.M. 1986. Classical conditioning rapidly induces specific changes in frequency receptive fields of single neurons in secondary and ventral ectosylvian auditory cortical fields. *Brain Res* 372:357–60.

Diamond, I.T. 1979. The subdivisions of the neocortex: A proposal to revise the traditional view of sensory, motor, and associational areas. *Progr Psychobiol Physiol Psychol* 8:1–43.

Farley, J., and Alkon, D.L. 1985. Cellular mechanisms of learning, memory, and information storage. *Annu Rev Psychol* 36:419–94.

Fearing, F. 1930. *Reflex Action: A Study in the History of Physiological Psychology*. Baltimore: William & Wilkins Co.

Fleshsig, P. 1901. Developmental (myelogenetic) localization of the cerebral cortex in the human subject. *Lancet* 2:1027–29.

Gao, E., and Suga, N. 1998. Experience-dependent corticofugal adjustment of midbrain frequency map in bat auditory system. *Proc Natl Acad Sci USA* 95:12663–70.

Gonzalez-Lima, F. 1989. Functional brain circuitry related to arousal and learning in rats. In *Visuomotor Coordination*, eds. J.-P. Ewert and M.A. Arbib, 729–765. New York: Plenum.

Gonzalez-Lima, F., and Scheich, H. 1986. Neural substrates for tone-conditioned bradycardia demonstrated with 2-deoxyglucose. II. Auditory cortex plasticity. *Behav Brain Res* 20:281–93.

Heil, P., and Irvine, D.R. 1998. The posterior field P of cat auditory cortex: Coding of envelope transients. *Cereb Cortex* 8:125–41.

Hui, G.K., Wong, K.L., Chavez, C.M., Leon, M.I., Robin, K.M., and Weinberger, N.M. 2009. Conditioned tone control of brain reward behavior produces highly specific representational gain in the primary auditory cortex. *Neurobiol Learn Mem* 92:27–34.

John, E.R. 1961. High nervous functions: Brain functions and learning. *Annu Rev Physiol* 23:451–84.

Kaas, J.H., and Hackett, T.A. 2000. Subdivisions of auditory cortex and processing streams in primates. *Proc Natl Acad Sci USA* 97:11793–99.

Kacelnik, O., Nodal, F.R., Parsons, C.H., and King, A.J. 2006. Training-induced plasticity of auditory localization in adult mammals. *PLoS Biol* 4:e71.

Kilgard, M.P., and Merzenich, M.M. 2002. Order-sensitive plasticity in adult primary auditory cortex. *Proc Natl Acad Sci USA* 99:3205–9.

Kisley, M.A., and Gerstein, G.L. 2001. Daily variation and appetitive conditioning-induced plasticity of auditory cortex receptive fields. *Eur J Neurosci* 13:1993–2003.

Masterton, R.B. 1993. Central auditory system. *ORL J Otorhinolaryngol Relat Spec* 55:159–63.

Miasnikov, A.A., Chen, J.C., and Weinberger, N.M. 2008. Specific auditory memory induced by nucleus basalis stimulation depends on intrinsic acetylcholine. *Neurobiol Learn Mem* 90:443–54.

Molchan, S.E., Sunderland, T., McIntosh, A.R., Herscovitch, P., and Schreurs, B.G. 1994. A functional anatomical study of associative learning in humans. *Proc Natl Acad Sci USA* 91:8122–26.

Morris, J.S., Friston, K.J., and Dolan, R.J. 1998. Experience-dependent modulation of tonotopic neural responses in human auditory cortex. *Proc Biol Sci* 265:649–57.

Nakahara, K., Ohbayashi, M., Tomita, H., and Miyashita, Y. 1998. The neuronal basis of visual memory and imagery in the primate: A neurophysiological approach. *Adv Biophys* 35:103–19.

Phillips, M.L., and Heining, M. 2002. Neural correlates of emotion perception: From faces to taste. In *Olfaction, Taste, and Cognition*, eds. C. Rouby and B. Schaal (trans. and ed.), 196–208. New York: Cambridge University Press.

Polley, D.B., Heiser, M.A., Blake, D.T., Schreiner, C.E., and Merzenich, M.M. 2004. Associative learning shapes the neural code for stimulus magnitude in primary auditory cortex. *Proc Natl Acad Sci USA* 101:16351–56.

Recanzone, G.H., Schreiner, C.E., and Merzenich, M.M. 1993. Plasticity in the frequency representation of primary auditory cortex following discrimination training in adult owl monkeys. *J Neurosci* 13:87–103.

Scheich, H., Brechmann, A., Brosch, M., Budinger, E., and Ohl, F.W. 2007. The cognitive auditory cortex: Task-specificity of stimulus representations. *Hear Res* 229:213–24.

Schicknick, H., Schott, B.H., Budinger, E., Smalla, K.-H., Riedel, A., Seidenbecher, C.I., Scheich, H., Gundelfinger, E.D., and Tischmeyer, W. 2008. Dopaminergic modulation of auditory cortex-dependent memory consolidation through mTOR. *Cereb Cortex* 18:2646–58.

Schouenborg, J. 2004. Learning in sensorimotor circuits. *Curr Opin Neurobiol* 14:693–97.

Schreurs, B.G., McIntosh, A.R., Bahro, M., Herscovitch, P., Sunderland, T., and Molchan, S.E. 1997. Lateralization and behavioral correlation of changes in regional cerebral blood flow with classical conditioning of the human eyeblink response. *J Neurophysiol* 77:2153–63.

Suzuki, W.A., and Eichenbaum, H. 2000. The neurophysiology of memory. *Ann N Y Acad Sci* 911:175–91.

Thompson, R.F., Patterson, M.M., and Teyler, T.J. 1972. The neurophysiology of learning. *Annu Rev Psychol* 23:73–104.

Weinberger, N.M. 2007. Associative representational plasticity in the auditory cortex: A synthesis of two disciplines. *Learn Mem* 14:1–16.

———. 2010 in press. Reconceptualizing the primary auditory cortex: Learning, memory and specific plasticity. In *The Auditory Cortex*, eds. J.A. Winer and C.E. Schreiner. New York: Springer-Verlag.

Weinberger, N.M., and Diamond, D.M. 1987. Physiological plasticity in auditory cortex: Rapid induction by learning. *Prog Neurobiol* 29:1–55.

Young, R.M. 1970. *Mind, Brain and Adaptation in the Nineteenth Century*. Oxford: Clarendon Press.

2 A Brief History of Sensation and Reward

Lawrence E. Marks

CONTENTS

2.1 INTRODUCTION

The history of sensation and reward, the history of *the relation between sensation and reward*, is long and complex, sometimes to the point of being tortuous, with intertwined roots that begin in antiquity (see Figure 2.1 for some of the key individuals who embellish this history). The history of sensation

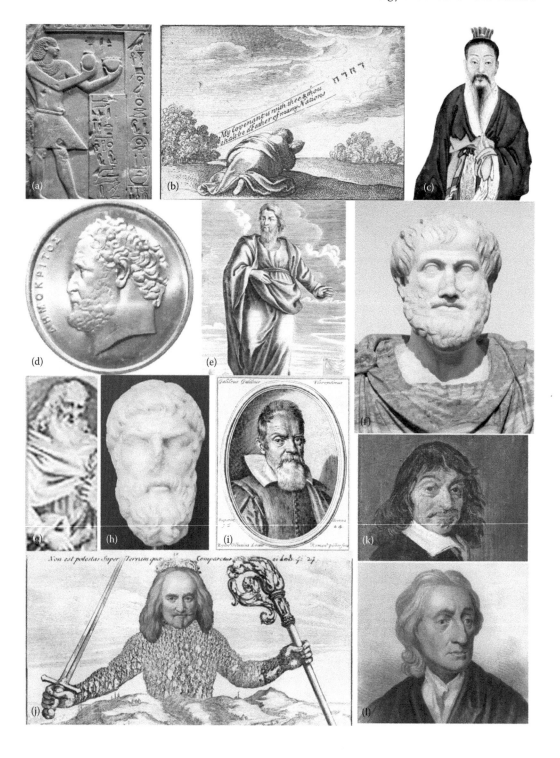

FIGURE 2.1 A rogues gallery of historical personages, appearing in rough chronological order of their date (or estimated date) of birth, from (a) to (z), who shaped modern scientific conceptions of sensation and reward. *Panel A:* (a) Egyptian King Intef II (ca. 2108–2069 BCE) funerary stele. On display at the Metropolitan Museum of Art. (From file "*Intef II*" by Wikimedia Commons user David Liam Moran, used under a Creative Commons Attribution-Share Alike 1.0 Generic license. File located at http://commons.wikimedia.org/wiki/File:Funerary_stele_of_Intef_II.jpg.) (b) Abram (Abraham) of the Old Testament. Etching, "*God's covenant with Abraham*, State 1," by Wenceslaus Hollar (1607–1677). (From Wikimedia Commons at http://commons.wikimedia.org/wiki/File:Wenceslas_Hollar_-_ God's_covenant_with_Abraham_(State_1).jpg.) (c) Guan Zhong (Kuan Chung; given name Yíwú), appointed Prime Minister of the Chinese state of Qi in 685 BCE. (From http://history.cultural-china.com/en/47History1499.html and also at http://www.xinfajia.net/english/4680.html. With permission.) (d) Democritus. Image depicted on the reverse of the Greek 10 drachma coin, 1976–2001. (e) Aristippus. Engraving appearing in *The History of Philosophy* by Thomas Stanley (1655). (f) Aristotle. Bust of Aristotle at the Natural Museum of Rome, Palazzo Altemps, Ludovisi Collection. Marble, Roman copy after a Greek bronze original by Lysippos from 330 BCE. (From Wikimedia Commons at http://commons.wikimedia.org/wiki/File:Aristotle_Altemps_Inv8575.jpg.) (g) Theophrastus. Engraving by John Payne, detail of title page from *The herball, or, Generall historie of plantes*, by John Gerard (From London: Richard Whitaker, 1633). (h) Epicurus. Bust of Epicurus at the National Museum of Rome, Palazzo Massimo alle Terme. Pentelic marble, Roman copy (first century CE) of a Greek original of the third century BCE. (From Wikimedia Commons at http://commons.wikimedia.org/wiki/ File:Epicurus_Massimo_Inv197306.jpg.) (i) Galileo Galilei. Galileo Galilei. Engraving. (From Wikimedia Commons at http://commons.wikimedia.org/ wiki/ File:Galileo_Galilei_4.jpg.) (j) Thomas Hobbes. Portrait detail (head) of Hobbes, overlaid on the frontispiece of his book, Leviathan, 1651. (Image of body taken from Wikimedia Commons at http://commons. wikimedia.org/wiki/ File:Leviathan.jpg.) (k) René Descartes. Detail from oil painting of Descartes, by Frans Hals (1649), National Gallery of Denmark. (From Wikimedia Commons at http://commons.wikimedia.org /wiki/File:Frans_Hals_111_WGA_ver-sion.jpg.) (l) John Locke. Lithograph from the Library of Congress, reproduction number LC-USZ62-59655: "by de Fonroug[...]? after H. Garnier. [No date found on item.]." (From Wikimedia Commons at http://commons.wikimedia. org/ wiki/File:Locke-John-LOC.jpg.) *Panel B:* (m) Jeremy Bentham. "Auto-Icon," with wax head. Acquired by University College London in 1850. (From file "Jeremy Bentham Auto-Icon.jpg" by Wikimedia Commons user Michael Reeve. File located at http://commons.wikimedia.org/wiki/ File:Jeremy_Bentham_Auto-Icon.jpg.) (n) Johannes Müller. Detail of oil painting by Pasquale Baroni. Museo di Anatomia Umana "Luigi Rolando," Torino, Italy. (From Wikimedia Commons at http://commons.wikimedia.org/wiki/File:Johannes_ Peter_Müller_by_Pasquale_Baroni.jpg.) (o) Gustav Fechner. Photograph. (From Wikimedia Commons at http://commons.wikimedia.org/wiki/File:Gustav_Fechner.jpg.) (p) Alexander Bain. Photograph. (From Wikimedia Commons at http://commons.wikimedia.org/wiki/File:AlexanderBain001.jpg.) (q) Hermann von Helmholtz. German postage stamp commemorating the 150th birthday of Helmholtz, issued 27 August 1971. (r) Wilhelm Wundt. Photograph, 1902. (Weltrundschau zu Reclams Universum. From Wikimedia Commons at http://commons.wikimediaorg/wiki/File:Wilhelm_Wundt.jpg.) (s) Ivan Pavlov. Portrait of Pavlov, 1920. Image taken from the Google-hosted LIFE Photo Archive, available under the filename 6bfe-762c14ea3e8d. (From Wikimedia Commons at http://commons.wikimedia.org/wiki/File:Ivan_Pavlov_LIFE. jpg.) (t) Sigmund Freud. Portrait of Freud, 1920, by Max Halberstadt. Image taken from the Google-hosted LIFE Photo Archive, available under the filename e45a47b1b422cca3. (From Wikimedia Commons at http://commons.wikimedia.org/wiki/File:Sigmund_Freud_LIFE.jpg.) (u) Walter Cannon. (Frontispiece portrait from the article by Mayer, *J. Nutr.,* 1965, reproduced with permission from the American Society for Nutrition.) (v) Edward Thorndike. Image courtesy of the National Library of Medicine. (w) Clark Hull. Anonymous. n.d. [Portrait of Clark L. Hull]. Photograph. (From the collection of Rand B. Evans. From http:// vlp.mpiwg-berlin.mpg.de/references?id=lit38366&page=p0001recto. With permission.) (x) Leonard Troland. Image from the Optical Society (OSA), Past Presidents 1922–1923. (Located at http://www.osa.org/about_ osa/leadership_and_volunteers/officers/past_presidents.aspx. With permission.) (y) Burrhus Frederic (B.F.) Skinner. Photograph. (From the file "B.F. Skinner at Harvard circa 1950.jpg" by Wikimedia Commons user Silly rabbit. From the file http://commons.wikimedia.org/wiki/File:B.F._Skinner_at_Harvard_circa_1950. jpg.) (z) James Olds. Photograph. (From the article by Thompson, Biographical Memoirs, 1999, with permission of the National Academy of Sciences, U.S.A.)

vis-à-vis reward taps into seminal themes in the history of Western thought—notably, empirical theories of the contents of mind and hedonistic theories of motivation; evolutionary biology and adaptation; and biological regulation and homeostasis. Because a brief history cannot possibly do justice to the interplay among these themes over a period of several thousand years, my goals here are as follows: to outline the main trends and developments (at the risk of oversimplification); to highlight a select number of important landmarks in the story of sensation and reward (at the risk of omission); and to do this as a temperate Aristotelian, navigating a middle course between, on the one hand, a Whiggish history that searches for old seeds to new flowers, with the virtue of providing a sense of the origins of current scientific discourse; and, on the other, an inclusive history that also considers issues that, from the perspective of the early twenty-first century, may appear to be weeds rather than blossoms. Because other chapters in this volume thoroughly review modern developments, the present chapter ends roughly a half century ago, with the discoveries by Olds, Milner, and others that spurred modern research into physiological and pharmacological mechanisms of reward.

The contemporary scientific conceptualization of *reward*, and its relation to sensory stimulation and sensory experience, finds its modern roots in the classical principle of learning, known as the *law of effect*, first named almost a century ago by Edward Lee Thorndike (1874–1949):

> The Law of Effect is that: Of several responses made to the same situation, those which are accompanied or closely followed by satisfaction to the animal will, other things being equal, be more firmly connected with the situation, so that, when it [the situation] recurs, they [the responses] will be more likely to recur; those which are accompanied or closely followed by discomfort to the animal will, other things being equal, have their connections with that situation weakened, so that, when it recurs, they will be less likely to occur. The greater the satisfaction or discomfort, the greater the strengthening or weakening of the bond. (Thorndike 1911, 244).

Note, however, that Thorndike (1898) had already more or less described the same general principle, without calling it a law, more than a decade earlier.

Thorndike's law of effect of 1911 had two parts: a law of reward (strengthening the bond between situation and response) and a law of punishment (weakening it). Two decades later, Thorndike (1931) would abandon the part of the law stating that punishments, or "annoyers" as he called them, act to weaken the bonds, while retaining the part stating that rewards, or "satisfiers," act to strengthen them. In brief, Thorndike's *revised law of effect* related a behavioral effect—an increase in the strength or probability of a particular response within a particular environment—to the satisfaction of the rewarding stimulus that follows the response.

Thorndike's law of effect spurred a new, scientific definition of reward: a reward is a stimulus that acts to strengthen the bond between the stimulus setting and the organism's instrumental response that led to the reward. This new definition was, to be sure, related to the older definitions of reward, definitions that had been and often still are used in general discourse, definitions that imply that reward serves as some kind of *recompense*. It is in this sense of reward that Silius Italicus wrote, in the first century CE, that "virtue herself is her own fairest reward."

Before there was "reward" in science, therefore, there was "reward" in other realms of discourse, realms in which the meaning of reward often had a distinctly ethical tone. Philosophical and religious writings were, and are, chock-full of allusions to rewards—and to punishments as well; indeed, punishment is often itself judged to be an appropriate reward for an inappropriate action. In religious and scriptural works especially, rewards and punishments are often characterized as just recompense for praiseworthy or blameworthy behaviors—though not always so, as even a casual reader of Job, for instance, readily discovers.

Prior to the law of effect, the connotations of the term reward were often ethical. And when the connotations were ethical, this ethical sense could contrast with or even contravene pleasure—as, for example, in the Epistle of Paul to the Hebrews 11:24–26, "Moses, when he was come to years, refused to be called the son of Pharaoh's daughter; choosing rather to suffer affliction with the people of God, than to enjoy the pleasures of sin for a season... for he had respect unto the recompense of the reward." Indeed, in the Hebrew Bible, the greatest reward is God: "After these things the word of the Lord came unto Abram* in a vision, saying, Fear not, Abram: I am thy shield, and thy exceeding great reward" (Gen. 1:15). Not that biblical authors were unaware of the pleasures of sensory experience; surely they were, as eloquently described, for example, in the Song of Songs 4:11, "Your lips drip flowing honey, O bride; honey and milk are under your tongue, and the fragrance of your garments is like the fragrance of Lebanon."

The present chapter is concerned with sensation vis-à-vis reward, but before Thorndike's law of effect, reward was not reward, which is to say, the use of the term reward lacked the scientific connotations of primary concern here. Consequently, the history of sensation and reward before Thorndike becomes largely the history of sensation in relation to *pleasure.*

Like many scientific terms, the meaning of reward has not stayed fixed since the time of Thorndike either. Indeed, for several decades, under the umbrella of a behaviorist movement that eschewed the use of subjectivist and mentalist terms, the term *reward*, with its hedonistic connotations of pleasure, gave way to *reinforcement.* Reinforcement clearly refers to effects on behavior, to the strengthening of responses, without implying any particular underlying mental process, indeed, without implying any particular process at all. In recent decades, however, the use of the term reward has become increasingly widespread, as various scientific paradigms seek to better understand reward and its mechanisms. The more we know about reward, and about the mental-behavioral processes and neural mechanisms underlying it, the more elaborate and nuanced its definition and meaning.

2.2   EMPIRICISM AND HEDONISM

To trace the history of sensation vis-à-vis reward is, therefore, not only to follow the developments after Thorndike's statement of the law of effect, but also to trace the roots of the law of effect itself. And to accomplish this means, in turn, examining two central themes in the history of Western thought: empiricism and hedonism.

First, empiricism: As the term is used here, empiricism refers to the notion that mind and behavior originate in experience and, especially pertinent to this chapter, to the notion that the contents of mental life derive from sensations.† Indeed, it is tempting to term this view *experientialism* or *sensationism*, to avoid confusion with another related sense of the word, empiricism, which is a method to uncover knowledge—commonly contrasted with intuitionism or with rationalism.

In the empirical theory of mental contents, mind is often called a *tabula rasa*, or unwritten slate, on which experience writes, through the senses. The Latin expression derived from Aristotle's phrase in Greek, *pinakis agraphos.* In *De Anima* (*On the Soul*), Aristotle (429b–430a) wrote, "What it [the mind] thinks must be in it [the mind] just as characters [letters, words] may be said to be on a writing-tablet on which as yet nothing actually stands written; this is exactly what happens with mind" (McKeon 1947, 683).

The second theme is hedonism, which refers to the notion that the goal of living is, or should be, the pursuit of pleasure and the avoidance of pain. Clearly, many pleasures and pains are

---

* At the end of his life, God gave to Abram a new name: "Neither shall thy name any more be called Abram, but thy name shall be Abraham; for a father of many nations have I made thee" (Gen. 17:5).

† Perhaps it would be better to define empiricism in terms of sensory experiences rather than sensations. At this moment, I perceive a computer in front of me. This perception is based on visual (sensory) processes. One view, which traces back through Hermann von Helmholtz (1867) to John Locke (1690), holds that perceptions are built on elementary sensations and constitute internal representations of the external world. Another view, which traces to Thomas Reid (1764), denies that perceptions derive from sensations, positing instead that perceptions directly provide information about the world. In this view, sensations come later, from subsequent, second-hand reflection on perceptions. Both of these views distinguish sensations from perceptions, albeit in different ways. Historically, however, the terms sensation and perception have, in fact, often been used interchangeably. Aristotle, for instance, had only a single term, *aesthesis*, to use for both sensation and perception (see Hamlyn 1959). For practical reasons, in the present chapter the terms sensation and perception will not be sharply distinguished, but will generally follow the context and language of the original authors.

sensory, linking hedonism to the study of the affective, or hedonic, qualities of sensory experience. Empiricism and hedonism have played prominent roles in the history of philosophy, empiricism in the theory of knowledge, or epistemology, and hedonism in ethics—anticipations of modern cognitive neuroscience and neuroethics.

Of the two themes, empiricism and hedonism, hedonism is probably the older. Inscribed four millennia ago on the tomb of the Egyptian King Intef was a harper song (Mackenzie 1907)—a harper song being, as the term suggests, a song that is sung while played on a harp. Like many harper songs, the *Lay of the Harper* expresses doubt about life after death, hence doubt about the value of deferring pleasures during life for possible rewards after death. Instead, the *Lay of the Harper* encourages its audience to live a life of what came to be called hedonism: "Revel in pleasure while your life endures/And deck your head with myrrh. Be richly clad/In white and perfumed linen; like the gods/Anointed be; and never weary grow/In eager quest of what your heart desires" (Mackenzie 1907, 246). Note the centrality of the sense of smell in the referents of these pleasures.

Egyptians of antiquity, no doubt like people living in other times and other places, readily fathomed the pleasures associated with sensory experiences, especially those associated with the savors of food and drink, and with sexual stimulation. A millennium and a half later, similar appeals to sensory pleasures appeared in the school of Charavaka in India and in the writings of Kuan-Yi-Wu (Guan Zhong) in China. Kuan's work is notable for its specific reference to sensory pleasures: "Allow the ear to hear anything that it likes to hear. Allow the eye to see anything that it likes to see. Allow the nose to smell whatever it likes to smell. Allow the body to enjoy whatever it likes to enjoy.… Allow the mind to do whatever it likes to do" (Riepe 1956, 551–52).

But science and philosophy require more than exhortations or recommendations, such as calls to enjoy pleasure. Science also requires an analytic stance, and the Western analytic stance originated in Greek thought.

2.3   SENSATION AND PLEASURE IN GREEK PHILOSOPHY

Within a period of only a few hundred years, from the fifth through the fourth centuries BCE, there was an astonishing growth of intellectual thought in the isles of the Aegean and on the western coast of what is now Turkey (for a provocative analysis of the Greek conception of mind, see Snell 1953, Ch. 10). During this period, a veritable pantheon of Greek philosophers, from Empedocles (ca. 495 to ca. 435 BCE) to Aristotle (384–322 BCE), tried to understand the relation between sensory experience and knowledge, and several of them, notably Aristotle, examined in detail the nature of sensory experience, especially in two of his works: *De Anima* (*On the Soul*) and *De Sensu* (*On the Senses*). Much of what other Greek philosophers knew or speculated about the senses appears in a single work, *De Sensibus* (*On the Senses*), written by Theophrastus (ca. 372 to ca. 287 BCE) (see Stratton 1917), a scholar at the Lyceum who became its head after Aristotle's death. A few of the Greek philosophers wrote on the affective or hedonic aspects of sensory experience, leading to two significant questions: First, are all sensory pleasures (indeed, are all pleasures) commensurable? Does the "pleasurableness" of sensory pleasures vary only in its intensity but not in the quality of the pleasures? And second, when are sensory pleasures absolute, and when are they relative? Which sensory pleasures are pleasurable in and of themselves, and which depend on what modern researchers call their "context," such as the body's internal state? Both sets of questions continue to reverberate in modern scientific discourse.

2.3.1   PLEASURES: COMMENSURABLE OR INCOMMENSURABLE?

Although the Greek philosopher Democritus (ca. 460 to ca. 370 BCE) averred, in the spirit of many subsequent ethical views, that higher intellectual pleasures exceed lower base ones, that "the great pleasures come from the contemplation of noble works" (Fragment 194: Freeman 1948, 110), he also remarked more pointedly [no pun intended] that "men get pleasure from scratching themselves: they feel an enjoyment like that of lovemaking" (Fragment 127: Freeman 1948, 104). To compare, as Democritus did, the pleasure of scratching an itch to the pleasure of lovemaking is to imply that the pleasurable qualities of these two classes of sensory experience are alike. This implication is consistent with the general tenor of Democritus's atomism, which asserted that the universe contains nothing but irreducible atoms in various sizes, shapes, masses, and numbers or clusters. As Democritus famously said, "Sweet exists by convention, bitter by convention, colour by convention; atoms and Void (*alone*) exist in reality.... We know nothing accurately in reality, but (*only*) as it changes according to the bodily condition, and the constitution of those things that flow upon (*the body*) and impinge upon it" (Fragment 9: Freeman 1948, 93).

The atomistic philosophy characterizes all variation in the universe as quantitative, none as qualitative. Sensory experiences of taste, smell, and color reflect quantitative variations in the properties of the atoms that strike the sense organs. And although sensations may vary in their degree of pleasurableness, pleasurableness varies only in degree but not in kind.

Commensurability was likely the view of Aristippus (ca. 435–366 BCE), a student of Socrates and one of the early proponents of hedonism, indeed, by tradition the founder of that doctrine. Commensurability was certainly the view of Aristippus's followers, the Cyrenaics. As Diogenes Laërtius (third century CE) recounted in his *Life of Aristippus*, "These men [the Cyrenaics]... adopted the following opinions. – They said that there were two emotions of the mind, pleasure and pain... that no one pleasure was different from or more pleasant than another; .... They also said that pleasure belonged to the body, and constituted its chief good..." (Diogenes Laërtius 1853, 89).

This is a principle of hedonic uniformity. It states that every object, event, or experience that produces pleasure is commensurable with every other object, event, or experience, at least with regard to pleasure itself. By implication, for example, the sensory pleasures experienced in eating a meal and in engaging in sexual intercourse may differ in magnitude—one may be greater than the other, hence more pleasurable—but there is no qualitative difference between pleasures as such. (From a modern, neural perspective, the implication is that all sensory pleasures involve the same reward circuitry.) Certainly, the experiences differ in their entirety, but they differ because other qualities aside from their pleasure differ; the experiences differ not because they vary in the quality of the pleasure but despite, speaking metaphorically, this *achromaticity of the pleasure*. Clearly, the principle of hedonic uniformity bears important implications for sensation and reward: Are rewards, *qua* rewards, all "of the same kind"? If they are, then, in one main respect, a reward is a reward is a reward.

More than 2000 years after Aristippus, Jeremy Bentham (1789) would take precisely this position, arguing that *utility*, his term for the measure of happiness or pleasure, varies only in amount and not in kind. Bentham's utilitarian theory deeply influenced both philosophical and scientific thought in the nineteenth century, and was one of the forerunners of later notions of reward and motivated behavior.

2.3.2 SENSORY PLEASURES: CONDITIONAL AND UNCONDITIONAL

The hedonism of Aristippus found several adherents. The philosophies of Epicurus (341–270 BCE) and his followers (the Epicureans) and of Zeno of Citium (ca. 333 to ca. 262 BCE) and his followers (the Stoics), like the philosophy of Aristippus, encouraged hedonistic goals: to maximize pleasure and minimize pain. But the most detailed and significant discussions among Greek philosophers of the relation between pleasure and sensation appear in the works of Aristotle.

Greek philosophers from Pythagoras in the sixth century BCE through Aristotle in the fourth century tried to determine whether, when, and how the senses can serve as sources of valid knowledge about the world. From the perspective of realism, all of the senses are, or can be, informative, but some of them—notably, pain, warmth, cold, taste, and smell—also play prominent roles in biological regulatory systems: in avoiding damage to body tissues, in controlling the body's thermal exchange with the environment, and in seeking and obtaining nourishment. A warm bath on a cold day can be especially pleasurable, and the pleasure of warming the skin is doubtless related to its homeostatic effect, as heating a chilled body helps maintain a stable core body temperature. Babies suck avidly at sweet flavors, and sweet tastes commonly signal the presence of calories.

Aristotle noted the contingency that can exist between pleasantness and internal biological states. Some sensations, Aristotle claimed, are unconditionally pleasant or unpleasant, pleasant or unpleasant in and of themselves. An example that Aristotle gave of an unconditionally pleasant sensation (and he may not be correct here) is the smell of certain flowers, which, he claimed, we enjoy in and of itself. By way of contrast, other sensations, Aristotle pointed out, are pleasant or unpleasant only conditionally. In these instances, whether the sensation is pleasant or unpleasant depends on the organism's internal biological state, and hence on the ability of the object that produces the sensation to meet a bodily need. With regard to conditionally pleasant or unpleasant smells, Aristotle wrote: "Their pleasantness and unpleasantness belong to them contingently, for, since they are qualities of that which forms our food, these smells are pleasant when we are hungry, but when we are sated and not requiring to eat, they are not pleasant; neither are they pleasant to those who dislike the food of which they are the odour" (Aristotle, *De Sensu*, 443b: Ross 1906, 75).

In pointing to the role of internal body states in sensory pleasure, Aristotle foreshadowed scientific research on the homeostatic role of internal states in reward and reinforcement. And a modern, expanded view of contingency would lead to research on the contextuality of pleasure and reward, for example, the ways that pleasure and reward depend on expectations and on previous pleasures and rewards (e.g., Solomon 1980; Flaherty 1982).

The smell and the flavor of a food provide knowledge of the food object, and these sensations are pleasant when we are hungry. The sensations signify nutritive value, and, when we have not eaten, they take on a positive affect. More than 2000 years after Aristotle, Cabanac (1971) would apply to this conditional relation the term *alliesthesia*. Cabanac (1971, 1105) noted the close connection, as Aristotle implied, between pleasure and the biological usefulness of the stimulus in maintaining homeostasis: "This relationship between pleasure and usefulness leads one to think that pleasure-displeasure is a determinant of an adapted behavior. A subject will seek all pleasant stimuli and try to avoid all unpleasant stimuli. Since pleasure is an indication of need or at least of usefulness ... this is a way that behavior can be adapted to its physiological aim. Indeed, it has been known for a long time that animal behavior, such as food or water intake, can be triggered by internal signals related to the 'milieu interieur.'" Internal body states also influence reward and reinforcement, for example, in what has been called *reinforcer devaluation* (e.g., Colwill and Rescorla 1985).

Finally, it is notable that Aristotle was explicit in his claim that unconditionally pleasant smells, such as the scents of flowers "have no influence either great or small in attracting us [humans] to our food nor do they contribute anything to the longing for it" (*De Sensu*, 443b: Ross 1906, 75). By implication, conditionally pleasant and unpleasant sensations of smell, by which Aristotle means the smells of food, are not only pleasurable, but may also motivate us to consume the food, presumably by creating a "longing" for it. This point anticipated recent claims that rewards have two components, one of "liking" (pleasantness) and the other of "wanting" (motivation) (Berridge 2003).

2.4 SENSATION AND THE EMERGENCE OF MODERN SCIENCE

The themes central to the present chapter made their first appearance more than two millennia ago, but would not be elaborated in important ways in Western science and philosophy until the seventeenth century, which saw the beginnings of modern science in the new physics of Galileo, Boyle, Newton, and others. This is not to say that nothing important happened in the previous 1500 years, but it is not possible in this chapter to treat the numerous developments that occurred in the West from the beginning of the Common Era through the end of the Renaissance roughly a millennium and a half later—except perhaps to note that, broadly speaking, under the impetus of early Christian religion, empirical observation as a method of inquiry gave way to Augustinian intuition, and hedonism to a cultivation of self-abnegation (see, e.g., Leahey 2004).

The empiricism of mental development implicit in Aristotle's notion of a *tabula rasa* did not, however, disappear. Also implicit in Aristotle's psychology is the maxim, "nihil est in intellectu quod non [prius] fuerit in sensu" (nothing in the intellect that does not arise in the senses). Although the maxim is sometimes attributed to John Locke in the seventeenth century, its origin is much older. Thomas Aquinas, in the thirteenth century, was among the first to use it or a variant of it (see Cranefield 1970).

Unfortunately, the Aristotelianism of Aquinas did not itself stimulate the study of the senses. Interest in the senses resumed in the seventeenth century, however, in conjunction with the physics of Galileo and Newton, the biology of Harvey, the chemistry of Boyle, and the philosophies of Descartes, Hobbes, and Locke. Three more or less concurrent developments were critical: the mechanization of the sciences, the distinction between sensation and stimulus, and the rebirth of empiricist and hedonistic philosophies.

2.4.1 PHYSICAL SCIENCE AND BIOLOGICAL SCIENCE

First was the "mechanization of the world picture" (Dijksterhuis 1961), the success of quantitative physical science, especially mechanics, both in understanding the inanimate world and in applying mechanical principles to biological processes of the animate world—well-known examples of the latter being the work, in the seventeenth century, of William Harvey on the circulation of blood and Giovanni Borelli on the flight of birds. This success suggested that physical science in general, and mechanics in particular, could serve as a model not only for physical processes, but also for biological processes and, yes, by implication, even for mental processes.

In the royal gardens of Germany, René Descartes (1596–1650) observed hydraulic automata, artificial animals that moved when powered by water, and these observations likely led him to conceive of behavior in terms of mechanics (e.g., Descartes 1664; for a detailed treatment, see Fearing 1929). To Descartes, animals were nothing but machines. So, too, were humans—albeit with the addition to mechanical human bodies of a rational soul, responsible for thought, language, and acts of the will. But even actions instigated by the Cartesian soul, through its hypothesized interactions with the body at the pineal gland, were constrained by the laws of physics. The Cartesian soul could not create anything material, including physical activity, out of nothing. It could influence the *direction* of bodily movement, but not its *quantity*; in the language of modern physics, the Cartesian soul could modify the vector quantity momentum, but not the scalar quantity energy. Indeed, from antiquity, nerve activity was often pictured in mechanical terms—some, such as Descartes, conceiving of nerves as hollow tubes filled with "animal spirits" that could, for instance, activate muscles by inflating them, whereas others, such as Newton, assuming that nerve activity consisted of mechanical vibrations of small particles within nerves.

Only after the discovery of electricity in the eighteenth century and research on the electrical properties of nerve transmission in the nineteenth century would the old mechanical theories of nerve function finally dissipate. Yet, even in the second half of the nineteenth century, Fechner (1860) wrote of brain activity in mechanical terms—so powerful was the continued conceptual hold of Galilean-Newtonian mechanics on the biological sciences. Indeed, a terminological vestige of this tradition continues, in that a goal of contemporary neuroscience, and other disciplines of biological science, is to understand *mechanisms* in terms that may be cellular, molecular, biochemical, biophysical—but, in a linguistic irony, rarely in terms that are *mechanical*.

2.4.2 STIMULUS AND SENSATION

A second critical development was the distinction between stimulus and sensation. Before Locke distinguished primary from secondary qualities (as discussed below), Galileo (1564–1642) had distinguished the physical properties of stimuli in the world from our sensory impressions of these stimuli. To many Greek thinkers, such as Empedocles in the fifth century BCE, perception was valid and veridical because, they believed, objects emit *eidola* or effluences, which are copies of the objects and which enter into our sense organs to create copies in our perception. To Galileo, however, and more in the spirit of Democritus, the realms of physics and sense perception were distinct, our sensory experience not a direct copy of the external world: "I do not believe that for exciting in us tastes, odors, and sounds there are required in external bodies anything but sizes, shapes, numbers, and slow or rapid movements; and I think that if ears, tongues, and noses were taken away, shapes, numbers, and motions would remain, but not odors or tastes or sounds" (Galileo 1623/1960, 311). Galileo himself discovered the relation between the subjective, sensory experience of pitch and the vibrations of a tone—a psychophysical relationship. If physics is to concern itself with shapes, numbers, and movements, then it followed that other disciplines could study the associated sensations.

2.4.3 HEDONISM AND EMPIRICISM

The third development was the return of hedonistic and empiricist theories of mind, invigorated, respectively, by Hobbes and Locke. Thomas Hobbes (1588–1679), a contemporary of Galileo, embedded his materialistic and mechanistic conception of the mind within a similarly material and mechanical body: Mind in general and sensation in particular, according to Hobbes (1650, 1651), represent physical processes in the brain—essentially motions in nerves. Just as importantly, he argued a hedonistic principle: the motivational and emotional roles of appetite or desire, the seeking of pleasure, and aversion, or the avoidance of pain. Motions in the nerves, which correspond to the sensations, can strengthen or weaken motions around the heart, which constitute the appetites and aversions: "And consequently all Appetite, Desire, and Love, is accompanied with some Delight more or lesse; and all Hatred, and Aversion, with more or lesse Displeasure and Offence" (Hobbes 1651, 25).

Four decades later, John Locke (1632–1704) would place sensations at the center of his philosophy of mind, in which, he asserted, "all of our ideas come from sensation or reflection"—reflection itself operating on ideas previously based in sensation (Locke 1690). Locke's distinction between primary and secondary qualities was a near cousin to Galileo's distinction between the physical properties of the world and the sensory properties of our experience, but differed from Galileo's in its terminology. Somewhat confusingly from a modern perspective, Locke followed a tradition that used the term "quality" to refer explicitly to characteristics of the physical world, not to attributes or features of sensations. To Locke, both primary and secondary qualities are *physical*. Both are properties of objects and events.

Primary and secondary qualities differ in that each kind of quality gives rise, according to Locke, to a different kind of perception: Primary qualities are perceived (more or less) as they "really" are—our perceptions of primary qualities, such as the shapes, sizes, textures, and numbers of objects, resemble the primary qualities that cause the sensations. In modern parlance, primary qualities refer to macroscopic features of objects. Secondary qualities, by contrast, are not perceived as they "really" are. Secondary qualities, such as the wavelength of light and the chemical structure of saccharides, give rise to sensations or perceptions of color and taste, perceptions that do not resemble the stimuli that cause the sensations. Perceptions of secondary qualities are caused by what are now called microscopic features, such as wavelength, molecular size, and molecular shape.

The flowering crabapple tree to the side of my house is considerably larger than the dogwood nearby, and the crabapple also looks larger. In Locke's terms, I perceive the primary quality of the size of the trees (more or less) accurately. But I don't perceive the colors of the trees' blossoms to differ in size, even though the crabapple's blossoms reflect much longer wavelengths of light. Instead, I see the crabapple's blossoms as deep red, the dogwood's as blue-white. In Locke's terms, I do not perceive the secondary quality of wavelength accurately. Perceptions of the qualities of objects are experiential; they are contents of mind. So, too, are pleasures and pains. According to Locke, the study of the mind's contents starts with its sensations, hence with the perceptions of primary and secondary qualities, including pleasures or pains, whose importance Locke emphasized.

2.5   A NEW HEDONISM: UTILITARIANISM

Philosophers of the eighteenth century—notably, Berkeley, Hume, Hartley, and Condillac—elaborated and extended the empirical, sensation-based theory of mind that Locke had promulgated, but by the nineteenth century, philosophers would also capitalize on a new version of hedonism. This new hedonism arrived in the utilitarian doctrine of Jeremy Bentham (1748–1842). Published in 1789, Bentham's *Introduction to the Principles of Morals and Legislation* argued that pleasure, happiness, or what Bentham called *utility* is and should be the goal of life (and society, through government). Bentham's own goals were largely normative—the utilitarian calculus makes it possible, Bentham believed, to quantify the overall utility of any course of action, equivalent to the overall resulting pleasure or good that is brought about, and therefore the calculus makes it possible to choose that course of action that will lead to "the greatest good for the greatest number."

Within this framework, Bentham aimed to consider everything that might contribute to utility, including sensory pleasures. Chapter 5 of his *Principles of Morals and Legislation* treated pleasures and pains, identifying sensory pleasure as one of the 14 classes that he called simple pleasures. Bentham then enumerated nine kinds of sensory pleasure: These included the sensory pleasures of the Aristotelian quintet of modalities—taste, smell, touch, hearing, and vision*—as well as pleasures of intoxication, of the sexual sense, of health, and, lastly, "of novelty: or, the pleasures derived from the gratification of the appetite of curiosity, by the application of new objects to any of the senses" (Bentham 1789, 31). This last of Bentham's sensory pleasures almost seems prescient, given the role that novelty and curiosity would come to play in the mid-twentieth century, in reward-based theories of associative learning.

Bentham's hedonistic utilitarianism followed the tradition of Democritus and Aristippus in supposing that all pleasures (utilities) are commensurable. That is, Bentham argued that all pleasures are qualitatively alike in being pleasant, with pleasures or utilities differing from one another only in their magnitude. In this spirit, the pleasure of scratching an itch would be qualitatively no different from the pleasure of sipping a fine Burgundy; as pleasures they would differ only in the extent that a person may find the one more pleasurable than the other.

Neither hedonism in general nor utilitarianism in particular need rely on the uniformity principle of Democritus, Aristippus, and Bentham, and, over the years, the principle has had its detractors—one being Plato and another being Bentham's godson, John Stuart Mill (1806–1873). Mill argued, contra Bentham, that if we *perceive* pleasures to be qualitatively different, then they must *be* qualitatively different: "Neither pains nor pleasures are homogenous, and pain is always heterogeneous with pleasure. What is there to decide whether a particular pleasure is worth purchasing at the cost of a particular pain, except the feelings and judgment of the experienced? When, therefore, those feelings and judgment declare the pleasures derived from the higher faculties to be preferable *in kind*, apart from the question of intensity, to those of which the animal nature, disjoined from the higher faculties, is susceptible, they are entitled on this subject to the same regard" (Mill 1863, 16). It is noteworthy that whereas Plato distinguished between lower and higher pleasures in an ethical sense, Mill argued (stimulated perhaps by his own ethical sensibility) that the distinction is one of perception and judgment, and so, by implication, one that is amenable to scientific inquiry.

2.6   EMPIRICISM, ASSOCIATIONISM, UTILITARIANISM

Sensation-based, empirical theories of mind were elaborated in the eighteenth century by David Hume, David Hartley, and Étienne Bonnot de Condillac and in the nineteenth century by James Mill and Alexander Bain, among many others. These theories were explicitly associationistic—they asserted that the organization of knowledge comes from the associations of sensations with

---

* Aristotle enumerated the special senses as five, even though he considered taste to be a species of (derivative from) touch (e.g., *De Sensu*, 439a: Ross 1906, 53).

other sensations, of sensations with ideas (themselves derived from sensations), or of ideas with ideas. Although Locke is widely acknowledged as the source of the principle of mental association and used the notion implicitly, his explicit contribution to associationism consisted of just one small chapter on the association of ideas, which he added to the fourth and last edition of his *Essay Concerning Human Understanding* (Locke 1700). In that chapter, Locke cautioned that, rather than being a source of knowledge, associations may distort or deceive.

Associationism hardly began with Locke; it can be found in Aristotle's writings on *Memory and Recollection* (*De Memoria*: Ross 1906)—although Locke apparently did not know Aristotle's work. In any case, the association of ideas, or mental connectionism, is pertinent to the present chapter because association serves as a historical antecedent for the stimulus-response (S-R), behavioral connectionism implicit in the law of effect: According to Thorndike's law of effect, a rewarding stimulus serves to strengthen the associative bond between the stimulus situation and the response leading to the reward. Thorndike's law conveniently marks the historical transition from mental associationism to behavioral associationism.

Several of the mental associationists of the eighteenth century, including Hume, Hartley, and Condillac, addressed the role of pleasure. de Condillac (1754) in particular sought to incorporate pleasantness and unpleasantness explicitly within an associationistic framework. But one of the most elegant statements on sensation and pleasure appeared a century later, in the popular textbook on *Mental Science*, written by the Scottish psychologist Alexander Bain (1818–1903): "The sensation of Warmth, on emerging from cold, is one of the greatest of physical enjoyments. It may be acute, as in drinking warm liquid, or massive, as in the bath, or other warm surrounding. Of passive physical pleasure, it is perhaps the typical form; other modes may be, and constantly are, illustrated by comparison with it; as are also the genial passive emotions – love, beauty, & c" (Bain 1868, 34).

Bain advocated both empirical and hedonic theories of mind: the empiricist notion that mind may effectively be analyzed into its components, at the heart of which are sensations, and a variant of the old hedonic notion that people try to maximize pleasure and minimize pain: "[T]he fundamental distinction of Pleasure and Pain, [involves] the sum of all human interests, the ends of all pursuit" (Bain 1868, 217). And, of course, many pleasures and pains are quintessentially sensory, for example, the pleasures of warmth sensations to which Bain alluded.

A follower of Bentham's utilitarianism, Bain elaborated the view that pleasure and pain serve as "the end of all pursuits." There is little new in this statement per se, which more or less follows the psychological (in contrast to the ethical) doctrines of hedonism from the time of Aristotle. But Bain went further, describing the role of pleasure (and pain) in learning, using terms similar to those that Thorndike would use less than half a century later: "We suppose movements spontaneously begun, and accidentally causing pleasure; we then assume that with the pleasure there will be an increase of vital energy, in which increase the fortunate movements will share, and thereby increase the pleasure.... A few repetitions of the fortuitous occurrence of pleasure and a certain movement, will lead to the forging of an acquired connection, under the law of Retentiveness or Contiguity [association], so that, at an after time, the pleasure or its idea shall evoke the proper movement at once" (Bain 1865, 310–11).

Similar views on what we now call associative learning were laid out by Herbert Spencer (1829–1903), notably, in the second edition to his *Principles of Psychology* (Spencer 1870; the discussion does not appear in the first edition of 1855). Spencer's explicitly adaptive, evolutionary account contains all the major components of Bain's: "In other words, those races of beings only can have survived in which, on the average, the agreeable or desired feelings [of pleasure] went along with activities conducive to the maintenance of life, while disagreeable and habitually-avoided feelings [of pain] went along with activities directly or indirectly destructive of life; and there must ever have been, other things being equal, the most numerous and long-continued survivals among races in which these adjustments of feelings to actions were the best, tending ever to bring about perfect adjustment" (Spencer 1870, 280). In brief, Spencer proposed that reward-based [pleasure-based] learning arose as an evolutionary adaptation. At the end of the nineteenth century, James Mark

Baldwin (1894) used the theories of Spencer and Bain to underpin his own version of pleasure-based learning (for an extensive review, see Cason 1932).

Just a few years later, Thorndike (1898) would first report his own, empirical studies of trial-and-error learning in animals, and in that report, he alluded to the law of effect, without yet naming it (this he would do in 1911). It is perhaps ironic, as Cason (1932) points out, that Thorndike did not cite (and perhaps did not know) the pleasure-based theories of learning of Bain, Spencer, and Baldwin, even though Thorndike's own formulation in 1898 was explicitly hedonistic: in that early formulation, it was pleasure that explicitly "stamped in" the learned responses (as discussed further in the chapters by White and by Balleine in this volume).

2.7   SENSATION AND BIOLOGICAL REGULATION: PHYSIOLOGY AND PSYCHOPHYSICS

Along with the threads of philosophical speculation in the nineteenth century about sensation, association, and pleasure, experimental studies of physiology, especially sensory physiology, led during that same period to remarkable scientific advances. The present summary largely follows Boring (1950). Early in the nineteenth century, anatomical and physiological studies revealed the distinction between sensory and motor nerves (Charles Bell, François Magendie), implying that conduction need not go in both directions through the same nerves. Slowly, the organizational structure of the brain became increasingly clear, including the organization of the sensory and motor regions of the cerebral cortex (Gustav Fritsch, Eduard Hitzig, David Ferrier). By the end of the century, the development of techniques to stain features of neurons (Ramón y Cajal, Camillo Golgi) helped define the microanatomy of the nervous system, leading to Cajal's discovery of synapses and thus the discontinuity of the nervous system. At the same time, the electrical nature of nerve propagation and its unidirectional transmission became better understood, as nerve activity was no longer conceived to consist of mechanical vibrations, as many earlier scientists and philosophers, including Newton and Hartley, had believed. By the early twentieth century, Lord Adrian would establish the all-or-none property of neural discharge.

2.7.1   THE DOCTRINE OF SPECIFIC NERVE ENERGIES

In synthesizing what was known of physiology just before the middle of the nineteenth century, the physiologist Johannes Müller (1801–1858) proposed the doctrine of "specific energies of nerve", which provided a neurophysiological underpinning to the empiricist theory of mind (Müller 1840). According to the empiricist theory, the mind consists of representations of the world, derived from sensations, but not necessarily *simulacra* of the world. In Locke's view, for instance, sensations of secondary qualities, such as colors, do not resemble their physical causes in the world, in this case, the wavelengths of light that produce the colors. Bishop George Berkeley (1710) would later point out that there is no basis for assuming that the perceptions of primary qualities resemble their physical causes either. By this token, the mind knows only itself and not the external world. Müller's doctrine proposes a way to describe how the brain mediates between the external world and the mental (sensory) representations of the world that arise from our interactions with it.

In the tradition of Locke and Müller, sensations represent only the mind's "acquaintance" with the nerves in the brain, and not with the external causes of sensation. The sensory experience in each modality—vision, hearing, touch, taste, and smell, again to use the Aristotelian quintet—is qualitatively different from the experience in any other modality because the experience in each modality depends on the properties of the sense-modality-specific nerves in the brain, not because of differences in external stimulation. Activity in visual nerves produces sensations of sight, regardless of whether the nerves are activated by light reflected from objects or from a mechanical blow to the head. Later in the nineteenth century, one of Müller's students, Hermann von Helmholtz, would extend the doctrine of specific nerve energies to account for different qualities within each modality—to different colors in vision, to different pitches in hearing, and so forth. By implication, sensory pleasures, too, are functions of the nerves in the brain.

2.7.2 PSYCHOLOGICAL AND PSYCHOPHYSICAL ANALYSIS OF SENSATION

In the nineteenth century, as in the twentieth and twenty-first centuries, many of the advances in understanding sensation came from psychophysical investigations, often carried out by sensory physiologists. Franciscus Donders and Helmholtz made important early measures of behavioral reaction times to sensory stimuli. Ernst Heinrich Weber established the basic properties of sensory discrimination, both intensity discrimination, for which he is best known, and spatial discrimination (two-point threshold), Ewald Hering and Helmholtz proposed physiological theories of color vision based on psychophysical evidence, and Hendrik Zwaardemaker constructed the first olfactometer to measure the basic properties of olfactory sensation and perception.

Experimental psychology arose formally in the second half of the nineteenth century, and psychophysical studies of sensation played a prominent role in this new discipline, in large measure because of the widely held view, empiricism, that the contents of the mind derive from sensations. Many of the new experimental psychologists recognized the significance of affective characteristics of experience, such as pleasantness. Still unclear, however, was how to relate pleasantness to other sensory qualities. Oswald Külpe (1893), for example, rejected the notion that pleasantness is another attribute of sensations, along with quality, intensity, and duration. According to Külpe, feelings are parallel to sensations; feelings may be associated with sensations, and when they are, they have their own attributes, including duration, intensity, and qualities such as pleasantness.

Wilhelm Wundt (1832–1920), by tradition the founder of experimental psychology, offered a general formulation to describe how the pleasantness of a sensory stimulus varies with the intensity of the stimulus: At very low levels, the sensory effect of stimulation is affectively neutral, but becomes increasingly pleasant with increasing stimulus level until pleasantness reaches a maximum, after which a further increase in stimulus level reduces pleasantness, eventually crossing through neutrality into unpleasantness (Wundt 1874). Unlike perceived intensity, which generally increases monotonically with increasing stimulus level, this hedonic function relating pleasantness-unpleasantness to stimulus intensity is distinctly non-monotonic. The inverted U-shaped function for hedonics has been called the *Wundt curve* (Berlyne 1971), and it can often describe not only affective judgments given by humans (e.g., Ekman and Åkeson 1965), but also behavioral responses of non-humans, such as rats (e.g., Young and Greene 1953). The ubiquity of the Wundt curve suggests that it may play a fundamental role in affective processes—an empirical counterpart to the old adage recommending "everything in moderation." To be sure, not every stimulus is pleasant even at very low levels, but among those that are, the Wundt curve implies that there is an intensity level at which pleasantness is maximal.

Pleasantness or unpleasantness, according to Wundt, is only one aspect of affect. After developing several mutually inconsistent proposals, he eventually settled on a three-dimensional theory of feeling (Wundt 1896). According to Wundt's tridimensional theory, all feelings, including those associated with sensory experiences, may be characterized by quantitative values on each of three bipolar dimensions: pleasantness-unpleasantness, strain-relaxation, and excitement-calm. The tridimensional theory continues to receive attention, largely as a result of a line of research on connotative meanings that began with Charles Osgood and colleagues (e.g., Osgood , Suci, and Tannenbaum 1957), showing that Wundt's tridimensional scheme applies across cultures and languages, to words and concepts as well as to percepts.

2.7.3 A BACK DOOR TO PLEASURE

While the new psychologists of the nineteenth century considered how to characterize pleasure within the framework of mental systems, other conceptualizations of pleasure emerged from a physiology steeped in physical science. Hermann von Helmholtz (1821–1894) famously put into mathematical form the principle of conservation of energy, formulated in 1842 by Robert Mayer. The principle states that energy can neither be created nor destroyed, meaning that the total amount of energy in the universe, or a closed system, remains constant. Consequently, as a person (or any other organism) exchanges energy with her or his environment, the conservation principle requires that the outflow of energy equals the inflow of energy, if the person is to maintain their state of energy balance. Applying it to biological systems, the principle of energy conservation indirectly—through a back door, so to speak—influenced theorizing about the physical sources of pleasure, and came to play a singular role in the history of sensation vis-à-vis reward.

Helmholtz was a thoroughgoing materialist, believing that biological systems can be fully explicated in terms of Newtonian mechanisms (unlike Helmholtz's teacher, Johannes Müller, who remained an avowed vitalist, believing that life required the presence of a vital force or *vis viva*, that is absent in non-living matter). Indeed, Helmholtz, together with other former students of Müller, helped found the Berliner Physikalische Gesellschaft (Berlin Physical Society), a group of physiologists dedicated to the proposition, as Emil DuBois-Reymond wrote in 1842 that, "No other forces than the common physical chemical ones are active within the organism. In those cases which cannot at the time be explained by these forces one has either to find the specific way or form of their action by means of physical mathematical method, or to assume new forces equal in dignity to the chemical physical forces inherent in matter, reducible to the force of attraction and repulsion" (quoted by Bernfeld 1944, 348).

Helmholtz read his paper on the conservation of energy to the Berlin Physical Society in 1847, and the conservation principle was clearly a linchpin of the society's reductionistic agenda. Another member of the society, also a former student of Müller, was Ernst Brücke, who later became director of the Physiological Institute in Vienna. In that position, Brücke would serve as an early mentor to a young medical student, Sigmund Freud. Bernfeld (1944) argued that Freud's psychoanalytic theory inherited, through Brücke, the reductionistic physiology of what Bernfeld called the "Helmholtz School." To be sure, Freud often fell back on the language of a physiology grounded in physical science, notably in an early work, the *Project for a Scientific Psychology*, written in 1895 but not published until more than a decade after Freud's death (Freud 1950/1966). It is noteworthy, however, that in all of Freud's writings, he mentioned Helmholtz only in broad, general ways. Although the principle of conservation of energy underlay Freud's psychoanalytic theory, it is plausible that a bridge between Helmholtz's materialistic physiology and the materialistic language of Freud's psychoanalysis came from the writings of the founder of psychophysics, Gustav Fechner (1801–1887) (see Marks 1992).

2.7.3.1 Fechner's Constancy Principle

Fechner's contributions to sensory psychophysics are well known: Fechner (1860) brought scientific attention to Weber's systematic studies of sensory discrimination and to the general principle, now known as Weber's law, that the just-noticeable-change in stimulus intensity, the discrimination threshold, is often (roughly) proportional to the intensity level from which the change is made. Fechner reported that it was only some time after he conceived his logarithmic law of sensation magnitude that he discovered Weber's work and recognized that the logarithmic law of sensation magnitude could be readily derived from Weber's law of sensory discrimination. Fechner even named the law of sensation magnitude after Weber, although tradition chose to name it after Fechner.

Less widely recognized are Fechner's (1873) provocative speculations on pleasure, which appear in his volume treating the *Origin and Evolutionary History of Organisms*. Fechner conceived of psychophysics more broadly than just the study of relations between stimulus and sensation. To Fechner, psychophysics referred to all the correlations between mind and the body, between the mental and the physical aspects of the world, especially the correlations between events in the mind and physical events in the brain; Fechner called the domain of these correlations *inner psychophysics* (he delegated the correlations between mind and external stimuli to the domain of *outer psychophysics*). Within this overarching framework, Fechner developed a theory of psychobiological function that is essentially "homeostatic"—"homeostatic" in quotes because the term did not yet exist when Fechner wrote. Although Walter Cannon (1926, cited in Cannon 1929) coined the term and then popularized it (Cannon 1932), the notion of biological regulation, which underlies homeostasis, was already clear in the work of Claude Bernard (1865; 1878–1879). Fechner based his own theory on a *principle of inertia* or *constancy principle*, perhaps borrowing the term *constancy* from Bernard. In the early formulation of psychoanalytic theory, Freud identified Fechner's constancy principle with his own pleasure principle.

Fechner's homeostatic theory started with physics (he was professor of physics at Leipzig University) and, more explicitly, with the Mayer-Helmholtz principle of conservation of energy. As a universal physical principle, conservation of energy applies to organisms as well as to inanimate matter. Thus, organisms cannot simply cause energy to disappear: If an organism takes in an excess of energy, the physical (biological) system becomes unstable, according to Fechner, and the excess energy must be dissipated in order to return the organism to energy balance, or stability: this is a homeostatic principle. All these exchanges of energy take place on the physical side of Fechner's psychophysical equation, and all of them have corresponding experiences on the mental side. To the presence of an excess of energy, with its resulting instability, corresponds the experience of displeasure. To the dissipation of energy, with its consequent return to balance and stability, corresponds the experience of pleasure.

2.7.3.2 Freud's Pleasure Principle

The correspondence between Fechner's constancy principle and the tenets of Freud's early psychoanalytic theory is striking. In *Beyond the Pleasure Principle*, Freud (1920/1955) acknowledged and lauded Fechner's anticipation of Freud's own work:

> We cannot, however, remain indifferent to the discovery that an investigator of such penetration as G.T. Fechner held a view on the subject of pleasure and unpleasure which coincides in all essentials with the one that has been forced upon us by psycho-analytic work. Fechner's statement ... reads as follows: 'In so far as conscious impulses always have some relation to pleasure or unpleasure, pleasure and unpleasure too can be regarded as having a psycho-physical relation to conditions of stability and instability.... According to this hypothesis, every psycho-physical motion rising above the threshold of consciousness is attended by pleasure in proportion as, beyond a certain limit, it approximates to complete stability, and is attended by unpleasure in proportion as, beyond a certain limit, it deviates from complete stability.' (Freud 1920/1955, 7–8)

A quarter century earlier, in his still unpublished *Project for a Scientific Psychology* of 1895, Freud had already proposed the existence of a mechanism for minimizing physical (neural, and hence also mental) energy or tension. This was his *principle of neuronal inertia*, the principle that "neurons tend to divest themselves of Q [quantity]" (Freud 1950/1966, 296). The physical-physiological principle of neuronal inertia and, later, the psychological pleasure principle were central to Freud's *economic view* of mental life: "The course taken by mental events is automatically regulated by the pleasure principle" (Freud 1920/1955, 7). Given the presence of an unpleasurable tension, reduction of that tension corresponds to "an avoidance of unpleasure or a production of pleasure" (ibid., 7). More specifically, Freud went on, "We have decided to relate pleasure and unpleasure to the quantity of excitation that is present in the mind but is not in any way 'bound'; and to relate them in such a manner that unpleasure (*Unlust*) corresponds to an *increase* in the quantity of excitation and pleasure (*Lust*) to a *diminution*...." Freud's theory, like many earlier ones including Fechner's, implies that pleasure consists of a release from pain.* The theory bears much in common with later

---

* For the sake of historical accuracy, it should be noted that Freud would subsequently identify Fechner's notion of stability with his own death drive (*Beyond the Pleasure Principle*: Freud 1920/1955), both of which represent a state of quiescence. Then, recognizing that some pleasures may involve increases in tension or energy rather than decreases, Freud once again revised his theory of the pleasure principle, in the end no longer identifying it with either the death drive or the principle of stability.

ones too—in particular, with the drive-reduction and drive-stimulus reduction theories of reward-based, S-R learning that would emerge under the rubric of behaviorism, as Clark Hull and his disciples and successors would try to understand the mechanistic basis to reward.

Finally, it is worth noting the close connection between the pleasure principle and the Freudian concept of *Trieb*, a term that is traditionally translated as instinct but is better designated as *drive* (see Bettelheim 1982). Drives reflect biological energy. A *Trieb*, or drive, arises biologically, from a bodily need, and "appears to us as a concept on the frontier between the mental and the somatic, as the psychical representative [psychischer Repräsentant] of the stimuli originating from within the organism and reaching into the mind" (Freud 1915/1967, 121–22). Not surprisingly, given the close correspondence between Freud's early formulation of the pleasure principle and Fechner's conservation principle, *Triebe* in Freud's theory, like utilities in Bentham's, differ from one another only with regard to quantity and not quality, as drives arise from physical processes within the body: "We do not know whether this process is invariably of a chemical nature or whether it may also correspond to the release of other, e.g., mechanical forces…. The [drives] are all qualitatively alike and owe the effect they make only to the amount of excitation they carry" (Freud 1915/1957, 123).

A decade later, John Paul Nafe (1924), working in the tradition of introspective experimentation, investigated the affective characteristics of sensory stimuli. Nafe asked his subjects to report their conscious experiences in response to stimuli presented to several modalities, including experiences of colors, musical chords, food flavors, scents, and tactually presented objects. The perceptual experiences often had affective qualities, and when they did, the qualities were virtually always pleasant or unpleasant. The experiences of pleasantness and unpleasantness, according to Nafe, invariably consisted of patterns of bright and dull pressure, respectively. As both Bentham and Freud might have predicted, neither pleasantness nor unpleasantness appeared to vary in quality, but could vary in intensity.

2.8   SENSATION, REWARD, AND MOTIVATION

Encompassing Fechner's constancy principle, Freud's pleasure principle, and Thorndike's law of effect is the central concept of motivation. Psychoanalytic theory is, of course, all about motivations, especially unconscious ones, and reward-based instrumental learning presumably would not occur without motivation. The word *motivation* appears in the title of Troland's (1928) important analysis of the topic (according to Herrnstein (1998), the first book to use "motivation" in its title).

2.8.1   SENSATION AND MOTIVATION

In his *Fundamentals of Human Motivation* (1928), the brilliant and eclectic Leonard Troland (1889–1932) offered another threefold classification, this one distinguishing sensory stimuli as *beneceptive* (those that are biologically helpful to the organism), *nociceptive* (those that are physiologically harmful), and *neutroceptive* (those that are physiologically neutral). The corresponding receptors he named *beneceptors*, *nociceptors*, and *neutroceptors*. Beneceptors include sweet-taste receptors, nociceptors include pain receptors, and neutroceptors include receptors of vision and hearing. This new terminology went hand in glow with Troland's attempt to define the receptor systems and their processes in functional rather than mental terms—although it is hard to ignore the connection between beneception and nociception, on the one hand, and pleasure and pain, on the other. Troland then went further, arguing in support of a principle that he called (with another new term) *retroflex action*, action that depends on the combined activity of beneceptors and nociceptors (a physiological cousin, perhaps, to Bentham's calculus of utility). In his conception of retroflex action, Troland incorporated both innate reflex acts and learned behaviors.

In Troland's system, beneception and nociception underpin reward-based and punishment-based learning (law of effect). Previously neutral stimuli are able to take on beneceptive or nociceptive capacities by being paired with primary beneception or nociception—this happening through a process of Pavlovian conditioning (see below). The broad outline of Troland's theory resembles the reward-based learning theory later developed by Clark Hull (1943).

As different as the theories are, both Freud and Troland grounded their theories—of pleasure and beneception, respectively—in what Cannon (1929) called homeostasis. Homeothermic organisms, for example, maintain their body temperature within narrow ranges, and Cannon described several physiological mechanisms, including sweating, shivering, and vascular constriction and relaxation, all of which help an organism maintain its body at a constant internal temperature. Because Cannon was concerned with automatic physiological mechanisms of regulation, however, he did not mention the central role of behavior in biological regulation.

Humans and non-humans regulate their body temperature through behavioral as well as physiological processes. Although Cannon conceived of homeostasis in terms of physiological mechanisms operating automatically, it is both sensible and appropriate to follow Curt Richter (e.g., 1947, 1954) and apply homeostasis to behavioral as well as physiological regulation. In humans, the behavioral regulation of internal body temperature includes both activities that operate over short periods of time, such as donning a coat on a cold day or moving from sun to shade on a hot one, and activities that operate over the long haul, such as building permanent shelters. Long-term adaptations capitalize on high-level cognitive processes of memory, planning, estimation, and calculation, and these can take place in the absence of immediate sources of pleasure or pain. Short-term adaptations, by contrast, tend to come about when peripheral or central signals provide the motivation (drive) to change one's microclimate, the changes in temperature serving as rewards, often accompanied by pleasurable thermal sensations.

2.8.2 SENSATION AND AFFECT

Whereas Troland's scheme for classifying sensations imputed a direct affective component to beneception and nociception, Paul Thomas Young (1892–1978) sought to distinguish more sharply between sensory qualities and affective qualities, between the informational and the hedonic. To Young (1961, 154), "Affective processes convey little or no information," by which he meant that affective dimensions of sensory response, such as the pleasantness of sugar, are distinct from non-affective dimensions, such as the intensity of the taste sensation. In what he called his "attempt to escape the limitations of a purely introspective study of the affective processes" (Young 1949, 98), Young aimed largely at understanding affective mechanisms of taste in rats, using behavioral measures of intake and choice (preference) (summarized in Young 1948, 1949, 1952). Despite his focus on objective measures, he had no qualms about inferring hedonic events taking place inside the animals' head: "To put the matter bluntly: Our work leads to the view that rats accept foods which they like (find enjoyable) and that foods differ in the degree to which they arouse immediate enjoyment" (Young 1949, 103).

Central to Young's theoretical stance were two main principles. On the one hand, Young endorsed the view that pleasantness affects the organization of behavior and performance, hence motivation. On the other, he was skeptical that pleasantness plays a role in reward or reinforcement per se. Defining learning as the modification of behavior and the underlying neural substrate, Young argued that affect was not necessary for learning to occur, but only for the organization of (learned) behaviors. As a result, he dismissed as useless the notion that learning itself involves processes of organization—a view likely to strike some as a semantic quibble.

2.9 SENSATION, REWARD, AND THE LAW OF EFFECT

In the early decades of the twentieth century, academic psychology evolved rapidly, especially in the United States. Functionalism begat behaviorism, which inherited from its parent an abiding interest in the ways that organisms adapt to their environment. This interest, in turn, impelled many behaviorists, including Clark Hull (1884–1932) and Burrhus F. Skinner (1904–1990) to study learning and the processes by which organisms come to respond adaptively to the world around them. Thirsty animals seek water, hungry ones seek food, in both cases learning to identify the environmental conditions that satisfy their biological needs. Laboratory experiments often simplify the conditions: A thirsty rat in a maze may explore until it stumbles on the path to water; a hungry rat in an operant chamber may sniff and paw until it depresses a lever that initiates the delivery of food; these are the kinds of behavioral activities used by Hull (1943) and Skinner (1938), respectively, as models for understanding reward-based learning (although Skinner denied that what he studied should be called learning). The behaviors are adaptive; animals need to learn where and how to find water and food. Skinner (1953, 90) argued that there is an analogy between the way natural selection acts on a species and the way reward acts on an individual: "Reflexes and other innate patterns of behavior evolve because they increase the chances of survival of the *species*. Operants [trial-and-error behaviors] grow strong because they are followed by important consequences in the life of the *individual*."

2.9.1 INSTRUMENTAL LEARNING AND PAVLOVIAN CONDITIONING: REWARDS AND REINFORCEMENTS

By the second decade of the twentieth century, there appeared to be two classes of learning, perhaps with separate sets of underlying mechanisms. One class was the trial-and-error learning that Thorndike had studied in animals. This kind of learning would come to be called *instrumental* (Hilgard and Marquis 1940), in that the organism's own behavior is instrumental in bringing about the reward, e.g., running a maze to its end, where water is located, or pressing a lever to obtain food. Skinner (1938) used the term *operant* to refer to behaviors that "operate on" the environment. Contrasting with instrumental learning and operant behavior was the kind of learning that came to be called *classical conditioning* (Hilgard and Marquis 1940). Classical conditioning, or Pavlovian conditioning, grew out of the work of Ivan Petrovich Pavlov (1849–1936). Pavlov's (1904/1967; see Pavlov 1927) research on conditioning was introduced to scientists in the United States and Europe by Yerkes and Morgulis (1909) and to a wide audience by John Watson (1916) through his presidential address to the American Psychological Association in December 1915.

In Pavlovian conditioning, an initially neutral stimulus, a conditioned stimulus (CS), is presented in close temporal proximity to an unconditioned stimulus (US), which is capable of eliciting some kind of automatic, unconditioned response (UR). An example is the pairing of a neutral light or buzzer (CS) with food placed in the mouth (US). According to a traditional interpretation of Pavlovian conditioning, pairing the CS with the US causes the CS to take on the properties of the US, so the CS can produce a conditioned response (CR) that resembles the UR. If placing food in the mouth increases salivation, then, after pairing a light with food, presenting the light alone comes to elicit saliva.

Modern interpretations of Pavlovian learning, discussed later, differ from this traditional interpretation. Nevertheless, it is important to keep in mind that throughout the first half of the twentieth century, Pavlovian conditioning was often contrasted with instrumental learning, with which it was seen to differ in important ways. This contrast often focused on the relation between the behavioral response and the reward or reinforcing stimulus. In instrumental learning, the reward is contingent on the response; rewards occur only after "correct responses," and learning takes place when a rewarding stimulus follows the response. In Skinner's terms, rewards select the behaviors that precede them. Pavlovian conditioning, by contrast, lacks this contingency between stimulus and response: Unconditioned stimuli, by traditional accounts, produce their responses automatically and non-contingently. In Pavlovian conditioning, the US was often called the reinforcement, because its presence was seen to reinforce or strengthen a connection between the CS and the CR. Skinner (1938), for one, distinguished sharply between operant (emitted) behaviors and respondent (elicited) behaviors—although other prominent behaviorists, notably Hull (1943), would follow Watson's lead in trying to adopt Pavlovian conditioning as a coherent, unifying principle for instrumental learning as well.

2.9.2   THE LAW OF EFFECT: REWARD AND REINFORCEMENT

The starting point for the modern scientific study of reward is Thorndike's revised law of effect—that rewards, or satisfiers, "stamp in" or strengthen the bond between the response that leads to the reward and the stimulus setting in which the response is made. But Thorndike's formulation posed a host of problems. Four are noted here. First of all, the formulation suggested mental causation: that satisfaction or pleasure acts to strengthen the S-R bonds. To many functionalists, behavioral research on animals made it possible to infer internal, mental events, the "fundamental utilities of consciousness" (Angell 1907, 85). Behaviorists, however, would have no part of this, eschewing explanations in mental terms, and often the very use of the terms. Second, there was ambiguity as to what was strengthened: the responses themselves, or S-R connections. Thorndike's (1911) definition of the law spoke directly of the strengthening of a bond, as did Hull's (1943); but Skinner's (1938) operant behaviorism implied changes in response probabilities, not in S-R connections. Third, Thorndike's formulation suggested "backward causation," with rewards acting back in time to affect previous S-R connections, a problem ameliorated by the assumption (made by Hull and others) that rewards act causally on currently active neural traces of previous stimuli and responses.

Fourth and lastly, a logical dilemma: How does one tell, independently or *a priori*, which stimuli will act as rewards? Thorndike suggested that satisfiers (rewards) are those stimuli that animals do not try to avoid and may seek out, but this attempt at an independent definition did not satisfy all his critics. One solution was simply to define rewards as those stimuli that strengthen S-R bonds or response probabilities. This solution tacitly acknowledges that the law of effect is circular: effects are produced by rewards, which in turn are defined by their effects (Postman 1947; but see Meehl (1950) for an alternative resolution). The upshot would be a restatement of the law of effect: if a response R is made in a given stimulus setting S and followed by certain other stimuli, then the S-R bond or probability of R will be strengthened. These response-strengthening stimuli continued often to be called rewards, but increasingly as *reinforcers*, thereby denoting the behavioral effect. One goal of behavioral learning theory became, therefore, to identify which stimuli under which conditions can act as reinforcers. To many behaviorists, especially those who adopted Skinner's theoretical stance, the law of effect assumed a distinctively descriptive cast, eschewing any attempt to determine the mechanisms responsible for the changes in instrumental behavior.

Over the next several decades, notwithstanding behaviorism's preeminence in academia, many academic researchers studying instrumental learning nevertheless continued to use the term reward, even in animal research (e.g., Cowles and Nissen 1937; Mowrer and Lamoreaux 1942; Kendler, Karasik, and Schrier 1954; Capaldi 1966). Others, however, used the term reinforcement, as early as the 1930s (e.g., Hull 1930; Skinner 1933). While it is obviously not possible to review all of the directions taken by reward- or reinforcement-based research on instrumental learning, the following section will review the significant line of behaviorist-inspired research initiated by Hull and his colleagues and students, as they tried to answer what was to many in the 1940s and 1950s the $64 question (by now, presumably inflated to $64,000): How do reinforcers reinforce? Equivalently, one may ask, how do rewards reward?

2.9.3   HOW DO REWARDS REWARD? REWARD AS BEHAVIORISTS SAW IT

Instrumental learning is adaptive. Rewards often, though not always, act through behavioral mechanisms to serve regulatory physiological processes. Thus, rewards can help maintain homeostasis. Just as Aristotle recognized that foods are most pleasant when we are hungry, food rewards or reinforcements are most effective in instrumental learning when organisms are deprived of food—behaviorism's "operational" definition of hunger.

After immersing himself in the writings of Isaac Newton, Clark Hull embarked on a mission to formalize learning theory within a mathematical system grounded in basic hypotheses and inferences drawn from these hypotheses. A central construct in Hull's system is habit strength, symbolized as sHR, which characterizes the strength of the connection between the stimulus S and the response R that led to reinforcement. The way that sHR increases with increasing numbers of rewarded trials reflects the operation of the law of effect: "Whenever an effector activity occurs in temporal contiguity with the afferent impulse, or the perseverative trace of such an impulse, resulting from the impact of a stimulus energy upon a receptor, and this conjunction is closely associated in time with the diminution in the receptor discharge characteristic of a need, there will result an increment to the tendency for that stimulus on subsequent occasions to evoke that reaction" (Hull 1943, 80).

Note Hull's qualification—that the effectiveness of a reinforcer (reward) comes through the "diminution in the receptor discharge characteristic of a need." Hull recognized a close, homeostatic connection between reward and the satisfaction of biological needs, but he also recognized that some rewards or reinforcements do not involve the reduction of a biological need; an example is the capacity of the non-nutritive sweetener saccharine to serve as a reward (e.g., Sheffield and Roby 1950). One possible explanation relies on the notion of secondary reinforcement. A stimulus that is repeatedly associated with a primary (need-reducing) reinforcer will take on the capacity to serve itself as a secondary reinforcer (Hull 1950). It is plausible that rewards that do not reduce needs act in a kind of *Ersatz* fashion as secondary reinforcers, through their history of Pavlovian association with primary reinforcers.

In the end, Hull (1952, 5) modified his implicit characterization of reward by changing "diminution in the receptor discharge characteristic of a need" to "diminution in the motivational stimulus." States of bodily need influence motivation, creating "drives" to satisfy those needs. Rewards may operate, then, by reducing drives rather than reducing needs—and if not by reducing primary drives, then by reducing secondary drives, created by Pavlovian conditioning of previously neutral stimuli to primary drives (or to other secondary drives). Need is gone, replaced by drive (Mowrer 1947; Wolpe 1950; Miller 1951a, 1951b)—or, perhaps more accurately, by the sensory stimulation that is associated with a drive, "the principle [being] that the prompt reduction in the strength of a strong drive stimulus acts as a reinforcement" (Dollard and Miller 1950, 40).

The attempt to specify the mechanism by which rewards reward—by satisfying needs, reducing drives, or reducing the stimuli associated with drives—expanded in scope in the 1950s with the research on curiosity, exploration, and manipulation, much of it conducted by behaviorists who couldn't, or wouldn't, look inside organisms. A central figure in this movement was Daniel Berlyne (1924–1976), who sought to place the concepts of curiosity and exploration into a neo-Hullian framework and, by doing so, contributed importantly to an expansion of the notion of reward. In his two-part theory, Berlyne (1950, 1955) postulated, first, that novel stimuli arouse a drive—or, more precisely, drive stimuli, which he identified with curiosity—and, second, that, when circumstances permit, these drive stimuli lead to exploration, which serves to reduce the curiosity drive (and the drive stimuli). Importantly, the opportunity to explore can subserve learning. To give two examples, Butler (1953) showed that rhesus monkeys could learn a color discrimination when the reward was the opportunity to explore and handle an apparatus, and Butler and Harlow (1957) showed that rhesus monkeys could learn a color discrimination when the reward was the opportunity to explore the environment visually.

Research on curiosity and exploration marked the end of an evolution in the conceptualization of reward: from reward as reduction in a biological need to reduction in a biological drive to reduction in internal stimuli aroused by a drive. In a sense, the drive-stimulus reduction theory of reward served as a behavioristic analog to the homeostatic constancy principle of Fechner and to the early formulation of the pleasure principle of Freud. Perhaps it is not surprising that Neal Miller (1961) was, at one time, a major proponent of the drive-stimulus reduction theory of reward, for Miller, a behaviorist who also trained in the Vienna Psychoanalytic Institute, had sought to assimilate Freudian theory to mid-twentieth century behaviorism (e.g., Dollard and Miller 1950).

At the risk of oversimplifying, we may conveniently divide into two groups the behaviorists concerned with the role of reward in learning. One group included those working in the tradition of Hull, many of whom sought to specify the mechanisms of reward or reinforcement, as just discussed. Another group included those working in the tradition of Skinner, many of whom seemed content with an empirical statement of the law of effect. A notable exception in this second group was David Premack (1925–), who sought to characterize reinforcers, or reinforcing stimuli, in terms of the probabilities of the behavioral responses associated with the stimuli: A food stimulus will reinforce other behaviors, according to Premack's (1959) formulation, when eating the food has a higher independent probability than the behavior that eating reinforces. Because a hungry rat will eat pellets of food more often than it will randomly press the lever in an operant chamber, presenting food pellets will reinforce lever pressing. If a monkey opens a window to look outside more frequently than the monkey pulls a chain, then, according to Premack, visual exploration will reinforce chain pulling. Presumably, one can use the independently determined rates of behaviors occurring in the presence of various stimuli in order to rank order the stimuli with regard to their reinforcing capacity. In fact, Premack's principle is reminiscent of Thorndike's attempt to define "satisfiers" as stimuli that organisms seek out. Although Premack's empirical, functional approach may have good predictive value, it leaves open the mechanistic questions as to why and how the relative probabilities of different behaviors determine their reinforcing capacity. Nevertheless, the findings reiterate and expand an important ontological point: that reinforcing capacity (like pleasure) is not absolute but is relative and contextually determined.

2.9.4 Reward-based Learning Redux

Several prominent theories of learning of the mid-twentieth century, notably those of Hull and his followers, aimed to bridge the seeming gap between Thorndikean learning and Pavlovian conditioning. In the paradigm of Thorndike and others, a particular response R emitted in a particular stimulus setting S is followed by a rewarding or reinforcing stimulus S*, which, depending on one's theoretical predilection, either increases the probability of R or increases the strength of a connection between S and R. In the traditional paradigm of Pavlovian conditioning, a previously neutral stimulus S, which becomes the CS, is presented together with, or just before, the reinforcing S*, or US. The US is chosen because it more or less automatically produces a UR. By pairing a CS with a US, the CS comes to elicit a CR that resembles the UR. To use a classic example from Pavlov, placing food in a dog's mouth (S*) elicits salivation (R*). After pairing a previously neutral stimulus, such as a light (S) with food in the mouth, presenting the light alone (CS) leads to salivation (CR).

From the late 1960s onward, however, researchers in learning came increasingly to interpret Pavlovian conditioning in a new way (see Rescorla 1988). The traditional view saw Pavlovian conditioning as the mechanical forging of new S-R connections, for example, between a flash of light and salivation. But studies by Rescorla (1968), Kamin (1969), and others cast doubt on this interpretation. In brief, these studies showed that mere contiguity between CS and US is not always sufficient to produce Pavlovian learning; what appears to be both necessary and sufficient is that there is a predictive relation: the CS must be able to predict the occurrence of the US. Over the last few decades, therefore, Pavlovian learning has come increasingly to be seen as involving the acquisition of information about the US, typically, about the prediction of rewards (e.g., Schultz, Dayan, and Montague 1997; Day and Carelli 2007; and see chapters in Part III of this volume). Within this framework, Pavlovian learning is interpreted as more of a "cognitive" process of predicting stimuli than an automatic acquisition or strengthening of either S-S or S-R connections. The appearance of a CS "means" the US is likely on its way. It is perhaps not surprising that this more cognitive interpretation of Pavlovian learning paralleled the growth of the cognitive sciences, including cognitive neuroscience.

Conditioned responses, by this interpretation, provide behavioral evidence that the organism has learned that a previously neutral stimulus can predict the occurrence of a US. Importantly, appetitive unconditioned stimuli (but, obviously, not aversive ones) can readily be described as rewards. In fact, as pointed out by one skeptic of the modern view (Bitterman 2006), Pavlov (1904/1967) himself recognized that conditioning is not simply a matter of contiguity, but also involves contingency.

What an organism—human or animal—learns is that the CS *predicts* the occurrence of the US. To a dog in Pavlov's laboratory, the onset of a light means that food is on its way.

This new interpretation of Pavlovian learning, as learning to predict rewards, points to the important role that Pavlovian processes may play in instrumental learning. As Day and Carelli (2007, 149) wrote, "In real life, organisms use environmental cues to update expectancies and allocate behavioral resources in a way that maximizes value and minimizes energy expenditure. Therefore, Pavlovian relationships may be embedded within virtually all situations involving operant behavior or instrumental learning. For example, general contextual stimuli (e.g., a place where rewards are consumed) may come to be explicitly associated with reward delivery and operate as conditioned stimuli." It is likely that most food odors, for example, are not "natural" rewards, but assume their rewarding capacity through Pavlovian association with food being consumed; that is, food odors not only come to predict the presence of food, but also may serve as conditioned reinforcers or rewards. Modern views of reward-based learning have come to incorporate Pavlovian as well as Thorndikean principles, and these views are prominent in modern research into brain mechanisms of reward.

2.10   HOW DO REWARDS REWARD? BRAIN MECHANISMS OF REWARD

In 1954, James Olds (1922–1976) and Peter Milner reported the results of what became a milestone in research on mechanisms of reward. Olds and Milner inserted electrodes into the brains of rats, then placed the rats in operant chambers equipped with a lever that, when depressed, would deliver current to the electrodes. Under these conditions, when an electrode was implanted in certain regions of the brain, notably the septal area, the rats would press the lever "to stimulate itself in these places frequently and regularly for long periods of time if permitted to do so" (Olds and Milner 1954, 426). Not only will animals work for food when they're hungry or for water when they're thirsty, but, even when sated, rats will work for electrical stimulation of their brains.

The seminal work of Olds and Milner (1954) unleashed a barrage of research studies into brain mechanisms of reward, most immediately, a spate of studies on electrical brain reward. Subsequent papers by Olds and colleagues (e.g., Margules and Olds 1962; Olds 1958a, 1958b), as well as others (e.g., Delgado, Roberts, and Miller 1954; Routtenberg and Lindy 1956), showed the rewarding effects of stimulating a variety of subcortical sites in several species, including humans (Bishop, Elder, and Heath 1963; for review, see Olds 1969). Stimulating the brain can also produce aversive responses; as with aversive Pavlovian conditioning, however, research on aversive brain stimulation falls outside the scope of the present chapter.

Olds (1958b) argued that the findings on brain reward provide *ipso facto* evidence against drive-reduction and drive-stimulus reduction theories. These theories, in the tradition of Fechner and early Freud, maintain that reward results from a reduction in (unpleasant) internal stimulation. But in the experimental paradigm of Olds and Milner, reward comes from *adding* stimulation to the brain. Of course, it is possible that a natural stimulus such as food is rewarding when it reduces the level of an internal drive stimulus (associated with hunger), and that this reduction in turn generates a signal (added stimulation) to upstream reward mechanisms. Even so, it is possible to bypass the hypothesized stage of stimulus reduction and activate reward mechanisms directly. By implication, the reduction in a drive stimulus may be sufficient for reward but is not necessary.

2.10.1   Pleasure Centers in the Brain?

Spurred by Olds's findings, the locus of research on long-standing topics of reward, reinforcement, utility, and pleasure moved into the brain. At one point, Olds (1956) himself referred to sites in the brain that sustain electrical self-stimulation as "pleasure centers"—although he later backed off from this ascription (see Wise 1980). The notion of pleasure centers faced two main criticisms: First, the electrical stimulation of the brain may not necessarily arouse "pleasure." And second, it is not clear that the brain functions through the operation of discrete "centers." Roy Wise (1980, 92), however, later took up the banner, writing that Olds's view "may not have been far wrong; the synaptic junction where sensory impressions of rewarding environmental stimuli take on the subjective experience of pleasure may have been but a half-axon's length away from Olds' best self-stimulation sites." On the basis of anatomical, physiological, and pharmacological evidence, Wise suggested that dopamine-mediated synapses near the sites of appetitive electrical self-stimulation may be crucial, that "there is a motivational or affective role for dopamine in behavior, and that the dopamine junction represents a synaptic way station for messages signaling the rewarding impact of a variety of normally powerful rewarding events" (Wise 1980, 94).

2.10.2 THE ROLE OF PLEASURE IN REWARD

To be sure, people and other organisms must be able to distinguish reinforcers or rewards in order to choose among them. In making these choices possible, different rewarding stimuli may evoke qualitatively as well as quantitatively different reward values, perhaps even qualitatively as well as quantitatively different pleasures—contrary to the hypothesis that all pleasures are qualitatively commensurable. The pleasure or rewarding value associated with sexual intercourse may not only exceed in magnitude that associated with scratching an itch, but moreover, in line with the stances of Plato and John Stuart Mill, the two pleasures or rewarding values may themselves differ in kind. And different subsets of neurons, or different patterns of activity in neural networks, might correlate with, even mediate, qualitatively different rewards or pleasures. Alternatively, in line with the stance of Democritus and Bentham, sensory reward values and pleasures, perhaps all reward values and pleasures, may differ only in their magnitudes but not in their qualities. The overall experiences of sexual intercourse and of scratching an itch differ qualitatively, to be sure, but the qualitative difference between them need not inhere in the associated pleasures. Research in neuroscience continues to speak, directly or indirectly, to these long-standing issues, as other chapters of this volume attest.

2.11 CONCLUSION

The very title of this final section is a misnomer. History has no conclusion (I hope) and, Giambattista Vico and James Joyce notwithstanding, history is rarely cyclical. In the history of science, important themes appear and reappear, but when they do, like many of Homer's protean gods, they often assume new shapes, calling for a fresh examination of the critical issues. Although Democritus and Bentham assumed the commensurability of pleasures, including sensory pleasures, neither philosopher could have conceived of the ways that this notion would play itself out in the study of brain mechanisms of reward. Indeed, the very conception of reward itself has evolved. A century ago came Thorndike's law of effect, which eventually led to redefining reward in terms of the reinforcement of behaviors. The concept of reward subsequently evolved and differentiated—with respect to its hedonic, motivational, and informational properties—in the light of research into its behavioral and neural mechanisms. There are, of course, important questions that remain unanswered, significant problems still not wholly resolved: Are rewards commensurable? Or might sensory rewards be special? Are pleasures commensurable? Or might sensory pleasures be special? Answers remain hidden behind doors that an acquaintance with history alone surely cannot unlock. What history does provide, however, is a context for posing the questions within larger themes in Western thought, as well as a humbling regard for the creative struggles with these questions on the part of philosophers and scientists, recent and past.

ACKNOWLEDGMENTS

Preparation of this chapter was supported in part by grants DC006688 and DC009021 from the National Institute on Deafness and Other Communication Disorders to Lawrence E. Marks. I thank Jay Gottfried, Ivan de Araujo, and Mark Laubach for thoughtful comments on an earlier version of this chapter. I would also like to thank Jay Gottfried for creating the rogues gallery in Figure 2.1.

REFERENCES

Angell, J.R. 1907. The province of functional psychology. *Psychol Rev* 94:61–91.

Bain, A. 1865. *The Emotions and the Will*. 2nd ed. London: Longmans, Green.

———. 1868. *Mental Science: A Compendium of Psychology and the History of Philosophy*. New York: Appleton.

Baldwin, J.M. 1894. *Mental Development in the Child and the Race: Method and Processes*. New York: Macmillan.

Bentham, J. 1789. *An Introduction to the Principles of Morals and Legislation*. London: Payne.

Berkeley, G. 1710. *A Treatise Concerning the Principles of Human Knowledge. Wherein the Chief Causes of Error and Difficulty in the Sciences, with the Grounds of Scepticism, Atheism, and Irreligion, are Inquir'd into*. Dublin: Pepyat.

Berlyne, D.E. 1950. Novelty and curiosity as determinants of exploratory behavior. *Br J Psychol* 42:68–80.

———. 1955. The arousal and satiation of perceptual curiosity in the rat. *J Comp Physiol Psychol* 48:238–46.

———. 1971. *Aesthetics and Psychobiology*. New York: Appleton-Century-Crofts.

Bernard, C. 1865. *Introduction à l'Étude de la Médicine Expérimentale*. Paris: Baillière et fils.

———. 1878–1879. *Leçons sur les Phénomènes de la Vie Communs aux Animaux et aux Végétaux*. Paris: Baillière et fils.

Bernfeld, S. 1944. Freud's earliest theories and the School of Helmholtz. *Psychoanal Quart* 13:341–62.

Berridge, K.C. 2003. Pleasures of the brain. *Brain Cogn* 52:106–28.

Bettelheim, B. 1982. *Freud and Man's Soul*. New York: Knopf.

Bishop, M.P., Elder, S.T., and Heath, R.G. 1963. Intracranial self-stimulation in man. *Science* 140:394–96.

Bitterman, M.E. 2006. Classical conditioning since Pavlov. *Rev Gen Psychol* 10:365–76.

Boring, E.G. 1950. *A History of Experimental Psychology*. 2nd ed. New York: Appleton-Century-Crofts.

Butler, R.A. 1953. Discrimination learning by rhesus monkeys to visual-exploration motivation. *J Comp Physiol Psychol* 46:95–98.

Butler, R.A., and Harlow, H.F. 1957. Discrimination learning and learning sets to visual exploration incentives. *J Gen Psychol* 57:257–64.

Cabanac, M. 1971. Physiological role of pleasure. *Science* 173:1103–7.

Cannon, W.B. 1929. Organization for physiological homeostasis. *Physiol Rev* 9:399–430.

———. 1932. *The Wisdom of the Body*. New York: Norton.

Capaldi, E.J. 1966. Stimulus specificity: Nonreward. *J Exper Psychol* 72:410–14.

Cason, H. 1932. The pleasure-pain theory of learning. *Psychol Rev* 39:440–66.

Colwill, R.M., and Rescorla, R.A. 1985. Post-conditioning devaluation of a reinforcer affects instrumental responding. *J Exper Psychol Anim Behav Proc* 11:120–32.

Cowles, J.T., and Nissen, H.W. 1957. Reward-expectancy in delayed responses of chimpanzees. *J Comp Psychol* 24:345–58.

Cranefield, P.H. 1970. On the origin of phrase *nihil est in intellectu quod non prius fuerit in sensu*. *J Hist Med* 25:77–80.

Day, J.D., and Carelli, R.M. 2007. The nucleus accumbens and Pavlovian reward learning. *Neuroscientist* 13:148–59.

de Condillac, E.B. 1754. *Traité des Sensations*. Paris: de Bure.

Delgado, J.M.R., Roberts, W.W., and Miller, N.E. 1954. Learning motivated by electrical stimulation of the brain. *Am J Physiol* 179:587–93.

Descartes, R. 1664. *L'Homme*. Paris: Girard.

Dijksterhuis, E.J. 1961. *The Mechanization of the World Picture*. Oxford: Clarendon Press.

Diogenes Laërtius. 1853. Life of Aristippus. In *The Lives and Opinions of Eminent Philosophers*, trans. C.D. Yonge, 81–96. London: Bohn.

Dollard, J., and Miller, N.E. 1950. *Personality and Psychotherapy*. New York: McGraw-Hill.

Ekman, G., and Åkeson C. 1965. Saltness, sweetness, and preference: A study of quantitative relations in individual subjects. *Scand J Psychol* 6:241–63.

Fearing, F. 1929. René Descartes. A study in the history of the theories of reflex action. *Psychol Rev* 36:375–88.

Fechner, G.T. 1860. *Elemente der Psychophysik*. Leipzig: Breitkopf und Härtel.

———. 1873. *Einige Ideen zur Schöpfungs- und Entwicklungsgeschichte der Organismen*. Leipzig: Breitkopf und Härtel.

Flaherty, C.F. 1982. Incentive motivation: A review of behavioral changes following shifts in reward. *Anim Learn Behav* 10:409–40.

Freeman, K. 1948. *Ancilla to the Pre-Socratic Philosophers: A Complete Translation of the Fragments in Diels "Fragmente der Vorsokratiker"*. Oxford: Blackwell.

Freud, S. 1955. Beyond the pleasure principle. In *Standard Edition of the Complete Psychological Works of Sigmund Freud, Volume 18 (1920–1922): Beyond the Pleasure Principle, Group Psychology and Other Works*, trans. and ed. J. Strachey, 1–64. London: Hogarth Press. (Orig. pub. 1920.)

———. 1957. Instincts and their vicissitudes. In *Standard Edition of the Complete Psychological Works of Sigmund Freud, Volume 14 (1914–1916): On the History of the Psycho-Analytic Movement, Papers on Metapsychology and Other Works*, trans. and ed. J. Strachey, 109–40. London: Hogarth Press. (Orig. pub. 1915.)

———. 1966. Project for a scientific psychology. In *Standard Edition of the Complete Psychological Works of Sigmund Freud, Volume 1 (1886–1899): Pre-Psycho-Analytic Publications and Unpublished Drafts*, trans. and ed. J. Strachey, 281–397. London: Hogarth Press. (Orig. written 1895; orig. pub. 1950.)

Galileo Galilei. 1960. The assayer. In *The Controversy on the Comets of 1618*, trans. S. Drake and C.D. O'Malley, 151–336. Philadelphia: University of Pennsylvania Press. (Orig. pub. 1623.)

Hamlyn, D.W. 1959. Aristotle's account of aesthesis in the De Anima. *Classical Quart* 9:6–16.

Helmholtz, H.L.F. von. 1867. *Handbuch der Physiologischen Optik*. Leipzig: Voss.

Herrnstein, R.J. 1998. Nature as nurture: Behaviorism and the instinct doctrine. *Behav Philos* 26:73–107.

Hilgard, E.R., and Marquis, D.G. 1940. *Conditioning and Learning*. New York: Appleton-Century.

Hobbes, T. 1650. *Humane Nature, or, the Fundamental Elements of Policie. Being a Discoverie of the Faculties, Acts and Passions of the Soul of Man, from Their Original Causes, According to such Philosophical Principles as are Not Commonly Known or Asserted*. London: Bowman.

———. 1651. *Leviathan, or, The Matter, Form and Power of a Common-wealth Ecclesiasticall and Civill*. London: Ckooke [*sic*].

Hull, C.L. 1930. Simple trial and error learning: A study in psychological theory. *Psychol Rev* 37:241–56.

———. 1943. *Principles of Behavior*. New York: Appleton-Century.

———. 1950. Behavior postulates and corollaries – 1949. *Psychol Rev* 57:173–80.

———. 1952. *A Behavior System*. New Haven, CT: Yale University Press.

Kamin, L.J. 1969. Predictability, surprise, attention, and conditioning. In *Punishment and Aversive Behavior*, eds. B.A. Campbell and R.M. Church, 276–96. New York: Appleton-Century-Crofts.

Kendler, H.H., Karasik, A.D., and Schrier, A.M. 1954. Studies of the effect of change of drive. III. Amounts of switching produced by shifting drive from thirst to hunger and from hunger to thirst. *J Exper Psychol* 47:179–82.

Külpe, O. 1893. *Grundriss der Psychologie*. Lepizig: Engelmann.

Leahey, T.H. 2004. *A History of Psychology: Main Currents of Psychological Thought*. 6th ed. Upper Saddle River, NJ: Pearson Prentice Hall.

Locke, J. 1690. *An Essay Concerning Humane Understanding*. London: Bassett.

———. 1700. *An Essay Concerning Humane Understanding*. 4th ed. London: Awnsham and Churchill.

Mackenzie, D. 1907. *Egyptian Myth and Legend, with Historical Narrative, Notes on Race Problems, Comparative Beliefs, etc*. London: Gresham.

Margules, D.L., and Olds, J. 1962. Identical "feeding" and "rewarding" systems in the lateral hypothalamus of rats. *Science* 135:374–75.

Marks, L.E. 1992. Freud and Fechner, desire and energy, hermeneutics and psychophysics. In *Cognition, Information Processing, and Psychophysics*, eds. H.-G. Geissler, S.W. Link, and J.T. Townsend, 23–42. Hillsdale, NJ: Erlbaum.

Mayer, J. 1965. Walter Bradford Cannon-A biographical sketch (October 19, 1871 – October 11, 1945). *J Nutr* 87:1–8.

McKeon, R. 1947. *Introduction to Aristotle, with a General Introduction and Introduction to Particular Works*. New York: Random House.

Meehl, P.E. 1950. On the circularity of the law of effect. *Psychol Bull* 47:52–75.

Mill, J.S. 1863. *Utilitarianism*. London: Parker, Son, and Bourne.

Miller, N.E. 1951a. Comments on multiple-process conceptions of learning. *Psychol Rev* 58:375–81.

———. 1951b. Learnable drives and rewards. In *Handbook of Experimental Psychology*, ed. S.S. Stevens, 435–72. New York: Wiley.

Mowrer, O.H. 1947. On the dual nature of learning—a re-interpretation of "conditioning" and "problem-solving". *Harv Educ Rev* 17:102–48.

Mowrer, O.H., and Lamoreaux, R.R. 1942. Avoidance conditioning and signal duration – a study of secondary motivation and reward. *Psychol Monogr* 54: Whole No. 34.

Müller, J. 1840. *Handbuch der Physiologie des Menschen für Vorlesungen, Zweiter Band,* 249–75. Coblenz: Hölscher.

Nafe, J.P. 1924. An experimental study of the affective qualities. *Am J Psychol* 35:507–44.

Olds, J. 1956. Pleasure centers in the brain. *Sci Am* 195:105–16.

———. 1958a. Effects of hunger and male sex hormone on self-stimulation of the brain. *J Comp Physiol Psychol* 51:320–24.

———. 1958b. Self-stimulation of the brain: Its use to study local effects of hunger, sex, and drugs. *Science* 127:315–24.

———. 1969. The central nervous system and the reinforcement of behavior. *Am Psychol* 24:114–32.

Olds, J., and Milner, P. 1954. Positive reinforcement produced by electrical stimulation of septal area and other regions of rat brain. *J Comp Physiol Psychol* 47:419–27.

Osgood, C.E., Suci, G.J., and Tannenbaum, P.H. 1957. *The Measurement of Meaning*. Urbana, IL: University of Illinois Press.

Pavlov, I.P. 1927. *Conditioned Reflexes*. Oxford: Oxford University Press.

———. 1967. Nobel lecture: The physiology of digestion, December 12, 1904. In *Nobel Lectures, Physiology or Medicine 1901–1921*, 141–55. Amsterdam: Elsevier. (Orig. pub. 1904.)

Postman, L. 1947. The history and present status of the Law of Effect. *Psychol Bull* 44:489–563.

Premack, D. 1959. Toward empirical behavioral laws. I. Positive reinforcement. *Psychol Rev* 66:219–33.

Reid, T. 1764. *An Inquiry into the Human Mind on the Principles of Common Sense*. Edinburgh: Kincaid and Bell.

Rescorla, R.A. 1968. Probability of shock in the presence and absence of CS in fear conditioning. *J Comp Physiol Psychol* 66:1–5.

———. 1988. Pavlovian conditioning: It's not what you think. *Am Psychol* 43:151–60.

Richter, C.P. 1947. Biology of drives. *J Comp Physiol Psychol* 40:129–34.

———. 1954. Behavioral regulators of carbohydrate homeostasis. *Acta Neuroveg* 9:247–59.

Riepe, D. 1956. Early Indian hedonism. *Philos Phenomenol Res* 16:551–55.

Ross, G.R.T., ed. 1906. *Aristotle: "De Sensu and De Memoria"*. Text and translation with Introduction and Commentary. Cambridge: Cambridge University Press.

Routtenberg, A., and Lindy, J. 1956. Effects of the availability of rewarding septal and hypothalamic stimulation on bar-pressing for food under conditions of deprivation. *J Comp Physiol Psychol* 60:158–61.

Schultz, W., Dayan, P., and Montague, P.R. 1997. A neural substrate of prediction and reward. *Science* 275:1593–99.

Sheffield, F.D., and Roby, T.B. 1950. Reward value of a non-nutritive sweet taste. *J Comp Physiol Psychol* 43:471–81.

Skinner, B.F. 1933. The rate of establishment of a discrimination. *J Gen Psychol* 9:302–50.

———. 1938. *The Behavior of Organisms*. New York: Appleton-Century.

———. 1953. *Science and Human Behavior*. New York: Macmillan.

Snell, B. 1953. *The Discovery of the Mind: The Greek Origins of European Thought*. Cambridge, MA: Harvard University Press.

Solomon, R.L. 1980. The opponent-process theory of acquired motivation: The costs of pleasure and the benefits of pain. *Am Psychol* 35:691–712.

Spencer, H. 1870. *The Principles of Psychology*. Vol. I, 2nd ed. London: Williams and Norgate.

Stratton, G.M. 1917. *Theophrastus and the Greek Physiological Psychology before Aristotle* London: Allen & Unwin.

Thompson, R.F. 1999. JAMES OLDS: May 30, 1922 – August 21, 1976. In *Biographical Memoirs, National Academy of Sciences U.S.A.*, Volume 77, 246–63. Washington: National Academy Press.

Thorndike, E.L. 1898. *Animal Intelligence: An Experimental Study of the Associative Processes in Animals*. New York: Macmillan.

———. 1911. *Animal Intelligence: Experimental Studies*. New York: Macmillan.

———. 1931. *Human Learning*. New York: Century.

Troland, L.T. 1928. *The Fundamentals of Human Motivation*. New York: van Nostrand.

Watson, J.B. 1916. The place of the conditioned-reflex in psychology. *Psychol Rev* 23:89–116.

Wise, R.A. 1980. The dopamine synapse and the notion of 'pleasure centers' in the brain. *Trends Neurosci* 3:91–95.

Wolpe, J. 1950. Need-reduction, drive-reduction, and reinforcement: A neurophysiological view. *Psychol Rev* 57:19–26.

Wundt, W. 1874. *Grundzüge der Physiologischen Psychologie*. Leipzig: Engelmann.

———. 1896. *Grundriss der Psychologie*. Leipzig: Engelmann.

Yerkes, R.M., and Morgulis, S. 1909. The method of Pawlow in animal psychology. *Psychol Bull* 6:257–73.

Young, P.T. 1948. Appetite, palatability, and feeding habit: A critical review. *Psychol Bull* 45:289–320.

————. 1949. Food-seeking drive, affective process, and learning. *Psychol Rev* 56:98–121.

————. 1952. The role of hedonic processes in the organization of behavior. *Psychol Rev* 59:249–62.

————. 1961. *Motivation and Emotion: A Survey of the Determinants of Human and Animal Activity.* New York: Wiley.

Young, P.T., and Greene, J.T. 1953. Relative acceptability of saccharine solutions as revealed by different methods. *J Comp Physiol Psychol* 46:295–98.

3 Reward: What Is It? How Can It Be Inferred from Behavior?

Norman M. White

CONTENTS

3.1 INTRODUCTION

In everyday use the word "reward" describes an event that produces a pleasant or positive affective experience. Among behavior scientists, reward is often used to describe an event that increases the probability or rate of a behavior when the event is contingent on the behavior. In this usage reward is a synonym of reinforcement. At best these common usages create ambiguity. At worst the two meanings of reward are conflated, leading to the assumption that reinforcement is always the result of positive affect produced by rewarding events. Although reward certainly influences behavior, its influence is not as straightforward as is often assumed, nor is reward the only reinforcement process that can influence behavior.

In the present analysis, "reinforcement" is the term used to describe any process that promotes learning: a change in behavior as the result of experience. The event (or stimulus) that initiates the process is called the reinforcer. Since both the reinforcer and its behavioral effects are observable and can be fully described, this can be taken as an operational definition. However, this definition is uninformative with respect to the processes that underlie the behavioral effects of reinforcement and tends to obscure the fact that there are several such effects, all of which result in behavioral change. This chapter discusses evidence for the existence and independent function of three reinforcement processes.

3.2 THREE REINFORCEMENT PROCESSES

Reinforcers are events that elicit several types of responses without prior experience. They are usually grouped into two broad types based on one class of these responses. Reinforcers that elicit approach responses are usually called positive; reinforcers that elicit withdrawal are called aversive or negative. These attributions are based on the assumption that approach-eliciting reinforcers also elicit some array of internal perturbations that constitute a pleasant or rewarding experience, and that withdrawal-eliciting reinforcers produce an aversive experience. Although these internal affective responses cannot be directly observed, their existence can be inferred from behavior in certain situations, making them a second type of response elicited by reinforcers. In addition to producing approach or withdrawal and reward or aversion, both positive and negative reinforcers also produce a third type of internal response that strengthens, or modulates, memories. Each of these three kinds of responses (approach/withdrawal, reward/aversion, and memory modulation) is a reinforcement process because each affects learning, albeit in different ways. The three processes are illustrated in Figure 3.1a.

An important feature of the responses elicited by reinforcers is that they are all subject to Pavlovian conditioning (Pavlov 1927). In Pavlovian terms the reinforcer is an unconditioned stimulus (US) and the responses it evokes are unconditioned responses (URs). Neutral stimuli present when such responses occur acquire the property of evoking very similar responses, thereby becoming conditioned stimuli (CS) that evoke conditioned responses (CRs). These CSs function as conditioned reinforcers with effects similar to those produced by the USs that generated them, so there are three kinds of conditioned reinforcement that parallel the three kinds of reinforcement (see Figure 3.1b). Importantly, the conditioned reinforcers function in the absence of the reinforcers.

Following a more detailed description of the three reinforcement processes, the main content of this chapter reviews evidence from specific learning situations, showing how the existence of each of these rewarding processes can be deduced.

3.2.1 Approach/Withdrawal

Certain naturally occurring stimuli such as food, water, or a sexual partner can elicit observable, unlearned approach responses; other events that cause injury or fear of injury elicit withdrawal responses (Craig 1918; Maier and Schnierla 1964). Similar responses are elicited by CSs in rewarding (Kesner 1992; Schroeder and Packard 2004; White and Hiroi 1993; Koob 1992) and aversive (Fendt and Fanselow 1999; Davis 1990) situations. In the present analysis the approach and withdrawal (motor) responses are independent of the rewarding and aversive (affective) responses that usually accompany them. This assertion is primarily based on data showing that the two types of responses are impaired by lesions to different parts of the brain (Corbit and Balleine 2005; Balleine, Killcross, and Dickinson 2003; Dayan and Balleine 2002). These studies will not be reviewed here although they are addressed in other chapters in Part III of this volume.

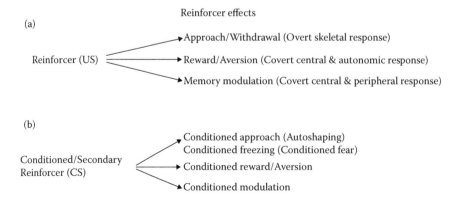

FIGURE 3.1 The three unconditioned (a) and conditioned (b) reinforcement processes.

3.2.2 AFFECTIVE STATES

Reward and aversion are internal affective states elicited by reinforcing events (Young 1959; Glickman and Schiff 1967; White 1989; Cabanac 1992). When affective states are considered independently of other reinforcement processes that may or may not accompany them, two things about them become clear. First, they must be consciously experienced in order to influence behavior (one of the basic ideas of affective neuroscience: Burgdorf and Panksepp 2006; Panksepp 1998). This differentiates affective states from the other reinforcement processes of approach/withdrawal and memory modulation, which function unconsciously. Second, the mere experience of an affective state has no influence on behavior. An affective state affects behavior only when an individual learns what to do to maintain or re-initiate a situation that the individual likes and wants, or what behavior leads to the termination of a state the individual does not like.

This kind of learning is necessarily a cognitive process that results in the representation of a contingent, or predictive, relationship between behavior and its consequences. This relationship has been described using the term expectancy (Tolman 1932), in the sense that an individual learns what behaviors or other events lead to a rewarding event. The behaviors are not fixed in form but remain oriented towards the desired change in affective state. This process has also been called instrumental learning (Thorndike 1933b) or operant conditioning (Skinner 1963). More recently, the term action-outcome learning has been used to describe learning about affective states (Everitt and Robbins 2005). This theme is taken up in greater detail in Chapter 13.

Action-outcome learning also occurs with artificially produced rewards such as electrical stimulation of certain brain areas (Olds 1956) or injection of an addictive drug (Weeks 1962; Pickens and Thompson 1968; Pickens and Harris 1968). Since these events lack external reference points, they may produce a disorganized increase or decrease in activity. Little can be inferred from such diffuse behavioral changes. Organized behavior that can be the basis of inferences about internal reward processing appears only when some form of action-outcome learning about the behaviors required to obtain the reward has occurred (White 1996). This illustrates the general point that in non-verbal animals there is no way to infer the existence of an affective state without action-outcome learning involving the reinforcer that produces the state.

3.2.3 MEMORY MODULATION

Memory modulation is a general process whereby central (McGaugh 2000; McGaugh and Petrinovitch 1965) and peripheral (Gold 1995; McGaugh and Gold 1989) processes initiated by reinforcers act in the brain to strengthen the neural representations of memories acquired around the same time as the reinforcer occurs. The effect has been demonstrated using a variety of memory tasks, from inhibitory avoidance in rats (Gold, McCarty, and Sternberg 1982; McGaugh 1988) to verbal learning in humans (Gold 1992; Watson and Craft 2004). Memory modulation is content-free: its occurrence does not involve learning anything about the reinforcer that modulates the representation of a memory. Accordingly, neither the approach/withdrawal nor the affective properties of a reinforcer are involved in its modulatory action. Both rewarding and aversive reinforcers modulate memories (Huston, Mueller, and Mondadori 1977). As described below, the modulatory action of reinforcers is inferred from experiments designed to eliminate the possibility that either the affective or approach processes can influence the learned behavior.

3.3 ANALYSIS OF REINFORCER ACTIONS IN LEARNING TASKS

3.3.1 INSTRUMENTAL LEARNING

Instrumental learning occurs when an individual acquires a new behavior that leads to a positive reinforcer (or avoids a negative reinforcer). However, in many learning tasks that fit the operational definition of instrumental learning, it is difficult to be sure which reinforcement process produced the behavioral change. This problem can be illustrated by examining how a food reinforcer acts to increase the running speed of a hungry rat over trials in which a rat is placed at one end of the runway and the food is at the other end.

One possibility is that running speed increases because the food reinforcer *modulates (strengthens)* a stimulus-response (S-R) association between the stimuli in the runway and the running response. An effect of this kind will be described below in the section on win-stay learning.

Another possibility is that the stimuli in the runway become CS that elicit *conditioned approach* responses which increase running speed. A mechanism very similar to this was proposed by Spence and Lippit (1946) as a modification of Hull's (1943) original S-R learning theory. An example will be described below in the section on conditioned cue preference (CCP) learning.

Finally, it is possible that the change in behavior is due to the acquisition of a neural representation of an *action-outcome* association (Everitt and Robbins 2005). Since consuming the food leads to a rewarding state, the rat may run faster because it learns that this behavior will lead to that state as soon as it reaches the end of the runway.

What kind of evidence would permit us to conclude that behavior is due to action-outcome learning? The partial reinforcement extinction effect (Skinner 1938; Capaldi 1966) is an experimental paradigm that was used by Tolman (1948) for this purpose. Two groups of rats were trained to run down a runway for food. One group found food at the end on every trial; the other found food on only 50% of the trials. After running speeds increased to similar levels in both groups, the food was eliminated completely for both groups and their extinction rates were compared. Results indicated that the group reinforced on 50% of the trials took significantly longer to extinguish than the group reinforced on 100% of the trials.

Both the modulation of S-R associations and conditioned approach are based on the promotion of specific behaviors by the reinforcer, leading to the prediction that resistance to extinction should be stronger in the 100% group because the response was reinforced twice as many times as in the 50% group. However, the behavior observed does not support these predictions.

Instead, the rats' behavior is consistent with the idea that the 100% rats learned to expect food on every trial while the 50% rats learned to expect food on only some of the trials. Therefore, fewer trials with no food were required to disconfirm the expectancy of the 100% rats than were required to disconfirm the 50% rats' expectancies. This interpretation suggests that the rats' behavior was determined by learned information (or knowledge) about the rewarding consequences of running down the alley (an action-outcome association). This information maps the relationships among the stimuli, one or more possible responses, and events, such as eating food, that lead to a rewarding state. Rather than promoting a specific behavior, the learned information contributes to determining behavior based on current conditions. Some authors have emphasized the flexibility of the behavior produced by this type of learning about reinforcers (Eichenbaum 1996).

Bar pressing for a reinforcer, possibly the most-used form of instrumental learning, is equally subject to influence by all three reinforcement processes. One of the most popular methods for studying the effects of addictive drugs in animals is the self-administration paradigm in which rats press a bar to deliver doses of a drug to themselves through implanted intravenous catheters. It is often claimed that the similarities between instrumental bar pressing and human drug self-administration make bar pressing a good model of addiction. However, given evidence that addictive drugs initiate the same array of reinforcement processes as naturally occurring reinforcers (White 1996), the difficulty of disambiguating these processes using an instrumental learning paradigm such as bar pressing raises the issue of whether instrumental learning is the best way to study the underlying processes of addiction. The use of reinforcement schedules such as progressive ratio (e.g., Roberts, Loh and Vickers. 1989) may help to sort out the processes, but it is arguable that other methods such as the CCP paradigm (see below) have revealed more about the reinforcing actions of drugs such as morphine than have self-administration studies (Jaeger and van der Kooy 1996; Bechara, Martin, Pridgar and van der Kooy. 1993; White, Chai, and Hamdani 2005).

Although the partial reinforcement extinction effect and other paradigms, such as reward contrast (McGaugh et al. 1995; Salinas and McGaugh 1996; Kesner and Gilbert 2007), are consistent with the idea that action-outcome learning is a major influence on behavior in the conditions of those experiments, they do not rule out the possibility that the other processes discussed also influence behavior in these and other learning situations. The following discussion focuses on an analysis of reinforcer actions in several learning paradigms: the CCP, post-training administration of reinforcers, and win-stay learning on the radial maze. In each case the goal is to deduce which of the three reinforcer processes produces the learned behaviors observed.

These are all paradigms for studying behavior in rats. It seems a fair assumption that the same processes apply to humans, although the interaction among them and with learned behaviors is undoubtedly more complex. The discussion is almost completely confined to the behavioral level of analysis and does not attempt to describe in detail evidence from studies using lesions, brain stimulation, or drugs, all of which contribute to understanding the physiological bases of reinforcement. However, it is important to note that the interpretation of physiological and neuroscientific information hinges on the way in which the *psychological* processes of reinforcement are understood to function, and therefore this analysis is intended as a contribution to that understanding.

3.3.2 Conditioned Cue Preference

This learning task (also known as conditioned place preference) uses an apparatus with two discriminable compartments and an area that connects them. Rats are confined in the two compartments on alternate days. One compartment always contains a reinforcer (e.g., food); the other is always empty. After several such training trials the rats are placed into the connecting area and allowed to move freely between the two compartments, neither of which contains food. Rats choose to spend more time in their food-paired compartments than in their unpaired compartments (Spyraki, Fibiger, and Phillips 1982; Everitt et al. 1991; White and McDonald 1993), an effect known as the conditioned cue preference because it indicates that the cues in the reinforcer-paired compartment have acquired conditioned stimulus properties. Addictive drugs also produce CCPs (Reicher and Holman 1977; Sherman et al. 1980; Mucha et al. 1982; Phillips, Spyraki, and Fibiger 1982; Tzschentke 1998, 2007).

In this task the reinforcer promotes a learned behavior: a preference for the stimuli in the reinforcer-paired compartment. Because the assignment of the reinforcer to the two compartments is counterbalanced, the preference cannot be attributed to any URs to the stimuli in either one. Since the rats are always confined in their reinforcer-paired compartments during training, they never have an opportunity to learn the response of entering them from the connecting area, nor are they ever required to perform any other new behavior to obtain the reinforcer, eliminating the possibility that an S-R association could be formed. This means that the preference cannot be due to instrumental learning about how to obtain the food reward.

The remaining possibility is that the preference is solely due to the CS properties acquired by the distinctive stimuli in the reinforcer-paired compartment during the training trials. During the test trial a CS in the paired compartment could elicit two different CRs in the absence of the reinforcer (US) (see Figure 3.1). One is a conditioned approach response from the point in the apparatus at which the rat can see the stimuli in the paired compartment. The other is conditioned reward, experienced when the rat has entered the compartment, resulting in action-outcome learning during the test trial (in contrast to the impossibility of such learning during the training trial when the rat is confined in the compartment with the reinforcer). Both of these CRs would increase the rat's tendency to enter and remain in the reinforcer-paired compartment, resulting in the observed preference. The CCP procedure and the influence of these CRs are illustrated in Figure 3.2, and the effect of the two CRs on behavior is discussed in more detail here.

3.3.2.1 Conditioned Approach

Conditioned approach behaviors have been demonstrated in a paradigm called autoshaping, first developed as a method of training pigeons to peck at a disc on a cage wall when the disc was lit. The standard method was to "shape" the pigeon's behavior by manually reinforcing successively

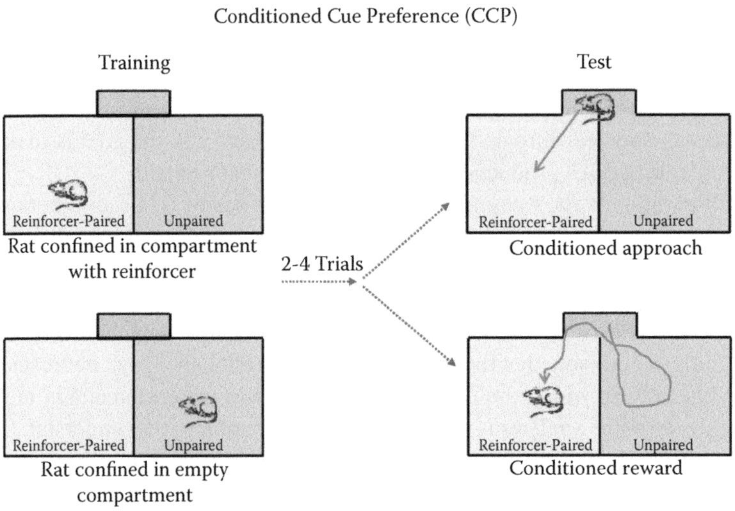

Conditioned Cue Preference (CCP)

FIGURE 3.2 Illustration of two conditioned responses that could produce conditioned cue preference. Rats are trained by placing them into one compartment with a reinforcer (e.g., food, drug injection) and into the other compartment with no reinforcer an equal number of times. The doors to the connecting tunnel are closed, confining the rats to the compartments during these trials. After several training trials the rats are given a preference test by placing them in the connecting tunnel with the doors to the compartments open. The gray arrows illustrate the two conditioned responses. Conditioned approach would attract the rat into the paired compartment when it sees the conditioned stimuli in that compartment from the tunnel. In the absence of this response the rat explores the apparatus, eventually entering the paired compartment, bringing it into contact with the conditioned stimuli. Conditioned reward elicited by those stimuli would instrumentally reinforce the response of entering the compartment. Both of these conditioned responses would increase a rat's tendency to enter and remain in the reinforcer-paired compartment, resulting in the observed preference. See text for further discussion.

closer approximations of the birds' responses until they pecked at the disc on their own. Brown and Jenkins (1968) attempted to eliminate experimenter involvement in shaping by simply illuminating the disc and providing access to the food when the illumination went off, regardless of what responses the pigeon made or did not make while the disc was lit. After a mean of about 50 such paired presentations the pigeons responded reliably to the light by pecking the disc, even though the reinforcement was not contingent on those responses. The pigeons had shaped themselves, hence "autoshaping." The sequence of events in the autoshaping paradigm is illustrated in Figure 3.3.

The increase in responding was attributed to Pavlovian conditioning (Gamzu and Williams 1971; Jenkins and Moore 1973). The conditioning process involved the transfer of a response (pecking) elicited by the food (the US) to the illuminated disc (the CS) due to the contiguity of the two stimuli.

One problem with this interpretation is that giving access to the reinforcer immediately after the disc light went off reinforced any responses made while it was on. It could therefore be argued that the increase in responding was due to adventitious response reinforcement without invoking Pavlovian conditioning. Evidence supporting the conditioning interpretation was obtained with a paradigm called reinforcer omission (Williams and Williams 1969), in which a response to the lit disc cancelled the access to the food that usually followed, completely eliminating response reinforcement (see Figure 3.3). Pigeons did not acquire the response in the reinforcer omission condition, but if they were trained on autoshaping first and then switched to reinforcer omission they maintained a higher rate of responding than birds trained in a control condition, such as random presentations of CS and US or CS presentations with no US at all. This supports the idea that increased responding to the CS observed in autoshaping is due to Pavlovian conditioning in which the response elicited by the reinforcer is transferred to the CS.

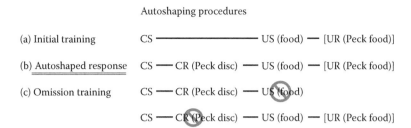

FIGURE 3.3 Autoshaping paradigm. (a) Initial training: CS (lit disc) is followed by access to food (US). Response to food (UR) may or may not occur (square brackets). (b) Autoshaped response: gradually the bird begins pecking the lit disc (the CR) reliably. Response to food may or may not occur. When it does occur it may reinforce the disc-pecking response. (c) Omission training: to prevent response reinforcement the US is omitted on trials when the CR (peck disc) occurs. The US is presented only when the rat does not emit the CR.

A conditioned approach response (see Figure 3.1) similar to the one that is acquired in the autoshaping procedure (see Figure 3.2) can explain the preference for the food-paired environment seen in the CCP paradigm. Exposure to the stimuli in that environment together with a reinforcer could result in a transfer of the approach response (UR) elicited by the food (US) to the environmental stimuli (CS). On the test trial the CS in the food-paired environment elicit the conditioned approach response (CR) resulting in a preference for that environment.

3.3.2.2 Conditioned Reward

In addition to producing conditioned approach, pairing the stimuli in the food-paired environment will also result in conditioned reward (see Figure 3.1). If the stimuli in the reinforcer-paired environment acquire conditioned rewarding properties during the training trials, a rat moving freely during the test trial would experience conditioned reward when it entered the reinforcer-paired environment (see Figure 3.2). This experience might induce the rat to remain in the environment but, perhaps more importantly, the rat would learn that entering the environment from the connecting area leads to a rewarding experience. This is action-outcome learning about the conditioned rewarding stimuli and is another way Pavlovian conditioning can produce a CCP.

This explanation of how the CCP can be produced by conditioned reward is an appetitive learning application of Mowrer's (Mowrer 1947; McAllister et al. 1986) two-factor theory of avoidance learning. According to this idea, when rats are shocked on one side of an apparatus with two distinct compartments they acquire conditioned aversive responses to the stimuli on the side where the shock is given. These CRs make the shock side of the apparatus aversive even if the shock is not given. When the rats run to the other (no-shock) side of the apparatus they learn how to escape from the aversive conditioned cues, an instance of instrumental learning.

As already mentioned, addictive drugs also produce CCPs, and it is usually assumed that they do so because of their rewarding effects. However, it is also possible that conditioned approach responses to initially neutral stimuli are acquired when the stimuli are paired with the effects of an addictive drug in a CCP or other paradigm. Although this issue is under investigation (Ito, Robbins, and Everitt 2004; Di Ciano and Everitt 2003; White, Chai, and Hamdani 2005), there is at present no clear evidence that drug-induced CCPs can be produced by conditioned approach responses.

3.3.3 POST-TRAINING ADMINISTRATION OF REINFORCERS

Memory modulation (McGaugh 1966, 2000) is a reinforcement process that improves, or strengthens, memory for an event or response. The reinforcer must be contemporaneous with the acquisition of the memory, but does not have to be related to it in any other way. A popular example of this effect is the observation that nearly everyone can remember where they were and what they were doing when they first heard about the destruction of the World Trade Center in New York on September 11, 2001 (Ferre 2006; Davidson, Cook, and Glisky 2006). This emotional (reinforcing) experience was presumably unrelated to the memory it strengthened in most of us but nevertheless acted to modulate our diverse individual memories.

The basis of this effect is the phenomenon of memory consolidation, first demonstrated by Muller and Pilzecker (1900), who showed that memory for a list of words was disrupted when a second list was learned a short time after the first one, but not when the second list was learned some time later. This suggested that the neural representation of the memory of the first list was relatively fragile immediately after it was learned, but became more robust with the passage of time (see McGaugh 1966 for review and discussion). This generalization has since been confirmed for a wide variety of memory tasks using a similarly large number of different post-learning events, including head trauma (Dana 1894; Russell and Nathan 1946) and electroconvulsive shock (Zubin and Barrera 1941; Duncan 1949). Other modulatory events improve rather than disrupt recall when they occur around the time of acquisition, but have no effect hours later. The earliest demonstrations of memory improvement used stimulant drugs such as strychnine and amphetamine (Breen and McGaugh 1961; Westbrook and McGaugh 1964; Krivanek and McGaugh 1969). These observations led to the notion that modulatory events accelerate consolidation by strengthening the representation of the memory (Bloch 1970).

A study by Huston, Mondadori, and Waser (1974) illustrates evidence supporting the claim that the memory modulation process is independent of the affective value of the reinforcer. Hungry mice were shocked when they stepped down from a platform (see Figure 3.4). When tested the next day, the shocked mice remained on the platform longer than unshocked controls. This behavior was probably due to action-outcome learning: the mice recalled the aversive experience produced by the shock when they stepped off the platform and altered their behavior accordingly; although it could also have been due to conditioned withdrawal (freezing).

In a subsequent phase of the study by Huston et al., one group of mice was fed immediately following their initial step-down–shock experience. On the test trial these mice remained on the

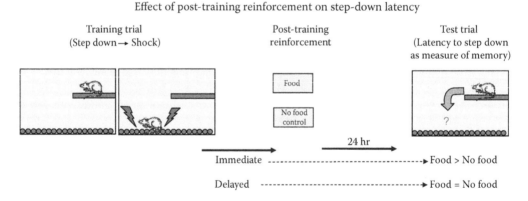

FIGURE 3.4 Post-training reinforcement effects of food reward on conditioned withdrawal. Hungry rats were placed on the platform and shocked when they stepped down onto the grid floor. Rats in the post-training reinforcement group were then placed in a separate cage where they were given food to eat; rats in the control group were placed in the same cage with no food. Rats in one experimental and one control group were placed in the food cage immediately after the shock; those in another group were given the same access to food after a short delay. The next day all rats were placed on the platform and their step-down latencies were measured. For the rats in the immediate group, latencies for the food group were longer than latencies for the no-food group, suggesting that the post-training treatment (eating) improved the memory of the immediate-group rats for the consequences of stepping down. No such difference was found for the delayed group, suggesting that the food treatment had no effect on memory. See text for discussion of why this result shows that post-training reinforcement improves memory independently of any rewarding properties of the reinforcer (and cf. Huston, Mondadori and Waser 1974).

platform longer than mice that had not been fed after the training trial (see Figure 3.4). Since the mice were hungry, the food had rewarding properties. If action-outcome learning about this reward had occurred, the fed mice would have stepped down more quickly than the unfed mice in order to obtain the food. Since the opposite was observed, the change in behavior (longer step-down times) produced by the food cannot be attributed to learning about its rewarding properties. Rather, the effect of the food suggests that it modulated, or strengthened, the memory for the relationship between stepping-down and shock. Consistent with this interpretation and with consolidation theory, mice that were fed after a delay did not show increased step-down times during the test.

Another demonstration (Huston, Mueller, and Mondadori 1977) showed that rewarding electrical stimulation of the lateral hypothalamus also produces memory modulation. Rats were trained to turn left for food in a T-maze. Every time they made an error (by turning right) they were removed from the goal box and placed in a cage where they received rewarding lateral hypothalamic stimulation for several minutes. The rats that received this treatment learned to make the correct response (the one that led to food, but away from rewarding stimulation) in fewer trials than rats in a control group that were not stimulated after either response. Since the rewarding stimulation improved memory for the location of food when it was given only after the rats made errors, its effect cannot be attributed to action-outcome learning about responses that led to its rewarding properties. The effect can be explained as a modulation of memory for the food location, for the no-food location, for the correct response, or for all three of these (Packard and Cahill 2001).

The electrical stimulation in the T-maze experiment was delivered by the experimenters, but self-stimulation also produces memory modulation (Major and White 1978; Coulombe and White, 1980, 1982). Although electrodes in several brain areas support self-stimulation, suggesting that the stimulation in all of them is rewarding, post-training stimulation in only some of these areas produces memory modulation. This suggests that reward is not a sufficient condition to produce modulation.

Another series of experiments leading to the same conclusion compared the modulatory effects of post-training consumption of sucrose and saccharin solutions (Messier and White 1984; Messier and Destrade 1988). The solutions were both preferred over water but were equally preferred to each other, suggesting that they had equivalent reward properties. The memory modulation effects of these two solutions were tested by allowing rats to drink them after paired presentations of a tone and a shock. The strength of the memory for the tone-shock association was estimated by measuring the pause in drinking by thirsty rats produced by presentation of the tone alone. The tone suppressed drinking more effectively in the rats that had drunk sucrose than in the rats that had drunk saccharin after the tone-shock pairings. Since the solutions had equivalent rewarding properties, the modulatory effect of sucrose cannot be attributed to its rewarding properties.

Post-training injections of glucose but not of saccharin, in amounts comparable to those ingested in the consumption experiments, also improved memory (Messier and White 1984), suggesting that the effect of sucrose was due to some post-ingestional effect, but not to its rewarding taste. The memory-modulating action of glucose has been studied in detail in rats and humans (Messier 2004; Korol 2002; Gold 1991, 1992, 1995). For further details see Chapter 12.

In human studies glucose is often consumed before acquisition of the memories it strengthens, with no diminution in its effect. This emphasizes that the post-training paradigm is simply a method for demonstrating the modulatory actions of reinforcers, not a requirement for them to have this effect. In fact, reminiscent of Thorndike's (1933a, 1933b) original studies on spread of effect (cf. Chapter 2), mere temporal contiguity of a reinforcing event with the acquisition of a memory is sufficient for modulation to occur. The fact that reinforcers improve performance of learned behavior when they occur before the behavior has been learned is also further evidence that modulation is not due to action-outcome learning about reward.

A related finding is that aversive post-training events also produce memory modulation (Mondadori, Waser, and Huston 1977; Huston, Mueller, and Mondadori 1977; Holahan and White 2002). In one study (White and Legree 1984) rats were given tone-shock pairings and immediately placed into a different cage where they were given a short, strong shock (instead of drinking a sweet solution). When tested for suppression of drinking 2 days later, the shocked rats exhibited stronger memory for the tone-shock association than rats that were not shocked and also stronger memory than rats that were shocked two hours after training. These findings are all consistent with the assertion that memory modulation is independent of the affective properties of reinforcers.

3.3.3.1 Conditioned Modulation

Memory modulation by a conditioned aversive stimulus has also been demonstrated (Holahan and White 2002, 2004) (see Figure 3.5). Rats were placed into a small cage where they received several shocks and into a second distinguishable cage where they did not receive shocks. They were then trained to find food in a Y-maze by forced entries into both the food and no-food arms in a predetermined order. On the last training trial all rats were forced to enter the food arm after which they were placed into either the shock or no-shock cage (no shocks were given at this time). Two days later the rats were tested on the Y-maze with no food in either arm. The rats in the group that had been placed into the shock cage after maze training made significantly more correct responses (to the arm that previously contained food) than the rats that had been placed into the no-shock cage (see Figure 3.5).

This increased resistance to extinction cannot be attributed to action-outcome learning about the shock because the rats were forced to run to the food-paired arm immediately before exposure to the shock cage. Action-outcome learning about the shock would have decreased their tendency to run to the food arm. Since the rats had an increased tendency to run to the food arm, the effect of exposure to the shock-paired cage is attributable to a memory modulation effect produced by exposure to the conditioned aversive cues in the shock cage.

In another control group, rapid extinction was observed in a group that was exposed to the shock cage 2 hrs after the last Y-maze training trial. This result is consistent with the conclusion that the conditioned contextual cues in the shock cage evoked a conditioned memory modulation response.

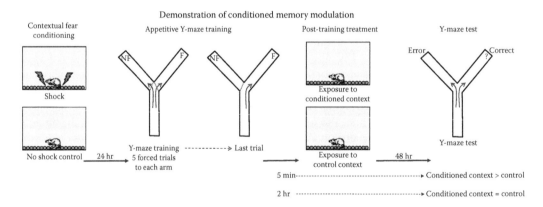

FIGURE 3.5 (See Color Insert) Conditioned modulation of approach behaviour by aversive stimulation. Rats were placed in a cage with a grid floor and shocked, and alternately into a discriminable cage and not shocked. On the next 2 days the rats were given a total of 10 training trials in a Y-maze. On each trial one arm was blocked and the rats were forced to run to the other arm. They ran to the food and no-food arms five times each; the last run was always to the food arm. After a 5 min delay, the rats in the experimental group were placed into the shock cage without receiving shock (the conditioned context) for 10 min. The rats in the control group were placed into the no-shock context. Another group of each type was placed into the contexts after a 2 hr delay. Then next day all rats were given a Y-maze test with no barriers or food on the maze in order to compare their memory for the location of the food. The rats in the 5 min delay group that had been exposed to the conditioned context made significantly more correct responses than the rats that had been placed into the control context. Context placement had no effect in the 2 hr delay group. See text for discussion of how this finding demonstrates post-training modulation by a conditioned reinforcer (instead of a reinforcer) and how it shows that modulation does not depend on the affective properties of the post-training treatment. (Results based on Holahan, M. R., White, N. M, *Behavioral Neuroscience* 118, 24, 2004.)

3.3.4 WIN-STAY LEARNING

Post-training reinforcers have been shown to modulate several different kinds of memory, including cognitive instrumental responses (Williams, Packard, and McGaugh 1994; Gonder-Frederick et al. 1987; Manning, Hall, and Gold 1990; Messier 2004), CCP learning (White and Carr 1985), and simple S-R associations (Packard and White 1991; Packard and Cahill 2001; Packard, Cahill, and McGaugh 1994).

In an example of S-R learning, the win-stay task (Packard, Hirsh, and White 1989), rats were placed on the center platform of an eight-arm radial maze. Four maze arms had lights at their entrances and only those arms contained food pellets. The other four arms were dark and did not contain food. Entries into dark arms with no food were scored as errors. On each daily trial a different set of four arms was lit and baited with food. Since the food was in a different spatial location on each trial the lights were the only information about the location of the food available to a rat on the center platform (see Figure 3.6). Rats acquired this S-R behavior slowly, achieving a rate of 80%–85% correct responses after approximately 30 trials.

Which reinforcement process produced this learned behavior? There are two possibilities. First, since the rats repeatedly run into the lit maze arms from the center platform, and since this behavior is followed by food reward, the increase in frequency of lit arm entries could be due to action-outcome learning about how the response leads to the food reward. Second, if a rat randomly enters an arm, passing the light at the entrance, and then eats food at the end of the arm, the memory modulation property of the food reinforcer would strengthen the association between the light stimulus and the arm-entering response. This type of learning has been described as the acquisition of a habit (Mishkin and Petri 1984; Mishkin, Malamut, and Bachevalier 1984), because nothing is learned about the relationship of either the stimulus or the response to the reinforcer.

Evidence against the action-outcome learning hypothesis was obtained using a devaluation procedure (Dickinson, Nicholas, and Adams 1983; Balleine and Dickinson 1992) in which consumption of a reinforcer is paired with injections of lithium chloride, which produces gastric malaise. The conditioned aversive response produced by the food CS, known as conditioned taste aversion (Garcia, Kimeldorf, and Koelling 1955; Garcia, Hankins, and Rusiniak 1976) reduces or eliminates consumption of the reinforcer. In the instrumental learning context this is known as devaluation of

Win-stay task

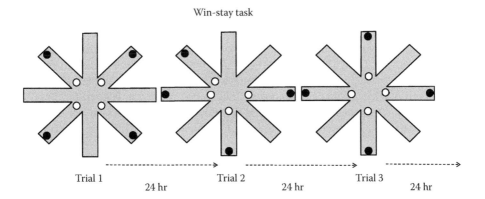

Trial 1   Trial 2   Trial 3
        24 hr   24 hr   24 hr

FIGURE 3.6 Schematic diagram of win-stay task. White dots are lights at the entrances to arms from the center platform. Black dots are food. On each daily trial four different arms are lit and baited with food. The other four arms are dark and contain no food. The diagram shows three possible configurations of lit/baited arms; other configurations are used on subsequent trials. On each trial the rats retrieve two pellets from each lit arm; when the first pellet in an arm has been eaten it is replaced. When the second pellet in an arm has been eaten the light is extinguished and no further pellets are placed there. Entries into unlit arms are errors. The trial ends when all lights are extinguished. See text for discussion of the kind of information used by the rats to locate the food in this task and the way in which the food reinforcer affects that information.

the reinforcer. Sage and Knowlton (2000) trained rats on the win-stay task and then devalued the food reinforcer by giving lithium injections following consumption in the rats' home cages. On subsequent win-stay trials the rats continued to enter lit arms on the maze, but stopped eating the food pellets.

If the pellets rewarded the response to the lit arms, the change in their affective value should have attenuated the rats' tendency to enter those arms. The fact that this did not happen suggests that reward was not the basis of the response, but that it was due to a modulated or strengthened memory for the S-R association. This occurred when the reinforcer was consumed shortly after the rats entered each lit arm. Since neither S-R learning nor its modulation involves information about the affective properties of the reinforcer, devaluation of the reinforcer did not affect the rats' behavior.

3.4 SUMMARY

Reinforcement can be operationally defined as the process that occurs when the presence of some object or event promotes observable behavioral changes. The present analysis argues that these new behaviors are due to at least three different, independently acting reinforcement processes: action-outcome learning about rewarding or aversive consequences of behavior, conditioned approach or withdrawal, and memory modulation.

Although reward is often assumed to be the only reinforcement process, evidence shown here suggests that this is not the case. Furthermore, when reward does influence behavior it can do so only as the result of either action-outcome learning or the sequential occurrence of Pavlovian conditioning and action-outcome learning.

Conditioned approach is another reinforcement process that can influence behavior if the learning conditions allow a view of the CS from a distance. Although approach and reward are normally produced simultaneously by naturally occurring reinforcers, there is evidence in the drug literature that approach and aversion co-occur, suggesting that affective states and approach-withdrawal behaviors result from independent processes (Wise, Yokel, and deWit 1976; White, Sklar, and Amit 1977; Reicher and Holman 1977; Sherman et al. 1980; Bechara and van der Kooy 1985; Carr and White 1986; Corrigall et al. 1986; Lett 1988; Brockwell, Eikelboom, and Beninger 1991).

Memory modulation is the third reinforcement process. It is continuously produced by contact with reinforcers and conditioned reinforcers, and affects all forms of learning. Situations in which no other form of reinforcement can operate provide evidence that modulation is an independently occurring process.

The major difficulty involved in distinguishing among these reinforcement processes is that they can all act simultaneously on different kinds of memory (White and McDonald 2002; McDonald, Devan, and Hong 2004). Analyses of a number of common situations used to study reward, such as bar pressing or running in a runway, suggest that it is difficult or impossible to show that reward is the only process that affects the learned behavior, suggesting that these may not be ideal for many purposes. When studying the effects of reinforcers, careful selection of the memory task—which, in turn, determines the type of learning from which the reinforcer action will be inferred—is critical.

REFERENCES

Balleine, B.W., and Dickinson, A. 1992. Signalling and incentive processes in instrumental reinforcer devaluation. *Quarterly Journal of Experimental Psychology* 45b:285–301.

Balleine, B.W., Killcross, A.S., and Dickinson, A. 2003. The effect of lesions of the basolateral amygdala on instrumental conditioning. *Journal of Neuroscience* 23:666–75.

Bechara, A., Martin, G. M., Pridgar, A., and van der Kooy, D. 1993. The parabrachial nucleus: A brain stem substrate critical for mediating the aversive motivational effects of morphine. *Behavioral Neuroscience* 107:147–60.

Bechara, A., and van der Kooy, D. 1985. Opposite motivational effects of endogenous opioids in brain and periphery. *Nature* 314:533–34.

Bloch, V. 1970. Facts and hypotheses concerning memory consolidation processes. *Brain Research* 24:561–75.

Breen, R.A., and McGaugh, J.L. 1961. Facilitation of maze learning with posttrial injections of picrotoxin. *Journal of Comparative and Physiological Psychology* 54:498–501.

Brockwell, N.T., Eikelboom, R., and Beninger, R.J. 1991. Caffeine-induced place and taste conditioning: Production of dose-dependent preference and aversion. *Pharmacology, Biochemistry and Behavior* 38:513–17.

Brown, P.L., and Jenkins, H.M. 1968. Autoshaping of the pigeon's keypeck. *Journal of the Experimental Analysis of Behavior* 11:1–8.

Burgdorf, J., and Panksepp, J. 2006. The neurobiology of positive emotions. *Neuroscience and Biobehavioral Reviews* 30:173–87.

Cabanac, M. 1992. Pleasure: The common currency. *Journal of Theoretical Biology* 155:173–200.

Capaldi, E.J. 1966. Partial reinforcement: A hypothesis of sequential effects. *Psychological Review* 73:459–77.

Carr, G.D., and White, N.M. 1986. Anatomical dissociation of amphetamine's rewarding and aversive effects: An intracranial microinjection study. *Psychopharmacology* 89:340–46.

Corbit, L.H., and Balleine, B. 2005. Double dissociation of basolateral and central amygdala lesions on the general and outcome-specific forms of Pavlovian-instrumental transfer. *Journal of Neuroscience* 25:962–70.

Corrigall, W.A., Linseman, M.A., D'Onofrio, R.M., and Lei, H. 1986. An analysis of the paradoxical effect of morphine on runway speed and food consumption. *Psychopharmacology* 89:327–33.

Coulombe, D., and White, N.M. 1980. The effect of post-training lateral hypothalamic self-stimulation on aversive and appetitive classical conditioning. *Physiology and Behavior* 25:267–72.

———. 1982. The effect of post-training lateral hypothalamic self-stimulation on sensory pre-conditioning in rats. *Canadian Journal of Psychology* 36:57–66.

Craig, W. 1918. Appetites and aversions as constituents of instincts. *Biological Bulletin* 34:91–107.

Dana, C.L. 1894. The study of a case of amnesia or "double consciousness." *Psychological Review* 1:570–80.

Davidson, P.S., Cook, S.P., and Glisky, E.L. 2006. Flashbulb memories for September 11th can be preserved in older adults. *Neuropsychology Development and Cognition. B: Aging Neuropsychology and Cognition* 13:196–206.

Davis, M. 1990. Animal models of anxiety based on classical conditioning: The conditioned emotional response (CER) and the fear-potentiated startle effect. *Pharmacology and Therapeutics* 47:147–65.

Dayan, P., and Balleine, B.W. 2002. Reward, motivation, and reinforcement learning. *Neuron* 36:285–98.

Di Ciano, P., and Everitt, B.J. 2003. Differential control over drug-seeking behavior by drug-associated conditioned reinforcers and discriminative stimuli predictive of drug availability. *Behavioral Neuroscience* 117:952–60.

Dickinson, A., Nicholas, D.J., and Adams, C.D. 1983. The effect of the instrumental training contingency on susceptibility to reinforcer devaluation. *Quarterly Journal of Experimental Psychology B* 35B:35–51.

Duncan, C.P. 1949. The retroactive effect of electroshock on learning. *Journal of Comparative and Physiological Psychology* 42:32–44.

Eichenbaum, H. 1996. Is the rodent hippocampus just for "place"? *Current Opinion in Neurobiology* 6:187–95.

Everitt, B.J., Morris, K.A., O'Brien, A., and Robbins, T.W. 1991. The basolateral amygdala-ventral striatal system and conditioned place preference: Further evidence of limbic-striatal interactions underlying reward-related processes. *Neuroscience* 42:1–18.

Everitt, B.J., and Robbins, T.W. 2005. Neural systems of reinforcement for drug addiction: From actions to habits to compulsion. *Nature Neuroscience* 8:1481–89.

Fendt, M., and Fanselow, M.S. 1999. The neuroanatomical and neurochemical basis of conditioned fear. *Neuroscience and Biobehavioral Reviews* 23:743–60.

Ferre, R.P. 2006. Memories of the terrorist attacks of September 11, 2001: A study of the consistency and phenomenal characteristics of flashbulb memories. *Spanish Journal of Psychology* 9:52–60.

Gamzu, E., and Williams, D.R. 1971. Classical conditioning of a complex skeletal response. *Science* 171:923–25.

Garcia, J., Hankins, W.G., and Rusiniak, K.W. 1976. Flavor aversion studies. *Science* 192:265–66.

Garcia, J., Kimeldorf, D.J., and Koelling, R.A. 1955. Conditioned aversion to saccharin resulting from exposure to gamma radiation. *Science* 122:157–58.

Glickman, S.E., and Schiff, B.B. 1967. A biological theory of reinforcement. *Psychological Review* 74:81–109.

Gold, P.E. 1991. An integrated memory regulation system: from blood to brain. In *Peripheral Signalling of the Brain: Neural, Immune and Cognitive Function*, eds. R.C.A. Frederickson, J.L. McGaugh, and D.L. Felten, 391–419. Toronto: Hogrefe and Huber.

———. 1992. Modulation of memory processing: enhancement of memory in rodents and humans. In *Neuropsychology of Memory*, 2 ed., eds. L.R. Squire and N. Butters, 402–14. New York: Guilford Press.

————. 1995. Role of glucose in regulating the brain and cognition. *American Journal of Clinical Nutrition* 61:987S–95S.

Gold, P.E., McCarty, R., and Sternberg, D.B. 1982. Peripheral catecholamines and memory modulation. In *Neuronal Plasticity and Memory Formation*, eds. C. Ajimone-Marsan and H. Matthies, 327-338. New York: Raven Press.

Gonder-Frederick, L., Hall, J.L., Vogt, J., Cox, D.J., Green, J., and Gold, P.E. 1987. Memory enhancement in elderly humans: effects of glucose ingestion. *Physiology and Behavior* 41:503–4.

Holahan, M.R., and White, N.M. 2002. Effects of lesions of amygdala subnuclei on conditioned memory consolidation, freezing and avoidance responses. *Neurobiology of Learning and Memory* 77:250–75.

————. 2004. Amygdala inactivation blocks expression of conditioned memory modulation and the promotion of avoidance and freezing. *Behavioral Neuroscience* 118:24–35.

Hull, C.L. 1943. *Principles of Behavior*. New York: Appleton-Century-Crofts.

Huston, J.P., Mondadori, C., and Waser, P.G. 1974. Facilitation of learning by reward of post-trial memory processes. *Experientia* 30:1038–40.

Huston, J.P., Mueller, C.C., and Mondadori, C. 1977. Memory facilitation by posttrial hypothalamic stimulation and other reinforcers: A central theory of reinforcement. *Biobehavioral Reviews* 1:143–50.

Ito, R., Robbins, T.W., and Everitt, B.J. 2004. Differential control over cocaine-seeking behavior by nucleus accumbens core and shell. *Nature Neuroscience* 7:389–97.

Jaeger, T. V., and van der Kooy, D. 1996. Separate neural substrates mediate the motivating and discriminative properties of morphine. *Behavioral Neuroscience* 110:181–201.

Jenkins, H.M., and Moore, B.R. 1973. The form of the auto-shaped response with food or water reinforcers. *Journal of the Experimental Analysis of Behavior* 20:163–81.

Kesner, R.P. 1992. Learning and memory in rats with an emphasis on the role of the amygdala. In *The Amygdala: Neurobiological Aspects of Emotion, Memory and Mental Dysfunction*, ed. J.P. Aggleton, 379–99. New York: Wiley-Liss.

Kesner, R.P., and Gilbert, P.E. 2007. The role of the agranular insular cortex in anticipation of reward contrast. *Neurobiology of Learning and Memory* 88:82–86.

Koob, G.F. 1992. Drugs of abuse: anatomy, pharmacology and function of reward pathways. *Trends in Pharmacological Sciences* 13:177–84.

Korol, D.L. 2002. Enhancing cognitive function across the life span. *Annals of the New York Academy of Sciences* 959:167–79.

Krivanek, J., and McGaugh, J.L. 1969. Facilitating effects of pre- and post-training amphetamine administration on discrimination learning in mice. *Agents and Actions* 1:36–42.

Lett, B.T. 1988. Enhancement of conditioned preference for a place paired with amphetamine produced by blocking the association between place and amphetamine-induced sickness. *Psychopharmacology* 95:390–94.

Maier, N.R.F., and Schnierla, T.C. 1964. *Principles of Animal Psychology*. New York: Dover.

Major, R., and White, N.M. 1978. Memory facilitation by self-stimulation reinforcement mediated by the nigro-neostriatal bundle. *Physiology and Behavior* 20:723–33.

Manning, C.A., Hall, J.L., and Gold, P.E. 1990. Glucose effects on memory and other neuropsychological tests in elderly humans. *Psychological Science* 1:307–11.

McAllister, W.R., McAllister, D.E., Scoles, M.T., and Hampton, S.R. 1986. Persistence of fear-reducing behavior: Relevance for conditioning theory of neurosis. *Journal of Abnormal Psychology* 95:365–72.

McDonald, R.J., Devan, B.D., and Hong, N.S. 2004. Multiple memory systems: The power of interactions. *Neurobiology of Learning and Memory* 82:333–46.

McGaugh, J.L. 1966. Time dependent processes in memory storage. *Science* 153:1351–58.

————. 1988. Modulation of memory storage processes. In *Memory—An Interdisciplinary Approach*, eds. P. R. Solomon, G.R. Goethals, C.M. Kelley, and B.R. Stephens, 33–64. New York: Spring Verlag.

————. 2000. Memory – a century of consolidation. *Science* 287:248–51.

McGaugh, J.L., Cahill, L.F., Parent, M.B., Mesches, M.H., Coleman-Mesches, K., and Salinas, J.A. 1995. Involvement of the amygdala in the regulation of memory storage. In *Plasticity in the Central Nervous System – Learning and Memory*, eds. J. L. McGaugh, F. Bermudez-Rattoni, and R.A. Prado-Alcala, 18-39. Hillsdale, NJ: Lawrence Earlbaum.

McGaugh, J.L., and Gold, P.E. 1989. Hormonal modulation of memory. In *Psychoendocrinology*, eds. R. B. Brush and S. Levine, 305–39. New York: Academic Press.

McGaugh, J.L., and Petrinovitch, L.F. 1965. Effects of drugs on learning and memory. *International Review of Neurobiology* 8:139–96.

Messier, C. 2004. Glucose improvement of memory: A review. *European Journal of Pharmacology* 490:33–57.

Messier, C., and Destrade, C. 1988. Improvement of memory for an operant response by post-training glucose in mice. *Behavioural Brain Research* 31:185–91.

Messier, C., and White, N.M. 1984. Contingent and non-contingent actions of sucrose and saccharin reinforcers: Effects on taste preference and memory. *Physiology and Behavior* 32:195–203.

Mishkin, M., Malamut, B., and Bachevalier, J. 1984. Memories and habits: Two neural systems. In *Neurobiology of Human Memory and Learning*, eds. G. Lynch, J.L. McGaugh, and N.M. Weinberger, 65–77. New York: Guilford Press.

Mishkin, M., and Petri, H.L. 1984. Memories and habits: Some implications for the analysis of learning and retention. In *Neuropsychology of Memory*, eds. L.R. Squire and N. Butters, 287–96. New York: Guilford Press.

Mondadori, C., Waser, P.G., and Huston, J.P. 1977. Time-dependent effects of post-trial reinforcement, punishment or ECS on passive avoidance learning. *Physiology and Behavior* 18:1103–9.

Mowrer, O.H. 1947. On the dual nature of learning – A reinterpretation of "conditioning" and "problem solving." *Harvard Educational Review* 17:102–48.

Mucha, R.F., van der Kooy, D., O'Shaughnessy, M., and Bucenieks, P. 1982. Drug reinforcement studied by the use of place conditioning in rat. *Brain Research* 243:91–105.

Muller, G.E., and Pilzecker, A. 1900. Experimentelle Beitrage zur Lehre vom Gedachtnis. *Zeitschrift fur Psychologie und Physiologie der Sennesorgane ergamzungsband* 1:1–288.

Olds, J. 1956. Pleasure center in the brain. *Scientific American* 195:105–16.

Packard, M.G., and Cahill, L.F. 2001. Affective modulation of multiple memory systems. *Current Opinion in Neurobiology* 11:752–56.

Packard, M.G., Cahill, L.F., and McGaugh, J.L. 1994. Amygdala modulation of hippocampal-dependent and caudate nucleus-dependent memory processes. *Proceedings of the National Academy of Sciences U.S.A.* 91:8477–81.

Packard, M.G., Hirsh, R., and White, N.M. 1989. Differential effects of fornix and caudate nucleus lesions on two radial maze tasks: evidence for multiple memory systems. *Journal of Neuroscience* 9:1465–72.

Packard, M.G., and White, N.M. 1991. Dissociation of hippocampal and caudate nucleus memory systems by post-training intracerebral injection of dopamine agonists. *Behavioral Neuroscience* 105:295–306.

Panksepp, J. 1998. *Affective Neuroscience*. New York: Oxford.

Pavlov, I.P. 1927. *Conditioned Reflexes*. Oxford: Oxford University Press.

Phillips, A.G., Spyraki, C., and Fibiger, H.C. 1982. Conditioned place preference with amphetamine and opiates as reward stimuli: attenuation by haloperidol. In *The Neural Basis of Feeding and Reward*, eds. B.G. Hoebel and D. Novin, 455–64. Brunswick, ME: Haer Institute.

Pickens, R., and Harris, W.C. 1968. Self-administration of d-amphetamine by rats. *Psychopharmacologia* 12:158–63.

Pickens, R., and Thompson, T. 1968. Cocaine reinforced behavior in rats: Effects of reinforcement magnitude and fixed ratio size. *Journal of Pharmacology and Experimental Therapeutics* 161:122–29.

Reicher, M.A., and Holman, E.W. 1977. Location preference and flavor aversion reinforced by amphetamine in rats. *Animal Learning and Behavior* 5:343–46.

Roberts, D.C.S., Loh, E.A., and Vickers, G. 1989. Self-administration of cocaine on a progressive ratio schedule in rats: Dose-response relationship and effect of haloperidol pretreatment. *Psychopharmacol* 97:535–38.

Russell, W.R., and Nathan, P.W. 1946. Traumatic amnesia. *Brain* 69:280–300.

Sage, J.R., and Knowlton, B.J. 2000. Effects of US devaluation on win-stay and win-shift radial maze performance in rats. *Behavioral Neuroscience* 114:295–306.

Salinas, J.A., and McGaugh, J.L. 1996. The amygdala modulates memory for changes in reward magnitude – involvement of the amygdaloid GABAergic system. *Behavioural Brain Research* 80:87–98.

Schroeder, J.P., and Packard, M.G. 2004. Facilitation of memory for extinction of drug-induced conditioned reward: Role of amygdala and acetylcholine. *Learning Memory* 11:641–47.

Sherman, J.E., Pickman, C., Rice, A., Liebeskind, J.C., and Holman, E.W. 1980. Rewarding and aversive effects of morphine: Temporal and pharmacological properties. *Pharmacology, Biochemistry and Behavior* 13:501–15.

Skinner, B.F. 1938. *The Behavior of Organisms*. New York: Appleton-Century-Crofts.

———. 1963. Operant behavior. *American Psychologist* 18:503–15.

Spence, K.W., and Lippitt, R. 1946. An experimental test of the sign-gestalt theory of trial and error learning. *Journal of Experimental Psychology* 36:491–502.

Spyraki, C., Fibiger, H.C., and Phillips, A.G. 1982. Attenuation by haloperidol of place preference conditioning using food reinforcement. *Psychopharmacology* 77:379–82.

Thorndike, E.L. 1933a. A proof of the law of effect. *Science* 77:173–75.

————. 1933b. A theory of the action of the after-effects of a connection upon it. *Psychological Review* 40:434–39.

Tolman, E.C. 1932. *Purposive Behavior in Animals and Men*. New York: Century.

————. 1948. Cognitive maps in rats and men. *Psychological Review* 56:144–55.

Tzschentke, T.M. 1998. Measuring reward with the conditioned place preference paradigm: A comprehensive review of drug effects, recent progress and new issues. *Progress in Neurobiology* 56:613–72.

————. 2007. Measuring reward with the conditioned place preference (CPP) paradigm: Update of the last decade. *Addiction Biology* 12:227–462.

Watson, G.S., and Craft, S. 2004. Modulation of memory by insulin and glucose: Neuropsychological observations in Alzheimer's disease. *European Journal of Pharmacology* 490:97–113.

Weeks, J.R. 1962. Experimental morphine addiction: Method for automatic intravenous injections in unrestrained rats. *Science* 138:143–44.

Westbrook, W.H., and McGaugh, J.L. 1964. Drug facilitation of latent learning. *Psychopharmacologia* 5:440–46.

White, N.M. 1989. Reward or reinforcement: What's the difference? *Neuroscience and Biobehavioral Reviews* 13:181–86.

————. 1996. Addictive drugs as reinforcers: Multiple partial actions on memory systems. *Addiction* 91:921–49.

White, N.M., and Carr, G.D. 1985. The conditioned place preference is affected by two independent reinforcement processes. *Pharmacology, Biochemistry and Behavior* 23:37–42.

White, N.M., Chai, S.-C., and Hamdani, S. 2005. Learning the morphine conditioned cue preference: Cue configuration determines effects of lesions. *Pharmacology Biochemistry and Behavior* 81:786–96.

White, N.M., and Hiroi, N. 1993. Amphetamine conditioned cue preference and the neurobiology of drug seeking. *Seminars in the Neurosciences* 5:329–36.

White, N.M., and Legree, P. 1984. Effect of post-training exposure to an aversive stimulus on retention. *Physiological Psychology* 12:233–36.

White, N.M., and McDonald, R.J. 1993. Acquisition of a spatial conditioned place preference is impaired by amygdala lesions and improved by fornix lesions. *Behavioural Brain Research* 55:269–81.

————. 2002. Multiple parallel memory systems in the brain of the rat. *Neurobiology of Learning and Memory* 77:125–84.

White, N.M., Sklar, L., and Amit, Z. 1977. The reinforcing action of morphine and its paradoxical side effect. *Psychopharmacology* 52:63–66.

Williams, C.L., Packard, M.G., and McGaugh, J.L. 1994. Amphetamine facilitation of win-shift radial-arm maze retention: The involvement of peripheral adrenergic and central dopaminergic systems. *Psychobiology* 22:141–48.

Williams, D.R., and Williams, H. 1969. Auto-maintenance in the pigeon: Sustained pecking despite contingent non-reinforcement. *Journal of the Experimental Analysis of Behavior* 12:511–20.

Wise, R.A., Yokel, R.A., and deWit, H. 1976. Both positive reinforcement and conditioned aversion from amphetamine and from apomorphine in rats. *Science* 191:1273–76.

Young, P.T. 1959. The role of affective processes in learning and motivation. *Psychological Review* 66:104–25.

Zubin, J., and Barrera, S.E. 1941. Effect of electric convulsive therapy on memory. *Proceedings of the Society for Experimental Biology and Medicine* 48:596–97.

4 What Can Different Brains Do with Reward?

Elisabeth A. Murray, Steven P. Wise, and Sarah E.V. Rhodes

CONTENTS

4.1   INTRODUCTION

In this chapter we ask: "What can brains do with reward?" Because different brains can do different things with reward, we focus on three kinds of brains: those of vertebrates, mammals, and primates. We do so not only because we are all of those things, but because about 530 million years ago the early vertebrates evolved a brain that we can recognize as reasonably like our own. That brain helps us, as it helped our ancient ancestors, deal with an intricate calculus of costs and benefits, some of which we call rewards. So at one level the answer to our question is relatively simple: Brains cause vertebrates to move as a coherent whole in order to obtain reward benefits at minimal cost. The complexity of cost-benefit calculations has multiplied manyfold since the early vertebrates first evolved, but they have bequeathed to us their life—our life—of decisions and goal-directed movement.

Although rewards can be understood in terms of costs and benefits, there is more to them than that. Along with vertebrates and other animals, plants endure costs and receive benefits, but these benefits do not count as rewards. Producing certain kinds of flowers and fruits, for example, incurs a metabolic and energetic cost while yielding a reproductive benefit. Most plants require water to produce these benefits. So why is fluid not a "reward" for a plant in need of turgor? The answer can be found in the word *animal* itself, which refers to a life of movement. Animals first appeared approximately 700–900 million years ago (Vermeij 1996), during a time of increasing atmospheric oxygen pressure, a contribution of our chloroplast-containing cousins. Higher oxygen pressure allowed organisms to use more energy, permitting the evolution of multicellular animals from unicellular ancestors. More energy, however, required more nutrients, and animals evolved the neural mechanisms for coordinated movements to obtain them. Movement, however, is both expensive and dangerous. A life of movement necessitates decisions about where to move, as well as when and if to do so. Animals must decide whether to abandon the relative safety that often comes from staying still in order to forage for nutrients and other rewards. When these decisions lead to actions having a beneficial outcome, we call that outcome reward (see Chapter 3). In short, animals evolved as reward seekers, and reward-seeking behavior (foraging) entails serious costs as well as benefits.

In psychology, a reward is defined operationally as anything that increases the behavior that leads to obtaining it. When reward acts in this way, psychologists also call it a positive reinforcer because it reinforces or strengthens the underlying associations in the brain that are said to "control" the reward-seeking behavior. The concept of reward can be divided into primary and secondary rewards, with primary rewards being those that directly meet biological needs (e.g., food, water, salt, and sex) and secondary rewards (also known as conditioned reinforcers) being stimuli that have acquired rewarding properties through their association with primary rewards.

Two important aspects of reward reflect its evolutionary history. First, evolution did not develop dedicated reward receptors. Instead, animals must learn the sensory properties of reward for both reward-seeking and reward-consuming behaviors (Hall, Arnold, and Myers 2000; Changizi, McGehee, and Hall 2002), and these sensory signals come from every modality. Second, although every sensory modality contributes to reward processing, the brain deals with a variety of rewards in similar ways. So, for example, even a uniquely human kind of reward, such as listening to a favorite piece of music, activates many of the same brain regions as a primary reward such as food (Blood and Zatorre 2001; and see Chapter 19 in this volume).

In what follows, we first sketch some current thinking on the evolution of some of the brain structures that deal with reward, with emphasis on those that likely appeared at three times during the history of animals (Figure 4.1): with early vertebrates, with early mammals, and during primate evolution. Next we use clues from laboratory research, mainly from studies of rats and rhesus monkeys, to paint a picture of structure-function relationships in each of these three clades. We recognize the hazards of this approach. Rats and rhesus monkeys are crown species, each with a long, separate evolutionary history and a complex suite of adaptations and specializations. The preferred way to understand the role of structures in early vertebrates or early mammals would involve the study of a diversity of animals in each group, selected according to the principles of evolutionary biology. Although a limited amount of such research is available, the literature from comparative psychology has not yet reached a point where we can use the preferred approach effectively. Accordingly, we use what hints come from rat and monkey research, in the context of comparative neuroanatomy, to explore what we have inherited from various ancestors. In doing so, we assume that structure-function relationships do not change haphazardly during evolution. Our approach will perhaps seem most peculiar when we use insights gained from the study of macaque monkeys and

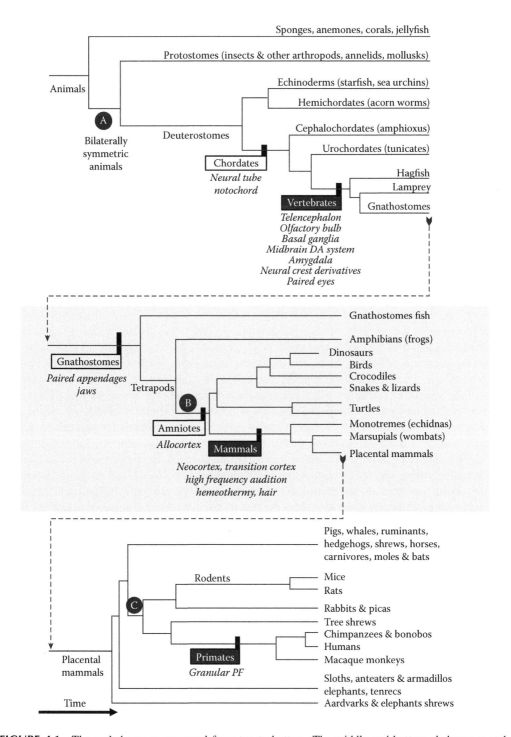

FIGURE 4.1 Three cladograms, arranged from top to bottom. The middle and bottom cladograms each develop one of the lineages from the cladogram above, as shown by the connecting arrows. The circled letters, A, B, and C, reference common ancestors referred to in the text. Beneath the names of selected clades, which are outlined in boxes, shared derived traits appear in italics. Abbreviation: DA, dopamine.

humans to explore the function of dopamine neurons and the basal ganglia in Section 4.3: "What Can Vertebrate Brains Do With Reward?" We adopt this approach because the original work was based on monkey research, and because we believe that the principles discerned from that research apply to other vertebrates as well. Our assumption is that such principles are highly conserved, although future work on a broad array of vertebrates is needed to confirm that view.

4.2 WHAT KINDS OF BRAINS DEAL WITH REWARD?

4.2.1 Brain History and Historiography

The field of brain evolution seeks to discern the history of the brain, but that field itself has a history, and not always an attractive one. Readers might wonder why the outline of brain evolution provided here is not widely known and why statements about brain evolution have an aura of controversy. In part, this feeling results from the bad reputation that brain evolution, as a field, has developed in certain circles. Nevertheless, progress has been made in understanding the history of brains, and we attempt to summarize some of that progress as it relates to reward processing.

Unfortunately, neuroscientists have often written about brain evolution in erroneous ways and in what follows we hope to avoid the mistakes of the past. In the mid-1980s, for example, one eminent neuroscientist (Jones 1985, 763) claimed that "no writer in the field today would seriously propose that the brains of existing nonmammalian vertebrates formed a scala naturae," while another did exactly that by speculating about how a certain behavioral capacity applied "across the entire phyletic scale" (Mishkin, Malamut, and Bachevalier 1984, 73). Evolutionary biologists long ago rejected the concept of a phyletic scale, also called the scale of nature or the *scala naturae*, in favor of the science of cladistics, which uses the evolutionary innovations of particular lineages to discern "evolutionary trees" through the principle of parsimony. Instead of ranking species as a linear progression from "lower" to "higher" forms—e.g., rats-cats-monkeys-chimpanzees-humans in the mammalian clade—most neuroscientists now understand that phylogenetic relationships have a branching architecture (Figure 4.1), with any two existing lineages sharing one most recent common ancestor. No existing species represents that common ancestor perfectly, and most, like laboratory rats and rhesus monkeys, do so rather poorly. Rats, for example, are highly specialized "gnawers," with a host of evolutionary innovations that the rodent-primate common ancestor lacked. Although we draw on data from rats and rhesus monkeys to explore what we have inherited from ancestral vertebrates, mammals, and primates, these groups do not make up a "phyletic scale" or anything of the sort. We discuss these groups in that order simply because we evolved from all of them and because they first appeared in that order.

We also discuss structures that evolved before or after others in evolutionary history, a topic that has many pitfalls. For example, one of the neuroscientists quoted above claimed that "the basal ganglia antedates both the cerebral cortex and the limbic system in phylogenesis" (Mishkin, Malamut, and Bachevalier 1984, 74). To the contrary, comparative neuroanatomy shows that homologues of the basal ganglia and many limbic structures, including parts of the cerebral cortex such as the hippocampus, evolved at about the same time, in relatively early vertebrates (Northcutt 1996; Striedter 2005).

Other discredited doctrines have also muddied the waters considerably. Some, such as MacLean's triune brain theory, have spawned nonsense such as "reptilian brains" inside human heads (MacLean 1985). In this chapter, we assume that structures such as the basal ganglia perform some conserved function, not that humans or other mammals have the same basal ganglia that some ancestral species or group had. (Also, technically, we did not evolve from "reptiles" but instead descended from ancestral amniotes.) Another discredited doctrine invokes the notion of "evolutionary trends" and schemes of connectional and functional hierarchies based on these imaginary trends. Sanides (1970) imagined that a group of medial cortical areas "arose" from the hippocampus, whereas a group of lateral cortical areas arose from the piriform cortex. He claimed that he could spot "trends" in the cytoarchitecture of these two groups of areas from phylogenetically older to newer parts of the cortex. Unfortunately, comparative evidence shows that many of his claims are wrong, while others are plausible but wholly lacking in evidentiary support. For example, whereas Sanides imagined that primary sensory areas appeared in later stages of mammalian evolution, comparative evidence shows that they were among the earliest areas to appear in mammals (Kaas 1995, 2008). Although Sanides's views were purely speculative, some neuroanatomists and others have adopted his opinions and interpreted their data according to those views (Barbas and Pandya 1989; Mesulam 1990; Cummings 1993). Nothing here should be construed as agreeing with any of that. When we talk about medial and lateral parts of the frontal cortex, we do not do so in the context of imaginary "evolutionary trends" of the sort popularized by Sanides and his followers.

To sketch the history of the brain, three concepts (homology, analogy, and homoplasy) and just five taxonomic terms (protostome, deuterostome, gnathostome, tetrapod, and amniote) clarify matters considerably. We explain the five taxonomic terms below, but take up the three evolutionary concepts here. *Homology* is a statement about ancestry, not function. A structure or behavior is homologous to another if two or more descendant species have inherited it from their most recent common ancestor. *Analogy* is a statement about function, not ancestry. A structure or behavior is analogous to another if it subserves the same function. The classic example involves wings. Insects, birds, and bats have wings, and they all perform the same function: producing lift for flight. We can be certain that they are only analogous, and not homologous, because the last common ancestor of each pair (birds and bats, birds and insects, bats and insects) lacked wings of any kind. Furthermore, insect wings evolved about 390 million years ago, hundreds of millions of years before the origin of either birds or bats (Grimaldi and Engel 2005), in a lineage that produced neither bats nor birds as descendants (Figure 4.1). A recent common ancestor of birds and insects is marked with an A in Figure 4.1, which also applies to bats and insects. This extinct species was a worm-like animal with no capacity for flight whatsoever. It had no wings or, indeed, appendages of any kind. Similarly, the last common ancestor of birds and bats is marked with a B. These ancient amniotes, which lived roughly 320 million years ago, also lacked wings. So like bird and insect wings, bird and bat wings are analogous, not homologous. This fact leads to the concept of *homoplasy*: something similar that has evolved in different lineages through parallel or convergent evolution. For example, the green lizard *Lacerta* adopts the habit of resting on green leaves. The color of lizard and leaf resemble each other, but this trait is neither analogous nor homologous. The trait (of being green) is not analogous because the color of the lizard performs a camouflage function, but the leaf has no interest in camouflage. The trait is not homologous because the lizard and plant did not inherit these characters from their nearest common ancestor, which was a single-cell eukaryote that lacked color in this sense. Instead, lizard and leaf exhibit homoplasy in being green. Some similarities stem from common developmental constraints and others result from the necessity of dealing with the physical and physiological realities of the world. For example, wings must produce lift from muscle power by interacting with air, so insect, bat, and bird wings (not to mention those of pterodactyls) inevitably shared certain features once they independently evolved.

Because of the prevalence of homoplasy, when we make statements such as "the amygdala and the dopamine system evolved in the early vertebrates," these conclusions are based on the identification of homologues of those structures in a diverse set of vertebrates, according to the principles of cladistics. Such statements do not imply that vertebrates are the only animals with neurons analogous to those in the mammalian amygdala or that they are the only animals with dopaminergic neurons.

4.2.2 VERTEBRATE BRAINS

A key development in the history of animals occurred when vertebrates evolved from invertebrate ancestors (Figure 4.1). Note that the term "invertebrates" refers to an arbitrary grouping of animals rather than a clade (a progenitor species and all of its descendants, living and extinct). The animals that most people think of as "invertebrates" are the protostomes, a group that includes insects, mollusks (such as squid and octopus), and segmented worms. Additional kinds of "invertebrates" are found among the deuterostomes. These two lineages, protostomes and deuterostomes, diverged more than 600 million years ago, and their last common ancestor was a worm-like animal with a diffuse nerve net (marked by an A in Figure 4.1). Both groups formed their nervous systems from ancestral conditions characterized by such nerve nets, which during parallel evolution condensed into concentrated neural structures independently (Valentine 2004; Grimaldi and Engel 2005). Many protostomes evolved to a highly advanced level, producing intelligent animals with extraordinary sensory sensitivity, motor capacity, and social organization. Nevertheless, the brains in these animals evolved independently from those in vertebrates and are not homologous to vertebrate brains. (This conclusion is not without controversy, but the analysis of the options proposed by Schmidt-Rhaesa (2007) in the context of the recent advances in understanding animal phylogeny (Dunn et al. 2008) supports this view.) The brains of protostomes differ in so many crucial respects from vertebrate brains that it is not particularly useful to apply the same name to them, although it is probably unavoidable. Most important for the present purposes, protostome brains lack homologues of the midbrain dopamine system, amygdala, hippocampus, and striatum; they lack anything remotely resembling the cerebral cortex of mammals; and to consider any part of their brains homologous to the prefrontal cortex of primates is preposterous. The brains that protostome invertebrates use to deal with reward have a long evolutionary history, one entirely separate from ours. Their brains are not only unlike those of vertebrates, they have virtually no resemblance to our common ancestral condition either.

Instead, as shown in Figure 4.1, vertebrates and their brains evolved in the deuterostome lineage, long after diverging from the line leading to modern protostomes. With these animals came a complex central nervous system, an ability to swim and eat, and to direct those actions with a brain and sensory organs on and in the head (Northcutt 1996; Holland and Holland 1999; Butler 2000; Nieuwenhuys 2002). (Protostomes developed similar features, too, but did so independently.) For the present purposes, the most important evolutionary developments involved the invention of the telencephalon and the vertebrate dopamine system. The early telencephalon included the olfactory bulb and a homologue of the piriform cortex of mammals (Wicht and Northcutt 1992, 1998; Northcutt 1996, 2001; Striedter 2005). Crucially, the telencephalon of early vertebrates contained homologues of the basal ganglia and, with somewhat less certainty, the amygdala and hippocampus, to use their mammalian names. (The strongest evidence traces all of these structures back to the early tetrapods (see Figure 4.1), but suggestive anatomical evidence indicates that homologues exist in various extant fish groups that make these structures likely to be innovations of relatively early vertebrates, if not the earliest ones (Neary 1990; Ulinski 1990; Northcutt 1996, 2008; Smeets, Marín, and Ganzález 2000; Wullimann and Mueller 2002; Wullimann and Rink 2002; Medina et al. 2005; Striedter 2005; Moreno and Gonzalez 2007).) Vertebrates also innovated paired eyes as part of their central nervous system, which, along with their olfactory system, supplied their brains with information about distant features of their environment, and these organs make crucial contributions to reward-seeking behavior. Thus, roughly 530 million years ago, early vertebrates developed the brain, several remote sensors on the head, and a motor apparatus, and they used these newly evolved systems to guide decisions and goal-directed movements. These evolutionary developments changed forever what brains—our brains by descent—can do with reward.

4.2.3 MAMMALIAN BRAINS

Mammals deal with reward with a mixture of old and new features, including those inherited from the early vertebrates: the dopamine system, amygdala, hippocampus, basal ganglia, olfactory system (with its olfactory bulb and piriform cortex), paired eyes, and so forth. In addition to these traits, shared with other vertebrates, the early mammals developed a new structure that deals with reward: the neocortex.

The top part of Figure 4.1 shows something about how this came about. Fairly early in vertebrate history, jawed fish (gnathostomes) evolved from jawless ancestors. Their descendants include all of the vertebrates in the bottom two parts of Figure 4.1, including the first land vertebrates, the ancestral tetrapods, which evolved 370 million years ago or so (Clack 2002). With amniotes, a group that appeared about 320 million years ago and includes all reptiles, birds, and mammals, the three-layered cortex known as the allocortex appeared. Parts of the allocortex have homologues in other vertebrates, but without its characteristic laminated structure (which was lost secondarily in birds).

Mammals evolved about 220 million years ago and have retained the allocortex as two substantial parts of their cerebral cortex, the piriform cortex and the hippocampus, along with some smaller, obscure allocortical areas. A different kind of cerebral cortex evolved in the earliest mammals or their immediate ancestors (Kaas 1995, 2008; Northcutt and Kaas 1995; Krubitzer and Kaas 2005). We can be certain of this conclusion because these areas occur only in mammals, and they can be found in all mammals, including monotremes and marsupials (see Figure 4.1). These mammalian innovations are often called the *neocortex* (meaning new cortex, which is true) or *isocortex* (meaning homogeneous cortex, which is false), but these terms are used inconsistently. Sometimes the terms neocortex and isocortex exclude mammalian innovations such as the so-called transition areas between the allocortex and the remainder of the neocortex. Here we use the term neocortex to include these transition areas. As noted above, rodent species, including rats, have their own set of adaptations and specializations, but the architecture of their frontal cortex represents the ancestral mammalian condition reasonably well. Rodents, like other mammals, possess several distinct neocortical fields: these include medial frontal areas, specifically the anterior cingulate (AC), infralimbic (IL), and prelimbic (PL) areas, as well as orbital frontal areas, specifically the agranular insular (Ia) and agranular orbitofrontal cortex. We use standard names and abbreviations for most of these areas, the ones noted in the previous sentence. We also use a new one, OFa, which stands for agranular orbitofrontal cortex. Some of these areas, specifically the medial frontal and orbitofrontal areas, go by the name "prefrontal cortex" in rodents and other nonprimate mammals (Kolb 2007), but their homologues in primates are not called prefrontal, so we use a different term. Regardless of their names, all of these specifically mammalian areas likely contribute to what mammalian brains can do with reward.

4.2.4 PRIMATE BRAINS

Primates first appeared about 60 million years ago. Early in this long history, primate brains evolved a new kind of frontal cortex, as well as some additional sensory areas. Figure 4.2 shows the basic, architectonic types of cerebral cortex in two species of primate and a rodent (Carmichael and Price 1994; Öngür, Ferry, and Price 2003; Palomero-Gallagher and Zilles 2004). In the primate frontal lobe, the main types of cortex are the agranular cortex, which lacks an internal granular layer (layer 4), and areas that have either a conspicuous or subtle layer 4, collectively called the granular prefrontal cortex (PFg). The PFg occurs in all primates and only in primates, and therefore represents a primate innovation (Preuss 1995, 2007). In appreciation of this historical fact, we use another new abbreviation, PFg, for these areas collectively. Figure 4.2 illustrates the parsimonious idea that rodents and primates, among other mammals, share the agranular parts of the frontal cortex, but primates alone have the granular parts. This idea has been controversial, but cortical topology (the spatial relationships of the frontal areas to each other and to the allocortex) and corticostriatal connectivity (from the agranular frontal areas to and near the ventral striatum, in particular) point to the same conclusions (Wise 2008). Crucially, none of the arguments brought forward to dispute the idea that PFg is a primate innovation—dopamine inputs, inputs from the mediodorsal nucleus of the thalamus, behavioral effects of brain lesions—stand up to scrutiny (Wise 2008). The lack of granular prefrontal areas in rodents should not be surprising given the antiquity of the most recent common ancestor of primates and rodents (marked with a C in Figure 4.1). Some of the primate innovations, such as granular parts of the orbitofrontal cortex (PFo), surely influence what primate brains can do with reward.

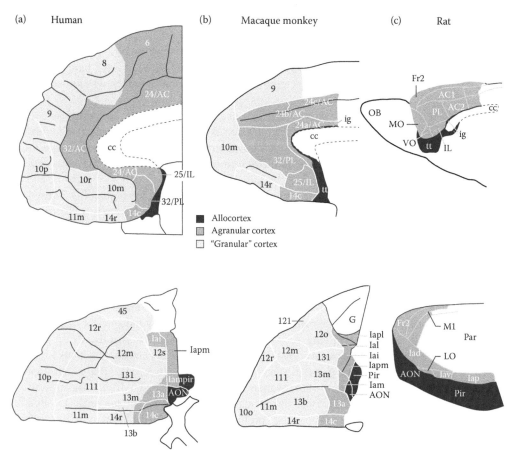

FIGURE 4.2 Basic types of frontal cortex. The key indicates three basic types of cerebral cortex by gray-scale code. (a) human, (b) macaque monkey, (c) rat. Architectonics from Öngür, Ferry, and Price (2003) (a), Carmichael and Price (1994) (b), and Palomero-Gallagher and Zilles (2004) (c). Abbreviations: AC, anterior cingulate area; AON, anterior olfactory "nucleus"; cc, corpus callosum; Fr2, second frontal area; I, insula; IL, infralimbic cortex; LO, lateral orbital area; MO, medial orbital area; M1, primary motor area; Par, parietal cortex; Pir, Piriform cortex; PL, prelimbic cortex; TT, tenia tectum; VO, ventral orbital area; l, lateral; m, medial; o, orbital; r, rostral; c, caudal; i, inferior; p, posterior; s, sulcal; v, ventral. "a" has two meanings: in Ia, it means agranular; in 13a, it distinguishes that area from area 13b. Otherwise, numbers indicate cortical fields.

4.3   WHAT CAN VERTEBRATE BRAINS DO WITH REWARD?

4.3.1   WHAT NONVERTEBRATE BRAINS CAN DO WITH REWARD

The vertebrate brain evolved a dramatically new central nervous system, one not shared by nonvertebrate species. Nevertheless, nonvertebrates deal with reward effectively. They have the ability to seek specific rewards (e.g., salt and water) and demonstrate the ability to learn about rewards via both associative mechanisms (e.g., Pavlovian and instrumental conditioning) and nonassociative ones (e.g., habituation and sensitization). As in vertebrates, associative learning about rewards in protostomes concerns how the animal learns to predict and respond to different events in their environment, and it involves the formation of associations between representations of different stimuli (S), responses (R), and rewards (often referred to as outcomes, or O), as formalized by the concepts of Pavlovian and instrumental conditioning. The stored knowledge acquired via these mechanisms guides reward-seeking behavior in protostomes as it does in vertebrates, and it is likely that almost all animals have the capacity for Pavlovian and instrumental conditioning. In Pavlovian conditioning, rewards are referred to as unconditioned stimuli (USs). Through experience (such as temporal contiguity), an originally neutral stimulus, called a conditioned stimulus (CS), becomes associated with a US and therefore comes to elicit the same innate responses (unconditioned responses, or URs) that are produced in response to the US. It is generally accepted that underlying Pavlovian conditioning is the formation of CS-US (stimulus-outcome or S-O) associations. Activation of the US representation via this association triggers the associated URs, which become known as conditioned responses (CRs) when evoked by presentations of the CS. In addition, Pavlovian conditioning may lead to the formation of direct CS-CR associations (also known as S-R associations) under certain circumstances (such as extinction, as discussed in Section 4.4.2.3). Learning about the predictive nature of environmental events allows the animal to act in anticipation of the US. When the US is the reward, this process increases the likelihood of obtaining it.

In instrumental conditioning, associations develop when an animal learns about the effect of its actions on the environment. Therefore, unlike Pavlovian conditioning, in instrumental conditioning the availability of reward depends on an animal's actions: its actions are *instrumental* in producing the reward. This type of learning involves the formation of associations between the instrumental response and reward (R-O and O-R associations). Guidance of instrumental behavior by R-O associations is conceptualized as reflecting *goal-directed* responding. If responding is followed by a reward in the presence of a particular stimulus, this stimulus comes to modulate the R-O association and is termed the discriminative stimulus. Furthermore, stimulus-response (S-R) associations also form between representations of the response and any stimuli present when the reward is delivered. Behavior predominantly guided by S-R associations is referred to as *habitual*. Chapter 14 explains these concepts in more detail.

Neither Pavlovian nor instrumental learning requires a vertebrate brain. Marine mollusks such as *Aplysia*, for example, demonstrate the ability to learn via both of these mechanisms (Carew, Walters, and kandel 1981; Carew and Sahley 1986; Brembs and Heisenberg 2000; Brembs et al. 2002). The garden slug, a terrestrial mollusk, shows evidence for more complex Pavlovian phenomena such as second-order conditioning, in which previously rewarded stimuli reinforce a new response. They also show the phenomena of blocking and extinction (Sahley, Gelperin, and Rudy 1981; Carew and Sahley 1986), which are described in more detail below. And nematodes (roundworms) can learn to discriminate between good and toxic food (Zhang, Lu, and Bargmann 2005). Given that nonvertebrates can do all that with reward, what is so special about what vertebrates can do with rewards?

4.3.2 Midbrain Dopamine Nucleus

The evolution of the midbrain dopamine system must have had a profound impact on what vertebrate brains can do with reward. Their function has been thoroughly reviewed (Schultz 2006, 2007) and so only selected aspects are mentioned here. In general, a reward acts as a reinforcer to strengthen associations between representations that are active in its presence. One consequence of this property is that initially neutral stimuli take on reward value, leading to the reinforcing value of secondary rewards, as mentioned in the previous section. Schultz showed in monkeys that as a stimulus becomes associated with reward, dopamine neurons increase their discharge rates when that stimulus appears. This finding explains why conditioned reinforcers work: dopamine cells respond to CSs much like they do to unconditioned reinforcers such as primary reward. Schultz also showed that dopamine-cell activity reflects an error signal that conveys the difference between predicted and received reward. The sources of the prediction and error remain unknown, but one possibility is that striatal GABAergic neurons provide a reward-prediction signal to the dopamine cells and that other inputs to the dopamine cells, such as those from the amygdala, hypothalamus, and peripeduncular nucleus or bed nucleus of the stria terminalis, provide the received reward signal. In this view, the dopamine cells compute the difference between these two inputs and their output serves as a teaching signal to its targets, among which the striatum figures especially prominently.

Receipt of reward at an unexpected time or in larger-than-expected amounts leads to strengthening of associations between representations active at the time of the teaching signal, which makes it more likely that the action made in that context will be repeated when the context recurs. Failure to receive a reward at the expected time and in the expected amount leads to a form of extinction, and puts the system in a state that promotes exploration of alternative actions. Comparison of predicted and received reward is the fundamental mechanism by which vertebrate brains deal with reward.

4.3.3 BASAL GANGLIA

One of the most vexing problems in understanding the general role of the midbrain dopamine system, especially in relation to basal ganglia function, is that one line of research has focused on its role in motor control, while another has focused on its role in reward-seeking behavior, as exemplified by Pavlovian approach behavior (S-O associations), Pavlovian-to-instrumental transfer (PIT), and instrumental conditioning. The first line of research has, after a period of uncertainty, confirmed earlier opinions (Northcutt 1996) that the largely diencephalic group of dopamine neurons found in jawless fish (lamprey) is homologous to the midbrain dopamine system in tetrapods (Smeets, Marín, and Ganzález 2000). Therefore, it now appears even more likely that the dopamine system evolved in the earliest vertebrates (Figure 4.1). The relatively new evidence is that the same toxin that produces Parkinson's disease (PD) in people causes similar symptoms in these jawless fish (Thompson et al. 2008). An equally ancient structure, the lateral habenula, appears to provide a signal to dopamine neurons that inhibits these neurons when rewards are not predicted (Matsumoto and Hikosaka 2007). It appears, then, that the evolution of this dopamine system in early vertebrates had profound implications for what their brains could do with reward, but it remains difficult to reconcile the motor functions of dopamine-basal ganglia interactions with those involving primary and secondary rewards. One approach toward this reconciliation has been to consider several different time courses for the actions of dopamine, extending several orders of magnitude from fractions of a second, as described above, to many hours (Schultz 2007). Alternatively, the dopamine system may compute the cost–benefit analyses that are fundamental to all decisions.

Recent work has pointed to a common function for both motor and nonmotor parts of the basal ganglia. According to Shadmehr and Krakauer (2008), the interaction between midbrain dopamine cells and the basal ganglia modifies the assessment of costs and benefits in a very general sense, one not limited to reward as traditionally construed. They based their view on a computational model and on recent psychophysical evidence that PD, which is caused by degeneration of the midbrain dopaminergic neurons, leads to a form of effort intolerance (Mazzoni, Hristova, and Krakauer 2007; Shadmehr and Krakauer 2008). Apart from tremor, two principal symptoms of PD are bradykinesia (slow movement) and akinesia (no movement). Shadmehr and Krakhauer have related these symptoms to a change in how the brain computes the value of a certain amount of energy expenditure. In a psychophysical experiment (Mazzoni, Hristova, and Krakauer 2007), human subjects made a series of voluntary reaching movements at varying speeds and distances. The investigators measured, among other things, the number of reaches needed to produce 20 reaches that fell within a given velocity range, such as 37–57 cm/s. Healthy people show a reluctance to make relatively fast reaching movements, measured by their need to attempt ~35 trials needed to make 20 within the fastest movement range. PD patients show even more reluctance to make fast movements than do healthy people, requiring ~45 trials to make the same number of correct ones. Yet their reaching movements maintain normal spatial accuracy. Similar conclusions come from the work of Desmurget and Turner (2008), who inactivated the internal segment of the globus pallidus, the output structure of the "motor" basal ganglia, in rhesus monkeys. This inactivation causes abnormally short movements (hypometria), which are slower than normal (bradykinetic), but have the normal trajectory, endpoint accuracy, and reaction times. Like Shadmehr and Krakauer, Desmurget and Turner concluded that the basal ganglia "may regulate energetic expenditures during movement."

Thus, the dopamine system—through its interaction with the basal ganglia—appears to lead to the minimization of a cost: the increased energy expenditure (effort) needed to make faster movements. In PD, the decrease in dopamine input to the striatum alters the sensitivity to such effort costs. According to Mazzoni, Hristova, and Krakauer (2007, 7105), "bradykinesia represents an implicit decision not to move fast because of a shift in the cost/benefit ratio of the energy expenditure needed to move at normal speed," which is an exaggerated form of a similar shift that occurs in healthy people. The basal ganglia and its dopamine inputs thus play a key role in computing the cost of reaching movements, and this function likely generalizes to volitional, skeletal movements of many kinds, including those involved in seeking and consuming primary rewards.

This kind of cost–benefit analysis could link the well-known motor functions of parts of the dorsal basal ganglia to the standard psychological framework implicating the ventral basal ganglia in decisions based on predicted reward outcome (Schultz 2006). This computational view of reward generalizes its definition to include benefits other than typical primary and secondary reinforcers, including benefits that come in the form of reduced energy costs. As Mazzoni, Hristova, and Krakauer (2007, 7105) put it, a role for dopamine in minimizing effort cost is "analogous to the established role of dopamine in explicit reward-seeking behavior." Note, however, that when Mazzoni et al. use the term "explicit" in this way, they do not do so in the sense usually meant in psychology, which implies conscious awareness and the ability to report on a decision and its basis. Instead, they mean to distinguish two levels of implicit decisions: one that leads to behaviors that have produced benefits in terms of primary and secondary rewards, and another that produces benefits in terms of reduced effort costs. It appears that the basal ganglia and its dopamine inputs play a key role in computing the costs and benefits in a very general sense, for use in neural functions that minimize costs and maximize benefits of many kinds.

We can only speculate about the common mechanisms between these two kinds of "reward" computations. One possibility is that dopamine cells do for effort costs more or less what they do for primary rewards. When vertebrates make a movement that uses more than the optimal amount of energy, i.e., they expend more effort than necessary to reach a movement target, the mismatch between predicted and actual effort would cause the dopamine cells to decrease their activity, and the behavior would be "extinguished." Conversely, if they reached the target with less effort than expected, the dopamine cells would increase their activity and this would teach the system to make the movement requiring this lesser effort in the future, thus optimizing the amount of energy expended to reach a movement target.

4.3.4 Amygdala

As they learn the biological value of rewards, vertebrates tend to select the behavior that yields the highest reward value. Not only can vertebrates learn the value of particular rewards, they can rapidly update the valuation of a reward based on their current state. The *reinforcer devaluation* procedure assesses this updating ability. When an animal has recently consumed one kind of food to satiety, that food becomes less valued, and animals will reduce a behavior that produced the devalued food. This happens even when their behavioral choices no longer yield the devalued reward (i.e., under extinction), which implies that they can recall the reward (outcome) value in its updated form.

In contrast to intact animals, rats or macaque monkeys with amygdala lesions fail to adjust their responses after reinforcer devaluation (Hatfield et al. 1996; Malkova, Gaffan, and Murray 1997; Balleine, Killcross, and Dickinson 2003; Izquierdo and Murray 2007). Although the experiments in monkeys involved the whole amygdala, work in rodents indicates that the basolateral amygdala (BLA), not the central nucleus, subserves this ability (Hatfield et al. 1996; Balleine, Killcross, and Dickinson 2003). Importantly, rats and monkeys with amygdala lesions have intact satiety mechanisms and intact sensorimotor discrimination abilities. The deficit in shifting choices after amygdala lesions thus results from an impairment either in updating food value or in linking stimuli and responses with this new value to guide behavior.

Wellman, Gale, and Malkova (2005), also studying macaque monkeys, disrupted reinforcer-devaluation effects by temporarily inactivating the BLA with the $GABA_A$ agonist muscimol. They replicated the result described above for amygdala lesions, but only when the inactivation occurred *during* the selective satiation procedure, not when inactivation occurred afterwards. Similarly, Wang et al. (2005) demonstrated in rats that, for updating to occur, protein synthesis is needed in the BLA immediately following devaluation. These findings show that the amygdala updates reward value, but has little role in guiding choices after the value of the food has been updated.

In addition to devaluation experiments, amygdala lesions in rodents and monkeys impair behavior on a number of tasks that require associating a stimulus with value, including second-order conditioning, conditioned cue preference, conditioned reinforcement, delay discounting, and PIT, among others (Everitt et al. 2003; Murray 2007). For example, in *delay discounting*, a lever press produces a large reward at an increasingly longer delay, which devalues or *discounts* that reward. As the large reward is progressively delayed, rats gradually shift their choices to a lever that produces a smaller, more immediate reward. Rats with BLA lesions shift responses more quickly than intact rats, thereby exhibiting a steeper discounting function than intact rats (Winstanley et al. 2004). This relative intolerance for delay could result from difficulty in maintaining the representation of reward value in its absence (Winstanley et al. 2004), but the results of the next experiment suggest a different account.

Rudebeck and Murray (2008) found that monkeys with selective amygdala lesions use reward information better than controls in certain circumstances. Their monkeys chose between two stimuli: one choice produced positive feedback (reward) and the other produced negative feedback (effort cost with no benefit). Monkeys with amygdala lesions performed better than controls on the trial following a correctly performed trial, indicating they used reward feedback more efficiently than controls. Both groups used negative feedback equally well. As noted above, Winstanley et al. (2004) suggested the BLA lesions facilitated delay discounting because the rats could not represent the value of the large reward in its absence. The findings of Rudebeck and Murray (2008) suggest instead that the amygdala lesion disrupted an affective signal associated with the lever that produced large rewards. Indeed, the results of both Winstanley et al. (2004) and Rudebeck and Murray (2008), and perhaps others as well (Izquierdo and Murray 2005), could result from the lack of an affective signal that biases choices towards a previously rewarded stimulus. This account implies that the amygdala can impede reward-based decision making and, without its affective signals, animals can sometimes behave more "rationally," as assessed in terms of costs and benefits.

We focused here on the BLA, but other parts of the amygdala might produce different reward-related signals. The BLA appears to link stimuli with the specific sensory properties of reward, whereas the central nucleus of the amygdala links them to reward's general affective properties (see Balleine and Killcross 2006, for detailed evidence). As Balleine and Killcross noted, these two roles could contribute to different aspects of foraging, with general affect promoting the preparatory and approach phase, as in arousal, and signals related to specific rewards promoting their consumption.

In sum, the amygdala mediates several mechanisms linking stimuli and responses with the affective properties of rewards, and these mechanisms likely evolved in early vertebrates.

4.3.5 HIPPOCAMPUS

There is ample evidence for a role of the hippocampus in spatial navigation and spatial memory in many vertebrate species, including animals in groups as diverse as reptiles, ray-finned fishes, birds, rodents, and primates. The activity of hippocampal pyramidal neurons in rats reflects spatially discrete locations (place fields) and collectively produces a map-like representation of the environment (O'Keefe and Nadel 1978). A growing body of evidence indicates that septal and temporal sectors of the hippocampus, often referred to in rats as the dorsal and ventral hippocampus, respectively, have dissociable functions. The septal region is essential for accurate spatial navigation and the temporal region subserves appropriate behavior in anxiety-inducing environments, as well as the contextual retrieval of internal cues based on motivational state (Kjelstrup et al. 2002; Bannerman et al. 2003). Anatomical relations of the septal and temporal portions of the hippocampus mirror their functional specializations. The temporal hippocampus, unlike the septal hippocampus, has extensive, reciprocal connections with the amygdala and hypothalamus, among other regions (Petrovich, Canteras, and Swanson 2001).

A recent finding places the septal-temporal dissociation of hippocampal function in a new perspective. Place fields within the hippocampus vary in size, with smaller place fields (<1 m) found septally and larger place fields (~10 m) located temporally (Kjelstrup et al. 2008). The large place fields in the temporal hippocampus may aid in identifying and representing familiar spatial contexts, as a form of generalization, and in identifying novel contexts. This ability to identify novel contexts could underlie both the reduction in exploratory behavior that occurs in anxiety-inducing environments and the reduction in eating that occurs in novel environments. These effects have been shown to depend on the temporal hippocampus (Kjelstrup et al. 2002; Bannerman et al. 2003; McHugh et al. 2004; Trivedi and Coover 2004) and likely reflect its interactions with the amygdala as well (Petrovich, Canteras, and Swanson 2001). Furthermore, the identification of familiar contexts is necessary for the contextual retrieval of cues based on internal signals regarding motivational state (Stouffer and White 2007). Thus, the functions of the temporal hippocampus may be an emergent property of its large place fields and specialized anatomical connections.

Table 4.1 summarizes the main points in Section 4.3, which outlines in broad terms what vertebrate brains can do with reward. Note that although Table 4.1 summarizes structure-function relationships in vertebrates, generally, it relies mainly on findings from mammals. Section 4.4 takes up the brain augmentations underlying what mammals can do with reward that their nonmammalian ancestors could not.

4.4 WHAT CAN MAMMALIAN BRAINS DO WITH REWARD?

4.4.1 MAMMAL BRAINS: BASIC VERTEBRATE MECHANISMS PLUS NEOCORTEX

As mammals evolved from our amniote ancestors (Figure 4.1), we came equipped with a brain that had come to include some new structures, compared to those found in ancestral, nonmammalian species. As outlined in Section 4.2.3, the frontal, parietal, occipital, and temporal areas that make up the neocortex changed forever how our brains could deal with rewards. This fact does not imply, however, that the systems that had developed earlier in vertebrate evolution became redundant; reward processing in mammalian brains still relies heavily on the midbrain dopamine system, basal ganglia, amygdala, and hippocampus. The phylogenetically older systems not only continue to operate alongside the mammalian innovations, but interact with them in a manner that transforms their

TABLE 4.1
Vertebrates

Midbrain dopamine nucleus	Reward prediction, comparison of prediction and received rewards
Basal ganglia	Effort cost analysis
Amygdala	
Central	General affective consequences of reward
Basolateral	Specific sensory properties of rewards
Hippocampus	
Temporal[a]	Coarse-grained representation of locations (contexts), association of contexts with specific rewards and affective states
Septal[b]	Fine-grained representation of locations for navigation during foraging

[a] Also known as ventral hippocampus in rodents and anterior hippocampus in primates.
[b] Also known as dorsal hippocampus in rodents and posterior hippocampus in primates.

function to an extent. Here we consider the adaptive advantages that the neocortex provides for reward-seeking behavior, in the context of cost-benefit analysis and foraging.

As discussed earlier in this chapter (and throughout other chapters in "Part III" of this volume), the central nervous system provides the means for an animal to learn about the relationships between its own actions, different aspects of its environment (including discrete stimuli and contexts), and the occurrence of reward, along with costs. The formation of internal representations of these stimuli, contexts, responses, and outcomes, as well as the association of these representations with each other, underpins learning. Although nonmammalian vertebrate brains form complex representations too, the evolution of neocortex in mammals enabled the formation of richer, more versatile internal representations of sensory inputs to associate with reward, including the sensory aspects of reward, its motivational aspects, and its hedonic properties. For example, sensory areas of the neocortex improve, relative to nonmammalian ancestors, the processing of the sensory aspects of rewards and stimuli associated with rewards. As we discuss in more detail in Section 4.5.2, sensory neocortical areas such as the perirhinal cortex, a feature of all mammalian brains, process sensory information in advanced ways. Accordingly, the mammalian brain does what the brain of its nonmammalian ancestors did with reward but does it better in certain ways.

Although reward processing involves all of the neocortex directly or indirectly, the agranular frontal areas seem to play a particularly large and specific role. Accordingly, this section focuses on these frontal areas. As indicated earlier, rats and other modern rodents do not replicate the traits of early mammals. They, like the other mammalian lineages, have many specializations that emerged over tens of millions of years of independent evolution. However, the organization of the agranular frontal areas in rats appears to be reasonably representative of the ancestral mammalian condition, notwithstanding the many anatomical and behavioral specializations of rats. The long, separate evolutionary history of the agranular frontal cortex in each mammalian lineage makes it likely that differences in both structure and function occur in each order and species of mammal. Nevertheless, we know of no reason to think that the fundamental functions of the agranular frontal areas differ among the mammals.

Learning, along with innate mechanisms, drives reward-seeking behavior in all vertebrates and, if left unsupervised, the most dominant associations ultimately control behavior. The agranular frontal cortex changes this state of affairs. We propose that among the most important innovations of early mammals was an improved "executive function," mediated by the agranular frontal cortex, including a top-down modulatory function that biases competition among different brain systems engaged in and competing for the control of behavior.

4.4.2 Agranular Frontal Cortex

The mammal innovations include the regions of the agranular frontal cortex called the anterior cingulate (AC), PL, IL, agranular orbital frontal (OFa), and agranular insular (Ia) cortex, which all mammals share (see Section 4.2.3). A growing body of evidence implicates these different parts of the frontal cortex in processes that, in the broadest sense, afford mammals a greater flexibility in their reward-seeking behavior and ultimately bestow upon them the ability to pick and choose between different behavioral options, selecting the one best suited for the exigencies of the moment. These areas have some unique inputs compared to other parts of the mammalian neocortex. The Ia cortex receives relatively direct inputs from parts of the gustatory cortex, located immediately caudal to it, as well as olfactory information from the piriform cortex and visceral related signals from the brainstem and thalamus (Ray and Price 1992). In addition, some "exotic" somatosensory inputs arrive there, leading to the view that it functions in interoception, including sensations conveying pain, itch, temperature, metabolic state (e.g., hypoxia or hypoglycemia), and inputs from the lungs, heart, baroreceptors, and digestive tract (Yaxley, Rolls, and Sienkiewicz 1990; Ray and Price 1992; Zhang, Dougherty, and Oppenheimer 1998; King et al. 1999; Craig 2002; Saper 2002; Drewes et al. 2006). In addition to these inputs, the agranular frontal cortex has reciprocal connections with the amygdala, hippocampus, and midbrain dopamine system, and projects to the basal ganglia. Through these and other pathways, the mammalian frontal cortex interacts with these phylogenetically older systems to promote flexibility in their function (see Sections 4.3.2–4.3.5). (See also Chapter 11 for an in-depth discussion of corticobasal ganglia circuitry and interactions.)

Aside from nonassociative processes such as sensitization and habituation, associative learning mechanisms account for much of a mammal's ability to obtain rewards. Through Pavlovian and instrumental conditioning that occurs during reward-related episodes, an associative network builds up between internal representations of rewards, actions, and stimuli; each representation has the potential to enter into associations with other representations. This vast associative structure involves several "memory systems," so called because different parts of the brain encode, represent, and store different kinds of information, which animals use to optimize their performance in terms of benefits and cost. Together with innate mechanisms, the associations mediated by these "memory systems" guide reward-seeking behavior, but do so in different ways. With no top-down modulation, the most dominant of these systems—perhaps based on the strength of its associations—will control behavior.

The behavior of nonmammalian vertebrates gives the impression of operating under the control of a dominant response system of the sort just described, sometimes called a prepotent response. Birds and reptiles often appear to have problems overriding their innate behavioral responses even when this prepotent response increases costs and reduces benefits. These relatively inflexible forms of behavior provide an advantage under many circumstances, especially routine ones. However, problems arise on the less frequent occasions when another behavioral choice would yield greater benefits or reduced costs. We present two examples of such inflexible behavior in nonmammals: one from a reptile, the other from a bird. Note from Figure 4.1 that both kinds of animals are crown species. Neither birds nor modern reptiles represent the ancestral vertebrate condition very well, but they do carry forward many traits from their most recent common ancestor and behavioral control by a dominant brain system may be one of them.

For instance, the green lizard *Lacerta* has an innate tendency to approach the color green, an instinctual response that presumably evolved to guide them towards leaves. Leaves are a good place for *Lacerta* to spend their time because they provide camouflage and house the insects and grubs that compose their diet. Wagner (1932) showed that it was almost impossible to train these lizards to override their innate response and reject green, even when in doing so they failed to obtain a palatable food reward. Lizards were trained on a task that required the choice of one of two panels: a red one with a palatable mealworm attached versus a green one with an unpalatable, saline-adulterated worm attached. The lizards consistently chose the green panel despite the cost of doing so: the loss of a tasty worm. Wagner (1932) reported that it took hundreds of trials for these lizards to override their innate response, if they could do so at all. Sensory and motor factors could not account for this result, as the lizards can easily distinguish the green and red panels and make the required choice. Instead, it appears that the prepotent response made it difficult or impossible for the brains of these lizards to use reward in a way that any mammal could easily do, at least any mammal that could see and discriminate red from green.

A frequently recounted study by Hershberger (1986) provides another clear demonstration of severe difficulty in using reward experience to override innate, prepotent responses, in this case from birds. Chicks (genus *Gallus*) were fed from a bowl with distinct visual characteristics and would subsequently approach the same food bowl if released at a distance from it. This consistent behavior shows that the bowl had acquired the properties of a Pavlovian CS, based on an association between the bowl and the reward outcome, which elicits an instinctive approach response. The chicks were then placed in a "looking-glass" scenario in which the bowl receded twice as fast as they ran towards it and approached them twice as fast as they ran away from it. They therefore had to inhibit their approach behavior in order to obtain the food. Even after a hundred minutes, the chicks continued to run toward the bowl. Apparently, they could not inhibit their conditioned approach response and alter their behavior, despite the fact that their actions cost them the reward every time. In short, the causal consequences of their behavior (i.e., the instrumental contingency) did not affect their behavior, either because they could not learn the experimentally contrived contingency or because they could not make use of it in the face of competition from the dominant response system, in this case one mediating an innate approach response.

Mammals, on the other hand, modify their reward-related behavior flexibly, adapting quickly to their changing needs and environment. This is not to say that instinct controls nonmammalian behavior but has no such control for mammalian behavior or that nonmammalian behavior lacks any flexibility. No such simplistic formulation could ever stand up to scrutiny. Rather, we suggest that mammals have brain areas—the agranular frontal areas—dedicated to modulating the balance among several behavioral-control systems. Nonmammals, lacking these areas, have more trouble doing that.

In rats, for example, the matching-to-position (MTP) task assesses the ability to override innate response tendencies. The MTP task pits a rat's innate foraging tendency against a newly learned behavior. In order to maximize the likelihood of finding food when foraging, rats have a natural tendency to explore locations that they have not visited recently (if at all) rather than places recently explored and exploited. This foraging strategy means that when given the choice between two arms of a maze, each of which might contain food, rats usually choose the arm visited least recently. The MTP task requires rats to override this innate tendency and learn to select the most recently visited arm in order to obtain a food reward. Normal rats can learn this task easily, successfully overriding their propensity to spontaneously alternate, but rats with prior removals of the PL/IL region of the agranular frontal cortex cannot do so (Dias and Aggleton 2000). The same lesion has no effect on the acquisition of the nonmatching-to-position task, which requires behavior consistent with the rat's innate tendencies.

The MTP task highlights the advantage of overriding innate responses, at least in a psychology laboratory. The ability to override this tendency must also have some value outside the laboratory. The tendency to vary foraging locations increases a rat's likelihood of finding fluid or nutrients, as it would have for ancestral mammals. However, when a single location provides a large benefit at low costs, the ability to suppress the prepotent foraging strategy ("look somewhere else") will likely reduce costs and increase benefits.

Sections 4.4.2.1–4.4.2.5 summarize some of the key research on subdivisions of the rat frontal cortex and their role in the flexible control of reward-seeking behavior, stressing two main themes: top-down biasing of behavioral control systems and flexible alterations of foraging strategies (Table 4.2).

4.4.2.1   Anterior Cingulate

The AC cortex lies along the medial wall of the rat frontal cortex (mainly AC1 in Figure 4.2c). At first glance, the results of AC lesions on reward-based tasks seem contradictory: some studies report AC lesion-induced disruption across a range of Pavlovian and instrumental tasks, and others report no effect (e.g., Bussey, Everitt, and Robbins 1997a, 1997b; Cardinal et al. 2003; Haddon and Killcross 2005; de Wit et al. 2006). However, once the number of stimuli present at any one time is considered, a fairly consistent pattern of results emerges. Rats with AC lesions show deficits when two or more stimuli simultaneously vie for control of performance (Cardinal et al. 2002, 2003). The AC cortex seems to play the same role for both Pavlovian CSs and instrumental discriminative stimuli: biasing behavioral control towards one among multiple competing stimuli (and their associations). This bias has the effect of ensuring that the most beneficial stimulus maintains control of behavior.

The AC cortex also influences the cost-benefit calculations that influence reward-seeking behavior. The amount of energy required to obtain a reward represents a major cost factor and the AC cortex contributes to weighing whether the benefit of acquiring a reward merits the effort required to obtain it (Walton, Bannerman, and Rushworth 2002; Rushworth et al. 2007). This conclusion is based on research that employs a specially designed task employing a T-maze, one that requires rats to choose between two arms of the maze, each containing food. If a small barrier in one arm requires the rat to climb in order to obtain food, all other things being equal the rat will prefer the no-barrier arm. A larger quantity of food at the end of the barrier arm can induce the rats to climb for their food. Compared to normal rats, those with AC lesions show less tolerance for this increased

TABLE 4.2
Mammals and Primates

AC	Anterior cingulate	Agranular frontal cortex: shared by rats, monkeys, and other mammals	Medial PFC in rodents	Biases competition among multiple stimuli and actions for dominant control over behavior
				Advantageous in the presence of multiple possible actions and outcomes
PL	Prelimbic			Biases behavior toward goal-directed choices (R-O associations) vs. habits (S-R associations)
				Advantageous in an dynamic environment, when the outcome of behavior is changing
				With IL, biases behavior toward repeated exploitation (based on outcome) vs. an innate tendency to avoid exploited places (independent of outcome)
IL	Infralimbic			Biases behavior toward habits (S-R associations) vs. goal-directed choices (R-O associations)
				Advantageous in a static environment, when the outcome of behavior is fixed
OFa	Agranular orbitofrontal		Orbital or lateral PFC in rodents	Biases control toward seeking transiently available, higher-value rewards ("impatient" foraging, increased delay discounting)
				Advantageous in the presence of a rich, available resource
				Improved stimulus-outcome (S-O) predictions, specific for sensory properties each kind of reward
Ia	Agranular insular			Biases control toward seeking rarely available, higher-value rewards ("patient" foraging, anticipatory contrast effect)
				Advantageous when a rich resource might be available later
				Improved storage or recall of sensory aspects of specific rewards
PFg	Granular prefrontal cortex	Absent in rodents and other nonprimate mammals		Advanced feature-value conjunctions, including "menu"-independent value, drive-independent value ("the common currency"), association of strategies and rules with reward, value-free use of rewards

Abbreviation: PFC, prefrontal cortex.

effort (Walton, Bannerman, and Rushworth 2002; Schweimer and Hauber 2005; Rudebeck et al. 2006). Although AC lesions disrupt effort-based performance in the T-maze, other tasks that involve manipulating the level of effort fail to show such effects. Schweimer and Hauber (2005) reported that when the number of lever-presses required to produce a reward progressively increased in stages (called a progressive ratio schedule), the performance of rats with AC lesions did not differ from that of intact rats. Together, these findings suggest that the AC cortex may only be required for cost-benefit computations when rats must choose between two or more response options.

Based on these findings, we propose that the AC cortex contributes to optimizing behavior under conditions of multiple, competing stimuli and actions. The AC cortex may provide the mammalian brain with the means to weigh more behavioral options than our nonmammalian ancestors could.

4.4.2.2 Prelimbic Cortex

The PL cortex lies along the medial wall of the rat frontal cortex (Figure 4.2c). Like the AC cortex, it contributes to behavioral flexibility; however, unlike the AC cortex, the PL cortex appears to be important specifically in situations involving competition between associations guiding instrumental behavior. Evidence from PL lesions indicates that it is required for the encoding of response-outcome (R-O) associations during learning. This means that PL plays a vital part in maintaining control of instrumental behavior in favor of *goal-directed* R-O associations over competing *habitual* stimulus-response (S-R) associations, which are built up simultaneously during learning. If a behavior is goal-directed, and therefore under the control of R-O associations, an animal shows sensitivity to both the current value of the reward and to the contingency between its response and reward. If a behavior is habitual, and therefore under the control of competing S-R associations, these factors do not influence behavior (see Chapter 14).

The level of behavioral control by R-O associations can be assessed by the reinforcer devaluation procedure described earlier in Section 4.3.4, in which the reward outcome value is reduced. Alternatively, behavioral control by R-O associations can be assessed by a *contingency degradation* procedure, in which the causal relationship between a response and an outcome is degraded (so that the reinforcement is just as likely if the animal does not respond as if it does), while the probability of a rat receiving reinforcement following a response is kept the same. In this situation, rats are able to detect the loss in contingency (providing evidence for sensitivity to R-O correlations) and so reduce responding. Several studies using these techniques show that PL lesions lead to an insensitivity to reinforcer devaluation and contingency degradation procedures. Thus, rats with PL lesions are left with habitual behavior controlled purely by S-R associations; they are blind to the causal relationship between their responses and reward, as well as to the value of the outcome that their actions produce (Balleine and Dickinson 1998; Corbit and Balleine 2003; Killcross and Coutureau 2003). The role of PL is, however, time-limited; lesions of this region that occur after training do not disrupt the aforementioned devaluation procedure in rats, which indicates that PL is involved in the encoding (but not the storage or utilization) of R-O associations (Ostlund and Balleine 2005).

Through the use of inactivation techniques, the PL cortex has also been implicated in switching between strategies used to solve the problem posed by certain tasks. It is important to note that in these experiments the inactivation encompassed parts of both the PL and IL areas, although the inhibitory injections were centered on the PL. For example, Ragozzino et al. (1999, 2003) found that inactivation of the PL/IL cortex selectively disrupted switching between two strategies in a plus maze (see Ragozzino 2007 for review). In one experiment (Ragozzino et al. 2003), rats had learned to use odor cues (independently of spatial location) to guide responses. They were later required to use spatial location (independently of odor cues) to guide responses. Other rats were trained in the opposite order. PL/IL inactivation had no effect on initial performance of the task irrespective of which strategy was correct. However, relative to controls, the rats with PL/IL inactivation were slow to acquire the new strategy, regardless of which of the two strategies was new. The impairment was quite selective: PL/IL inactivation had no effect on reversal of the odor-reward contingencies, even when that task was matched for difficulty to the strategy-switching task. Although the precise role of PL in these strategy-switching tasks remains unknown, the data are consistent with the idea that PL plays a role in the formation of R-O associations. According to this model, PL inactivation would disrupt acquisition of R-O associations while leaving the formation of S-R associations unaffected. If so, PL inactivation during learning of the first strategy might not affect performance because, under these noncompetitive conditions, S-R associations would be sufficient to support behavior. However, PL inactivation during the switch—as the rats attempted to learn the second strategy— would disrupt the formation of the R-O associations encoding the new strategy. As a result, only the newly formed S-R associations would be available to compete with the older S-R and R-O associations, which encoded the first strategy.

One additional clue about the nature of PL's contribution to strategy switching comes from another recent study (Rich and Shapiro 2007). Rats were required to switch between a response strategy (e.g., turn left) and a spatial or place strategy (e.g., turn north). Although PL/IL inactivation did not disrupt performance on the day of the switch, it did lead to increased use of the old strategy when rats were tested 24 hrs later. Interestingly, this effect was transient and did not persist over multiple switches between the two strategies, indicating that PL is not essential for switching between familiar, well-learned strategies. Thus, it appears that PL is not necessary for switching per se but rather for a mechanism that influences switches when the behaviors are not well learned. This conclusion makes sense if one considers that, across training, instrumental responding shifts from R-O to S-R control (Dickinson and Balleine 1994). In this view, early switches would rely heavily on R-O associations, whereas later switches would predominately involve S-R associations. Disruption of the formation of R-O associations by PL inactivation would therefore be expected to disrupt early switches more than later ones. Rich and Shapiro (personal communication) also examined the activity of neurons in the medial frontal cortex during a place-response strategy switching task. They found that the activity of PL neurons was correlated with the behavioral switch. In addition, although both PL and IL neurons were active during the switch phase, the changes in activity of PL neurons preceded those in IL neurons; the change in activity of IL neurons occurred when performance was close to being proficient once again. This finding probably reflects the passage of control of instrumental behavior from PL-dependent R-O associations to IL-dependent S-R associations (see Section 4.4.2.3). (Note that the neurophysiological data of Rich and Shapiro appeared in the *Journal of Neuroscience* after the completion of this chapter.)

In summary, a substantial body of work implicates PL in mediating goal-directed behavior by supporting the development of R-O associations. Recall our earlier discussion of the role played by the amygdala in updating the biological value of rewards (Section 4.3.4). In this context, PL could promote behavioral flexibility by allowing animals to take into account both the current value of a reward as well as the relationship between its response and the reward it produces. This role for PL cortex could also account for its contribution to strategy switching.

4.4.2.3 Infralimbic Cortex

The IL cortex lies along the medial wall of the frontal cortex and, in rats, is located immediately ventral to the PL cortex (Figure 4.2c). As in some of the rule-switching studies mentioned above, researchers have often considered the PL and IL cortex together. Lesions limited to the IL cortex, however, yield different behavioral effects than lesions limited to the PL cortex, especially for instrumental behavior and extinction.

Whereas the PL cortex supports the control of instrumental behavior by response-outcome (R-O) associations (see Section 4.4.2.2), the IL does the opposite, promoting behavioral control by S-R associations. Using satiety-specific devaluation of a reward associated with an instrumental response, Killcross and Coutureau (2003) demonstrated a double dissociation of function between the PL and IL cortex. As described in the previous section, PL (but not IL) lesions cause a rat's behavior to become insensitive to devaluation of its associated reward, leaving them stimulus-bound, literally creatures of habit. By contrast, IL (but not PL) lesions result in behavior that remains under the control of a predicted reward outcome and its biological value, even after a lengthy training period. Put in positive terms, as opposed to the negative descriptions that arise from describing the effects of lesions or inactivations, these findings imply that whereas PL supports behavioral control of instrumental responding by outcome-dependent, goal-directed R-O associations, IL biases behavior toward outcome-independent, habitual S-R associations (Table 4.2). Given the well-documented shift in behavioral control from goal-directed to habitual responding that occurs with lengthy periods of instrumental training (Dickinson and Balleine 1994), such a transition presumably reflects a change in the balance between PL and IL influences, with IL ultimately prevailing for situations of high predictability. Consistent with this idea, temporary inactivation of IL following extensive training reinstates outcome-dependent, goal-directed performance after behavior becomes habitual (Coutureau and Killcross 2003).

The IL cortex also plays a role in extinction, during which the performance of behaviors that previously led to reward no longer do so. Extinction differs from forgetting or unlearning in that it depends on new learning that overrides the original knowledge, and evidence suggests that this learning takes the form of inhibitory S-R associations (Rescorla 1993). Like the kinds of behavioral competition mentioned earlier, extinction training establishes a situation in which the original learning and the new learning compete for behavioral control. Under certain conditions, such as the passage of time, the original learning can once again prevail, and this leads to a transient restoration of the original response referred to as spontaneous recovery. IL has been shown to play a vital role in the expression of extinction learning. Although most of the work has focused on extinction of Pavlovian conditioned fear (see Quirk and Mueller 2008 for a full review), IL also contributes to the extinction of appetitive Pavlovian learning (Rhodes and Killcross 2004). Rats with IL lesions have no problem in acquiring extinction; they show both the normal reduction in responding on the first day of extinction training, as well as faster extinction on the second day compared to the first (i.e., savings). Rats with IL lesions do, however, show a deficit in retrieving the extinction memory at the start of the second extinction session; they display a greater level of spontaneous recovery of the original CR compared to normal rats (Rhodes and Killcross 2004). This finding indicates that IL biases behavioral control towards the newer, weaker extinction learning and that, without IL, the more dominant, original learning maintains behavioral control. This idea is supported by the finding that the IL-lesion-induced increase in spontaneous recovery reduces across successive extinction sessions as the extinction learning strengthens (Lebron, Milad, and Quirk 2004).

Although the two lines of research summarized here for the IL cortex involve different learning mechanisms—one Pavlovian, the other instrumental—both represent situations in which two associations involving the same stimulus vie for control over behavior. In Pavlovian extinction, the inhibitory stimulus-response (S-R) association competes with previous learning regarding the same stimulus. By contrast, in instrumental behavior, the habitual S-R association competes with the goal-directed S-R-O associations, which are also under the control of the same stimulus. The IL cortex contributes in much the same way to other behaviors that involve competition between two associations involving the same stimulus, such as Pavlovian conditioned inhibition (Rhodes and Killcross 2007) and discrimination reversal learning (Chudasama and Robbins 2003).

The IL-dependent ability to bias the control by one kind of association over another would confer an adaptive advantage for mammals over nonmammalian ancestors that lacked an IL cortex. These nonmammalian ancestors would not have the behavioral flexibility provided by the ability to learn and bias competing and sometimes contradictory associations, a contribution that comes to the fore when an otherwise dominant behavior no longer yields a sufficient benefit.

Taken together, the medial frontal areas of mammals—AC, PL, and IL cortex—show similarities in their function. In their specialized ways, each appears to allow rapid adaptability to a changeable environment in which experiences lead to competing responses, any of which might be appropriate in a given circumstance (Table 4.2). Sections 4.4.2.4 and 4.4.2.5 take up the orbital frontal areas, OFa and Ia.

4.4.2.4   Orbitofrontal Cortex

Of the many parts of the agranular frontal cortex, the agranular orbitofrontal (OFa) cortex has perhaps attracted the most attention. A major problem with this literature involves the assumption that the regions called the "orbitofrontal" cortex in rats correspond to the regions with the same name in primates. As noted above, comparative neuroanatomy indicates, to the contrary, that the rostral, *granular* parts of the orbitofrontal cortex in primates have no homologue in nonprimate species, including rats (Figure 4.2). Accordingly, we distinguish between the granular parts of the orbitofrontal cortex in primates, which rodents lack, and the agranular orbitofrontal cortex, which is homologous in rodents and primates. For convenience, we call the former PFo, for orbital prefrontal cortex, and the latter OFa, for agranular orbitofrontal cortex.

There is little question that the rat OFa plays an important role in the control of reward-seeking behavior. The activity of neurons in OFa reflects reward expectation, especially the sensory-specific properties of reward (Schoenbaum, Chiba, and Gallagher 1998). In addition, OFa lesions impair the ability to make decisions on the basis of reward expectations (Gallagher, McMahan, and Schoenbaum 1999). Current views emphasize two interpretations of this kind of result: one posits a contribution to reward valuation per se and the other emphasizes the learning of predictive relationships between stimuli and rewards (i.e., Pavlovian S-O associations). Detailed reviews on this topic have been published (Holland and Gallagher 2004; Delamater 2007; Ostlund and Balleine 2007; Rushworth et al. 2007) and are not recapitulated here.

There is general agreement that lesions of the OFa cause deficits in choices based on reward value. Rats with OFa lesions are more likely than control rats to choose or approach stimuli associated with a devalued reward (Gallagher, McMahan, and Schoenbaum 1999; Pickens et al. 2003, 2005). These results could, however, reflect either a general deficit in the computation of reward values, which would disrupt behaviors driven by both instrumental R-O and Pavlovian S-O associations, or a specific deficit in some aspect of S-O associations. A recent study supports the latter interpretation; OFa lesions fail to disrupt the effect of reinforcer devaluation on an instrumental response, which relies on the integrity of R-O associations (Ostlund and Balleine 2007). This finding indicates that the OFa does not compute (or represent) current reward value in general and, by default, likely plays a more specific role in some aspect of the relationship between stimuli and the rewards that they predict. Findings from other behavioral paradigms that rely on S-O mechanisms, but do not involve any manipulation of outcome value, provide further evidence in support of this notion. These procedures include Pavlovian contingency degradation, PIT and the differential-outcomes effect (McDannald et al. 2005; Ostlund and Balleine 2007). These studies, among others, show outcome-selective effects of OFa lesions, suggesting that the OFa plays an important role in the association between a stimulus and the sensory aspects of reward (as opposed to its motivational or temporal aspects).

Burke et al. (2008) provided direct evidence for this idea using a trans-reinforcer blocking design to control those aspects of a reward representation that become associated with a stimulus. Rats were first trained to associate one stimulus (S1) with a particular reward (O1) and were subsequently presented with S1 as part of a compound stimulus (S1+S2). The compound stimulus was paired with a different reward (O2). Importantly, O1 and O2 were equally palatable food pellets that shared the same general motivational and hedonic characteristics but differed in their sensory-specific characteristics (taste). The associations built up between S1 and O1 in the first stage blocked the formation of associations between S2 and the *general* aspects of O2 in the second stage because they were already predicted by O1. However, associations between S2 and the sensory-*specific* properties of O2 are not blocked, and their presence was identified by testing the ability of S2 to act as a conditioned reinforcer (i.e., whether it had acquired reinforcing properties via its association with the food reward O2). When rats were given the opportunity to press a lever to bring about S2 presentations, S2 was able to support conditioned lever pressing (compared to a fully blocked control stimulus) and this lever pressing decreased following satiety-specific devaluation of O2. These findings from intact rats indicate that the conditioned responding was mediated by the associations between S2 and the specific taste of O2. Rats with OFa lesions, however, would not work for presentations of S2, demonstrating that behavioral control was unaffected by the relationship between a stimulus and the specific sensory properties of reward. The results thus indicate that OFa contributes more to learning associations between CSs and the sensory aspects of reward (e.g., taste, smell, visceral, tactile, and visual concomitants, etc.) than to a role in computing biological value per se.

Lesions of OFa also cause changes in delay tolerance. Relative to control rats, rats with OFa lesions have a greater tendency to choose larger, delayed rewards versus smaller, more immediate rewards (Winstanley et al. 2004; Kheramin et al. 2002). This result surprised many experts, who predicted that rats with OFa lesions would become "impulsive" based on a loss of "behavioral inhibition." In that case, rats with OFa lesions should have become less tolerant of delay, the opposite of what was observed in the two papers cited above. Although we are aware that other experiments have yielded a different result, possibly because of differences in testing and training methods, let us assume for the sake of discussion that OFa lesions cause increased delay tolerance, as the two cited papers reported. Changes in delay tolerance probably reflect something about foraging strategies. For example, Stevens, Hallinan, and Hauser (2005) and Stevens, Rosati et al. (2005) have pointed to species-typical foraging strategies to account for different degrees of delay tolerance in two species of New World monkeys. Tamarins are more insectivorous than common marmosets, whereas marmosets eat more tree gum. Gum eating involves scratching tree bark and waiting for the sap to flow, whereas foraging for insects favors immediate acquisition of a widely dispersed, transiently available food source. Species that typically exploit lower value, but readily available, food resources tend to show considerable delay tolerance; they can wait a long time, their favored food will always be there. By contrast, species that favor high-value but less frequently available foods have little tolerance for delay; they must get what they want now or never. So how could S-O associations contribute to the increased delay tolerance shown by rats with OFa lesions? Perhaps, on the basis of environmental stimuli and their S-O associations, the OFa mediates a prediction of the specific sensory properties of a high-quality reward, such as its taste or smell. This prediction could lead to an affective response that biases rats toward seeking that higher-quality food source. Without this affective signal, the lower-value, more readily available reward predominates.

This view of OFa function stresses sensory processing and prediction, specifically about the sensory aspects of rewards, and complements the proposals put forward above for the medial frontal areas, AC, IL, and PL. Table 4.2 suggests that medial frontal areas bias behavioral control among competing stimuli, actions, and kinds of associations, such as R-O versus S-R associations. Perhaps the OFa, a more lateral part of the frontal cortex, instead biases behavioral control among competing foraging strategies.

4.4.2.5 Agranular Insular

The agranular insular (Ia) cortex comprises the lateral part of the rat's agranular frontal cortex and is an elongated structure, extending more caudally than any other structures considered part of the agranular frontal cortex (see Figure 4.2). Probably as a direct result of this caudal extension, there is considerable variability in the lesions used to study this region; many lesions encroach into neighboring regions, such as the overlying granular insular cortex or the OFa, and many studies employ lesions that differ in rostrocaudal extent (cf. Balleine and Dickinson 2000; Boulougouris, Dalley, and Robbins 2007). Because Ia is near to and connected with the gustatory cortex in the rat, considerable research has focused on its involvement in food reward processing. The Ia cortex appears to be involved in retrieving the representation of a food reward's incentive value when the food is absent. For example, Balleine and Dickinson (2000) used the reinforcer devaluation procedure to show that rats with lesions that probably included the caudal part of Ia, along with adjacent granular areas, have deficits when they need to activate the memory of a taste in order to access incentive-value information. In their task, the rats could press either of two levers to produce two different food rewards and subsequently underwent satiety-specific devaluation. When tested under conditions in which the lever presses resulted in the delivery of a reward, rats with Ia lesions, like controls, pressed the lever associated with the devalued reward less often. However, when tested under extinction (i.e., in the absence of reward), rats with Ia lesions pressed both levers equally, unlike controls. Balleine and Dickinson (2000) concluded that rats with Ia lesions could not recall the sensory properties of the food (e.g., its taste) in its absence and therefore could not activate a representation of the food's value to guide behavior. More research needs to be done to test this hypothesis directly, but if the Ia cortex is involved in recalling the sensory properties of an *instrumental* reward in its absence (based on R-O associations), then it would play a role complementary to that of the OFa. As discussed above, the OFa may be involved in recalling the sensory properties of a reward associated with a *Pavlovian* stimulus (based on S-O associations). However, Holland (1998) has suggested that the Ia cortex might also contribute to the latter function. This idea is based on results by Kiefer and Orr (1992), who showed that rats with Ia lesions—a lesion that spared OFa—failed to show conditioned orofacial responses (a normal result of the decrease in palatability or hedonic value) to a flavored drink that had been associated with illness. This deficit occurred despite the fact that conditioning did take place, as indicated by a reduced consumption of the drink due to the decrease in its motivational value. This finding seems to argue against a distinction between OFa and Ia function in terms of instrumental (R-O) associations versus Pavlovian (S-O) associations. Nevertheless, it appears that like the OFa, the Ia plays some role in the associative activation of some aspect of reward representations.

The concept of accessing the representation of reward value in a food's absence has explanatory value for other studies as well. Kesner and Gilbert (2007) found that rats with Ia lesions do not show the normal reduction in the consumption of a drink of lower-value reward (a 2% sucrose solution) when it is always followed by a drink of higher value (32% sucrose), a phenomenon called the *anticipatory contrast effect*. To perform this task well, rats need to recall the representation of the anticipated second solution when drinking the first, and based on their relative values defer consumption of the first solution pending the availability of the more beneficial reward. The Ia cortex could contribute to the process by which drinking the first solution leads to retrieval of the representation of the sensory characteristics of the second solution from memory. Once the representation of the second solution is activated in this way, its value can be accessed and compared to the value of the first solution to influence behavior.

According to the data summarized above, both Ia and OFa play a role in the associative activation of some aspect of reward, which enables the memory of that reward to guide behavior even when the reward is absent. Compared to nonmammalian ancestors that lack these frontal areas, the improved ability to remember and predict the specific sensations associated with each particular kind of reward would yield advantages when foraging choices need to be based on current needs. The results from Kesner and Gilbert (2007) indicate that Ia biases intact rats toward reducing their consumption of a lower-value reward when a higher-value one is expected later, in a sense promoting a more "patient" foraging strategy. The results from Winstanley et al. (2004) and Kheramin et al. (2002) indicate that the OFa biases rats toward obtaining lower-quality but readily available food rewards: in a sense, promoting a less "patient" foraging strategy aimed at rapid exploitation of a food resource (see Table 4.2).

Specific tests and refinement of these ideas remain for the future, but we propose that the core function—and adaptive advantage—conferred by the lateral parts of the rat's agranular frontal cortex (OFa and Ia) involves the ability to have the best of both worlds, depending on circumstances. When cost-benefit calculations indicate an advantage in patient foraging for a highly reliable and available food source, Ia learns the context for this circumstance. Later, when this circumstance occurs again, Ia biases behavioral control toward high levels of delay tolerance. When cost-benefit calculations indicate an advantage in the rapid exploitation of an unreliable but highly beneficial food source, the OFa learns the context for this circumstance and later biases behavioral control toward "impatient" foraging, as reflected in steep delay discounting functions (Table 4.2). According to this scheme, medial frontal cortex mediates different influences over foraging, based more on spatial, motor, and associative factors, such as R-O, S-O, and S-R associations. For example, the results from Dias and Aggleton (2000), discussed in Section 4.4.2, indicate that the medial frontal cortex biases rats toward repeated exploitation of a given location's resources, overcoming an innate tendency to avoid previously exploited foraging sites.

Stated generally, the idea presented here is that the agranular frontal areas, which evolved in early mammals, fine-tune the balance of behavioral control instantiated by phylogenetically older brain systems, which early mammals inherited from their amniote ancestors. Competing behavior-control systems all remain viable options, available simultaneously to contribute to the animal's fitness when the appropriate context arises. This form of top-down control allows a choice among various courses of action, often promoting a less frequently practiced behavior if it is the more appropriate option for the current situation, taking into account both long-term and short-term benefits and costs. The ability to learn contradictory lessons from experience and to choose the behavioral-control system best suited to the exigencies of the moment provides mammals with a substantial advantage over ancestral species that had less proficiency in learning such contradictions.

4.5 WHAT CAN PRIMATE BRAINS DO WITH REWARD?

4.5.1 Primate Brains: Basic Mammalian Mechanisms Plus Granular Prefrontal Cortex

In Section 4.4, we noted that as mammals emerged from our amniote ancestors (Figure 4.1), the systems that had developed earlier in vertebrate evolution remained in operation, albeit with modifications. The midbrain dopamine system, basal ganglia, amygdala, and hippocampus continued to do more or less what they had done for hundreds of millions of years. Likewise for the emergence of primates, the ancestral systems, including the agranular frontal cortex, continued to play a central role in what primate brains can do with reward.

As explained in Section 4.2.4, primates and nonprimate mammals share the agranular frontal areas, but do not share a PFg (Figure 4.2), which evolved in primates. Among the prefrontal areas, the granular parts of the orbitofrontal cortex (PFo) play a large role in what primate brains can do with reward. These areas include area 11 as well as the light gray parts of areas 13 and 14 in Figure 4.2. In addition, primates also innovated certain sensory areas. Given the importance of vision in primates, the inferotemporal cortex (IT) is notable among these primate innovations. (Note that certain carnivores, such as domestic cats, also evolved high-order visual areas, but these likely arose independently, through parallel evolution (Preuss 2007).) Their long, separate evolutionary history means that the shared, agranular frontal areas will also have diverged in both structure and function in primates versus nonprimates.

Assuming that the emergence of the granular prefrontal areas provided primates with an advantage over the ancestral condition, what could that advantage be? Most answers to this question fall into one of two categories: (1) the new cortical areas functioned much like older parts, but performed those functions better in some way; and (2) the new areas permitted a novel function. However, there is no need to choose between the two and the emergence of a PFg probably led to both kinds of advances.

4.5.2 New Visual Areas in the Context of Feature Conjunctions

Cost-benefit analysis improves when the stimuli that predict rewards and costs do so with more fidelity, and also when those stimuli can be combined or categorized in useful ways. Some of the mechanisms in the PFg, generally, and the granular orbitofrontal cortex (PFo), in particular, seem to support such an advance.

The PFo of primates is in a pivotal position to permit novel kinds of S-O associations, if only because it receives information from visual areas, such as the IT, that other mammals lack (Figure 4.3). The PFo receives strong projections from the IT and other posterior sensory areas of cortex, including the auditory cortex. It also receives inputs from the agranular frontal areas, in particular areas specialized for gustatory, olfactory, and visceral inputs (Rolls, Yaxley, and Sienkiewicz 1990; Webster, Bachevalier, and Ungerleider 1994; Carmichael and Price 1995; Cavada et al. 2000; Kondo, Saleem, and Price 2005). By virtue of its receipt of all these kinds of information, PFo is one of the earliest sites for convergence of visual information with gustatory, olfactory, and visceral inputs. Furthermore, PFo receives inputs from the perioral and tongue representations of the primary somatosensory cortex, as well as strong projections from the perirhinal cortex, a multimodal area that has been shown to operate as an extension of the ventral visual stream or object-processing pathway (Murray, Bussey, and Saksida 2007). Consistent with the anatomical evidence, individual neurons in the PFo of macaque monkeys respond to gustatory, olfactory, and visual stimuli presented separately or together (Rolls and Baylis 1994; Pritchard et al. 2008).

These findings suggest that primates may have a greater capacity to link visceral, olfactory, and gustatory inputs with high- and intermediate-order visual stimuli than our nonprimate ancestors. This new hypothesis originates, in part, from recent research on visual processing mechanisms in macaque monkeys. This work indicates that the perirhinal cortex represents complex conjunctions of visual features that promote object identification (Bussey and Saksida 2007; Murray, Bussey, and Saksida 2007). Other sensory modalities are also important to perirhinal cortex function, but given

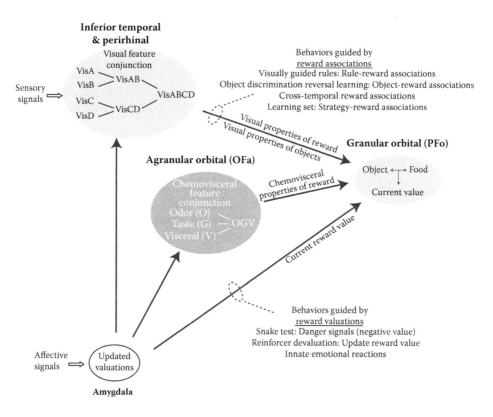

FIGURE 4.3 Convergent contributions to PFo function. PFo is depicted as receiving three inputs that influence what it can do with rewards: representations of visual feature conjunctions for objects and rewards; updated valuations of rewards; and conjunctions of chemosensory and visceral features of rewards. Abbreviations: VisA … VisD, visual features of an object or scene.

the predominance of vision in primates, we will focus on the visual aspects of scenes and objects. Bussey, Saksida, and Murray (2002, 2003) developed and tested this idea with a series of visual discrimination tasks in which combinations of features, rather than any individual feature, best predicted reward. In the context of discrimination learning, this objective can be attained by having individual features appear as part of both the correct (S+) and incorrect (S−) objects in a pair. In one experiment, for example, compound objects were comprised of two of four possible features: A, B, C, and D. Monkeys received rewards when they selected the compound stimuli AB and CD, but not when they selected stimuli CB and AD. This experimental design created a conflict, called feature ambiguity, because individual features were associated equally with both reward and non-reward. Accordingly, to obtain a reward, the monkeys were required to represent the combinations of features such as AB (depicted as VisAB in Figure 4.3, with the individual features depicted as VisA and VisB, respectively). Bussey, Saksida, and Murray (2002, 2003) showed that lesions of the perirhinal cortex disrupted performance in several tests based on this premise. Their results showed that the representation of feature conjunctions in the perirhinal cortex helped resolve feature ambiguity and promote object identification.

Homologues of the primate perirhinal cortex, along with lower-order visual areas such as the striate and near-striate cortex, can be found in rats and other mammals, but the evolution of additional areas such as the IT occurred in primates. What new capabilities could these new areas provide? Whereas perirhinal cortex is thought to represent the most complex (highest-order) visual features that compose objects, the IT cortex represents intermediate levels of feature conjunction. According to this view, the striate and near-striate cortex represent lower-order conjunctions, compared to the IT cortex, and the perirhinal cortex represents higher-order ones. Thus, it appears that during primate evolution, newly evolved areas emerged between the two extremes of conjunctive representations: mid-level conjunctions. Figure 4.3 depicts mid-level conjunctions as VisAB and VisCD, whereas high-level conjunctions are denoted as VisABCD. The existence of mid-level feature conjunctions changes what primate brains can do with reward, specifically by permitting categorization-based conjunctions that fall short of whole objects, but have greater complexity than their fundamental features.

4.5.3 ORBITAL PREFRONTAL CORTEX

As illustrated in Figure 4.3, the PFo appears to be the earliest site at which the high- and mid-level feature conjunctions of an object (from perirhinal and IT cortex, respectively) converge with olfactory, gustatory, and visceral inputs (relayed via the agranular orbitofrontal areas). This anatomical convergence enables feature conjunctions of a particularly important kind. Comparison to mammals that lack a PFo, which is to say all nonprimate mammals, indicates that the PFo may provide the neural substrate for finer distinctions among food and other rewards based on a combination of their visual attributes with their chemosensory and visceral sensory properties (collectively depicted as "chemovisceral" feature conjunctions in Figure 4.3). In addition to representing finer distinctions among reward items, the PFo could also represent more generalized categories of foods. We propose that these visual-chemovisceral feature conjunctions promote an enhanced flexibility in responding to rewards. The sight of a particular food, a fruit at a particular stage of ripeness for example, might evoke its taste and smell, along with the pleasant sensations and positive affect that follow its consumption. Similarly, the fruit's odor might evoke the whole, cross-modal configuration of features as well. This idea is much like that put forward in Section 4.4.2.4 for the OFa, but by making use of the richer analytical capacities of IT, for example, the PFo would provide an advantage over animals that lacked these areas.

So far, we have stressed ways in which the PFo functions much like the agranular frontal areas of rodents (and, presumably, the agranular frontal areas of primates as well). What about new functions? Some evidence points to a role for the PFo in representing value in a "common currency," one that is independent of the specific kind of reward (benefit) or a particular kind of cost. In many neurophysiological studies in monkeys, a particular visual stimulus has been associated with a specific reward through learning, primarily via Pavlovian (S-O) associations. At the time the monkey views the stimulus, the activity of many PFo neurons reflects the value of the associated reward, even though the reward itself arrives much later. This property is by no means unique to the PFo: in this regard it resembles phylogenetically older structures, such as the midbrain dopamine neurons and many other parts of the frontal cortex. Yet, the reward-expectancy signal that the PFo generates when elicited by the sight of a particular food might differ in a subtle but important way from similar signals in these other structures. Recent neurophysiological studies in monkeys have found that the activity of PFo neurons reflects the value of rewards independent of reward type (Wallis and Miller 2003; Padoa-Schioppa and Assad 2006). In addition, neurons in the PFo (along with neurons in other parts of the cortex) encode at least three factors that enter into the calculation of reward value: the magnitude of reward, the probability of reward, and the effort required to obtain it (Kennerley et al. 2009). Thus, it appears that the PFo's outcome representations extend beyond the particular substance (sucrose, quinine, juice) paired with a stimulus to signal a value that transcends particular types or amounts of reward (Padoa-Schioppa and Assad 2006).

To date, no neurophysiological study of the rodent frontal cortex has shown this property, although some have shown that neurons in the agranular frontal areas have activity that reflects *specific* expected reward outcomes and specific S-O associations (Schoenbaum, Chiba, and Gallagher 1998; Schoenbaum and Eichenbaum 1995). One possibility, then, is that the multiple levels of sensory convergence available in the PFo might enable the computation of both the *"menu-independent"* value signal observed by Padoa-Schioppa and Assad (2008), which indicates outcome values regardless of the reward options, and a *drive-independent* value signal (the "common currency"), which indicates values for rewards (and costs) of any kind (e.g., sex, food, fluid, exertion, etc.). This cost-benefit analysis can only be performed efficiently by reference to a common measure of value. It is possible that these features of PFo build on properties of the nearby agranular frontal areas, but do so in a particularly advanced way because of the extensive and varied visual inputs provided to the PFo by the perirhinal and IT cortex. Recall that the mid-order feature conjunctions provided by the IT permit a kind of generalization that would be much harder to come by in brains, such as rodent brains, that lack an IT cortex.

Autonomic signals or information about emotional state may also influence the calculation of value in the common currency described above. Experiments in monkeys have taken advantage of monkeys' innate fear of fake or real snakes to provide findings consistent with this idea. In an experiment by Izquierdo et al. (2005), monkeys were given the opportunity to reach over either a neutral object or a fake snake to retrieve a food reward. Thus, the benefits of obtaining the food reward were directly pitted against the costs (in terms of risk) of approaching the snake. When required to reach over the fake snake, intact monkeys displayed long food-retrieval latencies or refused to take the food altogether. In contrast, monkeys with PFo lesions exhibited short food-retrieval latencies, similar to when they reached over neutral objects. Similar findings have been reported by Rushworth et al. (2007). It seems that monkeys with PFo lesions cannot integrate the sensory signals, in this case arising from visual signals about a potential predator, with food value to guide their behavior. The inability to compute value in a common currency for risks and reward could account for these findings. Figure 4.3 depicts the results of the snake test as depending on interactions between the amygdala and PFo.

Like amygdala lesions (see Section 4.3.4), PFo lesions cause a deficit on the reinforcer devaluation task (Izquierdo, Suda, and Murray 2004). Compared to controls, monkeys with bilateral removals of PFo more often choose an object that will produce the devalued reward. Accordingly, Figure 4.3 depicts this test as depending on interactions between the amygdala and PFo (Baxter and Murray 2002), with the PFo mediating the associations between the visually sensed objects, the sight of a particular food, and its updated value.

4.5.4 Other Granular Prefrontal Areas

So far, this section has concentrated on the PFo, but primate brains have many granular frontal areas, including the dorsal, dorsolateral, ventrolateral, medial, and polar prefrontal cortex (Figure 4.2). These areas function in higher-order aspects of cognitive control, including those involved in monitoring actions (Petrides 1991) and implementing rules and strategies (Murray, Bussey, and Wise 2000; Wallis, Anderson, and Miller 2001; Mansouri, Buckley, and Tanaka 2007). Rules and strategies allow animals to gain rewards when faced with novel stimuli and contexts. Rules and strategies have many similarities, but a few differences as well. Rules are input-output algorithms that can be applied to solve a behavioral problem for novel as well as familiar stimuli; strategies are either one of many solutions to a problem or a partial solution to a problem.

4.5.4.1 Rule-Outcome and Strategy-Outcome Associations

Monkeys can learn abstract rules and strategies, and it has been proposed that these abstract rules arise from reinforcement of rules rather than stimuli (Gaffan 1985). That is, much as the representation of a stimulus can be associated with reward outcome, as in Pavlovian conditioning (which forms S-O associations), the cognitive representation of a rule can be associated with reward outcome. For example, one widely used test of object-reward association—called win-stay by some (e.g., see Chapter 3), but called *congruent recall* by Gaffan—has two parts: acquisition and test. In the acquisition phase, a monkey displaces one object that yields a reward and a second object that does not. After a delay, the same two objects appear, and the monkey can choose only one. The monkey can obtain another food reward by displacing the object that had been paired with food in the acquisition phase. Orthodox animal learning theory predicts that the stimulus-outcome contingency controls the monkey's choice. Gaffan's results contradicted that prediction. For a different group of monkeys, he applied the opposite rule, which some might call win-shift, but Gaffan called *incongruent recall*. This response rule requires the monkey to avoid the object that had been associated with reward during the acquisition phase and choose the other one instead. Macaque monkeys not only acquired the incongruent-recall rule as quickly as they learned the congruent-recall rule, but also performed equally well on subsequent tests requiring the application of the rule to lists of novel stimuli. Thus, even though in one case (the win-shift, incongruent-recall rule) the monkeys had to overcome behavioral "control" by S-O associations, experienced monkeys efficiently did so. This finding shows that although monkeys learn the association of individual stimuli with rewards, they also learn the association of rules with rewards and when these two aspects of their experience come into competition they can bias behavioral control toward the brain systems that deal with choosing rules rather than choosing stimuli.

This idea relates to the discussion of foraging strategies developed in Section 4.4, as well as the concept of biases among competing mechanisms for behavioral control. As for the latter, a prediction arising from this idea is that monkeys without a PFo should be deficient in implementing rules (based on rule-reward associations), especially when the rule must override S-O associations. Consistent with this idea, macaques with lesions that include the PFo lost the ability to use preoperatively acquired performance rules (Rushworth et al. 1997; Bussey, Wise, and Murray 2001; Buckley et al. 2007).

Similarly, Murray and Gaffan (2006) proposed that monkeys can associate reward with problem-solving strategies. In traditional tests of object (or stimulus) discrimination learning set, a single pair of stimuli is used repeatedly for several if not dozens of trials. New stimulus discrimination problems are introduced periodically. When monkeys have had experience with many problems of this type, they exhibit faster learning. More rapid acquisition of later problems relative to earlier ones demonstrates the development of a *learning set*. Because discrimination problems typically involve massed trials (i.e., many consecutive trials with one and the same pair of stimuli), monkeys presumably retain a memory of the correct response from one trial to the next that permits the development of object discrimination learning set. To test this idea, Murray and Gaffan trained monkeys on a large set of discrimination problems involving long (24-hr) intertrial intervals. If the short-term memory of the outcome of the previous trial was important for developing learning set, monkeys given 24-hr intertrial intervals should fail to develop a learning set. In the Murray and Gaffan study, monkeys learned 20 stimulus discrimination problems concurrently. They received one trial per problem per day, with the 20 pairs used for the 20 trials per daily test session. The same stimulus pairs were presented for choice each day, until the monkeys reliably approached and displaced the rewarded object of each pair. After mastering one set of 20 problems, the monkeys learned another set, and so on, until they had learned 10 consecutive sets of discrimination problems. Despite their extensive experience—each monkey had learned 200 individual problems—the monkeys with 24-hr intertrial intervals failed to develop a learning set. Murray and Gaffan (2006) concluded that monkeys only acquire a learning set when intertrial intervals are short enough to allow the events of the preceding trial to be retained in short-term memory. Specifically, they proposed that the rule learning underlying learning set involves laying down a prospective memory of what stimulus to choose on the next trial. When the monkey succeeds, the reward reinforces the laying down of the prospective memory from each trial, gradually leading to the formation of a learning set. Thus, rather than the learning set resulting from improved response rules at the time of the choice test, a kind of learning-to-learn that would occur regardless of the length of intertrial intervals employed, learning set instead depends on learning what to remember. In short, what is reinforced is a strategy about what to remember over the intertrial interval. This idea likely applies to a wide range of rule learning (e.g., congruent and incongruent recall tasks, delayed nonmatching-to-sample, object reversal learning set, etc.). Recent evidence shows that both learning set and reversal learning set depend on the interaction of the PFg and IT cortex (Browning , Easton, and Gaffan 2007; Wilson and Gaffan 2008).

4.5.4.2   Value-Free Reward Representations

Monkeys can use reward information in other ways as well. One underappreciated aspect of reward-based learning is that animals can process positive reinforcement much like they would any other sensory signal. That is, the role of food reward in many instances can derive from its informational value, as opposed to its reinforcing or emotional value. Evidence in support of this idea comes from neuropsychological experiments showing a dissociation between the neural structures required to support learning about food value versus those required for learning about the occurrence of food (through object-reward associations). For example, the amygdala is essential for updating food value and linking that value with object-food associations, as explained in Section 4.3.4. The amygdala is not necessary, however, for learning which stimuli are associated with food reward when object-reward associations change, as they do in the object discrimination reversal task (Izquierdo and Murray 2007). (Recall that in this task, a previously rewarded object changes to become unrewarded and vice versa.) It appears that, in these tasks, the occurrence or nonoccurrence of food guides the selection of a performance rule. Once a performance rule has been learned, the role of food in such tasks is largely limited to its informational value, as opposed to its reinforcing or emotional value, and this informational processing does not depend on the amygdala. Accordingly, Figure 4.3 depicts this process as depending on interactions between visual areas (perirhinal cortex and IT) and the PFo, although other prefrontal areas also play a role.

To sum up this idea about value-free food representations, we propose that reward can operate independently of biological value in primates. When monkeys implement rules efficiently, as is the case for experienced monkeys performing object discrimination learning (learning set) or object discrimination reversal tasks (reversal set), the outcome serves primarily to provide feedback as to the success of that trial, which might be a visual representation of a food among other possible types of feedback, rather than a representation of food value itself.

4.5.5   What Can Human Brains Do with Reward?

The age-old question of what separates us from other animals is beyond the scope of this article, but a couple of comments seem pertinent because we humans can do things with reward that other animals cannot. For example, we can talk about reward. We also have the ability, called mental time travel, to remember past events and to imagine future ones as if they were current, incorporating a representation of ourselves into the event (Suddendorf and Corballis 2007; Corballis 2009). This ability probably contributes to overcoming the deep devaluation of rewards that occurs at a long delay. Mental time travel allows low-frequency and temporally distant events to guide behavior and to override choices prompted by high-frequency and more proximate events. Episodic memory can provide the substrate for a single event to guide behavior decades later. The human capacity to both remember and to plan over an extraordinarily long time horizon, therefore, has to count as one of the things that human brains—and therefore primate brains—can do with rewards.

Finally, we note that humans, at least, also have the capacity to develop second-order intentions about rewards. For example, people may not want to want food rewards. Second-order intentions are another novel way in which the primate brain deals with reward. It is extraordinarily doubtful that such a capacity exists in nonprimates and it is probably absent in nonhuman primates as well.

4.6   SUMMARY

With the early vertebrates came a long list of neural innovations, including the "midbrain" dopamine system and the telencephalon (Northcutt 1996; Thompson et al. 2008). The dopamine system plays a role in predicting the benefits called rewards and computing the differences between such predictions and what ultimately occurs (Schultz 2006, 2007). In the telencephalon, the amygdala updates neural representations of biological value—for example, the value of a food—based on the current state of an animal (Murray 2007). The striatum also arose in early vertebrates (Northcutt 1996,

2008) and it plays a major role in cost-benefit analysis. These functions all support the decisions needed for optimal foraging. The hippocampus appears to play a role in the navigation involved in foraging, including the affective correlates of places and behavioral contexts. These structure-function relationships (Table 4.1) determined what these ancient animals could do with reward and their legacy continues in modified form in descendant animals, including us (Figure 4.1).

With the origin of mammals came the neocortex, which includes the orbital and medial frontal areas (Kaas 1995; Northcutt and Kaas 1995; Krubitzer and Kaas 2005). These agranular frontal areas include the AC, PL, IL, OFa, and Ia cortex (Table 4.2), which permit finer distinctions among specific rewards (based on olfactory, gustatory, and visceral signals). The most important advance in the mammalian way of using reward, however, may be the ability to bias one behavioral control system over another when they compete to guide responses. This "executive function" can have the effect of potentiating a weaker behavioral-control system, which would otherwise not gain behavioral expression, and thus optimize the cost-benefit tradeoffs inherent in competing foraging strategies. Table 4.2 summarizes the possible contributions of each part of the mammalian agranular frontal cortex in terms of associative memory and foraging strategies. Through top-down, biased competition, mammals can take advantage of parallel "memory systems," in which experience can lead to competing responses. This level of executive function allows mammals to explore and exploit a changeable environment, with sufficient behavioral flexibility to permit different systems to control behavior in different circumstances.

As for primates, we retain many of the traits of vertebrates and other mammals. But our primate ancestors evolved additional cortex areas collectively called the PFg (Figure 4.2) (Preuss 1995, 2007), which includes the PFo. These newly evolved prefrontal areas permit a primate way of using rewards, including the use of reward-specific sensory information that has been dissociated from its emotional, motivational, and affective attributes. This kind of value-free reward information could then be available for use in the same way that any other information is used by the primate brain, and it appears to be especially important for rule-guided behavior. Another primate way of using rewards is somewhat opposite to the first: the association of value with rules, strategies, and other cognitive constructs. In addition, the PFo is a site of convergence for visual feature conjunctions, which represent objects and categories of objects, and chemovisceral feature conjunctions, which represent the olfactory, gustatory, and visceral attributes of rewards. Finally, the brains of human primates use rewards in yet other ways, through mental time travel and second-order intentions. Mental time travel can overcome delay intolerance and produce behaviors aimed at both maximizing reward benefits and minimizing cost over the *very* long term. The ability to develop second-order intentions about reward, e.g., not wanting to want chocolate or wanting to pay the cost incurred by enervating exercise, probably depends on cognitive capacities that developed only during the evolution of modern humans.

REFERENCES

Balleine, B.W., and Dickinson, A. 1998. Goal-directed instrumental action: Contingency and incentive learning and their cortical substrates. *Neuropharmacology* 37:407–19.
———. 2000. The effect of lesions of the insular cortex on instrumental conditioning: Evidence for a role in incentive memory. *J Neurosci* 20:8954–64.
Balleine, B.W., and Killcross, S. 2006. Parallel incentive processing: An integrated view of amygdala function. *Trends Neurosci* 29:272–79.
Balleine, B.W., Killcross, A.S., and Dickinson, A. 2003. The effect of lesions of the basolateral amygdala on instrumental conditioning. *J Neurosci* 23:666–75.
Bannerman, D.M., Grubb, M., Deacon, R.M., Yee, B.K., Feldon, J., and Rawlins, J.N. 2003. Ventral hippocampal lesions affect anxiety but not spatial learning. *Behav Brain Res* 139:197–213.
Barbas, H., and Pandya, D.N. 1989. Architecture and intrinsic connections of the prefrontal cortex in the rhesus monkey. *J Comp Neurol* 286:353–75.

Baxter, M.G., and Murray, E.A. 2002. The amygdala and reward. *Nat Rev Neurosci* 3:563–73.

Blood, A.J., and Zatorre, R.J. 2001. Intensely pleasurable responses to music correlate with activity in brain regions implicated in reward and emotion. *Proc Natl Acad Sci USA* 98:11818–23.

Boulougouris, V., Dalley, J.W., and Robbins, T.W. 2007. Effects of orbitofrontal, infralimbic and prelimbic cortical lesions on serial spatial reversal learning in the rat. *Behav Brain Res* 179:219–28.

Brembs, B., and Heisenberg, M. 2000. The operant and the classical in conditioned orientation of *Drosophila melanogaster* at the flight simulator. *Learn Mem* 7:104–15.

Brembs, B., Lorenzetti, F.D., Reyes, F.D., Baxter, D.A., and Byrne, J.H. 2002. Operant reward learning in *Aplysia*: Neuronal correlates and mechanisms. *Science* 296:1706–9.

Browning, P.G.F., Easton, A., and Gaffan, D. 2007. Frontal-temporal disconnection abolishes object discrimination learning set in macaque monkeys. *Cereb Cortex* 17:859–64.

Buckley, M.J., Tanaka, K., Mahboubi, M., and Mansouri, F.A. 2007. Double dissociations in the effects of prefrontal cortex lesions in macaque monkeys on a Wisconsin card sorting task analogue. In *Linking Affect to Action: Critical Contributions of the Orbitofrontal Cortex*, eds. G. Schoenbaum, J.A. Gottfried, E.A. Murray, and S.J. Ramus, pp. 657–58. Boston: Blackwell.

Burke, K.A., Franz, T.M., Miller, D.N., and Schoenbaum, G. 2008. The role of the orbitofrontal cortex in the pursuit of happiness and more specific rewards. *Nature* 454:340–44.

Bussey, T.J., Everitt, B.J., and Robbins, T.W. 1997a. Dissociable effects of cingulate and medial frontal cortex lesions on stimulus-reward learning using a novel Pavlovian autoshaping procedure for the rat: Implications for the neurobiology of emotion. *Behav Neurosci* 111:908–19.

———. 1997b. Triple dissociation of anterior cingulate, posterior cingulate, and medial frontal cortices on visual discrimination tasks using a touchscreen testing procedure for the rat. *Behav Neurosci* 111:920–36.

Bussey, T.J., and Saksida, L.M. 2007. Memory, perception, and the ventral visual-perirhinal-hippocampal stream: Thinking outside of the boxes. *Hippocampus* 17:898–908.

Bussey, T.J., Saksida, L.M., and Murray, E.A. 2002. Perirhinal cortex resolves feature ambiguity in complex visual discriminations. *Eur J Neurosci* 15:365–74.

———. 2003. Impairments in visual discrimination after perirhinal cortex lesions: Testing "declarative" vs. "perceptual-mnemonic" views of perirhinal cortex function. *Eur J Neurosci* 17:649–60.

Bussey, T.J., Wise, S.P., and Murray, E.A. 2001. The role of ventral and orbital prefrontal cortex in conditional visuomotor learning and strategy use in rhesus monkeys (*Macaca mulatta*). *Behav Neurosci* 115:971–82.

Butler, A.B. 2000. Sensory system evolution at the origin of craniates. *Phil Trans Roy Soc Lond B* 355:1309–13.

Cardinal, R.N., Parkinson, J.A., Lachenal, G., Halkerston, K.M., Rudarakanchana, N., Hall, J., Morrison, C.H., Howes, S.R., Robbins, T.W., and Everitt B.J. 2002. Effects of selective excitotoxic lesions of the nucleus accumbens core, anterior cingulate cortex, and central nucleus of the amygdala on autoshaping performance in rats. *Behav Neurosci* 116:553–67.

Cardinal, R.N., Parkinson, J.A., Marbini, H.D., Toner, A.J., Bussey, T.J., Robbins, T.W., and Everitt, B.J. 2003. Role of the anterior cingulate cortex in the control over behavior by Pavlovian conditioned stimuli in rats. *Behav Neurosci* 117:566–87.

Carew, T.J., and Sahley, C.L. 1986. Invertebrate learning and memory: From behavior to molecules. *Annu Rev Neurosci* 9:435–87.

Carew, T.J., Walters, E.T., and Kandel, E.R. 1981. Associative learning in *Aplysia*: Cellular correlates supporting a conditioned fear hypothesis. *Science* 211:501–4.

Carmichael, S.T., and Price, J.L. 1994. Architectonic subdivision of the orbital and medial prefrontal cortex in the macaque monkey. *J Comp Neurol* 346:366–402.

———. 1995. Sensory and premotor connections of the orbital and medial prefrontal cortex of macaque monkeys. *J Comp Neurol* 363:642–64.

Cavada, C., Company, T., Tejedor, J., Cruz-Rizzolo, R.J., and Reinoso-Suarez, F. 2000. The anatomical connections of the macaque monkey orbitofrontal cortex. A review. *Cereb Cortex* 10:220–42.

Changizi, M.A., McGehee, R.M., and Hall, W.G. 2002. Evidence that appetitive responses for dehydration and food-deprivation are learned. *Physiol Behav* 75:295–304.

Chudasama, Y., and Robbins, T.W. 2003. Dissociable contributions of the orbitofrontal and infralimbic cortex to pavlovian autoshaping and discrimination reversal learning: Further evidence for the functional heterogeneity of the rodent frontal cortex. *J Neurosci* 23:8771–80.

Clack, J.A. 2002. *Gaining Ground: The Origin and Evolution of Tetrapods*. Bloomington: Indiana University Press.

Corballis, M.C. 2009. Mental time travel and the shaping of language. *Exp Brain Res* 192:553–60.

Corbit, L.H., and Balleine, B.W. 2003. Instrumental and Pavlovian incentive processes have dissociable effects on components of a heterogeneous instrumental chain. *J Exp Psychol Anim Behav Process* 29:99–106.

Coutureau, E., and Killcross, S. 2003. Inactivation of the infralimbic prefrontal cortex reinstates goal-directed responding in overtrained rats. *Behav Brain Res* 146:167–74.

Craig, A.D. 2002. How do you feel? Interoception: The sense of the physiological condition of the body. *Nat Rev Neurosci* 3:655–66.

Cummings, J.L. 1993. Frontal-subcortical circuits and human behavior. *Arch Neurol* 50:873–80.

de Wit, S., Kosaki, Y., Balleine, B.W., and Dickinson, A. 2006. Dorsomedial prefrontal cortex resolves response conflict in rats. *J Neurosci* 26:5224–29.

Delamater, A.R. 2007. The role of the orbitofrontal cortex in sensory-specific encoding of associations in pavlovian and instrumental conditioning. *Ann N Y Acad Sci* 1121:152–73.

Desmurget, M., and Turner, R.S. 2008. Testing basal ganglia motor functions through reversible inactivations in the posterior internal globus pallidus. *J Neurophysiol* 99:1057–76.

Dickinson, A., and Balleine, B. 1994. Motivational control of goal-directed action. *Anim Learn Behav* 22:1–18.

Drewes, A.M., Dimcevski, G., Sami, S.A., Funch-Jensen, P., Huynh, K.D., Le Pera, D., Arendt-Nielsen, L., and Valeriani M. 2006. The "human visceral homunculus" to pain evoked in the oesophagus, stomach, duodenum and sigmoid colon. *Exp Brain Res* 174:443–52.

Dunn, C.W., Hejnol, A., Matus, D.Q., Pang, K., Browne, W.E., Smith, S.A., Seaver, E., Giribet, G. et al. 2008. Broad phylogenomic sampling improves resolution of the animal tree of life. *Nature* 452:745–49.

Everitt, B.J., Cardinal, R.N., Parkinson, J.A., Robbins, T.W. 2003. Appetitive behavior: Impact of amygdala-dependent mechanisms of emotional learning. *Ann N Y Acad Sci* 985:233–50.

Gaffan, D. 1985. Hippocampus: Memory, habit and voluntary movement. *Phil Trans R Soc Lond B* 308:87–99.

Gallagher, M., McMahan, R.W., and Schoenbaum, G. 1999. Orbitofrontal cortex and representation of incentive value in associative learning. *J Neurosci* 19:6610–14.

Grimaldi, D., and Engel, M.S. 2005. *Evolution of the Insects*. Cambridge: Cambridge University Press.

Haddon, J.E., and Killcross, A.S. 2005. Medial prefrontal cortex lesions abolish contextual control of competing responses. *J Exp Anal Behav* 84:485–504.

Hall, W.G., Arnold, H.M., and Myers, K.P. 2000. The acquisition of an appetite. *Psychol Sci* 11:101–5.

Hatfield, T., Han, J., Conley, M., Gallagher, M., and Holland, P. 1996. Neurotoxic lesions of basolateral, but not central, amygdala interfere with Pavlovian second-order conditioning and reinforcer devaluation effects. *J Neurosci* 16:5256–65.

Hershberger, W.A. 1986. An approach through the looking-glass. *Anim Learn Behav* 14:443–51.

Holland, L.Z., and Holland, N.D. 1999. Chordate origins of the vertebrate central nervous system. *Curr Opin Neurobiol* 9:596–602.

Holland, P. 1998. Amount of training affects associatively-activated event representation. *Neuropharmacology* 37:461–69.

Holland, P.C., and Gallagher, M. 2004. Amygdala-frontal interactions and reward expectancy. *Curr Opin Neurobiol* 14:148–55.

Izquierdo, A., and Murray, E.A. 2005. Opposing effects of amygdala and orbital prefrontal cortex lesions on the extinction of instrumental responding in macaque monkeys. *Eur J Neurosci* 22:2341–46.

———. 2007. Selective bilateral amygdala lesions in rhesus monkeys fail to disrupt object reversal learning. *J Neurosci* 27:1054–62.

Izquierdo, A., Suda, R.K., and Murray, E.A. 2004. Bilateral orbital prefrontal cortex lesions in rhesus monkeys disrupt choices guided by both reward value and reward contingency. *J Neurosci* 24:7540–48.

———. 2005. Comparison of the effects of bilateral orbital prefrontal cortex lesions and amygdala lesions on emotional responses in rhesus monkeys. *J Neurosci* 25:8534–42.

Jones, E.G. 1985. *The Thalamus*. New York: Plenum Press.

Kaas, J.H. 1995. The evolution of isocortex. *Brain Behav Evol* 46:187–96.

———. 2008. The evolution of the complex sensory and motor systems of the human brain. *Brain Res Bull* 75:384–90.

Kennerley, S.W., Dahmubed, A.F., Lara, A.H., Wallis, J.D. 2009. Neurons in the frontal lobe encode the value of multiple decision variables. *J Cogn Neurosci* 21:1162–78.

Kesner, R.P., and Gilbert, P.E. 2007. The role of the agranular insular cortex in anticipation of reward contrast. *Neurobiol Learn Mem* 88:82–86.

Kheramin, S., Body, S., Mobini, S., Ho, M.Y., Velazquez-Martinez, D.N., Bradshaw, C.M., Szabadi, E., Deakin, J.F., and Anderson, I.M. 2002. Effects of quinolinic acid-induced lesions of the orbital prefrontal cortex on inter-temporal choice: A quantitative analysis. *Psychopharmacology (Berl)* 165:9–17.

Kiefer, S.W., and Orr, M.R. 1992. Taste avoidance, but not aversion, learning in rats lacking gustatory cortex. *Behav Neurosci* 106:140–46.

Killcross, S., and Coutureau, E. 2003. Coordination of actions and habits in the medial prefrontal cortex of rats. *Cereb Cortex* 13:400–8.

King, A.B., Menon, R.S., Hachinski, V., Cechetto, D.F. 1999. Human forebrain activation by visceral stimuli. *J Comp Neurol* 413:572–82.

Kjelstrup, K.B., Solstad, T., Brun, V.H., Hafting, T., Leutgeb, S., Witter, M.P., Moser, E.I., and Moser, M.B. 2008. Finite scale of spatial representation in the hippocampus. *Science* 321:140–43.

Kjelstrup, K.G., Tuvnes, F.A., Steffenach, H.A., Murison, R., Moser, E.I., and Moser, M.B. 2002. Reduced fear expression after lesions of the ventral hippocampus. *Proc Natl Acad Sci USA* 99:10825–30.

Kolb, B. 2007. Do all mammals have a prefrontal cortex? In *The Evolution of Primate Nervous Systems*, eds. J.H. Kaas and L. Krubitzer, pp. 443–50. Amsterdam: Elsevier.

Kondo, H., Saleem, K.S., and Price, J.L. 2005. Differential connections of the perirhinal and parahippocampal cortex with the orbital and medial prefrontal networks in macaque monkeys. *J Comp Neurol* 493:479–509.

Krubitzer, L., and Kaas, J. 2005. The evolution of the neocortex in mammals: How is phenotypic diversity generated? *Curr Opin Neurobiol* 15:444–53.

Lebron, K., Milad, M.R., and Quirk, G.J. 2004. Delayed recall of fear extinction in rats with lesions of ventral medial prefrontal cortex. *Learn Mem* 11:544–48.

MacLean, P.D. 1985. Evolutionary psychiatry and the triune brain. *Psychol Med* 15:219–21.

Malkova, L., Gaffan, D., and Murray, E.A. 1997. Excitotoxic lesions of the amygdala fail to produce impairment in visual learning for auditory secondary reinforcement but interfere with reinforcer devaluation effects in rhesus monkeys. *J Neurosci* 17:6011–20.

Mansouri, F.A., Buckley, M.J., and Tanaka, K. 2007. Mnemonic function of the dorsolateral prefrontal cortex in conflict-induced behavioral adjustment. *Science* 318:987–90.

Matsumoto, M., and Hikosaka, O. 2007. Lateral habenula as a source of negative reward signals in dopamine neurons. *Nature* 447:1111–15.

Mazzoni, P., Hristova, A., and Krakauer, J.W. 2007. Why don't we move faster? Parkinson's Disease, movement vigor, and implicit motivation. *J Neurosci* 27:7105–16.

McDannald, M.A., Saddoris, M.P., Gallagher, M., and Holland, P.C. 2005. Lesions of orbitofrontal cortex impair rats' differential outcome expectancy learning but not conditioned stimulus-potentiated feeding. *J Neurosci* 25:4626–32.

McHugh, S.B., Deacon, R.M., Rawlins, J.N., and Bannerman, D.M. 2004. Amygdala and ventral hippocampus contribute differentially to mechanisms of fear and anxiety. *Behav Neurosci* 118:63–78.

Medina, L., Brox, A., Legaz, I., Garcia-Lopez, M., and Puelles, L. 2005. Expression patterns of developmental regulatory genes show comparable divisions in the telencephalon of *Xenopus* and mouse: Insights into the evolution of the forebrain. *Brain Res Bull* 66:297–302.

Mesulam, M.M. 1990. Large-scale neurocognitive networks and distributed processing for attention, language, and memory. *Ann Neurol* 28:597–613.

Mishkin, M., Malamut, B., and Bachevalier, J. 1984. Memories and habits: Two neural systems. In *Neurobiology of Learning and Memory*, eds. G. Lynch, J. McGaugh, and N.M. Weinberger, pp. 65–77. New York: Guilford.

Moreno, N., and Gonzalez, A. 2007. Evolution of the amygdaloid complex in vertebrates, with special reference to the anamnio-amniotic transition. *J Anat* 211:151–63.

Murray, E.A. 2007. The amygdala, reward and emotion. *Trends Cogn Sci* 11:489–97.

Murray, E.A., Bussey, T.J., and Saksida, L.M. 2007. Visual perception and memory: A new view of medial temporal lobe function in primates and rodents. *Ann Rev Neurosci* 30:99–122.

Murray, E.A., Bussey, T.J., and Wise, S.P. 2000. Role of prefrontal cortex in a network for arbitrary visuomotor mapping. *Exp Brain Res* 133:114–29.

Murray, E.A., and Gaffan, D. 2006. Prospective memory in the formation of learning sets by rhesus monkeys (*Macaca mulatta*). *J Exp Psychol Anim Behav Process* 32:87–90.

Neary, T.J. 1990. The pallium of anuran amphibians. In *Cerebral Cortex: Comparative Structure and Evolution of Cerebral Cortex, Part I*, eds. E.G. Jones and A. Peters, pp. 107–38. New York: Plenum.

Nieuwenhuys, R. 2002. Deuterostome brains: Synopsis and commentary. *Brain Res Bull* 57:257–70.

Northcutt, R.G. 1996. The agnathan ark: The origin of craniate brains. *Brain Behav Evol* 48:237–47.

———. 2001. Changing views of brain evolution. *Brain Res Bull* 55:663–74.

———. 2008. Forebrain evolution in bony fishes. *Brain Res Bull* 75:191–205.

Northcutt, R.G., and Kaas, J.H. 1995. The emergence and evolution of mammalian neocortex. *Trends Neurosci* 18:373–79.

O'Keefe, J., and Nadel, L. 1978. *The Hippocampus as a Cognitive Map*. Oxford: Clarendon Press.

Öngür, D., Ferry, A.T., and Price, J.L. 2003. Architectonic subdivision of the human orbital and medial prefrontal cortex. *J Comp Neurol* 460:425–49.

Ostlund, S.B., and Balleine, B.W. 2005. Lesions of medial prefrontal cortex disrupt the acquisition but not the expression of goal-directed learning. *J Neurosci* 25:7763–70.

———. 2007. Orbitofrontal cortex mediates outcome encoding in pavlovian but not instrumental conditioning. *J Neurosci* 27:4819–25.

Padoa-Schioppa, C., and Assad, J.A. 2006. Neurons in the orbitofrontal cortex encode economic value. *Nature* 441:223–26.

———. 2008. The representation of economic value in the orbitofrontal cortex is invariant for changes of menu. *Nat Neurosci* 11:95–102.

Palomero-Gallagher, N., and Zilles, K. 2004. Isocortex. In *The Rat Nervous System*, ed. G. Paxinos, pp. 729–57. San Diego, CA: Elsevier Academic Press.

Petrides, M. 1991. Monitoring of selections of visual stimuli and the primate frontal cortex. *Proc R Soc Lond B Biol Sci* 246:293–98.

Petrovich, G.D., Canteras, N.S., and Swanson, L.W. 2001. Combinatorial amygdalar inputs to hippocampal domains and hypothalamic behavior systems. *Brain Res Rev* 38:247–89.

Pickens, C.L., Saddoris, M.P., Gallagher, M., and Holland, P.C. 2005. Orbitofrontal lesions impair use of cue-outcome associations in a devaluation task. *Behav Neurosci* 119:317–22.

Pickens, C.L., Saddoris, M.P., Setlow, B., Gallagher, M., Holland, P.C., and Schoenbaum, G. 2003. Different roles for orbitofrontal cortex and basolateral amygdala in a reinforcer devaluation task. *J Neurosci* 23:11078–84.

Preuss, T.M. 1995. Do rats have prefrontal cortex? The Rose-Woolsey-Akert program reconsidered. *J Cog Neurosci* 7:1–24.

———. 2007. Primate brain evolution in phylogenetic context. In *Evolution of Nervous Systems*, eds. J.H. Kaas and T.M. Preuss, pp. 2–34. Oxford: Elsevier.

Pritchard, T.C., Nedderman, E.N., Edwards, E.M., Petticoffer, A.C., Schwartz, G.J., and Scott, T.R. 2008. Satiety-responsive neurons in the medial orbitofrontal cortex of the macaque. *Behav Neurosci* 122:174–82.

Quirk, G.J., and Mueller, D. 2008. Neural mechanisms of extinction learning and retrieval. *Neuropsychopharmacology* 33:56–72.

Ragozzino, M.E. 2007. The contribution of the medial prefrontal cortex, orbitofrontal cortex, and dorsomedial striatum to behavioral flexibility. *Ann N Y Acad Sci* 1121:355–75.

Ragozzino, M.E., Kim, J., Hassert, D., Minniti, N., and Kiang, C. 2003. The contribution of the rat prelimbic-infralimbic areas to different forms of task switching. *Behav Neurosci* 117:1054–65.

Ragozzino, M.E., Wilcox, C., Raso, M., and Kesner, R.P. 1999. Involvement of rodent prefrontal cortex subregions in strategy switching. *Behav Neurosci* 113:32–41.

Ray, J.P., and Price, J.L. 1992. The organization of the thalamocortical connections of the mediodorsal thalamic nucleus in the rat, related to the ventral forebrain-prefrontal cortex topography. *J Comp Neurol* 323:167–97.

Rescorla, R.A. 1993. Inhibitory associations between S and R in extinction. *Learn Behav* 21:327–36.

Rhodes, S.E., and Killcross, S. 2004. Lesions of rat infralimbic cortex enhance recovery and reinstatement of an appetitive Pavlovian response. *Learn Mem* 11:611–16.

Rhodes, S.E., and Killcross, A.S. 2007. Lesions of rat infralimbic cortex result in disrupted retardation but normal summation test performance following training on a Pavlovian conditioned inhibition procedure. *Eur J Neurosci* 26:2654–60.

Rich, E.L., and Shapiro, M.L. 2007. Prelimbic/infralimbic inactivation impairs memory for multiple task switches, but not flexible selection of familiar tasks. *J Neurosci* 27:4747–55.

Rolls, E.T., and Baylis, L.L. 1994. Gustatory, olfactory, and visual convergence within the primate orbitofrontal cortex. *J Neurosci* 14:5437–52.

Rolls, E.T., Yaxley, S., and Sienkiewicz, Z.J. 1990. Gustatory responses of single neurons in the caudolateral orbitofrontal cortex of the macaque monkey. *J Neurophysiol* 64:1055–66.

Rudebeck, P.H., and Murray, E.A. 2008. Amygdala and orbitofrontal cortex lesions differentially influence choices during object reversal learning. *J Neurosci* 28:8338–43.

Rudebeck, P.H., Walton, M.E., Smyth, A.N., Bannerman, D.M., and Rushworth, M.F. 2006. Separate neural pathways process different decision costs. *Nat Neurosci* 9:1161–68.

Rushworth, M.F., Behrens, T.E., Rudebeck, P.H., and Walton, M.E. 2007. Contrasting roles for cingulate and orbitofrontal cortex in decisions and social behaviour. *Trends Cogn Sci* 11:168–76.

Rushworth, M.F., Nixon, P.D., Eacott, M.J., and Passingham, R.E. 1997. Ventral prefrontal cortex is not essential for working memory. *J Neurosci* 17:4829–38.

Sahley, C., Gelperin, A., and Rudy, J.W. 1981. One-trial associative learning modifies food odor preferences of a terrestrial mollusc. *Proc Natl Acad Sci USA* 78:640–42.

Sanides, F. 1970. Functional architecture of motor and sensory cortices in primates in the light of a new concept of neocortex evolution. In *The Primate Brain*, eds. C.R. Noback, W. Montagna, pp. 137–208. New York: Appleton-Century-Crofts.

Saper, C.B. 2002. The central autonomic nervous system: Conscious visceral perception and autonomic pattern generation. *Ann Rev Neurosci* 25:433–69.

Schmidt-Rhaesa, A. 2007. *The Evolution of Organ Systems*. Oxford: Oxford University Press.

Schoenbaum, G., Chiba, A.A., and Gallagher, M. 1998. Orbitofrontal cortex and basolateral amygdala encode expected outcomes during learning. *Nat Neurosci* 1:155–59.

Schoenbaum, G., and Eichenbaum, H. 1995. Information coding in the rodent prefrontal cortex. I. Single-neuron activity in orbitofrontal cortex compared with that in pyriform cortex. *J Neurophysiol* 74:733–50.

Schultz, W. 2006. Behavioral theories and the neurophysiology of reward. *Annu Rev Psychol* 57:87–115.

———. 2007. Multiple dopamine functions at different time courses. *Annu Rev Neurosci* 30:259–88.

Schweimer, J., and Hauber, W. 2005. Involvement of the rat anterior cingulate cortex in control of instrumental responses guided by reward expectancy. *Learn Mem* 12:334–42.

Shadmehr, R., and Krakauer, J.W. 2008. A computational neuroanatomy for motor control. *Exp Brain Res* 185:359–81.

Smeets, W.J.A.J., Marín, O., and Ganzález, A. 2000. Evolution of the basal ganglia: New perspectives through a comparative approach. *J Anat* 196:501–17.

Stevens, J.R., Hallinan, E.V., and Hauser, M.D. 2005. The ecology and evolution of patience in two New World monkeys. *Biol Lett* 1:223–26.

Stevens, J.R., Rosati, A.G., Ross, K.R., and Hauser, M.D. 2005. Will travel for food: Spatial discounting in two new world monkeys. *Curr Biol* 15:1855–60.

Stouffer, E.M., and White, N.M. 2007. Roles of learning and motivation in preference behavior: Mediation by entorhinal cortex, dorsal and ventral hippocampus. *Hippocampus* 17:147–60.

Striedter, G.F. 2005. *Principles of Brain Evolution*. Sunderland, MA: Sinauer.

Suddendorf, T., and Corballis, M.C. 2007. The evolution of foresight: What is mental time travel, and is it unique to humans? *Behav Brain Sci* 30:299–313.

Thompson, R.H., Menard, A., Pombal, M., and Grillner, S. 2008. Forebrain dopamine depletion impairs motor behavior in lamprey. *Eur J Neurosci* 27:1452–60.

Trivedi, M.A., and Coover, G.D. 2004. Lesions of the ventral hippocampus, but not the dorsal hippocampus, impair conditioned fear expression and inhibitory avoidance on the elevated T-maze. *Neurobiol Learn Mem* 81:172–84.

Ulinski, P.S. 1990. The cerebral cortex of reptiles. In *Cerebral Cortex: Comparative Structure and Evolution of Cerebral Cortex, Part I*, eds. E.G. Jones and A. Peters, pp. 139–215. New York: Plenum.

Valentine, J.W. 2004. *On the Origin of Phyla*. Chicago: University of Chicago Press.

Vermeij, G.J. 1996. Animal origins. *Science* 274:525–26.

Wagner, H. 1932. Uber den Farbensinn der Eidechsen. *Comp Physiol [A]: Neuroethol* 18:378–92.

Wallis, J.D., Anderson, K.C., and Miller, E.K. 2001. Single neurons in prefrontal cortex encode abstract rules. *Nature* 411:953–56.

Wallis, J.D., and Miller, E.K. 2003. Neuronal activity in primate dorsolateral and orbital prefrontal cortex during performance of a reward preference task. *Eur J Neurosci* 18:2069–81.

Walton, M.E., Bannerman, D.M., and Rushworth, M.F. 2002. The role of rat medial frontal cortex in effort-based decision making. *J Neurosci* 22:10996–11003.

Wang, S.H., Ostlund, S.B., Nader, K., and Balleine, B.W. 2005. Consolidation and reconsolidation of incentive learning in the amygdala. *J Neurosci* 25:830–35.

Webster, M.J., Bachevalier, J., and Ungerleider, L.G. 1994. Connections of inferior temporal areas TEO and TE with parietal and frontal cortex in macaque monkeys. *Cereb Cortex* 4:470–83.

Wellman, L.L., Gale, K., and Malkova, L. 2005. GABAA-mediated inhibition of basolateral amygdala blocks reward devaluation in macaques. *J Neurosci* 25:4577–86.

Wicht, H., and Northcutt, R.G. 1992. The forebrain of the Pacific hagfish: A cladistic reconstruction of the ancestral craniate forebrain. *Brain Behav Evol* 40:25–64.

———. 1998. Telencephalic connections in the Pacific hagfish (*Eptatretus stouti*), with special reference to the thalamopallial system. *J Comp Neurol* 395:245–60.

Wilson, C.R., and Gaffan, D. 2008. Prefrontal-inferotemporal interaction is not always necessary for reversal learning. *J Neurosci* 28:5529–38.

Winstanley, C.A., Theobald, D.E., Cardinal, and R.N., Robbins, T.W. 2004. Contrasting roles of basolateral amygdala and orbitofrontal cortex in impulsive choice. *J Neurosci* 24:4718–22.

Wise, S.P. 2008. Forward frontal fields: Phylogeny and fundamental function. *Trends Neurosci* 31:599–608.

Wullimann, M.F., and Mueller, T. 2002. Expression of Zash-1a in the postembryonic zebrafish brain allows comparison to mouse Mash1 domains. *Brain Res Gene Expr Patterns* 1:187–92.

Wullimann, M.F., and Rink, E. 2002. The teleostean forebrain: A comparative and developmental view based on early proliferation, Pax6 activity and catecholaminergic organization. *Brain Res Bull* 57:363–70.

Yaxley, S., Rolls, E.T., and Sienkiewicz, Z.J. 1990. Gustatory responses of single neurons in the insula of the macaque monkey. *J Neurophysiol* 63:689–700.

Zhang, Y., Lu, H., and Bargmann, C.I. 2005. Pathogenic bacteria induce aversive olfactory learning in *Caenorhabditis elegans*. *Nature* 438:179–84.

Zhang, Z.H., Dougherty, P.M., and Oppenheimer, S.M. 1998. Characterization of baroreceptor-related neurons in the monkey insular cortex. *Brain Res* 796:303–6.

Part II

A Systems Organization of the Senses

5 Smell

Jay A. Gottfried and Donald A. Wilson

CONTENTS

> To speak generally then, things that have been cooked, delicate things, and things which are least of an earthy nature have a good odour, (odour being a matter of exhalation), and it is obvious that those of an opposite character have an evil odour.
>
> **Theophrastus c. 300 BC**

5.1 INTRODUCTION

Smell is arguably the most ancient sense. Consider for example a bacterial prokaryote, monocellular, anuclear, flagella-rotating, as it tumbled through the Archaean (Archaeozoic) seas roughly 2–3 billion years ago. For such an organism, chemotaxis—the ability to redirect its movements in the presence of chemical gradients—was synonymous with survival. This most rudimentary sense of smell was critical for finding chemoattractants like organic nutrients (Figure 5.1a; Adler 1969) and evading chemorepellents like excreted waste (Figure 5.1b; Tso and Adler 1974; Adler 1978). In this manner, the sense of smell was rooted at the earliest evolutionary stages with the machinery of affective processing, to the extent that chemical sensing was indistinguishable from the sensing of biological imperatives.

Interestingly, the key features of bacterial chemotaxis presage many of the same principles guiding olfactory processes in higher animals (Koshland 1980; Kleene 1986; Baker, Wolanin, and Stock 2006). *First*, the bacterial chemoreceptor complex is highly localized into clusters at the cell

(a)   (b)

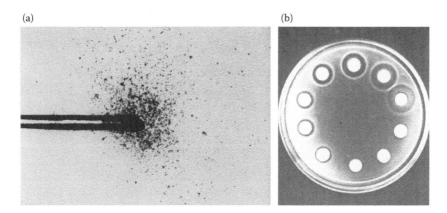

FIGURE 5.1 Olfactory hedonics in bacteria. (a) Chemosensory attraction. *E. coli* bacteria are using chemo-reception to migrate towards the open end of a capillary tube that has been filled with aspartate, an amino acid with nutritive value (Adler 1969). (Reprinted and modified from Adler, J., *Science* 166, 1588, 1969. Reprinted with permission of AAAS.) (b) Chemosensory aversion. In a microbiology assay of negative chemotaxis, agar plugs containing acetate, a harmful chemical compound, are placed around the perimeter of a petri dish, with acetate concentration in the plugs progressively increasing in a clockwise direction, starting at the four o'clock position. Higher concentrations of acetate induce greater repulsion in *E. coli* bacteria, indicated by progressive clearing at further distances from the plug. (Reprinted from *The Harvey Lectures: 1976–1977*, Series 72, Adler, J., Chemotaxis in bacteria, pp. 195–230, Copyright 1978, with permission from **Elsevier**.)

poles (Figure 5.2; Maddock and Shapiro 1993). This anatomical polarity effectively segregates sensory detection into spatially discrete regions of the organism, an arrangement that reaches its full eukaryotic expression in the form of antennae, noses, and olfactory epithelia.

Second, bacteria adapt to their chemical environment. Directional movement through a chemical gradient relies on concentration changes, rather than on absolute concentration per se, so that chemotaxis ceases once the bacterium finds itself in an isotropic solution. Reduced responsiveness to an unvarying chemical background closely parallels sensory adaptation in the mammalian olfactory system, which, as described later in this chapter, is an important mechanism underlying odor discrimination.

Third, bacteria learn from experience. Wild-type *Escherichia coli* show exuberant chemotaxis toward maltose when grown in a medium containing maltose, but respond minimally to the same attractant when grown in a maltose-free medium (Adler, Hazelbauer, and Dahl

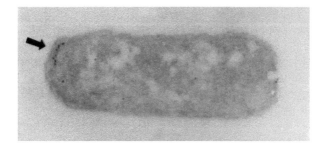

FIGURE 5.2 A primordial nose. Immunoelectron microscopy shows that the chemoreceptor complex of *E. coli* bacteria is clustered at a discrete polar location of the cell's inner membrane (arrow) (Maddock and Shapiro 1993). This cellular sequestration of chemoreceptive machinery represents an ancient forerunner of regional sensory specialization. (From Maddock, J.R., Shapiro, L. *Science*, 259, 1720, 1993. Reprinted with permission of AAAS.)

1973). As a basic example of sensory-specific plasticity, this experience-dependent gain in chemical sensitivity is also instantiated in vertebrate and invertebrate olfaction, whereby learning and experience robustly modify how smells are perceived. Remarkably, human neonates respond selectively to the odor of familiar (vs. unfamiliar) amniotic fluid (Schaal, Marlier, and Soussignan 1998) and they prefer food flavors that had been experienced prenatally by way of maternal consumption (Mennella, Jagnow, and Beauchamp 2001). That human fetuses "grown" in carrot-containing amniotic fluid are attracted to carrot flavor postnatally (Mennella, Jagnow, and Beauchamp 2001) curiously mirrors the growth-dependent inducibility of chemotaxis in prokaryote species. The role of experience in modulating odor coding is addressed in detail in Section 5.4.

Fourth, bacterial chemotaxis is synthetic. In other words, the decision to swim or tumble is based on the integration of chemical inputs. Simultaneous exposure to two different chemicals often elicits chemotactic behaviors that cannot be simply predicted from the mere sum of the individual components (Rubik and Koshland 1978). Even in *Salmonella typhimurium*, which contains a humble armament of five chemoreceptor genes (Wadhams and Armitage 2004; Hazelbauer, Falke, and Parkinson 2008), response potentiation (supra-additivity) to a combination of chemoattractants can be observed. The idea that behavioral non-linearities emerge in the presence of chemical mixtures has interesting implications for odor quality perception, multisensory integration, and flavor processing in rodents and humans, topics that are examined in Section 5.4, and throughout the sensory chapters of this book.

Fifth, and last, bacteria have a memory for smells, albeit on an extremely short timescale. The very fact that a 2-μm bacterium successfully swims through a chemical gradient to pinpoint a food source suggests that it must be able to retain a memory for what just came before (Macnab and Koshland 1972). Without the means to compare chemical concentrations from adjacent time frames (i.e., temporal integration), the bacterium would soon founder. This feature is not meant to imply that the olfactory systems of "higher" animals have adopted the same molecular cellular strategies to remember smells. Rather the point is that already from a very primordial moment in animal evolution, the same functional strategies were in place to maximize contact with behaviorally beneficial, or rewarding, chemical stimuli.

It is thus perhaps no accident that the human sense of smell remains so closely aligned with emotional processes, anatomically, physiologically, and psychologically. To our predecessors the bacteria, chemical sensation was fundamentally intertwined with the acquisition of rewards and the avoidance of threats. The operating principles by which bacteria achieved better living through chemistry—anatomical segregation of chemical detectors, response adaptation to static environments, sensory plasticity, signal integration, and chemical memory—continue to define and constrain the ways that the olfactory systems of higher organisms, including humans, contend with the odor landscape.

An olfactory enthusiast might plausibly go so far as to state that all biological subsystems are essentially a spin-off of the chemical detector apparatus. Since Buck and Axel first identified a large multigene family of G-protein-coupled receptor genes on rat olfactory sensory neurons (OSNs) (Buck and Axel 1991), it has become apparent that the G-protein-coupled conformation of the olfactory receptor shares much in common with receptors that bind a great many other critical biological ligands, including neurotransmitters, neuromodulators, tastants, light, hormones, growth factors, chemokines, cell adhesion molecules, and chemotactic peptides (Dryer 2000; Fredriksson et al. 2003). Laurence Dryer (2000) has nicely captured the idea that olfaction is a virtual microcosm of the human nervous system:

> This functional and structural diversity [within the olfactory receptor gene family] is not surprising if one bears in mind that there is no essential difference between neurotransmitter binding and odorant detection. Whether they are expressed at the synapse or in sensory cells, chemoreceptors serve the same function, that is, the chemical detection of a ligand carried by extracellular or external space.

5.2   THE ODOR STIMULUS SHAPES THE OLFACTORY SYSTEM

If a sensory system is to extract meaning from a sensory stimulus, what is the natural form of that stimulus in the real-world environment that a brain is likely to confront? For the olfactory system, a general answer to this question would be to state that olfactory percepts arise from physical and chemical properties of odor molecules, just as visual percepts arise from wavelengths of light.[*] This is correct, though troublingly vague, and indeed much of our current scientific understanding of olfactory systems in insects, rodents, and primates stems from research using monomolecular odorants (L-carvone, amyl acetate, etc.) with well-defined physical features, helping to ensure experimental control over the sensory input and to promote experimental replicability across different laboratories and species.

But it is critical to consider that the *ecological context* of an odor source defines what type of odor signal will be transmitted to a receiver. (See Dusenbery 1992 for an extensive and insightful overview of the sensory ecology of chemical signals, as well as thermal, light, sound, and mechanical signals.) The important biological implication is that, as an outcome of natural selection, how an odor is realistically encountered in the environment will ultimately shape how an olfactory nervous system evolves to optimize feature detection (Gottfried 2009).

The first point to keep in mind is that most real-world odors are *not* monomolecular compounds courtesy of the Sigma-Aldrich Flavors & Fragrances catalog. Single-molecule, single-meaning odorants are extremely rare in nature. The majority of emitted odors making contact with a chemical biosensor are mixtures of tens to hundreds of different odorant molecules. For example, microwave-popped popcorn contains at least 56 distinct volatile compounds (Walradt, Lindsay, and Libbey 1970), pressure-cooked pork liver liberates 179 volatile components (Mussinan and Walradt 1974), and the smell of chocolate is a blend of over 600 chemical constituents (Counet et al. 2002). Even the release of a potent moth sex attractant in the wild is often packaged together with many additional volatile components (Linn, Campbell, and Roelofs 1987; Dusenbery 1992). The compound nature of most environmental odors suggests that an olfactory system with the ability to encode information about the overall configuration of an emitted chemical stimulus would have a sensory discriminatory advantage. This issue is discussed at length in the final section of this chapter.

The second point is that odors are highly subject to the whims of their natural settings. The same odor may be experienced in the context of different background smells, potentially distorting how the odor is perceived. The direction or strength of the wind may make all the difference to an organism using airborne odor cues to locate its lunch. The length of time that has elapsed since the origination of an odor at its source will determine whether that odor cue still contains reliable predictive information, for example, if the odor source has run away (in the case of an animal), washed away (in the case of rain), or chemically transformed (in the case of a decomposing fruit). Each of these ecological scenarios means that an animal may have partial or corrupted access to an olfactory signal. Therefore, an olfactory system that reconstructs odor meaning—i.e., species, gender, sexual receptive state, social dominance, edibility—from sensory fragments would provide maximal adaptability to an organism.

The last point is that earthly transmission of an odor message markedly differs from the transmission of the other distance senses. In vision and audition, spatial patterns in the external world provide important sources of information that the brain can sample, extract, and codify in the form of spatial patterns of peripheral receptor activation. Critically, it is the spatial fidelity of visual and auditory stimuli—all the way from point source to receiver—that makes possible the use of spatial "codes" within these sensory modalities. However, in contrast to sights and sounds, smells diffuse

---

[*] The "building-blocks" comparison of odorant molecular components to visual light wavelength is commonly made but is really only appropriate at the receptor input stage of sensory information processing. Wavelength tells us about color per se, whereas it is the combination (and context) of many colors assembled into complex patterns that tell us what visual objects we see—just as it is the combination and context of many odorants assembled into complex patterns that tells us what odor objects we smell.

rather less predictably and less quickly from their sources. As mentioned above, air currents rapidly shear an odor message away from its spatial site of origin. Indeed, this carries certain advantages: olfactory signals can defy mundane physical obstacles, such as trees, bushes, hilltops, that would quickly extinguish a visual signal unable to bend around these foes. Thus, with the possible exception of localizing smells to one nostril or the other (Wilson 1997; Porter et al. 2007), spatial information is not a feature with which the olfactory system has evolved to contend. It is true that, as described in the next section, odorant-specific spatial patterns of neural activity are evoked within the olfactory bulb, but these patterns probably have more to do with encoding of odor identity, as opposed to odor space per se. Actually the spatial organization of receptor projections and bulb activity patterns are lost en route to olfactory cortex, and at this level temporal codes may be more critical for information processing (Haberly 2001).

In concluding this section, a proviso is in order. Tremendous advances in our neuroscientific understanding of the visual and auditory systems have been possible because the fundamental physical "primitives" underlying visual and auditory processing have been precisely identified. That is, receptive fields of color in vision or pitch in audition can be mapped precisely by varying the wavelength of light or sound along a linear dimension.[*] However, a precise characterization of odorant receptive fields remains elusive, given that the specific metric along which olfactory space is measured is not clear. For example, the physicochemical dimensions along which piperonyl acetate or 4-hydroxy-3-methoxy-benzaldehyde lie are still rather obscure; of equal perplexity is the lack of consensus regarding the *perceptual* dimensions along which the smell of cherry or vanilla lie. In spite of these uncertainties, much contemporary olfactory research, with modest successes, has placed an emphasis on hydrocarbon chain length and chemical functional group as two putative dimensions of odorant space, which will become evident from the following discussion.

5.3 THE OLFACTORY SYSTEM: ANATOMY AND PHYSIOLOGY

The olfactory system shares many functional properties with its counterpart sensory systems, yet is organizationally unique among mammalian sensory pathways. Broadly speaking, the functional anatomy of all sensory systems begins with an analysis of complex information streams into component features, local circuit enhancement of contrasts between features, and subsequent synthesis of those features into perceptual objects. Especially for complex organisms, the neurobiological assembly of sensory inputs into sensory "outputs" (i.e., perceptually meaningful objects) is optimized when sensory systems have an opportunity to integrate ascending and descending information flow, contextual and state-dependent modulatory effects, and multisensory interactions. Current evidence suggests that olfaction shares these functional properties with each of the other sensory systems.

A canonical mammalian sensory nervous system includes peripheral receptors, initial local circuit processing (e.g., retinal bipolar cells, auditory brainstem nuclei, dorsal root ganglia), a thalamic nucleus processing center, and a primary sensory cortex with reciprocal corticothalamic connections. Higher-order cortical areas may be specialized for specific, behaviorally relevant information content within one sensory modality, for example, faces, speech, or pain. Many sensory systems include parallel processing streams, wherein different aspects of the same stimulus are simultaneously captured and processed through anatomically distinct subsystems. Finally, it is becoming apparent that the opportunity for multimodal sensory convergence occurs at least as early as primary sensory cortex, an area traditionally considered to be a unimodal processing stream (e.g., Lakatos et al. 2007). Again, olfaction shares many of these anatomical characteristics with the other sensory systems, but there are important differences, as outlined below.

---

[*] Again, echoing the prior footnote, once we move from wavelengths to objects (faces, voices) and their cortical representations, the integrity of these fundamental dimensions breaks down both perceptually and topographically within sensory pathways. In area IT, for example, a neuron responsive to a garden gnome may be located adjacent to a neuron responsive to Greta Garbo, the objects of which share very little in the way of color, contrast, or curves.

The anatomy of the initial stages of the olfactory pathway suggests a hierarchical, combinatorial processing of odorants, with initial feature extraction in the periphery and subsequent convergence and blending of features through subsequent cortical stages. The following sections trace the encoding and transfiguration of an odor message as it moves through the olfactory system, beginning with the nose, olfactory epithelium, and OSNs, then pausing at the olfactory bulb, glomeruli, and mitral and tufted cells (the second-order neurons), and finishing at the olfactory cortex, including piriform cortex, and higher-order brain regions. The reader is referred to many comprehensive reviews (Price 1990; Carmichael, Clugnet, and Price 1994; Shipley and Ennis 1996; Haberly 1998; Cleland , Linster, and Doty 2003; Wilson, Sullivan, and Doty 2003; Zelano and Sobel 2005; Gottfried 2006; Gottfried, Small, and Zald 2006a) for in-depth discussions of olfactory system anatomy and physiology across different species.

5.3.1 PERIPHERAL CONSIDERATIONS

The mammalian olfactory system begins at the olfactory receptor sheet, where it lines the deeper recesses along the nasal passage. The sheltered location of the olfactory epithelium means that stimulus sampling is tied to airflow, if not absolutely dependent on it: in the "orthonasal" direction, inspiration draws in odorous molecules (odorants) from the outside environment and across the receptor layer; in the "retronasal" direction, expiration forces out odorant molecules from the inside environment (typically, food-based substances residing in the oral cavity) across the receptor layer. Thus, as Rozin (1982) has argued, olfaction is the only dual sensory system, operating both as a distance sense and as a contact sense.

The route of stimulus access can affect stimulus perception (e.g., King et al. 2006) and central brain circuit activation (Small et al. 2005). Detection thresholds and perceptual qualities for a given odorant are both dependent on the pathway by which that stimulus is delivered. These route-specific differences may arise because of variations in stimulus access to different zones within the olfactory receptor sheet (Scott and Brierley 1999; Scott et al. 2007) or because of gas chromatographic effects as the odorant molecules diffuse through the mucus overlying the sensory receptor cilia (Mozell 1991). Odor pleasantness, irritancy, and intensity, as well as state effects related to arousal and attention, can influence sniffing rate and volume (Mainland and Sobel 2006; Wesson et al. 2008), with further potential for modulating olfactory perception. Odor sampling behavior itself has an impact on olfactory coding: high-frequency sniffing vs. low-frequency passive respiration has been shown to filter out static background odors, enhancing an animal's ability to detect changes in the olfactory landscape (Verhagen et al. 2007).

5.3.2 CONTACT: OLFACTORY SENSORY NEURONS

Initial contact between an odor stimulus and the olfactory system occurs at the receptor endings of OSNs. Each OSN expresses one specific G-protein-coupled receptor gene (Buck and Axel 1991), of which there are approximately 1000 in rodents, 350 in humans, and 100 in fish (Firestein 2001). These receptors are studded along 20–30 long dendritic cilia that poke into the overlying mucus of the olfactory epithelium. Interestingly, OSNs expressing the same receptor gene do not segregate together within the epithelium, but instead are widely dispersed throughout different epithelial zones (Ressler, Sullivan, and Buck 1993). The lack of a systematic topographical organization within the olfactory receptor sheet already marks a fundamental difference between olfaction and the other distance senses (vision and audition).

The G-protein-coupled receptors may play a role not only in stimulus binding on the receptor cilia, but also in helping target the sensory neuron axons to discrete locations in the olfactory bulb called glomeruli (Vassalli et al. 2002). It is important to bear in mind that olfactory receptors do not bind and transduce "smells" (e.g., banana, rose), or even entire molecules (e.g., isoamyl acetate, phenyl ethyl alcohol), but sub-molecular moieties (Araneda, Kini, and Firestein 2000). Furthermore, an individual odorant molecule can serve as a full agonist of an olfactory receptor, a partial agonist, or even an antagonist (Gentilcore and Derby 1998). Thus, some interaction between odorant molecules occurs at the receptor sheet.

Once bound to receptor proteins, odorants evoke neural activity in OSNs, which transmit that activity into the central nervous system. Like neurons in other sensory systems, OSNs and their central targets respond to only a subset of all possible odorants (Duchamp-Viret, Chaput, and Duchamp 1999; Malnic et al. 1999), which is termed their molecular receptive range or, in the vernacular of sensory neuroscience, their odorant receptive field. The receptor ligand binding site on the OSN appears to be selective for molecular constituents of an odorant. In consequence, a given monomolecular odorant may bind to multiple different odorant receptors and a given receptor may bind multiple different odorants. Therefore, a natural scent, potentially composed of many different monomolecular odorants, will activate a unique combination of OSNs. Despite these guiding principles, recent work suggests that, at least in *Drosophila*, some individual receptor neurons may exhibit a highly narrow tuning profile for biologically salient odors such as pheromones (Schlief and Wilson 2007).

5.3.3 First Synapse: Olfactory Bulb Glomerulus

The OSNs send afferent projections directly into the central nervous system, terminating in a forebrain structure called the olfactory bulb. Olfactory glomeruli, lying near the surface of the olfactory bulb, are the principal functional units of information exchange between first-order neurons (OSNs) and second-order neurons (mitral and tufted cells). Individual glomeruli receive axons from OSNs all expressing the same receptor protein, which in turn form excitatory synapses onto mitral/tufted cell dendrites (see below). In the rodent olfactory system, the axons from several thousand OSNs (of the same receptor type) project to just one or two glomeruli (Firestein 2001) and a single mitral or tufted cell innervates just one glomerulus. This architecture ensures that each second-order neuron is targeted by a homogenous population of OSNs, providing an opportunity for dense anatomical convergence. On the other hand, in humans, with only 350–400 different olfactory receptor genes, this pattern breaks down, as there may be more than 10,000 glomeruli (Maresh et al. 2008). Nonetheless, given the homogenous receptor input to a given glomerulus, odorant stimulation evokes a stimulus-specific, combinatorial spatial pattern of glomerular activation across the olfactory bulb.

While there is debate over whether the glomerular spatial patterns of sensory input of odor-induced activity constitute a veridical map of odor perception (Laurent 1997), regional variation within the bulb in response to different molecular functional groups (e.g., alcohols or aldehydes; Johnson and Leon 2007) is certainly evident (though this chemotopy may not be preserved on a finer scale; Soucy et al. 2009). Furthermore, there may be regions that are more responsive to intrinsically meaningful odors (fox urine) and others more responsive to novel odors (Schaefer, Young, and Restrepo 2001; Kobayakawa et al. 2007). Recent work suggests that intact sensory input to the dorsal olfactory bulb is required for the expression of fear-based behavioral responses to innately aversive odors, whereas the ventral bulb appears sufficient to support learned odor aversions (Kobayakawa et al. 2007). This functional dissociation between innate and learned odors may be associated with differential output projections, and thus could comprise an important avenue of parallel processing.

5.3.4 Olfactory Bulb Output: Mitral and Tufted Cells

Mitral and tufted cells comprise the second-order olfactory neurons, whose cell bodies reside deep in the glomerular layer of the olfactory bulb. The dendrites arising from approximately 5–25 mitral/tufted cells typically innervate a single glomerulus (Firestein 2001), where they receive excitatory input from the OSN axons, as noted above. Given the precise spatial organization of odorant molecular information at the input (OSN) side of the glomerulus (Rubin and Katz 1999; Wachowiak et al. 2000; Johnson and Leon 2007), it is not surprising that the output side also shows a high degree of spatial patterning for odorant molecular features, with, for example, mitral/tufted cells in dorsome-dial olfactory bulb responsive to aldehydes (Imamura, Mataga, and Mori 1992) and cells in ventrolateral bulb responsive to aromatic (benzene-like) compounds (Katoh et al. 1993). Generally speaking, mitral and tufted cells are broadly responsive, or "tuned," to the presence of specific chain lengths or functional groups (Luo and Katz 2001; Fletcher and Wilson 2003; Igarashi and Mori 2005; Tan et al. 2010) with a specificity that resembles the receptive fields of the OSNs synapsing on them.

The hierarchical elaboration of odorant receptive fields from sensory neurons to central neurons shows broad similarities to processing hierarchies in the visual system. A few studies comparing odorant receptive fields across a variety of central olfactory pathway stages indicate that stimulus encoding becomes sparser, and receptive fields become more selective, as information ascends through the system (Tanabe et al. 1975a; Litaudon et al. 2003). For example, in monkeys, olfactory bulb neurons are broadly tuned, responding to an average of three to five stimuli within a fixed set of eight molecularly diverse odorants. To the same set of odorants, neurons in the amygdala and piriform cortex responded to an average of three odorants, and neurons in the orbitofrontal cortex (OFC) responded to one (Tanabe et al 1975a). Neurons within lateral hypothalamus are about as selective as olfactory bulb neurons (Scott and Pfaffmann 1972; Kogure and Onoda 1983).

An important organizational feature of the olfactory bulb is that second-order neurons located near each other express similar odorant receptive fields. As in other sensory systems, this arrangement provides an opportunity for lateral and feedback excitation and inhibition, which could contribute to lateral interactions, contrast enhancement, and gain control (Wilson and Leon 1987; Urban 2002; McGann et al. 2005; Olsen and Wilson 2008). A large population of GABAergic inhibitory interneurons, granule cells, forms dendrodendritic synapses with the output neurons in the bulb and probably helps mediate local interactions by serving as an inhibitory link among neighboring glomeruli. In fact, single-unit physiological processes resembling lateral inhibition have been demonstrated in mitral cell responses to odors and may even contribute to behavioral discrimination (Yokoi, Mori, and Nakanishi 1995; Luo and Katz 2001). Extensive mixture suppression is observed throughout the olfactory pathway (Kadohisa and Wilson 2006a; Lei, Mooney, and Katz 2006), which may reflect interactions at the sensory neuron level or through lateral inhibitory interactions within central circuits. Adaptation to one stimulus within the receptive field produces broad cross-adaptation to other odorants within the field, and stimuli evoking stronger initial responses induce greater cross-adaptation (Fletcher and Wilson 2003). These results again are consistent with a feature detection process, wherein the mitral or tufted cell responds to many odorants as long as they contain a particular feature.

A potential form of parallel processing may actually arise from the two main classes of glomerular output neurons. Mitral cell projections disseminate extensively throughout olfactory cortex and as far caudal as the entorhinal cortex, whereas tufted cell projections are limited largely to the most rostral portions of olfactory cortex (Scott 1981). Tufted cells have also been found to have lower thresholds for olfactory nerve activation than mitral cells (Harrison and Scott 1986; Nagayama et al. 2004).

5.3.5 OLFACTORY CORTEX

Afferent output from the olfactory bulb targets a large forebrain area called the olfactory cortex, a relatively simple trilaminar paleocortex that is devoid of classic cortical columns. The primary recipient of bulbar projections is called the piriform ("pear-shaped") cortex. In rodents this structure is found at the ventrolateral margin of the cortex and superficially borders the amygdala subnuclei for much of its length; in primates it is centered at the junction between the basal frontal and medial temporal lobes. Other direct recipients of olfactory bulb input include the anterior olfactory nucleus (AON), the olfactory tubercle, cortical nuclei of the amygdala, and entorhinal cortex. All of these regions, apart from the tubercle, project back to the bulb, allowing for descending modulation of bulb activity (Price 1990; Carmichael, Clugnet, and Price 1994; Shipley and Ennis 1996; Haberly 1998). The AON is the major intermediary of hemispheric cross-talk between olfactory structures, connecting the two olfactory bulbs through inhibitory (granule cell interneuron) relays (Yan et al. 2008), and linking left and right piriform cortices, perhaps as a critical link for transfer of odor information and memories during brief periods of unilateral blockade of nasal airflow (Kucharski and Hall 1987; Yeshurun, Dudai, and Sobel 2008). Whether these crossed pathways are functionally active in the human olfactory system is unclear.

Cortical afferents from the olfactory bulb project widely across piriform cortex, terminating in broad patches (Ojima, Mori, and Kishi 1984; Buonviso, Revial, and Jourdan 1991), in the absence of clear topographical organization. The relative lack of odor-specific spatial patterning within piriform cortex is distinctly different from the precise spatial arrangement of odor-specific activity within the olfactory bulb. It is hypothesized that the axon terminals from different populations of mitral cells, each conveying unique information about specific receptor activation, overlap in this region. In this manner, multiple features of an odorant, or of multiple odorants, extracted by disparate receptors may converge onto individual cortical pyramidal cells. Diverse techniques including c-*fos* immunohistochemistry (Datiche, Roullet, and Cattarelli 2001; Illig and Haberly 2003), voltage-sensitive dyes (Litaudon et al. 1997), microelectrode arrays (Rennaker et al. 2007), and optical imaging (Stettler and Axel 2009) have separately confirmed a diffuse pattern of odor-evoked activity throughout rodent anterior piriform cortex. The use of cortical flattening algorithms to generate two-dimensional "flat" maps of odor-evoked fMRI activity in the human brain has identified an equally diffuse, distributed projection pattern in posterior piriform cortex (Howard et al. 2009). Albeit at a more macroscopic (millimeter) level of resolution, these imaging findings show that the same piriform voxel may respond to more than one odor quality category, and neighboring voxels may respond to different odor categories, further reinforcing the idea of a non-topographical organization in piriform cortex.

Importantly, piriform cortex contains an extensive associational fiber system (Johnson et al. 2000; Yang et al. 2004). Individual cortical pyramidal cells make excitatory connections with several thousand other pyramidal cells, greatly expanding the associative convergence of information from different olfactory receptors. Further potential for information exchange occurs via extracortical connections that reciprocally join piriform cortex with higher-order areas such as amygdala, entorhinal cortex, and prefrontal cortex (Johnson et al. 2000). These association fiber synapses are highly plastic, allowing formation of templates of previously experienced patterns of afferent input in a content-addressable format (Haberly 2001). Computational models suggest that a key benefit of experience-dependent templates is efficient reconstruction of odor object percepts from patterned input, even if those inputs are degraded or fragmented (Hasselmo et al. 1990; Hopfield 1991; Haberly 2001).

The complex assortment of patchy mitral cell input and dense associational connections (both intracortical and extracortical) suggests that odorant receptive fields of neurons in the olfactory cortex may reflect an integrated, non-linear combination of features that cannot be predicted from afferent ("bottom-up") inputs alone (Lei, Mooney, and Katz 2006; Yoshida and Mori 2007; Barnes et al. 2008; Howard et al. 2009). For example, in some neurons of the anterior olfactory nucleus, responses to mixtures exceed the algebraic summation of the response to individual components (Lei, Mooney, and Katz 2006). Additionally, in both anterior olfactory nucleus (Lei, Mooney, and Katz 2006) and anterior piriform cortex (Yoshida and Mori 2007; Barnes et al. 2008), single neurons respond to a molecularly diverse range of odorants, while their mitral and tufted cell afferents respond to a more narrow range of molecular moieties (Lei, Mooney, and Katz 2006). Finally, olfactory cross-adaptation paradigms indicate that anterior piriform cortical neurons can learn to discriminate between mixtures and their elemental components, unlike olfactory bulb neurons that treat the parts and the whole much the same (Wilson 2000a, 2000b). These latter results suggest that, in contrast to mitral and tufted cells, piriform cortical neurons treat mixtures as unique objects, distinct from their components. Indeed, odor categorical perception can be estimated from distributed ensemble patterns of fMRI activity in human posterior piriform cortex (Howard et al. 2009), such that odorants more (or less) similar in perceptual quality exhibit more (or less) fMRI pattern overlap in this region, but not in anterior piriform cortex, OFC, or amygdala. Taken together these findings imply that piriform cortex is a critical repository for the encoding, retrieval, and modulation of odor objects (Wilson and Stevenson 2003; Gottfried 2010). This topic will be further examined in Sections 5.4 and 5.5.

5.3.6 HIGHER-ORDER PROJECTIONS

A rich array of regions receives input from olfactory cortex, including most of the limbic system—amygdala, hypothalamus, entorhinal cortex—as well as OFC, perirhinal neocortex, and mediodorsal thalamus. As with the olfactory bulb and cortex, many of these areas are reciprocally connected. Indirect connections include the insula, cingulate cortex, nucleus accumbens, ventral putamen, and hippocampus. Thus, the olfactory system has strong links with circuits involved in emotion, hedonics, memory, and decision making, and these systems in turn feed back to very early stages of olfactory processing.

The OFC is the principal olfactory neocortical projection site. The location of odor-responsive cells in OFC appears to vary across species: areas LO and VLO in rodent orbital cortex (Carmichael, Clugnet, and Price 1994); posterior orbital segments of Walker's area 13 in non-human primates (Tanabe et al. 1975b; Yarita et al. 1980; Carmichael, Clugnet, and Price 1994); and more anterior orbital segments adjacent to Walker's area 11 in humans (Ongur, Ferry, and Price 2003; Gottfried and Zald 2005). These anatomical differences may reflect biological differences in the way each of these species incorporates odors into their behavioral repertoires. Neurons in OFC respond not only to specific odors, but also to odor-contextual cues and odor cue-outcome contingencies (Schoenbaum and Eichenbaum 1995; Critchley and Rolls 1996a; Ramus and Eichenbaum 2000). In primate OFC both the identity of an odor and its reinforcement value can be extracted from the spike trains of individual neurons (Rolls et al. 1996), and similar results have been seen in rat OFC, leading to the idea that the OFC integrates sensory representations with associated reward value to help guide motivated behavior (Schoenbaum et al. 2003). This interpretation from single-unit recordings is supported by lesion studies in animals (Otto and Eichenbaum 1992; Schoenbaum et al. 2003) and functional neuroimaging studies in humans (O'Doherty et al. 2000; Gottfried, O'Doherty, and Dolan 2002; Gottfried, O'Doherty, and Dolan 2003; Gottfried and Dolan 2004; Gottfried 2007). To a lesser extent, similar profiles have been observed in piriform cortex (Gottfried, O'Doherty, and Dolan 2003).

One of the unusual characteristics of the mammalian olfactory system is the lack of an obligate thalamic relay between the periphery and the primary sensory neocortex. Anatomical and physiological data indicate that the major route of odor information into olfactory neocortex is a "direct" pathway between piriform cortex and OFC (Johnson et al. 2000; Cohen et al. 2008). However, an "indirect" trans-thalamic pathway does exist. The mediodorsal nucleus of the thalamus receives direct projections from the piriform cortex (Kuroda et al. 1992) and in turn projects to the OFC and to other prefrontal regions. Interestingly, the termination of direct piriform inputs into OFC appears to overlap with those arriving indirectly via mediodorsal thalamus, suggesting a convergent triangulation between these regions (Ray and Price 1992). The mediodorsal thalamus also receives input from the cortical and basal nuclei of the amygdala, potentially enriching the olfactory input to OFC with emotional context and hedonic valence (Pickens et al. 2003). The specific function of thalamocortical olfactory pathways in odor perception is not entirely known. Lesions of the mediodorsal thalamus in rats impair learning and memory of discriminative odor cues (Slotnick and Kaneko 1981), although this impairment appears to not be related to odor discrimination per se (Staubli , Schottler, and Nejat-Bina 1987b; Zhang, Schottler, and Nejat-Bina 1998). Recent work combining human olfactory fMRI and effective connectivity analysis (Plailly et al. 2008) shows that the transthalamic pathway is selectively recruited during olfactory attentional processes, perhaps helping to optimize conscious analysis of a smell. Other recent studies suggest involvement of human mediodorsal thalamus in both olfactory hedonic processing and associative learning (Zelano et al. 2007; Small et al. 2008; Sela et al. 2009).

It is worth noting that neurons within all central olfactory regions express some aspect of multimodal coding. Thus, the activity of even second-order neurons reflects not only receptor input, but also aspects of internal state (Pager et al. 1972), behavioral arousal (Gervais and Pager 1979; Kay and Laurent 1999) and past experience (Wilson and Leon 1987; Kendrick, Levy, and Keverne 1992; Fletcher and Wilson 2003). For example, mitral cell responses to food odors are enhanced in hungry animals (Pager et al. 1972). In awake rats performing an odor discrimination task, only 10%–20% of mitral cell activity differentially encodes odor information; most neurons respond to other aspects of the task (Kay and Laurent 1999; Rinberg, Koulakov, and Gelperin 2006). Similarly, piriform cortical neuron activity reflects not only odor stimulation but also a variety of non-olfactory task-related stimuli (Schoenbaum and Eichenbaum 1995), arousal (Murakami et al. 2005), predictive reward value (Li et al. 2008), and memory and past experience (Schoenbaum and Eichenbaum 1995; Dade et al. 1998; Kadohisa and Wilson 2006a; Li et al. 2006; Moriceau et al. 2006). Single neurons within the olfactory tubercle show bimodal activation, with activity driven by both odors and sounds (Wesson and Wilson 2010). Finally, the OFC is strongly multimodal, with activity reflecting olfactory, gustatory, somatosensory, and visual aspects of stimuli (Rolls and Baylis 1994; De Araujo et al. 2003; Gottfried and Dolan 2003; Kadohisa, Rolls, and Verhagen 2004; Small et al. 2004; Gottfried 2007).

Finally, the olfactory pathway is heavily innervated by neuromodulatory systems known to regulate cell excitability and plasticity in the setting of attention, arousal, and internal state. Both the olfactory bulb and cortex receive a strong cholinergic input from the horizontal limb of the diagonal band of Broca (Shipley and Ennis 1996), which itself is responsive to olfactory input (Linster and Hasselmo 2000). This creates an interesting feedback loop where cholinergic modulation of olfactory processing is itself partially under olfactory control. Norepinephrine from the locus coeruleus in the brainstem is another key neuromodulator (Shipley and Ennis 1996). Its release in the olfactory bulb is dependent on behavioral state and multimodal stimulus novelty, and it can modulate both mitral/tufted and piriform cortical neuron responses to odors (Jiang et al. 1996; Bouret and Sara 2002) as well as short-term and long-term olfactory plasticity (Gray, Freeman, and Skinner 1986; Sullivan, Wilson, and Leon 1989; Brennan, Kaba, and Keverne 1990; Shea, Katz, and Mooney 2008).

5.4 ODOR OBJECT SYNTHESIS AND PLASTICITY

Earlier sections of this chapter illustrate how ecological constraints and behavioral necessities have combined to guide the development of an olfactory brain system that can optimally extract meaningful information from odorous stimuli. As described above, the first step in the neural assembly of an odor object takes place with the progressive convergence and synthesis of molecular features present within an odorant (and across mixtures of odorants). This combinatorial scheme begins with receptor neuron axons projecting into olfactory bulb glomeruli and reaches its full expression in olfactory cortex, especially piriform cortex. Odor decorrelation (discrimination) improves as information moves from bulb to paleocortex to neocortex, though information flow to limbic regions such as amygdala and hypothalamus may be less specific.

The combinatorial processing of odors within a foreshortened sensory pathway (two synapses from sensory neuron to cortical neuron) and the strong odor-evoked activity within limbic brain areas suggest that cortical representations of odor are a strong echo of their stimulus inputs. However, the rapid and repeated convergence of non-olfactory information from higher-order areas into multiple levels of the information stream creates considerable opportunities for perceptual modulation and plasticity in the brain, such that odor representations become irreducibly connected to their meanings and associations. The goal of this section is to highlight some of the basic ecological pressures and physiological mechanisms driving sensory plasticity in the olfactory system.

Sensory perception and central processing of an odor are both strongly affected by past experience, current context, and behavioral state. The olfactory system shows remarkable plasticity in its response to odors, with rapid adjustments in sensitivity and acuity. These neural changes contribute to odor habituation, perception of odors against odorous backgrounds (background segmentation), and learning-induced changes in acuity as a result of perceptual and associative experience.

5.4.1 Reduced Input

The olfactory system regulates its sensitivity to odors through a variety of mechanisms and in reaction to different environmental settings. For example, a sustained period of *reduced* input, e.g., nasal obstruction due to rhinitis or (in the lab) due to nostril occlusion, results in an *increase* in odor sensitivity at the olfactory bulb (Wilson and Sullivan 1995). Such a mechanism of gain control helps to maintain bulb responsiveness in the face of diminished afferent information. However, this increase in gain comes at the cost of discriminability. Thus, spatial patterns of odor-evoked olfactory activity become blurred in the deprived hemisphere (Guthrie, Wilson, and Leon 1990), with increasing expansion and overlap of mitral cell receptive fields (Wilson and Sullivan 1995). A major component of the change in gain and loss of discrimination is an activity-dependent regulation of dopamine expression by juxtaglomerular interneurons (Baker et al. 1983; Baker 1990). Decreased sensory neuron input depresses synthesis and release of dopamine from juxtaglomerular neurons. Dopamine has a number of local circuit effects, but one is a D2 receptor-mediated pre-synaptic inhibition of glutamate release from sensory neuron terminals (Koster et al. 1999). In fact, the D2 receptor antagonist spiperone mimics the effects of sensory deprivation on mitral cell odor responses (Wilson and Sullivan 1995).

5.4.2 Prolonged Input

In contrast to the effects of reduced input, *prolonged* or repeated odor stimulation induces habituation and a *decrease* in odor sensitivity (Dalton and Wysocki 1996). Although OSNs (Kurahashi and Menini 1997; Zufall and Leinders-Zufall 2000) and mitral cells (Scott 1977; Wilson 1998b) both adapt to odors, piriform cortex undergoes the most robust adaptation, often exhibiting marked response suppression within one minute of stimulation (Wilson 1998b; Sobel et al. 2000; Poellinger et al. 2001; Li et al. 2006). As noted above, olfactory cortical adaptation is highly odor specific, allowing maintained responsiveness to novel odors (Wilson 2000a). Short-term cortical adaptation is mediated by activity-dependent depression of mitral/tufted cell cortical afferent synapses. These are glutamatergic synapses that express a pre-synaptic metabotropic glutamate receptor (group III). Following tens of seconds of normal odor exposure, these synapses are depressed (release less glutamate), reducing piriform pyramidal cell responses to afferent input and to odor stimuli (Wilson 1998a; Best et al. 2005). Blockade of these receptors prevents both cortical adaptation and behavioral odor habituation in animals (Best and Wilson 2004; Yadon and Wilson 2005). Both the synaptic depression and cortical odor adaptation recover within a few minutes (Best et al. 2005).

One important functional outcome of odor-specific cortical adaptation is its contribution to odor background segmentation (or "figure-ground" separation in the parlance of vision) (Kadohisa and Wilson 2006a; McNamara et al. 2008). Reduced piriform responsiveness to a steady background odor leaves intact the responses to novel odors appearing against that background olfactory landscape (Kadohisa and Wilson 2006a). This cortical mechanism allows for a separation of unvarying background information from dynamic inputs to optimize feature extraction of behaviorally relevant odor objects.

5.4.3 Perceptual (Non-Associative) Learning

It is important to bear in mind that cortical adaptation is more than just a high-pass filter of static information. To the extent that prolonged odor exposure implies an opportunity to acquire odor experience and familiarity, cortical adaptation is a critical component of olfactory perceptual learning. The consequence is that as an odor becomes more familiar, it becomes more distinct from other, similar odorants (Fletcher and Wilson 2002).

Olfactory perceptual learning is associated with changes in odor coding at the olfactory bulb (Fletcher and Wilson 2003; Moreno et al. 2009), piriform cortex (Wilson 2003; Kadohisa and

Wilson 2006b; Li et al. 2006), and OFC (Li et al. 2006). In the olfactory bulb, mitral and tufted cells express narrowed receptive fields to familiar odorants (Fletcher and Wilson 2003), perhaps due to fine-tuning of local inhibitory circuits (Moreno et al. 2009). These more acute receptive fields presumably encode the stimulus features more accurately. In the anterior piriform cortex, co-occurring features become synthesized into an odor object by experience, providing information about the identity of the stimulus (e.g., amyl acetate; Gottfried, Winston, and Dolan 2006b; Kadohisa and Wilson 2006b), whereas posterior piriform cortex encodes aspects of odor quality or category (e.g., fruity or banana; Gottfried, Winston, and Dolan 2006b; Kadohisa and Wilson 2006b; Howard et al. 2009), reflecting the influence of experience.

These experience-dependent changes can occur after simple exposure and familiarization (Fletcher and Wilson 2003; Wilson 2003; Moreno et al. 2009), even in the absence of specific training or associative learning, and thus represent an implicit perceptual learning process. In fact, this mode of learning may reflect the primary mechanism by which organisms compile their "vocabulary" of smells, with ever-increasing perceptual refinement and differentiation of odor objects. Based on the high-molecular dimensionality of odor stimuli and the roughly N-factorial combinations of N odorants that can be mixed together (Gottfried 2009), the number of unique discriminable smells is nearly limitless, highlighting a profound perceptual acuity that may largely be driven by mechanisms of implicit learning.

5.4.4 Associative Learning

Explicit training and associative experience can also modify receptive fields and odor processing, allowing learned important odors to be more discretely encoded and perceived (Rabin 1988; Cleland et al. 2002; Fletcher and Wilson 2002; Kadohisa and Wilson 2006a; Li et al. 2008). Thus, an organism can direct its behavior toward particular stimuli that have come to signal specific consequences. For example, in Pavlovian conditioning paradigms, learning that an odor predicts delivery of an aversive electric shock modifies mitral and tufted cell responses selectively to that odor (Sullivan and Wilson 1991), modulates piriform cortical ensemble responses selectively to that odor (Li et al. 2008), enhances odor-evoked activity within the basolateral amygdala (Sullivan et al. 2000; Rosenkranz and Grace 2002), and enhances behavioral discrimination of that odor from other very similar odors that previously could not be discriminated (Fletcher and Wilson 2002; Li et al. 2008). In rodents, the learned changes in behavior and cortical activity are acetylcholine-dependent (Wilson 2001; Fletcher and Wilson 2002).

In addition to changes in local circuit function and neural ensembles within specific brain regions, olfactory associative learning also modifies functional connectivity between different components of the olfactory and limbic systems. For example, associative conditioning enhances coherence of electrical activity between the olfactory bulb and piriform cortex (Martin et al. 2006), as well as between the olfactory bulb and hippocampus (Martin, Beshel, and Kay 2007). Associative conditioning can enhance synaptic strength of both afferent input to the piriform cortex (Truchet et al. 2002) and intracortical association fiber synapses (Saar, Grossman, and Barkai 2002), as well as strengthen connectivity between the OFC and the piriform cortex (Cohen et al. 2008). Through such changes, neurons in the piriform cortex (especially posterior regions) come to encode not only odor identity or quality, but also hedonic valence (e.g., Calu et al. 2007).

5.4.5 A Critical Interface in the Olfactory Orbitofrontal Cortex

The olfactory OFC appears to be a critical locus linking odor sensation, perception, and experience. Human olfactory fMRI studies increasingly show that OFC activity is highly plastic and can be updated by contextual, emotional, and cognitive factors. For example, odor-evoked activation in OFC is enhanced when a smell (e.g., scent of a rose) is presented with a semantically congruent picture (e.g., image of a flower), compared to a semantically incongruent picture (e.g., image of a bus), demonstrating the response sensitivity of OFC to familiar semantic contexts (Gottfried and Dolan 2003). Similar contextual findings in OFC have been observed with combinations of odors and tastes (Small et al. 2004) and odors and words (De Araujo et al. 2003). The idea that OFC helps integrate prior (learned) information about an odor receives further support from the perceptual learning study described above (Li et al. 2006). In this experiment, the magnitude of experience-induced response enhancement in OFC closely correlated with the degree of olfactory perceptual improvement, on a subject-by-subject basis, suggesting that OFC is instrumental in mediating behavioral changes in odor expertise.

The OFC also shows striking plasticity to manipulations of odor reward value. Human imaging studies of sensory-specific satiety (O'Doherty et al. 2000; Small et al. 2001; Gottfried, O'Doherty, and Dolan 2003; Kringelbach et al. 2003) indicate that when subjects consume a food until it is no longer palatable, the odor or flavor corresponding to that sated food elicits reduced activity in OFC, suggesting that this brain region provides a dynamic index of food motivational state. These latter findings closely accord with neurophysiological work in monkeys (Critchley and Rolls 1996b), which reveals similar orbitofrontal response profiles as a function of current olfactory reward value. It is interesting to speculate that the tendency of food rewards, and their olfactory cues, to lose their "rewarding-ness" upon consumption (i.e., satiety) might perhaps be based on the biological design of olfactory sensory systems that already had a tendency to adapt with prolonged and/or intense stimulation. The net result of odor-specific neuronal adaptation in a sensory-motor interface area like OFC would be to dampen consummatory behavior and perhaps to encourage a new search for food items containing different nutritional constituents.

Finally, a recent lesion study provides new evidence to suggest that the materialization of olfactory conscious awareness relies on an intact OFC (Li et al. 2010). A previously healthy 33-year-old man without any prior history of smell or taste problems developed anosmia (complete smell loss) following focal traumatic brain injury to the right OFC. Despite a total inability to perceive smells presented to either nostril, the patient nevertheless demonstrated preserved odor-evoked autonomic (skin conductance) responses to unpleasant vs. neutral smells, with concomitant odor-evoked fMRI activity in piriform cortex bilaterally as well as in left OFC. These findings implicate a central role of the right OFC in facilitating the transformation of an upstream olfactory message into a conscious percept, and at the same time suggest that the left olfactory pathway is not sufficient to sustain conscious olfaction.

5.5   FROM SENSATION TO REWARD: ARE ODORS REWARDING?

The main function of sensory systems is to extract behaviorally relevant information (meaning) from sensory signals in the environment. For an olfactory system, the natural world abounds with meaningful odor information: these are the smells of mother, child, and kin; home and territory; predator and prey; suitable and unsuitable mates. That survival throughout much of the animal kingdom hinges so substantially on odor sensations begs important questions. Is odor meaning an intrinsic or acquired feature of the physical stimulus? Are there innately pleasant (rewarding) and unpleasant (aversive) smells? Like food, water, and sex, do smells qualify as primary unconditioned reinforcers of behavior, an end unto themselves? That is, will an animal work to obtain more smell?

For aquatic microscopic species the answer is probably yes, because in the water there may be very little functional difference between olfaction and gustation. Inasmuch as a "smell" signifies a distant beacon or cue for an upstream reward (such as a source of food), a unicellular bacterium like *E. coli*, or a protozoan like the *Amoeba* (Figure 5.3), can *ingest* the cue—diverse soups of amino acids, peptides, disaccharides, algae particles—every bit as easily as it can the source. Thus the "smell" cue itself represents a consummatory object, a primary reinforcer. The fact that these water-bound organisms utilize a common chemosensory system for smell (i.e., food search, localization, detection) and taste (i.e., food reception, ingestion, phagocytosis) underscores the idea that there was little selective pressure for distinguishing cue from source in the earliest stages of animal evolution.

(a) (b) (c)

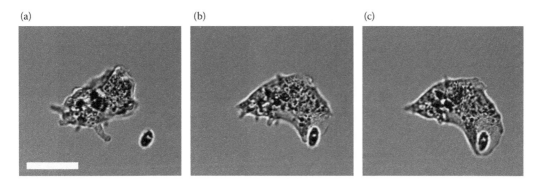

FIGURE 5.3 The functional boundaries between smell and taste are obscured in an aquatic protozoan like this amoeba. Standard bright-field microscopy depicts the sequence of events surrounding amoeboid food search and consumption. (a) An amoeba, moving slowly towards the right, uses chemotaxis to locate a microscopic aquatic organism, *Chilomonas* (small ovoid particle), which at this stage effectively acts as a chemosensory (olfactory) distance cue. Preparation for ingestion commences. (b) The advancing pseudopodia of the amoeba begin to surround the organism. (c) The amoeba has now completely engulfed the organism in a food vacuole, thus concluding consumption of the self-same cue that initially served as chemotactic stimulus. White scale bar, 50 μm. (Images reproduced and modified with permission of Tristan Ursell, PhD, Caltech.)

Further along the evolutionary time-line, a direct link between odor stimulus and innate value becomes slightly more tenuous. In multicellular organisms, and particularly in terrestrial species, the progressive increase in body size, metabolic requirements, and biological complexity meant that food smells alone could no longer satisfy the alimentary want. The scent of ripe fruit or young rabbit may be a potent food cue for locating fruits or rabbits, but the act of inhaling these aromas has no positive impact on nutritional status. Interestingly, the idea that smells do not a meal make (what would today appear to be an indisputable fact) was lost on some of history's eminent philosophers and clergymen. In his seventeenth-century quarto on traveling to the moon (Figure 5.4), Dr. John Wilkins (1614–1672), scientist, bishop, and visionary, proposed that the smell of food alone might be sufficient to sustain the nutritional needs of lunar voyagers (Wilkins 1684), with the following plausible rationale:

> Or, if we must needs feed upon something else, why may not smells nourish us? Plutarch and Pliny and divers other ancients, tell us of a nation in India that lived only upon pleasing odors. And 'tis the common opinion of physicians, that these do strangely both strengthen and repair the spirits. Hence was it that Democritus was able for divers days together to feed himself with the meer smel of hot bread.

The above viewpoint aside, it is clear that certain odor stimuli provoke hard-wired, unlearned behavioral responses. The predator smell of cat urine (Apfelbach et al. 2005; Takahashi et al. 2005), the moth sex pheromone bombykol (Carde et al. 1997), the alarm and recruitment pheromones of fire ants (Vander Meer, Slowik, and Thorvilson 2002), and the ink-opaline chemorepellent of the sea hare (Kicklighter et al. 2005) are just a few of the many examples of odor signals endowed with innate biological significance. In fact in rodents, cat odor can serve as an unconditioned stimulus in conditioning paradigms (Blanchard et al. 2001). Through natural selection, olfactory systems evolved to optimize detection and processing of these vital chemical messages in a species-specific manner. In invertebrates and vertebrates (with the exception of humans), specialized systems including vomeronasal organs and accessory olfactory bulbs may have evolved to handle these intrinsically meaningful odors—though to what extent the main olfactory system may also support such functions remains unclear.

Is there any reason to believe that the reinforcing properties of an odor might lie within the odor itself? A sensory ecology approach to the chemistry of pheromones (discussed in Dusenbery

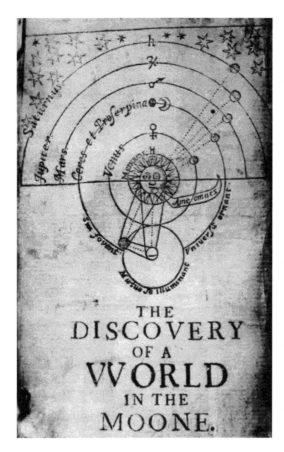 

FIGURE 5.4 Frontispieces of the first (1638) and fourth (1684) editions of a book by John Wilkins, the late Lord Bishop of Chester, in which he surmised that the smell of food might provide sufficient nourishment to ensure a healthy, comfortable sojourn and nutritionally adequate lifestyle on the moon. (From Wilkins, J. 1684. *A Discovery of a New World, or, A Discourse Tending to prove, that 'tis Probable there may be another Habitable World in the Moon.* 4th ed. corrected and amended, p. 172. London: T.M & J.A. for John Gillibrand at the Golden-Ball in St. Pauls Church-Yard.)

1992) suggests that molecular size of the physical stimulus is a key determinant of behavior. Small molecules (100–200 daltons; 6–15 carbons) make good alarm signals and repellants, because they diffuse rapidly into the air and dissipate as soon as the danger has passed. If the receiver has the appropriate receptors for that molecule and if those receptors are connected to the appropriate central motor control circuits, information eliciting escape can be quickly transmitted and acted upon. In turn, large molecules (200–300 daltons; 15–20 carbons) make good territorial markers because they stay in place for a long period of time due to their lower volatility. Large molecules also make good attractants, because their structural complexity provides numerous opportunities for chemical substitutions to the odorant molecule, helping to ensure species specificity (for example, a female silk moth would be most dismayed to find a male gypsy moth following her plume).

It is therefore plausible that basic appetitive or aversive features of a smell are methodically linked to the physical composition of an odorant (Yeshurun and Sobel 2010). Theories propounding a molecular basis for odor pleasantness have been present since antiquity (Cain 1978; Finger 1994). Democritus and Epicurus, and later Lucretius, championed the idea that the atoms of pleasant smells were smooth and round, whereas those of unpleasant smells were hooked, rending the nasal membrane full of nasty holes. Of the many odor classification schemes dating back to Linnaeus

(1707–1748), most of them contain at least one category for unpleasant smells, and even as recently as the 1960s "putrid" smells formed one of seven "primary odors" with its own unique electro-chemical configuration (Amoore 1952, 1970). Although empirical evidence in support of these ideas has historically been in short supply, multidimensional scaling studies indicate that valence is the primary dimension along which humans categorize odors (Berglund et al. 1973; Schiffman 1974), and a recent study using principal component analysis suggests that the subjective pleasantness of a smell can be roughly estimated from a set of >1500 chemical and molecular properties describing a given odorant stimulus (Khan et al. 2007).

Finally, it is worth noting that the meaning of a smell (its reinforcing property) depends importantly on the receiver. The fragrance of a red fox will simultaneously evoke dread in a rabbit and drool in a hound dog. In humans the earthy scent of Époisses cheese, or the brackish scent of sea urchin roe, may evoke strongly contrasting emotional responses even in two members of the same family. Both animal and human data (Critchley and Rolls 1996b; Rolls and Rolls 1997; O'Doherty et al. 2000) make it abundantly clear that through the mechanism of sensory-specific satiety, the pleasurable smell and flavor of a delectable food become unbearable after a gluttonous surfeit of said food. These examples underscore the hedonic relativity of odors, between species, within species, and even within individuals, suggesting that a "universal grammar" of chemical determinants to define the affective meaning of a given smell is unlikely to be identified.

5.6 THE NEUROBIOLOGY OF SENSATION AND REWARD: CONVERGENCE AND CODA

Faced with the natural complexities of an airborne odor, different organisms (and sometimes even the same organism) may adopt either analytical (elemental) or configural (synthetic) strategies for contending with an olfactory stimulus (Staubli et al. 1987a; Livermore et al. 1997; Kay, Lowry, and Jacobs 2003; Wiltrout, Dogra, and Linster 2003; Kay, Crk, and Thorngate 2005). In analytical odor processing, the whole stimulus (odor blend) is perceived to smell the *same* as the sum of the parts (odor elements)—or, in non-human animals, to evoke the same behavioral response, or to carry the same meaning. For example: $[A + B + C] = [A] + [B] + [C]$. Such mechanisms would promote stimulus generalization, by optimizing sensitivity to a biologically salient odor component, regardless of whether it appears alone or in a mixture (see Derby et al. 1996). Analytical processing may also help prevent signal corruption of the original input, preserving stimulus fidelity of the chemical message. However, these gains come at the potential expense of response over-generalization and a loss of odor specificity. An organism may respond indiscriminately to any compound stimulus that contains one particular "salient" odor X (e.g., $[A + B + X]$ and $[C + D + X]$), though that response may be impulsive or inappropriate to the present context.

In configural odor processing, the whole stimulus is perceived to smell *different* from the sum of the parts, such that a novel percept is generated, with a different meaning from the components themselves. In this instance, $[A + B + C] \neq [A] + [B] + [C]$. The capacity to generate new perceptual configurations adds substantially to stimulus discrimination (Livermore et al. 1997), with the possibility of emergent interactions and non-linearities among odor inputs. Synthetic processing may also be a more efficient mechanism of odor information storage (i.e., memory), because it reduces the computational burden that an olfactory brain would otherwise require to encode each and every element of a complex odor stimulus. At the same time, the integration of odor inputs is a useful way for assigning a unique perceptual stamp or signature to every odor-emanating object in the environment. In such cases, odor meaning is extracted from identifying the gestalt of the mixture, which is merged into a unique odor object. The main drawback of a synthetic system is the inevitable loss of signal input integrity, which could become problematic if detection of salient odor signal X decreases as a consequence of being incorporated into a new perceptual entity (where X is being suppressed) (Staubli et al. 1987a).

Biologically it appears that these two mechanisms can be engaged simultaneously. As discussed in Section 5.3, electrophysiological work by Wilson and colleagues (Wilson 2000b; Barnes et al. 2008) has demonstrated a functional double dissociation in rodents, whereby odor mixtures are encoded analytically in mitral/tufted cells of the olfactory bulb, and synthetically in the pyramidal cells of anterior piriform cortex. Complementary behavioral studies actually indicate that the ratio of components, or their perceived similarity, in a binary mixture can determine whether a rat will react to the odor blend analytically or synthetically (Kay, Lowry, and Jacobs 2003; Wiltrout, Dogra, and Linster 2003). That different levels of information processing in the rodent olfactory brain are organized to handle odor blends as elements or wholes is compatible with the notion that the rat can adopt different olfactory behavior strategies depending on environmental contingencies and task necessities.

Traditionally, the dichotomy between elemental and configural representations has strongly informed conceptual models of learning and memory (Rescorla 1972; Pearce 1987; Rudy and Sutherland 1992; Pearce and Bouton 2001). Since Pavlov it has been evident that an animal conditioned to a complex stimulus may not respond when the constituent stimuli are presented separately. Recognizing a behavioral distinction between the parts and the whole, Pavlov wrote that "a definite interaction takes place between the different cells of the cortex, resulting in a fusion or synthesis of their physiological activities on simultaneous excitation" (Pavlov 1927, 144). Woodbury (1943), evaluating these ideas more systematically in a reinforcement learning paradigm, found that when dogs were trained to respond to a *combination* of a high-pitched buzzer and a low-pitched buzzer (HL), they failed to react to the sound of either buzzer alone (Figure 5.5). Such observations led to the idea that a compound conditioned cue (AB) could gain access to a representation of an unconditioned stimulus (US) in either of two ways: elements A and B might each form a separate associative link with the US, such that either A or B would elicit a conditioned response; or, A and B might be conjoined into a unique configuration (AB), which itself becomes linked with the US. In this latter instance, only AB would elicit a response.

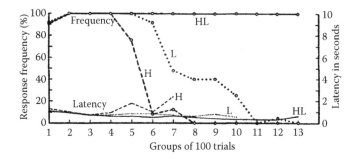

FIGURE 5.5 An early example of configural learning. In an instrumental conditioning task, dogs were trained to lift a wooden bar with the nose in order to receive positive reinforcement in the form of food pellets (Woodbury 1943). Each trial began with sounding of a high-pitched buzzer (H), a low-pitched buzzer (L), or a combination of both (HL), but critically, food reinforcement was provided only when bar-lifting behavior was preceded by HL. The figure shows that as training progressed, the animal successfully discriminated between the component and compound stimuli, such that response frequency (bar-lifting) remained high for the compound HL stimulus, but not for the component stimuli. In separate animals (data not shown), when responses to either auditory component alone (H or L) were reinforced, but responses to the compound (HL) were not, a similar profile was seen, with persistence of behavior for the component stimuli only. The discriminative enhancement between a stimulus combination and its components suggests that compound conditioning induces a fundamental qualitative change in the way that the combination is perceived, in keeping with modern tenets of configural learning. (Reprinted with permission of APA (American Psychological Association) from: Woodbury, C.B. 1943. The learning of stimulus patterns by dogs. *J Comp Psychol* 35:29–40.)

In all likelihood, elemental and configural mechanisms operate in tandem, though there is reason to believe that configural learning should confer much greater behavioral flexibility upon an organism. Rudy and Sutherland (1992) elegantly illustrated this point in a review article in which they considered two retrieval cues (A and B) that gain access to different reinforcers depending on stimulus context (Figure 5.6). In context 1 (C1), A activates a representation of the US, and B activates a representation of the absent US ("no-US"). In context 2 (C2), these contingencies are reversed: A activates a representation of the no-US, and B activates a representation of the US.

Using a simple model Rudy and Sutherland show that elemental representations cannot easily sustain context-dependent associative switching (the so-called trans-switching discrimination problem). With the formation of elemental associations (Figure 5.6, left), representational cues a and b and contexts c1 and c2 are by definition fully connected both to US and to no-US. As a result, there are no unique combinations of retrieval cues and contextual states that can selectively activate the full range of representational outcomes. However, with the formation of configural associations (Figure 5.6, right), unique conjunctions of stimulus and context information (c1a, c1b, c2a, c2b) ensure associative flexibility between a given retrieval cue and a behavioral reinforcer. The specificity of such an arrangement will also help to disambiguate potentially conflicting associations.

This marks a perfect example of systems convergence between sensation and reward, in keeping with a central theme of this book. For many animals the sensitivity of odor discrimination and the efficacy of reward learning both rely on an ability to forge novel associations between physically distinct stimuli. It is tempting to speculate that the neurobiology of (olfactory) sensation and the neurobiology of reward must have co-evolved, to the extent that many of the same anatomical circuits and physiological mechanisms are employed to achieve the same basic end. The potential co-dependence of these systems would have important implications for how sensory systems and reward systems operate. More complex organisms with a greater capacity for configural learning (odor-to-odor in the case of olfaction; cue-to-context in the case of reward) will be better equipped to adapt their behavior to changing environmental contingencies and homeostatic states.

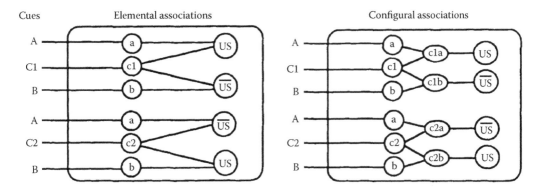

FIGURE 5.6 A schematic diagram contrasting the different internal representations that might arise during the formation of elemental (left) or configural (right) associations. A richer, more complex layer of associations can be generated during configural learning, allowing for greater discriminative capacity and memory retrieval for a given set of sensory cues (A, B), contexts (C1, C2), and outcomes (US, $\overline{US}$). See main text for further details. (Reproduced and modified with permission from Rudy, J.W., and Sutherland, R.J. 1992. Configural and elemental associations and the memory coherence problem. *J Cogn Neurosci* 4:208–16.)

REFERENCES

Adler, J. 1969. Chemoreceptors in bacteria. *Science* 166:1588–97.

————. 1978. Chemotaxis in bacteria. In *The Harvey Lectures – 1976–1977, Series 72*, eds. J. Adler et al., pp. 195–230. New York: Academic Press.

Adler, J., Hazelbauer, G.L., and Dahl, M.M. 1973. Chemotaxis toward sugars in *Escherichia coli. J Bacteriol* 115:824–47.

Amoore, J.E. 1952. The stereochemical specificities of human olfactory receptors. *Perfum Essent Oil Rec* 43:321–23, 330.

————. 1970. *Molecular Basis of Odor.* Springfield, IL: Charles C. Thomas.

Apfelbach, R., Blanchard, C.D., Blanchard, R.J., Hayes, R.A., and McGregor, I.S. 2005. The effects of predator odors in mammalian prey species: A review of field and laboratory studies. *Neurosci Biobehav Rev* 29:1123–44.

Araneda, R.C., Kini, A.D., and Firestein, S. 2000. The molecular receptive range of an odorant receptor. *Nat Neurosci* 3:1248–55.

Baker, H. 1990. Unilateral, neonatal olfactory deprivation alters tyrosine hydroxylase expression but not aromatic amino acid decarboxylase or GABA immunoreactivity. *Neuroscience* 36:761–71.

Baker, H., Kawano, T., Margolis, F.L., and Joh, T.H. 1983. Transneuronal regulation of tyrosine hydroxylase expression in olfactory bulb of mouse and rat. *J Neurosci* 3:69–78.

Baker, M.D., Wolanin, P.M., and Stock, J.B. 2006. Signal transduction in bacterial chemotaxis. *Bioessays* 28:9–22.

Barnes, D.C., Hofacer, R.D., Zaman, A.R., Rennaker, R.L., and Wilson, D.A. 2008. Olfactory perceptual stability and discrimination. *Nat Neurosci* 11:1378–80.

Berglund, B., Berglund, U., Engen, T., and Ekman, G. 1973. Multidimensional analysis of twenty-one odors. *Scand J Psychol* 14:131–37.

Best, A.R., Thompson, J.V., Fletcher, M.L., and Wilson, D.A. 2005. Cortical metabotropic glutamate receptors contribute to habituation of a simple odor-evoked behavior. *J Neurosci* 25:2513–17.

Best, A.R., and Wilson, D.A. 2004. Coordinate synaptic mechanisms contributing to olfactory cortical adaptation. *J Neurosci* 24:652–60.

Blanchard, R.J., Yang, M., Li, C.I., Gervacio, A., and Blanchard, D.C. 2001. Cue and context conditioning of defensive behaviors to cat odor stimuli. *Neurosci Biobehav Rev* 25:587–95.

Bouret, S., and Sara, S.J. 2002. Locus coeruleus activation modulates firing rate and temporal organization of odour-induced single-cell responses in rat piriform cortex. *Eur J Neurosci* 16:2371–82.

Brennan, P., Kaba, H., and Keverne, E.B. 1990. Olfactory recognition: A simple memory system. *Science* 250:1223–26.

Buck, L., and Axel, R. 1991. A novel multigene family may encode odorant receptors: A molecular basis for odor recognition. *Cell* 65:175–87.

Buonviso, N., Revial, M.F., and Jourdan, F. 1991. The projections of mitral cells from small local regions of the olfactory bulb: An anterograde tracing study using PHA-L (Phaseolus vulgaris Leucoagglutinin). *Eur J Neurosci* 3:493–500.

Cain, W.S. 1978. History of research on smell. In *Handbook of Perception: Vol. VIA. Tasting and Smelling*, eds. E.C. Carterette and M.P. Friedman, 197–229. New York: Academic Press.

Calu, D.J., Roesch, M.R., Stalnaker, T.A., and Schoenbaum, G. 2007. Associative encoding in posterior piriform cortex during odor discrimination and reversal learning. *Cereb Cortex* 17:1342–49.

Carde, R.T., Mafra-Neto, A., Carde, R.T., and Minks, A.K. 1997. Mechanisms of flight of male moths to pheromone. In *Insect Pheromone Research: New Directions*, eds. R.T. Carde and A.K. Minks, 275–90. New York: Chapman and Hall.

Carmichael, S.T., Clugnet, M.C., and Price, J.L. 1994. Central olfactory connections in the macaque monkey. *J Comp Neurol* 346:403–34.

Cleland, T.A., Linster, C., and Doty, R.L. 2003. Central olfactory structures. In *Handbook of Olfaction and Gustation*, ed. R.L. Doty, 165–80. New York: Marcel Dekker.

Cleland, T.A., Morse, A., Yue, E.L., and Linster, C. 2002. Behavioral models of odor similarity. *Behav Neurosci* 116:222–31.

Cohen, Y., Reuveni, I., Barkai, E., and Maroun, M. 2008. Olfactory learning-induced long-lasting enhancement of descending and ascending synaptic transmission to the piriform cortex. *J Neurosci* 28:6664–69.

Counet, C., Callemien, D., Ouwerx, C., and Collin, S. 2002. Use of gas chromatography-olfactometry to identify key odorant compounds in dark chocolate. Comparison of samples before and after conching. *J Agric Food Chem* 50:2385–91.

Critchley, H.D., and Rolls, E.T. 1996a. Olfactory neuronal responses in the primate orbitofrontal cortex: Analysis in an olfactory discrimination task. *J Neurophysiol* 75:1659–72.

———. 1996b. Hunger and satiety modify the responses of olfactory and visual neurons in the primate orbitofrontal cortex. *J Neurophysiol* 75:1673–86.

Dade, L.A., Jones-Gotman, M., Zatorre, R.J., and Evans, A.C. 1998. Human brain function during odor encoding and recognition. A PET activation study. *Ann N Y Acad Sci* 855:572–74.

Dalton, P., and Wysocki, C.J. 1996. The nature and duration of adaptation following long-term odor exposure. *Percept Psychophys* 58:781–92.

Datiche, F., Roullet, F., and Cattarelli, M. 2001. Expression of Fos in the piriform cortex after acquisition of olfactory learning: An immunohistochemical study in the rat. *Brain Res Bull* 55:95–99.

De Araujo, I.E., Rolls, E.T., Kringelbach, M.L., McGlone, F., and Phillips, N. 2003. Taste-olfactory convergence, and the representation of the pleasantness of flavour, in the human brain. *Eur J Neurosci* 18:2059–68.

Derby, C.D., Hutson, M., Livermore, B.A., and Lynn, W.H. 1996. Generalization among related complex odorant mixtures and their components: Analysis of olfactory perception in the spiny lobster. *Physiol Behav* 60:87–95.

Dryer, L. 2000. Evolution of odorant receptors. *Bioessays* 22:803–10.

Duchamp-Viret, P., Chaput, M.A., and Duchamp, A. 1999. Odor response properties of rat olfactory receptor neurons. *Science* 284:2171–74.

Dusenbery, D.B. 1992. *Sensory Ecology: How Organisms Acquire and Respond to Information.* New York: W.H. Freeman.

Finger, S. 1994. *Origins of Neuroscience: A History of Explorations into Brain Function.* New York: Oxford University Press.

Firestein, S. 2001. How the olfactory system makes sense of scents. *Nature* 413:211–18.

Fletcher, M.L., and Wilson, D.A. 2002. Experience modifies olfactory acuity: Acetylcholine-dependent learning decreases behavioral generalization between similar odorants. *J Neurosci* 22:RC201.

———. 2003. Olfactory bulb mitral-tufted cell plasticity: Odorant-specific tuning reflects previous odorant exposure. *J Neurosci* 23:6946–55.

Fredriksson, R., Lagerstrom, M.C., Lundin, L.G., and Schioth, H.B. 2003. The G-protein-coupled receptors in the human genome form five main families. Phylogenetic analysis, paralogon groups, and fingerprints. *Mol Pharmacol* 63:1256–72.

Gentilcore, L.R., and Derby, C.D. 1998. Complex binding interactions between multicomponent mixtures and odorant receptors in the olfactory organ of the Caribbean spiny lobster Panulirus argus. *Chem Senses* 23:269–81.

Gervais, R., and Pager, J. 1979. Combined modulating effects of the general arousal and the specific hunger arousal on the olfactory bulb responses in the rat. *Electroencephalogr Clin Neurophysiol* 46:87–94.

Gottfried, J.A. 2006. Smell: Central nervous processing. *Adv Otorhinolaryngol* 63:44–69.

———. 2007. What can an orbitofrontal cortex-endowed animal do with smells? *Ann N Y Acad Sci* 1121:102–20.

———. 2009. Function follows form: Ecological constraints on odor codes and olfactory percepts. *Curr Opin Neurobiol* 19:422–29.

———. 2010. Central mechanisms of odour object perception. *Nat Rev Neurosci* 11:628–41.

Gottfried, J.A., and Dolan, R.J. 2003. The nose smells what the eye sees: Crossmodal visual facilitation of human olfactory perception. *Neuron* 39:375–86.

———. 2004. Human orbitofrontal cortex mediates extinction learning while accessing conditioned representations of value *Nat Neurosci* 7:1144–52.

Gottfried, J.A., O'Doherty, J., and Dolan, R.J. 2002. Appetitive and aversive olfactory learning in humans studied using event-related functional magnetic resonance imaging. *J Neurosci* 22:10829–37.

———. 2003. Encoding predictive reward value in human amygdala and orbitofrontal cortex. *Science* 301:1104–7.

Gottfried, J.A., Small, D.M., and Zald, D.H. 2006a. The chemical senses. In *The Orbitofrontal Cortex*, eds. D.H. Zald, and S.L. Rauch, 125–79. New York: Oxford University Press.

Gottfried, J.A., Winston, J.S., and Dolan, R.J. 2006b. Dissociable codes of odor quality and odorant structure in human piriform cortex. *Neuron* 49:467–79.

Gottfried, J.A., and Zald, D.H. 2005. On the scent of human olfactory orbitofrontal cortex: Meta-analysis and comparison to non-human primates. *Brain Res Brain Res Rev* 50:287–304.

Gray, C.M., Freeman, W.J., and Skinner, J.E. 1986. Chemical dependencies of learning in the rabbit olfactory bulb: Acquisition of the transient spatial pattern change depends on norepinephrine. *Behav Neurosci* 100:585–96.

Guthrie, K.M., Wilson, D.A., and Leon, M. 1990. Early unilateral deprivation modifies olfactory bulb function. *J Neurosci* 10:3402–12.

Haberly, L.B. 1998. Olfactory cortex. In *The Synaptic Organization of the Brain*, ed. G.M. Shepherd, 377–416. New York: Oxford University Press.

———. 2001. Parallel-distributed processing in olfactory cortex: New insights from morphological and physiological analysis of neuronal circuitry. *Chem Senses* 26:551–76.

Harrison, T.A., and Scott, J.W. 1986. Olfactory bulb responses to odor stimulation: Analysis of response pattern and intensity relationships. *J Neurophysiol* 56:1571–89.

Hasselmo, M.E., Wilson, M.A., Anderson, B.P., and Bower, J.M. 1990. Associative memory function in piriform (olfactory) cortex: Computational modeling and neuropharmacology. *Cold Spring Harb Symp Quant Biol* 55:599–610.

Hazelbauer, G.L., Falke, J.J., and Parkinson, J.S. 2008. Bacterial chemoreceptors: Hgh-performance signaling in networked arrays. *Trends Biochem Sci* 33:9–19.

Hopfield, J.J. 1991. Olfactory computation and object perception. *Proc Natl Acad Sci USA* 88:6462–66.

Howard, J.D., Plailly, J., Grueschow, M., Haynes, J.D., and Gottfried, J.A. 2009. Odor quality coding and categorization in human posterior piriform cortex. *Nat Neurosci* 12:932–38.

Igarashi, K.M., and Mori, K. 2005. Spatial representation of hydrocarbon odorants in the ventrolateral zones of the rat olfactory bulb. *J Neurophysiol* 93:1007–19.

Illig, K.R., and Haberly, L.B. 2003. Odor–evoked activity is spatially distributed in piriform cortex. *J Comp Neurol* 457:361–73.

Imamura, K., Mataga, N., and Mori, K. 1992. Coding of odor molecules by mitral/tufted cells in rabbit olfactory bulb. I. Aliphatic compounds. *J Neurophysiol* 68:1986–2002.

Jiang, M., Griff, E.R., Ennis, M., Zimmer, L.A., and Shipley, M.T. 1996. Activation of locus coeruleus enhances the responses of olfactory bulb mitral cells to weak olfactory nerve input. *J Neurosci* 16:6319–29.

Johnson, B.A., and Leon, M. 2007. Chemotopic odorant coding in a mammalian olfactory system. *J Comp Neurol* 503:1–34.

Johnson, D.M., Illig, K.R., Behan, M., and Haberly, L.B. 2000. New features of connectivity in piriform cortex visualized by intracellular injection of pyramidal cells suggest that "primary" olfactory cortex functions like "association" cortex in other sensory systems. *J Neurosci* 20:6974–82.

Kadohisa, M., Rolls, E.T., and Verhagen, J.V. 2004. Orbitofrontal cortex: Neuronal representation of oral temperature and capsaicin in addition to taste and texture. *Neuroscience* 127:207–21.

Kadohisa, M., and Wilson, D.A. 2006a. Olfactory cortical adaptation facilitates detection of odors against background. *J Neurophysiol* 95:1888–96.

———. 2006b. Separate encoding of identity and similarity of complex familiar odors in piriform cortex. *Proc Natl Acad Sci USA* 103:15206–11.

Katoh, K., Koshimoto, H., Tani, A., and Mori, K. 1993. Coding of odor molecules by mitral/tufted cells in rabbit olfactory bulb. II. Aromatic compounds. *J Neurophysiol* 70:2161–75.

Kay, L.M., Crk, T., and Thorngate, J. 2005. A redefinition of odor mixture quality. *Behav Neurosci* 119:726–33.

Kay, L.M., and Laurent, G. 1999. Odor- and context-dependent modulation of mitral cell activity in behaving rats. *Nat Neurosci* 2:1003–9.

Kay, L.M., Lowry, C.A., and Jacobs, H.A. 2003. Receptor contributions to configural and elemental odor mixture perception. *Behav Neurosci* 117:1108–14.

Kendrick, K.M., Levy, F., and Keverne, E.B. 1992. Changes in the sensory processing of olfactory signals induced by birth in sleep. *Science* 256:833–36.

Khan, R.M., Luk, C.H., Flinker, A., Aggarwal, A., Lapid, H., Haddad, R., and Sobel, N. 2007. Predicting odor pleasantness from odorant structure: Pleasantness as a reflection of the physical world. *J Neurosci* 27:10015–23.

Kicklighter, C.E., Shabani, S., Johnson, P.M., and Derby, C.D. 2005. Sea hares use novel antipredatory chemical defenses. *Curr Biol* 15:549–54.

King, B.M., Arents, P., Duineveld, C.A., Meyners, M., Schroff, S.I., and Soekhai, S.T. 2006. Orthonasal and retronasal perception of some green leaf volatiles used in beverage flavors. *J Agric Food Chem* 54:2664–70.

Kleene, S.J. 1986. Bacterial chemotaxis and vertebrate olfaction. *Experientia* 42:241–50.

Kobayakawa, K., Kobayakawa, R., Matsumoto, H., Oka, Y., Imai, T., Ikawa, M., Okabe, M., et al. 2007. Innate versus learned odour processing in the mouse olfactory bulb. *Nature* 450:503–8.

Kogure, S., and Onoda, N. 1983. Response characteristics of lateral hypothalamic neurons to odors in unanesthetized rabbits. *J Neurophysiol* 50:609–17.

Koshland, Jr. D.E., 1980. Bacterial chemotaxis in relation to neurobiology. *Annu Rev Neurosci* 3:43–75.

Koster, N.L., Norman, A.B., Richtand, N.M., Nickell, W.T., Puche, A.C., Pixley, S.K., and Shipley, M.T. 1999. Olfactory receptor neurons express D2 dopamine receptors. *J Comp Neurol* 411:666–73.

Kringelbach, M.L., O'Doherty, J., Rolls, E.T., and Andrews, C. 2003. Activation of the human orbitofrontal cortex to a liquid food stimulus is correlated with subjective pleasantness. *Cereb Cortex* 13:1064–71.

Kucharski, D., and Hall, W.G. 1987. New routes to early memories. *Science* 238:786–88.

Kurahashi, T., and Menini, A. 1997. Mechanism of odorant adaptation in the olfactory receptor cell. *Nature* 385:725–29.

Kuroda, M., Murakami, K., Kishi, K., and Price, J.L. 1992. Distribution of the piriform cortical terminals to cells in the central segment of the mediodorsal thalamic nucleus of the rat. *Brain Res* 595:159–63.

Lakatos, P., Chen, C.M., O'Connell, M.N., Mills, A., and Schroeder, C.E. 2007. Neuronal oscillations and multisensory interaction in primary auditory cortex. *Neuron* 53:279–92.

Laurent, G. 1997. Olfactory processing: Maps, time and codes. *Curr Opin Neurobiol* 7:547–53.

Lei, H., Mooney, R., and Katz, L.C. 2006. Synaptic integration of olfactory information in mouse anterior olfactory nucleus. *J Neurosci* 26:12023–32.

Li, W., Howard, J.D., Parrish, T.B., and Gottfried, J.A. 2008. Aversive learning enhances perceptual and cortical discrimination of indiscriminable odor cues. *Science* 319:1842–45.

Li, W., Lopez, L., Osher, J., Howard, J.D., Parrish, T.B., and Gottfried, J.A. 2010. Right orbitofrontal cortex mediates conscious olfactory perception. *Psychol Sci* 21:1454–63.

Li, W., Luxenberg, E., Parrish, T., and Gottfried, J.A. 2006. Learning to smell the roses: Experience-dependent neural plasticity in human piriform and orbitofrontal cortices. *Neuron* 52:1097–1108.

Linn, Jr., C.E., Campbell, M.G., and Roelofs, W.L. 1987. Pheromone components and active spaces: What do moths smell and where do they smell it? *Science* 237:650–52.

Linster, C., and Hasselmo, M.E. 2000. Neural activity in the horizontal limb of the diagonal band of broca can be modulated by electrical stimulation of the olfactory bulb and cortex in rats. *Neurosci Lett* 282:157–60.

Litaudon, P., Amat, C., Bertrand, B., Vigouroux, M., and Buonviso, N. 2003. Piriform cortex functional heterogeneity revealed by cellular responses to odours. *Eur J Neurosci* 17:2457–61.

Litaudon, P., Mouly, A.M., Sullivan, R., Gervais, R., and Cattarelli, M. 1997. Learning-induced changes in rat piriform cortex activity mapped using multisite recording with voltage sensitive dye. *Eur J Neurosci* 9:1593–1602.

Livermore, A., Hutson, M., Ngo, V., Hadjisimos, R., and Derby, C.D. 1997. Elemental and configural learning and the perception of odorant mixtures by the spiny lobster Panulirus argus. *Physiol Behav* 62:169–74.

Luo, M., and Katz, L.C. 2001. Response correlation maps of neurons in the mammalian olfactory bulb. *Neuron* 32:1165–79.

Macnab, R.M., and Koshland, Jr. D.E., 1972. The gradient-sensing mechanism in bacterial chemotaxis. *Proc Natl Acad Sci USA* 69:2509–12.

Maddock, J.R., and Shapiro, L. 1993. Polar location of the chemoreceptor complex in the Escherichia coli cell. *Science* 259:1717–23.

Mainland, J., and Sobel, N. 2006. The sniff is part of the olfactory percept. *Chem Senses* 31:181–96.

Malnic, B., Hirono, J., Sato, T., and Buck, L.B. 1999. Combinatorial receptor codes for odors. *Cell* 96:713–23.

Maresh, A., Rodriguez Gil, D., Whitman, M.C., and Greer, C.A. 2008. Principles of glomerular organization in the human olfactory bulb – implications for odor processing. *PLoS ONE* 3:e2640.

Martin, C., Beshel, J., and Kay, L.M. 2007. An olfacto-hippocampal network is dynamically involved in odor-discrimination learning. *J Neurophysiol* 98:2196–2205.

Martin, C., Gervais, R., Messaoudi, B., and Ravel, N. 2006. Learning-induced oscillatory activities correlated to odour recognition: A network activity. *Eur J Neurosci* 23:1801–10.

McGann, J.P., Pirez, N., Gainey, M.A., Muratore, C., Elias, A.S., and Wachowiak, M. 2005. Odorant representations are modulated by intra- but not interglomerular presynaptic inhibition of olfactory sensory neurons. *Neuron* 48:1039–53.

McNamara, A.M., Magidson, P.D., Linster, C., Wilson, D.A., and Cleland, T.A. 2008. Distinct neural mechanisms mediate olfactory memory formation at different timescales. *Learn Mem* 15:117–25.

Mennella, J.A., Jagnow, C.P., and Beauchamp, G.K. 2001. Prenatal and postnatal flavor learning by human infants. *Pediatrics* 107:E88.

Moreno, M.M., Linster, C., Escanilla, O., Sacquet, J., Didier, A., and Mandairon, N. 2009. Olfactory perceptual learning requires adult neurogenesis. *Proc Natl Acad Sci USA* 106:17980–85.

Moriceau, S., Wilson, D.A., Levine, S., and Sullivan, R.M. 2006. Dual circuitry for odor-shock conditioning during infancy: Corticosterone switches between fear and attraction via amygdala. *J Neurosci* 26:6737–48.

Mozell, M.M., Kent, P.F., and Murphy, S.J. 1991. The effect of flow rate upon the magnitude of the olfactory response differs for different odorants. *Chem Senses* 16:631–49.

Murakami, M., Kashiwadani, H., Kirino, Y., and Mori, K. 2005. State-dependent sensory gating in olfactory cortex. *Neuron* 46:285–96.

Mussinan, C.J., and Walradt, J.P. 1974. Volatile constituents of pressure cooked pork liver. *J Agric Food Chem* 22:827–31.

Nagayama, S., Takahashi, Y.K., Yoshihara, Y., and Mori, K. 2004. Mitral and tufted cells differ in the decoding manner of odor maps in the rat olfactory bulb. *J Neurophysiol* 91:2532–40.

O'Doherty, J., Rolls, E.T., Francis, S., Bowtell, R., McGlone, F., Kobal, G., Renner, B., and Ahne, G. 2000. Sensory-specific satiety-related olfactory activation of the human orbitofrontal cortex. *Neuroreport* 11:893–97.

Ojima, H., Mori, K., and Kishi, K. 1984. The trajectory of mitral cell axons in the rabbit olfactory cortex revealed by intracellular HRP injection. *J Comp Neurol* 230:77–87.

Olsen, S.R., and Wilson, R.I. 2008. Lateral presynaptic inhibition mediates gain control in an olfactory circuit. *Nature* 452:956–60.

Ongur, D., Ferry, A.T., and Price, J.L. 2003. Architectonic subdivision of the human orbital and medial prefrontal cortex. *J Comp Neurol* 460:425–49.

Otto, T., and Eichenbaum, H. 1992. Complementary roles of the orbital prefrontal cortex and the perirhinal-entorhinal cortices in an odor-guided delayed-nonmatching-to-sample task. *Behav Neurosci* 106:762–75.

Pager, J., Giachetti, I., Holley, A., and Le Magnen, J. 1972. A selective control of olfactory bulb electrical activity in relation to food deprivation and satiety in rats. *Physiol Behav* 9:573–79.

Pavlov, I.P. 1927. *Conditioned Reflexes: An Investigation of the Physiological Activity of the Cerebral Cortex*, trans. and ed. G.V. Anrep. London: Oxford University Press.

Pearce, J.M. 1987. A model for stimulus generalization in Pavlovian conditioning. *Psychol Rev* 94:61–73.

Pearce, J.M., and Bouton, M.E. 2001. Theories of associative learning in animals. *Annu Rev Psychol* 52:111–39.

Pickens, C.L., Saddoris, M.P., Setlow, B., Gallagher, M., Holland, P.C., and Schoenbaum, G. 2003. Different roles for orbitofrontal cortex and basolateral amygdala in a reinforcer devaluation task. *J Neurosci* 23:11078–84.

Plailly, J., Howard, J.D., Gitelman, D.R., and Gottfried, J.A. 2008. Attention to odor modulates thalamocortical connectivity in the human brain. *J Neurosci* 28:5257–67.

Poellinger, A., Thomas, R., Lio, P., Lee, A., Makris, N., Rosen, B.R., and Kwong, K.K. 2001. Activation and habituation in olfaction – an fMRI study. *Neuroimage* 13:547–60.

Porter, J., Craven, B., Khan, R.M., Chang, S.J., Kang, I., Judkewicz, B., Volpe, J., Settles, G., and Sobel, N. 2007. Mechanisms of scent-tracking in humans. *Nat Neurosci* 10:27–29.

Price, J.L. 1990. Olfactory system. In *The Human Nervous System*, ed. G. Paxinos, 979–1001. San Diego: Academic Press.

Rabin, M.D. 1988. Experience facilitates olfactory quality discrimination. *Percept Psychophys* 44:532–40.

Ramus, S.J., and Eichenbaum, H. 2000. Neural correlates of olfactory recognition memory in the rat orbitofrontal cortex. *J Neurosci* 20:8199–8208.

Ray, J.P., and Price, J.L. 1992. The organization of the thalamocortical connections of the mediodorsal thalamic nucleus in the rat, related to the ventral forebrain-prefrontal cortex topography. *J Comp Neurol* 323:167–97.

Rennaker, R.L., Chen, C.F., Ruyle, A.M., Sloan, A.M., and Wilson, D.A. 2007. Spatial and temporal distribution of odorant-evoked activity in the piriform cortex. *J Neurosci* 27:1534–42.

Rescorla, R.A. 1972. "Configural" conditioning in discrete-trial bar pressing. *J Comp Physiol Psychol* 79:307–17.

Ressler, K.J., Sullivan, S.L., and Buck, L.B. 1993. A zonal organization of odorant receptor gene expression in the olfactory epithelium. *Cell* 73:597–609.

Rinberg, D., Koulakov, A., and Gelperin, A. 2006. Sparse odor coding in awake behaving mice. *J Neurosci* 26:8857–65.

Rolls, E.T., and Baylis, L.L. 1994. Gustatory, olfactory, and visual convergence within the primate orbitofrontal cortex. *J Neurosci* 14:5437–52.

Rolls, E.T., Critchley, H.D., Mason, R., and Wakeman, E.A. 1996. Orbitofrontal cortex neurons: Role in olfactory and visual association learning. *J Neurophysiol* 75:1970–81.

Rolls, E.T., and Rolls, J.H. 1997. Olfactory sensory-specific satiety in humans. *Physiol Behav* 61:461–73.

Rosenkranz, J.A., and Grace, A.A. 2002. Dopamine-mediated modulation of odour-evoked amygdala potentials during pavlovian conditioning. *Nature* 417:282–87.

Rozin, P. 1982. "Taste-smell confusions" and the duality of the olfactory sense. *Percept Psychophys* 31:397–401.

Rubik, B.A., and Koshland, Jr. D.E., 1978. Potentiation, desensitization, and inversion of response in bacterial sensing of chemical stimuli. *Proc Natl Acad Sci USA* 75:2820–24.

Rubin, B.D., and Katz, L.C. 1999. Optical imaging of odorant representations in the mammalian olfactory bulb. *Neuron* 23:499–511.

Rudy, J.W., and Sutherland, R.J. 1992. Configural and elemental associations and the memory coherence problem. *J Cogn Neurosci* 4:208–16.

Saar, D., Grossman, Y., and Barkai, E. 2002. Learning-induced enhancement of postsynaptic potentials in pyramidal neurons. *J Neurophysiol* 87:2358–63.

Schaal, B., Marlier, L., and Soussignan, R. 1998. Olfactory function in the human fetus: Evidence from selective neonatal responsiveness to the odor of amniotic fluid. *Behav Neurosci* 112:1438–49.

Schaefer, M.L., Young, D.A., and Restrepo, D. 2001. Olfactory fingerprints for major histocompatibility complex-determined body odors. *J Neurosci* 21:2481–87.

Schiffman, S.S. 1974. Physicochemical correlates of olfactory quality. *Science* 185:112–17.

Schlief, M.L., and Wilson, R.I. 2007. Olfactory processing and behavior downstream from highly selective receptor neurons. *Nat Neurosci* 10:623–30.

Schoenbaum, G., and Eichenbaum, H. 1995. Information coding in the rodent prefrontal cortex. I. Single-neuron activity in orbitofrontal cortex compared with that in pyriform cortex. *J Neurophysiol* 74:733–50.

Schoenbaum, G., Setlow, B., Saddoris, M.P., and Gallagher, M. 2003. Encoding predicted outcome and acquired value in orbitofrontal cortex during cue sampling depends upon input from basolateral amygdala. *Neuron* 39:855–67.

Scott, J.W. 1977. A measure of extracellular unit responses to repeated stimulation applied to observations of the time course of olfactory responses. *Brain Res* 132:247–58.

———. 1981. Electrophysiological identification of mitral and tufted cells and distributions of their axons in olfactory system of the rat. *J Neurophysiol* 46:918–31.

Scott, J.W., Acevedo, H.P., Sherrill, L., and Phan, M. 2007. Responses of the rat olfactory epithelium to retronasal air flow. *J Neurophysiol* 97:1941–50.

Scott, J.W., and Brierley, T. 1999. A functional map in rat olfactory epithelium. *Chem Senses* 24:679–90.

Scott, J.W., and Pfaffmann, C. 1972. Characteristics of responses of lateral hypothalamic neurons to stimulation of the olfactory system. *Brain Res* 48:251–64.

Sela, L., Sacher, Y., Serfaty, C., Yeshurun, Y., Soroker, N., and Sobel, N. 2009. Spared and impaired olfactory abilities after thalamic lesions. *J Neurosci* 29:12059–69.

Shea, S.D., Katz, L.C., and Mooney, R. 2008. Noradrenergic induction of odor-specific neural habituation and olfactory memories. *J Neurosci* 28:10711–19.

Shipley, M.T., and Ennis, M. 1996. Functional organization of olfactory system. *J Neurobiol* 30:123–76.

Slotnick, B.M., and Kaneko, N. 1981. Role of mediodorsal thalamic nucleus in olfactory discrimination learning in rats. *Science* 214:91–92.

Small, D.M., Gerber, J.C., Mak, Y.E., and Hummel, T. 2005. Differential neural responses evoked by orthonasal versus retronasal odorant perception in humans. *Neuron* 47:593–605.

Small, D.M., Veldhuizen, M.G., Felsted, J., Mak, Y.E., and McGlone, F. 2008. Separable substrates for anticipatory and consummatory food chemosensation. *Neuron* 57:786–97.

Small, D.M., Voss, J., Mak, Y.E., Simmons, K.B., Parrish, T., and Gitelman, D. 2004. Experience-dependent neural integration of taste and smell in the human brain. *J Neurophysiol* 92:1892–1903.

Small, D.M., Zatorre, R.J., Dagher, A., Evans, A.C., and Jones-Gotman, M. 2001. Changes in brain activity related to eating chocolate: From pleasure to aversion. *Brain* 124:1720–33.

Sobel, N., Prabhakaran, V., Zhao, Z., Desmond, J.E., Glover, G.H., Sullivan, E.V., and Gabrieli, J.D. 2000. Time course of odorant-induced activation in the human primary olfactory cortex. *J Neurophysiol* 83:537–51.

Soucy, E.R., Albeanu, D.F., Fantana, A.L., Murthy, V.N., and Meister, M. 2009. Precision and diversity in an odor map on the olfactory bulb. *Nat Neurosci* 12:210–20.

Staubli, U., Fraser, D., Faraday, R., and Lynch, G. 1987a. Olfaction and the "data" memory system in rats. *Behav Neurosci* 101:757–65.

Staubli, U., Schottler, F., and Nejat-Bina, D. 1987b. Role of dorsomedial thalamic nucleus and piriform cortex in processing olfactory information. *Behav Brain Res* 25:117–29.

Sullivan, R.M., Landers, M., Yeaman, B., and Wilson, D.A. 2000. Good memories of bad events in infancy. *Nature* 407:38–39.

Sullivan, R.M., and Wilson, D.A. 1991. Neural correlates of conditioned odor avoidance in infant rats. *Behav Neurosci* 105:307–12.

Sullivan, R.M., Wilson, D.A., and Leon, M. 1989. Norepinephrine and learning-induced plasticity in infant rat olfactory system. *J Neurosci* 9:3998–4006.

Takahashi, L.K., Nakashima, B.R., Hong, H., and Watanabe, K. 2005. The smell of danger: A behavioral and neural analysis of predator odor-induced fear. *Neurosci Biobehav Rev* 29:1157–67.

Tan, J., Savigner, A., Ma, M., and Luo, M. 2010. Odor information processing by the olfactory bulb analyzed in gene-targeted mice. *Neuron* 65:912–26.

Tanabe, T., Iino, M., and Takagi, S.F. 1975a. Discrimination of odors in olfactory bulb, pyriform-amygdaloid areas, and orbitofrontal cortex of the monkey. *J Neurophysiol* 38:1284–96.

Tanabe, T., Yarita, H., Iino, M., Ooshima, Y., and Takagi, S.F. 1975b. An olfactory projection area in orbitofrontal cortex of the monkey. *J Neurophysiol* 38:1269–83.

Theophrastus c. 300 BC. *Treatise on Odours*; Vol. II, *Books 6–9: Enquiry into Plants*. Loeb Classical Library, 1916 ed., trans. Sir A.F. Hort. Cambridge, MA: Harvard University Press.

Truchet, B., Chaillan, F.A., Soumireu-Mourat, B., and Roman, F.S. 2002. Learning and memory of cue-reward association meaning by modifications of synaptic efficacy in dentate gyrus and piriform cortex. *Hippocampus* 12:600–8.

Tso, W.W., and Adler, J. 1974. Negative chemotaxis in Escherichia coli. *J Bacteriol* 118:560–76.

Urban, N.N. 2002. Lateral inhibition in the olfactory bulb and in olfaction. *Physiol Behav* 77:607–12.

Vander Meer, R.K., Slowik, T.J., and Thorvilson, H.G. 2002. Semiochemicals released by electrically stimulated red imported fire ants, Solenopsis invicta. *J Chem Ecol* 28:2585–2600.

Vassalli, A., Rothman, A., Feinstein, P., Zapotocky, M., and Mombaerts, P. 2002. Minigenes impart odorant receptor-specific axon guidance in the olfactory bulb. *Neuron* 35:681–96.

Verhagen, J.V., Wesson, D.W., Netoff, T.I., White, J.A., and Wachowiak, M. 2007. Sniffing controls an adaptive filter of sensory input to the olfactory bulb. *Nat Neurosci* 10:631–39.

Wachowiak, M., Zochowski, M., Cohen, L.B., and Falk, C.X. 2000. The spatial representation of odors by olfactory receptor neuron input to the olfactory bulb is concentration invariant. *Biol Bull* 199:162–63.

Wadhams, G.H., and Armitage, J.P. 2004. Making sense of it all: Bacterial chemotaxis. *Nat Rev Mol Cell Biol* 5:1024–37.

Walradt, J.P., Lindsay, R.C., and Libbey, L.M. 1970. Popcorn flavor: Identification of volatile compounds. *J Agric Food Chem* 18:926–28.

Wesson, D.W., Donahou, T.N., Johnson, M.O., and Wachowiak, M. 2008. Sniffing behavior of mice during performance in odor-guided tasks. *Chem Senses* 33:581–96.

Wesson, D.W., and Wilson, D.A. 2010. Smelling sounds: Olfactory-auditory sensory convergence in the olfactory tubercle. *J Neurosci* 30:3013–21.

Wilkins, J. 1684. *A Discovery of a New World, or, A Discourse Tending to prove, that 'tis Probable there may be another Habitable World in the Moon*. 4th ed. corrected and amended, p. 172. London: T.M & J.A. for John Gillibrand at the Golden-Ball in St. Pauls Church-Yard.

Wilson, D.A. 1997. Binaral interactions in the rat piriform cortex. *J Neurophysiol* 78:160–69.

———. 1998a. Synaptic correlates of odor habituation in the rat anterior piriform cortex. *J Neurophysiol* 80:998–1001.

———. 1998b. Habituation of odor responses in the rat anterior piriform cortex. *J Neurophysiol* 79:1425–40.

———. 2000a. Odor specificity of habituation in the rat anterior piriform cortex. *J Neurophysiol* 83:139–45.

———. 2000b. Comparison of odor receptive field plasticity in the rat olfactory bulb and anterior piriform cortex. *J Neurophysiol* 84:3036–42.

———. 2001. Scopolamine enhances generalization between odor representations in rat olfactory cortex. *Learn Mem* 8:279–85.

———. 2003. Rapid, experience-induced enhancement in odorant discrimination by anterior piriform cortex neurons. *J Neurophysiol* 90:65–72.

Wilson, D.A., and Leon, M. 1987. Evidence of lateral synaptic interactions in olfactory bulb output cell responses to odors. *Brain Res* 417:175–80.

Wilson, D.A., and Stevenson, R.J. 2003. The fundamental role of memory in olfactory perception. *Trends Neurosci* 26:243–47.

Wilson, D.A., and Sullivan, R.M. 1995. The D2 antagonist spiperone mimics the effects of olfactory deprivation on mitral/tufted cell odor response patterns. *J Neurosci* 15:5574–81.

Wilson, D.A., Sullivan, R.M., and Doty, R.L. 2003. Sensory physiology of central olfactory pathways. In *Handbook of Olfaction and Gustation*, ed. R.L. Doty, 181–201. New York: Marcel Dekker.

Wiltrout, C., Dogra, S., and Linster, C. 2003. Configurational and nonconfigurational interactions between odorants in binary mixtures. *Behav Neurosci* 117:236–45.

Woodbury, C.B. 1943. The learning of stimulus patterns by dogs. *J Comp Psychol* 35:29–40.

Yadon, C.A., and Wilson, D.A. 2005. The role of metabotropic glutamate receptors and cortical adaptation in habituation of odor-guided behavior. *Learn Mem* 12:601–5.

Yan, Z., Tan, J., Qin, C., Lu, Y., Ding, C., and Luo, M. 2008. Precise circuitry links bilaterally symmetric olfactory maps. *Neuron* 58:613–24.

Yang, J., Ul Quraish, A., Murakami, K., Ishikawa, Y., Takayanagi, M., Kakuta, S., and Kishi, K. 2004. Quantitative analysis of axon collaterals of single neurons in layer IIa of the piriform cortex of the guinea pig. *J Comp Neurol* 473:30–42.

Yarita, H., Iino, M., Tanabe, T., Kogure, S., and Takagi, S.F. 1980. A transthalamic olfactory pathway to orbitofrontal cortex in the monkey. *J Neurophysiol* 43:69–85.

Yeshurun, Y., Dudai, Y., and Sobel, N. 2008. Working memory across nostrils. *Behav Neurosci* 122:1031–37.

Yeshurun, Y., and Sobel, N. 2010. An odor is not worth a thousand words: From multidimensional odors to unidimensional odor objects. *Annu Rev Psychol* 61:219–41, C211–15.

Yokoi, M., Mori, K., and Nakanishi, S. 1995. Refinement of odor molecule tuning by dendrodentric synaptic inhibition in the olfactory bulb. *Proc Natl Acad Sci USA* 92:3371–75.

Yoshida, I., and Mori, K. 2007. Odorant category profile selectivity of olfactory cortex neurons. *J Neurosci* 27:9105–14.

Zelano, C., Montag, J., Johnson, B., Khan, R., and Sobel, N. 2007. Dissociated representations of irritation and valence in human primary olfactory cortex. *J Neurophysiol* 97:1969–76.

Zelano, C., and Sobel, N. 2005. Humans as an animal model for systems-level organization of olfaction. *Neuron* 48:431–54.

Zhang, Y., Burk, J.A., Glode, B.M., and Mair, R.G. 1998. Effects of thalamic and olfactory cortical lesions on continuous olfactory delayed nonmatching-to-sample and olfactory discrimination in rats (Rattus norvegicus). *Behav Neurosci* 112:39–53.

Zufall, F., and Leinders-Zufall, T. 2000. The cellular and molecular basis of odor adaptation. *Chem Senses* 25:473–81.

6 Taste

Donald B. Katz and Brian F. Sadacca

CONTENTS

6.1 INTRODUCTION

Taste stimuli are unique in the world of recognizable objects, in that they are perceived only after being selected and engulfed (Figure 6.1). The physical sources of your current visual, auditory, somatosensory, and olfactory percepts are essentially external—you see what's in front of your face, hear what's within earshot, feel what's within reach, and smell bits of external objects carried to you in the airstream (it is usually possible to determine from whence it was dealt once it has been smelt)—but a stimulus activates the gustatory system only after it has been purposefully removed from view. Organisms make a deliberate decision to have a taste experience, choosing an external object in their environment for consumption and experiencing the taste percept only after sending that object down the path toward digestion.

This concrete, physical difference between taste and other stimuli has important biopsychological implications: stimuli providing taste sensations affect well-being with a reliability that other stimuli do not. The five major categories of taste quality are all tuned to identify a specific nutrient or physiological threat, namely ensuring energy reserves (sweet), maintaining water balance (salty), guarding pH (sour), motivating protein intake (savory), and avoiding toxins (bitter, see Bartoshuk 1991). A smell can be noxious, a sound too loud, a touch too rough, or an image too dangerous/titillating for safe viewing, but the vast majority of such stimuli leave little physiological trace once they've passed by. Because tasty stimuli are already in the body when perceived (at least for mammals), however, they directly impact our health and happiness. They feed and fatten us, poison, pickle, or please us, replenish or repulse us, but they seldom leave us unaffected.

It is perhaps the fact that tastes invariably impact what Garcia referred to as the internal *milieu* (Garcia, Hankins, and Rusiniak 1974) that makes them by far the most directly rewarding (or punishing) of the sensory stimuli. Taste stimuli act as primary reinforcers in a wide variety of contexts (Berridge 1996; O'Doherty et al. 2002; Hajnal and Norgren 2005); in fact, most other stimuli reinforce on the basis of temporal association with basic tastes (Gallagher and Schoenbaum 1999; Balleine and Dickinson 2000; O'Doherty et al. 2003). Although it is true that both human and nonhuman primate males will pay to access naked images of female conspecifics (Deaner, Khera, and Platt 2005), visual, auditory, somatosensory, and olfactory stimuli are seldom as tightly linked to

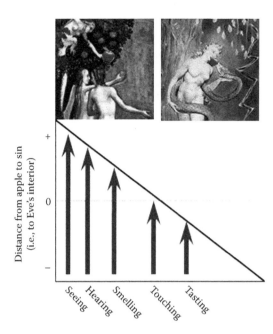

FIGURE 6.1 Egocentric distances of the five senses (units in Biblical proportions). At the top are two paintings showing Eve in the Garden of Eden. At the left is a detail from H. Bosch's *Last Judgment* and at the right is a detail from W. Blake's *The Temptation and Fall of Eve*. Below is a schematic diagram showing how the various attributes of the apple, proffered by the serpent, were communicated to her: the visual, the auditory (the sound of the apple falling or being plucked from the tree, perhaps), and the olfactory stimuli arrive while the object is still relatively far (i.e., >0 units distant) from her, and the somatosensory stimulus arrives when the apple is at her skin (i.e., 0 units distant); only the taste stimulus arrives after it is already too late—the apple is inside her mouth (i.e., <0 units distant from her).

reward as are stimuli imbued with taste attributes. Put succinctly, most sensory stimuli are used to attain reward, but thanks to their privileged access to the internal *milieu*, tastes *are* rewards.[*]

In the essay that follows, we will discuss what is known about taste in this context. We will first provide a brief anatomical overview of the gustatory system, and then describe taste-related behaviors and their underlying neurobiology, with particular consideration of the intimate links between taste and reward (for a comprehensive discussion of the link between post-ingestive effects and reward, see Chapter 12). We will go on to describe incentive learning research, which, while seldom discussed in these terms, demonstrates the ease with which intrinsic reward values of tastes can be transferred to temporally associated stimuli. We will suggest that this ease reflects an evolutionary imperative: the animal that survives will most likely be the one that is easily and strongly motivated by the sight, sound, and smell of an external object associated with this taste reward; thus, animals come equipped with strong connections between taste and the other sensory systems. This line of thinking—all of which follows more or less directly from the basic physical fact that taste sensations are unique in emanating from inside the body—leads us to a new conception of "flavor" as a byproduct of an evolutionary necessity, that of connecting all stimuli with the rewarding properties of taste.

---

[*] This is not to deny the reality of post-ingestive effects, which are a necessary and important consequence of a taste object's access to the digestive system. While data summarized in this chapter demonstrate that tastes have reward value in the absence of either experience or post-ingestive effects, it remains likely that taste becomes synonymous with reward only via "learning" about the post-ingestive effects of eating that occurred across evolutionary time (and, in certain circumstances, early in post-natal life, see Changizi, McGehee, and Hall 2002).

6.2 THE TASTE SYSTEM ITSELF

Tastes activate receptor cells in taste buds. The taste receptor cells appear to respond quite specifically to particular taste types (Yarmolinsky, Zuker, and Ryba 2009), although some evidence suggests that the "output" cells in the taste buds (a subset entirely distinct from those carrying transductive machinery, see Roper and Chaudhari 2009) may integrate information from multiple types of receptor cells (Tomchik et al. 2007). Cranial nerves VII, IX, and X carry this information to the nucleus of the solitary tract (NTS), which in turn projects (in rodents) to the parabrachial nuclei (PbN) of the pons, and to medullary motor nuclei (Norgren 1978). Neurons in both regions respond to the quality and concentration of tastes (Di Lorenzo 1988; Nishijo and Norgren 1990; Nakamura and Norgren 1991; Di Lorenzo and Victor 2003); these responses, which vary from narrowly to broadly tuned,[*] are sufficient to drive taste-specific ingestive and defensive behaviors in the absence of a forebrain (Grill and Norgren 1978b), but the behavioral repertoire may be abnormal (Grill and Norgren 1978b; Flynn and Grill 1988).

Beyond this brainstem circuit, the taste system becomes quite complex. Axons carry taste information from the pons (again, in the rodent) to the forebrain along three separate paths: one to the insular gustatory cortex (GC) via the parvicellular division of the ventroposteromedial (VPM) nucleus of the thalamus (Kosar, Grill, and Norgren 1986), one to the amygdala (Ottersen 1981), and one to the hypothalamus (Norgren 1974). While the cortical and limbic projections have been suggested to have distinct functions (Norgren, Hajnal, and Mungarndee 2006), they are closely related in a number of ways. In each, for instance, taste inputs are integrated with those from other sensory systems: GC receives somatosensory inputs from the oral cavity (Barnett et al. 1995) and olfactory inputs from the endopiriform nucleus (Fu et al. 2004), and amygdala receives auditory and visual cortical inputs (McDonald 1988). The hypothalamus receives interoceptive input from areas without a blood-brain barrier (Camargo, Saad, and Camargo 2000) and has a direct influence on sympathetic and parasympathetic outputs (van den Pol 1999). Of note, and as discussed later in this essay, the relay between NTS and PbN has not been identified in the primate gustatory system (Beckstead, Morse, and Norgren 1980; Norgren 1990; Pritchard, Hamilton, and Norgren 2000); instead, NTS projects directly to the VPM thalamus.

Cross-talk renders the relationships among the three parallel taste pathways explicit: the basolateral nucleus of the amygdala (BLA) is reciprocally connected with GC (McDonald and Jackson 1987; Shi and Cassell 1998), and the central nucleus of the amygdala (CeA) receives information from both GC (Turner and Zimmer 1984; McDonald 1988) and BLA (Savander et al. 1995); the lateral hypothalamus is reciprocally connected with CeA (Ottersen 1981) and GC (Allen et al. 1991). Finally, each of these structures feed back to the pons and medulla, ensuring that all but the first few milliseconds of taste responses are informed by processing in each pathway (Di Lorenzo 1990; Lundy and Norgren 2001, 2004).

6.3 TASTE-RELATED BEHAVIORS ARE INEVITABLE EXPRESSIONS OF INHERENT REWARD VALUE

Researchers routinely perform experiments in which conscious animals watch, feel, or listen to stimuli without making discernible behavioral responses. Although part of the explanation for this fact lies in the frequent use of stimuli lacking ethological validity (Weliky et al. 2003; Felsen and Dan 2005), even stimuli that provoke a specific response in an animal's natural environment (e.g., species-specific calls) often fail to do so in laboratory contexts (Ghazanfar and Santos 2004).

---

[*] The data alluded to in this paragraph form the substrate for the most heated debate in taste science. Researchers who focus on the function of taste receptor cells (Scott 2004), or who are impressed by narrowly tuned brainstem neurons (Frank 2000), favor a "labeled line" theory of coding, whereby tastes are identified via activation of specific subsets of highly responsive neurons (e.g., "sucrose-best neurons"). Researchers who focus on output cells in the taste buds (Roper and Chaudhari 2009), or who are impressed by broadly tuned brainstem neurons (Lemon and Katz 2007), instead argue an "across-neuron pattern" theory, whereby tastes are identified via decoding of the pattern of activation across responding and non-responding neurons.

Activation of the above-described taste system, in comparison, can seldom be observed passively: to wit, ingestion requires the active hand-to-mouth (paw-to-maw) participation of the subject. The most basic taste-related behaviors—orofacial responses referred to as "taste reactivity" (Grill and Norgren 1978a)—are reflex-like in their inevitability and reliability. Put a sample of something bitter into the mouth of a rodent, primate (Rosenstein and Oster 1988), or even amphibian (Liversedge 2003), and it will produce yawning "gapes" indicative of aversion. A drop of sweet fluid, meanwhile, induces lateral licks that indicate pleasure, and a mildly acidic taste causes a mixture of the two responses. These responses are evaluative in that they reveal a taste's hedonic properties (i.e., how much that particular animal "likes" the taste; Berridge 2000) and have even been used to suggest that taste palatability comprises a relatively simple continuum (Breslin, Spector, and Grill 1992).

Gapes and licks reflect the animal's desire to have less of or gain more of the substance affiliated with the taste, respectively. They inform an observant experimenter of the animal's taste preferences. Similarly, various measurements of an animal's consumption (Flynn, Culver, and Newton 2003; Caras et al. 2008), including the number and speed of licks at briefly available lick spouts (Davis 1973; Boughter et al. 2002; Zhang et al. 2003), and relative consumption of two simultaneously or sequentially available bottles (Touzani, Taghzouti, and Velley 1997; Curtis et al. 2004; Danilova and Hellekant 2004), reveal ubiquitous preference behavior generated by taste delivery.

While any one of these tasks can in some circumstances be dissociated from the others (Berridge 1996; Caras et al. 2008), what is reliable is that tastes have an intrinsic *value* that is far less obvious for stimuli in other modalities. As just one example, preferential looking tasks are frequently described in the literature, but with the exception of a few powerful visual stimuli (Deaner, Khera, and Platt 2005), preferences in infants typically reflect familiarity rather than inherent value judgment (for two of many examples, see Weizmann, Cohen, and Pratt 1971; Slater 2004); in taste, familiarity is just one variable that modulates the inevitable preference judgment, and its influence on preference is complex (De la Casa, Diaz, and Lubow 2003; Reilly and Bornovalova 2005; Rubin et al. 2009). Taste values can change with increasing familiarity, with taste (Garcia et al. 1985) and non-taste experience (Galef et al. 1997; Plassmann et al. 2008; Fortis-Santiago et al. 2010), and with physiological needs (Prakash and Norgren 1991), but value itself is an unavoidable part of taste perception.

6.4   THE INTERCONNECTEDNESS OF TASTE- AND REWARD-PROCESSING SYSTEMS IN THE BRAIN

Given that value is an intrinsic part of taste behaviors, it is unsurprising that value also turns out to be an intrinsic part of activity in the neural taste system. The ubiquity of value judgments in taste perception is reflected in the strong palatability-related information contained in neural taste responses from these regions. In awake rodents, palatability information has been observed in five- to ten-second averages of single neuron spike trains in the central amygdala (Nishijo et al. 1998), one of the primary targets of brainstem taste relays (Norgren 1976). Similar results have been reported for the pontine and medullary sources of these projections themselves (Scott and Mark 1987; Nishijo and Norgren 1997). Recent intrinsic signal imaging data, meanwhile, suggest that the spatial distribution of taste responses in primary GC is also palatability-related (Accolla et al. 2007), and while scant information is specifically available on taste responses in rodent orbitofrontal cortex (OFC, see de Araujo et al. 2006; Gutierrez et al. 2006), the large literature on incentive learning suggests that OFC codes tastes in a value-related manner (Gallagher, McMahan, and Schoenbaum 1999). Recent studies further suggest that GC and amygdala contain epochs within the time courses of their responses in which palatability-related information dwells (Katz, Simon, and Nicolelis 2001; Fontanini and Katz 2006; Fontanini et al. 2009). Clearly, value is a central facet of taste responses across the gustatory neuroaxis of rodents.

In primates, which again seem to possess neither a pontine taste relay (Pritchard, Hamilton, and Norgren 2000; Topolovec et al. 2004) nor direct brainstem-to-amygdala connections (Beckstead,

Morse, and Norgren 1980), the picture is less clear. OFC may play a privileged role in encoding value in primates. Imaging studies suggest that OFC responds differentially to palatable and aversive tastes, and that the valence-specific taste responses are independent of intensity (Small et al. 2003) and quality (Small et al. 2001; Kringelbach et al. 2003; Small et al. 2008). In addition, when subjects are asked to evaluate the pleasantness of a taste, the OFC is selectively engaged (Small et al. 2007; Grabenhorst and Rolls 2008).

Still, there is substantial evidence that primate GC and amygdala also respond in a palatability-specific manner to taste input. In GC, early reports suggested that satiety-induced changes in taste pleasantness did not affect GC responses (Rolls, Sienkiewicz, and Yaxley 1989), but fMRI data in humans suggests that many manipulations of palatability do in fact modulate GC responsiveness (Berns et al. 2001; Small et al. 2001; Nitschke et al. 2006; Smeets et al. 2006). The data from amygdala are more mixed, with electrophysiology from non-human primates (Yan and Scott 1996) converging with some human imaging studies (Zald et al. 1998) to suggest that amygdalar neurons respond differentially to palatable and aversive tastes (for analogous data regarding visual stimuli, see Paton et al. 2006), while other fMRI data suggest that the amygdala is more responsive to taste intensity than taste hedonics (Small et al. 2003). This latter result may reflect genuine interspecies differences; on the other hand, if neurons responding to either pleasant or unpleasant taste are highly intermingled in amygdala (as the rodent electrophysiology suggests, see Fontanini et al. 2009), then these differences might be beneath the spatial resolution of the fMRI technique.

The fact that taste-driven neural activity and behaviors are intrinsically value-laden suggests that tastes must also activate brain regions directly involved in the coding of stimulus value, and that these regions must feed back into the main taste neuroaxis. The obvious candidate is of course the dopamine system, widely agreed to be centrally involved in tagging experiences and stimuli as rewards (Schultz et al. 1995; Schultz 2001). In fact, recent research confirms the suspicion that taste stimuli activate neurons within the rodent nucleus accumbens (Roitman, Wheeler, and Carelli 2005) and ventral pallidum (Tindell et al. 2006), regions known to be involved in dopamine-linked reward processes (Berridge 1996; Cardinal et al. 2002; Kelley et al. 2002; Pecina and Berridge 2005). Furthermore, direct measurements show that palatable tastes are potent stimulators of dopamine release (Ahn and Phillips 1999; Hajnal and Norgren 2001), and that both systemic administration of dopamine antagonists and dopamine depletion of the ventral tegmental area (VTA, the primary source of accumbens dopamine, see Oades and Halliday 1987) inhibit consumption of such tastes (Roitman et al. 1997; Martinez-Hernandez, Lanuza, and Martinez-Garcia 2006). Taste-induced dopamine activity is increased even by sham feeding (Frankmann et al. 1994; Liang, Hajnal, and Norgren 2006), a fact that squarely implicates orosensory stimulation as the activity's cause.

These data indicate that the dopamine system is intimately intertwined with the taste system. A combination of anatomical and behavioral/lesion data reveal the nature of this relationship. In mouse, the pontine taste relay is directly and reciprocally connected to the VTA (Tokita, Inoue, and Boughter 2009). Taste-related dopamine activity in accumbens (and elsewhere) is not driven by this direct connection, however. Studies indicate that it is the amygdalar feedback pathway that is vital for taste-induced accumbal dopamine release (Ahn and Phillips 2002; Hajnal and Norgren 2005). Thus, it appears likely that the mingling of the taste and reward systems occurs similarly in rodents and primates, even though the latter lacks a direct ponto-VTA-accumbens pathway: tastes activate the nucleus accumbens upon reaching the forebrain taste relays.

6.5   THE LINK BETWEEN TASTE AND REWARD ACTIVITY DRIVES BEHAVIOR

The literature reviewed thus far suggests that taste stimuli are unavoidably and intrinsically laden with reward value. This linkage is seen in the behaviors that reveal each taste's current palatability, in the physiology of neural taste responses that intrinsically contain information concerning reward value, and in the anatomy of the taste system, which is reciprocally connected to basic reward centers in the brain. Based on such data, we suggest that tastes are, for all intents and purposes, rewards.

But perhaps the most telling evidence that tastes are rewards lies not in any particular taste data themselves, but in a meta-analysis of the use of tastes in studies of reward learning. Reward or incentive learning is currently an area of great interest in neuroscience (cf. Chapter 13). Many researchers are currently involved in studying the way in which arbitrary, otherwise neutral, stimuli become imbued with value when linked to a more intrinsically rewarding stimulus. At the most general level, these researchers seek to explain the nature of the association produced via this pairing (e.g., is it specific to the particular reward, or general to "rewardness?") and the involvement of the dopamine system in this process. But it all starts with placing an intrinsic reward into close temporal proximity with some arbitrary or neutral stimulus (or action).

The neutral stimuli used in primate and rodent incentive learning experiments are variously drawn from the visual, auditory, or olfactory domains, but in the vast majority of these studies, the intrinsically rewarding stimulus to which the neutral stimulus is linked is a taste—tastes are rewards for animals as diverse as monkeys and the flies that bother them. For example, when a researcher desires to imbue an otherwise neutral, innocuous olfactory stimulus with the ability to drive dopamine release in a rat, s/he delivers that odorant in close association to a sweet solution (Schoenbaum, Chiba, and Gallagher 1998; Schoenbaum 2001). This association "works" (once it is recognized by the rat) regardless of whether the sweet taste is caloric (e.g., Sheffield and Roby 1950; Dufour and Arnold 1966) and regardless of whether the animal is allowed to experience any post-ingestive effects of consuming the substance (e.g., Hull 1951).[*] It is clear, therefore, that it is the taste itself that carries reward value to be attached to the odor. Intracranial stimulation of the VTA also works well to drive reward learning (Fibiger et al. 1987; Garris et al. 1999), but the functionality of these physiologically unusual stimuli likely reflects the strong dopaminergic action of taste administration itself.

Further evidence that tastes are truly the rewards driving incentive learning to other stimuli comes from studies making use of the fact that a taste's reward value, while intrinsic, is plastic. For instance, when a taste is quickly "devalued" (when the experimenter induces a reduction of the taste's reward value) through either conditioned taste aversion (e.g., Colwill and Rescorla 1985) or feeding to satiation (e.g., Balleine and Dickinson 1998), associated non-taste stimuli are similarly robbed of reward value (this once again points to the importance of post-ingestive effects on calculation of reward; see Chapter 12). So effective is devaluation at interfering with the function of incentive learning mechanisms that the opposite of satiation—inducement of hunger via restriction of food access prior to training—is a nearly ubiquitous part of the preparation of rats for learning experiments.

In summary, an entire field of research involving both rodents and primates has been founded upon the intrinsic "rewardness" of taste, or more specifically on the ease with which any non-taste stimulus that is temporally linked to a taste stimulus can gain access to the reward system via that taste.[†] This associative access to reward networks makes perfect, even inevitable, evolutionary sense: as noted in the introduction, the fact that tastes are natural rewards is inextricably linked to the fact that the activation of reward pathways by taste stimulation occurs only late in the game (assuming that the game is diet procurement), after the potentially nourishing or poisoning food has been placed into the mouth. Reward value must subsequently be transferred from the taste of the food to the sight and smell of the food if the animal is to optimize its survival odds—a distinct advantage is conferred upon the animal that is able to predict nourishment or poison while the potential food object is still external. That is, incentive learning is evolutionarily adaptive, in much the same way that conditioned taste aversions and preferences are, because it helps an animal figure out what it should eat.

---

[*] Although, as discussed by de Araujo in this volume, the long-term maintenance of taste-reinforced associations may require the use of caloric, energy-rich tastes that elicit post-ingestive effects.

[†] Of course, the mere fact that taste research has been founded on taste "rewardness" is not iron-clad proof of taste rewardness; for much of history, a great deal of otherwise rigorous physics research was founded on the mistaken idea that the Earth is the center of the universe.

6.6 RETHINKING "FLAVOR" IN THIS FRAMEWORK

Put another way, the key to making the rewarding properties of taste "useful" to an animal involves having strong connections between the taste and other sensory systems. Taste input interacts with visual, somatosensory, auditory, and above all olfactory information to optimize an animal's success. Thus, it would be reasonable to expect that information in these other sensory systems had an impact on taste system function.

In fact, this impact is a much-studied one: it is well known that the "flavor" of a food item, while appearing to the taster to emerge from the tongue, is in fact a holistic, multi-modal perceptual construct. Each and every sensory modality appears to be involved in the construction of a flavor percept (Verhagen and Engelen 2006). Not only involved, however: flavor is the apparent enrichment and outright changing of the response to a relatively impoverished taste stimulus by these other sensory systems. While taste can modulate the intensity of olfactory and somatosensory percepts (Verhagen and Engelen 2006), the most common conclusion reached on the basis of flavor research is that taste is a relatively weak player in the determination of flavor.

But this work, which has been done in relative isolation from the research discussed in the previous section, begs the question "why does flavor exist?" Where is the survival value in the apparent taste of an object being so dependent on input from other sensory systems, when other percepts seem (at first blush, see below) to be largely determined by unimodal input? Why should stimulation of the tongue not provide a rich and reliable description of the food? If the answer is that taste information is too impoverished to identify the complex range of potential foods in the world (most believe that taste consists entirely of four or five basic qualities, see Smith and St John 1999), why then is that so? Why didn't we evolve to have a tongue that, like our noses, is outfitted with hundreds of different receptors (Kay and Stopfer 2006), each tuned to a subtle subset of the myriad molecules that make up food? What is adaptive about the loveliness of the gourmet restaurant meal?

We propose, perhaps provocatively, and in full awareness of the burgeoning field of flavor research, that the answer may well be "nothing." Flavor as a rich sensory experience may be a spandrel (Gould 1997), a side effect of natural selection for the deeply adaptive trait of incentive learning already described. Specifically, if the probability of survival is enhanced with proper recognition of stimuli that will ultimately reward and punish when in the mouth, such that animals benefit greatly from being able to "map" a food's taste properties onto stimuli impinging on other sensory systems, then what is probably selected for is strong connections between taste and the other sensory systems for the purpose of driving action in response to those stimuli. Flavor could simply be something that we get "for free" from a system designed to ensure the functioning of incentive learning in natural situations.

Note that we are not suggesting that feeding, an obvious target of selective pressure, relies on taste in the absence of influence from the other senses. Diet selection is clearly more a function of olfaction and vision than of taste. A basis of the argument made here is in fact that diet selection is almost exclusively under the control of olfaction and vision—the diet is selected before the food reaches the mouth—but that the effective use of olfaction and vision for these purposes is the result of incentive learning, for which simple taste properties are the primary rewards. We are simply unaware of a strong argument as to how appropriate feeding requires flavor in a manner that cannot be more easily explained under these terms (although one possibility is that multi-sensory stimulation allows for the detection of otherwise undetectable substances, see for instance Dalton et al. [2000]).*

If this reasoning is correct—if the purpose of intersensory interactions involving taste is not the emergence of flavor but the transmission of reward from taste to other sensory systems—then the

---

* Part of the attractiveness of this characterization lies in the fact that it also provides a simple, parsimonious explanation for the mysterious impoverishment of taste itself—i.e., the fact that the tongue seems to come equipped to detect only combinations of four to five tastes: if complex flavor perception isn't the *raison d'etre* of the system, then there was no selective pressure to increase the complexity of the input.

following prediction becomes reasonable: it should be possible, in a situation involving a rewarding non-taste stimulus, to demonstrate the same process that is evident in flavor working "in reverse," namely, passage of this reward to a consumed taste, and modulation of perception of the originally rewarding stimulus via perturbation of the taste system. We have recently published a study confirming this prediction, using social transmission of food preference. This task takes advantage of an olfactory stimulus that for the rat appears to have intrinsic, almost taste-like, reward value: carbon disulfide (CS_2), the smell of another rat's breath. A rat that smells a food odor mixed with CS_2 develops a preference for that food in a subsequent taste test. We first directly demonstrated that olfactory input is necessary and sufficient for proper preference development and then demonstrated the preference to be dependent on taste cortex using a temporary inactivation technique. Finally, by inactivating taste cortex during both training and testing sessions, we revealed a classic state-dependency: an odor is changed by taste cortex inactivation, such that proper recognition of that odor only occurs when taste cortex is again inactivated (Fortis-Santiago et al. 2010, Figure 6.2).

6.7 CONCLUSIONS

This research makes the argument that sensory system function has not been "specialized" for flavor. The linkage between taste and olfaction is a mutual one, allowing each to take advantage of the reward information inherent in the other. Each modulates the other as a byproduct of this optimization. Likely, the relationships between taste and the other sensory systems are to at least some degree similar. None of these relationships is 100% symmetric, for reasons laid out in the first sections of this essay, but the purpose of each intersensory interaction is probably the same. In fact, this finding allows us to more easily view flavor through the general lens of multi-sensory interaction, in

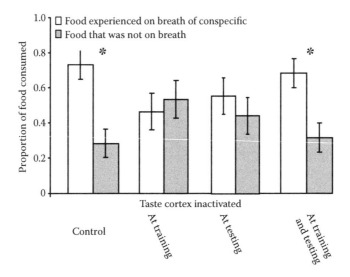

FIGURE 6.2 The taste system alters olfactory perception. The results of socially transmitted food preference tests, which rely wholly on olfactory stimulation (Fortis-Santiago et al. 2010). The left-most pair of bars show what happen when control rats are trained to prefer food with a particular odor—they eat significantly more (y-axis) of a food that had been on the breath of a conspecific (open bar) than one that had not (gray bar). If taste cortex was inactivated during either the training (2nd pair of bars) or testing sessions (3rd pair of bars), no evidence of preference emerged, but inactivation during *both* sessions (right-most pair of bars) resulted in learned preferences remaining intact. This double-inactivation effect proves that taste cortex was not necessary for learning and the testing session effect proves that taste cortical inactivation blocks expression of already-learned preferences; the most parsimonious explanation, therefore, is that taste cortical inactivation changed the nature of the olfactory percept, such that an odor altered by cortical inactivation could be recognized when cortex was inactivated once more.

which such reciprocal interactions are increasingly recognized to be commonplace (Ghazanfar and Schroeder 2006). The McGurk effect, wherein the sight and sound of a speaker mutually impact perception of each other, is the most well known result of these interactions, but in the service of effective communication—another prominent source of reward—vision and audition constantly update and change each other (Ghazanfar and Hauser 1999; Ghazanfar et al. 2005; Campbell 2008).

This broader view, which follows directly from the fact that tastes are the only stimuli that must be ingested to be sensed, has the power to change the way we think about, and do research studying, flavor. It suggests, for example, that we would be well advised to *not* look for circuitry that is dedicated to the production of flavor percepts, nor to think of flavor as a "one-way street" toward rich perceptual experience. At the most general level, this essay puts the study of taste and affiliated stimulus attributes into a more Gibsonian framework (Gibson 1966, 1982), in which the multi-modal flux of incoming stimulation is mined to reveal the important intrinsic properties of food objects in the environment; in this framework, the purpose of sensory input is to drive action toward such objects. The "flavor" of these objects is secondary to the divination of whether they afford consumption—an affordance that is directly ascertained through their taste attributes (which signal reward), and re-mapped to their visual, olfactory, auditory, and somatosensory attributes.

ACKNOWLEDGMENTS

The time spent musing on these issues was generously funded by the NIDCD (grants DC 006666 and DC 007102) and the Swartz Foundation. We are deeply indebted to the members of the Katz lab and Asif Ghazanfar for endless discussions, and to Dana Small for a little text and a lot of tolerance.

REFERENCES

Accolla, R., Bathellier, B., Petersen, C.C., and Carleton, A. 2007. Differential spatial representation of taste modalities in the rat gustatory cortex. *J Neurosci* 27:1396–1404.
Ahn, S., and Phillips, A.G. 1999. Dopaminergic correlates of sensory-specific satiety in the medial prefrontal cortex and nucleus accumbens of the rat. *J Neurosci* 19:RC29.
———. 2002. Modulation by central and basolateral amygdalar nuclei of dopaminergic correlates of feeding to satiety in the rat nucleus accumbens and medial prefrontal cortex. *J Neurosci* 22:10958–65.
Allen, G.V., Saper, C.B., Hurley, K.M., and Cechetto, D.F. 1991. Organization of visceral and limbic connections in the insular cortex of the rat. *J Comp Neurol* 311:1–16.
Balleine, B.W., and Dickinson, A. 1998. Goal-directed instrumental action: Contingency and incentive learning and their cortical substrates. *Neuropharmacology* 37:407–19.
———. 2000. The effect of lesions of the insular cortex on instrumental conditioning: Evidence for a role in incentive memory. *J Neurosci* 20:8954–64.
Barnett, E.M., Evans, G.D., Sun, N., Perlman, S., and Cassell, M.D. 1995. Anterograde tracing of trigeminal afferent pathways from the murine tooth pulp to cortex using herpes simplex virus type 1. *J Neurosci* 15:2972–84.
Bartoshuk, L.M. 1991. Taste, smell, and pleasure. In *The Hedonics of Taste and Smell*, ed. R.C. Bolles, pp. 15–28. Hillsdale, NJ: Erlbaum Associates.
Beckstead, R.M., Morse, J.R., and Norgren, R. 1980. The nucleus of the solitary tract in the monkey: Projections to the thalamus and brain stem nuclei. *J Comp Neurol* 190:259–82.
Berns, G.S., McClure, S.M., Pagnoni, G., and Montague, P.R. 2001. Predictability modulates human brain response to reward. *J Neurosci* 21:2793–98.
Berridge, K.C. 1996. Food reward: Brain substrates of wanting and liking. *Neurosci Biobehav Rev* 20:1–25.
———. 2000. Measuring hedonic impact in animals and infants: Microstructure of affective taste reactivity patterns. *Neurosci Biobehav Rev* 24:173–98.
Boughter Jr., J.D., John, S.J., Noel, D.T., Ndubuizu, O., and Smith, D.V. 2002. A Brief-access Test for Bitter Taste in Mice. *Chem Senses* 27:133–42.
Breslin, P.A.S., Spector, A.C., and Grill, H.J. 1992. A quantitative comparison of taste reactivity behaviors to sucrose before and after lithium chloride pairings: A unidimensional account of palatability. *Behav Neurosci* 106:820–36.

Camargo, L.A., Saad, W.A., and Camargo, G.P. 2000. Effects of subtypes alpha- and beta-adrenoceptors of the lateral hypothalamus on the water and sodium intake induced by angiotensin II injected into the subfornical organ. *Brain Res* 881:176–81.

Campbell, R. 2008. The processing of audio-visual speech: Empirical and neural bases. *Philos Trans R Soc Lond B Biol Sci* 363:1001–10.

Caras, M.L., Mackenzie, K., Rodwin, B., and Katz, D.B. 2008. Investigating the motivational mechanism of altered saline consumption following 5-HT-sub(1a) manipulation. *Behav Neurosci* 122:407–15.

Cardinal, R.N., Parkinson, J.A., Hall, J., and Everitt, B.J. 2002. Emotion and motivation: The role of the amygdala, ventral striatum, and prefrontal cortex. *Neurosci Biobehav Rev* 26:321–52.

Changizi, M.A., McGehee, R.M., and Hall, W.G. 2002. Evidence that appetitive responses for dehydration and food-deprivation are learned. *Physiol Behav* 75:295–304.

Colwill, R.M., and Rescorla, R.A. 1985. Postconditioning devaluation of a reinforcer affects instrumental responding. *J Exp Psychol Anim Behav Process* 11:120–32.

Curtis, K.S., Davis, L.M., Johnson, A.L., Therrien, K.L., and Contreras, R.J. 2004. Sex differences in behavioral taste responses to and ingestion of sucrose and NaCl solutions by rats. *Physiol Behav* 80:657–64.

Dalton, P., Doolittle, N., Nagata, H., and Breslin, P.A. 2000. The merging of the senses: Integration of subthreshold taste and smell. *Nat Neurosci* 3:431–32.

Danilova, V., and Hellekant, G. 2004. Sense of taste in a New World monkey, the common marmoset. II. Link between behavior and nerve activity. *J Neurophysiol* 92:1067–76.

Davis, J.D. 1973. The effectiveness of some sugars in stimulating licking behavior in the rat. *Physiol Behav* 11:39–45.

de Araujo, I.E., Gutierrez, R., Oliveira-Maia, A.J., Pereira Jr., A., Nicolelis, M.A., and Simon, S.A. 2006. Neural ensemble coding of satiety states. *Neuron* 51:483–94.

De la Casa, L.G., Diaz, E., and Lubow, R.E. 2003. Effects of post-treatment retention interval and context on neophobia and conditioned taste aversion. *Behav Processes* 63:159–70.

Deaner, R.O., Khera, A.V., and Platt, M.L. 2005. Monkeys pay per view: Adaptive valuation of social images by rhesus macaques. *Curr Biol* 15:543–48.

Di Lorenzo, P.M. 1988. Taste responses in the parabrachial pons of decerebrate rats. *J Neurophysiol* 59:1871–87.

———. 1990. Corticofugal influence on taste responses in the parabrachial pons of the rat. *Brain Res* 530:73–84.

Di Lorenzo, P.M., and Victor, J.D. 2003. Taste response variability and temporal coding in the nucleus of the solitary tract of the rat. *J Neurophysiol* 90:1418–31.

Dufour, V.L., and Arnold, M.B. 1966. Taste of saccharin as sufficient reward for performance. *Psychol Rep* 19:1293–94.

Felsen, G., and Dan, Y. 2005. A natural approach to studying vision. *Nat Neurosci* 8:1643–46.

Fibiger, H.C., LePiane, F.G., Jakubovic, A., and Phillips, A.G. 1987. The role of dopamine in intracranial self-stimulation of the ventral tegmental area. *J Neurosci* 7:3888–96.

Flynn, F.W., Culver, B., and Newton, S.V. 2003. Salt intake by normotensive and spontaneously hypertensive rats: Two-bottle and lick rate analyses. *Physiol Behav* 78:689–96.

Flynn, F.W., and Grill, H.J. 1988. Intraoral intake and taste reactivity responses elicited by sucrose and sodium chloride in chronic decerebrate rats. *Behav Neurosci* 102:934–41.

Fontanini, A., Grossman, S.E., Figueroa, J.A., and Katz, D.B. 2009. Distinct subtypes of basolateral amygdala taste neurons reflect palatability and reward. *J Neurosci* 29:2486–95.

Fontanini, A., and Katz, D.B. 2006. State-dependent modulation of time-varying gustatory responses. *J Neurophysiol* 96:3183–93.

Fortis-Santiago, Y., Rodwin, B.A., Neseliler, S., Piette, C.E., and Katz, D.B. 2010. State dependence of olfactory perception as a function of taste cortical inactivation. *Nat Neurosci* 13:158–59.

Frank, M.E. 2000. Neuron types, receptors, behavior, and taste quality. *Physiol Behav* 69:53–62.

Frankmann, S.P., Broder, L., Dokko, J.H., and Smith, G.P. 1994. Differential changes in central monoaminergic metabolism during first and multiple sodium depletions in rats. *Pharmacol Biochem Behav* 47:617–24.

Fu, W., Sugai, T., Yoshimura, H., and Onoda, N. 2004. Convergence of olfactory and gustatory connections onto the endopiriform nucleus in the rat. *Neuroscience* 126:1033–41.

Gallagher, M., McMahan, R.W., and Schoenbaum, G. 1999. Orbitofrontal cortex and representation of incentive value in associative learning. *J Neurosci* 19:6610–14.

Gallagher, M., and Schoenbaum, G. 1999. Functions of the amygdala and related forebrain areas in attention and cognition. *Ann N Y Acad Sci* 877:397–411.

Garcia, J., Hankins, W.G., and Rusiniak, K.W. 1974. Behavioral regulation of the milieu interne in man and rat. *Science* 185:824–31.

Garcia, J., Lasiter, P.S., Bermudez-Rattoni, F., and Deems, D.A. 1985. A general theory of aversion learning. *Ann N Y Acad Sci* 443:8–21.

Garris, P.A., Kilpatrick, M., Bunin, M.A., Michael, D., Walker, Q.D., and Wightman, R.M. 1999. Dissociation of dopamine release in the nucleus accumbens from intracranial self-stimulation. *Nature* 398:67–69.

Ghazanfar, A.A., and Hauser, M.D. 1999. The neuroethology of primate vocal communication: Substrates for the evolution of speech. *Trends Cogn Sci* 3:377–84.

Ghazanfar, A.A., Maier, J.X., Hoffman, K.L., and Logothetis, N.K. 2005. Multisensory integration of dynamic faces and voices in rhesus monkey auditory cortex. *J Neurosci* 25:5004–12.

Ghazanfar, A.A., and Santos, L.R. 2004. Primate brains in the wild: The sensory bases for social interactions. *Nat Rev Neurosci* 5:603–16.

Ghazanfar, A.A., and Schroeder, C.E. 2006. Is neocortex essentially multisensory? *Trends Cogn Sci* 10:278–85.

Gibson, J.J. 1966. *The Senses Considered as Perceptual Systems*. Boston: Houghton Mifflin.

———. 1982. The problem of temporal order in stimulation and perception. In *Reasons for Realism*, eds. E.S. Reed, and R. Jones, pp. 170–79. Hillsdale, NJ: LEA.

Gould, S.J. 1997. The exaptive excellence of spandrels as a term and prototype. *Proc Nat Acad Sci USA* 94:10750–55.

Grabenhorst, F., and Rolls, E.T. 2008. Selective attention to affective value alters how the brain processes taste stimuli. *Eur J Neurosci* 27:723–29.

Grill, H.J., and Norgren, R. 1978a. The taste reactivity test. I. Mimetic responses to gustatory stimuli in neurologically normal rats. *Brain Res* 143:263–79.

———. 1978b. The taste reactivity test. II. Mimetic responses to gustatory stimuli in chronic thalamic and chronic decerebrate rats. *Brain Res* 143:281–97.

Gutierrez, R., Carmena, J.M., Nicolelis, M.A., and Simon, S.A. 2006. Orbitofrontal ensemble activity monitors licking and distinguishes among natural rewards. *J Neurophysiol* 95:119–33.

Hajnal, A., and Norgren, R. 2001. Accumbens dopamine mechanisms in sucrose intake. *Brain Res* 904:76–84.

———. 2005. Taste pathways that mediate accumbens dopamine release by sapid sucrose. *Physiol Behav* 84:363–69.

Hull, C.L. 1951. *Essentials of Behavior*. New Haven, CT: Yale University Press.

Katz, D.B., Simon, S.A., and Nicolelis, M.A. 2001. Dynamic and multimodal responses of gustatory cortical neurons in awake rats. *J Neurosci* 21:4478–89.

Kay, L.M., and Stopfer, M. 2006. Information processing in the olfactory systems of insects and vertebrates. *Semin Cell Dev Biol* 17:433–42.

Kelley, A.E., Bakshi, V.P., Haber, S.N., Steininger, T.L., Will, M.J., and Zhang, M. 2002. Opioid modulation of taste hedonics within the ventral striatum. *Physiol Behav* 76:365–77.

Kosar, E., Grill, H.J., and Norgren, R. 1986. Gustatory cortex in the rat. I. Physiological properties and cytoarchitecture. *Brain Res* 379:329–41.

Kringelbach, M.L., O'Doherty, J., Rolls, E.T., and Andrews, C. 2003. Activation of the human orbitofrontal cortex to a liquid food stimulus is correlated with its subjective pleasantness. *Cereb Cortex* 13:1064–71.

Lemon, C.H., and Katz, D.B. 2007. The neural processing of taste. *BMC Neurosci* 8 Suppl 3:S5.

Liang, N.C., Hajnal, A., and Norgren, R. 2006. Sham feeding corn oil increases accumbens dopamine in the rat. *Am J Physiol Regul Integr Comp Physiol* 291:R1236–39.

Liversedge, T. 2003. Kalahari: The Great Thirstland. In Nature USA, Thirteen/WNET New York.

Lundy Jr., R.F., and Norgren, R. 2001. Pontine gustatory activity is altered by electrical stimulation in the central nucleus of the amygdala. *J Neurophysiol* 85:770–83.

———. 2004. Activity in the hypothalamus, amygdala, and cortex generates bilateral and convergent modulation of pontine gustatory neurons. *J Neurophysiol* 91:1143–57.

Martinez-Hernandez, J., Lanuza, E., and Martinez-Garcia, F. 2006. Selective dopaminergic lesions of the ventral tegmental area impair preference for sucrose but not for male sexual pheromones in female mice. *Eur J Neurosci* 24:885–93.

McDonald, A.J. 1988. Cortical pathways to the mammalian amygdala. *Prog Neurobiol* 55:257–332.

McDonald, A.J., and Jackson, T.R. 1987. Amygdaloid connections with posterior insular and temporal cortical areas in the rat. *J Comp Neurol* 262:59–77.

Nakamura, K., and Norgren, R. 1991. Gustatory responses of neurons in the nucleus of the solitary tract of behaving rats. *J Neurophysiol* 66:1232–48.

Nishijo, H., and Norgren, R. 1990. Responses from parabrachial gustatory neurons in behaving rats. *J Neurophysiol* 63:707–24.

————. 1997. Parabrachial neural coding of taste stimuli in awake rats. *J Neurophysiol* 78:2254–68.

Nishijo, H., Uwano, T., Tamura, R., and Ono, T. 1998. Gustatory and multimodal neuronal responses in the amygdala during licking and discrimination of sensory stimuli in awake rats. *J Neurophysiol* 79:21–36.

Nitschke, J.B., Dixon, G.E., Sarinopoulos, I., Short, S.J., Cohen, J.D., Smith, E.E., Kosslyn, S.M., Rose, R.M., and Davidson, R.J. 2006. Altering expectancy dampens neural response to aversive taste in primary taste cortex. *Nat Neurosci* 9:435–42.

Norgren, R. 1974. Gustatory afferents to ventral forebrain. *Brain Res* 81:285–95.

————. Taste pathways to hypothalamus and amygdala. *J Comp Neurol* 166:17–30.

————. 1978. Projections from the nucleus of the solitary tract in the rat. *Neuroscience* 3:207–18.

————. 1990. Gustatory system. In *The Human Nervous System*, ed. G. Paxinos, pp. 845–61. New York: Academic Press.

Norgren, R., Hajnal, A., and Mungarndee, S.S. 2006. Gustatory reward and the nucleus accumbens. *Physiol Behav* 89:531–35.

O'Doherty, J.P., Dayan, P., Friston, K., Critchley, H., and Dolan, R.J. 2003. Temporal difference models and reward-related learning in the human brain. *Neuron* 38:329–37.

O'Doherty, J.P., Deichmann, R., Critchley, H.D., and Dolan, R.J. 2002. Neural Responses during Anticipation of a Primary Taste Reward. *Neuron* 33:815–26.

Oades, R.D., and Halliday, G.M. 1987. Ventral tegmental (A10) system: Neurobiology. 1. Anatomy and connectivity. *Brain Res* 434:117–65.

Ottersen, O.P. 1981. Afferent connections to the amygdaloid complex of the rat with some observations in the cat. III. Afferents from the lower brain stem. *J Comp Neurol* 202:335–56.

Paton, J.J., Belova, M.A., Morrison, S.E., and Salzman, C.D. 2006. The primate amygdala represents the positive and negative value of visual stimuli during learning. *Nature* 439:865–70.

Pecina, S., and Berridge, K.C. 2005. Hedonic hot spot in nucleus accumbens shell: Where do mu-opioids cause increased hedonic impact of sweetness? *J Neurosci* 25:11777–86.

Plassmann, H., O'Doherty, J., Shiv, B., and Rangel, A. 2008. Marketing actions can modulate neural representations of experienced pleasantness. *Proc Natl Acad Sci USA* 105:1050–54.

Prakash, M.R., and Norgren, R. 1991. Comparing salt appetites: Induction with intracranial hormones or dietary sodium restriction. *Brain Res Bull* 27:397–401.

Pritchard, T.C., Hamilton, R.B., and Norgren, R. 2000. Projections of the parabrachial nucleus in the old world monkey. *Exp Neurol* 165:101–17.

Reilly, S., and Bornovalova, M.A. 2005. Conditioned taste aversion and amygdala lesions in the rat: A critical review. *Neurosci Biobehav Rev* 29:1067–88.

Roitman, M.F., Schafe, G.E., Thiele, T.E., and Bernstein, I.L. 1997. Dopamine and sodium appetite: Antagonists suppress sham drinking of NaCl solutions in the rat. *Behav Neurosci* 111:606–11.

Roitman, M.F., Wheeler, R.A., and Carelli, R.M. 2005. Nucleus accumbens neurons are innately tuned for rewarding and aversive taste stimuli, encode their predictors, and are linked to motor output. *Neuron* 45:587–97.

Rolls, E.T., Sienkiewicz, Z.J., and Yaxley, S. 1989. Hunger modulates the responses to gustatory stimuli of single neurons in the caudolateral orbitofrontal cortex of the macaque monkey. *Eur J Neurosci* 1:53–60.

Roper, S.D., and Chaudhari, N. 2009. Processing umami and other tastes in mammalian taste buds. *Ann N Y Acad Sci* 1170:60–65.

Rosenstein, D., and Oster, H. 1988. Differential facial responses to four basic tastes in newborns. *Child Dev* 59:1555–68.

Rubin, B.D., Sadacca, B.F., Keene, J.C., Chao, C., and Katz, D.B. 2009. Fear of the new: Neophobia and processing of novel vs. familiar tastes in rat insular cortex. In Neuroscience 2009 Abstracts Program No. 846.8, Society for Neuroscience, online.

Savander, V., Go, C.G., LeDoux, J.E., and Pitkanen, A. 1995. Intrinsic connections of the rat amygdaloid complex: Projections originating in the basal nucleus. *J Comp Neurol* 361:345–68.

Schoenbaum, G. 2001. Olfactory learning and the neurophysiological study of rat prefrontal function. In *Methods in Chemosensory Research*, eds. S.A. Simon, and M.A.L. Nicolelis, pp. 371–427. Boca Raton, FL: CRC.

Schoenbaum, G., Chiba, A.A., and Gallagher, M. 1998. Orbitofrontal cortex and basolateral amygdala encode expected outcomes during learning. *Nature Neurosci* 1:155–59.

Schultz, W. 2001. Reward signaling by dopamine neurons. *Neuroscientist* 7:293–302.

Schultz, W., Apicella, P., Romo, R., and Scarnati, E. 1995. Context-dependent activity in primate striatum reflecting past and future behavioral events. In *Models of Information Processing in the Basal Ganglia*, eds. J.C. Houk, J.L. Davis, and D.G. Beiser, pp. 11–27. Cambridge, MA: MIT Press.

Scott, K. 2004. The sweet and the bitter of mammalian taste. *Curr Opin Neurobiol* 14:423–27.

Scott, T.R., and Mark, G.P. 1987. The taste system encodes stimulus toxicity. *Brain Res* 414:197–203.

Sheffield, F.D., and Roby, T.B. 1950. Reward value of a non-nutritive sweet-taste. *J Comp Physiol Psychol* 43:471–81.

Shi, C.J., and Cassell, M.D. 1998. Cortical, thalamic, and amygdaloid connections of the anterior and posterior insular cortices. *J Comp Neurol* 399:440–68.

Slater, A. 2004. Novelty, familiarity, and infant reasoning. *Infant Child Dev* 13:353–55.

Small, D.M., Bender, G., Veldhuizen, M.G., Rudenga, K., Nachtigal, D., and Felsted, J. 2007. The role of the human orbitofrontal cortex in taste and flavor processing. *Ann N Y Acad Sci* 1121:136–51.

Small, D.M., Gregory, M.D., Mak, Y.E., Gitelman, D., Mesulam, M.M., and Parrish, T. 2003. Dissociation of neural representation of intensity and affective valuation in human gustation. *Neuron* 39:701–11.

Small, D.M., Veldhuizen, M.G., Felsted, J., Mak, Y.E., and McGlone, F. 2008. Separable substrates for anticipatory and consummatory food chemosensation. *Neuron* 57:786–97.

Small, D.M., Zatorre, R.J., Dagher, A., Evans, A.C., and Jones-Gotman, M. 2001. Changes in brain activity related to eating chocolate: From pleasure to aversion. *Brain* 124:1720–33.

Smeets, P.A., de Graaf, C., Stafleu, A., van Osch, M.J., Nievelstein, R.A., and van der Grond, J. 2006. Effect of satiety on brain activation during chocolate tasting in men and women. *Am J Clin Nutr* 83:1297–1305.

Smith, D.V., and St John, S.J. 1999. Neural coding of gustatory information. *Curr Opin Neurobiol* 9:427–35.

Tindell, A.J., Smith, K.S., Pecina, S., Berridge, K.C., and Aldridge, J.W. 2006. Ventral pallidum firing codes hedonic reward: When a bad taste turns good. *J Neurophysiol* 96:2399–2409.

Tokita, K., Inoue, T., and Boughter Jr., J.D. 2009. Afferent connections of the parabrachial nucleus in C57BL/6J mice. *Neuroscience* 161:475–88.

Tomchik, S.M., Berg, S., Kim, J.W., Chaudhari, N., and Roper, S.D. 2007. Breadth of tuning and taste coding in mammalian taste buds. *J Neurosci* 27:10840–48.

Topolovec, J.C., Gati, J.S., Menon, R.S., Shoemaker, J.K., and Cechetto, D.F. 2004. Human cardiovascular and gustatory brainstem sites observed by functional magnetic resonance imaging. *J Comp Neurol* 471:446–61.

Touzani, K., Taghzouti, K., and Velley, L. 1997. Increase of the aversive value of taste stimuli following ibotenic acid lesion of the central amygdaloid nucleus in the rat. *Behavioural Brain Research* 88:133–42.

Turner, B.H., and Zimmer, J. 1984. The architecture and some of the interconnections of the rat's amygdala and lateral periallocortex. *J Comp Neurol* 227:540–57.

van den Pol, A.N. 1999. Hypothalamic hypocretin (orexin): Robust innervation of the spinal cord. *J Neurosci* 19:3171–82.

Verhagen, J.V., and Engelen, L. 2006. The neurocognitive bases of human multimodal food perception: Sensory integration. *Neurosci Biobehav Rev* 30:613–50.

Weizmann, F., Cohen, L.B., and Pratt, R.J. 1971. Novelty, familiarity, and the development of infant attention. *Dev Psych* 4:149–54.

Weliky, M., Fiser, J., Hunt, R.H., and Wagner, D.N. 2003. Coding of natural scenes in primary visual cortex. *Neuron* 37:703–18.

Yan, J., and Scott, T.R. 1996. The effect of satiety on responses of gustatory neurons in the amygdala of alert cynomolgus macaques. *Brain Res* 740:193–200.

Yarmolinsky, D.A., Zuker, C.S., and Ryba, N.J. 2009. Common sense about taste: From mammals to insects. *Cell* 139:234–44.

Zald, D.H., Lee, J.T., Fluegel, K.W., and Pardo, J.V. 1998. Aversive gustatory stimulation activates limbic circuits in humans. *Brain* 121 (Pt 6):1143–54.

Zhang, Y., Hoon, M.A., Chandrashekar, J., Mueller, K.L., Cook, B., Wu, D., Zuker, C.S., and Ryba, N.J. 2003. Coding of sweet, bitter, and umami tastes: Different receptor cells sharing similar signaling pathways. *Cell* 112:293–301.

7 Touch

Steven Hsiao and Manuel Gomez-Ramirez

CONTENTS

7.1 INTRODUCTION

The somatosensory system, much like the gustatory system, processes information about events that are in direct contact with the animal. This feature critically distinguishes the sense of touch from the senses of sight, sound, and (to a certain extent) smell, all of which rely on receptors that infer events at a distance from the organism. Another defining characteristic of the somatosensory system is its ecological versatility, with an ability to detect and process information about a wide-ranging set of tactile perturbations (e.g., temperature, pain, itch, light touch, and joint position). Viewed in this manner, the somatosensory system is best conceptualized as a multimodal, rather than a unimodal, processor, comprised of multiple parallel systems carrying information about numerous aspects of environmental stimuli. To the extent that a given object encountered in the real world may simultaneously generate multiple tactile impressions, what is remarkable is that the somatosensory system unites these disparate channels into a unified percept.

All sensory systems face the same challenge: how to extract salience, or meaning, from afferent input. The potential complexity of sensory processing, however, varies across organisms. That is, while simple animals can only react to simple stimulation, higher animals like humans produce elaborate neural representations of the external world that require the integration of inputs that give higher animals more information about the world they live in and thus enhance their chances for survival. Thus, organisms that can extract more information from afferent inputs than simply "an event occurred at a particular location on my body" have a better chance of greeting another day.

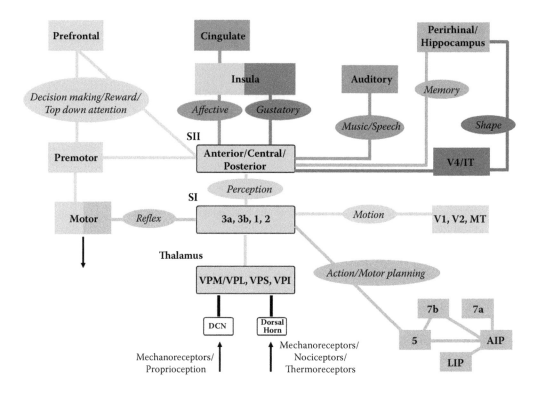

FIGURE 7.1   Block diagram illustrating the somatosensory system and its integration with other brain systems.

As discussed later in this chapter, the initial peripheral representation of spatial form in touch is isomorphic,* as individual receptors on the skin have small localized receptive fields. While these point sensors indicate the strength of a stimulus at a particular location on the skin, they do not individually convey salient information about features of the stimuli (e.g., whether the point is part of a spider creeping up your arm). It is possible that the central representation is also isomorphic. However, if this were the case, then even minor changes in the position, orientation, or size of the stimulus would require the activation of a different set of central neurons for perception of the object to occur—essentially requiring a unique complement of cells to represent each and every possible version of every tactile stimulus. This is clearly not reasonable since it would require an infinite set of stored representations of every object for perception to occur. A more realistic alternative is to transform the sensory input into a central representation that maintains perceptual constancy of object size, orientation, and position. Understanding how the brain transforms the sensory inputs into an *invariant representation* is a major thrust of contemporary neurophysiological research and in our opinion is key to deciphering the neural bases of sensation, reward, and behavior (Phillips et al. 1988).

7.1.1   SOMATOSENSORY CORTEX AND ITS INTERACTIONS WITH OTHER NEURAL SYSTEMS

The transformation of tactile sensory inputs into perceptual representations marks only the first step in a set of complex interactions with many neural systems (Figure 7.1). Somatosensory inputs are initially divided into two processing streams. One input is derived from the mechanoreceptors and proprioceptors, and the other is derived from the pain or nociceptors, and thermoreceptors. Briefly, these inputs are first processed in subcortical neural circuits in the brainstem and thalamus, and then processed in the primary (SI) and secondary somatosensory (SII) cortices to produce cortical

* The term "isomorphic" refers to the output having the same form or shape as the input.

representations of touch, which mediate tactile perception. Information is subsequently relayed to other brain structures where it interacts with inputs from other sensory modalities to carry out specific high-level functions. In what follows, we discuss the functional role of major sensory, motor and cognitive systems that interact with the somatosensory system to mediate complex behaviors. These interactions are not arranged in any rank order.

One major projection is to brain systems that are involved in action and motor planning. The role of the somatosensory input in this pathway is to provide information about planning to make movements. This is sometimes called the "how" pathway and is closely tied to inputs from the dorsal visual system. Evidence suggests that these inputs provide sensory feedback for how to move and position your hands, for example, when reaching out to grasp and manipulate objects (Jeannerod 1994). This pathway may also be involved in directing where attention should be allocat in space.

A second major set of projections is to brain systems involved in the sensory processing of functions such as motion, shape, audition, and taste. Tactile inputs must interact with what are traditionally called modality-specific areas that are related to vision or audition in order to integrate the common features of an object across different sensory channels. An example is music, which is traditionally considered to be an auditory function, but employs both auditory and somatosensory inputs (i.e., conveyance of temporal rhythm). Another example is taste, or food flavor, which has both thermal and tactile (e.g., texture) components in addition to flavor and smell. These crossmodal associations will be discussed in detail later in this chapter.

A third major projection is to brain systems involved in processing affective behaviors. There is a big difference, for example, when you are touching, or being touched by, another person—even though the sensory inputs in the two instances are identical. Further, the sense of touch has a sense of aesthetics, with some tactile inputs feeling more pleasurable to touch than others. This can easily be appreciated when performing tasks such as feeling a rough surface, which usually feels aesthetically bad, versus feeling a smooth surface, which usually feels good. We know very little about how and where the affective aspects of tactile perception are processed in the nervous system.

A fourth projection is to the frontal cortex. Numerous studies have shown that neurons in these areas are associated with behaviors related to executive functions such as decision making and reward (Romo et al. 1999; Hernandez, Zainos, and Romo 2002) and to top-down cognitive effects associated with selective attention. Little is known about how these pathways are associated with the somatosensory system, though it is evident that attentional effects on tactile information processing increase as one progresses from the thalamus to the SI cortex, and from SI to SII. One possible explanation for this systematic increase in attentional effects is that attention is selective for objects with the greatest behavioral relevance. In the case of most vertebrates, this selectivity would take place in areas like SII where meaningful holistic information about tactile objects is coded, rather than in area 3b where only basic stimulus features with less immediate relevance are represented.

Finally, the fifth major projection is to areas associated with direct feedback to the motor system. This system is made up of short and long reflex pathways in the motor system that use proprioceptive afferents as feedback signals necessary for controlling the positions and movements of our bodies and limbs.

7.1.2 Species Differences

Numerous studies have demonstrated significant differences in the organization of the somatosensory system across different species (Krubitzer 2000; Krubitzer and Kahn 2003). These include large differences in the size of the body representation and differences in the number of somatosensory areas (note Figure 7.1 pertains specifically to primates). There are even differences in the types of somatosensory receptors that are found in species that are otherwise evolutionarily adjacent (e.g., humans and monkeys). Figure 7.2 compares the body representation for a rodent (the naked mole-rat) and a primate. Note that the body map for the rodent in the SI has a large representation of the face, with a specialized area that is devoted to processing information from the whiskers and teeth, while the size of the rest of the body is relatively small. By contrast, the cortical magnification

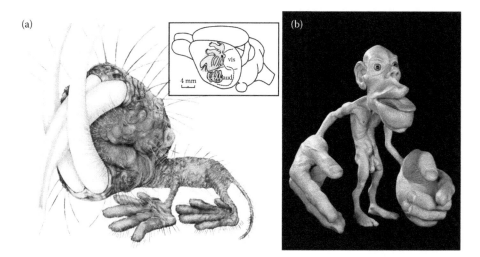

FIGURE 7.2   (See Color Insert) Illustration of how the body is represented in the rodent and human somato-sensory cortex. (a) Naked mole-ratunculus (Adapted from Catania and Remple, *Proc Natl Acad Sci USA* 99, 5692, 2002) and (b) human homunculus (Image from Natural History Museum London, reference no. 1914). *Inset*, mouse brain depicting somatosensory mouse-unculus representations in areas S1 (blue) and S2 (yellow); vis, visual cortex; aud, auditory cortex (Reprinted from Woolsey and Van der Loos, *Brain Res* 17, 205, 1970, with permission from Elsevier).

(i.e., cortical area/body area) in primates is very different. In primates, both the face and the hand representations are very large and there is no specialized region for processing information from whiskers (even for the bearded primates among us). Further, there are four areas that make up the SI in primates, while only a single area has been reported in rodents. The reason for these differences likely relates to differences in the specific goals and abilities of the two species, as well as to sensory ecological differences in the real-world form of species-relevant tactile objects.

The basis for these differences in the somatotopic representation of rodents and primates is not because the relative sizes of the body parts are different in the two species, but depends greatly on how the two species interact with their environment. The primary way that rodents explore their environment is through the whisker barrel system, which is devoted to whisking the hairs on the face back-and-forth, while sensors at the base of the whiskers infer information about the texture and shape of objects that contact the whiskers (Arabzadeh, Zorzin, and Diamond 2005; Neimark et al. 2003). The importance of this task to the animal has led to the evolution of specialized mod-ules, or barrels, in the rodent somatosensory cortex that are devoted to processing information from the whiskers. By contrast, although primates do have facial hair, the primary tool for exploring and manipulating objects in the environment is the hand. Yet this alone does not explain the increased representation of the hands in the primates. In primates, the areas of the body with high spatial acu-ity correspond to the hands and face. Both of these body parts have a dense innervation of cutaneous receptors that convey a spatial image of stimuli, providing primates with the ability to discriminate fine spatial details of surfaces and objects. High innervation densities translate to more afferent fibers and larger representations in the cortex. These examples demonstrate that the somatosensory system differs between species, and the representation and processing that is performed is dependent on the needs and behavioral goals of the animal and the environment that the animal must operate in.

The rest of this chapter will focus on what is currently known about the organization of the pri-mate somatosensory system. The main point to appreciate is the rich diversity of the afferent inputs from this system and how information about the cutaneous inputs are represented and transformed in the nervous system. We also discuss how these representations are integrated with other sensory systems from other modalities.

7.2 ANATOMICAL ORGANIZATION AND FUNCTION OF THE SOMATOSENSORY SYSTEM

As discussed above, the somatosensory system mediates multiple functions essential to human behavior. There are thirteen different kinds of afferent fibers that innervate the glabrous skin and at least three additional kinds of afferent fibers that innervate the hairy skin (for further details see: Hendry and Hsiao 2002; Hsiao, Johnson, and Yoshioka 2003). Briefly, the afferents can be divided into four main classes based on the kind of information that they convey. (1) Thermoreceptors come in two forms. These receptors respond to either warm or cold stimulation. (2) The nociceptive afferents convey information about sharp and dull pain, and a specialized receptor type conveys information about skin irritants and itch. All of these afferent fiber types have slow conduction velocities with information being carried to the CNS via small-diameter axons that can be myelinated or unmyelinated. These small-diameter afferents achieve their functional specificity to the environment via highly feature selective receptor channels that are located at the ends of the axon terminals that innervate the skin. The two remaining classes of afferents are (3) the mechanoreceptors and (4) the proprioceptors, which together convey information to the central nervous system about two- and three-dimensional form and shape, texture, vibration, motion, muscle force, joint angle, and body movement.

There are four distinct kinds of mechanoreceptors that innervate the glabrous skin of the human hand (three in the monkey), three kinds of mechanoreceptors that are only found in hairy skin, and four kinds of proprioceptors that are found in the muscles, tendons, and joints. The four kinds of mechanoreceptors found in humans are the slowly adapting type 1 (SA1) afferents, the slowly adapting type 2 (SA2) afferents, the rapidly adapting afferents (RA), and the Pacinian afferents (PC). The SA2 afferents are only found in humans and have never been observed in monkeys. Numerous studies that have combined psychophysical studies in humans with neurophysiological recording studies in monkeys have shown that each of these afferent systems is selective for different aspects of mechanical stimulation and is responsible for mediating different aspects of tactile perception (Johnson 2001).

The SA1 system is referred to as the tactile spatial system and receives its inputs from a specialized receptor called the Merkel/neurite complex. These afferents densely innervate the skin and have small punctate receptive fields (RF), and, as the name suggests, adapt slowly to sustained indentations of the skin. Individual SA1 afferents are capable of discriminating between two points on the skin of the distal fingerpad that are spaced about 1 mm apart, which is the spatial limit of acuity of the somatosensory system. Each SA1 afferent can be considered to be a single pixel of a spatial image that is conveyed to the cortex by a population of afferents that have similar RF properties. The SA1 system is responsible for processing information about two-dimensional form and texture. This system is analogous to the parvocellular system in the visual system, which is also slowly adapting and carries visual form information (see Chapter 8 in this volume; also cf. Hsiao 1998).

The RA system is referred to as the motion system. These afferents receive their inputs from a specialized ending called a Meissner corpuscle and only respond to stimulus transients during the indentation and retraction phases of stimuli. Like the SA1 afferents, they densely innervate the skin with a density that is a bit higher than the SA1 afferents. However, RA afferents have larger receptive fields with uniform response profiles (Figure 7.3). That is, the firing rates are basically independent of where the stimuli are indented within each neuron's RF. The spatial acuity of the RA afferents is five times worse than the SA1 afferents. These afferents respond well to low-frequency vibrations, which are called flutter (Talbot et al. 1968) and motion across the skin, and play a critical role in signaling when objects begin to slip when being lifted (Westling and Johansson 1987). The RA system is analogous to the magnocellular system in vision, which is also a rapidly adapting system that carries information about visual motion (Hsiao 1998).

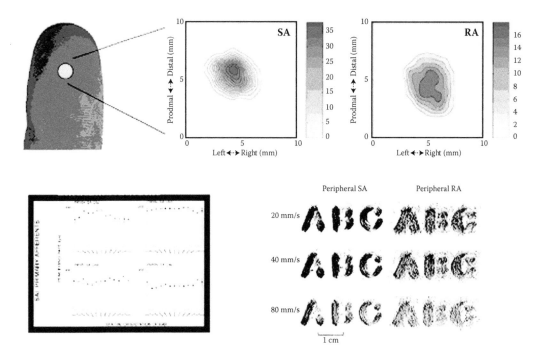

FIGURE 7.3 Typical receptive fields of SA1 and RA afferents. The bottom left figure shows that the SA1 afferents do not show orientation tuning. The bottom right figure shows the spatial event plots for the SA1 and RA afferents at three different velocities (Adapted from Hsiao, S.S., Bensmaia, S.J., in: Somatosensation, in *Handbook of the Senses*, Academic Press, Oxford, Vol. 6, pp. 55–66, 2008.)

As stated earlier, the SA2 afferents are not found in monkeys. However studies in humans show that these afferents have large RF that span 8 mm or more. While it was originally thought that the receptor ending for these afferents is the Ruffini corpuscle, recent studies suggest that this is not the case (Pare, Behets, and Cornu 2003). Thus, the receptor ending of these afferent fibers is currently unknown. Recordings from human SA2 afferents demonstrate that they respond poorly to punctuate stimuli but optimally to stretching the skin in a particular direction. It is curious why this system evolved. One possibility is that it codes for the direction that the skin is pulled for objects that are firmly grasped in the hand, though it is odd that a complex afferent system would have evolved for such an obscure function. The current working hypothesis is that this system evolved to signal joint angle and hand conformation (Edin and Johansson 1995). Evidence for this comes from a series of studies showing that the pattern of stretch on the back of the hand is unique for different hand postures and that the brain uses this neural signal to encode hand conformation (Edin and Johansson 1995; Edin 2004). It is interesting to note that monkeys, who don't have SA2 receptors, lack certain elements of fine hand coordination.

The fourth afferent mechanoreceptor afferent type is the Pacinian system. The PC system has been studied extensively. PC afferents end in a specialized structure called a Pacinian corpuscle. These afferents, like the RAs, are also rapidly adapting, but unlike the RA afferents have a low innervation density and are highly sensitive to high-frequency vibrations. The PC afferents can be considered to be a temporal processing system and are the somatosensory analog to the auditory system. The PC afferents play several important roles in somatosensory function. One function is simply to sense minute vibrations which provide information about mechanical events happening at a distance from the body. The other main function is to encode vibrations that are transmitted from the working end of tools. Tool use is an important component of tactile function. While the SA1 and RA afferents signal information about texture, shape and motion of the tool itself, they are unable to signal information about high-frequency vibrations that are transmitted through the tool to the hand. (Johnson and Hsiao 1992).

Thus, four types of mechanoreceptive afferents have evolved that respond to different ranges of spatial and temporal tactile inputs to the skin. The division of labor between the sensitivity ranges of these afferents corresponds directly to differences in the functional roles that each afferent system plays in perception.

Less is known about the four kinds of proprioceptive afferents. The Golgi tendon organs are located in the tendons of muscles and provide information about muscle force. The two kinds of muscle spindle afferents are located in the intrafusal muscles and provide information about muscle length and velocity. The joint afferents, which are located in the joint capsules, appear to signal information about when the joints are stretched to their extremes. How joint angle is coded is still unknown but it appears to involve inputs from both the muscle spindle and SA2 afferents (McCloskey et al. 1983; Moberg 1983).

7.3 CENTRAL PROCESSING OF SOMATOSENSORY INFORMATION

An organizational principle of the somatosensory system is modality segregation, which states that inputs from the different sensory afferents are sorted into segregated parallel processing streams. At the receptor level, receptors are distributed over the entire body surface. However by the time they reach cortex they are segregated and different cortical areas are specialized for processing information from particular sets of afferents. Modality segregation begins at the level of the spinal cord where small-diameter afferents that carry information about temperature, itch, and pain project directly onto neurons in the dorsal horn. These neurons in turn send their projections across the midline to form the spino-thalamic tract (see Figure 7.1). The large-diameter afferent fibers take a different route to the cortex. These afferents, which carry information related to mechanoreception and proprioception, enter the spinal cord and immediately ascend up the fiber tracts of the dorsal columns. It is not until they reach the brainstem that these fibers make their first synapse onto neurons in the dorsal column (cuneate and gracilis) nuclei. Neurons in the dorsal column nuclei then send their projections to the ventroposterior nuclei (VPL) of the contralateral thalamus (Figure 7.1). Current evidence suggests that, at least in primates, the information is not transformed along these ascending subcortical networks. A second level of modality segregation occurs at the thalamus where neurons in the central core of the VPL respond only to cutaneous inputs and have small SA1-like RF. Surrounding the core are neurons with cutaneous RFs that have larger receptive fields that are more RA-like. Finally there is a shell region that surrounds both of these regions with neurons that respond to proprioceptive input.

7.3.1 WHY MODALITY SEGREGATION IS NECESSARY

For cortical neurons to extract information about complex features of stimuli, it is critical that the afferents that are part of the neural transformation are in close proximity to each other. At the peripheral input level, neurons have small RFs that carry information about the intensity of stimuli at specific locations on the skin (Figure 7.3). However, in the cortex neurons must integrate inputs from afferent fibers that are of the same kind (e.g., SA1) to produce spatial filters that are selective to particular features of stimuli. In the next section we show that cortical neurons in area 3b have RFs composed of multiple subregions, much like cortical neurons in the primary visual cortex. And similar to neurons in the visual system, these neurons show specificity to different stimulus features. This type of integration can only be achieved if neurons of a like kind are in close proximity with each other and if the transformation is performed in a stepwise fashion (Bankman, Johnson, and Hsiao 1990). Modality segregation thus plays a critical role in allowing local neural circuits to integrate ascending inputs from neurons of the same modality into coherent spatial filters that are feature selective. These feature-selective neurons form the basis for the "higher" cortical areas to be selective to complex features of stimuli. Put simply, without modality segregation it would have been impossible for organisms to evolve the ability to perceive complex features of stimuli.

7.3.2 Primary Somatosensory Cortex: Tactile Integration

SI is the first place in the cortex where inputs from the different modalities are integrated to produce complex responses. In primates the integration is performed across four areas that individually process different aspects of the sensory input. We believe that the elaboration of function that occurs in primates must have been a direct consequence of the evolution of the hand as the primary sensory organ that primates use to explore their environment. Two of the four areas (areas 3a and 3b) are considered to be the true primary somatosensory areas, with area 3a devoted to processing information from the proprioceptive inputs and area 3b for cutaneous input.

While there have been few studies of area 3a, the general notion is that neurons in this area must play a role in converting *absolute* position information of individual joints into *relative* positions between the joints. Synergistic movements between the joints is critically important not only for making smooth movements of the hands (i.e., for producing actions) but also for producing sensory representations of tactile objects (a key role of area 2; see below).

Area 3b is the primary cortical area for processing cutaneous input. Animals that have selective lesions of this area are unable to perform tactile tasks that depend on cutaneous input (Randolph and Semmes 1974). Two key facts provide clues as to the function of area 3b in cortical function. The first is that neurons have small receptive fields confined to a single finger pad. While inputs from other fingers may modulate the responses of 3b neurons (Reed et al. 2008), it is clear that the main drive is from a single finger. The second clue comes from studies of the receptive fields and responses of these neurons to complex stimuli. As shown in Figure 7.4, the responses are more complex than what would be predicted if the RF were composed simply of an excitatory field (DiCarlo, Johnson, and Hsiao 1998; DiCarlo and Johnson 2000). Instead what is observed are neurons that show sensitivity to oriented bars and show complex responses to scanned embossed letter (see Hsiao 2008 for a review). The complexity of the responses

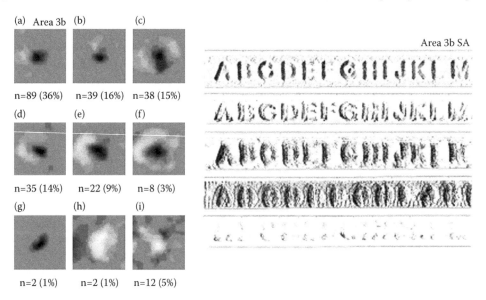

FIGURE 7.4 Receptive fields and responses to raised dots or embossed letters of neurons in area 3b. (Left) Receptive field profiles observed in area 3b in response to raised dots. Each panel (a-i) shows a typical example of the type, the total number of RFs fitting the description, and their percent of the total RF sample. Black/white shades represent excitatory/inhibitory regions of the RF; for example, the RF in panel (a) shows a single inhibitory region located on the trailing (distal) side of the excitatory region, and the RF in panel (f) shows a complete inhibitory surround. (Adapted from DiCarlo, Johnson, Hsiao, *J Neurosci* 18, 2626, 1998.) (Right) Responses of 5 SA neurons from area 3b to letters scanned across their visual fields; reconstructions arranged (from top to bottom) on the basis of decreasing subjective similarity to the stimulus letters, with greatest spatial resolution for the neuron in the top trace. (Adapted from Phillips, Johnson, Hsiao, *Proc Natl Acad Sci USA* 85, 1317, 1988.)

that are observed in 3b is similar to what is seen in the primary visual cortex, which supports the idea that vision and touch may have developed in parallel and use similar neural mechanisms (Hsiao 1998).

The responses of 3b neurons show invariant spatial responses to dynamically moving stimuli. Velocity invariance is explained by the spatio-temporal RFs of these cortical neurons, which are comprised of three main components (Sripati et al. 2006). The first is an excitatory input that is spatially flanked by a second inhibitory component that gives the neurons their spatial feature selectivity. These components are like simple cells in the V1 cortex. For example, some neurons in 3b have excitatory and inhibitory fields that provide them with orientation selectivity. The third component is a temporally delayed inhibitory field that lags the first two. This lagging inhibitory component is responsible for shutting off the excitatory response, which eliminates the effects of temporal smearing when stimuli are scanned at different velocities.

The two other areas that comprise the SI cortex are areas 1 and 2. While these areas are considered to be part of the SI cortex, they receive parallel inputs from areas 3a and 3b and from the thalamus, which suggests that they are at a higher stage of processing. Neurons in these areas have larger RFs, which implies that they must play a more integrative function. In fact, recent studies suggest that neurons in area 1 may be the tactile analog for area MT in vision and as such are specialized for processing tactile motion (Pei et al. 2009, 2010). While neurons in area 3b also respond to motion, their RFs are small and unable to capture the global motion signal in the entire pattern. Thus while 3b neurons respond to motion of one edge of plaid gratings, neurons in area 1 respond to the global motion of the plaid. Furthermore, the motion signals do not seem to pass from area 1 to area 2 since pattern motion neurons are not observed in area 2. The implication is that areas 1 and 2 are processing information in parallel rather than serially.

The fourth area that makes up the SI cortex is area 2, which receives both cutaneous and proprioceptive input. Neurons in this area have large RFs that integrate cutaneous input with hand conformation information. The integrative properties of these neurons make them an ideal candidate for coding information about global object shape, which must rely on integrating information about local cutaneous features of objects with information about where those contact points lie in 3D space (Hsiao 2008).

Neurons of the four areas of the SI project to several cortical systems (Figure 7.1). One projection is used to guide action. Another is to give feedback for movement and to signal when handheld objects are moving. The third main projection is to the SII cortex, which is responsible for integrating inputs from the SI to produce representations of complex objects, which are then integrated with systems related to higher cognitive functions.

7.3.3 Secondary Somatosensory Cortex: Tactile Shape

Recent studies show that the secondary somatosensory cortex (SII) is actually composed of several cortical fields. While the exact number and organization of these fields is not clear, it is now well established that it is responsible for higher-order tactile perception (Fitzgerald et al. 2004). Neurons in the SII have much larger receptive fields that span multiple fingers (Figure 7.5) or the entire hand. A large fraction of SII neurons even have RFs that include both hands. The RFs of neurons in the SII are highly heterogeneous, with the number of fingers that are driven by cutaneous inputs varying and the organization of the RF containing mixed excitatory and inhibitory regions. Further, a large fraction of the neurons appear to be modulated by proprioceptive input (Fitzgerald et al. 2004, 2006b). About 90% of the neurons in the SII are affected by the attentional state of the animal and many neurons seem to respond more effectively to the decision rather than to the stimulus (Hsiao, O'Shaughnessy, and Johnson 1993; Jiang, Tremblay, and Chapman 1997). Imaging studies in humans support the idea that this area is involved in higher cognitive functions related to tactile processing (Keysers et al. 2004). Furthermore, single-unit recordings show that a large fraction of the neurons show stimulus position invariance to oriented bars (Figure 7.5) (Thakur et al. 2006). Together these studies support the notion that the SII is related to tactile perception of complex objects.

Once the nervous system has constructed an internal representation of the somatosensory inputs, it can then integrate those representations for the higher functions that were discussed in the

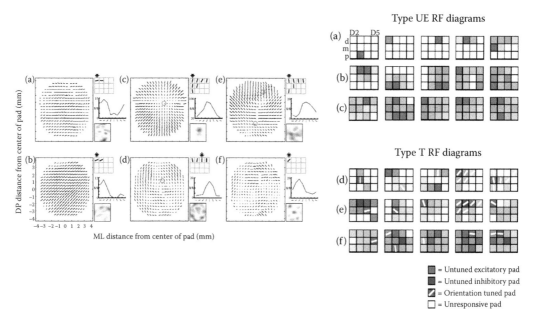

FIGURE 7.5 **(See Color Insert)** Receptive fields of S2 neurons. Left side: receptive fields on a single finger pad of six neurons with orientation-tuned responses. For each neuron a vector plot of the local orientation tuning, a map of the tuning across the distal, middle, and proximal pads of digits 2–5, an orientation tuning curve for the neuron, and the best fitting linear receptive field are shown. The left two neurons show position invariant responses that cannot be explained by the linear receptive field. Neurons labeled c, d, and e have orientation-tuned responses that are explained by having one or more inhibitory or excitatory zones. The neuron in panel f has a complex tuning response. (From Thakur, P.H., Fitzgerald, P.J., Lane, J.W., Hsiao, S.S., *J Neurosci* 26, 13571, 2006. With permission.) Right side: These maps show samples of the RFs of 30 neurons in the SII cortex. Each block of 12 panels represents the response from the distal, medial, and proximal pads of digits 2–5 (see top left panel for a key to which pad the blocks correspond to). Red colors indicate excitatory pads, blue colors indicate inhibitory pads, and pads containing a white bar were orientation tuned. (From Fitzgerald, P.J., Lane, J.W., Thakur, P.H., Hsiao, S.S. *J Neurosci* 26, 6490, 2006a.)

introduction (Figure 7.1). It should be noted that we do not mean to imply that there is a single hierarchical pathway leading to perception. Rather, information is processed in a distributed fashion, with perception being built up in parallel. Thus, perception of simple features such as cutaneous orientation or curvature (Thakur et al. 2006; Yau et al. 2009) most likely occurs in areas 3b and 2, whereas the perception of object size and shape occurs in SII. This scheme is not unlike the scheme that was originally proposed by Hubel and Wiesel in their studies of the visual system and the mechanisms of distributed representations proposed by Mountcastle (Mountcastle 1975; Hubel 1982).

7.4   CROSSMODAL INTERACTIONS IN TACTILE PERCEPTION

7.4.1   GENERAL PROPERTIES OF MULTISENSORY INTEGRATION

As different sensory systems specialize in encoding distinct features of objects and events, merging sensory inputs from different sensory modalities is fundamental for enhancing the efficacy of these computations. In touch, neocortical processing of the cutaneous and proprioceptive signals initiates at the levels of primary and secondary somatosensory cortices. However, recent evidence indicates that tactile signals are further resolved in non-somatosensory areas that pool complementary information from other sensory modalities such as vision and audition (Amedi et al. 2001; Bensmaia, Killebrew, and Craig 2006; Konkle et al. 2009; Lucan et al. 2010). This mechanism of multisensory integration is elemental to most if not all high-order animals (i.e., mammals) and perhaps to the earliest single-cell progenitors (Stein and Meredith 1993). Multisensory integration is particularly significant when inputs from one sensory modality are either weak and/or degraded.

The classical model of multisensory integration is that sensory inputs are initially processed in the sensory-specific areas and then combined in higher-order areas. This model is primarily derived from anatomical studies in animals, which showed limited connections between somatosensory, auditory, and visual areas (Kuypers et al. 1965). However, recent anatomical and physiological evidence indicates otherwise. These studies argue in favor of an alternative view of multisensory integration where interactions occur across many different levels of sensory processing, including primary sensory cortical areas. For example, tracing studies in non-human primates have revealed direct connections from the primary auditory cortex (A1) to the primary visual cortex (V1) and secondary visual cortex (V2) (Falchier et al. 2002), from areas 1 and 3b in SI to the medial temporal cortex (MT) and V2 (Cappe and Barone 2005; Cappe, Rouiller, and Barone 2009), and between SII and the core area in A1 (Coq et al. 2004; Cappe and Barone 2005). Furthermore, a recent study showed inter-hemispheric projections from V1 to SI in vole rodents (Campi et al. 2010). Furthermore, the effects in V1 appear to be specific to processing of form but not texture (Zangaladze et al. 1999; Merabet et al. 2004).

Neurophysiological findings in humans and non-human primates provide further evidence that integration occurs early in sensory processing. In particular, work from the Schroeder lab has revealed that auditory neurons in the caudo-medial area of the auditory cortex respond to somatosensory inputs (Schroeder et al. 2001; Fu et al. 2003). In addition, it was shown that "non-driving" somatosensory inputs in the supragranular layers of A1 modulate the firing response of neurons in the granular layers by resetting the phase of ongoing oscillations (Lakatos et al. 2007). Imaging and electrophysiological studies in humans demonstrated enhanced activation of auditory cortices to audio-tactile stimulation compared to the linear sum of the unisensory constituents (Foxe et al. 2002; Murray et al. 2005). These audio-tactile interactions occurred at the very early stages of auditory sensory processing (~50 ms and below). Zhou and his colleagues have described visual input to SI (Zhou and Fuster 2000). Notably, these multisensory interactions are not fully dependent on the attentional state of the animal since integration effects are also observed in anesthetized primates (Kayser et al. 2005).

The evidence that somatosensory signals interact with inputs from other sensory modalities is quite overwhelming. These interactions occur at many different levels of the sensory processing stream to give rise to meaningful percepts and sensations (Haenny, Maunsell, and Schiller 1988). However, the question of how these multisensory interactions occur is still the matter of much debate. In Figure 7.1, we outline a model that provides some insight into how and/or where cross-modal inputs become integrated to form coherent percepts of tactile perception. The model provides a general framework of a series of neural pathways, which we believe are dedicated to the analysis of distinct tactile functions. In one such pathway, cutaneous and proprioceptive signals are analyzed by SII, and further resolved in extrastriate areas such as V4 and inferotemporal (IT). This pathway is believed to mediate representations of shapes. These areas in turn interact with hippocampal structures to retrieve stored representations of tactile objects. We depict a separate pathway where somatosensory signals transmit cutaneous information (e.g., temporal rhythmic information) to auditory-related areas in order to complement perception of speech and music. In what follows we discuss recent data regarding the crossmodal interactions between the tactile and other sensory systems in low- and high-order perceptual functions.

7.4.2 Audio-Tactile Crossmodal Interactions

Neurophysiological recordings in awake-behaving non-human primates, as well as fMRI studies in humans, have shown that the belt region of the auditory cortex, which is considered part of the secondary auditory cortex (A2) (cf. Chapter 9), integrates different forms of somatosensory inputs across different body parts (Foxe et al. 2002; Fu et al. 2003). In particular, Fu et al. (2003) observed that neurons in A2 are strongly driven by cutaneous inputs from the head, neck, and hands, as well as by proprioceptive inputs from the elbow. Foxe et al. (2000) observed multisensory interactions between auditory and tactile inputs at very early stages of sensory processing (~50 ms). To investigate the behavioral significance of these multisensory interactions, Cappe, Rouiller, and Barone (2010) conducted a simple reaction time (SRT) task to auditory, tactile, and audio-tactile stimuli in humans. The authors found that early multisensory interactions (i.e., ~40–86 ms past stimulus onset) accounted for the fastest reaction times to audio-tactile stimuli and that these multisensory mechanisms occurred in the posterior sections of superior temporal cortices. Thus, it seems that these early integration mechanisms focus on global stimulus features to facilitate stimulus detection and integration.

Temporal frequency information is an important factor in the crossmodal integration of touch and audition. The spectral composition of vibratory signals is essential for tactile texture perception, while speech and music perception is partly encoded in the temporal frequency envelope of an auditory signal. It is in this area where significant audio-tactile interactions have been observed. Studies have shown that tactile texture perception is modulated by the presence of auditory inputs (Guest et al. 2002; Suzuki, Gyoba, and Sakamoto 2008) and that these effects might be frequency specific. For example, Guest et al. (2002) revealed that attenuating the high-frequency components in the auditory signal led to an increased perception of tactile smoothness. Yau et al. (2009) further characterized these audio-tactile dependencies in a tactile-frequency discrimination task. Subjects were instructed to make tactile-discrimination judgments in the presence of auditory distracters with different frequency values. The authors found that auditory stimuli interfered with tactile-frequency discrimination only when the frequencies of the auditory and tactile stimuli overlapped. In addition, the authors found that auditory distracters did not have an interference effect when subjects performed a tactile-intensity task using similar stimuli, thus suggesting that audio-tactile interactions for frequency discrimination are frequency and task dependent.

Additional temporal-frequency audio-tactile interactions have been observed in the area of speech and music perception. For instance, a recent study showed that, similar to vision, humans can utilize tactile information to enhance speech perception (Gick and Derrick 2009). Subjects were instructed to detect auditory syllables (e.g., *pa* or *ba*) in the presence of irrelevant tactile air puffs delivered to the hand or neck. The authors observed that syllables that were presented simultaneously with air puffs were more likely to be perceived as aspirated syllables (e.g., participants misperceived a *b* as a *p*). They suggested that speech is truly a multisensory phenomenon, and that tactile information can modulate speech perception similar to vision. This finding is further supported by work from Ito, Tiede, and Ostry (2009), who used a robotic device to create patterns of facial skin deformation that are normally produced during speech. They found that facial skin-stretching altered the perception of the words in a systematic way depending on the direction of skin stretch. Further, a perceptual change occurred only when the timing of skin stretch was comparable to that which occurs during speech production.

7.4.3 Visual-Tactile Crossmodal Interactions

Touch also has significant interactions with the visual system. A good example is the so-called rubber-hand illusion discovered by Botvinick and Cohen (1998). In this illusion, subjects have their hands visually hidden while an artificial arm model is placed directly in front of them. Two stimulating objects are then used to concomitantly stimulate one of the subject's hands as well as the artificial rubber hand. The typical finding is that subjects tend to perceive touch on the rubber hand instead of the hidden hand. This is a remarkable illusion as it illustrates that touch perception is partly reliant on visual elements. Furthermore, the study showed that proprioceptive inputs are also affected by this visual illusion. The authors ran a separate task where subjects displaced their right index finger along a straight line until it was judged to be in alignment with the left index finger. The data showed that subjects who were exposed to the illusion tended to displace their finger toward the rubber hand and that the magnitude of this displacement was proportional to the reported duration of the illusion. Vision also affects thermal sensation in the hands. In a task similar to the rubber-hand illusion, Durgin et al. (2007) used a beam of light from a laser pointer to produce somatic sensations and showed that subjects perceived thermal changes in the hand in the absence of any tactile stimulation. The authors also observed that this thermal illusion produced similar proprioceptive biases as in the Botvinick and Cohen (1998) study. Finally, Pavani, Spence, and Driver (2000) showed that visual distracting stimuli influenced subjects' ability to discriminate the location of vibrotactile stimuli delivered to digits D2 or D1 randomly. However, this visual-tactile interaction only occurred when the visual and tactile stimuli were in spatial accordance.

7.4.4 TACTILE CROSSMODAL INTERACTIONS IN MOTION PERCEPTION

Crossmodal interactions between touch and the other senses are also highly prevalent in higher-level perceptual tasks. However, while multisensory integration effects in low-level perceptual tasks typically focus on the global features of objects, the effects in higher-level tasks seem to be much more specific. That is, they seem to operate on the individual features of the object or event (e.g., shape or motion flow).

Several imaging studies suggest that the medial-temporal (MT) cortex is not solely devoted to analyzing visual inputs. Although the MT cortex is considered to be the core region of visual motion analysis, it has been shown to be activated by tactile and auditory motion stimuli (Hagen et al. 2002; Blake, Sobel, and James 2004; Beauchamp et al. 2007; Saenz et al. 2008; Summers et al. 2009). In one such study, Beauchamp et al. (2007) investigated whether tactile motion activated the MT and/or medial superior temporal (MST) cortex in humans. Further, they investigated whether activation in these areas was a result of tactile stimulation or visual imagery. Subjects were stimulated on different parts of the body (e.g., ipsilateral and contralateral hands and feet). They reasoned that if visual imagery is driving the activations in these areas then one should observe similar activation patterns regardless of which site was stimulated. However, if the activations are a result of tactile stimulation then MT and/or MST should display different patterns of activity depending on the body site of stimulation. The authors observed that tactile stimulation activated the MST cortex and not MT, which led them to conclude that the MST cortex is the human analogue of the MT cortex in non-human primates. However, a more parsimonious interpretation would be MST is a supramodal region of motion perception in humans. Furthermore, the authors reported that stimulating the hand resulted in more activation of the MST cortex than did foot stimulation, indicating that responses in this area were not due to visual imagery.

Psychophysical studies provide further evidence that tactile motion perception interacts with the visual modality (Bensmaia, Killebrew, and Craig 2006; Pei, Hsiao, and Bensmaia 2008; Konkle et al. 2009). Specifically, Bensmaia, Killebrew, and Craig (2006) revealed that tactile motion perception is affected by the presence of visual motion-distracting stimuli. They showed that when visual and tactile motion stimuli drifted in the same direction, speed perception of the tactile gratings increased. However, when both stimuli moved in opposite directions, tactile speed perception was either reduced or sometimes even reversed. Interestingly, they showed that when the stimuli were presented with temporal asynchrony the perceived effects were abolished. In a separate study, Konkle et al. (2009) implemented a motion aftereffect paradigm and revealed that repeated exposure to visual motion in a given direction produced a motion aftereffect in the tactile modality (i.e., the participant perceived tactile motion in the opposite direction). Similarly, they found that repeated exposure to tactile motion induced a visual motion aftereffect. Taken together, these patterns of effects provide strong evidence that motion perception might be achieved in a shared cortical area, perhaps MT or MST, with access to information from the auditory, tactile, and visual senses. In support of this idea, tracing studies have shown that the MT cortex in non-human primates has reciprocal connections with the hand and face regions of areas 1 and 3b (Cappe and Barone 2005) as well as V1 (Maunsell and Van Essen 1983).

7.4.5 CROSSMODAL INTERACTIONS IN SHAPE PERCEPTION

Similar to the findings in motion, several studies report that the lateral occipital complex (LOC), which is considered to be the human analogue of IT (Grill-Spector, Kourtzi, and Kanwisher 2001;

and Chapter 8, this volume), is involved in coding of both tactile and visual objects (Beauchamp 2005). In one study, Amedi et al. (2001) instructed subjects to judge the shape (e.g., syringes, forks) or the texture (e.g., sandpaper, fur) of objects. The authors found that fMRI activity in the LOC showed a preference for object-shapes compared to textures. The same was true when subjects were tested on an analogous task in the visual modality. In addition, there was considerable overlap between the voxels that responded to objects in the tactile and visual modalities. The results were interpreted to suggest that neurons in the LOC are bimodal (i.e., responsive to both tactile and visual objects), rather than the LOC having segregated maps for each sensory modality. These findings have been replicated in several other human imaging studies (e.g., Pietrini et al. 2004; Reed, Shoham, and Halgren 2004; Zhang et al. 2004; Lucan et al. 2010), though it is important to note that the limited spatial resolution of fMRI and surface EEG techniques makes it difficult to confirm whether the macroscopic LOC effects are relevant at the level of individual neurons.

While it is fairly evident that the LOC is activated in response to tactile objects, it is still unclear what type of neural mechanism is giving rise to these activations. For example, are these activations related to visual imagery or are they a function of tactile perception? To resolve this question, Lucan et al. (2010) conducted an event-related potential (ERP) study where subjects were asked to discriminate the shape or duration of a two-dimensional object presented on a single finger pad. The authors reasoned that if activity over the LOC is due to visual imagery then one should observe ERP activity occurring over the late stages of stimulus processing. However, if activity in this area was due to perceptual effects, then one should observe the LOC responding early in the stimulus processing stream. In support of the perceptual-based hypothesis, the authors found that the LOC began to respond approximately 140 ms (considered early stages of sensory processing) after stimulus presentation, leading them to conclude that object recognition regions in the visual cortex (i.e., LOC) serve a more general multisensory object recognition function. It should be noted, however, that several studies from our lab and others indicate that the SII cortex is highly involved in the coding of form and curvature of tactile stimuli (Fitzgerald et al. 2004, 2006a; Haggard 2006; Yau et al. 2009). Thus, it is likely that both the LOC and SII cortices are nodes of a multisensory neural network system whose functions is to build central representations of shape.

7.5   SOMATOSENSORY "REWARDS" AS BEHAVIORAL REINFORCERS

In keeping with the spirit of the present book, this final section will briefly consider the role of somatosensory events as rewards. At first pass, the value of tactile objects as positive reinforcers is not immediately apparent, especially when surveying the research literature on reward and reward learning: food and water are the predominant stimulus incentives used to motivate behavior, with monetary incentives having an additional exclusive effect on the motivation of human behavior. Food, water, and money are not necessarily "better" rewards than other stimuli, but they are certainly easy to dispense in research paradigms, and their sensitivity to the drives of hunger, thirst, and wealth make it relatively straightforward to modulate stimulus incentive value in an experimental setting. However, there is significant evidence that somatic inputs in mammals provide animals with extremely important rewards.

In addition to the avoidance of pain, which we do not discuss in this chapter, perhaps the most obvious tactile reward relates to sexual reproduction. There is no doubt that reproduction is axiomatic for survival of the species. In many invertebrates and vertebrates, sex is a major driving force of behavior, with sensory gratification or reinforcement arriving in the form of somatosensory stimulation.* In humans the affective perception of touch is extremely powerful in shaping behavior, for example a massage, a kiss, or a gentle caress (from the right person). Not surprisingly, there is an abundance of receptors in the genital regions sensitive to motion and vibration (Malinovsky

---

* For better or worse, protists and protozoa rely on asexual reproductive acts like fission in order to perpetuate their species, and almost all fish are oviparous, meaning that fertilization takes place outside of the mother's body. Thus, any tactile basis of reproductive reward for such organisms is probably moot.

et al. 1975; Halata and Munger 1986), highlighting the biological intimacy between somatosensory inputs and reproduction.

Interestingly, tactile stimulation, of the non-sexual variety, has been used in animal paradigms of classical conditioning. For example, in neonatal rat pups, stroking—with a sable hairbrush!—serves as an effective unconditioned stimulus (US), based on the idea that stroking mimics the grooming activity naturally performed by the pup's dam. In a series of studies by Sullivan, Wilson, and Leon (Sullivan and Leon 1986; Wilson, Sullivan, and Leon 1987; Sullivan and Hall 1988; Sullivan et al. 1988), pups who underwent repetitive pairing between a peppermint odor (the conditioned stimulus; CS) and tactile stroking exhibited a behavioral preference for the CS odor, in comparison to pups exposed to the same odor in the absence of stroking, or in comparison to stroking only. At the same time, odor-tactile learning selectively modulated single-unit activity and 2-deoxy-glucose (2-DG) response patterns in mitral/tufted cells in the olfactory bulb.

7.6   CONCLUSIONS

In this chapter we have reviewed the neurobiology of the somatosensory system and discussed the singular features of the sense of touch that distinguish it from all of the other senses. As a model system, somatosensation nicely embodies many of the key elements essential for integrating information both within and across sensory and motor domains. The great diversity of sensations, including thermoreception, nociception, itch, proprioception, and mechanoreception, elicits a diverse profile of somatosensory inputs, which in complex organisms are woven together into holistic tactile-perceptual representations that play important roles in other brain systems. At the same time, crossmodal interactions that link somatosensation with the other sensory modalities create central representations that are independent of a single modality. An example is the emergent representation of an apple, which is represented in several modalities. The internal percept of an apple can be evoked through several sensory channels that could not have been achieved from single sensory channels.

Based on the observations discussed above, it seems appropriate to suggest that the classical view of neocortical function, in which sensory inputs are selectively processed in separate pathways devoted to a single modality, and later converge in "higher-order association areas," is ripe for revision. As described here, the somatosensory system interacts with other senses at many different stages of the sensory processing stream, even within traditional "unimodal" cortices. The fact that multisensory integration occurs in unimodal sensory areas, and that neural processing in unimodal areas is heavily influenced by information from other senses, reflects a paradigm shift that has already begun to change the direction of neuroscientific research. This general theme is also taken up in Chapter 1 of this volume.

Indeed, given the prevalence and promiscuity of crossmodal interactions, it is likely that multisensory integration is the rule rather than the exception of sensory systems. From this perspective, it is reasonable to consider multisensory integration as the necessary and sufficient instrument of associative learning that links not only sensory interactions but also relationships between sensation and movement and sensation and reward.

REFERENCES

Amedi, A., Malach, R., Hendler, T., Peled, S., and Zohary, E. 2001. Visuo-haptic object-related activation in the ventral visual pathway. *Nat Neurosci* 4:324–30.

Arabzadeh, E., Zorzin, E., and Diamond, M.E. 2005. Neuronal encoding of texture in the whisker sensory pathway. *PLoS Biol* 3:e17.

Bankman, I.N., Johnson, K.O., and Hsiao, S.S. 1990. Neural image transformation in the somatosensory system of the monkey: Comparison of neurophysiological observations with responses in a neural network model. *Cold Spring Harb Symp Quant Biol* 55:611–20.

Beauchamp, M.S. 2005. See me, hear me, touch me: Multisensory integration in lateral occipital-temporal cortex. *Curr Opin Neurobiol* 15:145–53.

Beauchamp, M.S., Yasar, N.E., Kishan, N., and Ro, T. 2007. Human MST but not MT responds to tactile stimulation. *J Neurosci* 27:8261–67.

Bensmaia, S.J., Killebrew, J.H., and Craig, J.C. 2006. The influence of visual motion on tactile motion perception. *J Neurophysiol 96*, 1625–1637.

Blake, R., Sobel, K.V., and James, T.W. 2004. Neural synergy between kinetic vision and touch. *Psychol Sci* 15:397–402.

Botvinick, M., and Cohen, J. 1998. Rubber hands "feel" touch that eyes see. *Nature* 391:756.

Campi, K.L., Bales, K.L., Grunewald, R., and Krubitzer, L. 2010. Connections of auditory and visual cortex in the prairie vole (Microtus ochrogaster): Evidence for multisensory processing in primary sensory areas. *Cereb Cortex* 20:89–108.

Cappe, C., and Barone, P. 2005. Heteromodal connections supporting multisensory integration at low levels of cortical processing in the monkey. *Eur J Neurosci* 22:2886–902.

Cappe, C., Rouiller, E.M., and Barone, P. 2009. Multisensory anatomical pathways. *Hear Res* 258:28–36.

Catania, K.C., and Remple, M.S. 2002. Somatosensory cortex dominated by the representation of teeth in the naked mole-rat brain. *Proc Natl Acad Sci USA* 99:5692-7.

Coq, J.O., Qi, H., Collins, C.E., and Kaas, J.H. 2004. Anatomical and functional organization of somatosensory areas of the lateral fissure of the New World titi monkey (Callicebus moloch). *J Comp Neurol* 476:363–87.

DiCarlo, J.J., and Johnson, K.O. 2000. Spatial and temporal structure of receptive fields in primate somatosensory area 3b: Effects of stimulus scanning direction and orientation. *J Neurosci* 20:495–510.

DiCarlo, J.J., Johnson, K.O., and Hsiao, S.S. 1998. Structure of receptive fields in area 3b of primary somatosensory cortex in the alert monkey. *J Neurosci* 18:2626–45.

Durgin, F.H., Evans, L., Dunphy, N., Klostermann, S., and Simmons, K. 2007. Rubber hands feel the touch of light. *Psychol Sci* 18:152–57.

Edin, B.B. 2004. Quantitative analyses of dynamic strain sensitivity in human skin mechanoreceptors.

Edin, B.B., and Johansson, N. 1995. Skin strain patterns provide kinaesthetic information to the human central nervous system. *J Physiol* 487:243–51.

Falchier, A., Clavagnier, S., Barone, P., and Kennedy, H. 2002. Anatomical evidence of multimodal integration in primate striate cortex. *J Neurosci* 22:5749–59.

Fitzgerald, P.J., Lane, J.W., Thakur, P.H., and Hsiao, S.S. 2004. Receptive field properties of the macaque second somatosensory cortex: Evidence for multiple functional representations. *J Neurosci* 24:11193–204.

Fitzgerald, P.J., Lane, J.W., Thakur, P.H., and Hsiao, S.S. 2006a. Receptive field (RF) properties of the macaque second somatosensory cortex: RF size, shape, and somatotopic organization. *J Neurosci* 26:6485–95.

Fitzgerald, P.J., Lane, J.W., Thakur, P.H., and Hsiao, S.S. 2006b. Receptive field properties of the macaque second somatosensory cortex: Representation of orientation on different finger pads. *J Neurosci* 26:6473–84.

Foxe, J. J., Morocz, I. A., Murray, M. M., Higgins, B. A., Javitt, D. C., & Schroeder, C. E. 2000. Multisensory auditory-somatosensory interactions in early cortical processing revealed by high-density electrical mapping. *Brain Res Cogn Brain Res* 10 (1–2): 77–83. Retrieved from http://www.ncbi.nlm.nih.gov/pubmed/10978694.

Foxe, J.J., Wylie, G.R., Martinez, A., Schroeder, C.E., Javitt, D.C., Guilfoyle, D., Ritter, W., and Murray, M.M. 2002. Auditory-somatosensory multisensory processing in auditory association cortex: An fMRI study. *J Neurophysiol* 88:540–43.

Fu, K.M., Johnston, T.A., Shah, A.S., Arnold, L., Smiley, J., Hackett, T.A., Garraghty, P.E., and Schroeder, C.E. 2003. Auditory cortical neurons respond to somatosensory stimulation. *J Neurosci* 23:7510–15.

Gick, B., and Derrick, D. 2009. Aero-tactile integration in speech perception. *Nature* 462:502–4.

Grill-Spector, K., Kourtzi, Z., and Kanwisher, N. 2001. The lateral occipital complex and its role in object recognition. *Vision Res* 41:1409–22.

Guest, S., Catmur, C., Lloyd, D., and Spence, C. 2002. Audiotactile interactions in roughness perception. *Exp Brain Res* 146:161–71.

Haenny, P.E., Maunsell, J.H., and Schiller, P.H. 1988. State dependent activity in monkey visual cortex II. Retinal and extraretinal factors in V4. *Exp Brain Res* 69:245–59.

Hagen, M.C., Franzen, O., McGlone, F., Essick, G., Dancer, C., and Pardo, J.V. 2002. Tactile motion activates the human middle temporal/V5 (MT/V5) complex. *Eur J Neurosci* 16:957–64.

Haggard, P. 2006. Sensory neuroscience: From skin to object in the somatosensory cortex. *Curr Biol* 16:R884–86.

Halata, Z., and Munger, B.L. 1986. The neuroanatomical basis for the protopathic sensibility of the human glans penis. *Brain Res* 371:205–30.

Hendry, S.H.C., and Hsiao, S.S. 2002. The somatosensory system. In *Fundamental Neuroscience*, eds. L.R. Squire, F.E. Bloom, S.K. McConnell, J.L. Roberts, N.C. Spitzer, and M.J. Zigmond, pp. 668–96. San Diego: Academic Press.

Hernandez, A., Zainos, A., and Romo, R. 2002. Temporal evolution of a decision-making process in medial premotor cortex. *Neuron* 33:959–72.

Hsiao, S.S. 1998. Similarities between touch and vision. In *Neural Aspects of Tactile Sensation*, ed. J.W. Morley, pp. 131–65. Amsterdam: Elsevier.

Hsiao, S.S. 2008. Central mechanisms of tactile shape perception. *Curr Opin Neurobiol* 18:418–24.

Hsiao, S.S., Johnson, K.O., and Yoshioka, T. 2003. Processing of tactile information in the primate brain. In *Comprehensive Handbook of Psychology*, Vol. 3: *Biological Psychology*, eds. M. Gallagher and R.J. Nelson, pp. 211–36. New York: Wiley.

Hsiao, S.S., O'Shaughnessy, D.M., and Johnson, K.O. 1993. Effects of selective attention on spatial form processing in monkey primary and secondary somatosensory cortex. *J Neurophysiol* 70:444–47.

Hubel, D.H. 1982. Exploration of the primary visual cortex, 1955–78. *Nature* 299:515–24.

Ito, T., Tiede, M., and Ostry, D.J. 2009. Somatosensory function in speech perception. *Proc Natl Acad Sci USA* 106:1245–48.

Jeannerod, M. 1994. The hand and the object: The role of posterior parietal cortex in forming motor representations. *Can J Physiol Pharmacol* 72:535–41.

Jiang, W., Tremblay, F., and Chapman, C.E. 1997. Neuronal encoding of texture changes in the primary and the secondary somatosensory cortical areas of monkeys during passive texture discrimination. *J Neurophysiol* 77:1656–62.

Johnson, K.O. 2001. The roles and functions of cutaneous mechanoreceptors. *Curr Opin Neurobiol* 11:455–61.

Johnson, K.O., and Hsiao, S.S. 1992. Neural mechanisms of tactual form and texture perception. *Annu Rev Neurosci* 15:227–50.

Kayser, C., Petkov, C.I., Augath, M., and Logothetis, N.K. 2005. Integration of touch and sound in auditory cortex. *Neuron* 48:373–84.

Keysers, C., Wicker, B., Gazzola, V., Anton, J.L., Fogassi, L., and Gallese, V. 2004. A touching sight: SII/PV activation during the observation and experience of touch. *Neuron* 42:335–46.

Konkle, T., Wang, Q., Hayward, V., and Moore, C.I. 2009. Motion aftereffects transfer between touch and vision. *Curr Biol* 19:745–50.

Krubitzer, L., and Kahn, D.M. 2003. Nature versus nurture revisited: An old idea with a new twist. *Prog Neurobiol* 70:33–52.

Krubitzer, L.A. 2000. How does evolution build a complex brain? [In Process Citation]. *Novartis Found Symp* 228:206–20.

Kuypers, H.G., Szwarcbart, M.K., Mishkin, M., and Rosvold, H.E. 1965. Occipitotemporal corticocortical connections in the Rhesus monkey. *Exp Neurol* 11:245–62.

Lakatos, P., Chen, C.M., O'Connell, M.N., Mills, A., and Schroeder, C.E. 2007. Neuronal oscillations and multisensory interaction in primary auditory cortex. *Neuron* 53:279–92.

Lucan, J.N., Foxe, J.J., Gomez-Ramirez, M., Sathian, K., and Molholm, S. 2010. Tactile shape discrimination recruits human lateral occipital complex during early perceptual processing. *Hum Brain Mapp. 31*(11) 1813–1821.

Malinovsky, L., Sommerova, J., and Martincik, J. 1975. Quantitative evaluation of sensory nerve endings in hypertrophy of labia minora pudendi in women. *Acta Anat (Basel)* 92:129–44.

Maunsell, J.H., and Van Essen, D.C. 1983. Functional properties of neurons in middle temporal visual area of the macaque monkey. I. Selectivity for stimulus direction, speed, and orientation. *J Neurophysiol* 49:1127–47.

McCloskey, D.I., Cross, M.J., Honner, R., and Potter, E.K. 1983. Sensory effects of pulling or vibrating exposed tendons in man. *Brain* 106:21–37.

Merabet, L., Thut, G., Murray, B., Andrews, J., Hsiao, S., and Pascual-Leone, A. 2004. Feeling by sight or seeing by touch? *Neuron* 42:173–79.

Moberg, E. 1983. The role of cutaneous afferents in position sense, kinaesthesia, and motor function of the hand. *Brain* 106:1–19.

Mountcastle, V.B. 1975. The view from within: Pathways to the study of perception. *Johns Hopkins Med J* 136:109–31.

Murray, M.M., Molholm, S., Michel, C.M., Heslenfeld, D.J., Ritter, W., Javitt, D.C., Schroeder, C.E., and Foxe, J.J. 2005. Grabbing your ear: Rapid auditory-somatosensory multisensory interactions in low-level sensory cortices are not constrained by stimulus alignment. *Cereb Cortex* 15:963–74.

Neimark, M.A., Andermann, M.L., Hopfield, J.J., and Moore, C.I. 2003. Vibrissa resonance as a transduction mechanism for tactile encoding. *J Neurosci* 23:6499–509.

Pare, M., Behets, C., and Cornu, O. 2003. Paucity of presumptive ruffini corpuscles in the index finger pad of humans. *J Comp Neurol* 456:260–66.

Pavani, F., Spence, C., and Driver, J. 2000. Visual capture of touch: Out-of-the-body experiences with rubber gloves. *Psychol Sci* 11:353–59.

Pei, Y.C., Denchev, P.V., Hsiao, S.S., Craig, J.C., and Bensmaia, S.J. 2009. Convergence of submodality-specific input onto neurons in primary somatosensory cortex. *J Neurophysiol* 102:1843–53.

Pei, Y.C., Hsiao, S.S., and Bensmaia, S.J. 2008. The tactile integration of local motion cues is analogous to its visual counterpart. *Proc Natl Acad Sci USA* 105:8130–35.

Pei, Y.C., Hsiao, S.S., Craig, J.C., and Bensmaia, S.J. 2010. Shape invariant coding of motion direction in somatosensory cortex. *PLoS Biol* 8:e1000305.

Phillips, J.R., Johnson, K.O., and Hsiao, S.S. 1988. Spatial pattern representation and transformation in monkey somatosensory cortex. *Proc Natl Acad Sci USA* 85:1317–21.

Pietrini, P., Furey, M.L., Ricciardi, E., Gobbini, M.I., Wu, W.H., Cohen, L., Guazzelli, M., and Haxby, J.V. 2004. Beyond sensory images: Object-based representation in the human ventral pathway. *Proc Natl Acad Sci USA* 101:5658–63.

Randolph, M., and Semmes, J. 1974. Behavioral consequences of selective ablations in the postcentral gyrus of Macaca mulatta. *Brain Res* 70:55–70.

Reed, C.L., Shoham, S., and Halgren, E. 2004. Neural substrates of tactile object recognition: An fMRI study. *Hum Brain Mapp* 21:236–46.

Reed, J.L., Pouget, P., Qi, H.X., Zhou, Z., Bernard, M.R., Burish, M.J., Haitas, J., Bonds, A.B., and Kaas, J.H. 2008. Widespread spatial integration in primary somatosensory cortex. *Proc Natl Acad Sci USA* 105:10233–37.

Romo, R., Brody, C.D., Hernández A., and Lemus, L. 1999. Neuronal correlates of parametric working memory in the prefrontal cortex. *Nature* 399:470–73.

Saenz, M., Lewis, L.B., Huth, A.G., Fine, I., and Koch, C. 2008. Visual Motion Area MT+/V5 Responds to Auditory Motion in Human Sight-Recovery Subjects. *J Neurosci* 28:5141–18.

Schroeder, C.E., Lindsley, R.W., Specht, C., Marcovici, A., Smiley, J.F., and Javitt, D.C. 2001. Somatosensory input to auditory association cortex in the macaque monkey. *J Neurophysiol* 85:1322–27.

Sripati, A.P., Yoshioka, T., Denchev, P., Hsiao, S.S., and Johnson, K.O. 2006. Spatiotemporal receptive fields of peripheral afferents and cortical area 3b and 1 neurons in the primate somatosensory system. *J Neurosci* 26:2101–14.

Stein, B.E., and Meredith, M.A. 1993. *The Merging of the Senses*. Cambridge, MA: MIT Press.

Sullivan, R.M., and Hall, W.G. 1988. Reinforcers in infancy: classical conditioning using stroking or intra-oral infusions of milk as UCS. *Dev Psychobiol* 21:215–23.

Sullivan, R.M., and Leon, M. 1986. Early olfactory learning induces an enhanced olfactory bulb response in young rats. *Brain Res* 392:278–82.

Sullivan, R.M., Wilson, D.A., Kim, M.H., and Leon, M. 1988. Behavioral and neural correlates of postnatal olfactory conditioning: I. Effect of respiration on conditioned neural responses. *Physiol Behav* 44:85–90.

Summers, I.R., Francis, S.T., Bowtell, R.W., McGlone, F.P., and Clemence, M. 2009. A functional-magnetic-resonance-imaging investigation of cortical activation from moving vibrotactile stimuli on the fingertip. *J Acoust Soc Am* 125:1033–39.

Suzuki, Y., Gyoba, J., and Sakamoto, S. 2008. Selective effects of auditory stimuli on tactile roughness perception. *Brain Res* 1242:87–94.

Talbot, W.H., Darian-Smith, I., Kornhuber, H.H., and Mountcastle, V.B. 1968. The sense of flutter-vibration: Comparison of the human capacity with response patterns of mechanoreceptive afferents from the monkey hand. *J Neurophysiol* 31:301–34.

Thakur, P.H., Fitzgerald, P.J., Lane, J.W., and Hsiao, S.S. 2006. Receptive field properties of the Macaque second somatosensory cortex: Nonlinear mechanisms underlying the representation of orientation within a finger pad. *J Neurosci* 26:13567–75.

Westling, G., and Johansson, R.S. 1987. Responses in glabrous skin mechanoreceptors during precision grip in humans. *Exp Brain Res* 66:128–40.

Wilson, P., and Snow, P.J. 1987. Reorganization of the receptive fields of spinocervical tract neurons following denervation of a single digit in the cat. *J Neurophysiol* 57:803–18.

Wilson, D.A., Sullivan, R.M., and Leon, M. 1987. Single-unit analysis of postnatal olfactory learning: Modified olfactory bulb output response patterns to learned attractive odors. *J Neurosci* 7:3154–62.

Woolsey, T.A., and Van der Loos, H. 1970. The structural organization of layer IV in the somatosensory region (SI) of mouse cerebral cortex. The description of a cortical field composed of discrete cytoarchitectonic units. *Brain Res* 17:205-42.

Yau, J. M., Olenczak, J. B., Dammann, J. F., & Bensmaia, S. J. 2009. Temporal frequency channels are linked across audition and touch. *Current Biology: CB, 19*(7), 561–6. Elsevier Ltd. doi: 10.1016/j.cub.2009.02.013.

Yau, J.M., Pasupathy, A., Fitzgerald, P.J., Hsiao, S.S., and Connor, C.E. 2009. Analogous intermediate shape coding in vision and touch. *Proc Natl Acad Sci USA* 106:16457–62.

Zangaladze, A., Epstein, C.M., Grafton, S.T., and Sathian, K. 1999. Involvement of visual cortex in tactile discrimination of orientation. *Nature* 401:587–90.

Zhang, M., Weisser, V.D., Stilla, R., Prather, S.C., and Sathian, K. 2004. Multisensory cortical processing of object shape and its relation to mental imagery. *Cogn Affect Behav Neurosci* 4:251–59.

Zhou, Y.D., and Fuster, J.M. 2000. Visuo-tactile cross-modal associations in cortical somatosensory cells. *Proc Natl Acad Sci USA* 97:9777–82.

8 Sight

Christina S. Konen and Sabine Kastner

CONTENTS

8.1 INTRODUCTION

Vision is the quintessential "distance" sense.[*] Barring any physical obstructions or opacities, the eye can detect light radiating from stars that loom billions of miles away—on a clear moonless night the Andromeda galaxy can be viewed from 2.5 million light years, or 15 million-trillion miles, away. These cosmic numbers also highlight the speed of vision: at light-speed rates of transmission there is no faster way to communicate information about objects in the environment. These sensory attributes easily outclass the senses of smell and hearing, which operate over shorter spatial ranges and slower temporal scales (see chapters by Gottfried and Wilson and by Camalier and Kaas for further ecological considerations of these other senses).

It is important to note that light is reflected, not emitted, from visual objects (Dusenbery 1992). This stands in contrast to the other distance senses: odor is emitted from olfactory objects and sound is emitted from auditory objects.[†] It follows that there is no visual object to be seen unless a light source is available to shine on that object. The fundamental fact that it is hard to see in the dark has had particular impact on visual system development of nocturnal animals, which found themselves at an evolutionary crossroads. Some of these animals stayed true to the visual path, developing a keen night-vision sense, but others veered off the visual path, instead developing greater reliance on olfactory and auditory channels for locating and identifying behaviorally salient objects.

For virtually all animals on planet earth living a cage-free life, the Sun is *the* origin of light that reflects off of natural objects, lending them a salient visual aspect until nightfall (Dusenbery 1992). The sun not only brings visual objects "to life" so to speak, but also represents the source energy of

---

[*] Note: much of the following discussion draws from these key resources: Kirschfeld 1976; Land 1981; Dusenbery 1992.

[†] A few organisms such as bioluminescent protozoa and fireflies produce light as a means of communication or courtship, but these are rare instances.

life itself. Nourishment and survival ultimately depend on energy provided by the sun, transformed by plants into consumable packets that ascend the food chain of herbivores and carnivores. As a consequence, animals favor sunny habitats, since these environments offer the best prospects of energy/food.

The natural environment of earth substantially constrains the spectral distribution of sunlight available to visual systems. Light scattering by the water content in the atmosphere and light absorption by ozone, carbon dioxide, and oxygen effectively restrict the frequency range between 300 and 1000 nm. Aqueous environments further narrow the spectral content, particularly in the deep sea, and in an elegant example of biology conforming to ecology, the rod pigments in deep-sea animals have shifted to absorb shorter (blue) wavelengths of light that maximally penetrate these underwater depths.

In general, the operating range for vertebrate visual systems is between 400 nm (violet) and 800 nm (red), so-called "visible light." Insects such as honeybees can "see" ultraviolet light, enabling them to identify and pollinate preferred species of flowers that would be indistinguishable from non-preferred flowers when viewed under the visible spectrum of light (Eisner et al. 1969). However, the eyes of most animals filter out UV light before it can reach the retina, probably because it has such deleterious effects on cells and tissues. At the opposite end of the spectrum, snakes such as the pit viper can detect infrared light, though the mechanism of detection involves converting the infrared radiation into heat, permitting measurement of temperature changes.

The biological form of a visual detection system can vary widely, according to an organism's needs and complexity: from the eye-spot apparatus of *Euglena* to the pinhole eye of Nautilus, from the eight-eye arrangement of spiders to the numerous reflector eyes of scallops lying just inside the mantle of the shell (Figure 8.1). For insects and other small animals, compound eyes—which are really an assembly of multiple lens eyes—are well suited for motion sensitivity and elemental pattern discrimination (Horridge 2009; Srinivasan 2010). These perceptual gains occur at the cost of absolute spatial resolution and visual acuity, though an insect equipped with a single lens eye (as exists in larger animals) would suffer even greater loss of visual sensitivity, due to physical aberrations in the lens with increasing angular distance from the fovea. On the other hand, if humans were equipped with compound eyes capable of achieving the same visual resolution afforded by single lens eyes, the grotesque size of such an organ would be physically untenable (Kirschfeld 1976) (Figure 8.2).

All mammals possess paired single-lens eyes, but not all mammals house their eyes in a side-by-side anterior orientation on the head. The 360° field-of-view that a rabbit gains by having its eyes at opposite positions on the head is sacrificed in primates for keen three-dimensional perception of visual objects, which requires that the two eyes receive sensory input from highly overlapping areas of visual space (Dusenbery 1992). The subsequent task of the visual system is to integrate monocular information into binocular representations of size, shape, and depth, and to construct visual objects that can be perceptually separated from each other. The mobility of the viewer, not to mention the mobility of the object and the mobility of the light source, present further challenges to a visual system seeking to recognize the same object from multiple different sizes and perspectives.

The ability to achieve object constancy under dynamic environmental conditions perhaps reaches its apogee in the primate visual system. The remainder of this chapter will focus on the neural and perceptual mechanisms of visual object perception in human and non-human primate brains, with an emphasis on the role of attention in the designation of visual objects that have particular relevance for reward-based behavior.

8.2   THE VISUAL SYSTEM

Most of our knowledge about the organization of visual cortex has been derived from invasive studies in non-human primates. These studies have shown that monkey cortex contains more than 32 separate areas that are implicated in visual processing (Felleman and Van Essen 1991). These areas have been

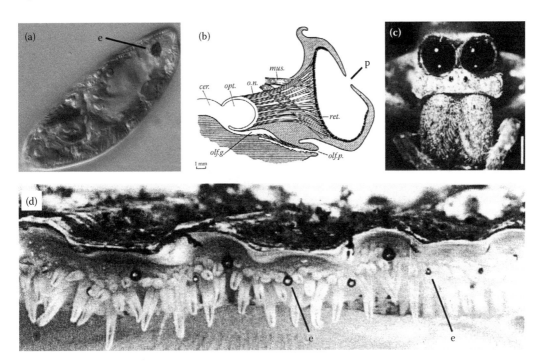

FIGURE 8.1 A myriad of eyes across the animal kingdom. (a) The eyespot apparatus (e) of *Euglena*, a protozoan with features common to both animal and plant life. (From National Oceanic & Atmospheric Administration [NOAA]. With permission.) (b) The pinhole eye (p) of *Nautilus*, a marine mollusk related to squid, octopus, and cuttlefish. (c) Frontal view showing four of the eight eyes of *Dinopis subrufus*, or the "Ogre-Faced Spider," a nocturnal spider found in eastern Australia. The two large eyes are the postero-medians; the two dots below them are the principal eyes. (d) In the scallop, *Pecten maximus*, approximately 60 image-forming reflector eyes (e) are found distributed around the edge of the mantle. (Images B–D, with kind permission of **Springer Science+Business Media** and the author, modified from: Land, M.F., Optics and vision in invertebrates, In *Handbook of Sensory Physiology: Comparative Physiology and Evolution of Vision in Invertebrates*, ed. Autrum, H., Vol. 7(6B), 1981, pp. 471–593.)

defined on the basis of anatomical organization, as revealed by staining techniques, physiological properties, in particular their receptive field (RF) architecture, and their connectivity to other areas. By contrast, in the human brain only a limited number of functional criteria are available to define distinct areas within the visual system. The criterion most commonly used is topographic organization, that is, the representation of the visual field in an orderly fashion within an area, also referred to as a visual "map" (Sereno et al. 1995; DeYoe et al. 1996; Engel, Glover, and Wandell 1997). Typically, these maps are retinotopically organized, i.e., the RFs of adjacent cortical neurons represent adjacent portions of the visual field. In the human visual system, at least 16 such maps have been identified using non-invasive functional magnetic resonance imaging (fMRI) in combination with phase-encoded retinotopic mapping (for reviews see Wandell, Dumoulin, and Brewer 2007). In these studies, a rotating wedge is presented at different angular positions to measure the representation of polar angle across the visual field (Figure 8.3a). Similarly, an expanding or contracting annulus is presented to measure the representations of different eccentricity bands within the visual field (Figure 8.3b).

Figure 8.3c shows topographically organized areas within ventral visual cortex of the right hemisphere in a representative subject. Since the responses of many areas are lateralized, the left visual hemifield is represented in the right hemisphere, whereas the right visual hemifield is represented in the left hemisphere. The color code indicates the phase of the fMRI response that corresponds to a given position in the visual field. Polar angle mapping reveals that primary visual cortex (V1) contains a continuous representation of the contralateral hemifield. Areas V2 and V3, which

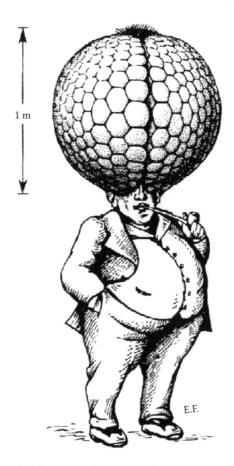

FIGURE 8.2 A human equipped with compound eyes, scaled to the size necessary to achieve a visual resolution matching that of single-lens eyes. (With kind permission of **Springer Science+Business Media** and the author, from: Kirschfeld, K., The resolution of lens and compound eyes, In *Neural Principles in Vision*, eds. Zettler, F., Weiler, R., 1976, pp. 354–70, Figure 7c.)

are both characterized by a discontinuous representation of a hemifield, surround area V1. The boundary between V1 and ventral V2 is formed by a representation of the upper vertical meridian (Figure 8.3c), whereas the boundary between V1 and dorsal V2 is formed by a lower vertical meridian representation. Accordingly, V2 is divided into quarterfield representations of the visual space with ventral V2 representing the upper quadrant and dorsal V2 representing the lower quadrant. Area V3 abuts on V2 and is similarly divided into ventral and dorsal quarterfield representations of visual space. The visual areas anterior to areas V1/V2/V3 are characterized by full hemifield representations and separated by upper or lower vertical meridian representations.

Eccentricity mapping reveals that V1, V2, V3, and adjacent V4 share one confluent foveal representation on the ventral-lateral surface near the occipital pole (Figure 8.3d). As one moves from posterior to anterior in visual cortex, the visual field representations shift from the center to the periphery, that is, increasingly peripheral stimuli are represented at increasingly anterior positions along the medial surface. Visual maps that share one common eccentricity representation appear to exhibit similar computational functions (VO-1/VO-2, Brewer et al. 2005; PHC-1/PHC-2, Arcaro, McMains, and Kastner 2009).

Retinotopic mapping has not only revealed the topography of areas in visual cortex, but also in subcortical nuclei. Using retinotopic mapping and high-resolution fMRI ($1.5 \times 1.5 \times 2$ mm^3), Schneider et al. (2004) found bilateral activations in the posterior thalamus, in the anatomical location of the lateral geniculate nucleus (LGN). The activations were strictly confined to

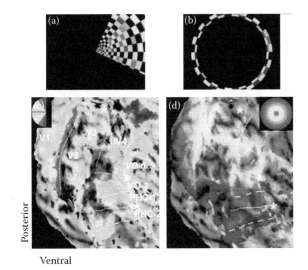

Ventral

FIGURE 8.3 (**See Color Insert**) (a, b) Examples of visual stimuli used for retinotopic mapping (Swisher et al. 2007). The stimuli consist of chromatic, flickering checkerboards. (a) The rotating wedge is typically used for polar angle mapping. (b) The expanding/contracting ring is typically used for eccentricity mapping. (c, d) Polar angle and eccentricity maps in human ventral visual cortex (Arcaro, McMains, and Kastner 2009). Flattened surface reconstruction of the right hemisphere depicting early (V1, V2, V3) and ventral visual cortex (hV4, VO-1, VO-2, PHC-1, PHC-2) of one representative subject. The color code indicates the phase of the fMRI response and labels the region of the visual field to which the surface node is most responsive to, as depicted in the visual field color legends. The dotted and dashed white lines indicate the upper and lower vertical meridians, respectively. The asterisks indicate foveal representations. (c) Polar angle map. (d) Eccentricity map. hV4 = human V4; VO = ventral occipital; PHC = parahippocampal cortex. (From Arcaro, M.J., McMains, S.A., Singer, B.D., Kastner, S., *J Neurosci* 29, 10638, 2009. With permission.)

stimulation of the contralateral visual field in each LGN. The lower visual field was represented in the medial-superior part and the upper field was represented in the posterior and superior part. The fovea was represented in posterior and superior portions, with increasing eccentricities represented more anteriorly. Furthermore, Schneider et al. (2004) attempted to segregate the magnocellular and parvocellular divisions of the LGN by using functional criteria. The magnocellular regions of the LGN were distinguished based on their sensitivity to low stimulus contrast and tended to be located in its inferior and medial portions. The results showed striking similarities in the topographic organization of human and monkey LGN.

The recent introduction of cognitive mapping has revealed topographically organized areas in higher-order cortex (for reviews see Wandell et al. 2007). For example, visual field maps in human posterior parietal cortex (PPC) can be identified with a memory-guided saccade paradigm, in which subjects perform memory-guided saccades to multiple peripheral locations arranged clockwise around a central fixation point (Sereno, Pitzalis, and Martinez 2001). In contrast to passive viewing during traditional retinotopic mapping, subjects have to attend to the location of the stimulus, store the location in working memory, and perform a saccadic eye movement to that location. This cognitive mapping approach revealed a contiguous band of topographically organized areas in PPC, consisting of five areas along and around the intraparietal sulcus (IPS; from posterior to anterior: IPS1–IPS5), and one area branching off into the superior parietal lobule (SPL1; Konen and Kastner 2008b). Figure 8.4 shows that each area contains a representation of the contralateral visual field and is separated towards neighboring areas by reversals in the visual field orientation. The identification of topographically organized areas is key to studying their functional properties and the large-scale functional organization of topographic units. In the following sections, we will review the main principles of organization of visual cortex.

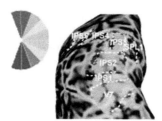
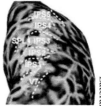

Posterior

Left hemisphere   Right hemisphere

FIGURE 8.4 (See Color Insert) Topographic maps in posterior parietal cortex obtained with the memory-guided saccade task. Briefly, targets appeared at successive locations "around the clock" and were followed by a ring of distracters. The disappearance of the distracters signaled subjects to perform saccadic eye movements towards the remembered target locations. The figure shows activation obtained from one subject overlaid on inflated left and right hemispheres. The color code is shown for voxels whose responses were correlated with the fundamental frequency of saccade direction, $r < .2$, indicating the phase of the response. The responses evoked in the memory-guided saccade task were lateralized such that the right visual field was represented in the left hemisphere, whereas the left visual field was represented in the right hemisphere. Area boundaries were formed by the alternating representation of either the upper (red) or lower (blue) vertical meridian. IPS = intraparietal sulcus. (From Konen, C.S., and Kastner, S. 2008. Representation of eye movements and stimulus motion in topographically organized areas of human posterior parietal cortex. *J Neurosci* 28:8361–75. With permission.)

8.2.1 HIERARCHICAL PROCESSING OF VISUAL INFORMATION

Neurons at successive stages of the visual system encode progressively more complex features of visual information, which is first represented in a localized and simple form and is then, through stagewise processing, transformed into more global and complex representations (DeYoe and Van Essen 1988). For example, a neuron in area V1 may respond to a contour such as a horizontal line presented in its RF, whereas neurons in V2 will respond to illusory contours (von der Heydt and Peterhans 1989), and neurons in ventral occipital and temporal cortex will respond selectively to combinations of contours such as shapes. No doubt the ability to encode visual objects holistically, with a capacity for perceptual generalization and categorization, marked a critical evolutionary advancement in visual system processing.

Likewise, the average RF size increases as one moves forward within the visual hierarchy. For example, the RFs of neurons in area V1 are about $1.5°$ and about $4°$ in intermediate area V4, whereas neurons in higher-order area TE have a median RF size of $26° \times 26°$ (Gattass and Gross 1981; Gattass, Sousa, and Gross 1988; Kastner et al. 2001). It thus appears that large RFs in higher-order areas are built from smaller RFs in lower visual areas, implying that the initial neural representation of visual scenes is the result of a bottom-up process that is driven by the physical properties of the visual input.

Anatomical studies have revealed that the successive stages in the hierarchically organized visual processing stream are reciprocally connected, i.e., projections from one area to the next are reciprocated by projections from the second area back onto the first (Felleman and Van Essen 1991). Reciprocal connections are important for higher visual areas to influence the activation in lower visual areas and vice versa. These feedback connections may constitute local networks that mediate complex neural computations across neighboring areas that require the integration of information and, importantly, also form the neural basis for cognitive functions such as selective attention, which will be discussed below.

8.2.2 ORGANIZATION INTO ANATOMICALLY DISTINCT AND
FUNCTIONALLY SPECIALIZED PATHWAYS

Visual areas appear to be divided into two major corticocortical processing pathways, each of which begins in primary visual cortex. The ventral pathway is directed into temporal cortex and is important for visual object recognition, or "what" an object is; the dorsal pathway is directed into the PPC and is important for spatial perception, or "where" an object is (Ungerleider and Mishkin 1982). The original evidence for separate processing pathways for object and spatial vision was derived from lesions of IT and PPC in monkeys (Ungerleider and Mishkin 1982). Inferior temporal lesions were shown to impair performance on tasks requiring discrimination of visual objects, but left performance on tasks requiring visuo-spatial judgments intact. By contrast, lesions of PPC did not impair object discrimination performance, but led to visuo-spatial deficits. Goodale and Milner (1992) provided an alternative interpretation of the two-pathways hypothesis by suggesting that the dorsal pathway is also involved in visually guided action, or "how" to interact with an object.

8.2.2.1 The Ventral Pathway

Monkey physiology studies have shown that areas at successive stages of the ventral pathway encode progressively more complex attributes of an object. Neurons in area V4 respond to a variety of simple and complex object features such as color and shape (Desimone et al. 1985; Gallant et al. 1996; Pasupathy and Connor 2002), while neurons in IT respond selectively to more sophisticated combinations of such features including specific objects such as hands or faces (Gross, Rocha-Miranda, and Bender 1972). Furthermore, it has been shown that these object representations are invariant to certain image transformations including size and viewpoint (Gross, Rocha-Miranda, and Bender 1972; Desimone et al. 1984; Booth and Rolls 1998).

The ventral pathway in humans seems to be similarly organized. Area V4 responds selectively to a variety of object stimuli such as line drawings of objects, 2D- and 3D-objects, but these responses depend on the size and viewpoint of an object (Konen and Kastner 2008a). The putative homologue to monkey IT cortex may be human lateral occipital complex (LOC), which responds more strongly to pictures of objects than to their scrambled counterparts (Malach et al. 1995; Grill-Spector et al. 1999; Kourtzi and Kanwisher 2001) and exhibits similar responses to objects defined by luminance, texture, motion, and other cues, thus representing objects independent of the precise physical cues that define an object (Grill-Spector et al. 1998; Kourtzi and Kanwisher 2000). Importantly, LOC represents objects invariant to changing external viewing conditions such as viewpoint or transformation of object size (Grill-Spector et al. 1999; Vuilleumier et al. 2002; Konen and Kastner 2008a). Thus, there is converging evidence for the hierarchical organization of the ventral pathway in both humans and non-human primates. At the top of the ventral pathway, both monkey IT and human LOC represent objects independent of image transformations. Accordingly, lesions of LOC in humans have devastating effects on object recognition, leading to object agnosia (Riddoch and Humphreys 1987; and see Chapter 10 in this volume). Object agnosia is a neuropsychological disorder characterized by the inability to recognize visual objects, in the absence of other perceptual impairments.

Within LOC, fMRI studies have reported discrete cortical regions in ventral temporal cortex that responded preferentially to biologically relevant objects such as faces (fusiform face area [FFA], Kanwisher, McDermott, and Chun 1997), places (parahippocampal place area [PPA], Epstein and Kanwisher 1998), and body parts (extrastriate body area [EBA], Downing et al. 2001). Each of these regions was anatomically consistent within and across subjects (Peelen and Downing 2005), and each was relatively spatially circumscribed (Spiridon, Fischl, and Kanwisher 2006). These findings suggest that one organizational principle of the object vision pathway is the segregation into category-specific modules for faces, houses, and places. The evolution of cortical modules related to these specific categories is likely due to their biological relevance for primates in order to interact with and orient themselves in their environment (Downing et al. 2006).

A major debate has centered on the issue of the specific neural representation of the large number of object categories that we can effortlessly discriminate. One hypothesis is that visual objects in the ventral visual pathway are represented in a domain-specific fashion by groups of highly selective neurons and that these neurons are clustered near each other in cortex (Kanwisher 2000; Spiridon and Kanwisher 2002). Key evidence in support of domain-specific organization has originated from fMRI-guided neurophysiology in non-human primates (Tsao et al. 2003). Tsao and colleagues found an area in monkey superior temporal sulcus in which 97% of neurons responded selectively to faces, thereby demonstrating the selectivity of one cortical area for a single high-level function—face perception—and providing the strongest evidence to date for both domain specificity and clustering of visual object representations in the ventral visual pathway.

By contrast, Tarr and Gauthier (2000) have argued for a different functional role of face processing modules, particularly the FFA. The authors showed that FFA activation was strongly influenced by the expertise of subjects. For example, FFA responded selectively not only to faces, but also to cars in car experts and to birds in bird experts. According to this account, activity in FFA reflected expertise in recognizing certain object categories, a function that was not restricted to faces. Thus, the specific face activation in this part of cortex may result from experience and the fact that we are all "face experts" due to the biological significance of face stimuli.

In an important challenge to the idea that the functional architecture of the ventral pathway is a mosaic of domain-specific modules, Haxby and coworkers (2001) showed that the representations of faces and objects were widely distributed and overlapping. According to their view, each object category is associated with its own differential pattern of response and represented not merely by a strong response in a small region of cortex, but by the entire distributed pattern of activation across the ventral visual pathway. The fact that all objects activate a broad expanse of ventral temporal cortex suggests that its functional architecture is based on a continuous representation of object features, such that features shared by members of a category tend to cluster together (Ishai et al. 1999). Such large-scale coding across distributed populations of neurons might be another important functional organizational principle of the visual pathway.

While theories of domain specificity versus distributed categorical information are often portrayed as conflicting, they are not mutually exclusive. A plausible scenario is that ventral temporal cortex has a feature-space organization distributed across the cortical sheet, but particular object categories that are especially relevant to the organism (either evolutionarily or experientially) will occupy a disproportionally large amount of neighboring neurons for computational efficiency. In other words, while there may be neurons that contribute to the representation of one category spread across a large span of cortex (i.e., as revealed by large-scale fMRI activity patterns), the bulk of the most relevant information processing for a few object categories will occur in localized neural modules.

Finally, Malach, Levy, and Hasson (2002) have proposed retinal eccentricity as another organizing principle of human object-related areas. According to the "eccentricity bias" account, brain regions that represent object categories depending on detailed scrutiny such as faces are more strongly associated with central information, as compared to representations of objects that may be recognized by integrating more peripheral information such as places. Thus, face-selective regions are associated with central visual field representations, whereas place-related regions are associated with peripheral visual field representations. This principle, however, has been called in question by recent findings of foveal representations within the region of the PPA and multiple topographic maps within ventral temporal cortex with systematic representation of both polar angle and eccentricity (Arcaro, McMains, and Kastner 2009).

8.2.2.2 The Dorsal Pathway

The IPS in monkeys contains several regions that can be distinguished on the basis of structural and functional criteria and exhibit a variety of different response properties related to the encoding of spatial information (Andersen 1997; Colby and Goldberg 1999). For example, the ventral intraparietal area (VIP), the lateral intraparietal area (LIP), and area 7a are reciprocally connected to adjacent areas along the IPS (Van Essen et al. 1990). LIP and area 7a have been shown to be involved in the encoding of saccadic eye movements (Andersen et al. 1990), while the majority of VIP neurons respond during smooth pursuit eye movements (Schlack, Hoffmann, and Bremmer 2003). Furthermore, VIP and area 7a at the apex of the dorsal stream contain neurons that respond selectively to radial motion (Steinmetz et al. 1987; Schaafsma and Duysens 1996; Siegel and Read 1997; Bremmer et al. 2002; Schlack, Hoffmann, and Bremmer 2002).

The recent discovery of topographically organized areas along the IPS provides a fruitful avenue for probing the functional characteristics of areas in human PPC and comparing them to those in monkey PPC (Sereno and Huang 2006; Levy et al. 2007; Konen and Kastner 2008a,b). A recent fMRI study investigated the core functions of the dorsal pathway, that is, the representation of eye movements and motion-selectivity in human PPC (Konen and Kastner 2008b). The results showed that the preferential responses during saccadic or smooth pursuit eye movements changed gradually across areas of the IPS with IPS1–IPS2 and the medial SPL1 preferring saccadic eye movements and IPS3–IPS5 preferring smooth pursuit eye movements. Accordingly, patients with lesions in more posterior regions of the PPC exhibit deficits in the execution of visually guided saccadic eye movements (Pierrot-Deseilligny et al. 1991), while patients with lesions in more anterior regions of the PPC show impairments in the performance of smooth pursuit eye movements (Heide, Kurzidim, and Kompf 1996). Interestingly, areas in close anatomical proximity such as IPS1–IPS2 and SPL1 in the posterior/medial PPC and IPS3–IPS5 in the anterior/lateral PPC showed similar response characteristics. This principle of functional organization is similar to the one in non-human primates, in which boundaries between areas in PPC are blurred, leading to systematic shifts of response characteristics from one area to the next and thus several functional gradients along the IPS (Colby and Duhamel 1996). Importantly, the direct comparison between both species revealed that saccade-related activity decreased while smooth pursuit eye movement-related activity increased from posterior to anterior/lateral in human PPC and from lateral/dorsal to ventral in monkey PPC. Thus, both human and monkey PPC exhibit a similar gradient organization in the representation of eye movements.

Furthermore, all topographically organized areas in human PPC responded to different types of motion including planar, circular, and radial optic flow (Konen and Kastner 2008b). IPS1–IPS3 preferred radial optic flow over planar motion, whereas areas in anterior PPC showed no preference for a particular motion type. These results are in agreement with monkey PPC, in which VIP and area 7a contain neurons that respond selectively to radial motion (Steinmetz et al. 1987; Schaafsma and Duysens 1996; Siegel and Read 1997; Bremmer et al. 2002; Schlack, Hoffmann, and Bremmer 2002). Thus, visual motion appears to be similarly represented in human and monkey PPC as well.

These findings contribute to a growing body of recent work that has related functional characteristics to underlying topography in human PPC (see Silver and Kastner 2009 for review). IPS1–IPS2 and SPL1 in humans exhibit similar response properties compared to LIP and area 7a in non-human primates. Physiology studies in monkeys have shown that both LIP and area 7a, which are located adjacent to visual cortex, occupy the highest position within the visual hierarchy of the dorsal processing stream (Felleman and Van Essen 1991). Likewise, IPS1–IPS2 and SPL1 are located just anterior to visual cortex in the posterior IPS. These areas as well as macaque LIP and area 7a have also been shown to be involved in spatial attention and working memory (Gnadt and Andersen 1988; Colby, Duhamel, and Goldberg 1996; Constantinidis and Steinmetz 1996, 2001; Schluppeck, Glimcher, and Heeger 2005; Silver, Ress, and Heeger 2005). IPS1–IPS2 and SPL1 showed a preferred representation of saccades relative to smooth pursuit eye movements, which is similar in neurons in LIP and area 7a (Andersen et al. 1990; Barash et al. 1991; Colby, Duhamel, and Goldberg 1996; Bremmer, Distler, and Hoffmann 1997; Konen and Kastner 2008b). Evidence for functional homology between IPS1–IPS2 and LIP comes also from neuroimaging findings that IPS1 and IPS2 exhibit both saccade- and reach-related activity (Hagler, Riecke, and Sereno 2007; Levy et al. 2007). Both effector-specific responses have been found in LIP neurons at the single-cell level (Snyder, Batista, and Andersen 1997). Furthermore, SPL1 shows a pattern of motion-selective responses that is similar to those found in area 7a neurons (Merchant, Battaglia-Mayer, and Georgopoulos 2001; Konen and Kastner 2008b). There is also converging evidence that IPS5 may be functionally similar to VIP in monkeys (Sereno and Huang 2006). The preference in responses to smooth pursuit eye movements and optic flow patterns in IPS5 is in agreement with the functional properties of VIP neurons (Schaafsma, Duysens, and Gielen 1997; Schlack, Hoffmann, and Bremmer 2003). Furthermore, the majority of VIP neurons are bimodal and respond both to tactile and visual stimulation (Colby, Duhamel, and Goldberg 1993; Duhamel, Colby, and Goldberg 1998). This characteristic is in agreement with coaligned representations of tactile and visual space in IPS5 (Sereno and Huang 2006). Taken together, human IPS1–IPS2, SPL1, and IPS5 exhibit similar functional characteristics compared to monkey LIP, 7a, and VIP, respectively.

8.2.2.3 Object Representations in the Dorsal Pathway

In monkeys, object-related responses in PPC have typically been reported in association with action planning, particularly in tasks that require the execution of grasping movements (Sakata 2003). For example, area AIP contains both visual- and motor-related neurons for the neural coding of finger shaping in monkeys trained to grasp objects (Murata et al. 2000). In humans, activations in the anterior IPS have been found while subjects viewed graspable objects such as tools, which appear to be automatically linked to actions that are executed during their usage, but not to non-graspable objects (Chao and Martin 2000; Culham and Valyear 2006).

However, physiology studies in monkeys have shown shape-selective responses in area LIP (Sereno and Maunsell 1998; Sereno et al. 2002; Lehky and Sereno 2007) related to simple geometric objects that were not associated with grasping movements. The finding of shape-selective responses suggests that the cortical representation of objects is not exclusively restricted to the ventral pathway. Recently, object-selective responses in human PPC were investigated under passive viewing conditions and when attention was drawn away from the stimuli (Konen and Kastner 2008a). Specifically, neural representations of different types of object stimuli (2D and 3D objects; line drawings of objects and tools) as well as size- and viewpoint-invariance were investigated in topographically defined areas of the ventral and dorsal pathways. It was shown that intermediate processing stages—V4 in the ventral stream and V3A, MT, and V7 in the dorsal stream—exhibited object-selective responses (Figure 8.5a). However, these object representations were not invariant to image transformations, indicating that object representations at these processing stages depend on low-level visual features. By contrast, advanced processing stages of both pathways—LOC in the ventral stream and IPS1 and IPS2 in the dorsal stream—exhibited not only similar object-selectivity, but also size- and viewpoint-invariant response properties. Figure 8.5b summarizes the invariant responses of areas along the visual hierarchy. Taken together, these findings suggest the existence of two parallel hierarchical neural systems for object information in the human ventral and dorsal pathways.

The classical view of the two pathways hypothesis, however, suggests the existence of two anatomically distinct and functionally specialized pathways: a ventral stream for object vision and a dorsal stream for spatial vision (Ungerleider and Mishkin 1982). However, the two pathways hypothesis was originally proposed based on lesion studies in monkeys. These apparent discrepancies may be reconciled with the hypothesis that object information is represented differently in the human and monkey dorsal pathways due to evolutionary pressures that necessitate a more complex representation in the human dorsal system. These object representations might play a role in the human-specific parietal network for sophisticated tool-use (Peeters et al. 2009). Comparative studies in humans and non-human primates might clarify inter-species differences of object representations in the dorsal pathway.

8.3   TOP-DOWN INFLUENCES ON VISUAL PROCESSING

Even though the neural representation of natural scenes in visual cortex is ultimately constrained by the physical properties of the input, i.e., its feature content in terms of contours, colors, moving items, and so on, these representations are constantly modulated by top-down influences that reflect the behavioral goals of the individual. Therefore, what we see and how we perceive it is often determined by our inner states. In this section, we will review such top-down influences related to visual attention.

8.3.1   VISUAL ATTENTION

Natural visual scenes are cluttered and contain many different objects. However, the capacity of the visual system to process information about multiple objects at any given moment in time is limited.

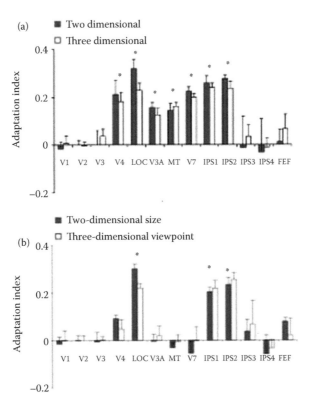

FIGURE 8.5 Object selectivity in the visual system. In these fMRI-adaptation experiments, object-selectivity (a) and invariance properties (b) of early visual areas, areas along the ventral pathway, and areas along the dorsal pathway were probed (Konen and Kastner 2008a). FMRI adaptation is a robust phenomenon in which repeated presentations of the same visual stimulus lead to gradual response reductions as a function of repetition frequency. The adaptation effects were quantified by an adaptation index. Positive index values indicate object-selective, size-, or viewpoint-invariant responses; negative values and values around zero indicate no selective responses. (a) The adapted condition consisted of the repeated presentations of identical 2D or 3D objects, while the non-adapted condition consisted of 16 different objects. The results showed object-selective responses not only in areas along the ventral pathway (V4, LOC), but also in several areas along the dorsal pathway, including V3A, MT, V7, IPS1, and IPS2. Importantly, LOC and the posterior IPS consistently showed the strongest adaptation effects as compared to the other areas of the visual system. (b) The adapted condition consisted of the repeated presentations of identical objects in 16 different sizes or viewpoints, while the non-adapted condition consisted of 16 different objects. The results showed that early visual areas and intermediate areas along both pathways exhibited no adaptation effects, suggesting processing of low-level aspects of object information. Adaptation in LOC, IPS1, and IPS2 remained under different viewing conditions of an object, indicating the representation of global rather than local aspects of object information at advanced processing stages of both pathways. (Reprinted by permission from Macmillan Publishers Ltd. [Nature Neuroscience] Konen, C.S., Kastner, S., 11, 224, copyright 2008.)

Hence, attentional mechanisms are needed to select relevant information and to filter out irrelevant information from cluttered visual scenes. Selective visual attention is a broad term that refers to a variety of different behavioral phenomena. Directing attention to a spatial location has been shown to improve the accuracy and speed of subjects' responses to target stimuli that occur in that location (Posner, Snyder, and Davidson 1980). Attention also increases the perceptual sensitivity for the discrimination of target stimuli (Lu 1998), increases contrast sensitivity (Cameron, Tai, and Carrasco 2002), reduces the interference caused by distracters (Shiu and Pashler 1995), and improves acuity (Yeshurun and Carrasco 1998). There is converging evidence from single-cell physiology studies in non-human primates and functional brain mapping studies in humans that selective attention modulates neural activity in the visual system in several ways. Here, we will briefly review the attentional effects on visual processing that may underlie these and other behavioral effects of visual attention.

8.3.2 ATTENTIONAL RESPONSE MODULATION IN THE VISUAL SYSTEM

In a typical fMRI study, the effects of space-based selection on neural responses have been investigated by presenting simple stimuli that activate the visual system well, such as flickering checkerboards, to the left or right visual hemifield, while subjects directed attention to the stimulus (attended condition) or away from the stimulus (unattended condition) (e.g., O'Connor et al. 2002). In the unattended condition, attention was directed away from the stimulus by having subjects count letters at fixation. The letter-counting task ensured proper fixation and effectively prevented subjects from covertly attending to the checkerboard stimuli. In the attended condition, subjects were instructed to covertly direct attention to the checkerboard stimulus and to detect luminance changes that occurred randomly in time at a peripheral stimulus location. Relative to the unattended condition, the mean fMRI signals evoked by a high-contrast checkerboard stimulus increased significantly in the attended condition in the LGN, and in visual cortex (Figure 8.6a,d). In particular, attentional response enhancement was found in striate cortex, and in each extrastriate area along the ventral and dorsal pathways (Figure 8.7a). Similar attentional response enhancement was obtained with activity evoked by a low-contrast checkerboard stimulus (Figure 8.6a,d). Notably, these attention effects were shown to be spatially specific in other studies, in which identical stimuli were presented simultaneously to the right and left of fixation, while subjects were instructed to direct attention covertly to one or the other side (Heinze et al. 1994; Tootell et al. 1998; Brefczynski and DeYoe 1999; O'Connor et al. 2002). Taken together, these findings suggest that selective attention facilitates visual processing at thalamic and cortical stages by enhancing neural responses to an attended stimulus relative to those evoked by the same stimulus when ignored. Attentional response enhancement may be a neural correlate for behavioral attention effects such as increased accuracy and response speed, or improved target discriminability (e.g., Posner 1980; Lu 1998).

Spatial attention affects not only the processing of the selected information, but also the processing of the unattended information, which is typically the vast majority of incoming information. The neural fate of unattended stimuli was investigated in an fMRI experiment, in which the attentional load of a task at fixation was varied (O'Connor et al. 2002). According to attentional load theory (Lavie and Tsal 1994), the degree to which ignored stimuli are processed is determined by the amount of attentional capacity that is not dedicated to the selection process. This account predicts that neural responses to unattended stimuli should be attenuated depending on the attentional load necessary to process the attended stimulus. This idea was tested by using the checkerboard paradigm described above while subjects performed either an easy (low load) attention task or a hard (high load) attention task at fixation and ignored the peripheral checkerboard stimuli. Relative to the easy task condition, mean fMRI signals evoked by the high-contrast and by the low-contrast stimuli decreased significantly in the hard task condition across the visual system with the smallest effects in early visual cortex and the largest effects in LGN and extrastriate cortex (Figure 8.6b,e). Taken together, these findings suggest that neural activity evoked by ignored stimuli is attenuated at several stages of visual processing as a function of the load of attentional resources engaged elsewhere (Rees, Frith, and Lavie 1997; O'Connor et al. 2002; Schwartz et al. 2005). Attentional-load-dependent suppression of unattended stimuli may be a neural correlate for behavioral effects such as reduction of interference caused by distracters (Shiu and Pashler 1995). Further, there is evidence that the enhancement of activity at an attended location and the suppression of activity at unattended locations operate in a push-pull fashion and thus represent co-dependent mechanisms (Pinsk, Doniger, and Kastner 2004; Schwartz et al. 2005).

A third attentional effect has been observed during periods when subjects deploy attention to a location in space at which visual stimuli are expected to occur. A neural correlate of cue-related activity has been found in physiology studies demonstrating that spontaneous (baseline) firing rates were 30%–40% higher for neurons in areas V2 and V4 when the animal was cued to attend covertly to a location within the neuron's RF before the stimulus was presented there; that is, in the absence

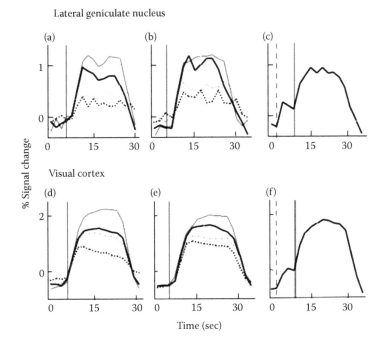

FIGURE 8.6 Attentional response enhancement, suppression, and increases in baseline activity in the LGN and in visual cortex. Time series of fMRI signals in the LGN and visual cortex were combined across left and right hemispheres. Activity in visual cortex was pooled across areas V1, V2, V3/VP, V4, TEO, V3A, and MT/MST. (a, d) Attentional enhancement. During directed attention to the stimuli (gray curves), responses to both the high-contrast stimulus (100%, solid curves) and low-contrast stimulus (5%, dashed curves) were enhanced relative to an unattended condition (black curves). (b, e) Attentional suppression. During an attentionally demanding "hard" fixation task (black curves), responses evoked by both the high-contrast stimulus (100%, solid curves) and low-contrast stimulus (10%, dashed curves) were attenuated relative to an easy attention task at fixation (gray curves). (c, f) Baseline increases. Baseline activity was elevated during directed attention to the periphery of the visual hemifield in expectation of the stimulus onset; the beginning of the expectation period is indicated by the dashed vertical line. Gray vertical lines indicate the beginning of checkerboard presentation periods. (Reprinted by permission from Macmillan Publishers Ltd. [Nature Neuroscience] O'Connor, D.H., Fukui, M.M., Pinsk, M.A., Kastner, S., 5, 1203, copyright 2002.)

of visual stimulation (Luck et al. 1997; Lee, Williford, and Maunsell 2007; but see McAdams and Maunsell 1999). This increased baseline activity has been interpreted as a direct demonstration of a top-down signal that feeds back from higher-order to lower-order areas. In the latter areas, this feedback signal appears to bias neurons representing the attended location, thereby favoring stimuli that will appear there at the expense of those appearing at unattended locations.

To investigate attention-related baseline increases in the human visual system in the absence of visual stimulation, fMRI activity was measured while subjects were cued to covertly direct attention to the periphery of the left or right visual hemifield and to expect the onset of a stimulus (Kastner et al. 1999; O'Connor et al. 2002; McMains et al. 2007; Sylvester et al. 2007). The expectation period, during which subjects were attending to the periphery without receiving visual input, was followed by attended presentations of a high-contrast checkerboard. During the attended presentations, subjects counted the occurrences of luminance changes. Relative to the preceding blank period in which subjects maintained fixation at the center of the screen and did not attend to the periphery, fMRI signals increased during the expectation period in the LGN, striate, and extrastriate cortex (Figure 8.6c,f and Figure 8.7c). This elevation of baseline activity was followed by a further response increase evoked by the visual stimuli (Figure 8.6c).

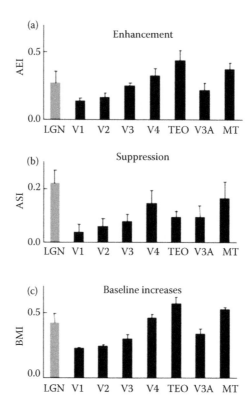

FIGURE 8.7 Attentional response modulation in the visual system. Attention effects that were obtained in the experiments presented in Figure 8.6 were quantified by defining several indices: (a) attentional enhancement index (AEI), (b) attentional suppression index (ASI), (c) baseline modulation index (BMI). For all indices, larger values indicate larger effects of attention. Index values were computed for each subject based on normalized and averaged signals obtained in the different attention conditions and are presented as averaged index values from 4 subjects (for index definitions, see O'Connor et al. 2002). In visual cortex, attention effects increased from early to later processing stages. Attention effects in the LGN were larger than in V1. Vertical bars indicate S.E.M. across subjects. (Reprinted by permission from Macmillan Publishers Ltd. [Nature Neuroscience] O'Connor, D.H., Fukui, M.M., Pinsk, M.A., Kastner, S., 5, 1203, copyright 2002.)

8.4 COMPARISON OF ATTENTION EFFECTS ACROSS THE VISUAL SYSTEM

With fMRI, neural responses can be investigated at the population level and across a wide range of different processing stages, allowing for quantitative comparisons of attentional modulatory effects across the visual system. For example, the attention effects of enhancement, suppression, and baseline increases can be quantified in each visual area by calculating an index value. Large index values indicate large effects of attention (for further details, see O'Connor et al. 2002). As shown in Figure 8.7, the magnitude of all attention effects increased from early to more advanced processing stages along both the ventral and dorsal pathways of visual cortex (Kastner et al. 1998; Martinez et al. 1999; Mehta, Ulbert, and Schroeder 2000; Cook and Maunsell 2002). (A similar profile has been observed in the somatosensory system: see Chapter 7.) This is consistent with the idea that attention operates through top-down signals that are transmitted via corticocortical feedback connections in a hierarchical fashion. Thereby, areas at advanced levels of visual cortical processing are more strongly controlled by attentional mechanisms than are early processing levels. This idea is supported by single-cell recording studies, which have shown that attentional effects in area TE of inferior temporal cortex have a latency of approximately 150 ms (Chelazzi et al. 1998), whereas attentional effects in V1 have a longer latency, approximately 230 ms (Roelfsema, Lamme, and Spekreijse 1998).

According to this account, one would predict smaller attention effects in the LGN than in striate cortex. Surprisingly, it was found that all attention effects tended to be larger in the LGN than in striate cortex (Figure 8.7a–c). This finding raises the possibility that attentional modulation in the LGN may not be exclusively attributable to corticothalamic feedback from striate cortex, but may also reflect additional modulatory influences from other sources. In addition to corticothalamic feedback projections from V1, the LGN receives inputs from the superior colliculus, which is part of a distributed network of areas controlling eye movements, and the thalamic reticular nucleus (TRN), which has long been implicated in theoretical accounts of selective attention (Crick 1984; Sherman and Guillery 2001). Evidence in support of this notion has been found in a recent physiology study demonstrating strong response modulation in both LGN and TRN (McAlonan, Cavanaugh, and Wurtz 2008). Thus, modulatory BOLD signals in the visual system may reflect the summed feedback input to an area rather than top-down processing that reverses the visual hierarchy.

8.5   SOURCES OF ATTENTIONAL MODULATORY SIGNALS

There is evidence from studies in patients suffering from attentional deficits due to brain damage and from functional brain imaging studies in healthy subjects performing attention tasks that attention-related modulatory signals are not generated within the visual system, but rather derive from higher-order areas in parietal and frontal cortex and are transmitted via feedback projections to the visual system (Kastner and Ungerleider 2000). This network includes the frontal eye field (FEF), supplementary eye field (SEF), and regions in the SPL, and has been found to be activated in a variety of visuo-spatial attention tasks, suggesting that it constitutes a rather general attention network that operates independent of the specific requirements of the visuo-spatial task (Figure 8.8a,b).

Neuroimaging and physiology studies suggest that this network reflects the top-down sources that generate attentional feedback signals, which in turn modulate visual processing at "sites" in sensory cortex. In fMRI studies, neural activations related to the attentional operations themselves were dissociated from modulatory effects on visual stimuli by probing the effects of spatially directed attention in the presence and in the absence of visual stimulation (Kastner et al. 1999). As described above, subjects were cued to direct attention to a peripheral target location and to expect the onset of visual stimuli, which occurred with a delay of several seconds after which subjects indicated the detection of target stimuli at the attended location. The increase of baseline activity in visual cortical areas was followed by a further increase of activity evoked by the onset of the stimulus presentations (gray shaded epochs). In parietal and frontal cortex, the same distributed network for spatial attention was activated during directed attention in the absence of visual stimulation as during directed attention in the presence of visual stimulation, consisting of the FEF, the SEF, and the SPL (Figure 8.8a,b). As in visual cortical areas, there was an increase in activity in these frontal and parietal areas due to directed attention in the absence of visual input. However, first, this increase in activity was stronger in FEF, SEF, and SPL than the increase in activity seen in visual cortex (as exemplified for FEF in Figure 8.8d), and second, there was no further increase in activity evoked by the attended stimulus presentations in these parietal and frontal areas. Rather, there was sustained activity throughout the expectation period and the attended presentations (Figure 8.8d). These results from parietal and frontal areas suggest that this activity reflected the attentional operations of the task and not visual processing per se. These findings therefore provide the first evidence that these parietal and frontal areas may be the sources of feedback that generated the top-down biasing signals seen in visual cortex. This notion has been strongly corroborated by physiology studies in monkeys showing that microstimulation of the FEF can mimic the modulatory effects of attention in downstream area V4 (Moore and Armstrong 2003).

In summary, visual attention has been shown to engage a widely distributed network of brain areas in the thalamus and cortex that cooperate to mediate the selection of behaviorally relevant information from the cluttered visual environment that we are constantly faced with.

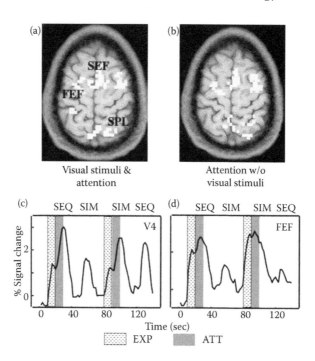

FIGURE 8.8 A fronto-parietal network for spatial attention. Axial slice through frontal and parietal cortex. (a) When the subject directed attention to a peripheral target location and performed a discrimination task, a distributed fronto-parietal network was activated including the SEF, the FEF, and the SPL. (b) The same network of frontal and parietal areas was activated when the subject directed attention to the peripheral target location in expectation of the stimulus onset. L indicates left hemisphere. Increases of baseline activity in the absence of visual stimulation. (c) Time series of fMRI signals in V4. Directing attention to a peripheral target location in the absence of visual stimulation led to an increase of baseline activity (textured blocks), which was followed by a further increase after the onset of the stimuli (gray-shaded blocks). Baseline increases were found in both striate and extrastriate visual cortex. Sequentially presented stimuli (SEQ) evoked more activity in V4 than did simultaneously presented stimuli (SIM). (d) Time series of fMRI signals in FEF. Directing attention to the peripheral target location in the absence of visual stimulation led to a stronger increase in baseline activity than in visual cortex; the further increase of activity after the onset of the stimuli was not significant. Sustained activity was seen in a distributed network of areas outside visual cortex, including SPL, FEF, and SEF, suggesting that these areas may provide the source for the attentional top-down signals seen in visual cortex. (From Kastner, S., and Ungerleider, L.G. 2000. Mechanisms of visual attention in the human cortex. *Annu Rev Neurosci* 23:315–41.)

8.6 EMOTIONAL RESPONSE MODULATION IN THE VISUAL SYSTEM

The previous sections have shown that selective attention provides an important mechanism for the filtering of distracting information in cluttered visual scenes in order to enhance the influence of behaviorally relevant stimuli at the expense of irrelevant ones. An exception to the critical role of attention in the processing of visual information, however, seems to be the neural processing of emotional stimuli. A widespread view is that the amygdala, a critical node in the neural circuit mediating the processing of stimulus valence, functions in a largely automatic fashion that is independent of top-down factors such as attention (Dolan and Vuilleumier 2003). Consistent with this assumption, amygdala responses have been observed when attention was allocated elsewhere (Vuilleumier et al. 2001; Whalen et al. 2004).

Pessoa et al. (2002) challenged this view by systematically investigating the effects of attention on the neural processing of emotional stimuli. FMRI responses evoked by pictures of faces with fearful, happy, or neutral expressions were measured when attention was directed to them (attended condition) or when attention was directed to oriented bars (unattended condition). The results showed that attended as compared to unattended faces evoked greater activation in the amygdala for all facial expressions. The amygdala responded differentially to faces with emotional content only in the attended condition, indicating a significant interaction between stimulus valence and attention. When all attentional resources were directed to the oriented bars, responses to faces were diminished. Taken together, the results showed that amygdala responses to emotional stimuli were not entirely automatic and instead required some degree of attention. Furthermore, the contrast between emotional and neutral faces evoked stronger activation not only in the amygdala, but also in visual areas including V1 and V2 (Moll et al. 2002; Mourao-Miranda et al. 2003). The increased activation in visual cortex evoked by emotional stimuli reflected emotional modulation, in which the neuronal processing was biased towards emotional faces as compared to neutral faces. Because of its strong projections to areas in visual cortex (Amaral and Insausti 1992), the amygdala might be the source of emotional modulation in visual areas.

Indeed, recent studies have shown that amygdala signals correlate with signals from visual areas (Morris, Ohman, and Dolan 1999; Pessoa et al. 2002). Further evidence for the correlation of signals between the amygdala and visual cortex comes from patients with lesions of the amygdala, who showed attenuated activation in visual cortex (Vuilleumier et al. 2004). Taken together, the amygdala might underlie a form of emotional modulation of information that in many ways parallels the attentional effects observed in visual cortex.

8.7 SUMMARY

In this chapter, we reviewed organizational principles of the visual system in humans and non-human primates. Decades of electrophysiological studies in non-human primates have yielded detailed response properties of visual neurons. Using fMRI, various mapping techniques appear to be a fruitful avenue to chart the visual system in humans and to systematically test functional characteristics of a multitude of distinct areas. This region of interest approach allows the direct comparison of the anatomical and functional organization of the visual system between species, which in turn yields valuable information about evolutionary processes. For example, it has been shown that the encoding of eye movements and visual motion appears to be similarly represented in human and monkey PPC. By contrast, object information is represented differently in the dorsal pathways of both species, possibly due to evolutionary pressures that necessitate a more complex representation in the human dorsal system in order to enable sophisticated tool-use.

Furthermore, we reviewed human fMRI studies of top-down influences related to visual attention. Selective attention can affect neural processing in several ways, which include: (1) the enhancement of neural responses to attended stimuli; (2) the filtering of unwanted information by counteracting the suppression induced by distracters; and (3) the biasing of signals in favor of an attended location by increases of baseline activity in expectation of a visual stimulus. Finally, we have shown that these forms of attentional modulation in visual cortex are similar to emotional modulation at subcortical stages, that is, in the amygdala.

REFERENCES

Amaral, D.G., and Insausti, R. 1992. Retrograde transport of D-[3H]-aspartate injected into the monkey amygdaloid complex. *Exp Brain Res Exp Hirnforsch* 88:375–88.

Andersen, R.A. 1997. Multimodal integration for the representation of space in the posterior parietal cortex. *Philos Trans R Soc Lond B Biol Sci* 352:1421–28.

Andersen, R.A., Bracewell, R.M., Barash, S., Gnadt, J.W., and Fogassi, L. 1990. Eye position effects on visual, memory, and saccade-related activity in areas LIP and 7a of macaque. *J Neurosci* 10:1176–96.

Arcaro, M., McMains, S.A., and Kastner, S. 2009. Retinotopic organization of human ventral visual cortex. *J Neurosci* 29:10638–52.

Barash, S., Bracewell, R.M., Fogassi, L., Gnadt, J.W., and Andersen, R.A. 1991. Saccade-related activity in the lateral intraparietal area. I. Temporal properties; comparison with area 7a. *J Neurophysiol* 66:1095–1108.

Booth, M.C., and Rolls, E.T. 1998. View-invariant representations of familiar objects by neurons in the inferior temporal visual cortex. *Cereb Cortex* 8:510–23.

Brefczynski, J.A., and DeYoe, E.A. 1999. A physiological correlate of the "spotlight" of visual attention. *Nat Neurosci* 2:370–74.

Bremmer, F., Distler, C., and Hoffmann, K.P. 1997. Eye position effects in monkey cortex. II. Pursuit- and fixation-related activity in posterior parietal areas LIP and 7A. *J Neurophysiol* 77:962–77.

Bremmer, F., Klam, F., Duhamel, J.R., Ben Hamed, S., and Graf, W. 2002. Visual-vestibular interactive responses in the macaque ventral intraparietal area (VIP). *Eur J Neurosci* 16:1569–86.

Brewer, A.A., Liu, J., Wade, A.R., and Wandell, B.A. 2005. Visual field maps and stimulus selectivity in human ventral occipital cortex. *Nat Neurosci* 8:1102–9.

Cameron, E.L., Tai, J.C., and Carrasco, M. 2002. Covert attention affects the psychometric function of contrast sensitivity. *Vision Res* 42:949–67.

Chao, L.L., and Martin, A. 2000. Representation of manipulable man-made objects in the dorsal stream. *NeuroImage* 12:478–84.

Chelazzi, L., Duncan, J., Miller, E.K., and Desimone, R. 1998. Responses of neurons in inferior temporal cortex during memory-guided visual search. *J Neurophysiol* 80:2918–40.

Colby, C.L., and Duhamel, J.R. 1996. Spatial representations for action in parietal cortex. *Brain Res* 5:105–15.

Colby, C.L., Duhamel, J.R., and Goldberg, M.E. 1993. Ventral intraparietal area of the macaque: anatomic location and visual response properties. *J Neurophysiol* 69:902–14.

———. 1996. Visual, presaccadic, and cognitive activation of single neurons in monkey lateral intraparietal area. *J Neurophysiol* 76:2841–52.

Colby, C.L., and Goldberg, M.E. 1999. Space and attention in parietal cortex. *Annu Rev Neurosci* 22:319–49.

Constantinidis, C., and Steinmetz, M.A. 1996. Neuronal activity in posterior parietal area 7a during the delay periods of a spatial memory task. *J Neurophysiol* 76:1352–55.

———. 2001. Neuronal responses in area 7a to multiple stimulus displays: II. responses are suppressed at the cued location. *Cereb Cortex* 11:592–97.

Cook, E.P., and Maunsell, J.H. 2002. Attentional modulation of behavioral performance and neuronal responses in middle temporal and ventral intraparietal areas of macaque monkey. *J Neurosci* 22:1994–2004.

Crick, F. 1984. Function of the thalamic reticular complex: the searchlight hypothesis. *Proc Nat Acad Sci USA* 81:4586–90.

Culham, J.C., and Valyear, K.F. 2006. Human parietal cortex in action. *Curr Opin Neurobiol* 16:205–12.

Desimone, R., Albright, T.D., Gross, C.G., and Bruce, C. 1984. Stimulus-selective properties of inferior temporal neurons in the macaque. *J Neurosci* 4:2051–62.

Desimone, R., Schein, S.J., Moran, J., and Ungerleider, L.G. 1985. Contour, color and shape analysis beyond the striate cortex. *Vis Res* 25:441–52.

DeYoe, E.A., Carman, G.J., Bandettini, P., Glickman, S., Wieser, J., Cox, R., Miller, D., and Neitz, J. 1996. Mapping striate and extrastriate visual areas in human cerebral cortex. *Proc Nat Acad Sci USA* 93:2382–86.

DeYoe, E.A., and Van Essen, D.C. 1988. Concurrent processing streams in monkey visual cortex. *Trends Neurosci* 11:219–26.

Dolan, R.J., and Vuilleumier, P. 2003. Amygdala automaticity in emotional processing. *Ann N Y Acad Sci* 985:348–55.

Downing, P.E., Chan, A.W., Peelen, M.V., Dodds, C.M., and Kanwisher, N. 2006. Domain specificity in visual cortex. *Cereb Cortex* 16:1453–61.

Downing, P.E., Jiang, Y., Shuman, M., and Kanwisher, N. 2001. A cortical area selective for visual processing of the human body. *Science* 293:2470–73.

Duhamel, J.R., Colby, C.L., and Goldberg, M.E. 1998. Ventral intraparietal area of the macaque: congruent visual and somatic response properties. *J Neurophysiol* 79:126–36.

Dusenbery, D.B. 1992. *Sensory Ecology: How Organisms Acquire and Respond to Information*. New York: W.H. Freeman.

Eisner, T., Silberglied, R.E., Aneshansley, D., Carrel, J.E., and Howland, H.C. 1969. Ultraviolet video-viewing: The television camera as an insect eye. *Science* 166:1172–74.

Engel, S.A., Glover, G.H., and Wandell, B.A. 1997. Retinotopic organization in human visual cortex and the spatial precision of functional MRI. *Cereb Cortex* 7:181–92.

Epstein, R., and Kanwisher, N. 1998. A cortical representation of the local visual environment. *Nature* 392:598–601.

Felleman, D.J., and Van Essen, D.C. 1991. Distributed hierarchical processing in the primate cerebral cortex. *Cereb Cortex* 1:1–47.

Gallant, J.L., Connor, C.E., Rakshit, S., Lewis, J.W., and Van Essen, D.C. 1996. Neural responses to polar, hyperbolic, and Cartesian gratings in area V4 of the macaque monkey. *J Neurophysiol* 76:2718–39.

Gattass, R., and Gross, C.G. 1981. Visual topography of striate projection zone (MT) in posterior superior temporal sulcus of the macaque. *J Neurophysiol* 46:621–38.

Gattass, R., Sousa, A.P., and Gross, C.G. 1988. Visuotopic organization and extent of V3 and V4 of the macaque. *J Neurosci* 8:1831–45.

Gnadt, J.W., and Andersen, R.A. 1988. Memory related motor planning activity in posterior parietal cortex of macaque. *Exp Brain Res Exp Hirnforsch* 70:216–20.

Goodale, M.A., and Milner, A.D. 1992. Separate visual pathways for perception and action. *Trends Neurosci* 15:20–25.

Grill-Spector, K., Kushnir, T., Edelman, S., Avidan, G., Itzchak, Y., and Malach, R. 1999. Differential processing of objects under various viewing conditions in the human lateral occipital complex. *Neuron* 24:187–203.

Grill-Spector, K., Kushnir, T., Edelman, S., Itzchak, Y., and Malach, R. 1998. Cue-invariant activation in object-related areas of the human occipital lobe. *Neuron* 21:191–202.

Gross, C.G., Rocha-Miranda, C.E., and Bender, D.B. 1972. Visual properties of neurons in inferotemporal cortex of the Macaque. *J Neurophysiol* 35:96–111.

Hagler, Jr., D.J., Riecke, L., and Sereno, M.I. 2007. Parietal and superior frontal visuospatial maps activated by pointing and saccades. *NeuroImage* 35:1562–77.

Haxby, J.V., Gobbini, M.I., Furey, M.L., Ishai, A., Schouten, J.L., and Pietrini, P. 2001. Distributed and overlapping representations of faces and objects in ventral temporal cortex. *Science* 293:2425–30.

Heide, W., Kurzidim, K., and Kompf, D. 1996. Deficits of smooth pursuit eye movements after frontal and parietal lesions. *Brain* 119 (6):1951–69.

Heinze, H.J., Luck, S.J., Munte, T.F., Gos, A., Mangun, G.R., and Hillyard, S.A. 1994. Attention to adjacent and separate positions in space: an electrophysiological analysis. *Percept Psychophys* 56:42–52.

Horridge, A. 2009. What does an insect see? *J Exp Biol* 212:2721–29.

Ishai, A., Ungerleider, L.G., Martin, A., Schouten, J.L., and Haxby, J.V. 1999. Distributed representation of objects in the human ventral visual pathway. *Proc Nat Acad Sci USA* 96:9379–84.

Kanwisher, N. 2000. Domain specificity in face perception. *Nat Neurosci* 3:759–63.

Kanwisher, N., McDermott, J., and Chun, M.M. 1997. The fusiform face area: A module in human extrastriate cortex specialized for face perception. *J Neurosci* 17:4302–11.

Kastner, S., De Weerd, P., Desimone, R., and Ungerleider, L.G. 1998. Mechanisms of directed attention in the human extrastriate cortex as revealed by functional MRI. *Science* 282:108–11.

Kastner, S., De Weerd, P., Pinsk, M.A., Elizondo, M.I., Desimone, R., and Ungerleider, L.G. 2001. Modulation of sensory suppression: implications for receptive field sizes in the human visual cortex. *J Neurophysiol* 86:1398–1411.

Kastner, S., Pinsk, M.A., De Weerd, P., Desimone, R., and Ungerleider, L.G. 1999. Increased activity in human visual cortex during directed attention in the absence of visual stimulation. *Neuron* 22:751–61.

Kastner, S., and Ungerleider, L.G. 2000. Mechanisms of visual attention in the human cortex. *Annu Rev Neurosci* 23:315–41.

Kirschfeld, K. 1976. The resolution of lens and compound eyes. In *Neural Principles in Vision*, eds. F. Zettler, and R. Weiler, pp. 354–70. Berlin: Springer-Verlag.

Konen, C.S., and Kastner, S. 2008a. Two hierarchically organized neural systems for object information in human visual cortex. *Nat Neurosci* 11:224–31.

———. 2008b. Representation of eye movements and stimulus motion in topographically organized areas of human posterior parietal cortex. *J Neurosci* 28:8361–75.

Kourtzi, Z., and Kanwisher, N. 2000. Cortical regions involved in perceiving object shape. *J Neurosci* 20:3310–18.

———. 2001. Representation of perceived object shape by the human lateral occipital complex. *Science* 293:1506–9.

Land, M.F. 1981. Optics and vision in invertebrates. In *Handbook of Sensory Physiology: Comparative Physiology and Evolution of Vision in Invertebrates*, Vol. 7 6B, ed. H. Autrum, pp. 471–593. Berlin: Springer-Verlag.

Lavie, N., and Tsal, Y. 1994. Perceptual load as a major determinant of the locus of selection in visual attention. *Percept Psychophys* 56:183–97.

Lee, J., Williford, T., and Maunsell, J.H. 2007. Spatial attention and the latency of neuronal responses in macaque area V4. *J Neurosci* 27:9632–37.

Levy, I., Schluppeck, D., Heeger, D.J., and Glimcher, P.W. 2007. Specificity of human cortical areas for reaches and saccades. *J Neurosci* 27:4687–96.

Lehky, S.R., and Sereno, A.B. 2007. Comparison of shape encoding in primate dorsal and ventral visual pathways. *J Neurophysiol* 97:307–19.

Lu, Z.L.D.B.A. 1998. External noise distinguishes attention mechanisms. *Vis Res* 38:1183–98.

Luck, S.J., Chelazzi, L., Hillyard, S.A., and Desimone, R. 1997. Neural mechanisms of spatial selective attention in areas V1, V2, and V4 of macaque visual cortex. *J Neurophysiol* 77:24–42.

Malach, R., Levy, I., and Hasson, U. 2002. The topography of high-order human object areas. *Trends Cogn Sci* 6:176–84.

Malach, R., Reppas, J.B., Benson, R.R., Kwong, K.K., Jiang, H., Kennedy, W.A., Ledden, P.J., Brady, T.J., Rosen, B.R., and Tootell, R.B. 1995. Object-related activity revealed by functional magnetic resonance imaging in human occipital cortex. *Proc Nat Acad Sci USA* 92:8135–39.

Martinez, A., Anllo-Vento, L., Sereno, M.I., Frank, L.R., Buxton, R.B., Dubowitz, D.J., Wong, E.C., Hinrichs, H., Heinze, H.J., and Hillyard, S.A. 1999. Involvement of striate and extrastriate visual cortical areas in spatial attention. *Nat Neurosci* 2:364–69.

McAdams, C.J., and Maunsell, J.H. 1999. Effects of attention on orientation-tuning functions of single neurons in macaque cortical area V4. *J Neurosci* 19:431–41.

McAlonan, K., Cavanaugh, J., and Wurtz, R.H. 2008. Guarding the gateway to cortex with attention in visual thalamus. *Nature* 456:391–94.

McMains, S.A., Fehd, H.M., Emmanouil, T.A., and Kastner, S. 2007. Mechanisms of feature- and space-based attention: response modulation and baseline increases. *J Neurophysiol* 98:2110–21.

Mehta, A.D., Ulbert, I., and Schroeder, C.E. 2000. Intermodal selective attention in monkeys. I: distribution and timing of effects across visual areas. *Cereb Cortex* 10:343–58.

Merchant, H., Battaglia-Mayer, A., and Georgopoulos, A.P. 2001. Effects of optic flow in motor cortex and area 7a. *J Neurophysiol* 86:1937–54.

Moll, J., de Oliveira-Souza, R., Eslinger, P.J., Bramati, I.E., Mourao-Miranda, J., Andreiuolo, P.A., and Pessoa, L. 2002. The neural correlates of moral sensitivity: a functional magnetic resonance imaging investigation of basic and moral emotions. *J Neurosci* 22:2730–36.

Moore, T., and Armstrong, K.M. 2003. Selective gating of visual signals by microstimulation of frontal cortex. *Nature* 421:370–73.

Morris, J.S., Ohman, A., and Dolan, R.J. 1999. A subcortical pathway to the right amygdala mediating "unseen" fear. *Proc Nat Acad Sci USA* 96:1680–85.

Mourao-Miranda, J., Volchan, E., Moll, J., de Oliveira-Souza, R., Oliveira, L., Bramati, I., Gattass, R., and Pessoa, L. 2003. Contributions of stimulus valence and arousal to visual activation during emotional perception. *NeuroImage* 20:1955–63.

Murata, A., Gallese, V., Luppino, G., Kaseda, M., and Sakata, H. 2000. Selectivity for the shape, size, and orientation of objects for grasping in neurons of monkey parietal area AIP. *J Neurophysiol* 83:2580–2601.

O'Connor, D.H., Fukui, M.M., Pinsk, M.A., and Kastner, S. 2002. Attention modulates responses in the human lateral geniculate nucleus. *Nat Neurosci* 5:1203–9.

Pasupathy, A., and Connor, C.E. 2002. Population coding of shape in area V4. *Nat Neurosci* 5:1332–38.

Peelen, M.V., and Downing, P.E. 2005. Within-subject reproducibility of category-specific visual activation with functional MRI. *Hum Brain Mapp* 25:402–8.

Peeters, R., Simone, L., Nelissen, K., Fabbri-Destro, M., Vanduffel, W., Rizzolatti, G., and Orban, G.A. 2009. The representation of tool use in humans and monkeys: Common and uniquely human features. *J Neurosci* 29:11523–39.

Pessoa, L., McKenna, M., Gutierrez, E., and Ungerleider, L.G. 2002. Neural processing of emotional faces requires attention. *Proc Nat Acad Sci USA* 99:11458–63.

Pierrot-Deseilligny, C., Rivaud, S., Gaymard, B., and Agid, Y. 1991. Cortical control of reflexive visually-guided saccades. *Brain* 114 (3):1473–85.

Pinsk, M.A., Doniger, G.M., and Kastner, S. 2004. Push-pull mechanism of selective attention in human extrastriate cortex. *J Neurophysiol* 92:622–29.

Posner, M.I. 1980. Orienting of attention. *Q J Exp Psychol* 32:3–25.

Posner, M.I., Snyder, C.R., and Davidson, B.J. 1980. Attention and the detection of signals. *J Exp Psychol* 109:160–74.

Rees, G., Frith, C.D., and Lavie, N. 1997. Modulating irrelevant motion perception by varying attentional load in an unrelated task. *Science* 278:1616–19.

Roelfsema, P.R., Lamme, V.A., and Spekreijse, H. 1998. Object-based attention in the primary visual cortex of the macaque monkey. *Nature* 395:376–81.

Sakata, H. 2003. The role of the parietal cortex in grasping. *Adv Neurol* 93:121–39.

Schaafsma, S.J., and Duysens, J. 1996. Neurons in the ventral intraparietal area of awake macaque monkey closely resemble neurons in the dorsal part of the medial superior temporal area in their responses to optic flow patterns. *J Neurophysiol* 76:4056–68.

Schaafsma, S.J., Duysens, J., and Gielen, C.C. 1997. Responses in ventral intraparietal area of awake macaque monkey to optic flow patterns corresponding to rotation of planes in depth can be explained by translation and expansion effects. *Vis Neurosci* 14:633–46.

Schlack, A., Hoffmann, K.P., and Bremmer, F. 2002. Interaction of linear vestibular and visual stimulation in the macaque ventral intraparietal area (VIP). *Eur J Neurosci* 16:1877–86.

———. 2003. Selectivity of macaque ventral intraparietal area (area VIP) for smooth pursuit eye movements. *J Physiol* 551:551–61.

Schluppeck, D., Glimcher, P.W., and Heeger, D.J. 2005. Topographic organization for delayed saccades in human posterior parietal cortex. *J Neurophysiol* 94:1372–84.

Schwartz, S., Vuilleumier, P., Hutton, C., Maravita, A., Dolan, R.J., and Driver, J. 2005. Attentional load and sensory competition in human vision: modulation of fMRI responses by load at fixation during task-irrelevant stimulation in the peripheral visual field. *Cereb Cortex* 15:770–86. Epub 2004 Sep 2030.

Sereno, A.B., and Maunsell, J.H. 1998. Shape selectivity in primate lateral intraparietal cortex. *Nature* 395:500–3.

Sereno, M.E., Trinath, T., Augath, M., and Logothetis, N.K. 2002. Three-dimensional shape representation in monkey cortex. *Neuron* 33:635–52.

Sereno, M.I., Dale, A.M., Reppas, J.B., Kwong, K.K., Belliveau, J.W., Brady, T.J., Rosen, B.R., and Tootell, R.B. 1995. Borders of multiple visual areas in humans revealed by functional magnetic resonance imaging. *Science* 268:889–93.

Sereno, M.I., and Huang, R.S. 2006. A human parietal face area contains aligned head-centered visual and tactile maps. *Nat Neurosci* 9:1337–43.

Sereno, M.I., Pitzalis, S., and Martinez, A. 2001. Mapping of contralateral space in retinotopic coordinates by a parietal cortical area in humans. *Science* 294:1350–54.

Sherman, S.M., and Guillery, R.W. 2001. *Exploring the Thalamus*. San Diego: Academic Press.

Shiu, L.P., and Pashler, H. 1995. Spatial attention and vernier acuity. *Vis Res* 35:337–43.

Siegel, R.M., and Read, H.L. 1997. Analysis of optic flow in the monkey parietal area 7a. *Cereb Cortex* 7:327–46.

Silver, M.A., Ress, D., and Heeger, D.J. 2005. Topographic maps of visual spatial attention in human parietal cortex. *J Neurophysiol* 94:1358–71.

Snyder, L.H., Batista, A.P., and Andersen, R.A. 1997. Coding of intention in the posterior parietal cortex. *Nature* 386:167–70.

Spiridon, M., and Kanwisher, N. 2002. How distributed is visual category information in human occipito-temporal cortex? An fMRI study. *Neuron* 35:1157–65.

Spiridon, M., Fischl, B., and Kanwisher, N. 2006. Location and spatial profile of category-specific regions in human extrastriate cortex. *Hum Brain Mapp* 27:77–89.

Srinivasan, M.V. 2010. Honey bees as a model for vision, perception, and cognition. *Annu Rev Entomol* 55:267–84.

Steinmetz, M.A., Motter, B.C., Duffy, C.J., and Mountcastle, V.B. 1987. Functional properties of parietal visual neurons: radial organization of directionalities within the visual field. *J Neurosci* 7:177–91.

Swisher, J.D., Halko, M.A., Merabet, L.B., McMains, S.A., and Somers, D.C. 2007. Visual topography of human intraparietal sulcus. *J Neurosci* 27:5326–37.

Sylvester, R., Josephs, O., Driver, J., and Rees, G. 2007. Visual FMRI responses in human superior colliculus show a temporal-nasal asymmetry that is absent in lateral geniculate and visual cortex. *J Neurophysiol* 97:1495–1502.

Tarr, M.J., and Gauthier, I. 2000. FFA: A flexible fusiform area for subordinate-level visual processing automatized by expertise. *Nat Neurosci* 3:764–69.

Tootell, R.B., Hadjikhani, N., Hall, E.K., Marrett, S., Vanduffel, W., Vaughan, J.T., and Dale, A.M. 1998. The retinotopy of visual spatial attention. *Neuron* 21:1409–22.

Tsao, D.Y., Freiwald, W.A., Knutsen, T.A., Mandeville, J.B., and Tootell, R.B. 2003. Faces and objects in macaque cerebral cortex. *Nat Neurosci* 6:989–95.

Ungerleider, L.G., and Mishkin, M. 1982. Two cortical visual systems. In *Analysis of Visual Behavior*, eds. D.J. Ingle, M.A. Goodale, and R.J.W. Mansfield, pp. 549–86. Cambridge, MA: MIT.

Van Essen, D.C., Felleman, D.J., DeYoe, E.A., Olavarria, J., and Knierim, J. 1990. Modular and hierarchical organization of extrastriate visual cortex in the macaque monkey. *Cold Spring Harb Symp Quant Biol* 55:679–96.

von der Heydt, R., and Peterhans, E. 1989. Mechanisms of contour perception in monkey visual cortex. I. Lines of pattern discontinuity. *J Neurosci* 9:1731–48.

Vuilleumier, P., Armony, J.L., Driver, J., and Dolan, R.J. 2001. Effects of attention and emotion on face processing in the human brain: An event-related fMRI study. *Neuron* 30:829–41.

Vuilleumier, P., Henson, R.N., Driver, J., and Dolan, R.J. 2002. Multiple levels of visual object constancy revealed by event-related fMRI of repetition priming. *Nat Neurosci* 5:491–99.

Vuilleumier, P., Richardson, M.P., Armony, J.L., Driver, J., and Dolan, R.J. 2004. Distant influences of amygdala lesion on visual cortical activation during emotional face processing. *Nat Neurosci* 7:1271–78.

Wandell, B.A., Dumoulin, S.O., and Brewer, A.A. 2007. Visual field maps in human cortex. *Neuron* 56:366–83.

Whalen, P.J., Kagan, J., Cook, R.G., Davis, F.C., Kim, H., Polis, S., McLaren, D.G., et al. 2004. Human amygdala responsivity to masked fearful eye whites. *Science* 306:2061.

Yeshurun, Y., and Carrasco, M. 1998. Attention improves or impairs visual performance by enhancing spatial resolution. *Nature* 396:72–75

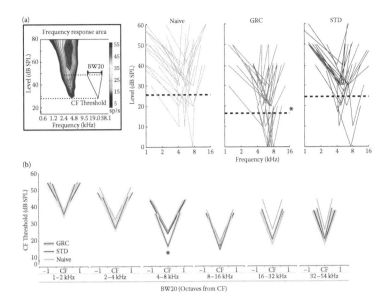

FIGURE 1.2 Tone-onset-to-error learning strategy results in reduced threshold and bandwidth in A1. Frequency response areas (FRA) were constructed for each recording site in A1 to determine threshold and bandwidth across A1 tonotopy (inset). (a) Individual threshold and bandwidth 20 dB SPL above threshold (BW20) at the characteristic frequency (CF) of the cortical site is represented by a "V" shape. Dashed lines show group mean CF threshold of the A1 population tuned within ± 0.5 octaves of the 5.0 kHz signal frequency. Only the GRC group using the TOTE strategy develops frequency-specific increases in neural sensitivity and selectivity. (b) Representational plasticity in threshold and BW20 is evident in the GRC group as the changes are specific only to A1 sites tuned near the signal frequency. Solid lines show group means with shaded areas showing ± standard error. Asterisks mark significant differences from a naive and STD group means. (Reprinted from *Neurobiology of Learning and Memory*, 89, Berlau, K.M., and Weinberger, N.M., Learning strategy determines auditory cortical plasticity, 153–66, 2008, with permission from **Elsevier**.)

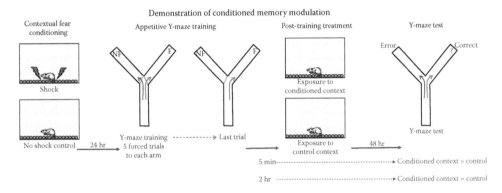

FIGURE 3.5 Conditioned modulation of approach behaviour by aversive stimulation. Rats were placed in a cage with a grid floor and shocked, and alternately into a discriminable cage and not shocked. On the next 2 days the rats were given a total of 10 training trials in a Y-maze. On each trial one arm was blocked and the rats were forced to run to the other arm. They ran to the food and no-food arms five times each; the last run was always to the food arm. After a 5 min delay, the rats in the experimental group were placed into the shock cage without receiving shock (the conditioned context) for 10 min. The rats in the control group were placed into the no-shock context. Another group of each type was placed into the contexts after a 2 hr delay. Then next day all rats were given a Y-maze test with no barriers or food on the maze in order to compare their memory for the location of the food. The rats in the 5 min delay group that had been exposed to the conditioned context made significantly more correct responses than the rats that had been placed into the control context. Context placement had no effect in the 2 hr delay group. See text for discussion of how this finding demonstrates post-training modulation by a conditioned reinforcer (instead of a reinforcer) and how it shows that modulation does not depend on the affective properties of the post-training treatment. (Results based on Holahan, M. R., White, N. M, *Behavioral Neuroscience* 118, 24, 2004.)

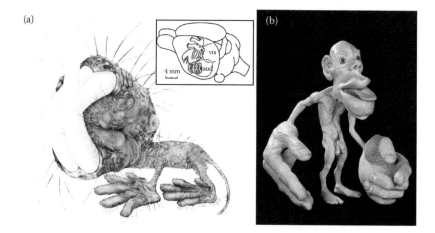

FIGURE 7.2 Illustration of how the body is represented in the rodent and human somatosensory cortex. (a) Naked mole-ratunculus (Adapted from Catania and Remple, *Proc Natl Acad Sci USA* 99, 5692, 2002) and (b) human homunculus (Image from Natural History Museum London, reference no. 1914). *Inset*, mouse brain depicting somatosensory mouse-unculus representations in areas S1 (blue) and S2 (yellow); vis, visual cortex; aud, auditory cortex (Reprinted from Woolsey and Van der Loos, *Brain Res* 17, 205, 1970, with permission from Elsevier).

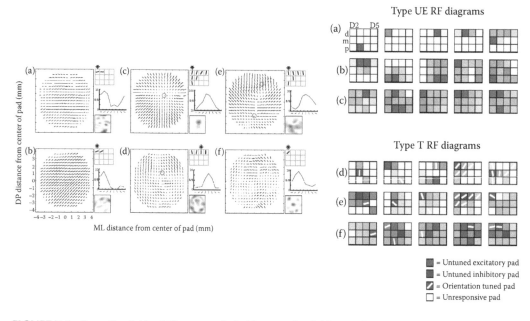

FIGURE 7.5 Receptive fields of S2 neurons. Left side: receptive fields on a single finger pad of six neurons with orientation-tuned responses. For each neuron a vector plot of the local orientation tuning, a map of the tuning across the distal, middle, and proximal pads of digits 2–5, an orientation tuning curve for the neuron, and the best fitting linear receptive field are shown. The left two neurons show position invariant responses that cannot be explained by the linear receptive field. Neurons labeled c, d, and e have orientation-tuned responses that are explained by having one or more inhibitory or excitatory zones. The neuron in panel f has a complex tuning response. (From Thakur, P.H., Fitzgerald, P.J., Lane, J.W., Hsiao, S.S., *J Neurosci* 26, 13571, 2006. With permission.) Right side: These maps show samples of the RFs of 30 neurons in the SII cortex. Each block of 12 panels represents the response from the distal, medial, and proximal pads of digits 2–5 (see top left panel for a key to which pad the blocks correspond to). Red colors indicate excitatory pads, blue colors indicate inhibitory pads, and pads containing a white bar were orientation tuned. (From Fitzgerald, P.J., Lane, J.W., Thakur, P.H., Hsiao, S.S. *J Neurosci* 26, 6490, 2006a.)

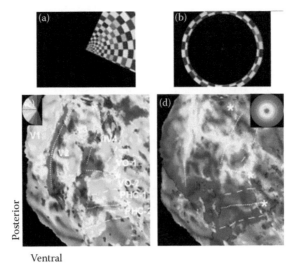

Posterior

Ventral

FIGURE 8.3 (a, b) Examples of visual stimuli used for retinotopic mapping (Swisher et al. 2007). The stimuli consist of chromatic, flickering checkerboards. (a) The rotating wedge is typically used for polar angle mapping. (b) The expanding/contracting ring is typically used for eccentricity mapping. (c, d) Polar angle and eccentricity maps in human ventral visual cortex (Arcaro, McMains, and Kastner 2009). Flattened surface reconstruction of the right hemisphere depicting early (V1, V2, V3) and ventral visual cortex (hV4, VO-1, VO-2, PHC-1, PHC-2) of one representative subject. The color code indicates the phase of the fMRI response and labels the region of the visual field to which the surface node is most responsive to, as depicted in the visual field color legends. The dotted and dashed white lines indicate the upper and lower vertical meridians, respectively. The asterisks indicate foveal representations. (c) Polar angle map. (d) Eccentricity map. hV4 = human V4; VO = ventral occipital; PHC = parahippocampal cortex. (From Arcaro, M.J., McMains, S.A., Singer, B.D., Kastner, S., *J Neurosci* 29, 10638, 2009. With permission.)

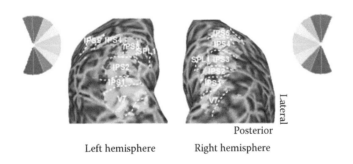

Lateral

Posterior

Left hemisphere   Right hemisphere

FIGURE 8.4 Topographic maps in posterior parietal cortex obtained with the memory-guided saccade task. Briefly, targets appeared at successive locations "around the clock" and were followed by a ring of distracters. The disappearance of the distracters signaled subjects to perform saccadic eye movements towards the remembered target locations. The figure shows activation obtained from one subject overlaid on inflated left and right hemispheres. The color code is shown for voxels whose responses were correlated with the fundamental frequency of saccade direction, $r < .2$, indicating the phase of the response. The responses evoked in the memory-guided saccade task were lateralized such that the right visual field was represented in the left hemisphere, whereas the left visual field was represented in the right hemisphere. Area boundaries were formed by the alternating representation of either the upper (red) or lower (blue) vertical meridian. IPS = intraparietal sulcus. (From Konen, C.S., and Kastner, S. 2008. Representation of eye movements and stimulus motion in topographically organized areas of human posterior parietal cortex. *J Neurosci* 28:8361–75. With permission.)

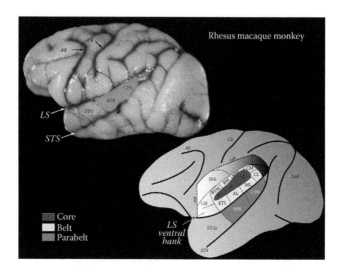

FIGURE 9.2 Location of primate auditory cortex (macaque). Upper figure: location of auditory cortex in macaque. Note that only the parabelt (blue: RPB, CPB) is exposed on the surface of the brain. Lower figure: the parietal and frontal cortices have been graphically cut away to reveal approximate location of core and belt (red and yellow). Abbreviations: LS, lateral sulcus; STS, superior temporal sulcus; AS, arcuate sulcus; CS, central sulcus; STG, superior temporal gyrus; STS, superior temporal sulcus; CIS, circular sulcus; INS, insula; LuS, lunate sulcus. See text for other abbreviations.

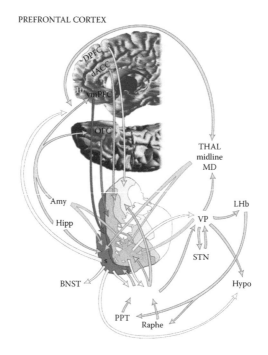

FIGURE 11.1 Schematic illustrating key structures and pathways of the reward circuit, with a focus on the connections of the ventral striatum. Red arrow = striatal input from the vmPFC; dark orange arrow = striatal input from the OFC; light orange arrow = striatal input from the dACC; yellow arrow = striatal input from the dPFC; white arrows = inputs into ventral striatum; gray arrows = outputs from ventral striatum; brown arrows = other non-striatal connections of the reward circuit. Amy = amygdala; BNST = bed nucleus of the stria terminalis; dACC = dorsal anterior cingulate cortex; DPFC = dorsal prefrontal cortex; Hipp = hippocampus; LHb = lateral habenula; Hypo = hypothalamus; MD = mediodorsal nucleus of the thalamus; OFC = orbital frontal cortex; PPT = pedunculopontine nucleus; Raphe = dorsal raphe; S = shell; SN = substantia nigra; STN = subthalamic nucleus; THAL = thalamus; VP ventral pallidum; VTA = ventral tegmental area; vmPFC = ventral medial prefrontal cortex. (Reprinted from Haber, S.N. and Knutson, B., *Neuropsychopharmacology*, 35, 4, 2010.)

FIGURE 11.6 Schematic illustrating the complex connections between the striatum and substantia nigra. The arrows illustrate how the ventral striatum can influence the dorsal striatum through the midbrain dopamine cells. Colors indicate functional regions of the striatum, based on cortical inputs. Midbrain projections from the shell target both the VTA and ventromedial SNc. Projections from the VTA to the shell form a "closed" reciprocal loop, but also project more laterally to impact on dopamine cells projecting the rest of the ventral striatum forming the first part of a feed-forward loop (or spiral). The spiral continues through the striato-nigro-striatal projections through which the ventral striatum impacts on cognitive and motor striatal areas via the midbrain dopamine cells. Red=inputs from the vmPFC; orange=inputs from the OFC and dACC; yellow=inputs from the dPFC; green and blue=inputs from motor control areas. (Reprinted from Haber, S.N., Fudge, J.L. and McFarland, N.R. *J Neurosci*, 20, 2369, 2000.)

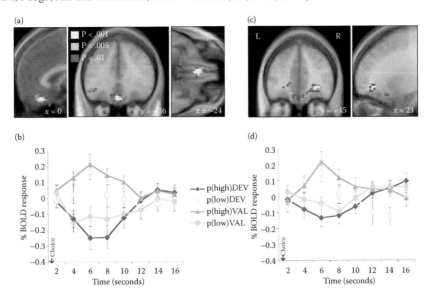

FIGURE 14.3 Regions of vmPFC and OFC showing response properties consistent with action-outcome learning. Neural activity during action selection for reward shows a change in response properties as a function of the value of the outcome with each action. Choice of an action leading to a high probability of obtaining an outcome that had been devalued (*p*(high)DEV) led to a decrease in activity in these areas whereas choice of an action leading to a high probability of obtaining an outcome that was still valued led to an increase in activity in the same areas. Devaluation was accomplished by means of feeding the subject to satiety on that outcome prior to the test period. (a) A region of medial OFC showing a significant modulation in its activity during instrumental action selection as a function of the value of the associated outcome. (b) Time-course plots derived from the peak voxel (from each individual subject) in the mOFC during trials in which subjects chose each one of the four different actions (choice of the high- vs. low-probability action in either the valued or devalued conditions). (c) A region of the right central OFC also showed significant modulation during instrumental conditioning. (d) Accompanying time-course plots from central OFC are shown. (Data from Valentin, V.V., Dickinson, A., and O'Doherty, J.P., *J Neurosci*, 27, 4019, 2007.)

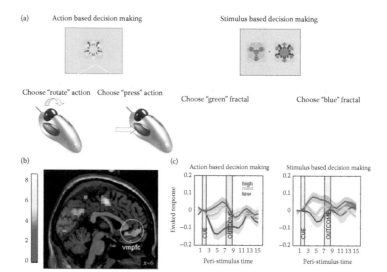

FIGURE 14.4 Expected reward representations in vmPFC during action-based and stimulus-based decision making. (a) Illustration of action-based decision-making task in which subjects must choose between one of two different physical actions that yield monetary rewards with differing probabilities, and a stimulus-based decision-making task in which subjects must choose between one of two different fractal stimuli randomly assigned to either the left or right of the screen, which yield rewards with differing probabilities. (b) A region of vmPFC correlating with expected reward during *both* action- and stimulus-based decision making. (c) Time courses from vmPFC during the action- and stimulus-based decision-making tasks plotted as a function of model-predicted expected reward from low (blue) to high (red). (From Glascher, J., Hampton, A.N., and O'Doherty, J.P., *Cereb Cortex*, 19, 483, 2009. With permission.)

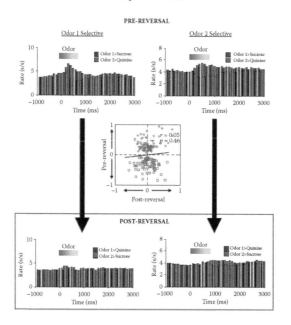

FIGURE 15.2 Flexibility of associative encoding in orbitofrontal cortex. Population response of neurons in orbitofrontal cortex identified as cue-selective during learning. Average activity per neuron is shown, synchronized to odor onset, during and after reversal. Inset scatter plot (middle part of figure) compares the cue-selectivity indices before (*y* axis) and after (*x* axis) reversal for all the cue-selective neurons used to construct the population histograms. Blue and red symbols show data for "Odor 1 Selective" neurons and "Odor 2 Selective" neurons, respectively. Unlike the single-unit example in Figure 15.1, the population responses failed to reverse and the cue-selectivity indices showed no correlation. (Adapted from Stalnaker, T.A. et al.: Abnormal associative encoding in orbitofrontal neurons in cocaine-experienced rats during decision-making. *Eur J Neurosci* 2006, 24, 2643, Copyright Wiley-VCH Verlag GmbH & Co. KGaA. With permission.)

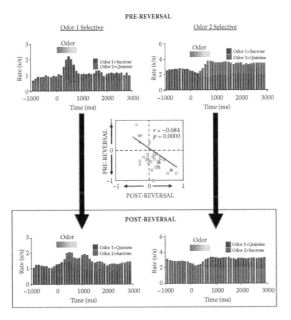

FIGURE 15.3 Flexibility of associative encoding in basolateral amygdala. Population response of neurons in basolateral amygdala identified as cue-selective during learning. Average activity per neuron is shown, synchronized to odor onset, during and after reversal. Inset scatter plot compares the cue-selectivity indices before (y axis) and after (x axis) reversal for all the cue-selective neurons used to construct the population histograms. Blue and red symbols show data for "Odor 1 Selective" neurons and "Odor 2 Selective" neurons, respectively. In contrast to similar data from orbitofrontal cortex in Figure 15.2, the population responses reversed, and the cue-selectivity indices showed a strongly significant inverse correlation. (Adapted from Stalnaker, T.A. et al., *Nat Neurosci,* 10, 949, 2007. With permission.)

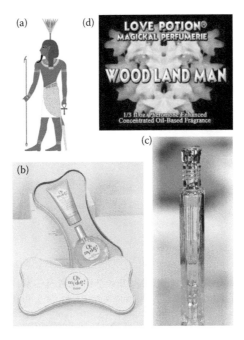

FIGURE 17.1 A limited pictorial survey of perfume and its vessels. (a) Nefertum, the Egyptian God of perfume and luck. (From file "Nefertem.svg" by Wikimedia Commons user Jeff Dahl. File at http://commons.wikimedia.org/wiki/File:Nefertum.svg.) (b) *Oh My Dog!* cologne for dogs by Etienne de Swardt. (From Elizabeth duPar at www.ohmydogstore.com. With permission.) (c) Marie Antoinette's perfume bottle *Eau de Sillage*. (Permission grated by original author. For copyright details of this image please contact chapter author Rachel Herz.) (d) *Love Potion* perfume explicitly marketed as a potent human sex pheromone. (From Love Potion Woodland Man; artwork by Ravon, Mara Fox at LovePotionPerfume.com. With permission.)

FIGURE 18.1 Visual aesthetics recapitulates visual processing: a hierarchical progression as documented through the brushstrokes of three renowned artists. Each painting depicts the female figure, through successive levels of representational complexity, from Malevich (early) and Picasso (intermediate) to Titian (late). (a) *Torso*, 1928–32 (oil on canvas) by Kazimir Severinovich Malevich (1878–1935). (Reproduced from *Torso*, 1928–32 (oil on canvas) by Kazimir Severinovich Malevich (1878–1935) State Russian Museum, St. Petersburg, Russia/Bridgeman Art Library. With permission.) (b) *Weeping Woman*, 1937 (oil on canvas) by Pablo Picasso (1881–1973). © 2010 Estate of Pablo Picasso/Artists Rights Society (ARS). (With permission: Reproduction, including downloading of Picasso works, is prohibited by copyright laws and international conventions without the express written permission of Artist Rights Society (ARS), New York.) (c) *Venus of Urbino*, 1538 (oil on canvas) by Titian (Tiziano Vecellio) (c. 1488–1576).

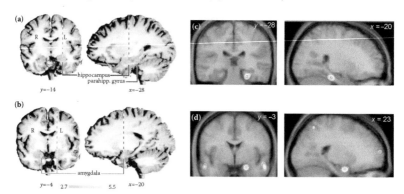

FIGURE 19.1 Changes in regional cerebral blood flow during unpleasant (dissonant) and pleasant music. (a and b) Blood oxygenation signal increases associated with listening to unpleasant dissonant music: activation was detected in the hippocampus, parahippocampal gyrus, and amygdala. (Modified from Koelsch, S. et al.: Investigating emotion with music: An fMRI study. *Human Brain Mapping*. 27, 239, 2006. Copyright Wiley-VCH Verlag GmbH & Co. KGaA. With permission.) (c) Increases in cerebral blood flow in the parahippocampal area associated with unpleasant dissonant music. (Modified by permission from Macmillan Publishers Ltd. [*Nature Neuroscience*] Blood, A.J. et al. 2, 382, copyright 1999.) (d) Decreases in cerebral blood flow in amygdala associated with highly pleasurable music. (Modified from Blood, A.J., Zatorre, R.J., *Proc Natl Acad Sci*, 98, 11818, 2001.) Note similarity across studies in recruitment of parahippocampal region (a and c) to negatively valenced music; and in activation or deactivation of amygdala to negatively or positively valenced music, respectively (b and d).

9 Sound

Corrie R. Camalier and Jon H. Kaas

CONTENTS

9.1 INTRODUCTION

From the piercing wail of an ambulance to the soothing melody of a Mozart sonata, it is clear that sound carries enormous meaning in daily life. Sound can be a reinforcing stimulus used to guide flexible behavior—as such, its meaning is often dependent on contextual cues. For example, the ringing of a bell at different times in one's career may mean you are late for class or that your speaker time at a conference is over. Also, what we believe to be the inherent pleasing or aversive properties of sound are often counterintuitive. For example, psychophysical work suggests that it is the low-frequency spectral components that lead to the "chilling" effect of fingernails scratching on a blackboard, not the high frequencies that most people associate with the unpleasant nature of that sound (Halpern et al. 1986).

This chapter will focus on cortical aspects of sound processing in primates. However, due to the incomplete nature of primate literature, conclusions based on data from other species will be discussed. Insights from other mammalian species (i.e., cats, ferrets, rats) will be used to shape hypotheses. It is worth noting that important species differences exist in the auditory system, even within primates. For example, many species of Old and New World monkeys (e.g., macaques, capuchins) prefer silence over music (McDermott and Hauser 2007) (also see Chapter 19 in this volume). Another example is that language comprehension is something at which humans uniquely excel.

However, as the specialized anatomy is not well understood, we will not review the pathways dedicated to language processing and production in humans.

9.2 EVOLUTIONARY CONSIDERATIONS: COMPARISON ACROSS MAMMALIAN TAXA

Sound has a necessarily complex role in guiding behavior and the systems that subserve this role have had a long time to evolve. To give some context for understanding the functions of the auditory system, we first briefly cover the evolutionary development of this system in mammals, and particularly primates.

The part of the auditory system that is most varied in mammals is auditory cortex (for review see Kaas and Hackett 2008). While reptiles have a dorsal cortex that is homologous to neocortex, this dorsal cortex does not have auditory inputs. Instead, the projections of the auditory thalamus are subcortical. Yet, all studied mammals have a region of temporal cortex that gets inputs from the auditory thalamus and is responsive to auditory stimuli. Thus, early mammals or the ancestors of mammals somehow acquired direct thalamic auditory projection to cortex. Most studied mammals have several areas of auditory cortex, including two or three primary or primary-like areas that are characterized by direct inputs from the tonotopically organized ventral nucleus of the medial geniculate complex (MGC), MGv, and are in turn also tonotopically organized. In addition, these primary fields are surrounded by a belt of secondary auditory areas, and additional, higher-level areas of auditory or multisensory processing may be present. For now, conclusions about how auditory cortex varies in organization across mammals need to be limited, as species in some of the major branches of mammalian evolution have not been studied, and studies have been few and incomplete for species in other major branches. Thus, we do not yet know how auditory cortex is organized in the major clades of Monotremata (platypus and echidna), Afrotheria (elephant, tenrec, etc.), and Xenarthra (armadillo, sloth, and anteater), and little is known about Marsupialia, except that possums have at least one primary area. Only one primary area has been demonstrated in Scandentia (tree shrews), which are close relatives of primates.

More progress has been made in studies of Carnivora (cats, dogs, ferrets), where at least two primary areas exist: an anterior auditory field (AAF) and the classical primary field (A1). A posterior field (P or PAF) has some of the characteristics of a primary field, and seven or eight secondary fields have been described. With more than one field having primary area features, identifying the same (homologous) areas across members of different orders of mammals has been challenging, but the characteristics of relative position, connections, architecture, and tonotopic organization have been helpful. While terms have been applied inconsistently, it now appears that rodents have primary AAF and A1 areas, and possibly a posterior field, that are homologous to those in carnivores, and comparable AAF and A1 areas have been identified in bats.

Primates also have three primary fields: a posterior A1, an anterior "rostral" area, R, and a more anterior rostrotemporal area, RT. Areas A1 and R share a common border representing low frequencies in primates, and A1 and AAF adjoin along a high-frequency representation in carnivores, rodents, and bats, so it is not clear that A1 and R in primates correspond to A1 and AAF in other mammals. One possibility is that if the expansion of the temporal lobe in primates rotated R from the posterior to the anterior border of A1, then AAF in carnivores could correspond to primate belt area CM. Correspondingly, both A1/AAF and A1/CM share a high-frequency border. Given these uncertainties about the identities of primary areas and the limited comparative evidence, there is little understanding of what secondary areas may be homologous, if any, across mammalian taxa. For now, we can surmise that early mammals had at least one primary area and a bordering secondary area or areas, and this organization has been partly retained, but variously elaborated, in the major lines leading to extant mammals. A likely common feature is that some of the secondary areas are bi- or multisensory.

In addition to taxa of mammals varying in the arrangement and number of areas of auditory cortex, auditory systems are sometimes obviously specialized in order to mediate unusual abilities.

For example, echolocating bats have auditory systems that have expanded representations of the sound frequencies used in echolocation as well as specialized areas of auditory cortex for analyzing specific aspects of the echoes to provide information about target distance, size, and nature. The auditory systems of mice have specialized sectors devoted to the ultra-high frequencies used in social communication. Moreover, humans have greatly specialized auditory cortex of the left cerebral hemisphere for the processing of language.

9.3 PROPERTIES OF SOUND AND ETHOLOGICAL CONSIDERATIONS

To understand some basic properties of sound, consider a marmoset monkey twittering a "love song" to its mate in the dense tree cover of a Brazilian rain forest. It is plausible to suggest that this call is a rewarding stimulus, and it serves as a useful example. The first important question one has to ask (if one is a marmoset): Who is she? Identity cues, such as spectral frequency structure (i.e., which frequencies are in the call) and temporal modulation rates (i.e., how the amplitudes of the frequencies change in time), give rise to complex percepts such as pitch and harmonicity. These percepts, combined with other systems such as emotion and memory systems, in turn give rise to meaning and speaker identity (Moore 1997).

The second important question is: Where is she? Location cues include loudness (is she getting louder/approaching?), frequency structure (the outer ear, or pinna, filters sound in particular ways depending on their vertical location), and differences between the two ears in intensity (interaural intensity differences: IID) and time (interaural time differences: ITD). These features give rise to cues about direction of motion, vertical location, and horizontal location, respectively.

No animal processes sound outside of the confines of its surrounding environment, so ethology must be taken into consideration. Any forest, grassland, or classroom is like an auditory hall of mirrors, absorbing, reflecting, and distorting sound in characteristic ways (reviewed in Fitch 2002; Hauser 1996). The acoustic environment presents numerous challenges to the detection and discrimination of sounds, such as degradation (frequency-dependent attenuation, reverberation, and irregular amplitude fluctuations), and the levels and quality of ambient noise in the surroundings.

These degradations and distortions affect sound differently in various environments. For example, in a forest biome, reverberation off objects (such as trees) is more severe than in an open habitat, and reverberation effects are stronger for higher temporal modulation frequencies. Thus, the higher temporal frequencies of amplitude- and frequency-modulated sounds will be masked in a closed environment. In contrast, in open environments amplitude fluctuations from atmospheric inhomogeneities are more likely to be a factor. These inhomogeneities are generally less than 50 Hz and so will affect low frequencies of temporal modulation of the sound (reviewed in Brown and Handford 2000; Wiley and Richards 1978).

In both environments, spectral frequency-dependent degradation gets worse with increasing spectral frequency, but for the forest environments, there appears to be a low-frequency pass-band window for which sound passes with less attenuation (Brown 1989; Morton 1975). This may be due to lower-frequency sound bouncing off the canopy or off the canopy's midday thermal gradient. Thus, open environments propagate sound best with low spectral frequencies and high temporal modulation, and closed environments propagate sound best with low spectral frequencies (especially in the pass-band window) with slower rates of temporal modulation. Frequency-dependent ground attenuation is also a factor, but is similar in both environments. At 1 m above the earth, sounds in the range of 300–3000 Hz are attenuated the greatest. The higher the source is, the less attenuation occurs, especially at higher frequencies (Wiley and Richards 1978). Lastly, the source and frequency content of background masking noise is different for different habitats (Brown and Waser 1988).

The sound transmission effects described above are considerations in the design of communication sounds. The constraints of ecology on communication sounds have been well explored in fields such as birdsong. Songbirds are largely believed to have characteristic, highly structured calls. Numerous studies have looked at the correlation of habitat on birdsong and have shown numerous species whose calls operate within restricted ranges of spectral and temporal frequency. Acoustic environmental effects seem to show selection pressures, even at surprisingly short timescales. For example, the habitat for the California White-crowned Sparrow went from grassland to scrub over a 35-year period. In this time, the birds' trill modulation rates also decreased (Derryberry 2009), presumably reflecting adaptation to increased selection pressure for better penetrating sound.

What can ethological considerations tell us about the constraints on the vocal behavior of primates? An influential hypothesis is that animals that do not have a complex social structure will communicate in a nongraded system (e.g., Marler 1975). A nongraded system is characterized by large feature distances between exemplars, which make it easier to distinguish between calls, even degraded ones. In contrast, the hypothesis predicts that animals that have close contact and easy visual access, such as in grassland, will develop graded vocalizations, where the calls have variability in a given exemplar. For these animals, there may be only subtle differences between different calls. This hypothesis also makes a second prediction that most long-range communication calls should exhibit a nongraded structure to counteract environmental degradation.

Consistent with this hypothesis, certain species of primates with complex social structures, such as macaques, exhibit graded vocalizations. However, even long-range calls, such as shrill barks, are graded when they are predicted to be nongraded (Fischer and Hammerschmidt 2002). What explains this discrepancy? Macaques live in highly variable habitats, from forest to semidesert, and many live in villages and towns. Because they need to be highly adaptable, we can conclude that ecological acoustics are not a primary factor in the evolution of production of the macaque vocal repertoire. However, there is at least one example where graded macaque calls appear to be optimized for long-range communication. Lost calls in toque macaques come in two basic types: one that is long duration and low frequency, better for long-distance propagation, and the other that is higher frequency with shorter repetitions, better to transmit location cues based on differences between the two ears (Dittus 1988).

For macaques it appears that Marler's predictions are of limited usefulness. This is also true for other species of primates, both in Old World monkeys such as baboons (Fischer et al. 2001) and for New World arboreal species such as marmosets, squirrel monkeys, and tamarins (Hauser 1996; Schrader and Todt 1993). Given the risk of ambiguities introduced by sound degradation in a graded repertoire, why communicate in such a system? Precisely because they are graded, these repertoires may be able to carry more information (Hauser 1996). It appears that graded differences between similar calls are often the mark of individual voices (Hammerschmidt and Todt 1995). In monkeys, spectral peak patterns or differences in spectral composition in certain vocalizations help identify individual voices (Hauser and Fowler 1992; Rendall, Owren, and Rodman 1998; Rendall, Rodman, and Emond 1996). These are likened to human vocal tract resonances and are supposed to cue individual identity and morphology (e.g., size, gender) (Ghazanfar et al. 2007).

Following the school of thought pioneered by Barlow, the auditory system can be understood as a system that evolved to process behaviorally relevant sounds (Barlow 1961). To understand the mechanisms by which we process complex sounds to guide behavior, a natural place to start is by understanding the neural mechanisms of sound processing, including how these pathways interact with reward and emotional systems in the brain. From the earlier marmoset example we can see that both auditory object identity and location rely on partially overlapping sets of cues. The spectral frequency structure of the sound is important for both, and the changes of amplitude and frequency content over time are important for both. Auditory cues are critically time dependent, so a hallmark of the auditory system is its highly parallel nature. Thus, the architecture of auditory processing is characterized by multiple interacting streams even at its earliest levels. Also, context plays a large role in the relative importance of processing identity or space. Sometimes you need to know if it is your mate. At other times it is more important to know that an object is on a collision path with you than to specifically identify what it is you are dodging!

This chapter aims to review the main features and pathways of the auditory system in order to provide a foundation for the other chapters on reward. It describes auditory processing as sound waves hit the cochlea, travel up through the nuclei of the brainstem, further disseminate in multiple cortical streams, and finally arrive at associative areas such as the prefrontal and orbitofrontal cortices. Processing streams that have been identified anatomically are described, and physiological properties and possible functions are described where data are available. We then discuss how this system could interact with reward and limbic systems, providing examples where reward-based information influences auditory processing, and explore possible anatomical underpinnings of such activity. Because perceptual plasticity is so often based on reward, we will also briefly touch on the effects reward has on the processing of auditory stimuli in the context of learning. Lastly, we cover areas where auditory information is processed in some reward-based and limbic structures.

9.4 AUDITORY PROCESSING STREAMS AND PATHWAYS IN THE PRIMATE BRAIN

9.4.1 EARLY AUDITORY PROCESSING: THE PATH TO CORTEX

9.4.1.1 Cochlea to the Inferior Colliculus

The most obvious and possibly most fundamental organizing principle of the auditory system is tonotopy, an orderly representation of sound frequency across a one-dimensional space. Tonotopy is first established at the level of the sensory epithelium (the cochlea). When sound waves hit the spiraled structure of the mammalian cochlea, the basilar membrane splits the sound into its frequency components. The basilar membrane has graded stiffness along its length, so wave amplitude changes in a frequency-dependent manner along the basilar membrane. Higher frequencies stimulate inner hair cells at the closest portion of the membrane (the base) and lower frequencies stimulate inner hair cells at the farthest portion (the apex). In mammals, the length of the basilar membrane is related to the range of high and low frequencies an animal can hear (West 1985). Thus, the cochlea establishes tonotopy, an organizational principle preserved though the levels of auditory processing through auditory cortex.

Responses from the cochlea project via the eighth nerve to the cochlear nucleus (Figure 9.1). The cochlear nucleus projects to structures in the superior olivary complex (SOC) such as the medial superior olive (MSO), the lateral superior olive (LSO), and the medial nucleus of the trapezoid body (MNTB) (reviewed in Pickles 1988; Rouiller 1997). It also projects to the nuclei of the lateral lemniscus. Again, each of these structures maintains a basic tonotopy established by the cochlea.

Response properties of these structures show that neurons still faithfully represent sound by encoding spatial frequency at very high resolution. Neurons in these structures also demonstrate temporal modulation rate tuning (or how fast the neuron can synchronize with the temporal structure of the sound) at high rates. Nuclei of the superior olivary complex are especially important for the encoding of sound location, as they are the first place where ascending information from the two ears is combined (for review, see Kelly et al. 2002). From these structures and nuclei of the lateral lemniscus, responses reach the inferior colliculus (IC) of the midbrain. The inferior colliculus can be divided into two major portions: the central and external nuclei. Investigators also commonly distinguish the dorsal cortex, the dorsoventral nucleus, and the pericentral nucleus of the inferior colliculus.

The central nucleus (ICc) is considered to be the main relay nucleus of the inferior colliculus. It is tonotopically organized and receives a direct projection from the lateral lemniscus. Responses are tightly tuned to tones, modulation rate encoding shows synchronization up to 120 Hz, and response latencies are generally short (Langner and Schreiner 1988; Ryan and Miller 1978). In contrast, the external nucleus (ICx) is not tonotopically organized, its neurons have longer latencies, and it receives most of its inputs from sources other than the lateral lemniscus. Thus, there are two pathways: a fast, direct, tonotopically organized pathway (lemniscal pathway through the ICc) and a slower, indirect, nontonotopically organized pathway (nonlemniscal pathways through the ICx).

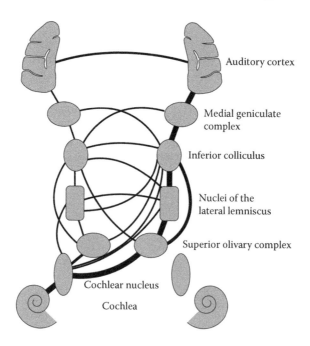

Auditory cortex

Medial geniculate
complex

Inferior colliculus

Nuclei of the
lateral lemniscus

Superior olivary complex

Cochlear nucleus

Cochlea

FIGURE 9.1 Major ascending connections of the subcortical auditory system. Selected ascending pathways from the cochlea to auditory cortex; major pathways are shown in thick lines. Divisions of subcortical nuclei are indicated in text.

When compared to other sensory systems, the proliferation of brainstem nuclei in the auditory system is striking. The exact significance of this is not clear, but it may allow for more parallel processing, allowing greater and faster stimulus processing early to transform a simple one-dimensional representation of spectral frequency in time into complex percepts in space and time. This element of parallel processing is one of the hallmarks of the auditory system and is an architecture best suited to processing stimuli that occur on a very fast timescale, as in audition.

9.4.1.2 The Auditory Thalamus

Lying just medial and posterior to the lateral geniculate of the visual system, the MGC is a small and heterogeneous thalamic structure. The MGC is characterized as the primary feedforward auditory division because its inputs are dominated by the inferior colliculus (Jones 2003; Winer, Wenstrup, and Larue 1992). There is a multiplicity of pathways from the cochlea to the thalamus, but the MGC is an obligatory relay of auditory information into auditory cortex. Thus, it is useful to spend some time describing the organization and response properties of neurons in the MGC, since cortical responses can best be understood in the light of their thalamic inputs.

While the functional organization of the MGC has not been extensively explored, especially in primates, a general picture based on connectivity and microelectrode studies is emerging (reviewed in de Ribaupierre 1997). The MGC of primates consists of at least three main divisions: ventral (MGv), dorsal (MGd), and medial (MGm) (n.b., more than three divisions are distinguished in other mammalian species, such as cats). Based partly on the paucity of data, the nature and specialization of these divisions has been a matter of speculation for some time. Very early, Poljak (1926) posited that the MGv aided in localization and the MGd was involved in the discrimination of sounds. Later, Evans advanced a similar idea that the MGv was involved in localization and the MGd was involved in pattern recognition (Evans 1974). The current understanding of the MGC is that the divisions perhaps do not divide function so cleanly. What has become clear is that these divisions have different input connections and internal architecture, leading to neurons with different spectral frequency tuning properties, modulation rate tuning, response latencies, and sometimes multisensory properties.

The first division of the MGC, the MGv, receives tonotopically organized projections from both ipsi- and contralateral ICc, but ipsilateral input is stronger. This leads to a structure that is itself tonotopically organized. Neurons respond well to pure tones and are generally narrowly tuned to spectral frequency–they respond best to a small range of frequencies even at high intensities, perhaps only a quarter of an octave (Allon, Yeshurun, and Wollberg 1981; Calford 1983). Response latencies are quite short. In addition, temporal modulation rate tuning indicates that the responses of these neurons can follow and distinguish very rapid rates of stimulation (Allon, Yeshurun, and Wollberg 1981; Wang et al. 2008). In terms of selectivity to sound identity or location, the majority are primarily excited by sound coming from the contralateral ear, and sensitive to difference cues such as interaural intensity and time differences (Barone et al. 1996; Calford 1983; Starr and Don 1972). These MGv neurons are not selective for particular vocal stimuli (Symmes, Alexander, and Newman 1980), evidence that complex sound identity cues, such as call type, are not distinguished at this level.

A second division of the auditory thalamus, the MGd, receives most of its input from noncentral portions of the inferior colliculus. There is little evidence of tonotopy in MGd (but see Gross, Lifschitz, and Anderson 1974) and its neurons are generally poorly responsive to pure tones, with broad or multipeaked spectral frequency tuning. These neurons have long response latencies, consistent with their inputs from noncentral collicular nuclei. However, MGd neurons exhibit robust responses to complex sounds (Allon, Yeshurun, and Wollberg 1981; Calford 1983; He and Hu 2002). An important caveat is that MGd can be subdivided further in some species. It has been suggested that this region has two divisions in primates: anterior (MGad) and posterior (MGpd) (reviewed in Jones 2003). It is possible that the response properties differ between the two subdivisions, and it is suspected that the MGad may in fact be tonotopically organized and have neurons with short latencies, resembling MGv neurons.

A third division of the auditory thalamus, the MGm, receives inputs from both the central and external divisions of the inferior colliculus. MGm also receives significant projections from vestibular nuclei, the spinal cord, and the superior colliculus (SC) (Calford and Aitkin 1983; Rouiller 1997; Winer, Wenstrup, and Larue 1992). Connectivity of neurons within this nucleus may be highly variable, as MGm neurons project to all three core belt and parabelt regions of auditory cortex, as well as to other regions. There is also evidence that different cell classes within MGm also project to different cortical layers (Hashikawa et al. 1995; Molinari et al. 1995). There may be tonotopy in the rostral division of the MGm (Rouiller et al. 1989), but for the most part tonotopy through the entire structure is lacking. Much like the MGd, neurons are broadly spectrally tuned and are often multipeaked to tone stimuli. Response latencies in the MGm are also variable (Allon, Yeshurun, and Wollberg 1981; Calford 1983) and, consistent with its heterogeneous inputs, there is evidence for neurons with multisensory responses in at least some nonprimate species (e.g., Calford and Aitkin 1983; LeDoux et al. 1987; Rouiller et al. 1989).

There are other auditory-related nuclei in the primate thalamus, but their primary inputs are from structures such as cortical and nonprimary subcortical auditory structures, and cortical multisensory and brainstem nuclei. These include the posterior nuclear group (PO), medial pulvinar (PM), suprageniculate (SG), and limitans (Lim) (de la Mothe et al. 2006b; de Ribaupierre 1997; Rouiller and Durif 2004). The posterolateral section of the thalamic reticular nucleus is heavily implicated in mediating feedback from cortical structures. The posterior nuclear group lies dorsal and medial to the MGC. The medial pulvinar is the auditory-responsive region of the pulvinar and receives inputs from the superior colliculus, but whether it receives inputs from the inferior colliculus is not known. The medial pulvinar projects broadly to temporal, frontal, and cingulate cortex (Gutierrez et al. 2000). The thalamic reticular nucleus can be broken down into three parts: an anterior division that responds primarily to somatosensory inputs, a dorsal division that responds primarily to visual inputs, and a ventral division that responds primarily to auditory inputs. Neurons in the auditory sector are broadly tuned to tones, but can act in a frequency-specific manner mediated by connections with the MGC (Crabtree 1998).

The importance of the MGC cannot be overestimated in understanding auditory cortical processing: it is an obligatory relay to the cortex. In primates, each of the three divisions of the MGC transforms and modulates auditory neural responses in a different way. These three divisions project to different parts of auditory cortex in different degrees (see Figure 9.4), creating the firmament of the organization and response properties seen there.

9.4.2 Auditory Cortex

Auditory cortex includes cortex that receives preferential projections from the MGC and is highly responsive to auditory stimuli. In humans, auditory cortex corresponds to Brodmann's areas 41 and 42 located in the vicinity of Heschl's gyrus on the superior temporal plane (Hackett et al. 2001). In macaques, auditory cortex is located on the caudal portion of the lower bank of the lateral sulcus and the superior temporal gyrus (Figure 9.2). Since only a small portion is visible on the surface of the macaque brain (upper brain), the parietal and frontal cortex has been "cut" away to reveal the areas of auditory cortex hidden deep in the lateral and circular sulcus (lower brain). From the figure, it is easy to appreciate one difficulty of studying auditory cortex: this part of cortex is almost completely covered by the parietal lobe in Old World primates such as macaques, chimpanzees, and humans.

In this chapter, we emphasize a primate model of auditory cortical organization based on decades of anatomical and physiological research (Kaas and Hackett 1998, 2000; Hackett 2010). According to this working model, auditory cortex is first divided into three **regions**, which can be thought of as levels of processing (Figure 9.3). These regions are further subdivided into thirteen **areas**. Regions are subdivisions of auditory cortex and areas are subdivisions of regions.

Distinctions between regions and areas are based on three features: *connections* (i.e., from thalamus and to other cortical areas), the cellular and histochemical *architecture* of cortical tissue, and *functional organization*, presumably reflected by specificity of neural response properties (such as patterns of tonotopic organization and differences in response properties). The anatomical

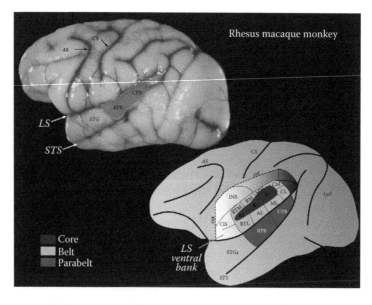

FIGURE 9.2 (**See Color Insert**) Location of primate auditory cortex (macaque). Upper figure: location of auditory cortex in macaque. Note that only the parabelt (blue: RPB, CPB) is exposed on the surface of the brain. Lower figure: the parietal and frontal cortices have been graphically cut away to reveal approximate location of core and belt (red and yellow). Abbreviations: LS, lateral sulcus; STS, superior temporal sulcus; AS, arcuate sulcus; CS, central sulcus; STG, superior temporal gyrus; STS, superior temporal sulcus; CIS, circular sulcus; INS, insula; LuS, lunate sulcus. See text for other abbreviations.

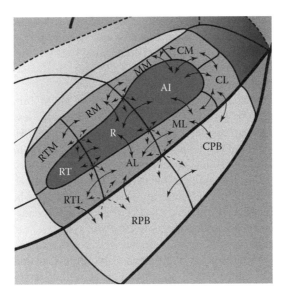

FIGURE 9.3 Organization of primate auditory cortex. A schematic of primate auditory cortex showing core, belt, and parabelt regions with areal subdivisions, and some short-range connections. For clarity, medial belt projections to parabelt are not pictured.

differences in connections and architecture are thought to subserve the differences in function. In the following section, we will first describe general regional characteristics and then fill in details, where known, of areal characteristics.

In auditory cortex, three major regions are defined: **core**, a **belt** that wraps around the core, and a **parabelt** region lying lateral to the belt. Distinguishing architectonic features of auditory cortex can include markers for cellular and molecular features such as cytochrome oxidase (CO), acetylcholinesterase (AChE), parvalbumin, vesicular glutamate transporter 2 (vGluT2), and density of myelination. These markers change roughly stepwise as one progresses across regions in a medial to lateral direction (de la Mothe et al. 2006a; Hackett and de la Mothe 2009).

The core region receives its primary input from the MGv, and also receives a projection from the MGm (Figure 9.4). The core is densely myelinated, has a broad layer IV, and exhibits a high expression of CO, AChE, parvalbumin, and vGluT2, all consistent with receiving a dense and rapidly conducting projection from the thalamus (de la Mothe et al. 2006a, 2006b). This core region is densely connected within itself (between areas) and with areas in the adjoining belt region, but not with the

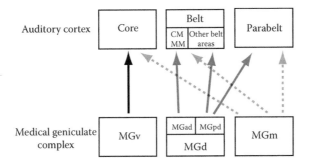

FIGURE 9.4 Connections of the MGC with regions of auditory cortex. A schematic of connections of the three subdivisions of the MGC (MGv, MGd, MGm) with the three regions of auditory cortex (core, belt, parabelt). For simplicity, only major connections are shown. Arrow type denotes different thalamic sources, not quantity of projections.

parabelt (we will come back to this later). Compared to other regions, neurons in the core tend to have short response latencies (though this varies across areas, see below), narrow spectral frequency tuning functions, and relatively fast modulation frequency tuning (e.g., Bendor and Wang 2008; Kajikawa et al. 2008; Kusmierek and Rauschecker 2009; Merzenich and Brugge 1973; Recanzone 2000a; Recanzone, Guard, and Phan 2000a).

The belt region receives thalamic projections from the MGd and MGm, but not MGv (Figure 9.4). The belt has a less pronounced layer IV than the core, and it is also less myelinated and exhibits reduced expressions of the markers described above, consistent with a less robust projection from the MGC. The belt is divided into a medial and a lateral region, relative to its position in relation to the core. The medial belt region is most connected within itself and the adjoining core regions, and has additional connections to the parabelt (Figure 9.3). The lateral belt is most connected to itself, adjoining core, and adjoining parabelt. Neural responses in the belt have been less well studied electrophysiologically than in the core because most of the region responds poorly under anesthesia. Compared to core neurons, belt neurons appear to have wider spectral tuning functions, often with broad or multipeaked frequency tuning, and respond well to spectrally complex sounds (Kajikawa et al. 2008; Kusmierek and Rauschecker 2009; Recanzone 2008; Tian and Rauschecker 2004). Belt neurons also exhibit longer latencies and do not entrain as well to temporally modulated stimuli. Instead, temporal modulation rate may be encoded by firing rate (see Wang et al. 2008).

The third region, the parabelt, also receives thalamic projections from the MGd and MGm (Figure 9.4). The parabelt has a less pronounced layer IV than core or belt, and it is also less myelinated and exhibits further reduced expression of the markers described above. The parabelt region is most connected within itself and with the medial and lateral belt region (Figure 9.3). It may have a weak feedback projection to core, but no direct projection from it. The response properties of neurons have not been well studied, but based on patterns of connections and responses known thus far, parabelt neurons are expected to exhibit long latencies and respond extremely poorly to tones, with wide and probably complex frequency tuning functions. This region will probably be more responsive to sounds that are both spectrally and temporally complex, such as vocalizations.

Architectural distinctions between regions exist roughly stepwise in a medial to lateral direction, but it is also important to note that there is a distinct rostral to caudal gradient of the same molecular markers described above (i.e., markers for differences in cellular and molecular features, such as CO, AChE, parvalbumin, vGluT2, and density of myelination). In general, these features change gradually, where the strongest expression of these markers in a given region is caudal and the weakest expression is rostral.

Thus, upon this regional organization, regions are divided into areas (Figures 9.3 and 9.4) (Kaas and Hackett 1998, 2000).

The core region contains three areas, from caudal to rostral: the "primary" area A1, the rostral area R, and the rostral temporal area RT. The medial belt contains four areas, also named by location: the caudal medial area CM, the middle medial area MM, the rostral medial area RM, and the rostrotemporal medial area RTM. The lateral belt also has four areas: the caudal lateral area CL, the middle lateral area ML, the anterolateral area AL, and the rostrotemporal lateral area RTL. Lastly, the parabelt has at least two areas: the caudal parabelt area CPB and the rostral parabelt area RPB. Most, if not all, of these areas have their own, often crude, tonotopic map which flips representational order along caudorostral borders (see Petkov et al. 2006). Core area A1 has been best explored, followed next by belt area CM, and the rest are under active investigation. The functions of any of these areas have not been fully elucidated, partly because they must be interpreted in the context of the others. We do know something about specificities of neural responses in some areas and can make educated predictions about the rest.

Given these subdivisions of regions and areas, how does auditory information flow between and across regions and areas? Based on connectional anatomy and physiology, a picture of graded hierarchy of informational flow between regions is emerging. The MGv is the only subdivision of the MGC that has demonstrated strong tonotopy, so the tonotopy exhibited in the belt and parabelt (which do not receive projections from the MGv) is probably inherited from the core (Kaas and Hackett 2000; Rauschecker et al. 1997). Additionally, there is no direct projection from core to para-belt, further suggesting a high degree of serial processing from core to belt to parabelt. There is a convergence of inputs from each region to the next, which presumably leads to the wider frequency tuning and altered response specificity as one progresses across regions.

In further support of this direction of flow, latencies increase from core to lateral belt at the same rostrocaudal level (Issa and Wang 2008; Kusmierek and Rauschecker 2009; Rauschecker and Tian 2004; Recanzone 2008; Woods et al. 2006). Other support comes from sound-level functions. At lower levels, sound level (loudness) is encoded as a monotonic function (as sound level rises, so does neural firing rate). As one ascends the hierarchy of sound processing, one sees more complex, nonmonotonic cells, where cells reach peak firing at a certain sound level and then are less responsive at higher sound levels. Also, neural response thresholds (lowest sound level to elicit a response) get higher from core to belt. Tuning widths for tone frequencies also become wider for neurons from core to belt, presumably reflecting convergence in the belt of more tightly tuned inputs originating in the core or MGC (Rauschecker and Tian 2004; Woods et al. 2006). Temporal modulation tuning also progressively decreases and this is thought to be due to timing imprecision introduced by successive synaptic delays. Presumably to compensate for this, a nonsynchronized firing rate code for modulation rate also occurs at the level of auditory cortex (Bartlett and Wang 2007; reviewed in Wang et al. 2008). It should be noted, however, that while information flow across regions may be roughly serial, informational processing is not thought to occur in a strictly staged process. Instead, it is likely that many perceptual processes are occurring in parallel with each other in a graded serial fashion.

As mentioned before, the regions express dramatic medial to lateral stepwise changes in architecture and connectivity. Within each region, the architecture and response properties follow a less dramatic, but distinct, caudal to rostral decrease in molecular marker expression. Also, feedforward connectivity across areas within a region seems to have a preferential caudal to rostral flow, where connections within a region seem to be more robust caudal to rostral (e.g., A1 to R) than rostral to caudal (e.g., R to A1) (see Hackett 2010).

Rostrocaudal changes in architecture and differences in connectivity indicate that each area within a region has a unique profile of architecture and connectivity, which probably creates differences in functions between areas. For example, the most caudal core area (A1) is more myelinated than the most rostral core area (RT) (i.e., de la Mothe et al. 2006a) and demonstrates faster latencies than other core areas (Bendor and Wang 2008). Current evidence suggests that most of the belt areas exhibit slower latencies than the adjacent core area (i.e., ML slower than A1). Due to this rostral to caudal gradient it is not as easy to make predictions regarding response latencies between nonadjacent areas. There is growing evidence that the most caudal belt region, CM, has many neurons with response latencies that are as fast or faster than those in the most caudal core region, A1 (Kajikawa et al. 2005; Lakatos et al. 2005; Rauschecker and Tian 2004). However, CM receives most of its MGd inputs from the MGad nuclei, which, as discussed earlier, may exhibit tonotopy and fast latencies. Yet, physiological and lesion evidence seems to suggest that CM appears to depend completely on A1 inputs for its tone responses (Rauschecker et al. 1997). Clearly, these and other findings are presenting challenges to the present model in terms of information flow within auditory cortex.

As discussed in the introduction, auditory object location and identity share partially overlapping cues, whose coding is described above. To date, there has been little electrophysiological evidence of strong feature identity selectivity (e.g., for different calls) for neurons in core and belt areas of auditory cortex (Kusmierek and Rauschecker 2009; Recanzone 2008; Wang et al. 1995). There is a bias for recording in the larger and more accessible caudal core and belt areas, so the lack of selectivity found thus far may be simply due to this. One aspect of object identity is the subjective perception of pitch (irrespective of whether the frequency is actually present). A module of neurons that appears to be selective for pitch has been described on the lateral low frequency border of A1 with RT (Bendor and Wang 2005) and there seems to be converging evidence from functional imaging for a similar processing zone in human core auditory cortex, though it is often not as tightly localized (reviewed by Bendor and Wang 2006; but see Hall and Plack 2009).

The coding of object location in auditory cortex has been an area of intense interest. Many neurons appear to be sensitive to the spatial location of freefield sounds as well as headphone-based interaural intensity and time differences (IID and ITD), especially in the caudal belt fields (Ahissar et al. 1992; Miller and Recanzone 2009; Recanzone 2000b; Recanzone et al. 2000b; Scott, Malone, and Semple 2007; Woods et al. 2006). How this spatial selectivity is propagated is not well understood, as spatial selectivity in the MGC is virtually unknown. There has been little evidence for an ordered spatiotopic map in auditory cortex—instead, location in space is probably represented across a distributed population of neurons (Miller and Recanzone 2009). A co-registration of auditory information within an ordered spatiotopic map could occur with the lower layers of the superior colliculus (SC), which have significant inputs from auditory and multisensory areas of neocortex (e.g., Collins, Lyon, and Kaas 2005). Higher-order spatial perception such as the perception of auditory motion is also poorly understood, but belt areas have been shown to be sensitive to the presentation of approaching "looming" stimuli (Maier and Ghazanfar 2007). Areas of temporal and posterior parietal cortex that are sensitive to visual motion may be also sensitive to or modulated by auditory motion (Alink, Singer, and Muckli 2008).

9.4.3 Auditory-Responsive Cortex beyond Classical Auditory Cortex

9.4.3.1 Superior Temporal Gyrus

Areas of the superior temporal gyrus (STG) include Ts1, Ts2, the superior temporal polysensory region (STP), and parts of the temporoparietal temporal area (Tpt) (Figure 9.5). These areas have been shown to have dense reciprocal connections with belt and parabelt, and some have weaker auditory thalamic inputs from the MGC and multisensory nuclei of the posterior thalamus (Galaburda and Pandya 1983; Hackett et al. 2007a, 1998a, 1998b; Kosmal et al. 1997; Markowitsch et al. 1985; Pandya and Rosene 1993; Pandya, Rosene, and Doolittle 1994; Trojanowski and Jacobson 1975). Evidence from fMRI and PET studies in primates shows responsiveness to auditory stimuli, as well as to other modalities (Gil-da-Costa et al. 2006; Leinonen, Hyvarinen, and Sovijarvi 1980; Petkov et al. 2008; Poremba et al. 2004, 2003). Rostral STG has been shown to be particularly responsive to vocalizations and there is growing evidence for a voice identity processing area in the superior

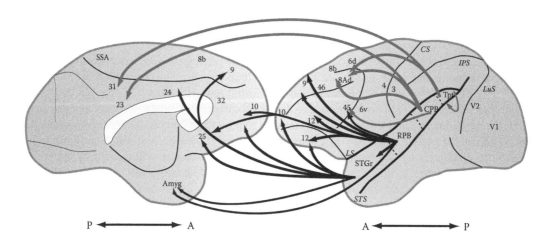

FIGURE 9.5 Auditory cortical projections to prefrontal, limbic structures. Long-range connections of auditory and auditory-related cortical fields to prefrontal and limbic structures projected on a macaque brain. Note the dorsoventral topography. For clarity, projections from core and belt of auditory cortex, insular cortex, basal ganglia, and basal nuclei projections are not shown.

temporal region—one that responds preferentially to the vocal identity of particular callers (see also review of human literature by Belin 2006; Petkov et al. 2008; Poremba et al. 2004).

9.4.3.2 Prefrontal Cortex and Ventral Premotor Cortex

Auditory cortical belt and parabelt project to areas in the prefrontal cortex, orbitofrontal cortex, and cingulate cortices in a topographic manner. Caudal parabelt primarily projects to dorsal prefrontal cortex and rostral parabelt primarily projects to ventral prefrontal cortex (Figure 9.5). Auditory-related areas on the STG described above project in a similar topographic manner (Barbas 2007; Barbas et al. 2005; Cavada et al. 2000; Morecraft et al. 2004; Pandya, Hallett, and Kmukherjee 1969; Petrides and Pandya 2002, 2007; Roberts et al. 2007; Saleem, Kondo, and Price 2008). These areas show auditory responsiveness to complex stimuli such as vocalizations or auditory responsiveness in a task-specific manner (Artchakov et al. 2007; Azuma and Suzuki 1984; Bodner, Kroger, and Fuster 1996; Cohen, Hauser, and Russ 2006; Cohen et al. 2007; Fuster, Bodner, and Kroger 2000; Ito 1982; Kikuchi-Yorioka and Sawaguchi 2000; Romanski 2007; Romanski, Averbeck, and Diltz 2005; Sugihara et al. 2006). For example, the dorsal-most portion of the frontal eye field responds to auditory stimuli and it is thought to mediate auditory-guided saccades. This is consistent with projection patterns described above showing a caudal projection from the caudal belt and adjoining association cortices to the portion of the frontal cortex containing the frontal eye fields. Auditory responsiveness has also been explored in the ventrolateral prefrontal cortex. Neurons in this part of cortex are not responsive to tones or noise, but are selectively responsive to different vocalizations (Cohen, Hauser, and Russ 2006, 2007; reviewed in Romanski and Averbeck 2009; Romanski, Averbeck, and Diltz 2005; Romanski and Goldman-Rakic 2002; Russ et al. 2008).

This dorsoventral topography of auditory belt and parabelt projections to the frontal cortex has led to the proposal that there exists a domain specificity of auditory processing in the auditory cortex (but see Recanzone and Cohen 2009; Romanski, Bates, and Goldman-Rakic 1999; Romanski, Tian, et al. 1999), much like the domain specificity described in the visual cortex (Ungerleider and Mishkin 1982). It is hypothesized that the dorsal prefrontal cortex receives projections from the dorsocaudal "where" stream of auditory processing and the ventral prefrontal cortex receives projections from the ventrocaudal "what" stream of auditory processing (Romanski, Tian, et al. 1999). This can be interpreted as functional specialization of prefrontal and auditory cortices (e.g., Rauschecker and Tian 2000; Romanski and Goldman-Rakic 2002; Tian et al. 2001).

In addition to prefrontal cortex being involved in auditory functions, ventral premotor cortex in monkeys contains neurons that respond to sounds signifying hand or mouth actions (Keysers et al. 2003). This region of cortex is also activated in humans when they listen to the sound of an action (Gazzola, Aziz-Zadeh, and Keysers 2006).

9.4.3.3 Insular Cortex

Another auditory-responsive region is insular cortex, lying medial to the medial belt of auditory cortex. This area is connected to the medial belt, and to a lesser extent lateral belt and parabelt, as well as areas of the superior temporal gyrus and prefrontal cortex (de la Mothe et al. 2006a; Hackett et al. 2007b). Early electrophysiological studies indicated it was responsive to both simple auditory stimuli (i.e., tones and clicks) and more complex stimuli (i.e., vocalizations) (Bieser 1998; Bieser and Muller-Preuss 1996; Sudakov et al. 1971) (see also Remedios, Logothetis, and Kayser 2009). Insular cortex also has been implicated in auditory functions in humans (Griffiths et al. 1997; Zatorre, Evans, and Meyer 1994). In a recent study, insular neurons responded preferentially to conspecific vocalizations over sounds with similar spectral or envelope structure, indicating that they are responding preferentially to the vocalization (Remedios, Logothetis, and Kayser 2009). How insular cortex fits into the "what vs. where" stream hypothesis has yet to be determined. The region of the anterior insula has been implicated in the ability to understand the emotional experiences of others (e.g. Peyron, Laurent, and Garcia-Larrea 2000), and auditory input to insular cortex may have an affective component and relate to emotional appreciation of music in humans (Craig 2008).

9.4.3.4 Corticofugal Projections

There are extensive corticofugal projections from auditory and auditory-related cortex (reviewed in Winer 2005), presumed to play a major role in top-down modulatory and learning effects. These connections have been most extensively explored in cats, but they have been demonstrated in primates as well (de la Mothe et al. 2006b; FitzPatrick and Imig 1978; Luethke, Krubitzer, and Kaas 1989; Morel and Kaas 1992). These corticofugal projections target primarily ipsilateral nuclei and structures. There are massive projections to the MGC (de la Mothe et al. 2006b; Winer et al. 2002), inferior colliculus (mostly outside of the central nucleus) (Winer et al. 1998), superior olivary complex (SOC), cochlear nucleus (CoN), pons (Brodal 1972), and basal ganglia (Reale and Imig 1983). Input to claustrum and endopiriform nucleus also arises from areas of auditory cortex (Beneyto and Prieto 2001). The dorsal putamen and caudate nucleus receive topographic projections from areas of auditory cortex (Reale and Imig 1983), and appear to have a role in sensory processing. Motor behavior could be modulated by inputs from auditory cortex to the basal ganglia (Beneyto and Prieto 2001). Auditory cortex also is positioned to affect autonomic function via its projections to amygdala (e.g., Romanski and LeDoux 1993) and central gray (Winer et al. 1998). Other inputs to the amygdala come directly from the auditory thalamus, at least in rats (Iwata et al. 1986).

9.5   REWARD-RELATED ACTIVITY IN AUDITORY CORTEX AND OTHER AUDITORY AREAS

9.5.1   Early Lesion Studies of Auditory Cortex

Having reviewed the basic framework for auditory processing, we now focus on roles of reward and emotion in auditory perception as well as connections that could be mediating these interactions. There have been a number of studies of the effects of cortical lesions on conditioned auditory discriminative behavior in cats (e.g., Butler, Diamond, and Neff 1957; Diamond, Goldberg, and Neff 1962; Jenkins and Merzenich 1984) and to a lesser extent in monkeys (e.g., Heffner and Heffner 1990a, 1990b). Lesions generally involved all of the primary core auditory cortex and much of the secondary auditory cortex of both hemispheres. The usual finding was that learned discriminations were lost immediately after such lesions. Simple discriminations, such as responding to a frequency change of a pulsating tone, were rapidly relearned, while more complex auditory tasks, such as responding to a change in a pattern of two pulsing tones, were not. Sound localization tasks were permanently impaired, while reflexive head-orienting responses remained, but were less accurate (Beitel and Kaas 1993; Jenkins and Merzenich 1984). Based on these studies, extensive bilateral lesions of auditory cortex did not seem to interfere with associating sounds with reward or punishment, but did produce deficits in complex auditory discriminations.

In rats, extensive bilateral lesions of the auditory region, including all known primary and secondary areas, did not abolish the ability to acquire auditory fear conditioning, but lesions of the auditory thalamus, particularly those that include MGm, did (Romanski and LeDoux 1992). The authors concluded that fear conditioning can be mediated by projections from the auditory thalamus to the amygdala (see below, and see also Edeline and Weinberger 1991, 1992). Quite possibly, auditory stimuli can be associated with reward or punishment using such subcortical pathways if the relevant auditory stimuli do not depend on cortical mechanisms for analysis.

9.5.2   Reward-Related Changes in the Responses of Neurons and Sensory Map Plasticity

In a developing animal, passive exposure to a sensory stimulus will increase that stimulus' representation in cortex (Sanes and Bao 2009), but in intact adult animals the stimulus must be paired with a task to create the same changes. Large-scale reorganization of sensory topographic maps is highly reward dependent (Blake et al. 2006). In owl monkeys, training on a target frequency will expand the representation of that frequency in A1 (Blake et al. 2002; Recanzone, Schreiner, and Merzenich 1993), which is similar to results in rats (Blake et al. 2006; Polley, Steinberg, and Merzenich 2006). This phenomenon is commonly referred to as perceptual learning.

There have been many examples where pairing an auditory stimulus with reinforcement alters the receptive field organization of auditory cortical fields (e.g., Atiani et al. 2009; Beitel et al. 2003; Brosch, Selezneva, and Scheich 2005; Durif, Jouffrais, and Rouiller 2003; Hocherman, Itzhaki, and Gilat 1981), as well as the auditory thalamus (e.g., Edeline, Neuenschwander-el Massioui, and Dutrieux 1990; Edeline and Weinberger 1992; Gabriel, Saltwick, and Miller 1975; Halas et al. 1970; Komura et al. 2001, 2005; Ryugo and Weinberger 1978), the inferior colliculus (e.g., Disterhoft and Stuart 1977; Metzger et al. 2006; Nienhuis and Olds 1978; Olds, Nienhuis, and Olds 1978), and even the cochlear nucleus (e.g., Oleson, Ashe, and Weinberger 1975). When a stimulus (such as a tone) is paired with a reward, the neurons activated by that auditory stimulus generally increase their response rate, and/or auditory areas and nuclei devote larger proportions of their neurons to the representation of that rewarded sound.

Such behaviorally mediated plasticity is regulated by the basal cholinergic system that projects broadly to auditory and other forebrain structures (Kilgard and Merzenich 1998; Ma and Suga 2005; Weinberger 2003). Furthermore, lesions of the basal forebrain cholinergic system prevent the changes in sensory neuron responsiveness with reward (Conner, Chiba, and Tuszynski 2005; Juliano, Ma, and Eslin 1991; Prakash et al. 2004).

There are at least two proposed models for the basis of this plasticity (reviewed by Edeline 1999; Edeline and Weinberger 1992). The first model proposes that long-lasting effects in the nonlemniscal pathway contribute to cortical plasticity. This model postulates that plasticity effects are mainly at the level of the MGm, where acetylcholine released from the nucleus basalis creates long-lasting receptive field changes at the MGm, and possibly the MGv. These changes are perpetuated up through A1 (see Weinberger 1998, 2004; and see see Chapter 1 by Weinberger and Bieszczad in this volume). The second model proposes that cortical activity activates the amygdala, which activates the nucleus basalis, which in turn affects change in cortex. It implicates an association cortex-amygdala-nucleus basalis-auditory cortex loop (Gao and Suga 1998; Ma and Suga 2005). In the second model, changes in corticofugal pathways promote long-lasting thalamic plasticity (Weinberger et al. 1990). Regardless of the mechanism, plasticity is affected by a number of other neurotransmitters besides acetylcholine. Dopamine from the ventral tegmental area and norepinephrine from the locus coeruleus have been shown to induce auditory receptive field plasticity (see Bao, Chan, and Merzenich 2001; Edeline 1999; Seitz and Watanabe 2005).

Reward-based plasticity can reflect an increased processing of stimuli that show a predictive relationship to the reinforcing signal. For example, in a tone-conditioning task in rats, cortical expansion occurred at the representation corresponding to the target frequency, and also that corresponding to the low-frequency sounds emitted by the reward delivery system (Rutkowski and Weinberger 2005). Similarly, somatosensory and visual stimuli can evoke responses in auditory cortex when they are associated with an auditory stimulus and reward. After intensive training on an auditory task, cells in A1 of macaque responded to the visual cue that preceded the auditory stimulus and also responded to the touch of the bar that was used to indicate a response (Brosch, Selezneva, and Scheich 2005). Rather than propose that A1 is multisensory, it may be that in the context of overtraining A1 is sensitive to signals that predict reward. Lastly, there is also evidence to suggest that different topographic maps can reorganize independently without disturbing one another (Polley, Steinberg, and Merzenich 2006).

9.5.3 CONNECTIONS WITH THE REWARD SYSTEM

In order for auditory signals to motivate behavior, connections with brain structures involved in reward would appear to be necessary (reviewed in Schultz 2000). Thus, the next section seeks to identify connections of the auditory pathway with areas associated with the reward system, such as the basal ganglia, amygdala, basal forebrain, and frontal limbic cortices.

In monkeys, the proportion of auditory cortical projections to basal ganglia generally increases as one ascends from core to belt to parabelt to auditory-related cortices (Borgmann and Jurgens 1999; de la Mothe et al. 2006a; Selemon and Goldman-Rakic 1985; Smiley et al. 2007; Yeterian and Pandya 1998). The densest connections with the most striking topography emerge at the level of parabelt and auditory-related areas of the temporal lobe. Caudal areas project to dorsal parts of the caudate and putamen, and lateral and rostral areas project to the ventral portions of the caudate and putamen (Borgmann and Jurgens 1999; Forbes and Moskowitz 1974; Nauta and Whitlock 1956; Yeterian and Pandya 1998). Consistent with these projections, auditory-responsive neurons have been recorded in caudate, putamen, globus pallidus, and substantia nigra (Aosaki, Kimura, and Graybiel 1995; Chudler, Sugiyama, and Dong 1995; Hikosaka, Sakamoto, and Usui 1989; Hikosaka and Wurtz 1983; Nagy et al. 2005; Steinfels et al. 1983; Strecker et al. 1985).

The lateral nucleus of the amygdala (AL) receives highly processed visual and auditory information. It can receive auditory information from the thalamus (mainly, but not limited to, the MGm) or cortex (belt and parabelt of auditory cortex, and auditory-related areas such as rostral temporal cortices) (reviewed in Amaral et al. 1992; Romanski and LeDoux 1993). Most of these cortical projections arise from the rostral putative "what" pathway, similar to the visual system (see Amaral et al. 1992). The thalamoamygdala and corticoamygdala pathways have different but overlapping functions. It is believed that both pathways can mediate acquisition of simple auditory discriminations in fear-conditioning tasks, such as the acoustic startle. However, the corticoamygdala projections appear to be necessary to mediate acquisition of more complex auditory discriminations (Jarrell et al. 1987; LeDoux, Sakaguchi, and Reis 1984; Romanski and LeDoux 1992). The amygdala is responsive to auditory stimuli, especially in the context of a task, and is also responsive to emotional vocalizations in monkeys (Kuraoka and Nakamura 2007).

The basal forebrain nuclei are the main extrinsic sources of acetylcholine to auditory cortex (Jones and Burton 1976), a neurotransmitter previously discussed regarding its implications with plasticity and learning. The heavy AChE banding in layers 3 and 4 of auditory cortex, particularly in core and caudomedial areas (de la Mothe et al. 2006a; Winer and Lee 2007), presumably arises from basal forebrain. Basal forebrain projections to auditory cortex may be useful to mediate plasticity and memory effects. Consistent with this, the basal forebrain has demonstrated auditory activity when an auditory stimulus cues a reward (Wilson and Rolls 1990).

The prefrontal limbic cortices are located on the ventral and medial surface of the most anterior portion of the frontal lobe. Based on intrinsic and extrinsic connection patterns, they are generally considered to consist of two networks, the orbital (or ventral) and medial networks (Barbas 2007; Carmichael and Price 1996; Cavada et al. 2000; Price 2006). Areas on the ventral and ventromedial surface of the frontal lobe (e.g., 12, 11) have robust connections with rostral parabelt and rostral superior temporal gyrus (Barbas 1993; Hackett et al. 1999; Romanski, Bates, and Goldman-Rakic 1999; Saleem, Kondo, and Price 2008). These areas are auditory responsive and show responses to complex auditory stimuli, demonstrated both electrophysiologically (Rolls et al. 2006) and metabolically via 2-deoxyglucose (Poremba et al. 2003; Wiley and Richards 1978). Furthermore, these areas are important for responding to social auditory stimuli. Macaques with lesions of the OFC show less appropriate responses to vocal social cues such as threat and affiliation calls (Machado and Bachevalier 2006).

In the second network, areas on the medial surface of the frontal lobe and cingulate cortex have robust connections with rostral auditory parabelt and rostral superior temporal gyrus (Hackett et al. 1999; Romanski, Bates, and Goldman-Rakic 1999; Vogt and Pandya 1987). This region of frontal and cingulate cortex is also connected to brainstem vocalization centers and appears to be involved in emotional vocal processing and production (Alheid and Heimer 1996; Holstege 1991; Holstege, Bandler, and Saper 1996). It has long been known that vocalizations (among other things) can be elicited by electrical stimulation of anterior cingulate cortex in monkeys (e.g., Smith 1945) and that lesions of this cortex reduce or eliminate the productions of isolation calls (MacLean and Newman 1988).

9.6 CONCLUSIONS

This chapter reviews the auditory processing pathways, with particular emphasis on cortical and subcortical structures implicated in reward and reward-based plasticity. To provide a foundation for the other chapters on rewards, we have described the functional organization of auditory processing pathways. Where possible, we have described examples of reward-based activity within this pathway and described pathways that likely mediate this activity.

Analysis of the pathways of reward in the context of sensory systems is illuminating because it reveals ways that reward-based information could guide behavior. Additionally, analysis of these pathways also informs our evolving ideas about the basic structure of auditory processing. For example, perhaps the parts of the auditory system that are preferentially connected to reward structures are participating in different computations from those that are not. Analysis of the pathways of auditory informational flow is also useful to illuminate ideas about sensory organization. For example, if what/where is a global organizational principle of many sensory systems, then it is reasonable to expect that this organization is perpetuated into nonsensory systems, such as reward.

These pathways are also interesting from a comparative evolutionary perspective. In the course of primate evolution, the human brain underwent dramatic expansion, and the auditory cortices are no exception (Hackett et al. 2001). Comparing evidence from different primates such as marmosets, macaques, and humans inspires a search for similarities and differences between species. As discussed previously, differences may be due to evolutionary selection pressures in their specific ecological niche. Differences may also be due to constraints that emerge to minimize the metabolic cost and connection length of a larger brain (reviewed in Kaas 2000). For example, laterality of function in larger brains is hypothesized to have arisen from constraints to limit the number of long-range interhemispheric connections (Ringo et al. 1994). These connections are slower, temporally imprecise, and metabolically costly. Given the auditory system's dependence on precise timing, it is no surprise that evidence for laterality in large brains is strongest in this sensory modality (e.g., language processing in humans).

There are still many fundamental questions left on the nature of auditory processing and reward processing within the auditory system. We need a more complete understanding of anatomical underpinnings, especially in close human relatives. We also need more studies of neuronal responses in the awake animal, especially in the context of task-guided behavior. Careful studies across different species will help elucidate global roles of reward and its specific interactions with the auditory system in primates.

ACKNOWLEDGMENTS

The authors would like to thank Drs. Troy Hackett and Lisa de la Mothe of Vanderbilt University and Dr. Daniel Polley of Harvard Medical School for helpful comments on this chapter.

REFERENCES

Ahissar, M., Ahissar, E., Bergman, H., and Vaadia, E. 1992. Encoding of sound-source location and movement: Activity of single neurons and interactions between adjacent neurons in the monkey auditory cortex. *J Neurophysiol* 67:203–15.

Alheid, G.F., and Heimer, L. 1996. Theories of basal forebrain organization and the "emotional motor system". *Prog Brain Res* 107:461–84.

Alink, A., Singer, W., and Muckli, L. 2008. Capture of auditory motion by vision is represented by an activation shift from auditory to visual motion cortex. *J Neurosci* 28:2690–97.

Allon, N., Yeshurun, Y., and Wollberg, Z. 1981. Responses of single cells in the medial geniculate body of awake squirrel monkeys. *Exp Brain Res* 41:222–32.

Amaral, D., Price, J., Pitkanen, A., and Carmichael, S. 1992. Anatomical organization of the primate amygdaloid complex. In *The Amygdala: Neurobiological Aspects of Emotion, Memory, and Mental Dysfunction*, ed. J. Aggleton, pp. 1–66. New York: Wiley-Liss.

Aosaki, T., Kimura, M., and Graybiel, A.M. 1995. Temporal and spatial characteristics of tonically active neurons of the primate's striatum. *J Neurophysiol* 73:1234–52.

Artchakov, D., Tikhonravov, D., Vuontela, V., Linnankoski, I., Korvenoja, A., and Carlson, S. 2007. Processing of auditory and visual location information in the monkey prefrontal cortex. *Exp Brain Res* 180:469–79.

Atiani, S., Elhilali, M., David, S.V., Fritz, J.B., and Shamma, S.A. 2009. Task difficulty and performance induce diverse adaptive patterns in gain and shape of primary auditory cortical receptive fields. *Neuron* 61:467–80.

Azuma, M., and Suzuki, H. 1984. Properties and distribution of auditory neurons in the dorsolateral prefrontal cortex of the alert monkey. *Brain Res* 298:343–46.

Bao, S., Chan, V.T., and Merzenich, M.M. 2001. Cortical remodelling induced by activity of ventral tegmental dopamine neurons. *Nature* 412:79–83.

Barbas, H. 1993. Organization of cortical afferent input to orbitofrontal areas in the rhesus monkey. *Neuroscience* 56:841–64.

———. 2007. Specialized elements of orbitofrontal cortex in primates. *Ann N Y Acad Sci* 1121:10–32.

Barbas, H., Hilgetag, C.C., Saha, S., Dermon, C.R., and Suski, J.L. 2005. Parallel organization of contralateral and ipsilateral prefrontal cortical projections in the rhesus monkey. *BMC Neurosci* 6:32.

Barlow, H. 1961. Possible principles underlying the transformation of sensory messages. In *Sensory Communication*, ed. W. Rosenblith, pp. 217–34. Cambridge, MIT Press.

Barone, P., Clarey, J.C., Irons, W.A., and Imig, T.J. 1996. Cortical synthesis of azimuth-sensitive single-unit responses with nonmonotonic level tuning: A thalamocortical comparison in the cat. *J Neurophysiol* 75:1206–20.

Bartlett, E.L., and Wang, X. 2007. Neural representations of temporally modulated signals in the auditory thalamus of awake primates. *J Neurophysiol* 97:1005–17.

Beitel, R.E., and Kaas, J.H. 1993. Effects of bilateral and unilateral ablation of auditory cortex in cats on the unconditioned head orienting response to acoustic stimuli. *J Neurophysiol* 70:351–69.

Beitel, R.E., Schreiner, C.E., Cheung, S.W., Wang, X., and Merzenich, M.M. 2003. Reward-dependent plasticity in the primary auditory cortex of adult monkeys trained to discriminate temporally modulated signals. *Proc Natl Acad Sci USA* 100:11070–75.

Belin, P. 2006. Voice processing in human and non-human primates. *Philos Trans R Soc Lond B Biol Sci* 361:2091–107.

Bendor, D., and Wang, X. 2005. The neuronal representation of pitch in primate auditory cortex. *Nature* 436:1161–65.

———. 2006. Cortical representations of pitch in monkeys and humans. *Curr Opin Neurobiol* 16:391–99.

———. 2008. Neural response properties of primary, rostral, and rostrotemporal core fields in the auditory cortex of marmoset monkeys. *J Neurophysiol* 100:888–906.

Beneyto, M., and Prieto, J.J. 2001. Connections of the auditory cortex with the claustrum and the endopiriform nucleus in the cat. *Brain Res Bull* 54:485–98.

Bieser, A. 1998. Processing of twitter-call fundamental frequencies in insula and auditory cortex of squirrel monkeys. *Exp Brain Res* 122:139–48.

Bieser, A., and Muller-Preuss, P. 1996. Auditory responsive cortex in the squirrel monkey: Neural responses to amplitude-modulated sounds. *Exp Brain Res* 108:273–84.

Blake, D.T., Heiser, M.A., Caywood, M., and Merzenich, M.M. 2006. Experience-dependent adult cortical plasticity requires cognitive association between sensation and reward. *Neuron* 52:371–81.

Blake, D.T., Strata, F., Churchland, A.K., and Merzenich, M.M. 2002. Neural correlates of instrumental learning in primary auditory cortex. *Proc Natl Acad Sci USA* 99:10114–19.

Bodner, M., Kroger, J., and Fuster, J.M. 1996. Auditory memory cells in dorsolateral prefrontal cortex. *Neuroreport* 7:1905–8.

Borgmann, S., and Jurgens, U. 1999. Lack of cortico-striatal projections from the primary auditory cortex in the squirrel monkey. *Brain Res* 836:225–28.

Brodal, P. 1972. The corticopontine projection in the cat. The projection from the auditory cortex. *Arch Ital Biol* 110:119–44.

Brosch, M., Selezneva, E., and Scheich, H. 2005. Nonauditory events of a behavioral procedure activate auditory cortex of highly trained monkeys. *J Neurosci* 25:6797–806.

Brown, C., and Waser, P. 1988. Environmental influences on the structure of primate vocalizations. In *Primate Vocal Communication*, eds. D. Todt, P. Goedeking, and D. Symmes, pp. 51–66. Berlin: Springer-Verlag.

Brown, C.H. 1989. The acoustic ecology of east African primates and the perception of vocal signals by grey-cheeked mangabeys and blue monkeys. In *The Comparative Psychology of Audition: Perceiving Complex Sounds*, eds. R. Dooling, and S. Hulse, pp. 201–39. Hillsdale, NJ: Erlbaum.

Brown, T.J., and Handford, P. 2000. Sound design for vocalizations: Quality in the woods, consistency in the fields. *Condor* 102:81–92.

Butler, R.A., Diamond, I.T., and Neff, W.D. 1957. Role of auditory cortex in discrimination of changes in frequency. *J Neurophysiol* 20:108–20.

Calford, M.B. 1983. The parcellation of the medial geniculate body of the cat defined by the auditory response properties of single units. *J Neurosci* 3:2350–64.

Calford, M.B., and Aitkin, L.M. 1983. Ascending projections to the medial geniculate body of the cat: Evidence for multiple, parallel auditory pathways through thalamus. *J Neurosci* 3:2365–80.

Carmichael, S.T., and Price, J.L. 1996. Connectional networks within the orbital and medial prefrontal cortex of macaque monkeys. *J Comp Neurol* 371:179–207.

Cavada, C., Company, T., Tejedor, J., Cruz-Rizzolo, R.J., and Reinoso-Suarez, F. 2000. The anatomical connections of the macaque monkey orbitofrontal cortex. A review. *Cereb Cortex* 10:220–42.

Chudler, E.H., Sugiyama, K., and Dong, W.K. 1995. Multisensory convergence and integration in the neostriatum and globus pallidus of the rat. *Brain Res* 674:33–45.

Cohen, Y.E., Hauser, M.D., and Russ, B.E. 2006. Spontaneous processing of abstract categorical information in the ventrolateral prefrontal cortex. *Biol Lett* 2:261–65.

Cohen, Y.E., Theunissen, F., Russ, B.E., and Gill, P. 2007. Acoustic features of rhesus vocalizations and their representation in the ventrolateral prefrontal cortex. *J Neurophysiol* 97:1470–84.

Collins, C.E., Lyon, D.C., and Kaas, J.H. 2005. Distribution across cortical areas of neurons projecting to the superior colliculus in new world monkeys. *Anat Rec A Discov Mol Cell Evol Biol* 285:619–27.

Conner, J.M., Chiba, A.A., and Tuszynski, M.H. 2005. The basal forebrain cholinergic system is essential for cortical plasticity and functional recovery following brain injury. *Neuron* 46:173–79.

Crabtree, J.W. 1998. Organization in the auditory sector of the cat's thalamic reticular nucleus. *J Comp Neurol* 390:167–82.

Craig, A. 2008. Interoception and emotion, a neuroanatomical perspective. In *Handbook of Emotions*, eds. M. Lewis, J. Haviland-Jones, and L. Barrett, pp. 272–88. New York: Guilford Press.

de la Mothe, L.A., Blumell, S., Kajikawa, Y., and Hackett, T.A. 2006a. Cortical connections of the auditory cortex in marmoset monkeys: Core and medial belt regions. *J Comp Neurol* 496:27–71.

———. 2006b. Thalamic connections of the auditory cortex in marmoset monkeys: Core and medial belt regions. *J Comp Neurol* 496:72–96.

de Ribaupierre, F. 1997. Acoustical information processing in the auditory thalamus and cerebral cortex. In *The Central Auditory System*, eds. G. Ehret, and R. Romand, pp. 317–98. New York: Oxford University Press.

Derryberry, E.P. 2009. Ecology shapes birdsong evolution: Variation in morphology and habitat explains variation in White-crowned Sparrow song. *Am Naturalist* 174:24–33.

Diamond, I.T., Goldberg, J.M., and Neff, W.D. 1962. Tonal discrimination after ablation of auditory cortex. *J Neurophysiol* 25:223–35.

Disterhoft, J.F., and Stuart, D.K. 1977. Differentiated short latency response increases after conditioning in inferior colliculus neurons of alert rat. *Brain Res* 130:315–33.

Dittus, W. 1988. An analysis of toque macaque cohesion calls from an ecological perspective. In *Primate Vocal Communication*, eds. D. Todt, P. Goedeking, and D. Symmes, pp. 31–49. Berlin: Springer-Verlag.

Durif, C., Jouffrais, C., and Rouiller, E.M. 2003. Single-unit responses in the auditory cortex of monkeys performing a conditional acousticomotor task. *Exp Brain Res* 153:614–27.

Edeline, J.M. 1999. Learning-induced physiological plasticity in the thalamo-cortical sensory systems: A critical evaluation of receptive field plasticity, map changes and their potential mechanisms. *Prog Neurobiol* 57:165–224.

Edeline, J.M., Neuenschwander-el Massioui, N., and Dutrieux, G. 1990. Discriminative long-term retention of rapidly induced multiunit changes in the hippocampus, medial geniculate and auditory cortex. *Behav Brain Res* 39:145–55.

Edeline, J.M., and Weinberger, N.M. 1991. Subcortical adaptive filtering in the auditory system: Associative receptive field plasticity in the dorsal medial geniculate body. *Behav Neurosci* 105:154–75.

———. 1992. Associative retuning in the thalamic source of input to the amygdala and auditory cortex: Receptive field plasticity in the medial division of the medial geniculate body. *Behav Neurosci* 106:81–105.

Evans, E. 1974. Neural processes for the detection of acoustic patterns and for sound localization. In *The Neurosciences, Third Study Program*, eds. F. Schmidt, and F. Wordern, pp. 131–45. Cambridge: MIT Press.

Fischer, J., and Hammerschmidt, K. 2002. An overview of the barbary macaque, Macaca sylvanus, vocal repertoire. *Folia Primatol (Basel)* 73:32–45.

Fischer, J., Hammerschmidt, K., Cheney, D.L., and Seyfarth, R.M. 2001. Acoustic features of female chacma baboon barks. *Ethology* 107:33–54.

Fitch, W.T.S. 2002. Primate vocal production and its implications for auditory research. In *Primate Audition: Ethology and Neurobiology*, ed. A. Ghazanfar, pp. 87–108. London: CRC Press.

FitzPatrick, K.A., and Imig, T.J. 1978. Projections of auditory cortex upon the thalamus and midbrain in the owl monkey. *J Comp Neurol* 177:573–55.

Forbes, B.F., and Moskowitz, N. 1974. Projections of auditory responsive cortex in the squirrel monkey. *Brain Res* 67:239–54.

Fuster, J.M., Bodner, M., and Kroger, J.K. 2000. Cross-modal and cross-temporal association in neurons of frontal cortex. *Nature* 405:347–51.

Gabriel, M., Saltwick, S.E., and Miller, J.D. 1975. Conditioning and reversal of short-latency multiple-unit responses in the rabbit medial geniculate nucleus. *Science* 189:1108–9.

Galaburda, A.M., and Pandya, D.N. 1983. The intrinsic architectonic and connectional organization of the superior temporal region of the rhesus monkey. *J Comp Neurol* 221:169–84.

Gao, E., and Suga, N. 1998. Experience-dependent corticofugal adjustment of midbrain frequency map in bat auditory system. *Proc Natl Acad Sci USA* 95:12663–70.

Gazzola, V., Aziz-Zadeh, L., and Keysers, C. 2006. Empathy and the somatotopic auditory mirror system in humans. *Curr Biol* 16:1824–29.

Ghazanfar, A.A., Turesson, H.K., Maier, J.X., van Dinther, R., Patterson, R.D., and Logothetis, N.K. 2007. Vocal-tract resonances as indexical cues in rhesus monkeys. *Curr Biol* 17:425–30.

Gil-da-Costa, R., Martin, A., Lopes, M.A., Munoz, M., Fritz, J.B., and Braun, A.R. 2006. Species-specific calls activate homologs of Broca's and Wernicke's areas in the macaque. *Nat Neurosci* 9:1064–70.

Griffiths, T.D., Rees, A., Witton, C., Cross, P.M., Shakir, R.A., and Green, G.G. 1997. Spatial and temporal auditory processing deficits following right hemisphere infarction. A psychophysical study. *Brain* 120 (5): 785–94.

Gross, N.B., Lifschitz, W.S., and Anderson, D.J. 1974. The tonotopic organization of the auditory thalamus of the squirrel monkey (Saimiri sciureus). *Brain Res* 65:323–32.

Gutierrez, C., Cola, M.G., Seltzer, B., and Cusick, C. 2000. Neurochemical and connectional organization of the dorsal pulvinar complex in monkeys. *J Comp Neurol* 419:61–86.

Hackett, T.A. 2010. Information flow in the auditory cortical network. *Hear Res*, doi:10.1016/j. heares. 2010.01.011.

Hackett, T.A., and de la Mothe, L.A. 2009. Regional and laminar distribution of the vesicular glutamate transporter, VGluT2, in the macaque monkey auditory cortex. *J Chem Neuroanat* 38:106–16.

Hackett, T.A., De La Mothe, L.A., Ulbert, I., Karmos, G., Smiley, J., and Schroeder, C.E. 2007a. Multisensory convergence in auditory cortex, II. Thalamocortical connections of the caudal superior temporal plane. *J Comp Neurol* 502:924–52.

Hackett, T.A., Preuss, T.M., and Kaas, J.H. 2001. Architectonic identification of the core region in auditory cortex of macaques, chimpanzees, and humans. *J Comp Neurol* 441:197–222.

Hackett, T.A., Smiley, J.F., Ulbert, I., Karmos, G., Lakatos, P., de la Mothe, L.A., and Schroeder, C.E. 2007b. Sources of somatosensory input to the caudal belt areas of auditory cortex. *Perception* 36:1419–30.

Hackett, T.A., Stepniewska, I., and Kaas, J.H. 1998a. Subdivisions of auditory cortex and ipsilateral cortical connections of the parabelt auditory cortex in macaque monkeys. *J Comp Neurol* 394:475–95.

———. 1998b. Thalamocortical connections of the parabelt auditory cortex in macaque monkeys. *J Comp Neurol* 400:271–86.

———. 1999. Prefrontal connections of the parabelt auditory cortex in macaque monkeys. *Brain Res* 817:45–58.

Halas, E.S., Beardsley, J.V., and Sandlie, M.E. 1970. Conditioned neuronal responses at various levels in conditioning paradigms. *Electroencephalogr Clin Neurophysiol* 28:468–77.

Hall, D.A., and Plack, C.J. 2009. Pitch processing sites in the human auditory brain. *Cereb Cortex* 19:576–85.

Halpern, D.L., Blake, R., and Hillenbrand, J. 1986. Psychoacoustics of a chilling sound. *Percept Psychophys* 39:77–80.

Hammerschmidt, K., and Todt, D. 1995. Individual-differences in vocalizations of young barbary macaques (macaca sylvanus) – a multi-parametric analysis to identify critical cues in acoustic signaling. *Behaviour* 132:381–99.

Hashikawa, T., Molinari, M., Rausell, E., and Jones, E.G. 1995. Patchy and laminar terminations of medial geniculate axons in monkey auditory cortex. *J Comp Neurol* 362:195–208.

Hauser, M.D. 1996. *The Evolution of Communication*. Cambridge, MA: MIT Press.

Hauser, M.D., and Fowler, C.A. 1992. Fundamental frequency declination is not unique to human speech: Evidence from nonhuman primates. *J Acoust Soc Am* 91:363–69.

He, J., and Hu, B. 2002. Differential distribution of burst and single-spike responses in auditory thalamus. *J Neurophysiol* 88:2152–56.

Heffner, H.E., and Heffner, R.S. 1990a. Effect of bilateral auditory cortex lesions on absolute thresholds in Japanese macaques. *J Neurophysiol* 64:191–205.

———. 1990b. Effect of bilateral auditory cortex lesions on sound localization in Japanese macaques. *J Neurophysiol* 64:915–31.

Hikosaka, O., Sakamoto, M., and Usui, S. 1989. Functional properties of monkey caudate neurons. II. Visual and auditory responses. *J Neurophysiol* 61:799–813.

Hikosaka, O., and Wurtz, R.H. 1983. Visual and oculomotor functions of monkey substantia nigra pars reticulata. I. Relation of visual and auditory responses to saccades. *J Neurophysiol* 49:1230–53.

Hocherman, S., Itzhaki, A., and Gilat, E. 1981. The response of single units in the auditory cortex of rhesus monkeys to predicted and to unpredicted sound stimuli. *Brain Res* 230:65–86.

Holstege, G. 1991. Descending motor pathways and the spinal motor system: Limbic and non-limbic components. *Prog Brain Res* 87:307–421.

Holstege, G., Bandler, R., and Saper, C.B. 1996. The emotional motor system. *Prog Brain Res* 107:3–6.

Issa, E.B., and Wang, X. 2008. Sensory responses during sleep in primate primary and secondary auditory cortex. *J Neurosci* 28:14467–80.

Ito, S.I. 1982. Prefrontal unit activity of macaque monkeys during auditory and visual reaction time tasks. *Brain Res* 247:39–47.

Iwata, J., LeDoux, J.E., Meeley, M.P., Arneric, S., and Reis, D.J. 1986. Intrinsic neurons in the amygdaloid field projected to by the medial geniculate body mediate emotional responses conditioned to acoustic stimuli. *Brain Res* 383:195–214.

Jarrell, T.W., Gentile, C.G., Romanski, L.M., McCabe, P.M., and Schneiderman, N. 1987. Involvement of cortical and thalamic auditory regions in retention of differential bradycardiac conditioning to acoustic conditioned stimuli in rabbits. *Brain Res* 412:285–94.

Jenkins, W.M., and Merzenich, M.M. 1984. Role of cat primary auditory cortex for sound-localization behavior. *J Neurophysiol* 52:819–47.

Jones, E.G. 2003. Chemically defined parallel pathways in the monkey auditory system. *Ann N Y Acad Sci* 999:218–33.

Jones, E.G., and Burton, H. 1976. Areal differences in the laminar distribution of thalamic afferents in cortical fields of the insular, parietal and temporal regions of primates. *J Comp Neurol* 168:197–247.

Juliano, S.L., Ma, W., and Eslin, D. 1991. Cholinergic depletion prevents expansion of topographic maps in somatosensory cortex. *Proc Natl Acad Sci USA* 88:780–84.

Kaas, J. 2000. Why is brain size so important: Design problems and solutions as neocortex gets bigger or smaller. *Brain Mind* 1:7–23.

Kaas, J., and Hackett, T. 2008. The functional neuroanatomy of the auditory cortex. In *The Senses*, eds. P. Doallos, and D. Oertel, pp. 765–80. Amsterdam: Elsevier.

Kaas, J.H., and Hackett, T.A. 1998. Subdivisions of auditory cortex and levels of processing in primates. *Audiol Neurootol* 3:73–85.

———. 2000. Subdivisions of auditory cortex and processing streams in primates. *Proc Natl Acad Sci USA* 97:11793–99.

Kajikawa, Y., de La Mothe, L., Blumell, S., and Hackett, T.A. 2005. A comparison of neuron response properties in areas A1 and CM of the marmoset monkey auditory cortex: Tones and broadband noise. *J Neurophysiol* 93:22–34.

Kajikawa, Y., de la Mothe, L.A., Blumell, S., Sterbing-D'Angelo, S.J., D'Angelo, W., Camalier, C.R., and Hackett, T.A. 2008. Coding of FM sweep trains and twitter calls in area CM of marmoset auditory cortex. *Hear Res* 239:107–25.

Kelly, K., Metzger, R., Mulette-Gillman, O., Werner-Reiss, U., and Groh, J. 2002. Representation of sound location in the primate brain. In *Primate Audition: Ethology and Neurobiology*, ed. A. Ghazanfar, pp. 177–98. London: CRC Press.

Keysers, C., Kohler, E., Umilta, M.A., Nanetti, L., Fogassi, L., and Gallese, V. 2003. Audiovisual mirror neurons and action recognition. *Exp Brain Res* 153:628–36.

Kikuchi-Yorioka, Y., and Sawaguchi, T. 2000. Parallel visuospatial and audiospatial working memory processes in the monkey dorsolateral prefrontal cortex. *Nat Neurosci* 3:1075–76.

Kilgard, M.P., and Merzenich, M.M. 1998. Cortical map reorganization enabled by nucleus basalis activity. *Science* 279:1714–18.

Komura, Y., Tamura, R., Uwano, T., Nishijo, H., Kaga, K., and Ono, T. 2001. Retrospective and prospective coding for predicted reward in the sensory thalamus. *Nature* 412:546–49.

Komura, Y., Tamura, R., Uwano, T., Nishijo, H., and Ono, T. 2005. Auditory thalamus integrates visual inputs into behavioral gains. *Nat Neurosci* 8:1203–9.

Kosmal, A., Malinowska, M., and Kowalska, D.M. 1997. Thalamic and amygdaloid connections of the auditory association cortex of the superior temporal gyrus in rhesus monkey (Macaca mulatta). *Acta Neurobiol Exp (Wars)* 57:165–88.

Kuraoka, K., and Nakamura, K. 2007. Responses of single neurons in monkey amygdala to facial and vocal emotions. *J Neurophysiol* 97:1379–87.

Kusmierek, P., and Rauschecker, J.P. 2009. Functional specialization of medial auditory belt cortex in the alert rhesus monkey. *J Neurophysiol* 102:1606–22.

Lakatos, P., Pincze, Z., Fu, K.M., Javitt, D.C., Karmos, G., and Schroeder, C.E. 2005. Timing of pure tone and noise-evoked responses in macaque auditory cortex. *Neuroreport* 16:933–37.

Langner, G., and Schreiner, C.E. 1988. Periodicity coding in the inferior colliculus of the cat. I. Neuronal mechanisms. *J Neurophysiol* 60:1799–1822.

LeDoux, J.E., Ruggiero, D.A., Forest, R., Stornetta, R., and Reis, D.J. 1987. Topographic organization of convergent projections to the thalamus from the inferior colliculus and spinal cord in the rat. *J Comp Neurol* 264:123–46.

LeDoux, J.E., Sakaguchi, A., and Reis, D.J. 1984. Subcortical efferent projections of the medial geniculate nucleus mediate emotional responses conditioned to acoustic stimuli. *J Neurosci* 4:683–98.

Leinonen, L., Hyvarinen, J., and Sovijarvi, A.R. 1980. Functional properties of neurons in the temporo-parietal association cortex of awake monkey. *Exp Brain Res* 39:203–15.

Luethke, L.E., Krubitzer, L.A., and Kaas, J.H. 1989. Connections of primary auditory cortex in the New World monkey, Saguinus. *J Comp Neurol* 285:487–513.

Ma, X., and Suga, N. 2005. Long-term cortical plasticity evoked by electric stimulation and acetylcholine applied to the auditory cortex. *Proc Natl Acad Sci USA* 102:9335–40.

Machado, C.J., and Bachevalier, J. 2006. The impact of selective amygdala, orbital frontal cortex, or hippocampal formation lesions on established social relationships in rhesus monkeys (Macaca mulatta). *Behav Neurosci* 120:761–86.

MacLean, P.D., and Newman, J.D. 1988. Role of midline frontolimbic cortex in production of the isolation call of squirrel monkeys. *Brain Res* 450:111–23.

Maier, J.X., and Ghazanfar, A.A. 2007. Looming biases in monkey auditory cortex. *J Neurosci* 27:4093–100.

Markowitsch, H.J., Emmans, D., Irle, E., Streicher, M., and Preilowski, B. 1985. Cortical and subcortical afferent connections of the primate's temporal pole: A study of rhesus monkeys, squirrel monkeys, and marmosets. *J Comp Neurol* 242:425–58.

Marler, P. 1975. On the origin of speech from animal sounds. In *The Role of Speech in Language*, eds. J. Kavanaugh, and J. Cutting, pp. 11–37. Cambridge: MIT Press.

McDermott, J., and Hauser, M.D. 2007. Nonhuman primates prefer slow tempos but dislike music overall. *Cognition* 104:654–68.

Merzenich, M.M., and Brugge, J.F. 1973. Representation of the cochlear partition of the superior temporal plane of the macaque monkey. *Brain Res* 50:275–96.

Metzger, R.R., Greene, N.T., Porter, K.K., and Groh, J.M. 2006. Effects of reward and behavioral context on neural activity in the primate inferior colliculus. *J Neurosci* 26:7468–76.

Miller, L.M., and Recanzone, G.H. 2009. Populations of auditory cortical neurons can accurately encode acoustic space across stimulus intensity. *Proc Natl Acad Sci USA* 106:5931–35.

Molinari, M., Dell'Anna, M.E., Rausell, E., Leggio, M.G., Hashikawa, T., and Jones, E.G. 1995. Auditory thalamocortical pathways defined in monkeys by calcium-binding protein immunoreactivity. *J Comp Neurol* 362:171–94.

Moore, B. 1997. The Psychology of Hearing, 4th edn. San Diego, CA: Academic Press.

Morecraft, R.J., Cipolloni, P.B., Stilwell-Morecraft, K.S., Gedney, M.T., and Pandya, D.N. 2004. Cytoarchitecture and cortical connections of the posterior cingulate and adjacent somatosensory fields in the rhesus monkey. *J Comp Neurol* 469:37–69.

Morel, A., and Kaas, J.H. 1992. Subdivisions and connections of auditory cortex in owl monkeys. *J Comp Neurol* 318:27–63.

Morton, E.S. 1975. Ecological sources of selection on avian sounds. *Am Naturalist* 109:17–34.

Nagy, A., Paroczy, Z., Norita, M., and Benedek, G. 2005. Multisensory responses and receptive field properties of neurons in the substantia nigra and in the caudate nucleus. Eur *J Neurosci* 22:419–24.

Nauta, W.J., and Whitlock, D.G. 1956. Subcortical projections from the temporal neocortex in Macaca mulatta. *J Comp Neurol* 106:183–212.

Nienhuis, R., and Olds, J. 1978. Changes in unit responses to tones after food reinforcement in the auditory pathway of the rat: Intertrial arousal. *Exp Neurol* 59:229–42.

Olds, J., Nienhuis, R., and Olds, M.E. 1978. Patterns of conditioned unit responses in the auditory system of the rat. *Exp Neurol* 59:209–28.

Oleson, T.D., Ashe, J.H., and Weinberger, N.M. 1975. Modification of auditory and somatosensory system activity during pupillary conditioning in the paralyzed cat. *J Neurophysiol* 38:1114–39.

Pandya, D.N., Hallett, M., and Kmukherjee, S.K. 1969. Intra- and interhemispheric connections of the neocortical auditory system in the rhesus monkey. *Brain Res* 14:49–65.

Pandya, D.N., and Rosene, D.L. 1993. Laminar termination patterns of thalamic, callosal, and association afferents in the primary auditory area of the rhesus monkey. *Exp Neurol* 119:220–34.

Pandya, D.N., Rosene, D.L., and Doolittle, A.M. 1994. Corticothalamic connections of auditory-related areas of the temporal lobe in the rhesus monkey. *J Comp Neurol* 345:447–71.

Petkov, C.I., Kayser, C., Augath, M., and Logothetis, N.K. 2006. Functional imaging reveals numerous fields in the monkey auditory cortex. *PLoS Biol* 4:e215.

Petkov, C.I., Kayser, C., Steudel, T., Whittingstall, K., Augath, M., and Logothetis, N.K. 2008. A voice region in the monkey brain. *Nat Neurosci* 11:367–74.

Petrides, M., and Pandya, D.N. 2002. Comparative cytoarchitectonic analysis of the human and the macaque ventrolateral prefrontal cortex and corticocortical connection patterns in the monkey. *Eur J Neurosci* 16:291–310.

———. 2007. Efferent association pathways from the rostral prefrontal cortex in the macaque monkey. *J Neurosci* 27:11573–86.

Peyron, R., Laurent, B., and Garcia-Larrea, L. 2000. Functional imaging of brain responses to pain. A review and meta-analysis. *Neurophysiol Clin* 30:263–88.

Pickles, J. 1988. *An Introduction to the Physiology of Hearing.* 2nd ed. San Diego, CA: Academic Press.

Poljak, S. 1926. The connections of the acoustic nerve. *J Anat* 60:465–69.

Polley, D.B., Steinberg, E.E., and Merzenich, M.M. 2006. Perceptual learning directs auditory cortical map reorganization through top-down influences. *J Neurosci* 26:4970–82.

Poremba, A., Malloy, M., Saunders, R.C., Carson, R.E., Herscovitch, P., and Mishkin, M. 2004. Species-specific calls evoke asymmetric activity in the monkey's temporal poles. *Nature* 427:448–51.

Poremba, A., Saunders, R.C., Crane, A.M., Cook, M., Sokoloff, L., and Mishkin, M. 2003. Functional mapping of the primate auditory system. *Science* 299:568–72.

Prakash, N., Cohen-Cory, S., Penschuck, S., and Frostig, R.D. 2004. Basal forebrain cholinergic system is involved in rapid nerve growth factor (NGF)-induced plasticity in the barrel cortex of adult rats. *J Neurophysiol* 91:424–37.

Price, J.L. 2006. Connections of orbital cortex. In *The Orbitofrontal Cortex*, eds. D. Zald, and S. Rauch, pp. 39–55. New York: Oxford University Press.

Rauschecker, J.P., and Tian, B. 2000. Mechanisms and streams for processing of "what" and "where" in auditory cortex. *Proc Natl Acad Sci USA* 97:11800–6.

———. 2004. Processing of band-passed noise in the lateral auditory belt cortex of the rhesus monkey. *J Neurophysiol* 91:2578–89.

Rauschecker, J.P., Tian, B., Pons, T., and Mishkin, M. 1997. Serial and parallel processing in rhesus monkey auditory cortex. *J Comp Neurol* 382:89–103.

Reale, R.A., and Imig, T.J. 1983. Auditory cortical field projections to the basal ganglia of the cat. *Neuroscience* 8:67–86.

Recanzone, G.H. 2000a. Response profiles of auditory cortical neurons to tones and noise in behaving macaque monkeys. *Hear Res* 150:104–18.

———. 2000b. Spatial processing in the auditory cortex of the macaque monkey. *Proc Natl Acad Sci USA* 97:11829–35.

———. 2008. Representation of con-specific vocalizations in the core and belt areas of the auditory cortex in the alert macaque monkey. *J Neurosci* 28:13184–93.

Recanzone, G.H., and Cohen, Y.E. 2009. Serial and parallel processing in the primate auditory cortex revisited. *Behav Brain Res* 206:1–7.

Recanzone, G.H., Guard, D.C., and Phan, M.L. 2000a. Frequency and intensity response properties of single neurons in the auditory cortex of the behaving macaque monkey. *J Neurophysiol* 83:2315–31.

Recanzone, G.H., Guard, D.C., Phan, M.L., and Su, T.K. 2000b. Correlation between the activity of single auditory cortical neurons and sound-localization behavior in the macaque monkey. *J Neurophysiol* 83:2723–39.

Recanzone, G.H., Schreiner, C.E., and Merzenich, M.M. 1993. Plasticity in the frequency representation of primary auditory cortex following discrimination training in adult owl monkeys. *J Neurosci* 13:87–103.

Remedios, R., Logothetis, N.K., and Kayser, C. 2009. An auditory region in the primate insular cortex responding preferentially to vocal communication sounds. *J Neurosci* 29:1034–45.

Rendall, D., Owren, M.J., and Rodman, P.S. 1998. The role of vocal tract filtering in identity cueing in rhesus monkey (Macaca mulatta) vocalizations. *J Acoust Soc Am* 103:602–14.

Rendall, D., Rodman, P.S., and Emond, R.E. 1996. Vocal recognition of individuals and kin in free-ranging rhesus monkeys. *Anim Behav* 51:1007–15.

Ringo, J.L., Doty, R.W., Demeter, S., and Simard, P.Y. 1994. Time is of the essence: A conjecture that hemispheric specialization arises from interhemispheric conduction delay. *Cereb Cortex* 4:331–43.

Roberts, A.C., Tomic, D.L., Parkinson, C.H., Roeling, T.A., Cutter, D.J., Robbins, T.W., and Everitt, B.J. 2007. Forebrain connectivity of the prefrontal cortex in the marmoset monkey (Callithrix jacchus): An anterograde and retrograde tract-tracing study. *J Comp Neurol* 502:86–112.

Rolls, E.T., Critchley, H.D., Browning, A.S., and Inoue, K. 2006. Face-selective and auditory neurons in the primate orbitofrontal cortex. *Exp Brain Res* 170:74–87.

Romanski, L.M. 2007. Representation and integration of auditory and visual stimuli in the primate ventral lateral prefrontal cortex. *Cereb Cortex* 17 (Suppl 1):i61–69.

Romanski, L.M., and Averbeck, B.B. 2009. The primate cortical auditory system and neural representation of conspecific vocalizations. *Annu Rev Neurosci* 32:315–46.

Romanski, L.M., Averbeck, B.B., and Diltz, M. 2005. Neural representation of vocalizations in the primate ventrolateral prefrontal cortex. *J Neurophysiol* 93:734–47.

Romanski, L.M., Bates, J.F., and Goldman-Rakic, P.S. 1999. Auditory belt and parabelt projections to the prefrontal cortex in the rhesus monkey. *J Comp Neurol* 403:141–57.

Romanski, L.M., and Goldman-Rakic, P.S. 2002. An auditory domain in primate prefrontal cortex. *Nat Neurosci* 5:15–16.

Romanski, L.M., and LeDoux, J.E. 1992. Equipotentiality of thalamo-amygdala and thalamo-cortico-amygdala circuits in auditory fear conditioning. *J Neurosci* 12:4501–9.

———. 1993. Information cascade from primary auditory cortex to the amygdala: Corticocortical and corticoamygdaloid projections of temporal cortex in the rat. *Cereb Cortex* 3:515–32.

Romanski, L.M., Tian, B., Fritz, J., Mishkin, M., Goldman-Rakic, P.S., and Rauschecker, J.P. 1999. Dual streams of auditory afferents target multiple domains in the primate prefrontal cortex. *Nat Neurosci* 2:1131–36.

Rouiller, E. 1997. Functional organization of the auditory pathways. In *The Central Auditory System*, eds. G. Ehret, and R. Romand, pp. 3–65. New York: Oxford University Press.

Rouiller, E.M., and Durif, C. 2004. The dual pattern of corticothalamic projection of the primary auditory cortex in macaque monkey. *Neurosci Lett* 358:49–52.

Rouiller, E.M., Rodrigues-Dagaeff, C., Simm, G., De Ribaupierre, Y., Villa, A., and De Ribaupierre, F. 1989. Functional organization of the medial division of the medial geniculate body of the cat: Tonotopic organization, spatial distribution of response properties and cortical connections. *Hear Res* 39:127–42.

Russ, B.E., Ackelson, A.L., Baker, A.E., and Cohen, Y.E. 2008. Coding of auditory-stimulus identity in the auditory non-spatial processing stream. *J Neurophysiol* 99:87–95.

Rutkowski, R.G., and Weinberger, N.M. 2005. Encoding of learned importance of sound by magnitude of representational area in primary auditory cortex. *Proc Natl Acad Sci USA* 102:13664–69.

Ryan, A., and Miller, J. 1978. Single unit responses in the inferior colliculus of the awake and performing rhesus monkey. *Exp Brain Res* 32:389–407.

Ryugo, D.K., and Weinberger, N.M. 1978. Differential plasticity of morphologically distinct neuron populations in the medical geniculate body of the cat during classical conditioning. *Behav Biol* 22:275–301.

Saleem, K.S., Kondo, H., and Price, J.L. 2008. Complementary circuits connecting the orbital and medial prefrontal networks with the temporal, insular, and opercular cortex in the macaque monkey. *J Comp Neurol* 506:659–93.

Sanes, D.H., and Bao, S. 2009. Tuning up the developing auditory CNS. *Curr Opin Neurobiol* 19:188–99.

Schrader, L., and Todt, D. 1993. Contact call parameters covary with social-context in common marmosets, Callithrix-j-jacchus. *Anim Behav* 46:1026–28.

Schultz, W. 2000. Multiple reward signals in the brain. *Nat Rev Neurosci* 1:199–207.

Scott, B.H., Malone, B.J., and Semple, M.N. 2007. Effect of behavioral context on representation of a spatial cue in core auditory cortex of awake macaques. *J Neurosci* 27:6489–99.

Seitz, A., and Watanabe, T. 2005. A unified model for perceptual learning. *Trends Cogn Sci* 9:329–34.

Selemon, L.D., and Goldman-Rakic, P.S. 1985. Longitudinal topography and interdigitation of corticostriatal projections in the rhesus monkey. *J Neurosci* 5:776–94.

Smiley, J.F., Hackett, T.A., Ulbert, I., Karmas, G., Lakatos, P., Javitt, D.C., and Schroeder, C.E. 2007. Multisensory convergence in auditory cortex, I. Cortical connections of the caudal superior temporal plane in macaque monkeys. *J Comp Neurol* 502:894–923.

Smith, W. 1945. The functional significance of the rostral cingular cortex as revealed by its responses to electrical excitation. *J Neurophysiol* 8:241–55.

Starr, A., and Don, M. 1972. Responses of squirrel monkey (Samiri sciureus) medial geniculate units to binaural click stimuli. *J Neurophysiol* 35:501–17.

Steinfels, G.F., Heym, J., Strecker, R.E., and Jacobs, B.L. 1983. Response of dopaminergic neurons in cat to auditory stimuli presented across the sleep-waking cycle. *Brain Res* 277:150–54.

Strecker, R.E., Steinfels, G.F., Abercrombie, E.D., and Jacobs, B.L. 1985. Caudate unit activity in freely moving cats: Effects of phasic auditory and visual stimuli. *Brain Res* 329:350–53.

Sudakov, K., MacLean, P.D., Reeves, A., and Marino, R. 1971. Unit study of exteroceptive inputs to claustrocortex in awake, sitting, squirrel monkey. *Brain Res* 28:19–34.

Sugihara, T., Diltz, M.D., Averbeck, B.B., and Romanski, L.M. 2006. Integration of auditory and visual communication information in the primate ventrolateral prefrontal cortex. *J Neurosci* 26:11138–47.

Symmes, D., Alexander, G.E., and Newman, J.D. 1980. Neural processing of vocalizations and artificial stimuli in the medial geniculate body of squirrel monkey. *Hear Res* 3:133–46.

Tian, B., and Rauschecker, J.P. 2004. Processing of frequency-modulated sounds in the lateral auditory belt cortex of the rhesus monkey. *J Neurophysiol* 92:2993–3013.

Tian, B., Reser, D., Durham, A., Kustov, A., and Rauschecker, J.P. 2001. Functional specialization in rhesus monkey auditory cortex. *Science* 292:290–93.

Trojanowski, J.Q., and Jacobson, S. 1975. A combined horseradish peroxidase-autoradiographic investigation of reciprocal connections between superior temporal gyrus and pulvinar in squirrel monkey. *Brain Res* 85:347–53.

Ungerleider, L., and Mishkin, M. 1982. Two cortical visual systems. In *Analysis of Visual Behavior*, eds. Ingle, D.J., Goodale, M.A., and Mansfield, R.J.W., pp. 549–86. Cambridge, MA: MIT Press.

Vogt, B.A., and Pandya, D.N. 1987. Cingulate cortex of the rhesus monkey: II. Cortical afferents. *J Comp Neurol* 262:271–89.

Wang, X., Lu, T., Bendor, D., and Bartlett, E. 2008. Neural coding of temporal information in auditory thalamus and cortex. *Neuroscience* 157:484–94.

Wang, X., Merzenich, M.M., Beitel, R., and Schreiner, C.E. 1995. Representation of a species-specific vocalization in the primary auditory cortex of the common marmoset: Temporal and spectral characteristics. *J Neurophysiol* 74:2685–706.

Weinberger, N., Ashe, J., Metherate, R., McKenna, T., Diamond, D., and Bakin, J. 1990. Retuning auditory cortex by learning: A preliminary model of receptive field plasticity. *Concepts Neurosci* 1:91–132.

Weinberger, N.M. 1998. Physiological memory in primary auditory cortex: Characteristics and mechanisms. *Neurobiol Learn Mem* 70:226–51.

———. 2003. The nucleus basalis and memory codes: Auditory cortical plasticity and the induction of specific, associative behavioral memory. *Neurobiol Learn Mem* 80:268–84.

———. 2004. Specific long-term memory traces in primary auditory cortex. *Nat Rev Neurosci* 5:279–90.

West, C.D. 1985. The relationship of the spiral turns of the cochlea and the length of the basilar membrane to the range of audible frequencies in ground dwelling mammals. *J Acoust Soc Am* 77:1091–1101.

Wiley, R.H., and Richards, D.G. 1978. Physical constraints on acoustic communication in atmosphere – implications for evolution of animal vocalizations. *Behav Ecol Sociobiol* 3:69–94.

Wilson, F.A., and Rolls, E.T. 1990. Neuronal responses related to reinforcement in the primate basal forebrain. *Brain Res* 509:213–31.

Winer, J.A. 2005. Decoding the auditory corticofugal systems. *Hear Res* 207:1–9.

Winer, J.A., Chernock, M.L., Larue, D.T., and Cheung, S.W. 2002. Descending projections to the inferior colliculus from the posterior thalamus and the auditory cortex in rat, cat, and monkey. *Hear Res* 168:181–95.

Winer, J.A., Larue, D.T., Diehl, J.J., and Hefti, B.J. 1998. Auditory cortical projections to the cat inferior colliculus. *J Comp Neurol* 400:147–74.

Winer, J.A., and Lee, C.C. 2007. The distributed auditory cortex. *Hear Res* 229:3–13.

Winer, J.A., Wenstrup, J.J., and Larue, D.T. 1992. Patterns of GABAergic immunoreactivity define subdivisions of the mustached bat's medial geniculate body. *J Comp Neurol* 319:172–90.

Woods, T.M., Lopez, S.E., Long, J.H., Rahman, J.E., and Recanzone, G.H. 2006. Effects of stimulus azimuth and intensity on the single-neuron activity in the auditory cortex of the alert macaque monkey. *J Neurophysiol* 96:3323–37.

Yeterian, E.H., and Pandya, D.N. 1998. Corticostriatal connections of the superior temporal region in rhesus monkeys. *J Comp Neurol* 399:384–402.

Zatorre, R.J., Evans, A.C., and Meyer, E. 1994. Neural mechanisms underlying melodic perception and memory for pitch. *J Neurosci* 14:1908–19.

10 Sensory Agnosias

H. Branch Coslett

CONTENTS

10.1 INTRODUCTION AND HISTORICAL OVERVIEW

Sensory agnosias are relatively uncommon clinical syndromes characterized by a failure of recognition that cannot be attributed to the loss of primary sensory function, inattentiveness, general mental impairment, or lack of familiarity with the stimulus (Fredericks 1969; Bauer 2003). In other words, sensory agnosias are disorders that are bracketed by failures of early sensory processing on the input side and inability to attend to or comprehend the output of high-level sensory processing on the output side. For example, a blind subject who fails to recognize a rose or a deaf subject who does not recognize the sound of a hammer driving a nail are not agnosic; similarly, a demented person who no longer knows what a hammer is would not be considered agnosic if he failed to recognize the object.

Although the term "agnosia" was coined by Freud (1891) in his discussion of aphasia and related disorders, descriptions of the disorder predate Freud. Finkelnburg (1870) described the syndrome of "asymbolia" in which a percept failed to contact the knowledge relevant to that percept and Hughlings Jackson (1876) described the phenomenon of "imperception" in the context of a patient with a large tumor of the posterior portion of the right hemisphere who was unable to navigate in familiar places and did not recognize people or places. He postulated that the posterior portion of the right hemisphere was crucial for visual memory and recognition. Munk (1881) provided an early and influential description of dogs with parieto-occipital lesions that were able to navigate about their surroundings without bumping into objects or getting lost yet didn't recognize objects of obvious relevance such as food; he termed this disorder "Seelenblindheit" or mindblindness. Interestingly, similar findings were subsequently reported in macaques after lesions of the occipital lobe and temporal projections (Horel and Keating 1972).

An important early theoretical contribution was made by Lissauer (1890), who distinguished between two forms of the disorder: "apperceptive" and "associative" agnosias. As this terminology continues to be widely employed in the clinical literature, it will be briefly reviewed. For Lissauer "apperception" referred to the "stage of conscious awareness of a sensory impression" (translated into English by Jackson [1932]). Apperceptive agnosias, in Lissauer's scheme, represented a disorder in which the early visual processing of a stimulus is disrupted, with a resulting failure to generate a fully specified perceptual representation. Associative agnosias, in contrast, were characterized by preserved ability to compute a representation of a visual stimulus but an inability to recognize the object, as indicated by the ability to name the object or produce verbal or non-verbal information that would unambiguously identify the stimulus. Presaging a number of contemporary accounts of the processes involved in recognition (e.g., Damasio 1989; Simmons and Barsalou 2003; Farah 2000; Humphreys, Riddoch, and Fortt 2006), for Lissauer recognition entailed the simultaneous activation of multiple attributes (e.g., sound, touch, smell, etc.) that were linked to the visual form. That is, for Lissauer as for a number of contemporary theorists, "recognition" of a telephone entailed the contemporaneous activation of the visual image, sound, heft, texture, manner of use, and function of a telephone (Allport 1985).

Although the boundaries between agnosia and primary visual loss on the input side and disorders of semantics at a higher level were hazy in Lissauer's account (and remain so today), he offered a clear distinction between apperceptive and associative visual agnosias: the ability to copy a visual stimulus. Reflecting their inability to generate an adequate sensory representation of the stimulus, apperceptive agnosics are unable to copy a stimulus whereas associative agnosics are able to copy a figure but remain unable to recognize what they have copied (Rubens and Benson 1971). As has been emphasized by Farah (2004) in her authoritative account of the visual agnosias, the reproductions of associative agnosics may be extremely detailed but appear to be slavish copies of the stimulus, uninformed by stored knowledge of the stimulus. As indicated in Figure 10.1, when asked to copy a drawing that has been deliberately distorted, associative agnosics may include the distortion in their drawing without appearing to be aware of the error.

Unlike other classical neurologic syndromes such as aphasia or neglect, the status of the concept of agnosia has varied substantially over the century since its description. In the early portion of the

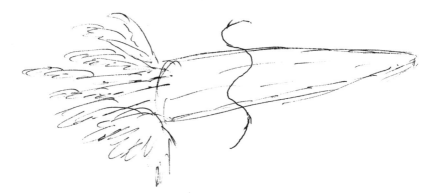

FIGURE 10.1   A copy of a drawing by a visual agnosic of a carrot with a line drawn through it. The patient failed to recognize the carrot; reflecting the lack of top-down processing from stored knowledge of the object, his copy incorporated the out of place line.

twentieth century, the discussion of the disorder was largely framed by the prevailing Gestalt theory of psychology; Poppelreuter (1923) and Goldstein (Goldstein 1943; Goldstein and Gelb 1918), for example, interpreted their subjects' behaviors with respect to such concepts such as "closure" and "invariance."

The very existence of the phenomenon of visual agnosia was questioned by a number of investigators. This was expressed by Pavlov (1927), for example, who, in response to Munk's phrase "the dog sees but does not understand," countered that "the dog understands but does not see well enough." In later years Bay (1953) and Bender and Feldman (1972) argued that apparent visual agnosias were attributable to perceptual impairment (e.g., "tunnel vision"), general cognitive impairment, or some combination thereof.

In recent years, there has been a resurgence of interest in the topic for several reasons. One is the development of more sophisticated and nuanced models of "recognition" according to which perception is taken to be a multi-stage interactive process; on such accounts, the distinction between perceptual disorders and agnosia is seen not as dichotomous but rather as a process in which sensory inputs give rise to progressively more elaborated representations in which different types of information (e.g., shape, color, location) may be emphasized (Heinke and Humphreys 2003; Ellis and Young 1988). Second, in light of the increasingly complex and interactive models of recognition, the heuristic value of data from patients with agnosia has proven to be substantial. As illustrated by the influential contributions of Farah (2004), Riddoch and Humphreys (1987), and others, data from agnosic subjects may serve to indicate the fault lines in the process of recognition and offer important constraints for accounts based on animal work, modeling, and studies of normal subjects.

10.2 TYPES OF AGNOSIA

As sensory agnosia is defined as a modality-specific disorder of recognition, the syndrome may be encountered in any sensory channel that permits entities to be identified. Reflecting the central role of vision in humans as well as the fact that vision has been studied more extensively than other sensory modalities, most work has focused on visual agnosia. Tactile, auditory, and even olfactory and gustatory agnosias have also been described. In this chapter we focus on different types of visual agnosia and present a theoretical framework for understanding them before discussing other types of agnosia.

10.2.1 Visual Agnosia

Most studies of visual agnosia have emphasized the recognition of man-made objects. More recently, however, investigations of agnosic subjects have demonstrated remarkable specificity with respect to the types of stimuli with which subjects may be impaired. Agnosias for objects, faces ("prosopagnosia"), words ("pure word blindness"), colors, and the environment (including landmarks) have all been described. Finally, the disorder of simultanagnosia—an inability to "see" more than one object at a time—is often regarded as a type of visual agnosia. Consistent with these behavioral dissociations, functional imaging in normal subjects demonstrates that different parts of the brain may, at least to some degree, be optimized for the processing of different classes of visual stimuli. We briefly discuss these types of visual agnosia in turn. Before considering the specific disorders of higher-level visual processing, however, we present an information-processing account of visual processing. A detailed account of visual processing is beyond the scope of this chapter but can be found in Chapter 8; the model described briefly below is intended only to provide a framework within which the complex and rich clinical literature can be understood.

10.2.1.1 Visual Object Recognition: A Theoretical Overview

The most important insight regarding visual processing during the last century has been the recognition that different types of visual information are segregated at the retina and that different types of information are processed in parallel. Although there are numerous and strong interactions between the processing streams, the segregation persists until late in visual processing, as indicated by the fundamental distinction between the ventral "what" stream and the dorsal "where" (Ungerleider and Mishkin 1982) or "how to" (Milner and Goodale 1995) streams. A cartoon depicting the basic architecture of the visual processing system is presented in Figure 10.2.

Physiologic, anatomic and, more recently, imaging studies have demonstrated conclusively that different visual attributes such as color, angle, motion, depth, orientation, and length are processed in parallel in different brain regions. Low-level visual routines such as "boundary marking" (e.g., Ullman 1984; Borenstein and Ullman 2008) serve to group different stimuli leading to visual forms. Although these processes were intensively investigated by Gestalt psychologists, the neural bases of the routines remain poorly understood. These processes, including grouping of visual features into candidate "objects" and marking the boundary of candidate objects, appear to be crucial for object recognition. These processes do not require attention or effort and are assumed to be "automatic."

Although there is convincing evidence that visual attributes are processed in parallel, the visual environment consists of stimuli or regions of space in which the attributes are integrated; that is, in order to generate an interpretable, coherent picture of the environment, information that is processed in parallel in different brain regions must be integrated so that, for example, the color red and oval shape are linked to generate the percept of an apple. Abundant experimental evidence demonstrates that this "binding" of visual feature information is mediated by a limited-capacity operation typically referred to as "visual attention" (the neural mechanisms of which are described in detail in Chapter 8). The binding function of visual attention is illustrated by the phenomenon of "illusory conjunctions." Treisman (Treisman and Gelade 1980; Treisman and Souther 1985) and others (Cave 1999; Prinzmetal et al. 2002) have demonstrated that when visual attention is "overloaded," visual attributes can miscombine. For example, when presented with an array containing red X's and green T's for 200 ms, normal subjects may report seeing a red T despite the fact that no such stimulus was present. These and a host of similar findings suggest that a limited-capacity, relatively fast but not infallible, "glue" links visual feature information computed in parallel. This glue is visual attention. As suggested by the commonly used "spotlight" metaphor (Eriksen and Hoffmann 1973; Posner

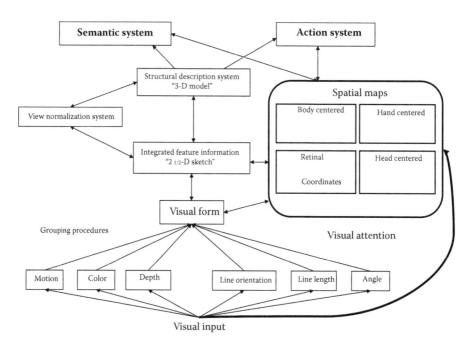

FIGURE 10.2 A cartoon of the processes involved in visual object recognition.

1980), visual attention appears to be spatially based under most circumstances. Experimental evidence suggests that attention can also be allocated to other visual attributes, including color (Cave 1999), motion, and even objects (Duncan 1984; Vecera and Farah 1994).

Limited-capacity operations that serve to select visual information for additional processing (including integration) have been demonstrated at many levels of neural processing, ranging from relatively high-level vision (Corbetta 1998), intermediate vision (Moran and Desimone 1985), and even the primary visual cortex (Vidyasagar 1998). Despite the fact that these operations differ somewhat from the procedure by which visual features are integrated, this process is also typically described as visual attention. As will be discussed below, disruption of visual attention at different levels of visual processing may give rise to distinctly different clinical syndromes.

The integration of visual feature information generates a viewer-centered representation of the orientations and depths of the surfaces of an object as well as the discontinuities between the surfaces. This representation is similar in most important respects to the "2 1/2D model" described by Marr (1982), which makes explicit the form, shape, and volume of the object as well as the hierarchical relationships between the parts. This type of representation has been termed a "structural description" (Riddoch and Humphreys 1987) and is similar to the 3-D representation of Marr (1982) in that it is assumed to be perspective independent; that is, at this level of processing, the representation computed by the brain specifies the relationships between the visual features in an "object-centered" fashion. The nature of the processes that mediate between the viewer-dependent, integrated feature representation and the perspective-independent or object-centered structural description remain unknown. As reviewed briefly by Farah (2000), a variety of accounts have been proposed, from connectionist architectures to template-matching procedures. Details of these proposals are beyond the scope of this chapter.

As will be discussed below, a large number of imaging studies suggest that the fusiform gyri and lateral occipital region are crucial anatomical substrates for the integrated feature representation and structural description systems (Grill-Spector, Knouf, and Kanwisher 2004; Haxby, Hoffman, and Gobbini 2000 for reviews).

Familiar objects are quickly and effortlessly recognized when viewed across a wide range of angles and perspectives. A bowl, for example, may be viewed from the side when sitting in a dishwasher, from below when stacked on a high shelf, or from above when clearing a table.

Although the low-level visual information regarding surfaces, color, form, etc. is remarkably different in these circumstances, under normal circumstances these stimuli are immediately recognized as the same. In the context of Figure 10.2, the mechanism that supports object constancy is termed the view normalization system.

The visual processing discussed to this point is in the service of object recognition—that is, knowledge of the form, function, name, and other attributes of entities in the environment. Following Lissauer (1890) as well as recent accounts of semantic representations (Allport 1985; Damasio 1989; Rogers et al. 2004; Saffran and Schwartz 1994; Warrington and Shallice 1984; Simmons and Barsalou 2003), I suggest that recognition entails the simultaneous activation of the many aspects of knowledge-specific knowledge; thus, as argued by Lissauer (1890), for example, recognition of a violin consists of the activation of stored knowledge regarding the sound, heft, feel, and manner of manipulation of the instrument.

With this overview of the series of processes underlying visual recognition in mind, we now turn to a consideration of the different types of visual agnosia. Rather than use the traditional "apperceptive–associative" distinction described above, we discuss the clinical phenomena with reference to the putative locus of the processing deficit exhibited by the patient.

10.2.1.2 Disorders of Low-Level Vision

A variety of clinical disorders have been described in which the primary deficit is a disruption of processing of different types of visual feature information. These are reviewed briefly in this section.

10.2.1.2.1 Achromatopsia

Achromatopsia is an acquired disorder of color perception characterized by a loss of the ability to distinguish color. The disorder is probably far more common than widely appreciated because it varies in severity from a mild loss of the richness of color (e.g., "red desaturation") to a complete loss of the sense of color. Milder forms are often not recognized by the patient whereas in more severe forms the patient may indicate that they see the world in black and white or shades of gray. The defect is often in one visual field or part thereof but may involve the entire visual field if both hemispheres are affected.

Since Verrey's initial report of the disorder in 1888, achromatopsia has been consistently associated with lesions involving the lingual or fusiform gyri. Subsequent studies with static brain imaging (Damasio et al. 1980) and functional brain imaging have yielded generally similar results. Reflecting this anatomic substrate, achromatopsia may be observed in isolation or in association with conditions such as prosopagnosia, alexia, or superior visual field deficits that are associated with lesions to nearby cortex (e.g., Pearlman, Birch, and Meadows 1979). Interestingly, the loss of color perception is not typically associated with visual object agnosia.

10.2.1.2.2 Impaired Motion Perception (Akinetopsia)

Reflecting its crucial role in vision, motion cells that appear to be optimized for motion perception may be identified at the retina, lateral geniculate, primary visual cortex, and a number of higher-level visual cortices, including MT, a cortical region that appears to be specialized for motion processing (see below).

Relatively pure disorders of motion perception are rare. The syndrome was first reported almost a century ago (Goldstein and Gelb 1918; Potzl and Redlich 1911). In recent years, a subject with this disorder, LM, has been extensively investigated (e.g., Zihl, Cramon, and Mai 1983). LM developed a profound impairment in the ability to detect motion after bilateral infarcts involving the posterior portions of the middle temporal gyri extending into the occipital lobe as well as adjacent subcortical white matter. The deficit had profound consequences for her ability to negotiate her environment. She did not perceive movement as a continuous process but stated that objects seemed to jump from one position to the next. When she poured water into a cup, the liquid appeared to be static, like a piece of ice. Although profoundly impaired in motion perception, LM performed well on other measures of visual processing. Her visual fields were full and she performed normally on tests of stereopsis, visual acuity, color perception, and critical flicker fusion. At least under most circumstances, LM exhibited no impairment in object recognition.

Several patients have been reported whose object recognition is influenced by motion. Botez and Serbanescu (1967) reported two patients with a "static form agnosia" characterized by a failure to recognize stationary stimuli but much improved performance when the same stimuli were moved. A similar but perhaps less striking facilitation of recognition with movement was exhibited by the "visual form agnosic" reported by Benson and Greenberg (1969) as well as a patient reported by Horner and Massey (1986). We have also observed this phenomenon in a number of patients with hypoxic brain injury or degenerative diseases preferentially involving the posterior portions of the hemispheres (that is, "posterior cortical atrophy"). In these patients, it appeared that patients were unable to reliably deploy visual attention to different stimuli in the array; movement seemed to permit the subjects to foveate the stimulus that was then recognized by normal procedures.

Data from patients with achromatopsia as well as abundant fMRI studies implicate the posterior portion of the middle temporal gyrus ("V5") as a cortical region specialized for motion processing (Kable, Lease-Spellmeyer, and Chatterjee 2002; Watson et al. 1993).

10.2.1.3 Visual Form Agnosia

One well-described but relatively rare syndrome, "visual form agnosia," may represent a deficit in the segregation of coherent regions of visual input. One such subject, a 24-year-old man who had suffered carbon monoxide poisoning (Mr. S), was reported by Benson and Greenberg (1969) and investigated in detail by Efron (1968). Mr. S. was relatively intact with respect to general cognitive function but was substantially impaired in recognizing visually presented objects, drawings, letters, or faces. Extensive testing of several low-level visual attributes revealed normal performance on tasks requiring the detection of differences in luminance, wavelength, area, and motion. Perhaps his most striking visual deficit was a profound impairment in the ability to discriminate shape; for example, if presented with a square and a tall, skinny rectangle (height:width ratio of 4:1), matched for area, color, and other visual qualities, he was unable to indicate if the stimuli were the same or different shapes.

More recently, Milner and Heywood (1989) reported a second visual form agnosic, DF; like LM, she had suffered carbon monoxide poisoning. Structural MRI demonstrated bilateral lesions involving the lateral occipital area. She exhibited profound visual recognition problems and performed poorly on tasks requiring that she discriminate between different shapes.

DF exhibited a finding of great theoretical interest that had not been described previously. Although unable to distinguish between visual forms or to name objects, she performed normally with respect to hand posture and shape when asked to pick up the objects; thus, when asked to pick up rectangles whose shape she was unable to describe, the distance between her thumb and index finger and timing of the movements of the fingers in the reach trajectory were normal. Thus, information regarding visual form that was not available for the purposes of object analysis was available to the motor system.

10.2.2 DISORDERS OF VISUAL ATTENTION

As described above, visual attention is a limited-capacity resource that is typically accorded a variety of roles in visual processing. Reflecting the diversity of functions attributed to visual attention, it is perhaps not surprising that disorders of visual attention have been implicated in a number of different clinical syndromes.

Perhaps the prototypical syndrome of this type is "simultanagnosia," an inability to "see" more than one object in an array (Wolpert 1924). The first detailed description of this syndrome was by Balint (1909), who described a patient with bilateral posterior parietal infarcts who was able to identify visually presented familiar objects when presented in isolation but exhibited a striking difficulty in the processing of visual arrays. For example, when shown a letter and a triangle, he reported seeing only the letter; when told that a second object was present, he reported the triangle but no longer saw the letter. As Balint's patient had normal visual fields and visual acuity, the disorder could not be attributed to low-level visual processing deficits. Similar patients have been reported by a number of investigators (Holmes 1918; Luria 1959; Coslett and Saffran 1991; Coslett and Lie 2008).

Simultanagnosic subjects are often impaired in recognizing single objects as well as arrays. The deficit with single objects usually consists of a failure to appreciate the entire stimulus. For example, when confronted with a complex object such as a car, subjects with this disorder may report only a tire; similarly, when shown the word "table," subjects may not see the entire word but report constituent letters, often identifying the word in a letter-by-letter fashion. Object size does not appreciably influence performance. When confronted with an array, simultanagnosics often report seeing only one item at a time.

Farah (2004) introduced the distinction between "dorsal" and "ventral" simultanagnosia. The former is associated with dominant hemisphere posterior lesions and is usually associated with a hemianopia. Dorsal simultanagnosia is associated with bilateral posterior parietal-occipital lesions (Rizzo and Hurtig 1987). We have recently suggested that dorsal simultanagnosia may be further subdivided (Coslett and Chatterjee 2003). On our analysis, one form of dorsal simultanagnosia may be attributable to an impairment in the process by which visual attention is allocated or serves to integrate visual feature information. As might be expected given the role accorded visual attention in the integration of visual feature information, patients with this disorder generate frequent "illusory conjunctions" characterized by the incorrect combination of visual features; these subjects may report, for example, a red *T* when shown an array of red *X*'s and green *T*'s (Pavese et al. 2002; Robertson et al. 1997). We suggested that simultanagnosia may also be attributable to an impairment in linking object location and identity (Coslett and Chatterjee 2003). Consistent with this perspective, one subject with simultanagnosia was impaired in reporting not only more than one object in an array, but also more than one attribute of a single object. For example, he was unable to report both the color of the ink in which a word was written as well as the word. For both this and a previously reported subject, performance was significantly influenced by semantic factors; for example, both subjects were able to report two items from an array on a significantly greater number of trials if the items were semantically related (e.g., both tools) as compared to trials on which they were unrelated (e.g., one tool, one animal). Thus, data from these subjects demonstrated that the patients were, in fact, processing visual information that they were unable to report.

Disorders of visual attention are a prominent feature of the "simultanagnosia-like" disturbance encountered in other conditions such as the syndrome of "posterior cortical atrophy" (Benson, Davis, and Snyder 1988). The most common cause of this syndrome is Alzheimer's disease. Like patients with simultanagnosia from focal parieto-occipital lesions, these patients often identify single objects relatively well but are substantially impaired in the processing of arrays. These patients are often severely impaired in everyday activities such as finding an item in the refrigerator or the butter dish on the table. They may be unable to find their way in their own home; we have encountered a number of patients whose visual world is so "confusing" that they close their eyes when walking or searching a complex array. The visual processing deficit in these patients typically differs in one telling way from that of simultanagnosic patients with focal parieto-occipital lesions: they exhibit a striking effect of object size (Coslett et al. 1995; Saffran, Fitzpatrick-DeSalme, and Coslett 1990). We have suggested that these and other findings exhibited by these patients are consistent with the hypothesis that the patients suffer from a pathologic restriction in their capacity to integrate visual feature information.

10.2.3 Impairments in the View Normalization System

Impairments in identifying objects seen from unusual perspectives or matching objects across different views (see Figure 10.2) is a relatively common but rarely diagnosed disorder. The first systematic investigations of the phenomenon were reported by Warrington and Taylor (1973, 1978). In one study they asked patients to name two sets of 20 objects, one depicting the object in a canonical or standard view and the other in an unusual view. They found that subjects with lesions involving the posterior portion of the right (non-dominant) hemisphere were selectively impaired in the recognition of the unusual views of these objects. Based on a series of elegant studies and tasks, Humphreys and Riddoch (1984, 2006) suggested that the deficit may be observed for different reasons. Thus, they reported one patient who appeared to rely on distinctive feature information; other patients, in contrast, have been found to be particularly sensitive to foreshortening (see also Warrington and James 1986), suggesting that these patients are unable to generate a representation of the object relative to its principal axis.

10.3 IMPAIRMENTS OF THE STRUCTURAL DESCRIPTION SYSTEM ("ASSOCIATIVE AGNOSIAS")

Failures of object recognition in the context of visual processing that is adequate to support copying of a stimulus have been reported on a number of occasions (Davidoff and Wilson 1985; Levine 1978; Levine and Calvanio 1989; see Farah 2004 for a comprehensive review). Rubens and Benson (1971) reported a patient who exhibited a striking form of associative agnosia after a hypotensive episode. Elementary visual function was normal except for right hemianopia; the subject was unable to name most objects yet drew accurate and quite detailed depictions of the very objects that he was unable to name.

From Lissauer's initial account to the present, the pattern of deficits displayed by Rubens and Benson's patient has been attributed to disruption of the structural description system. Associative visual agnosia is differentiated from loss of knowledge of the object (semantics) by the fact that, when queried verbally or by means of auditory or tactile input, the patients are able to provide appropriate information about objects that they are unable to recognize visually; thus, when asked to provide a verbal description of a hammer, for example, an associative agnosic would be expected to describe its function, heft, and sound as well as demonstrate its manner of manipulation. Similarly, the patient would be expected to be able to name the hammer after handling it.

Despite the fact that they are able to describe or copy visual stimuli quite adequately, it seems unlikely that they are contacting normal structural descriptions for several reasons. First, as emphasized by Farah (2000), copies generated by visual agnosics are typically produced in a "slavish" and painstaking fashion, presumably reflecting a reliance on the exact physical attributes of the stimulus with little or no "top-down" input from stored knowledge of object appearance. This point is illustrated by the copy depicted in Figure 10.2; this subject copied the stimulus—including the extraneous, sinusoidal line—without distinguishing between the irrelevant and object-specific information. Second, it is noteworthy that the quality of the copies generated by associative agnosics is typically strongly influenced by the richness of the visual image; subjects generally perform best with real objects, less well with pictures, and worst with line drawings. This hierarchy would not be expected if all three types of stimuli contacted the same structural description.

Finally, it should be noted that when asked to match or sort complex non-object figures, all associative agnosics of which we are aware exhibit impairments (Humphreys and Riddoch 1987; Ratcliff and Newcombe 1982; see Farah 2004 for discussion). This observation suggests that these subjects have some degree of impairment in the processes by which structural descriptions are accessed.

It should be emphasized that although the model depicted in Figure 10.1 provides a useful framework for considering the general principles involved in visual processing and their breakdown, it does not readily accommodate the full range of agnosic performance. One well-studied "integrative agnosic," HJA (Riddoch and Humphreys 1987), illustrates this fact. Although the details of the subject's performance are beyond the scope of this chapter, it is noteworthy that HJA appears to be an associative agnosic (Farah 2004), whereas in other regards his performance is more typical of apperceptive agnosia (Riddoch and Humphreys 2003).

10.3.1 Prosopagnosia

Prosopagnosia, or face blindness, is a condition in which subjects are unable to identify another person by viewing the person's face (Bodamer 1947). Although they are often capable of recognizing that a face is a face and, in many instances, are able to indicate the gender, age, and likely occupation of the person, they are unable to identify the person from vision alone. The deficit may be so profound that subjects are unable to identify their own face or that of immediate family members (Bauer and Verfaellie 1986).

Prosopagnosia may be asymptomatic in some cases. Many subjects rely on information regarding hairstyle, habitus, clothing, or voice to circumvent their visual deficits. In some instances, the deficit may only be apparent when subjects encounter a familiar person in an atypical setting. We have seen patients, for example, who were only noted to be prosopagnosic when they failed to recognize a close family member who was wearing a nursing uniform. In this context, the absence of cues from clothes, hairstyle, and context unmasked a deficit for which the patient typically compensated quite successfully. Many prosopagnosic subjects exhibit additional deficits in other aspects of visual processing such as achromatopsia, visual field deficit, topographical memory loss, or object agnosia.

Although "pure" prosopagnosics have been described, many patients with the disorder exhibit deficits in the recognition of other types of stimuli at a subordinate level. That is, subjects may be able to recognize dogs but not able to discriminate between their dog and others of the same species (see Borenstein, Sroka, and Munitz 1969).

A variety of hypotheses have been advanced to explain the phenomenon of prosopagnosia. On some accounts, the deficit is assumed to be an inability to discriminate individual exemplars within a class of visually similar items (Damasio, Damasio, and Hoesen 1982). Citing the double disso-ciation between visual object agnosia and prosopagnosia (Moscovitch, Winocur, and Behrmann 1997), other investigators have argued that faces are processed by a distinct system; Ellis and Young (1988), for example, have proposed that processing of faces is in some respects akin to that of objects, in that "early" disruptions of the face processing system are associated with "apperceptive" prosopagnosia whereas disorders later in the information processing cascade are associated with "associative" prosopagnosia. On these accounts, subjects who are unable to derive information about gender and other attributes from faces fall into the former category whereas patients who performed well on these tasks but who could not identify the subject would fall into the latter category.

A third account has been articulated by Farah (2004). She noted that whereas deficits restricted to words or faces have been described with some regularity, patients exhibiting deficits with objects in the context of normal recognition of faces or words are quite rare (but see Buxbaum, Glosser, and Coslett 1996). These findings led Farah (2004) to propose that two distinct but interactive modes of object processing subserve stimulus recognition; by her account, one mechanism is specialized for objects that are processed as a unit, that is, with relatively little decomposition into simpler parts; various lines of evidence support the view that faces are processed in this manner. A second mechanism is specialized for stimuli that undergo substantial decomposition; for these items, the parts into which they are segregated may be discrete objects themselves. As they are composed of a finite number of discrete objects and their identity is entirely determined by the constituent units, words may represent the prototypical stimulus for the decomposition procedure. Intense debate continues regarding the degree to which the processing of faces represents a distinct module or is best conceptualized as one end of a continuum along the dimension of holistic (configural) as opposed to decompositional (elemental) processing (for sensory-perceptual perspectives on this debate, see Chapter 5).

Prosopagnosia is usually associated with lesions of the temporo-occipital regions bilaterally (Damasio, Damasio, and Hoesen 1982). Several well-studied cases (e.g., DeRenzi 1986; Michel, Pernin, and Sieroff 1986) have been described with lesions involving only the right hemisphere. Additionally, a substantial functional imaging literature has identified a region of the fusiform gyrus bilaterally—the "fusiform face area"—that is reliably activated by faces (for a recent review see Harris and Aguirre 2008). Whether this should be considered a cortical module dedicated to the processing of faces or a part of the broader visual recognition system that is optimized for stimulus properties that characterize faces has been a topic of substantial debate (Haxby, Hoffman, and Gobbini 2000; McKone and Kanwisher 2004; Tarr and Gauthier 2000).

10.3.2 AGNOSIA FOR WORDS

This is also known as pure alexia, alexia without agraphia, or pure word blindness. Although this condition is usually discussed in the context of language impairments, it is an agnostic symptom, as the condition is restricted to a specific type of visual stimulus; most subjects with this condition do not exhibit impairment in language (e.g., speech, writing, auditory comprehension). As noted initially by Dejerine (1892) and confirmed by multiple investigators since (e.g., Saffran and Coslett 1996), the disorder is typically associated with lesions involving the dominant occipital lobe and the splenium of the corpus callosum that serve to deprive the left hemisphere of visual input while simultaneously disconnecting the right occipital lobe from the left peri-Sylvian cortex.

10.4   CATEGORY-SPECIFIC RECOGNITION DEFICITS

As noted previously, for some patients the ability to recognize visually presented stimuli is significantly influenced by the semantic category of the item. Thus, a number of patients have been reported who are able to name man-made objects but are severely impaired in naming naturally occurring items such as animals, fruits, and vegetables. A number of competing hypotheses have been proposed to explain these category-specific deficits. The modality-specific hypothesis (Warrington and

Shallice 1984), later named the Sensory-Functional Theory (SFT; Caramazza and Shelton 1998), postulates that the animate versus inanimate dissociation follows from a selective impairment of the sensory or functional attributes that subserve the processing of either of these two categories. By this account, the identification of animate objects relies more on sensory attributes and would be disproportionately impaired by damage to the processing of sensory features associated with these objects. In contrast, inanimate objects may be known primarily by virtue of their function and the manner in which they are used or manipulated. As a consequence, a deficit in the recognition of inanimate objects would be disrupted by loss of information regarding the function of an object or sensory-motor knowledge regarding the manner in which the object is manipulated. A competing hypothesis is that the semantic knowledge is organized categorically in the brain (Caramazza and Shelton 1998). Clearly, teleologic explanations for this organization can be made. Evolutionary pressures would favor an animal that could easily recognize and distinguish other animals that are potential predator or prey, or plants that are potential sources of food. Further, developmental data support the idea that infants as young as 3 months of age can differentiate living from non-living things.

We recently studied a 50-year-old patient who exhibited a category-specific visual agnosia. After suffering a stroke of the right temporal lobe in the context of hypotension, he developed amnesia, prosopagnosia, and an inability to recognize living things. He was significantly impaired naming animate as compared with inanimate items. Additional testing revealed that his ability to name objects was strongly predicted by the nature of his experience with the object. Those objects—whether animate or inanimate—that he knew by virtue of using or manipulating (e.g., hammer) were named relatively accurately whereas those items that he knew primarily by sight (e.g., elephant) were named poorly. We argued that he exhibited a mild visual agnosia but that "motor knowledge" could compensate at least in part for the deficits in the object recognition system.

10.5 FUNCTIONAL IMAGING CORRELATES OF OBJECT RECOGNITION

In the past few years, functional imaging in healthy subjects has helped to elucidate the anatomy of object recognition. Malach et al. (1995) described the lateral occipital cortex (LOC), which is located at the lateral and ventral aspects of the occipito-temporal cortex. This area is activated preferentially by objects compared with scrambled objects or textures, regardless of the nature of the object (e.g., faces, cars, common objects, and even unfamiliar abstract objects) (Malach, Levy, and Hasson 2002). Further processing of the visual stimuli occurs in specific brain areas according to stimulus category. For example, faces are processed in the fusiform face area (Kanwisher, McDermott, and Chun 1997) and in the occipital face area (Gauthier, Tarr, and Moylan 2000); the former is more selective for faces than the latter. Places and/or spatial layouts are processed in the parahippocampal place area (Epstein and Kanwisher 1998), whereas objects like animals and tools are processed in specific loci in the fusiform and the superior and middle temporal gyri (Chao, Haxby, and Martin 1999). Orthographic stimuli are processed in the left inferior occipito-temporal cortex on or near the left fusiform gyrus (Polk et al. 2002; Allison et al. 1994), although this localization is highly debated (Price and Devlin 2003). Thus, the specific perceptual categories of visual stimuli dictate very different ways of their processing in the brain and might give an anatomic basis for the different agnostic syndromes.

10.6 AUDITORY AGNOSIAS

Like the visual agnosias discussed above, auditory agnosias are characterized by an inability to recognize a stimulus—in this case, a sound—that cannot be explained by inadequate elementary sensory processing, generalized loss of knowledge (e.g., dementia), or inadequate attention to the task. Perhaps reflecting the fact that in the context of normal vision, audition is not usually needed for object recognition, auditory agnosias have been reported far less commonly than visual agnosias.

Indeed, in most naturalistic settings, one is rarely required to identify an object on the basis of sound alone. Even in those circumstances in which this is necessary—for example, identifying the sound of a telephone—the range of potential stimuli is narrow, limiting the complexity of the processing required. Perhaps because of these factors, complaints of poor recognition of objects from sound are uncommon. In our experience, when questioned about their auditory recognition deficit, most subjects will concede that sounds simply "aren't right" and, when pressed, attribute the disorder to hearing loss.

There is, however, reason to believe that auditory agnosias are under-reported. On a number of occasions, we have observed patients with auditory recognition deficits evident on formal testing who were utterly unaware of the deficit. Additionally, most patients with documented auditory sound agnosia do not complain of an inability to recognize sounds (Saygin et al. 2003; Vignolo 2003). For example, although aphasic subjects rarely complain of difficulty understanding non-speech sounds, acquired language disorders are frequently associated with a generalized disorder of auditory recognition (Vignolo 1982). Saygin et al. (2003), for example, recently reported an elegant study in which 30 left-hemisphere-damaged aphasic subjects were asked to match environmental sounds (sound of a cow mooing) or linguistic phrases ("cow mooing") to pictures. They found that the aphasic subjects were impaired with both word and sound stimuli; furthermore, there was a high performance correlation between severity of aphasia and accuracy on the environmental sound-matching task. Lesion overlay analysis demonstrated that damage to posterior regions in the left middle and superior temporal gyri and inferior parietal lobe was a predictor of impaired performance on both tasks.

Although accounts of auditory recognition are, in general, less well-developed than those in the visual domain, several investigators have argued that two forms of auditory sound agnosia may be identified (Kleist 1928). Echoing Lissauer's distinction between apperceptive and associative visual agnosias, Vignolo (1982) distinguished between a perceptual or discriminative agnosia linked to right hemisphere lesions and an associative agnosia observed with left hemisphere lesions. Although subjects with both conditions exhibit the same fundamental deficit, an inability to recognize objects from sound, support for the claim that they arise at different levels of processing comes from analyses of the errors (Vignolo 1982; Schnider, Benson, and Scharre 1994). In the right hemisphere "apperceptive" form of auditory agnosia, confusions are typically based on the features of the sound; for example, a car horn may be confused with a musical instrument. In left hemisphere "associative" auditory agnosia, errors appear to be semantically based; for example, subjects may confuse a police whistle with a siren.

Vignolo (2003) recently reported data that support this lateralization of function. Tasks assessing music and identification of environmental sounds were administered to 40 subjects with unilateral stroke; right hemisphere lesions tended to disrupt the apperception of environmental sounds whereas left hemisphere lesions disrupted the semantic identification of sounds. Finally, recent evidence from functional imaging studies in normal subjects is also consistent with this claim. Lewis et al. (2004) reported data from an fMRI study in which participants listened to a wide range of environmental sounds while undergoing BOLD imaging. The contrast between recognizable and unrecognizable stimuli revealed activity in a distributed network of brain regions previously associated with semantic processing, most of which were in the left hemisphere.

Auditory information is not only relevant to the recognition of words and entities in the environment but also to spatial processing; that is, just as vision may be used in parallel for the identification and localization of objects, sound may also be employed in the service of spatial processing. Although beyond the purview of this chapter, it should be noted that both functional imaging and lesion studies have suggested that distinct processing pathways may underlie these capacities (Griffiths et al. 1998). Clarke et al. (2000), for example, reported detailed investigations of four subjects with left hemisphere lesions who exhibited substantial dissociations in the ability to recognize and localize auditory stimuli, suggesting that the capacities to identify and localize auditory stimuli are, at least in part, distinct and dissociable.

In the following sections we briefly review the major syndromes of auditory agnosia starting with cortical deafness and culminating in disorders of auditory word recognition and recognition of auditory affect.

10.7   CORTICAL DEAFNESS AND GENERALIZED AUDITORY AGNOSIA

Cortical deafness is a rare disorder characterized by profound loss of awareness of sound; in its most profound form, patients appear to be deaf. Patients with this disorder have typically suffered extensive bihemispheric lesions that destroy superior temporal cortex crucial for the analysis of sound (Leicester 1980). More recently, several patients with extensive subcortical bilateral lesions that undercut and, presumably, disconnect the auditory cortex from input from the medial geniculate have been reported (e.g., Kazui et al. 1990; Kaga et al. 2005).

As in the visual domain, the distinction between a primary disorder of auditory processing and "higher-level" disorders of auditory recognition remains controversial. Michel et al. (1980) proposed a number of criteria to differentiate these conditions, including auditory evoked potentials as well as mapping of the lesions to koniocortex as opposed to pro- and parakoniocortex. Neither of these approaches has been demonstrated to provide an adequate account of the distinction.

Consistent with the animal literature demonstrating that bilateral ablations of auditory cortex do not consistently produce complete deafness (e.g., Neff 1961; see also Celesia 1976), most subjects who present with substantial bilateral temporal strokes also do not exhibit cortical deafness. Similarly, the rare subjects who present with cortical deafness typically regain awareness of sound but continue to exhibit profound auditory recognition deficits for all kinds of stimuli. With formal testing, the errors of subjects with generalized auditory agnosia typically represent confusions based on acoustic properties rather than meaning; thus, in the context of the two-stage model described above, the disorder may be described as an "apperceptive" disorder.

10.7.1   AUDITORY SOUND AGNOSIA (AUDITORY AGNOSIA FOR NON-SPEECH SOUNDS)

This disorder is characterized by apparently normal awareness of sound but an inability to identify non-verbal sounds. Although, as noted above, some patients with auditory sound agnosia may be impaired in recognizing speech (e.g., Saygin et al. 2003), in other subjects the disorder may be relatively "pure" in that speech may be largely preserved whereas recognition of auditory sounds is clearly impaired (e.g., Spreen, Benton, and Finchman 1965; Vignolo 1982; Fujii et al. 1990; Schnider, Benson, and Scharre 1994; Clarke et al. 2000). Auditory agnosia restricted to non-verbal sounds is typically associated with bilateral (Spreen, Benton, and Finchman 1965; Albert et al. 1972; Kazui et al. 1990) or right hemisphere lesions (Fujii et al. 1990).

10.7.2   PURE WORD DEAFNESS (AUDITORY AGNOSIA FOR SPEECH)

Pure word deafness is a rare and often striking disorder in which patients are unable to understand auditory language but exhibit no significant hearing loss and maintain the ability to recognize sounds. It is differentiated from aphasia by the preservation of language as manifested by speech production, naming, reading, and writing. Reflecting the central role of speech in human interaction, patients with pure word deafness are typically aware of and distressed by their difficulty.

Two major pathologic substrates of pure word deafness have been identified. One group of patients suffers from bitemporal lesions, usually involving the primary and secondary auditory cortices of the superior temporal gyrus (e.g., Coslett, Brashear, and Heilman 1984); another group suffers from single lesions of the left temporal lobe or temporal isthmus. From an anatomic perspective, both types of lesions are assumed to give rise to the same basic problem: a dissociation between intact language cortices in the left perisylvian region and auditory input.

Several investigators have proposed that distinct subtypes of pure word deafness may be identified. Some investigators (e.g., Albert and Bear 1974; Auerbach et al. 1982; Mendez and Geehan

1988; Coslett, Brashear, and Heilman 1984) have proposed that the disorder is attributable to a loss of temporal acuity; on this account, speech is selectively impaired because it is far more demanding with respect to temporal precision than auditory sounds, most of which do not require fine auditory distinctions (Albert and Bear 1974). An alternative interpretation is that the disorder is caused by a disruption of phonemic discrimination (Saffran, Schwartz, and Marin 1976; Denes and Semenza 1975). Auerbach et al. (1982) suggested that the former subtype of pure word deafness was associated with bilateral lesions whereas the latter was associated with left hemisphere lesions.

Although they may perform at chance on auditory word-to-picture matching tasks, patients with pure word deafness often perform surprisingly well in some naturalistic settings. For example, whereas patients with pure word deafness may be utterly unable to communicate by telephone, they may communicate relatively well in face-to-face conversations. Several factors may contribute to this. One possibility is that these patients may be adept at using a "top-down" process to narrow the possible range of candidate words from which the subject must choose; thus, in a conversation about politics, one may expect certain words to appear, and this knowledge may bias the impaired recognition system by priming likely words. Alternatively, some investigators have argued that lip-reading provides useful information regarding factors such as place of articulation that supplements the impoverished auditory input (Auerbach et al. 1982). Additionally, some patients may be able to use non-verbal auditory cues to supplement deficient word decoding.

Of note, we demonstrated that a patient with profound pure word deafness from bilateral lesions performed well on a task in which he was asked to identify the affect (or emotion) expressed in a verbal utterance (Coslett, Brashear, and Heilman 1984). These results suggested that patients with pure word deafness perform better in face-to-face communication because they are able to make use of affective and other non-verbal cues. The dissociation between object identification and affective discrimination underscores the possibility that patients with sensory agnosia, in the textbook sense, may yet have sensory-specific access to the emotional content of a stimulus. This provocative idea is taken up in Section 10.9.

10.7.3 Tactile Agnosias (Somatosensory Agnosia)

Disorders of object recognition from tactile input have received little attention and remain poorly understood. There is substantial terminologic confusion in this domain with a variety of different names being applied, often inconsistently, to the general class of phenomena. Some investigators have applied the general term "tactile agnosia" to all disorders characterized by poor identification of objects from palpation whereas others have suggested that this term be restricted to those cases in which subjects fail to recognize palpated objects yet have normal or near normal somatosensory processing (see Bauer 2003 and Bohlhalter, Fretz, and Weder 2002 for reviews). Similarly, some investigators have reserved the term "astereognosis" for disorders arising from a failure in low-level sensory processing. Here we refer to the entire range of phenomena as tactile agnosias with the understanding that this does not reflect a consensus view and that the reader must be attentive to terminology when exploring this literature.

As with the visual and auditory agnosias previously discussed, tactile agnosias are defined both by what they are and what they are not. Tactile agnosia is a disorder of object recognition from touch that cannot be explained by severe sensory-motor disturbance, inattentiveness, or general intellectual decline. Although there may be debate regarding the boundaries between primary sensory-motor disorders at the one end of the spectrum and cognitive disorders at the other (e.g., Bay 1944; Head and Holmes 1911), there is considerable support for the claim that tactile agnosia is more than either of these disorders. One strong argument for this point comes from the double dissociation between elementary somatosensory processing disorders and tactile agnosia. In an investigation of 84 subjects with tactile object recognition impairments, Caselli (1991) reported substantial deficits in patients with normal or only mild deficits in somatosensory processing; these and other data (see also Hecaen 1972; Corkin 1978) demonstrate that significant disorders of somatosensory processing do not necessarily cause tactile agnosia; other investigators (e.g., Bohlhalter, Fretz, and Weder 2002) have demonstrated that subjects with tactile agnosia may exhibit normal or near-normal processing across a range of somatosensory tasks.

Wernicke (1895) distinguished between tactile agnosias that involved a loss of a "tactile image" from those that were attributable to an inability to associate the tactile image to its meaning. This distinction is, of course, reminiscent of Lissauer's apperceptive/associative dichotomy. More recently, Wernicke's distinction has been advocated by a number of investigators (Mauguiere and Isnard 1995; Platz 1996; Reed, Caselli, and Farah 1996; Caselli 1997). On this classification, those subjects with at least relatively normal elementary somatosensory function but an impairment in object identification through palpation who are impaired in matching items on the basis of tactile input and whose errors appear to reflect deficits in form, shape, or texture discrimination are considered apperceptive tactile agnosics; those subjects who perform well on tasks such as matching items by palpation and whose errors appear to reflect imprecision at the level of meaning (e.g., responding "fork" when presented a spoon) are considered to be associative agnosics. Hecaen and David (1945), for example, reported a patient who could not name palpated objects but could draw the object with sufficient precision that he could then name the object (see also Newcombe and Ratcliff 1974).

Other investigators have attempted to classify the disorders on the basis of the putative type of sensory processing deficit. For example, Delay (1935), building on the work of von Frey (1895), Head (1918), and others, proposed that tactile agnosias could be divided into subtypes in which shape and size were predominantly affected, as compared to forms in which other qualities such as texture, weight, and temperature were affected. Consistent with such a view, Bohlhalter, Fretz, and Weder (2002) recently reported data from two subjects who differed with respect to their ability to process "microgeometrical" (e.g., texture) from "macrogeometrical" (e.g., length) information.

There is substantial agreement from the clinical literature regarding the anatomic bases of the tactile agnosias. The great majority of patients with these disorders for which imaging or autopsy data are available exhibit lesions involving the parietal lobe. Three regions of parietal cortex may be particularly relevant. First, the post-central gyrus or primary somatosensory cortex is involved in many subjects (Platz 1996). Second, the parietal operculum, often designated SII or the "secondary sensory cortex," has been lesioned in a number of well-studied subjects (e.g., Reed, Caselli, and Farah 1996; Caselli 1991; Bohlhalter, Fretz, and Weder 2002). Finally, the superior parietal lobule (BA 5 and 7) receives direct projections from the primary somatosensory cortex and has been postulated to be crucial for the integration of somatosensory information to generate a high-level representation of the body and objects in space (Platz 1996).

There is some data suggesting that the parietal operculum and the superior parietal lobule subserve different functions (Ledberg et al. 1995; Roland, Sullivan, and Kawashima 1998). Whereas both areas are implicated in tactile recognition in normal subjects undergoing fMRI scanning, Binkofski et al. (1999) suggested that the parietal operculum (SII) may be important for "microgeometrical" properties such as texture, while the superior parietal lobule may be more important for macrogeometrical properties such as size and shape. Chapter 7 contains further general details of somatosensory function.

Finally, an unresolved issue concerns the hemispheric basis of tactile agnosias. As reviewed by Bauer (2003), there is no clear evidence of a hemispheric difference in tactile agnosia. There is, however, suggestive evidence that disorders in the ability to appreciate shape, size, and other "spatial" attributes of a stimulus may be more marked in subjects with right hemisphere lesions. A second point concerns the observation that some patients exhibit tactile agnosia in both hands after a unilateral hemispheric lesion. For example, Corkin, Milner, and Rasmussen (1970) demonstrated that 20 of 50 patients with unilateral hemispheric cortical excisions exhibited bilateral sensory deficits; this was attributed to damage involving the parietal operculum, which receives prominent projections from both primary somatosensory cortices. Lesions of primary somatosensory cortex typically produce contralesional tactile agnosia.

10.8   OTHER AGNOSIAS

There has been an explosion of interest in the human neurobiology of taste and smell in recent years (Small 2006; Gottfried 2006). As the brain mechanisms underlying these disorders have been investigated, patients who meet traditional criteria for olfactory or gustatory agnosia have been identified. Small et al. (2005) reported a thorough investigation of a subject with longstanding bilateral temporal lobe dysfunction who developed a gustatory agnosia after surgical resection of the left anterior medial temporal lobe for control of seizures. Prior to surgery the patient performed normally on tasks assessing her ability to detect and estimate the intensity of taste; additionally, although recognition thresholds were elevated, she successfully named basic tastes (e.g., sweet, sour). After resection of the left anterior medial temporal lobe she retained the ability to detect, discriminate, and react hedonically to tastants but lost the ability to recognize the tastant. Based on these and other data (Small et al. 1997), the investigators argued that recognition of taste involves interactions between the primary gustatory centers located in the insula/opercular region and the regions of the anterior medial temporal lobe that are critical for taste recognition.

Olfactory agnosias have also been described. For example, Mendez and Ghajarnia (2001) reported a patient who was unable to recognize faces and odors in the context of right temporal lobe dysfunction. Jones-Gotman, Rouleau, and Snyder (1997) reported data from a smell recognition test from 70 subjects who had undergone temporal lobectomy for seizure control. They found that damage to the anterior temporal lobe of either hemisphere was associated with impairment of odor identification with impairment of odor identification. However, because basic smell detection was not formally assessed, one cannot state with certainty that these subjects met the criteria for olfactory agnosia. Eichenbaum et al. (1983) demonstrated that patient HM, who had undergone bilateral anterior temporal lobe resections, exhibited preserved early olfactory processing but substantial impairment in higher-order olfactory processing. Gottfried and Zald (2005) have recently reviewed the contributions of orbitofrontal cortex to olfactory processing and the implications of lesions in this region for odor recognition in humans and non-human primates.

10.9   AGNOSIA AND CONSCIOUS AWARENESS

As previously noted, agnosia is a disorder of stimulus recognition. Traditionally, evidence for "recognition" has come from explicit report: subjects name the object or provide a verbal description that is sufficiently detailed so that the object can be unambiguously identified. As in other domains in neuropsychology (e.g., Schacter 1987; McGlinchey-Berroth 1996), there is now abundant evidence that some agnosic subjects may derive substantial information from stimuli that they fail to "recognize." This phenomenon, which has been termed "covert recognition" or "processing without awareness," has been studied most extensively in prosopagnosics (see Bruyer 1991 for review). Bruyer et al. (1983) demonstrated that a prosopagnosic subject exhibited greater difficulty learning to associate familiar but unrecognized faces with fictitious as opposed to real names. Similar data from a face-name interference test were reported by DeHaan, Young, and Newcombe (1987) as well as Young et al. (1986). Additional evidence from psychophysical tasks was provided by Bauer (1984), who demonstrated that a prosopagnosic subject exhibited the largest galvanic skin ("lie detector") response to unrecognized faces when the face was paired with the correct name (see also Bauer and Verfaellie 1986; Tranel and Damasio 1985).

Similar findings have been reported in other agnosic syndromes. For example, Coslett and Saffran (1989; Coslett et al. 1993) reported that patients with written word agnosia ("pure word deafness") responded significantly better than chance on a variety of semantic tasks to words that they claimed not to have "seen." Several investigators have demonstrated that simultanagnosic subjects derive information from unreported stimuli (Robertson et al. 1997). Coslett and Lie (2008), for example, demonstrated that a simultanagnosic patient's report of two items in an array was significantly influenced by the semantic relationship between items in an array: like a previously reported subject (Coslett and Saffran 1993), this subject reported items from the same semantic class (e.g., hammer, pliers) more reliably than items from different categories (e.g., skirt, apple).

The interpretation of these and other data demonstrating dissociations between implicit and explicit knowledge remains controversial. It is clear from these and multiple other sources of data that patients with brain injury derive substantially more information than they realize. In light of this observation, one may question whether this implicit knowledge influences the behavior of these subjects. For example, is a visual agnosic who can't discriminate between a shotgun and fire-poker on formal testing likely to pick up a shotgun to stoke a fire? Although one might expect that implicit knowledge would constrain the behavior of agnosic subjects, we are unaware of any data that speak directly to this point.

Although beyond the scope of this chapter, there is a substantial (but controversial) body of literature suggesting that unconscious thought is an important and efficient tool for decision making, problem solving, and attitude formation (see Dijksterhuis et al. 2006). In a series of experiments, Dijksterhuis and colleagues have demonstrated that "unconscious" thought may, under some circumstances, be more rational and efficient than "conscious" thought. In one experiment (Dijksterhuis 2004), subjects were asked to rate the desirability of different apartments after being told of 12 dimensions (e.g., cost, size) on which the apartments differed. One group of subjects was asked to respond immediately upon hearing about the apartments, another group was given three minutes to explicitly consider the apartments, and a third group performed a distracting (2-back) task for three minutes before responding. Subjects who were distracted for three minutes and therefore not able to consciously consider the apartments performed significantly better in discriminating between the most and least desirable apartments than those subjects who responded immediately or carefully considered the options. Dijksterhuis and colleagues argue for a fundamental discrepancy between "conscious" and "unconscious" thought and claim that the latter may be more efficient for certain types of mental operations. The implications of "unconscious" thought for the processes underlying sensory agnosia are, at present, not clear, but they represent a fertile domain for future research.

10.10 CONCLUSION

Sensory agnosias are relatively uncommon but often striking disorders of recognition that are of considerable interest for several reasons. Not only are they important from a clinical perspective but also, as emphasized by a number of investigators (e.g., Farah 2004; Bauer 2003), by dissecting perceptual systems at the fault lines, the disorders may provide important insights regarding the basic mechanisms of information processing and their anatomic bases. Using visual agnosia as the prototype, we have attempted to demonstrate that the confusion that arises from a consideration of the phenomenology of the disorders can be mitigated by considering the disorders in the context of a theoretically motivated account of basic perceptual mechanisms. Such an approach is, in principle, applicable to agnosias in other modalities. Finally, there are several areas for future research that appear particularly promising. The study of high-level recognition deficits including the study of "category-specific" agnosias offers great promise in the exploration of semantics, including the role of "grounded cognition" (see Barsalou 2008).

Additionally, in light of recent demonstrations of the role of unconscious operations in perception and thought, the study of agnosic subjects may contribute to the understanding of the manner in which motivations, goals, and drives are influenced by factors about which subjects remain unaware. For example, the extent to which Pavlovian conditioning is present in agnosic patients is not clear. The demonstration that a prosopagnosic patient could form a conditioned response between a face that they could not identify and an unconditioned stimulus would suggest that reward-based conditioning does not require recognition but could be based on lower-level sensory processing. To this end, Chapter 16 explicitly considers the novel idea that many of the behavioral deficits observed in patients with frontal lobe dysfunction reflect an agnosia for object "value."

REFERENCES

Albert, M., and Bear, D. 1974. Time to understand: a case study of word deafness with reference to the role of time in auditory comprehension. *Brain* 97:373–84.

Albert, M., Sparks, R., von Stockert, T., and Sax, D. 1972. A case study of auditory agnosia: Linguistic and nonlinguistic processing. *Cortex* 8:427–33.

Allison, T., McCarthy, G., Nobre, A., Puce, A., and Belger, A. 1994. Human extrastriate visual cortex and the perception of faces, words, numbers, and colors. *Cerebral Cortex* 4:544–54.

Allport, D. 1985. Distributed memory, modular subsystems and dysphasia. In *Current Perspectives in Dysphasia*, eds. S.K. Newman and R. Epstein, pp. 32–60. Edinburgh: Churchill Livingstone.

Auerbach, S., Allard, T., Naeser, M., Alexander, M., and Albert, M. 1982. Pure word deafness. Analysis of a case with bilateral lesions and a defect at the prephonemic level. *Brain* 105:271–300.

Balint, R. 1909. Seelenlahmung des "Schauens", optische Ataxie, raumliche Storung der Aufmerksamkeit. *Monatschrift fur Psychiatrie und Neurologie* 25:51–81.

Barsalou, L. 2008. Grounded cognition. *Annual Review of Psychology* 59:617–45.

Bauer, R. 1984. Autonomic recognition of names and faces in prosopagnosia: A neuropsychological application of the Guilty Knowledge Test. *Neuropsychologia* 22:457–69.

———. 2003. Agnosia. In *Clinical Neuropsychology*, ed. K.H.a.E. Valenstein, pp. 236–95. Oxford: Oxford University Press.

Bauer, R., and Verfaellie, M. 1986. Electrodermal discrimination of familiar but not unfamiliar faces in prosopagnosia. *Brain Cognition* 8:240–52.

Bay, E. 1944. Zum problem der taktilen Agnosie. *Deutsche Zeitschrift für Nervenheilkunde* 156:1–3; 64–96.

———. 1953. Disturbances of visual perception and their examination. *Brain* 76:515–50.

Bender, M., and Feldman, M. 1972. The so called "visual agnosias". *Brain* 95:173–86.

Benson, D., Davis, J., and Snyder, B. 1988. Posterior cortical atrophy. *Archives of Neurology* 45:789–93.

Benson, D., and Greenberg, J. 1969. Visual form agnosia. A specific defect in visual discrimination. *Archives of Neurology* 20:82–89.

Binkofski, F., Buccino, G., Posse, S., Seitz, R., Rizzolatti, G., and Freund, H. 1999. A fronto-parietal circuit for object manipulation in man: Evidence from an fMRI-study. *European Journal of Neuroscience* 11:3276–86.

Bodamer, J. 1947. Die Prosop-Agnosie. *Archiv für Psychiatric und Zeitschrift Neurologie*. 179:6–54.

Bohlhalter, S., Fretz, C., and Weder, B. 2002. Hierarchical versus parallel processing in tactile object recognition: A behavioural-neuroanatomical study of aperceptive tactile agnosia. *Brain* 125:2537–48.

Borenstein, E., and Ullman, S. 2008. Combined top-down/bottom-up segmentation. *IEEE Transactions on Pattern Analysis & Machine Intelligence* 30:2109–25.

Bornstein, B., Sroka, H., and Munitz, H. 1969. Prosopagnosia with animal face agnosia. *Cortex* 5:164–69.

Botez, M., and Serbănescu, T. 1967. Course and outcome of visual static agnosia. *Journal of the Neurological Sciences* 4:289–97.

Bruyer, R. 1991. Covert face recognition in prosopagnosia: A review. *Brain Cognition* 15:223–35.

Bruyer, R., Laterre, C., Seron, X., Feyereisen, P., Strypstein, E., Pierrard, E., and Rectem, D. 1983. A case of prosopagnosia with some preserved covert remembrance of familiar faces. *Brain Cognition* 2:257–84.

Buxbaum, L., Glosser, G., and Coslett, H. 1996. Relative sparing of object recognition in alexia-prosopagnosia. *Brain and Cognition* 32:202–5.

Caramazza, A., and Shelton, J. 1998. Domain-specific knowledge systems in the brain the animate-inanimate distinction. *Journal of Cognitive Neuroscience* 10:1–34.

Caselli, R. 1991. Bilateral impairment of somesthetically mediated object recognition in humans. *Mayo Clinic Proceedings* 66:357–64.

———. 1997. Tactile agnosia and disorders of tactile perception. In *Behavioral Neurology and Neuropsychology*, eds. T. E. Feinberg and M.J. Farah, pp. 277–88. New York: McGraw-Hill.

Cave, K. 1999. The Feature Gate model of visual selection. *Psychological Research* 62:182–94.

Celesia, G. 1976. Organization of auditory cortical areas in man. *Brain* 99:403–14.

Chao, L., Haxby, J., and Martin, A. 1999. Attribute-based neural substrates in temporal cortex for perceiving and knowing about objects. *Nature Neuroscience* 2:913–19.

Clarke, S., Bellmanm, A., Meuli, R., Assal, G., and Steck, A. 2000. Auditory agnosia and auditory spatial deficits following left hemispheric lesions: Evidence for distinct processing pathways. *Neuropsychologia* 38:797–807.

Corbetta, M. 1998. Frontoparietal cortical networks for directing attention and the eye to visual locations: Identical, independent, or overlapping neural systems. *Proceedings of the National Academy of Science* 95:831–38.

Corkin, S. 1978. The role of different cerebral structures in somesthetic perception. In *Handbook of Perception*, ed. C.E.C.a.M.P. Friedman, pp. 105–55. New York: Academic Press.

Corkin, S., Milner, B., and Rasmussen, T. 1970. Somatosensory thresholds: Contrasting effects of postcentral-gyrus and posterior parietal-lobe excision. *Archives of Neurology* 23:41–58.

Coslett, H., Brashear, H., and Heilman, K. 1984. Pure word deafness after bilateral primary auditory cortex infarcts. *Neurology* 34:347–52.

Coslett, H., and Chatterjee, A. 2003. Balint's syndrome and related disorders. In *Behavioral Neurology and Neuropsychology*, eds. T. Feinberg and M. Farah, pp. 325–36. McGraw-Hill.

Coslett, H., and Lie, E. 2008. Simultanagnosia: Effects of semantic category and repetition blindness. *Neuropsychologia* 46:1853–63.

Coslett, H., and Saffran, E. 1989. Evidence for preserved reading in "pure alexia". *Brain* 112:327–59.

———. 1993. Simultanagnosia. To see but not two see. *Brain* 114:1523–45.

Coslett, H., Saffran, E., Greenbaum, S., and Schwartz, J. 1993. Reading in pure alexia. The effect of strategy. *Brain* 116:21–37.

Coslett, H., Stark, M., Rajaram, S., and Saffran, E. 1995. Narrowing the spotlight: A visual attentional disorder in presumed alzheimer's disease. *Neurocase* 1:305–18.

Damasio, A. 1989. Time-locked multiregional retroactivation: A system level proposal for the neural substrates of recall and recognition. *Cognition* 33:25–62.

Damasio, A., Damasio, H., and Hoesen, G.V. 1982. Prosopagnosia: Anatomic basis and behavioral mechanisms. *Neurology* 32:331–41.

Damasio, A., Yamada, T., Damasio, H., Corbet, J., and McKee, J. 1980. Central achromatopsia: Behavioral, anatomic, and physiologic aspects. *Neurology* 30:1064–71.

Davidoff, J., and Wilson, B. 1985. A case of visual agnosia showing a disorder of pre-semantic visual classification. *Cortex* 21:121–34.

DeHaan, E., Young, A., and Newcombe, F. 1987. Faces interfere with name classification in a prosopagnosic patient. *Cortex* 23:309–16.

Dejerine, J. 1892. Contribution a l'etude anatomo-pathologique et clinique des differentes varietes de cecite verbale. *Memoires de la Societe de Biologie* 4:61–90.

Delay, J. 1935. Les Astereognosies. Pathologie due Toucher. In *Clinique, Physiologie, Topographie*. Paris: Masson.

Denes, G., and Semenza, C. 1975. Auditory modality-specific anomia: Evidence from a case of pure word deafness. *Cortex* 11:401–11.

DeRenzi, E. 1986. Prosopagnosia in two patients with CT scan evidence of damage confined to the right hemisphere. *Neuropsychologia* 24:171–80.

Dijksterhuis, A. 2004. Think different: The merits of unconscious thought in preference development and decision making. *Journal of Personality and Social Psychology* 87:586–98.

Dijksterhuis, A., Bos, M., Nordgren, L., and Baaren, R.v. 2006. On making the right choice: The deliberation-without-attention effect. *Science* 311:1005–7.

Duncan, J. 1984. Selective attention and the organization of visual information. *Journal of Experimental Psychology: General* 113:501–17.

Efron, R. 1968. What is perception? *Boston Studies in Philosophy of Science* 4:137–73.

Eichenbaum, H., Morton, T., Potter, H., and Corkin, S. 1983. Selective olfactory deficits in case H.M. *Brain* 106:459–72.

Ellis, A., and Young, A. 1988. *Human Cognitive Neuropsychology*. Hillsdale, NJ: Lawrence Erlbaum Associates.

Epstein, R., and Kanwisher, N. 1998. A cortical representation of the local visual environment. *Nature* 392:598–601.

Eriksen, C., and Hoffman, J. 1973. The extent of processing noise elements during selective encoding from visual displays. *Perception and Psychophysics* 12:201–4.

Farah, M.J. 2000. *The Cognitive Neuroscience of Vision*. Malden: Blackwell Publishers.

———. 2004. *Visual Agnosia*. 2nd ed. Cambridge, MA: MIT Press.

Finkelnburg, F.C. 1870. Niederrheinische Gesellschaft in Bonn. Medicinische Section. *Berliner klinische Wochenschrift* 449–50, 460–61.

Fredericks, J.A.M. 1969. The agnosias. In *Handbook of Clinical Neurology*, ed. P.J.V.a.G.W. Bruyn. Amsterdam: Nother-Holland.

Freud, S. 1891. *Zur Auffasun der Aphasien. Eine Kritische Studie*. Vienna: Franz Deuticke.

Fujii, T., Fukatsu, R., Watabe, S., Ohnuma, A., Teramura, K., Kimura, I., Saso, S., and Kogure, K. 1990. Auditory sound agnosia without aphasia following a right temporal lobe lesion. *Cortex* 26:263–68.

Gauthier, I., Tarr, M., and Moylan, J. 2000. The fusiform "face area" is part of a network that processes faces at the individual level. *Journal of Cognitive Neuroscience* 12:495–504.

Goldstein, K. 1943. Some remarks on Russell Brain's article concerning visual object-agnosia. *Journal of Nervous and Mental Disease* 98:148–53.

Goldstein, K., and Gelb, A. 1918. Psychologische Analysen hirnpathologischer Falle auf Grund von Untersuchungen Hirnverletzter. *Zeitschrift fur die Neurologie and Psychiatrie* 41:1–42.

Gottfried, J. 2006. Central nervous processing. *Advances in Oto-rhino-laryngology* 63:44–69.

Gottfried, J., and Zald, D. 2005. On the scent of human olfactory orbitofrontal cortex: Meta-analysis and comparison to non-human primates. *Brain Research Reviews* 50:287–304.

Griffiths, T., Elliot, C., Coulthard, A., Cartlidge, N., and Green, G. 1998. A distinct low-level mechanism for interaural timing analysis in human hearing. *Neuroreport* 9:3383–86.

Grill-Spector, K., Knouf, N., and Kanwisher, N. 2004. The fusiform face area subserves face perception, not generic within-category identification. *Nature Neuroscience* 7:555–62.

Harris, A., and Aguirre, G. 2008. The representation of parts and wholes in face-selective cortex. *Journal of Cognitive Neuroscience* 20:863–78.

Haxby, J., Hoffman, E., and Gobbini, M. 2000. The distributed human neural system for face perception. *Trends in Cognitive Sciences* 4:223–33.

Head, H. 1918. Sensation and the cerebral cortex. *Brain* 41:57–253.

Head, H., and Holmes, G. 1911. Sensory disturbances from cerebral lesions. *Brain* 34:102–254.

Hecaen, H. 1972. *Introduction a la Neuropsychologie*. Paris: Larousse.

Hecaen, H., and David, M. 1945. Syndrome parietale traumatique: Asymbolie tactile et hemiasomatognosie paroxystique et douloureuse. *Revue Neurologique* 77:113–23.

Heinke, D., and Humphreys, G. 2003. Attention, spatial representation, and visual neglect: Simulating emergent attention and spatial memory in the selective attention for identification model SAIM. *Psychological Review* 110:29–87.

Holmes, G. 1918. Disturbances of visual orientation. *British Journal of Ophthalmology* 2:449–68.

Horel, J., and Keating, E. 1972. Recovery from a partial Kluver-Bucy syndrome induced by disconnection. *Journal of Comparative and Physiological Psychology* 79:105–14.

Horner, J., and Massey, E. 1986. Dynamic spelling alexia. *Journal of Neurology, Neurosurgery & Psychiatry* 49:455–57.

Humphreys, G., and Riddoch, M. 1984. Routes to object constancy: Implications from neurological impairments of object constancy. *Quarterly Journal of Experimental Psychology* 36A:385–415.

———. 1987. The fractionation of visual agnosia. In *Visual Object Processing: A Cognitive Neuropsychological Approach*, eds. G.W. Humphreys, and M.J. Riddoch. London: Lawrence Erlbaum Associates.

Humphreys, G., Riddoch, M., and Fortt, H. 2006. Action relations, semantic relations, and familiarity of spatial position in Balint's syndrome: Crossover effects on perceptual report and on localization. *Cognitive, Affective, & Behavioral Neuroscience* 6:236–45.

Jackson, J. 1876. Case of large cerebral tumor without optic neuritis and with left hemiplegia and imperception. *Royal London Ophthalmic Hospital Reports* 8:434–44.

———. 1932. *Selected Writings of John Hughlings Jackson*. London: Hodder and Stoughton.

Jones-Gotman, M., Rouleau, I., and Snyder, P. 1997. Clinical and research contributions of the intracarotid amobarbital procedure to neuropsychology. *Brain and Cognition* 33:1–6.

Kable, J., Lease-Spellmeyer, J., and Chatterjee, A. 2002. Neural substrates of action event knowledge. *Journal of Cognitive Neuroscience* 14:795–805.

Kaga, K., Kurauchi, T., Nakamura, M., Shindo, M., and Ishii, K. 2005. Magnetoencephalography and positron emission tomography studies of a patient with auditory agnosia caused by bilateral lesions confined to the auditory radiations. *Acta Oto-laryngologica* 125:1351–55.

Kanwisher, N., McDermott, J., and Chun, M. 1997. The fusiform face area: A module in human extrastriate cortex specialized for face perception. *Journal of Neuroscience* 17:4302–11.

Kazui, S., Naritomi, H., Sawada, T., Inoue, N., and Okuda, J. 1990. Subcortical auditory agnosia. *Brain and Language* 38:476–87.

Kleist, K. 1928. Gehirnpathologische und lokalisatorische Ergebnisse über Hörstörungen, Geräuschtaubheiten und Amusien. *Monatsschrift für Psychiatrie und Neurologie* 68:853–60.

Ledberg, A., O'Sullivan, B., Kinomura, S., and Roland, P. 1995. Somatosensory activations of the parietal operculum of man. A PET Study. *European Journal of Neuroscience* 7:1934–41.

Leicester, J. 1980. Central deafness and subcortical motor aphasia. *Brain and Language* 10:224–42.

Levine, D. 1978. Prosopagnosia and visual object agnosia: A behavioral study. *Brain and Language* 5:341–65.

Levine, D., and Calvanio, R. 1989. Prosopagnosia: A defect in visual configural processing. *Brain and Cognition* 10:149–70.

Lewis, J., Wightman, F., Brefczynski, J., Phinney, R., Binder, J., and DeYoe, E. 2004. Human brain regions involved in recognizing environmental sounds. *Cerebral Cortex* 14:1008–21.

Lissauer, H. 1890. Ein Fall von Seelenblindheit Nebst Einem Beitrage zur Theori derselben. *Archiv fur Psychiatrie und Nervenkrankheiten* 21:222–70.

Luria, A. 1959. Disorders of "simultaneous perception" in a case of bilateral occipito-parietal brain injury. *Brain* 83:437–49.

Malach, R., Levy, I., and Hasson, U. 2002. The topography of high-order human object areas. *Trends in Cognitive Sciences* 6:176–84.

Malach, R., Reppas, J., Benson, R., Kwong, K., Jiang, H., Kennedy, W., Ledden, P., Brady, T., Rosen, B., and Tootell, R. 1995. Object-related activity revealed by functional magnetic resonance imaging in human occipital cortex. *Proceedings of the National Academy of Sciences* 92:8135–39.

Marr, D. 1982. *Vision. A Computational Investigation into the Human Representation and Processing of Visual Information.* New York: W.H. Freeman.

Mauguiere, F., and Isnard, J. 1995. Tactile agnosia and dysfunction of the primary somatosensory area. Data of the study by somatosensory evoked potentials in patients with deficits of tactile object recognition. *Revue Neurologique* 151:518–27.

McGlinchey-Berroth, R., Bullis, D.P., Milberg, W.P., Verfaellie, M., Alexander, M., and D'Esposito, M. 1996. Assessment of neglect reveals dissociable behavioral but not neuroanatomic subtypes. *Journal of the International Neurospychological Society* 2:441–51.

McKone, E., and Kanwisher, N. 2004. Does the human brain process objects-of-expertise like faces: An update. In *From Monkey Brain to Human Brain*, eds. J.R.D.S. Dehaene, M. Hauser and G. Rizzolatti. Cambridge, MA: MIT Press.

Mendez, M., and Geehan, G. 1988. Cortical auditory disorders: Clinical and psychoacoustic features. *Journal of Neurology, Neurosurgery & Psychiatry* 51:1–9.

Mendez, M., and Ghajarnia, M. 2001. Agnosia for familiar faces and odors in a patient with right temporal lobe dysfunction. *Neurology* 57:519–21.

Michel, F., Pernin, M., and Sieroff, E. 1986. Porospagnosie sans hemianopsie apres lesion unilateralie occipito-temporale driote. *Revue Neurologique* 142:545–49.

Michel, J., Peronnet, F., and Schott, B. 1980. A case of cortical deafness: Clinical and electrophysiological data. *Brain Lang* 10:367–77.

Milner, A., and Goodale, M. 1995. *The Visual Brain in Action.* Oxford: Oxford University Press.

Milner, A., and Heywood, C. 1989. A disorder of lightness discrimination in a case of visual form agnosia. *Cortex* 25:489–94.

Moran, J., and Desimone, R. 1985. Selective attention gates visual processing in extrastriate cortex. *Science* 229:782–84.

Moscovitch, M., Winocur, G., and Behrmann, N. 1997. What is special about face recognition? Nineteen experiments on a person with visual object agnosia and dyslexia but normal face recognition. *Journal of Cognitive Neuroscience* 9:555–604.

Munk, H. 1881. *Ueber die Functionen der Grosshirnrinde. Gesammelte Mittheilungenaus den lahren.* Berlin: Hirschwald.

Neff, W. 1961. Neuronal mechanisms of auditory discrimination. In *Sensory Communication*, ed. N.A. Rosenblith. Cambridge, MA: MIT Press.

Newcombe, F., and Ratcliff, G. 1974. Agnosia: A disorder of object recognition. In *Les Syndromes de Disconnexion Calleuse chez L'homme*, ed. F. Michel and B. Schott. Lyon: Colloque International de Lyon.

Pavese, A., Coslett, H., Saffran, E., and Buxbaum, L. 2002. Limitations of attentional orienting. Effects of abrupt visual onsets and offsets on naming two objects in a patient with simultanagnosia. *Neuropsychologia* 40:1097–1103.

Pavlov, I.P. 1927. *Conditioned Reflexes; An Investigation of the Physiological Activity of the Cerebral Cortex.* Oxford: Humphrey Milford.

Pearlman, A., Birch, J., and Meadows, J. 1979. Cerebral color blindness: An acquired defect in hue discrimination. *Annals of Neurology* 5:253–61.

Platz, T. 1996. Tactile agnosia: Casuistic evidence and theoretical remarks on modality-specific meaning representations and sensorimotor integration. *Brain* 119:1565–74.

Polk, T., Stallcup, M., Aguirre, G.K., Alsop, D., D'Esposito, M., Detre, J., and Farah, M. 2002. Neural specialization for letter recognition. *Journal of Cognitive Neuroscience* 14:145–59.

Poppelreuter, W. 1923. Zur Psychologie und Pathologie der optischen Wahrehmung. *Zur Gesamtgebiete der Neurologie und Psychiatrie* 83:26–152.

Posner, M. 1980. Orienting of attention. *Quarterly Journal of Experimental Psychology* 32:3–25.

Potzl, O., and Redlich, E. 1911. Demonstration eines Falles von bilateraler Affektion beider Occipitallapen. *Weiner Klinsche Wochenschrift* 24:517–18.

Price, C., and Devlin, J. 2003. The myth of the visual word form area. *Neuroimage* 19:473–81.

Prinzmetal, W., Ivry, R., Beck, D., and Shimizu, N. 2002. A measurement theory of illusory conjunctions. *Journal of Experimental Psychology: Human Perception and Performance* 28:251–69.

Ratcliff, G., and Newcombe, F. 1982. Object recognition: Some deductions from the clinical evidence. In *Normality and Pathology in Cognitive Functions*, ed. A.W. Willis. New York: Academic Press.

Reed, C., Caselli, R., and Farah, M. 1996. Tactile agnosia. Underlying impairment and implications for normal tactile object recognition. *Brain* 119:875–88.

Riddoch, M., and Humphreys, G. 1987. A case of integrative visual agnosia. *Brain* 110:1431–62.

———. 2003. Visual agnosia. *Clinical Neurology* 21:501–20.

Rizzo, M., and Hurtig, R. 1987. Looking but not seeing: Attention, perception, and eye movements in simultanagnosia. *Neurology* 37:1642–48.

Robertson, L., Treisman, A., Freidman-Hill, S., and Grabowecky, M. 1997. The interaction of spatial and object pathways: Evidence from Balint's Syndrome. *Journal of Cognitive Neuroscience* 9:295–317.

Rogers, T., Lambon, R., Garrard, P., Bozeat, S., McClelland, J., Hodges, J., and Patterson, K. 2004. Structure and deterioration of semantic memory: A neuropsychological and computational investigation. *Psychological Review* 111:205–35.

Roland, P., Sullivan, B., and Kawashima, R. 1998. Shape and roughness activate different somatosensory areas in the human brain. *Proceedings of the National Academy of Sciences* 95:3295–3300.

Rubens, A., and Benson, D. 1971. Associative visual agnosia. *Archives of Neurology* 24:304–16.

Saffran, E., and Coslett, H. 1996. Attentional dyslexia in Alzheimer's disease: A case study. *Cognitive Neuropsychology* 13:205–28.

Saffran, E., Fitzpatrick-DeSalme, E., and Coslett, H. 1990. Visual disturbances in dementia. In *Modular Deficits in Alzheimer-type Dementia*, ed. M. Schwartz, pp. 297–327. Cambridge, MA: MIT Press.

Saffran, E., and Schwartz, M. 1994. Impairments of sentence comprehension. *Philosophical Transactions of the Royal Society B: Biological Sciences* 346:47–53.

Saffran, E., Schwartz, M., and Marin, O. 1976. An analysis of speech perception in word deafness. *Brain and Language* 3:209–28.

Saygin, A., Dick, F., Wilson, S., Dronkers, N., and Bates, E. 2003. Neural resources for processing language and environmental sounds. Evidence from aphasia. *Brain* 126:928–45.

Schacter, D. 1987. Memory, amnesia, and frontal lobe dysfunction. *Psychobiology* 15:21–36.

Schnider, A., Benson, D., and Scharre, D. 1994. Visual agnosia and optic aphasia: Are they anatomically distinct? *Cortex* 30:445–57.

Simmons, W., and Barsalou, L. 2003. The similarity-in-topography principle: Reconciling theories of conceptual deficits. *Cognitive Neuropsychology* 20:451–86.

Small, D. 2006. Gustatory processing in humans. *Advances in Oto-rhino-laryngology* 63:191–220.

Small, D., Bernasconi, N., Bernasconi, A., Sziklas, V., and Jones-Gotman, M. 2005. Gustatory agnosia. *Neurology* 64:311–17.

Small, D., Jones-Gotman, M., Zatorre, R., Petrides, M., and Evans, A. 1997. A role for the right anterior temporal lobe in taste quality recognition. *Journal of Neuroscience* 17:5136–42.

Spreen, O., Benton, A., and Finchman, R. 1965. Auditory agnosia without aphasia. *Archives of Neurology* 13:84–92.

Tarr, M., and Gauthier, I. 2000. FFA: A flexible fusiform area for subordinate-level visual processing automatized by expertise. *Nature Neuroscience* 3:764–69.

Tranel, D., and Damasio, A. 1985. Knowledge without awareness: An autonomic index of facial recognition by prosopagnosics. *Science* 228:1453–54.

Treisman, A., and Gelade, G. 1980. A Feature-Integration Theory of Attention. *Cognitive Psychology* 12:97–136.

Treisman, A., and Souther, J. 1985. Search asymmetry: A diagnostic for preattentive processing of separable features. *Journal of Experimental Psychology: General* 114:285–310.

Ullman, S. 1984. Visual routines. *Cognition* 18:97–159.

Ungerleider, L., and Mishkin, M. 1982. Two cortical visual systems. In *Analysis of Visual Behavior*, eds. D.J. Ingle, M.A. Goodale, and R.J.W. Mansfield, pp. 549–86. Cambridge: MIT Press.

Vecera, S., and Farah, M. 1994. Does visual attention select objects or locations? *Journal of Experimental Psychology: General* 123:146–60.

Verrey. 1888. Hé miachromatopsie droite absolue. *Arch Ophthalmol* (Paris) 14:422–34.

Vidyasagar, T. 1998. Gating of neuronal responses in macaque primary visual cortex by an attentional spotlight. *Neuroreport* 9:1947–52.

Vignolo, L. 1982. Auditory agnosia. *Philosophical Transactions of The Royal Society London B Biological Science* 298:49–57.

———. 2003. Music agnosia and auditory agnosia. Dissociations in stroke patients. *Annals of the New York Academy of Sciences* 999:50–57.

Von Frey, M. 1895. Beitrage zur Sinnesphysiologie des Haut. *Bericht uber die Verhandlungen der Sdchsischen Akademie der Wissenschaften zu Leipzig, Math.-Phys. Kl.* 47:166–84.

Warrington, E., and James, M. 1986. Visual object recognition in patients with right-hemisphere lesions: Axes or features? *Perception* 15:355–66.

Warrington, E., and Shallice, T. 1984. Category specific semantic impairments. *Brain* 107:829–54.

Warrington, E., and Taylor, A. 1973. Immediate memory for faces: Long- or short-term memory? *Quarterly Journal of Experimental Psychology* 25:316–22.

———. 1978. Two categorical stages in object recognition. *Perception* 7:695–705.

Watson, J., Myers, R., Frackowiak, R., Hanjal, J., Woods, R., Mazziotta, J., Shipp, S., and Zeki, S. 1993. Area V5 of the human brain: Evidence from a combined study using positron emission tomography and magnetic resonance imaging. *Cerebral Cortex* 3:79–94.

Wernicke, C. 1895. Zwei Falle von Rindenlasion. *Arbeiten Aus die Psychiatrische Klinic In Breslau* 2:35.

Wolpert, I. 1924. Die Simultanagnosie: Storung der gesamtauffassung. *Zeitschrift fur die gesamte Neurologie und Psychiatrie* 93:397–413.

Young, A., Ellis, A., Flude, B., McWeeny, K., and Hay, D. 1986. Face-name interference. *Journal of Experimental Psychology: Human Perception and Performance* 12:466–75.

Zihl, J., Cramon, D.V., and Mai, N. 1983. Selective disturbance of movement vision after bilateral brain damage. *Brain* 106:313–40.

Part III

From Sensation to Reward

11 Neuroanatomy of Reward: A View from the Ventral Striatum

Suzanne N. Haber

CONTENTS

11.1 INTRODUCTION

A key component to rational decision-making and appropriate goal-directed behaviors is the ability to evaluate reward value, predictability, and risk accurately. The reward circuit, central to mediating this information and to assessing the likely outcomes from different choices effectively, is a complex neural network. While the hypothalamus and other subcortical structures are involved in processing information about basic, or primary, rewards, higher-order cortical and subcortical forebrain structures are engaged when complex choices about these fundamental needs are required. Moreover, choices often involve secondary rewards, such as money, power, or challenge, that are more abstract (compared to primary needs). Although cells that respond to different aspects of reward such as

235

anticipation, value, etc., are found throughout the brain, at the center of this network is the cortico-ventral basal ganglia (BG) circuit. This includes the orbitofrontal cortex (OFC) and anterior cingulate cortex (ACC), the ventral striatum (VS), the ventral pallidum (VP), and the midbrain dopamine (DA) neurons. Overall, the BG works in concert with cortex to execute motivated, well-planned behaviors. The reward circuit is embedded within this system and is a key driving force for the development and monitoring of these behaviors (Figure 11.1). Other BG circuits, those associated with cognition and motor control, work in tandem with elements of the reward system to develop appropriate goal-directed actions.

An essential element in motivating drive is the interaction between sensory information and the reward pathways. The idea that goal-directed behavior relies on the combined interplay of sensory inputs, emotional information, and memories of prior outcomes is sometimes lost in the details of reward system circuitry and function. Given the fundamental role that the BG plays in reward processing, it is fair to ask: how do sensory inputs—namely, all of those conditioned stimuli eliciting goal-directed responses—make contact with the BG? While the BG does not receive direct sensory

PREFRONTAL CORTEX

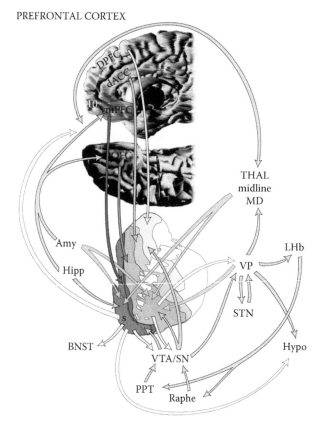

FIGURE 11.1 **(See Color Insert)** Schematic illustrating key structures and pathways of the reward circuit, with a focus on the connections of the ventral striatum. Red arrow = striatal input from the vmPFC; dark orange arrow = striatal input from the OFC; light orange arrow = striatal input from the dACC; yellow arrow = striatal input from the dPFC; white arrows = inputs into ventral striatum; gray arrows = outputs from ventral striatum; brown arrows = other non-striatal connections of the reward circuit. Amy = amygdala; BNST = bed nucleus of the stria terminalis; dACC = dorsal anterior cingulate cortex; DPFC = dorsal prefrontal cortex; Hipp = hippocampus; LHb = lateral habenula; Hypo = hypothalamus; MD = mediodorsal nucleus of the thalamus; OFC = orbital frontal cortex; PPT = pedunculopontine nucleus; Raphe = dorsal raphe; S = shell; SN = substantia nigra; STN = subthalamic nucleus; THAL = thalamus; VP ventral pallidum; VTA = ventral tegmental area; vmPFC = ventral medial prefrontal cortex. (Reprinted from Haber, S.N. and Knutson, B., *Neuropsychopharmacology*, 35, 4, 2010.)

inputs, it does receive processed sensory information. The main input structure of the BG is the striatum and its main input is derived from cortex. Most of cortex projects to the striatum, including sensory cortices. The VS is the striatal region most closely associated with reward. It receives its primary input from the orbital prefrontal cortex, insular cortex, and cingulate cortex. These cortical areas, particularly orbital and insular cortex, receive sensory information from all modalities. Moreover, the ventral striatal region receives a dense innervation from the amygdala, which is also tightly linked to sensory processing. In addition, recent evidence has shown that sensory-processing nuclei in the brainstem, particularly the superior colliculus, have a direct BG connection through an input to the substantia nigra. Finally, afferent projections from the olfactory bulb into the olfactory tubercle likely represent a rich source of direct olfactory input to the BG. Together these regions provide the main sensory input to the BG.

The ability to predict and evaluate reward value, and use that information to develop and execute an action plan efficiently, requires: first, integration of incoming sensory information with reward value, expectation, and memory; second, the incorporation of that information with cognition to develop the plan; and finally, the motor control to execute it. However, the BG are traditionally considered to process information in parallel and segregated functional streams consisting of reward (limbic), associative (cognitive), and motor control circuits (Alexander and Crutcher 1990). Moreover, microcircuits within each region are thought to mediate different aspects of each function (Middleton and Strick 2002). Nonetheless, expressed behaviors are the result of a combination of complex information processing that involves all of frontal cortex. Indeed, appropriate responses to environmental stimuli require continual updating and learning to adjust behaviors according to new data. This requires coordination between sensory, limbic, cognitive, and motor systems. While the anatomical pathways are generally topographic from cortex through BG circuits, a large body of growing evidence supports a dual processing system. Thus, information is not only processed in parallel streams, but also through integrative mechanisms through which information can be transferred between functional circuits (Bar-Gad et al. 2000; Belin and Everitt 2008; Bevan, Clarke, and Bolam 1997; Draganski et al. 2008; Haber et al. 2006; Haber, Fudge, and McFarland 2000; Kolomiets et al. 2001; McFarland and Haber 2002b; Mena-Segovia et al. 2005; Percheron and Filion 1991). This chapter will first discuss the place of the reward circuit in the BG; second, how the sensory systems interface within this circuit; and third, the anatomical basis for integrating the reward circuit with cognition and motor control systems.

11.2 THE PLACE OF THE REWARD CIRCUIT IN THE BASAL GANGLIA

Specific regions within the frontal-BG network play a unique role in different aspects of reward processing and evaluation of outcomes, including reward value, anticipation, predictability, and risk. The ACC and OFC prefrontal areas mediate different aspects of reward-based behaviors, error prediction, value, and the choice between short- and long-term gains (cf. other chapters in the Reward section of this volume). Cells in the VS and VP respond to anticipation of reward and reward detection. Reward prediction and error detection signals are generated, in part, from the midbrain dopamine cells. While the VS and the ventral tegmental area (VTA) dopamine neurons are the BG areas most commonly associated with reward, reward-responsive activation is not restricted to these, but found throughout the striatum and substantia nigra, pars compacta (SNc). Together, the frontal regions that mediate reward, motivation, and affect regulation project primarily to the rostral striatum, including the nucleus accumbens, the medial caudate nucleus, and the medial and ventral rostral putamen, collectively referred to as the VS. The area occupies over 20% of the striatum (Haber et al. 2006). This striatal region is also involved in various aspects of reward evaluation and incentive-based learning (Corlett et al. 2004; Elliott et al. 2003; Knutson et al. 2001; Schultz Tremblay, and Hollerman 2000; Tanaka et al. 2004), and is associated with pathological risk-taking and addictive behaviors (Kuhnen and Knutson 2005; Volkow et al. 2005). The VS projects to the VP and substantia nigra. From there information is transferred to the ACC and OFC via the mediodorsal nucleus of thalamus (MD) nucleus of the thalamus. While this ventral circuit is similar to the dorsal associative and motor circuits, there are also important differences (see Section 11.2.2.1).

The next section (Section 11.2.1) introduces the principal areas in prefrontal cortex that provide the bulk of afferent input into the VS. This overview will set the stage for an in-depth discussion of the anatomy and connectivity of the VS, which is the main focus of Section 11.2.2. Subsequent sections will consider other key components of the fronto-basal ganglia reward circuit (Sections 11.2.3 and 11.2.4) and how all of these elements are unified anatomically and functionally into an integrated reward circuit (Section 11.2.5 and 11.2.6).

11.2.1 ORBITAL PREFRONTAL CORTEX AND ANTERIOR CINGULATE CORTEX

Frontal cortex is organized in a hierarchical manner and can be divided into functional regions (Fuster 2001): the orbital (OFC) and anterior cingulate (ACC) prefrontal cortex are involved in reward, emotion, and motivation; the dorsal prefrontal cortex (DPFC) is involved in higher cognitive processes or "executive" functions; and the premotor and motor areas are involved in motor planning and the execution of those plans. Although cells throughout frontal cortex fire in response to various aspects of reward processing, the main components of evaluating reward value and outcome are the ACC and OFC. Each of these regions is comprised of several specific cortical areas: the ACC is divided into areas 24, 25, and 32; the orbital cortex is divided into areas 11, 12, 13, 14 and caudal regions referred to as either parts of insular cortex or periallo- and proiso-cortical areas (Barbas 1992; Carmichael and Price 1994). Based on specific roles for mediating different aspects of reward processing and emotional regulation, these regions can be functionally grouped into: (1) the dorsal ACC (dACC), which includes parts of areas 24 and 32; (2) the ventral, medial prefrontal cortex (vmPFC), which includes areas 25, 14, and subgenual area 32; and (3) the OFC, which includes areas 11, 13, and 12.

The vmPFC plays a role in monitoring correct responses based on previous experience and the internal milieu, and is engaged when previously learned responses are no longer appropriate and need to be suppressed. This region is a key player in the ability to extinguish previous negative associations and is positively correlated with the magnitude of extinction memory (Mayberg 2003; Milad et al. 2007). The OFC plays a central role in evaluation of value, magnitude, and probability of reward. However, responses to reward can transcend specific rewards, such as food or water, and seem also to code for general value (Padoa-Schioppa and Assad 2006; Roesch and Olson 2004; Tremblay and Schultz 2000; Wallis and Miller 2003). It is the OFC and insula that are most closely associated with the sensory systems (see below). The dACC is a unique part of frontal cortex, in that it contains within it a representation of many diverse frontal lobe functions, including motivation (areas 24a and b), cognition (area 24b), and motor control (area 24c). This is a complex area, but the overall role of the ACC appears to be monitoring these functions in conflict situations (Paus 2001; Vogt et al. 2005; Walton et al. 2003).

Anatomical relationships both within and between different PFC regions are complex. As such, several organizational schemes have been proposed based on combinations of cortical architecture and connectivity (Barbas 1992; Carmichael and Price 1994, 1996; Haber et al. 2006; Mesulam and Mufson 1993). In general, cortical areas within each prefrontal group are highly interconnected. However, these connections are quite specific in that a circumscribed region within each area projects to specific regions of other areas, but not throughout. Overall the vmPFC is primarily connected to other medial subgenual regions and to areas 24a and 12, with few connections to the dACC or areas 9 and 46. Area 24b of the dACC is interconnected with the different regions of the dACC areas and is also tightly linked to area 9. Different OFC regions are also highly interconnected and connected to areas 9 and 46.

In addition, both the vmPFC and OFC are connected to the hippocampus and amygdala (Barbas and Blatt 1995; Carmichael and Price 1995a; Ghashghaei and Barbas 2002). The hippocampus projects most densely to the vmPFC and less prominently to the OFC. By contrast, there are few projections to the dorsal and lateral PFC areas (dACC, areas 9 and 46). Amygdala projections to the PFC terminate primarily in different regions of the vmPFC, OFC, and dACC, with a particularly dense projection to the vmPFC and adjacent OFC. Unlike the hippocampal projections, the amygdalo-cortical projections are bidirectional. The primary target of these cortical areas is the basal and lateral nuclear complex. Here, the OFC-amygdalo projections target the intercalated masses, whereas terminals from the vmPFC and dACC are more diffuse. PFC projects to multiple subcortical brain regions, but their largest output is to the thalamus and striatum. Cortical connections to ventral BG output nuclei of the thalamus primarily target the mediodorsal nucleus and are bidirectional (see Section 11.4.3). The second largest subcortical PFC output is to the striatum.

11.2.2 The Ventral Striatum

The link between the BG (specifically, the nucleus accumbens) and reward was first demonstrated as part of the self-stimulation circuit originally described by Olds and Milner (1954). Since then, the nucleus accumbens (and the VS in general) has been a central site for studying reward and drug reinforcement and for the transition between drug use as a reward and as a habit (Kalivas, Volkow, and Seamans 2005; Taha and Fields 2006). The term VS, coined by Heimer, includes the nucleus accumbens and the broad continuity between the caudate nucleus and putamen ventral to the rostral internal capsule, the olfactory tubercle and the rostrolateral portion of the anterior perforated space adjacent to the lateral olfactory tract in primates (Heimer et al. 1999). From a connectional perspective, it also includes the medial caudate nucleus, rostral to the anterior commissure (Haber and McFarland 1999) (Section 11.2.2.2). Human imaging studies demonstrate the involvement of the VS in reward prediction and reward prediction errors (Knutson et al. 2001; O'Doherty et al. 2004; Pagnoni et al. 2002; Tanaka et al. 2004) and, consistent with physiological non-human primate studies, the region is activated during reward anticipation (Schultz 2000). Collectively, these studies demonstrate its key role in the acquisition and development of reward-based behaviors and its involvement in drug addiction and drug-seeking behaviors (Belin and Everitt 2008; Everitt and Robbins 2005; Porrino et al. 2007; Volkow et al. 2006).

11.2.2.1 Special Features of the Ventral Striatum

While the VS is similar to the dorsal striatum in most respects, there are also some unique features. The VS contains a subterritory, the shell,* which plays a particularly important role in the circuitry underlying goal-directed behaviors, behavioral sensitization, and changes in affective states (Carlezon and Wise 1996; Ito, Robbins, and Everitt 2004). The shell has some unique connections compared to the rest of the VS that are indicated below. Several other characteristics are unique to the VS. The dopamine transporter is relatively low throughout the VS. The cellular composition of the VS varies somewhat compared to the dorsal striatum (Bayer 1985; Chronister et al. 1981; Meyer et al. 1989). Of particular importance is the fact that, while both the dorsal and ventral striatum receive input from the cortex, thalamus, and brainstem, the VS alone also receives a dense projection from the amygdala and hippocampus. Collectively, these are important distinguishing features of the VS, but it is important to note that its dorsal and lateral border is continuous with the rest of the striatum, and neither cytoarchitectonic nor histochemical distinctions mark a clear boundary between it and the dorsal striatum. Indeed, the best way to define the VS is by its afferent projections from cortical areas that mediate different aspects of reward processing, the vmPFC, OFC, dACC, and the medial temporal lobe.

---

* In rodent models, the term "core" is often used to distinguish the "shell" from the rest of the nucleus accumbens. However, as discussed here, anatomical and physiological data indicate that reward-related processing extends well beyond the confines of nucleus accumbens. Therefore, in this chapter, the term ventral striatum (VS), which encompasses a much larger region, is favored. On the other hand, because VS does not have a clear set of boundaries, it is difficult to define a "core," so this term is not used here.

11.2.2.2 Connections of the Ventral Striatum (Figure 11.1)

11.2.2.2.1 Afferent Connections: Cortical Inputs

The VS is the main input structure of the ventral BG. Like the dorsal striatum, afferent projections to the VS are derived from three major sources: a massive, generally topographic input from cerebral cortex (as reviewed in Section 11.2.1); a large input from the thalamus; and a smaller but critical input from the brainstem, primarily from the midbrain dopaminergic cells. Cortico-striatal terminals are organized in two projection patterns: focal projection fields and diffuse projections (Calzavara, Mailly, and Haber 2007; Haber et al. 2006). Focal projection fields consist of dense clusters of terminals forming the well-known dense patches that can be visualized at relatively low magnification. The diffuse projections consist of clusters of terminal fibers that are widely distributed throughout the striatum, both expanding the borders of the focal terminal fields, but also extending widely throughout other regions of the striatum. We will return to the diffuse projections in Section 11.4.1.

It is the general distribution of the focal terminal fields that gives rise to the topography ascribed to the cortico-striatal projections. This organization is the foundation for the concept of parallel and segregated cortico-BG circuits (see Section 11.2.6). Together, these projections terminate primarily in the rostral, medial, and ventral parts of the striatum and define the ventral striatal territory (Haber et al. 1995a, 2006). The large extent of this region is consistent with the findings that diverse striatal areas are activated following reward-related behavioral paradigms (Apicella et al. 1991; Corlett et al. 2004; Delgado et al. 2003; Kuhnen and Knutson 2005; Tanaka et al. 2004). The focal projection field from the vmPFC is the most limited (particularly from area 25) and is concentrated within, and just lateral to, the shell. The innervation of the shell receives the densest input from area 25, although fibers from areas 14, 32, and from agranular insular cortex also terminate here. The vmPFC also projects to the medial wall of the caudate nucleus, adjacent to the ventricle. In contrast, the central and lateral parts of the VS (including the ventral caudate nucleus and putamen) receive inputs from the OFC. These terminals also extend dorsally, along the medial caudate nucleus, but lateral to those from the vmPFC. There is some medial-to-lateral and rostral-to-caudal topographic organization of the OFC terminal field. Projections from the dACC (area 24b) extend from the rostral pole of the striatum to the anterior commissure and are located in both the central caudate nucleus and putamen. They primarily avoid the shell region. These fibers terminate somewhat lateral and dorsal to those from the OFC. Thus, the OFC terminal fields are positioned between the vmPFC and dACC.

11.2.2.2.2 Afferent Connections: Amygdala and Hippocampal Input

The amygdala is a prominent limbic structure that also plays a key role in emotional coding of environmental stimuli and provides contextual information used for adjusting motivational level. Overall, the main source of amygdala inputs to the VS are the basal nucleus and the magnocellular division of the accessory basal nucleus (Fudge et al. 2002; Russchen et al. 1985). The lateral nucleus has a relatively minor input to the VS. The basal and accessory basal nuclei innervate both the shell and ventromedial striatum outside the shell. The amygdala sends few fibers to the dorsal striatum in primates. In contrast to the amygdala, the hippocampus projects to a more limited region of the VS and is essentially confined to the shell region, where fibers overlap with those from the amygdala (Friedman, Aggleton, and Saunders 2002).

11.2.2.2.3 Afferent Connections: Thalamic Inputs

The midline and medial intralaminar thalamic nuclei project to medial prefrontal areas, the amygdala, and hippocampus, and, as such, are considered the limbic-related thalamic nuclear groups. The VS receives dense projections from these thalamic nuclei, which are topographically organized (Giménez-Amaya et al. 1995). The shell of the nucleus accumbens receives the most limited projection, almost exclusively midline nuclei and the medial parafascicular nucleus. The medial wall of the caudate nucleus receives projections not only from these nuclei, but also from the central superior lateral nucleus. In contrast, the central and lateral parts of the VS receive their main input from the intralaminar nucleus, with a limited projection from the midline thalamic nuclei. In addition to the midline and intralaminar thalamostriatal projections, in primates there is a large input from the "specific" thalamic BG relay nuclei, the medial dorsalis nucleus (MD), and ventral anterior (VA) and ventral lateral (VL) nuclei (McFarland and Haber 2000, 2001). The VS receives these direct afferent projections primarily from the medial MD nucleus and a limited input from the magnocellular subdivision of the ventral anterior nucleus.

11.2.2.2.4 *Efferent Connections*

Efferent projections from the VS, like those from the dorsal striatum, project primarily to the pallidum and substantia nigra/VTA (Haber et al. 1990). Specifically, they terminate topographically in the subcommissural part of the globus pallidus (classically defined as the VP), the rostral pole of the external segment, and the rostromedial portion of the internal segment. The more central and caudal portions of the globus pallidus do not receive this input. VS projections to the substantia nigra are not as confined to a specific region as those to the globus pallidus. Although the densest terminal fields occur in the medial portion, numerous fibers also extend laterally to innervate a wide mediolateral expanse of the dopamine neurons (see Section 11.2.4). This projection extends throughout the rostral-caudal extent of the substantia nigra. In addition to projections to the typical BG output structures, the VS also projects to non-BG regions. The shell sends fibers caudally and medial into the lateral hypothalamus. Projections from the medial part of the VS also project more caudally, terminating in the pedunculopontine nucleus and to some extent in the medial central gray. Axons from the medial VS (including the shell) travel to and terminate in the bed nucleus of the stria terminalis, and parts of the ventral regions of the VS terminate in the nucleus basalis (Haber et al. 1990). This direct projection to the nucleus basalis in the basal forebrain is of particular interest, since it is the main source of cholinergic fibers to the cerebral cortex and the amygdala. Thus, the VS is in a position to influence cortex directly, without passing through the pallidal, thalamic loop (Beach, Tago, and McGeer 1987; Chang, Penny, and Kitai 1987; Haber 1987; Martinez-Murillo et al. 1988; Zaborszky and Cullinan 1992). Likewise, the projection to the bed nucleus of the stria terminalis indicates direct striatal influence on the extended amygdala.

11.2.3 THE VENTRAL PALLIDUM

The VP (Figure 11.2) is an important component of the reward circuit. These cells respond during the learning and performance of reward-incentive behaviors and are an area of focus in the study of addictive behaviors (Smith and Berridge 2007; Tindell et al. 2006). While the term VP typically refers to the region below the anterior commissure, in primates it is best defined by its input from the entire reward-related VS (Haber et al. 1990; Heimer 1978). As indicated above, that includes not only the subcommissural regions, but also the rostral pole of the external segment and the medial rostral internal segment of the globus pallidus. Like the dorsal pallidum, the VP contains two parts: a substance-P-positive component and an enkephalin-positive component, which project to thalamus and subthalamic nucleus (STN), respectively (Haber, Wolfe, and Groenewegen 1990; Haber, Lynd-Balta, and Mitchell 1993; Mai et al. 1986; Russchen, Amaral, and Price 1987). Pallidal neurons have a distinct morphology that is useful for determining the boundaries and extent of the VP (DiFiglia, Aronin, and Martin 1982; Fox et al. 1974; Haber and Nauta 1983; Haber and Watson 1985). The VP not only reaches ventrally, but also rostrally to invade the rostral and ventral portions of the VS. In addition to the GABAergic input from the VS, there is a glutamatergic input from the STN and a dopaminergic input from the midbrain (Klitenick et al. 1992; Turner et al. 2001).

Descending efferent projections from the enkephalin-positive VP terminate primarily in the medial subthalamic nucleus, extending into the adjacent lateral hypothalamus (Haber et al. 1985; Haber, Lynd-Balta, and Mitchell 1993; Zahm 1989). The VP also projects to the substantia nigra,

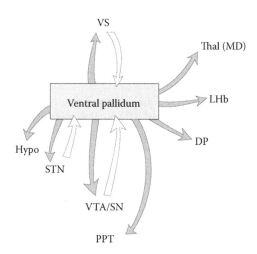

FIGURE 11.2 Schematic illustrating the connections of the ventral pallidum. Same abbreviations as in Figure 11.1. (Reprinted from Haber, S.N., Anatomy and connectivity of the reward circuit, pp. 3–27, in *Handbook of Reward and Decision Making*, Dreher, J.C. and Tremblay, L. (eds.), Copyright 2009, with permission from **Elsevier**.)

terminating medially in the SNc, SN pars reticulata (SNr), and the VTA. Projections from the VP to the subthalamic nucleus and the lateral hypothalamus are topographically arranged. By contrast, terminating fibers from the VP in the substantia nigra overlap extensively, suggesting convergence of terminals from different ventral pallidal regions. VP fibers also innervate the pedunculopontine nucleus. As with the dorsal pallidum, components of the substance P-positive VP project to the thalamus, terminating in the midline nuclei and medial MD. Pallidal fibers entering the thalamus give off several collaterals forming branches that terminate primarily onto the soma and proximal dendrites of thalamic projection cells. In addition, some synaptic contact is also made with local circuit neurons, indicating that, while pallidal projections are primarily inhibitory on thalamic relay cells, they may also function to disinhibit projection cells via the local circuit neurons (Arecchi-Bouchhioua et al. 1997; Ilinsky, Yi, and Kultas-Ilinsky 1997). In addition, the VP also projects to both the internal and external segments of the dorsal pallidum. This is a unique projection, in that the dorsal pallidum does not seem to project ventrally, into the VP.

Parts of the VP (along with the dorsal pallidum) project to the lateral habenular nucleus (LHb). Recent data have supported the role of the LHb in generating a negative reward signal. In particular, it provides the signal that inhibits dopamine activity when an expected reward does not occur (Lecourtier and Kelly 2007; Matsumoto and Hikosaka 2007; Ullsperger and von Cramon 2003). Thus, LHb cells are inhibited by a reward-predicting stimulus, but fire following an unexpected non-reward signal, thus providing a negative reward-related signal to the substantia nigra, pars compacta (SNc). Most pallidal cells that project to the lateral habenula are embedded within accessory medullary laminae that divide the lateral and medial portions of the dorsal internal pallidal segment. In addition, other LHb projecting cells are found circumventing the VP (Haber et al. 1985; Parent and De Bellefeuille 1982). Finally, part of the VP (as with the GPe) also projects to the striatum (Spooren et al. 1996). This pallidostriatal pathway is extensive in the monkey and is organized in a topographic manner preserving a general, but not strict, medial-to-lateral and ventral-to-dorsal organization.

11.2.4 THE MIDBRAIN DOPAMINE NEURONS

Dopamine neurons play a central role in the reward circuit (Schultz 2002; Wise 2002). While behavioral and pharmacological studies of dopamine pathways have led to the association of the mesolimbic pathway and nigrostriatal pathway with reward and motor activity, respectively, more recently both of these cell groups have been associated with reward. The midbrain dopamine neurons project widely throughout the brain. However, studies of the rapid signaling that is associated with incentive learning and habit formation focus on the dopamine striatal pathways. Before turning to its projections, it is important to understand the organization of the midbrain dopamine cells in primates.

The midbrain dopamine neurons are divided into the VTA and the substantia nigra, pars compacta (SNc) (Figure 11.3a). Based on projections and chemical signatures, these cells are also referred to as the dorsal and ventral tier neurons (Haber et al. 1995b). The dorsal tier includes the VTA and the dorsal part of the SNc (also referred to as the retrorubral cell group). The cells of the dorsal tier are calbindin-positive and contain relatively low levels of mRNA for the dopamine transporter and D2 receptor subtype. They project to the VS, cortex, hypothalamus, and amygdala. The ventral tier of dopamine cells are calbindin-negative, have relatively high levels of mRNA for the dopamine transporter and D2 receptor and project primarily to the dorsal striatum. Ventral tier cells (calbindin poor, but DAT and D2 receptor rich) are more vulnerable to degeneration in Parkinson's disease and to N-methyl-4-phenyl-1,2,3,6-tetrahydropyridine (MPTP)-induced toxicity, while the dorsal tier cells are selectively spared (Lavoie and Parent 1991). As mentioned above, despite these distinctions, both cell groups respond to unexpected rewards.

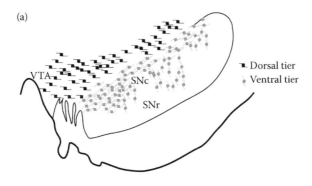

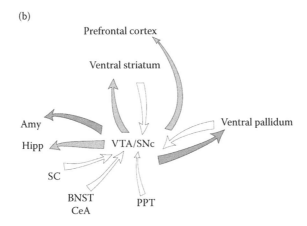

FIGURE 11.3 Schematic illustrating the organization (a) and connections (b) of the midbrain dopamine cells. CeA=central nucleus of the amygdala; SC=superior colliculus; SNc=substantia nigra, pars compacta; SNr=substantia nigra, pars reticulate. Other abbreviations as per Figure 11.1. (Reprinted from Haber, S.N., Anatomy and connectivity of the reward circuit, pp. 3–27, in *Handbook of Reward and Decision Making,* Dreher, J.C. and Tremblay, L. (eds.), Copyright 2009, with permission from **Elsevier**.)

11.2.4.1 Afferent Connections of the Dopamine Neurons

Input to the midbrain dopamine neurons is primarily from the striatum, both the external segment of the globus pallidus (GPe) and the VP, and the brainstem (Figure 11.3b). In addition, there are projections to the dorsal tier from the bed nucleus of the stria terminalis, the sublenticular substantia innominata, and the central amygdala nucleus (Fudge and Haber 2000, 2001; Haber, Lynd-Balta, and Mitchell 1993; Hedreen and DeLong 1991; Lynd-Balta and Haber 1994a).

As described above, the striatonigral projection is the most massive projection to the SN and terminates in both the VTA/SNc and the SNr (Hedreen and DeLong 1991; Lynd-Balta and Haber 1994a; Szabo 1979). The ventral (like the dorsal) striatonigral connection terminates throughout the rostro-caudal extent of the substantia nigra. There is an inverse ventral/dorsal topography to the striatonigral projections. The ventral striatonigral inputs terminate in the VTA, the dorsal part of the ventral tier, and in the medial and dorsal SNr. Thus the VS projects not only throughout the rostrocaudal extent of the substantia nigra, but also covers a wide mediolateral range. In contrast, the dorsolateral striatonigral inputs are concentrated in the ventrolateral SN. These striatal cells project primarily to the SNr, but also terminate on the cell columns of dopamine neurons that penetrate deep into the SNr (Haber, Fudge, and McFarland 2000).

Both the GPe and the VP project to the substantia nigra. The pallidal projection follows a similar inverse dorsal/ventral organization to the striatonigral projection. Thus, the VP projects dorsally, primarily to the dorsal tier and dorsal SNc. The pedunculopontine nucleus sends a major glutamatergic input to the dopaminergic cell bodies. In addition, there is a serotonergic innervation from the dorsal raphe nucleus, though there is disagreement regarding whether fibers terminate primarily in the pars compacta or pars reticulata. Other brain-stem inputs to the dopamine neurons include those from the superior colliculus, the parabrachial nucleus, and locus coeruleus. These inputs raise the interesting possibility that dopamine cells receive a direct sensory input (see Section 11.3.4). Finally, in primates, there is a small and limited projection from the PFC to the midbrain DA neurons, and to both the VTA and SNc. While considerable attention has been given to this projection, relative to the density of its other inputs, this projection is weak in primates (Frankle, Laruelle, and Haber 2006).

11.2.4.2 Efferent Projections

The midbrain dopamine neurons send their largest output to the striatum (Hedreen and DeLong 1991; Lynd-Balta and Haber 1994b; Szabo 1979). As with the descending striatonigral pathway, there is a mediolateral and an inverse dorsoventral topography arrangement to the projection. The ventral pars compacta neurons project to the dorsal striatum, and the dorsally located dopamine neurons project to the VS. The shell region receives the most limited input, primarily derived from the medial VTA (Lynd-Balta and Haber 1994c). The rest of the VS receives input from the entire dorsal tier. In contrast to the VS, the central striatal area (the region innervated by the DPFC) receives input from a wide region of the SNc. The dorsolateral (motor-related) striatum receives the largest midbrain projection from cells in the ventral tier. In contrast to the dorsolateral region of the striatum, the VS receives the most limited dopamine cell input. Thus, in addition to an inverse topography, there is also a differential ratio of dopamine projections to the different striatal areas (Haber, Fudge, and McFarland 2000).

The dorsal tier cells also project widely throughout the primate cortex and are found not only in granular areas but also in agranular frontal regions, parietal cortex, temporal cortex, and, albeit sparsely, in occipital cortex (Gaspar, Stepneiwska, and Kaas 1992; Lidow et al. 1991). The majority of DA cortical projections are from the parabrachial pigmented nucleus of the VTA and the dorsal part of the SNc. The VTA also projects to the hippocampus, though to a lesser extent than in neocortex. The dopamine cells that project to functionally different cortical regions are intermingled with each other, in that individual neurons send collateral axons to different cortical regions. Thus the nigrocortical projection is a more diffuse system compared to the nigrostriatal system and can modulate cortical activity at several levels. Dopamine fibers are located in superficial layers, including a prominent projection throughout layer I. This input therefore is in a position to modulate the distal apical dendrites. Dopamine fibers are also found in the deep layers in specific cortical areas (Goldman-Rakic et al. 1999; Lewis 1992). Projections to the amygdala arise primarily from the dorsal tier. These terminals form symmetric synapses primarily with spiny cells of specific subpopulations in the amygdala (Brinley-Reed and McDonald 1999). As indicated above, dopamine fibers also project to the VP.

11.2.5 COMPLETING THE CORTICO-BASAL GANGLIA REWARD CIRCUIT

The MD nucleus projects to the PFC and is the final link in the reward circuit (Haber, Lynd-Balta, and Mitchell 1993; McFarland and Haber 2002b; Ray and Price 1993). Projections from the VP terminate primarily in the medial mediodorsal nucleus (MD) and in the adjacent midline nuclei (Haber, Lynd-Balta, and Mitchell 1993) (see Figure 11.1). This projection is also topographic, such that the medial part of the VP, which receives its primary input from the shell, is connected mainly to the midline nuclei. The central parts of the VP that receive input from central regions of the VS project to the medial magnocellular MD, while more lateral regions project more laterally. These different MD thalamic areas are then connected to the vmPFC, OFC, and dACC, respectively (Ray and Price 1993).

11.2.6 THE PLACE OF THE REWARD CIRCUIT IN THE BASAL GANGLIA

Afferent projections to the striatum terminate in a general topographic manner. Different frontal cortical areas have corresponding striatal regions that are involved in various aspects of reward cognition and motor control. As indicated above, the vmPFC, OFC, and dACC project to the ventromedial striatum. The DPFC projects to the head of the caudate nucleus and to the putamen rostral to the anterior commissure. Caudal to the commissure, this projection is confined to the medial, central portion of the head of the caudate nucleus, with few terminals in the central and caudal putamen (Calzavara and Haber 2006; Haber et al. 2006). Physiological, imaging, and lesion studies support the idea that these areas are involved in working memory and strategic planning processes (Battig, Rosvold, and Mishkin 1960; Levy et al. 1997; Pasupathy and Miller 2005). Both caudal and rostral motor areas occupy much of the putamen caudal to the anterior commissure, a region that also receives overlapping projections from somatosensory cortex, resulting in a somatotopically organized sensory-motor area (Aldridge, Anderson, and Murphy 1980; Flaherty and Graybiel 1994; Kimura 1986). In summary, projections from frontal cortex form a functional gradient of inputs from the ventromedial to the dorsolateral striatum, with the medial and orbital prefrontal cortex terminating in the ventromedial part, and the motor cortex terminating in the dorsolateral region. As seen with the cortico-striatal projection, thalamostriatal projections are also organized in a general topographical manner, such that interconnected and functionally associated thalamic and cortical regions terminate in the same general striatal region (McFarland and Haber 2000).

The striatal projections to the pallidal complex and substantia nigra (pars reticulata) are also generally topographically organized, thus maintaining the functional organization of the striatum in these output nuclei (Haber et al. 1990; Hedreen and DeLong 1991; Lynd-Balta and Haber 1994a; Middleton and Strick 2002; Selemon and Goldman-Rakic 1990). The VS terminates in the VP and in the dorsal part of the midbrain. Terminals from the central striatum terminate more centrally in both the pallidum and the pars reticulata, while those from the sensorimotor areas of the striatum innervate the ventrolateral part of each pallidal segment and the ventrolateral SN. Finally, the pallidum and pars reticulata project to the different BG output nuclei of the thalamus, the mediodorsal, ventral anterior, and ventral lateral cell groups, which are connected respectively to limbic, associative, and motor control areas (Ilinsky, Jouandet, and Goldman-Rakic 1985; Kuo and Carpenter 1973; McFarland and Haber 2002a; Middleton and Strick 2002; Strick 1976). Thus, the organization of connections through the cortico-BG–cortical network preserves a general functional topography within each structure, from the cortex through the striatum, from the striatum to the pallidum/pars reticulata, from these output structures to the thalamus, and finally, back to cortex.

This organization has led to the concept that each functionally identified cortical region drives (and is driven by) a specific BG loop or circuit, leading, in turn, to the idea of parallel processing of cortical information through segregated BG circuits (Alexander and Crutcher 1990). This concept focuses on the role of the BG in the selection and implementation of an appropriate motor response, while inhibiting unwanted ones (Mink 1996). The model assumes, however, that the behavior has been learned and the role of the BG is to carry out a coordinated action. We now know that the cortico-BG network is critical in mediating the learning process to adapt and to accommodate past experiences to modify behavioral responses (Cools, Clark, and Robbins 2004; Hikosaka et al. 1998; Muhammad, Wallis, and Miller 2006; Pasupathy and Miller 2005; Wise, Murray, and Gerfen 1996). This requires communication links between circuits. However, before we discuss the integration between the reward circuit and cognitive and motor control circuits, we turn to where the sensory systems enter the cortico-BG reward network.

11.3 SENSORY INPUTS TO THE REWARD CIRCUIT

Sensory systems play a key role in initiating and developing reward responses. While the cortico-BG reward circuit does not receive direct sensory input (with the possible exception of olfactory input), highly processed sensory information does reach the VS indirectly via cortical, amygdala, and midbrain inputs. Moreover, the reward system has access to sensory modulation through its output, albeit limited, to the hypothalamus and brainstem.

11.3.1 ORBITAL AND INSULAR PREFRONTAL CORTEX

The main cortical sensory input to the VS is through the OFC and adjacent insula. The insula is divided into three cytoarchitectonic areas that are associated with different sensory functions: (1) a rostroventral agranular insula (Ia) that is related to olfactory and autonomic functions; (2) an intermediate dysgranular insula (Id) that is associated with gustatory and some visual and somatosensory functions; and (3) a caudodorsal granular insula (Ig) that is associated with somatosensory, auditory, and visual functions. These three areas are arranged in a radial manner around the piriform olfactory cortex (Friedman et al. 1986; Mesulam and Mufson 1993; Penfield and Faulk 1955; Schneider Friedman, and Mishkin 1993; Showers and Lauer 1961). Taste and visceral information from primary gustatory cortex and olfactory information from piriform cortex overlap in agranular insula. Moreover, direct connections from the visceral nucleus of the thalamus also converge in specific agranular areas (Carmichael and Price 1995b). Anatomical and physiological studies suggest that the anterior Ig and adjacent Id may contribute to tactile object recognition in the hand and mouth associated with feeding behavior (Friedman et al. 1986; Preuss and Goldman-Rakic 1989). The OFC receives input from all of the sensory modalities (Barbas 1993; Carmichael and Price 1995b). Sensory information arrives with different levels of processing, with visual, auditory, and somatosensory systems passing through several cortical areas before reaching the OFC and insula. By contrast, olfaction and gustatory information is derived from direct inputs from primary cortices.

Links between insular cortex, the vmPFC, and OFC are complex. Ia has a tight connection with the vmPFC (Carmichael and Price 1996). Area 13 of orbital cortex receives inputs from olfactory and gustatory areas that overlap with highly processed information from other modalities. Lateral area 12 receives a substantial input from visual association cortex, area TE. In addition, fibers from TE also project to specific regions of area 13. As mentioned above, OFC regions are tightly linked. Thus, the OFC, particularly the caudal regions, receive both primary and multimodal sensory input from high-order association cortical areas. Taken together, through interconnections between the OFC areas 12 and 13 appear to integrate input from multisensory regions (Barbas 1992; Barbas and Pandya 1989; Carmichael and Price 1995b; Morecraft, Geula, and Mesulam 1992).

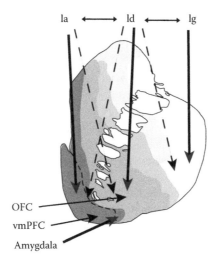

FIGURE 11.4   Schematic illustrating projections from the insula to the striatum. Ia = agranular insula; Id = dysgranular insula; Ig = granular insula; OFC = orbital prefrontal cortex; vmPFC = ventral medial prefrontal cortex.

The VS receives input from both the Ia and Id insular cortex, but not Ig[*] (Chikama et al. 1997) (Figure 11.4). The densest input from Ia terminates in the shell and the medial wall of the caudate. This ventral striatal region therefore receives convergent input from the olfactory and visceral-associated insula, and input from the vmPFC. The gustatory insular regions in the anterior and central portions of Id (Smith-Swintosky, Plata-Salaman, and Scott 1991; Yaxley, Rolls, and Sienkiewicz 1990) primarily project to the central VS. Moreover, there are additional inputs here from agranular areas associated with olfactory and visceral responses. Finally, the dorsal portion of Id that receives somatosensory information regarding the hand and face (Mesulam and Mufson 1993) also projects partially to VS. Overall, areas 13 and 12 appear to receive the most convergent input from processed multimodal sensory systems. Projections to the striatum from areas 12 and 13 are organized somewhat topographically in that, generally, area 13 projects more centrally and area 12 more laterally. As indicated above, both areas 12 and 13 are involved in multimodal sensory processing and are further divided into somewhat different combinations of sensory input (Carmichael and Price 1995b). However, their projections to the VS do not reflect these subdivisions. It is therefore difficult to match specific sensory modalities to the topographic terminal organization in the striatum. In general the striatal region that receives this input also receives input from the amygdala, Ia, and Id. This combination of inputs that link visual and somatosensory stimuli associated with feeding behaviors provides a wide spectrum of information regarding food acquisition centered in the VS. In addition, although more rostral OFC regions, areas 10 and 11, receive less direct sensory input, here via OFC connections, the above-mentioned modalities are combined with information from auditory areas (Barbas et al. 1999). The VS therefore appears to receive dual sensory inputs from both the OFC and insula. That is, Ia and Id project directly to the shell and the VS, respectively, and also to different OFC regions. OFC also sends inputs to the VS, overlapping with those from the insula.

11.3.2   Interface between Amygdala and Cortical Inputs to the Ventral Striatum

The three main amygdaloid regions are the basolateral nuclear group (BLNG), the cortico-medial region (including the periamygdaloid cortex [PAC]) and medial nucleus), and the CeA, which are

---

[*] The absence of Ig input to VS would be predicted on evolutionary grounds, since anatomical homologues of the basal ganglia and VS exist in non-mammalian vertebrates that lack granular isocortex, as discussed in detail in Chapter 4.

characterized by functional differences based on specific intrinsic and extrinsic connections (Amaral et al. 1992; Carmichael and Price 1995a; Fudge et al. 2002; Jolkkonen and Pitkanen 1998; Pitkanen and Amaral 1998; Saunders, Rosene, and Van Hoesen 1988). The BLNG includes the basal and accessory basal nuclei which process higher-order sensory inputs in all modalities except olfaction. This is the main input to the VS (Iwai and Yukie 1987; Nishijo, Ono, and Nishino 1988a, 1988b; Ono et al. 1989; Pitkanen and Amaral 1998; Turner, Mishkin, and Knapp 1980). The shell receives the most complex amygdala input. This is derived not only from the BLNG, but also from the cortico-medial region and the CeA. The CeA can be viewed as a site where the "drive" value of a stimulus is determined based on its converging inputs from the "external" milieu (via the BLNG) and "internal" milieu (via the lateral hypothalamus and brainstem) (Aggleton 1985; Amaral et al. 1992; Ricardo and Koh 1978; Saper and Loewy 1980). Thus, these inputs directed to the shell add important information about matching a stimulus with the animal's internal "drive" state, e.g., whether the animal is hungry when a food-associated stimulus is presented. These inputs are further strengthened by projections derived from similar functional regions, the vmPFC and agranular insular cortex. The BLNG is the source of all amygdaloid input to the VS outside of the shell. This projection is concentrated in the central part of the VS, in addition to the shell. However, the lateral VS also receives this input. The central and lateral striatum receive additional projections from multimodal sensory regions of the OFC, particularly areas 13 and 12, thus reinforcing multimodal sensory information derived from the amygdala. Overall, the amygdala has strong bilateral connections with the OFC. However, it is important to note that the connections between cortex and striatum and between amygdala and striatum are not bidirectional. Thus, these afferent projections to the VS from the cortex and amygdala leave the cortico-amygdala interface and enter another system, the BG.

11.3.3 FOLLOWING THE CORTICO-AMYGDALA SENSORY TOPOGRAPHY THROUGH THE REWARD CIRCUIT

The shell region of the VS is unique as it receives inputs that the rest of the VS does not. In particular this area contains overlapping sensory inputs from the vmPFC, Ia, and the CeA and PAC amygdala nuclei (as indicated in Figure 11.4). This information is then combined with inputs from the hippocampus. In addition, the shell, like the rest of the VS, receives input from the BLNG and the VTA. The shell also has some unique outputs. In addition to its projections to the pallidum and the substantia nigra, unlike the rest of the VS, the shell also projects to the hypothalamus and the bed nucleus of the stria terminalis (see Figure 11.1). This provides direct feedback into systems that monitor the internal milieu. Moreover, its projection terminates in the most medial aspect of the VP. This part of the VP projects not only to the subthalamic nucleus, but also to the adjacent lateral hypothalamus. Finally, this VP region projects primarily to the midline thalamic nuclei, with few terminals in the MD. While these features set the shell and medial VP apart from the rest of the VS/VP, they contribute to the overall integration of the cortico-BG network through their integrative connectivities with other BG elements (see Section 11.4).

The central VS receives input primarily from the OFC, particularly from multimodal sensory areas, particularly area 13, along with the amygdala. In addition, this region also receives input from the more anterior OFC, including areas 10 and 11. Collectively, these inputs converge with those from the dorsal tier dopamine neurons, including part of the VTA. Projections from the central VS terminate centrally in subcommissural VP, extending dorsal and medial, above the anterior commissure. In turn, this area projects to the medial subthalamic nucleus and to the medial parts of the MD. Few, if any, fibers reach the lateral hypothalamus, either directly from the VS or from the VP. More lateral VS receives similar cortical input to the central VS, with a great proportion of OFC projections derived from area 12. In addition, here there is convergence between inputs from OFC and those from dACC. Projections from the lateral VS mirror those from the central VS, but terminate somewhat more laterally. As will be discussed in Section 11.4.2, terminals from throughout the VS converge extensively in the substantia nigra.

11.3.4 Superior Colliculus Inputs to the Midbrain Dopamine Neurons

There is a direct efferent projection from the superior colliculus (SC) to the midbrain dopamine neurons. The SC projects to both the dorsal and ventral tier midbrain dopamine cells, and to a lesser extent to the substantia nigra, pars reticulata. These inputs terminate in close association with TH-positive cells, particularly within the dorsal tier of DA neurons. The SC-projecting cells are located in the intermediate and deep layers of the SC (May et al. 2009; McHaffie et al. 2006). Several SC cell types project to the dopamine neurons, suggesting a heterogeneous input as indicated by differences in DA cell responses to various categories of visual events, e.g., appearance, disappearance, movement (Wurtz and Albano 1980), and looming (Westby et al. 1990). Few of the tectonigral neurons are located in the exclusively visual, superficial layers of the SC, suggesting that the tectonigral cells are multisensory neurons (May et al. 2009). It has been suggested that these provide the multimodal sensory input that tunes the dopamine cells to the novel and salient stimuli in the environment (Stein and Meredith 1993; Stein and Stanford 2008). These cells may include those that increase the gain of their activity for reward-related responses but do not show motor-related activity (Ikeda and Hikosaka 2003).

The dorsal tier of DA neurons receives the densest tectonigral projection. As indicated above, the dorsal tier cells also receive unique inputs from the amygdala and BNST, along with those from the VS. Moreover, these cells preferentially innervate VS (Haber, Fudge, and McFarland 2000). Thus, the tectonigral projection provides a signal of biologically salient sensory stimuli (via a glutamatergic excitatory input) to elicit phasic DA modulation in the VS system. This could be of importance for reinforcing reward-related learning activity derived from cortical and amygdala inputs to the VS (Gruber et al. 2006).

The well-documented nigrotectal projection, derived from the SNr, mediates gaze-related activity of the SC (Graybiel 1978; Chevalier et al. 1981; Hikosaka and Wurtz 1983; Huerta et al. 1991). This pathway provides an output mechanism of the BG to influence sensory modulation of motor systems. The direct tectal projection to the pars reticulata may modulate the afferent signals it receives from the BG (Comoli et al. 2003). Thus, tectonigral input to SNr may modulate the disinhibitory output signals from the BG that gate SC saccade-related outputs (Chevalier and Deniau 1990; Hikosaka et al. 2000).

11.4 INTEGRATING THE REWARD CIRCUIT INTO COGNITIVE AND MOTOR BASAL GANGLIA PATHWAYS

The cortico-BG is a complex system through which prefrontal cortex exploits the BG for additional processing of reward to effectively modulate learning, leading to the development of goal-directed behaviors and action plans. To develop an appropriate behavioral response to external environmental stimuli, sensory information must be integrated into the reward circuit, to develop a strategy and an action plan for obtaining the goal. Thus action plans developed towards obtaining a goal require a combination of sensory and reward processing, cognition, and motor control. Although theories related to cortico-BG processing have traditionally emphasized the segregation of functions (including different reward circuits), highlighting separate and parallel pathways (Alexander and Crutcher 1990; Middleton and Strick 2002; Price, Carmichael, and Drevets 1996), it is now evident that the reward circuit does not work in isolation. As such, this complex circuitry interfaces with pathways that mediate cognitive function and motor planning. There are regions of integration and convergence linking together areas that are associated with different functional domains, both within and between each of the cortico-BG structures. First, there is convergence between terminals derived from different cortical areas in the striatum that permits cortical information to be integrated across multiple functional regions within the striatum. Second, there are interconnections between structures that have both reciprocal and non-reciprocal connections that link across functional domains. The two major ones are the striato-nigro-striatal pathway and the cortico-thalamo-cortical network. In addition, the VP-VS-VP network also has an important non-reciprocal component. Through these networks, sensory information enters the reward circuit, which then impacts on cognition and motor control through several different interactive routes, allowing information about reward to be channeled through cognitive and motor control circuits to mediate the development of appropriate action plans.

11.4.1 CONVERGENCE OF CORTICO-STRIATAL PROJECTIONS

Despite the general topography described above, focal terminal fields from the vmPFC, OFC, and dACC show a complex interweaving and convergence, providing an anatomical substrate for modulation between circuits within the reward network (Haber et al. 2006). Focal projections from the OFC extend into areas innervated by the dACC and the vmPFC (Figure 11.5a). Indeed, these terminal fields do not occupy completely separate territories in any part of the striatum, but converge most extensively at rostral levels. Regions of convergence between the focal terminal fields of the vmPFC, OFC, and dACC provide an anatomical substrate for integration between sensory inputs, which, along with different reward processing circuits within specific striatal areas, may represent "hot spots" of plasticity for integrating reward value, predictability, and salience. In addition to convergence between vmPFC, dACC, and OFC focal terminal fields, projections from dACC and OFC also converge with inputs from the DPFC, demonstrating that functionally diverse PFC projections also interface in the striatum. At rostral levels, DPFC terminals converge with those from both the dACC and OFC, although each cortical projection also occupies its own territory. Here, projections from all PFC areas occupy a central region, with the different cortical projection extending into non-overlapping zones. The anterior striatum is, therefore, a particularly critical place where sensory, emotional, and cognitive information intermingle. Coordinated activation of DPFC, dACC, and/or OFC terminals in the striatum could produce a unique combinatorial activation at the specific sites, enabling reward-based incentive drive to impact on long-term strategic planning. Convergence between these areas is less prominent caudally, with almost complete separation of the dense terminals from the DPFC and dACC/OFC just rostral to the anterior commissure.

In addition to the focal projections, each cortical region sends a diffuse fiber projection that extends outside of its focal terminal field (Figure 11.5b) (Haber et al. 2006). These axons can travel some distance, invading striatal regions that receive their focal input from other prefrontal cortex areas. For example, the diffuse projection from the OFC extends deep into the dorsal, central caudate and putamen, with extensive convergence with the focal and diffuse projections from both the dACC and the DPFC. Likewise, the diffuse projections from dACC overlap with focal projections from the vmPFC, OFC, and DPFC. Moreover, clusters of fibers are found in the dorsal lateral caudate nucleus and in the caudal ventral putamen, areas that do not receive a focal input from other prefrontal regions. Finally, clusters of DPFC fibers terminate throughout the rostral striatum, including the VS and lateral putamen. Although the focal projections do not reach into the ventromedial region, clusters of labeled fibers are located here.

Significant and extensive diffuse projections from each frontal cortical region are consistent with the demonstration that a single cortico-striatal axon can innervate 14% of the striatum (Zheng and Wilson 2002). However, activation of medium spiny neurons requires a large coordinated glutamatergic input from many cortical cells (Wilson 2004). Therefore, the invasions of relatively small fiber clusters from other functional regions are not considered to have much relevance for cortico-striatal information processing and, as a result, anatomical studies have focused on the large, dense focal projections (Ferry et al. 2000; Selemon and Goldman-Rakic 1985). While under normal conditions in which a routine behavior is executed these fibers may have little impact, this diffuse projection may serve a separate integrative or modulatory function. Collectively, these projections represent a large population of axons invading each focal projection field and, under certain conditions, if collectively activated, they may provide the recruitment strength necessary to modulate the focal signal. This would serve to broadly disseminate cortical activity to a wide striatal region, thus providing an anatomical substrate for cross-encoding cortical information to influence the future firing

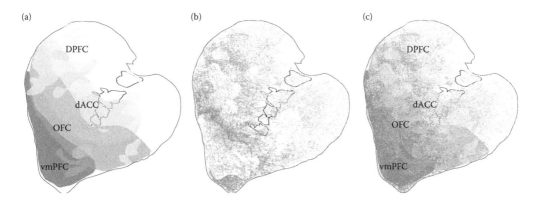

FIGURE 11.5 Schematic diagrams demonstrating convergence of cortical projections from different reward-related regions and dorsal prefrontal areas. (a) Convergence between focal projections from different prefrontal regions. (b) Distribution of diffuse fibers from different prefrontal regions. (c) Combination of focal and diffuse fibers. dACC = dorsal anterior cingulate cortex; DPFC = dorsal lateral prefrontal cortex; OFC = orbital prefrontal cortex; vmPFC = ventral medial prefrontal cortex. The four levels of gray shading indicate input from vmPFC (darkest gray), OFC, dACC, and DPFC (lightest gray).

of medium spiny neurons (Kasanetz et al. 2008). Taken together, the combination of focal and diffuse projections from frontal cortex occupies the rostral striatum and continues caudally through the caudate nucleus and putamen (Figure 11.5c). The fronto-striatal network, therefore, constitutes a dual system comprising both topographically organized terminal fields, along with subregions that contain convergent pathways derived from functionally discrete cortical areas (Draganski et al. 2008; Haber et al. 2006).

11.4.2 The Striato-Nigro-Striatal Network

While the role of dopamine and reward is well established, the latency between the presentation of the reward stimuli and the activity of the DA cells is too short to reflect higher cortical processing necessary for linking a stimulus with its rewarding properties. The fast, burst-firing activity is likely, therefore, to be generated from other inputs such as brainstem glutamatergic nuclei (see Section 11.3.4) (Dommett et al. 2005). An interesting issue is, then, how do the dopamine cells receive information concerning reward value? The largest forebrain input to the dopamine neurons is from the striatum. However, this is a relatively slow GABAergic inhibitory projection, unlikely to result in the immediate, fast burst-firing activity. Nonetheless, the collective complex network of PFC, amygdala, and hippocampal inputs to the VS integrate information related to reward processing and memory to modulate striatal activity. These striatal cells then impact directly on a subset of medial dopamine neurons, which, through a series of connections described below, can modulate the dorsal striatum.

As mentioned above, projections from the striatum to the midbrain are arranged in an inverse dorsal-ventral topography and there is also an inverse dorsal-ventral topographic organization to the midbrain striatal projection. When considered separately, each limb of the system creates a loose topographic organization: the VTA and medial SN being associated with the limbic system, and the central and ventral SN with the associative and motor striatal regions, respectively. However, each functional region differs in their proportional projections that significantly alter their relationship to each other. The VS receives a limited midbrain input, but projects to a large region. By contrast, the dorsolateral striatum receives a wide input, but projects to a limited region. In other words, the VS influences a wide range of dopamine neurons, but is itself influenced by a relatively limited group of dopamine cells. On the other hand, the dorsolateral striatum influences a limited midbrain region, but is affected by a relatively large midbrain region.

The proportional differences between inputs and outputs of the dopamine neurons, coupled with their topography, result in complex interweaving of functional pathways. For each striatal region, the afferent and efferent striato-nigro-striatal projection system contains three components in the midbrain. There is a reciprocal connection that is flanked by two non-reciprocal connections. The reciprocal component contains cells that project to a specific striatal area. These cells are embedded within terminals from that same striatal area. Dorsal to this region lies a group of cells that project to the same striatal region but do not lie within its reciprocal terminal field. In other words, these cells receive a striatal projection from a region to which they do not project. Finally, ventral to the reciprocal component are efferent terminals. However, there are no cells embedded in these terminals that project to that same specific striatal region. The cells located in this terminal field project to a different striatal area. These three components for each striato-nigro-striatal projection system occupy different positions within the midbrain. The VS system lies dorsomedially, the dorsolateral striatum system lies ventrolaterally, and the central striatal system is positioned between the two. Thus, the size and position of the afferent and efferent connections for each system, together with the arrangement into three components, allow information from the limbic system to reach the motor system through a series of connections (Haber, Fudge, and McFarland 2000) (Figure 11.6). With this arrangement, while the VS receives input from the vmPFC, OFC, dACC, and amygdala, its efferent projection to the midbrain extends beyond the tight VS/dorsal tier dopamine/VS circuit. It terminates also in the ventral tier, to influence the dorsal striatum. Moreover, this part of the ventral tier is reciprocally connected to the central (or associative) striatum. The central striatum also projects to a

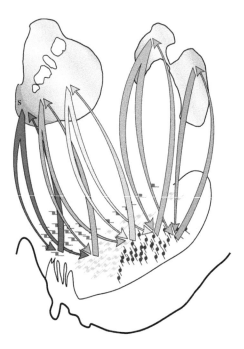

FIGURE 11.6 (See Color Insert) Schematic illustrating the complex connections between the striatum and substantia nigra. The arrows illustrate how the ventral striatum can influence the dorsal striatum through the midbrain dopamine cells. Colors indicate functional regions of the striatum, based on cortical inputs. Midbrain projections from the shell target both the VTA and ventromedial SNc. Projections from the VTA to the shell form a "closed" reciprocal loop, but also project more laterally to impact on dopamine cells projecting the rest of the ventral striatum forming the first part of a feed-forward loop (or spiral). The spiral continues through the striato-nigro-striatal projections through which the ventral striatum impacts on cognitive and motor striatal areas via the midbrain dopamine cells. Red=inputs from the vmPFC; orange=inputs from the OFC and dACC; yellow=inputs from the dPFC; green and blue=inputs from motor control areas. (Reprinted from Haber, S.N., Fudge, J.L. and McFarland, N.R. *J Neurosci*, 20, 2369, 2000.)

more ventral region than it receives input from. This region, in turn, projects to the dorsolateral (or motor) striatum. Taken together, the interface between different striatal regions via the midbrain DA cells is organized in an ascending spiral, interconnecting different functional regions of the striatum and creating a feed-forward organization, from reward-related regions of the striatum to cognitive and motor areas (Figure 11.6). Thus, although the short-latency burst-firing activity of dopamine that signals immediate reinforcement is likely to be triggered from brainstem nuclei, the cortico-striato-midbrain pathway is in the position to influence dopamine cells to distinguish rewards and modify responses to incoming salient stimuli over time. This pathway is further reinforced via the nigro-striatal pathway, placing the striato-nigro-striatal pathway in a pivotal position for transferring information from the VS to the dorsal striatum during learning and habit formation. Indeed, cells in the dorsal striatum are progressively recruited during different types of learning, from simple motor tasks to drug self-administration (Everitt and Robbins 2005; Lehericy et al. 2005; Pasupathy and Miller 2005; Porrino et al. 2004; Volkow et al. 2006). Moreover, when the striato-nigro-striatal circuit is interrupted, information transfer from Pavlovian (stimulus-based) to instrumental (action-based) learning does not take place (Belin and Everitt 2008).

11.4.3 THE PLACE OF THE THALAMUS IN BASAL GANGLIA CIRCUITRY

The thalamocortical pathway is the last link in the circuit and is often treated as a simple "one-way relay" back to cortex. However, this pathway does not transfer information passively, but rather plays a key role in regulating cortical ensembles of neurons through its non-reciprocal connections with cortex. This occurs in two ways. First, the thalamus projects to different cortical layers. Therefore, while the thalamus receives input from the deep cortical layers, the thalamic projection to cortex, from the BG relay nuclei, terminates in superficial, middle, and deep layers (layers I/II, III/IV, and V, respectively) (Erickson and Lewis 2004; McFarland and Haber 2002b). Projections that terminate in layer V form both direct thalamo-cortico-thalamic and thalamo-cortico-striatal loops, thus sustaining information processing from the thalamus through each specific cortico-BG circuit. However, projections to the superficial layers play a key role in cortico-cortical processing. These are particularly interesting in that they have a more global recruiting action response affecting wide networks of cortical activity. In contrast to the topographically specific thalamocortical projections to middle layers, the more widespread, diffuse terminals to layer I are in a position to modulate neuronal activity from all cortical layers, with apical dendrites ascending into layer I. Moreover, this projection can provide an important mechanism for cross-communication between BG circuits. Projections to superficial layers also interface with cortico-cortical connections. These cortical regions, in turn, send axons to the striatum, thereby potentially modulating a different loop.

Second, while cortico-thalamic projections to specific relay nuclei are thought to follow a general rule of reciprocity, cortico-thalamic projections to VA/VL and central MD sites, as seen in other thalamocortical systems, are more extensive than thalamocortical projections (Catsman-Berrevoets and Kuypers 1978; Darian-Smith, Tan, and Edwards 1999; Deschenes, Veinante, and Zhang 1998; Hoogland, Welker, and Van der Loos 1987; Jones 1998; McFarland and Haber 2002b; Sherman and Guillery 1996). Furthermore, they are derived from areas not innervated by the same thalamic region, indicating non-reciprocal cortico-thalamic projections to specific BG relay nuclei (McFarland and Haber 2002b). Although each thalamic nucleus completes the cortico-BG segregated circuit, the non-reciprocal component is derived from a functionally distinct frontal cortical area. For example, the central MD has reciprocal connections with the lateral and orbital prefrontal areas and also a non-reciprocal input from medial prefrontal areas; VA has reciprocal connections with dorsal premotor areas and caudal DLPFC, and also a non-reciprocal connection from medial prefrontal areas; and VL has reciprocal connections with caudal motor areas along with a non-reciprocal connection from rostral motor regions. The potential for relaying information between circuits through thalamic connections, therefore, is accomplished both through the organization of projections to different layers and through the non-reciprocal cortico-thalamic pathways.

Thus, similar to the striato-nigro-striatal project system, the thalamic relay nuclei from the BG also appear to mediate information flow from higher cortical "association" areas of the prefrontal cortex to rostral motor areas involved in "cognitive" or integrative aspects of motor control to primary motor areas that direct movement execution.

11.5 CONCLUSIONS

A key component for developing appropriate goal-directed behaviors is the ability to correctly evaluate different aspects of reward and to select an appropriate action based on previous experience. These calculations rely on integration of sensory input with different aspects of reward processing and cognition to develop and execute appropriate action plans. While parallel networks that mediate different functions are critical to maintaining coordinated behaviors, cross talk between functional circuits during learning and adaptation is critical. Indeed, reward, associative, and motor control functions are not clearly and completely separated within the striatum. For example, consistent with human imaging studies, reward-responsive neurons are not restricted to the VS, but rather are found throughout the striatum. Moreover, cells responding in working memory tasks are often found also in the VS (Apicella et al. 1991; Cromwell and Schultz 2003; Delgado et al. 2005; Hassani, Cromwell, and Schultz 2001; Levy et al. 1997; Takikawa, Kawagoe, and Hikosaka 2002; Tanaka et al. 2004; Watanabe, Lauwereyns, and Hikosaka 2003).

As described above, embedded within limbic, associative, and motor-control striatal territories are subregions containing convergent terminals between different reward-processing cortical areas, and between these projections and those from the DPFC. These nodes of converging terminals may represent "hot spots" that may be particularly sensitive to synchronizing information across functional areas to impact on long-term strategic planning and habit formation (Kasanetz et al. 2008). Indeed, cells in the dorsal striatum are progressively recruited during different types of learning, from simple motor tasks to drug self-administration (Lehericy et al. 2005; Pasupathy and Miller 2005; Porrino et al. 2004; Volkow et al. 2006). The existence of convergent fibers from cortex within the VS, taken together with hippocampal and amygdalo-striatal projections, places the VS as a key entry port for the processing of sensory, emotional, and motivational information that, in turn, drives BG-mediated action selection and output. The ventral, reward-based striatal region, and the associative, central striatal region can impact on motor output circuits, not only through convergent terminal fields within the striatum, but also through the striato-nigro-striatal pathways. One can hypothesize that initially the nodal points of interface between the reward and associative circuits, for example, send a coordinated signal to dopamine cells. This pathway is in a pivotal position for temporal "training" of dopamine cells. In turn, these nodal points may be further reinforced through the burst-firing activity of the nigro-striatal pathway, thus transferring that impact back to the striatum. Moreover, through the striato-nigro-striatal system, information is transferred to other functional regions, during learning and habit formation (Belin et al. 2008; Everitt and Robbins 2005; Porrino et al. 2007; Volkow et al. 2006). This signal then enters the parallel system and, via the pallidum and thalamus, carries an integrated signal back to cortex. Indeed, when the striato-nigro-striatal circuit is interrupted, information transfer from Pavlovian to instrumental learning does not take place (Belin and Everitt 2008).

REFERENCES

Aggleton, J.P. 1985. A description of intra-amygdaloid connections in old world monkeys. *Exp Brain Res* 57:390–99.
Aldridge, J.W., Anderson, R.J., and Murphy, J.T. 1980. Sensory-motor processing in the caudate nuleus and globus pallidus: A single-unit study in behaving primates. *Can J Physiol Pharmacol* 58:1192–201.
Alexander, G.E., and Crutcher, M.D. 1990. Functional architecture of basal ganglia circuits: Neural substrates of parallel processing. *Trends Neurosci* 13:266–71.

Amaral, D.G., Price, J.L., Pitkanen, A., and Carmichael, S.T. 1992. Anatomical organization of the primate amygdaloid complex. In *The Amygdala: Neurobiological Aspects of Emotion, Memory, and Mental Dysfunction*, pp. 1–66. Wiley-Liss.

Apicella, P., Ljungberg, T., Scarnati, E., and Schultz, W. 1991. Responses to reward in monkey dorsal and ventral striatum. *Exp Brain Res* 85:491–500.

Arecchi-Bouchhioua, P., Yelnik, J., Francois, C., Percheron, G., and Tande, D. 1997. Three-dimensional morphology and distribution of pallidal axons projecting to both the lateral region of the thalamus and the central complex in primates. *Brain Res* 754:311–14.

Bar-Gad, I., Havazelet-Heimer, G., Goldberg, J.A., Ruppin, E., and Bergman, H. 2000. Reinforcement-driven dimensionality reduction—a model for information processing in the basal ganglia. *J Basic Clin Physiol Pharmacol* 11:305–20.

Barbas, H. 1992. Architecture and cortical connections of the prefrontal cortex in the rhesus monkey. In *Advances in Neurology*, eds. P. Chauvel and A.V. Delgado-Escueta, pp. 91–115. New York: Raven Press.

———. 1993. Organization of cortical afferent input to orbitofrontal areas in the Rhesus monkey. *Neuroscience* 56:841–64.

Barbas, H., and Blatt, G.J. 1995. Topographically specific hippocampal projections target functionally distinct prefrontal areas in the rhesus monkey. *Hippocampus* 5:511–33.

Barbas, H., Ghashghaei, H., Dombrowski, S.M., and Rempel-Clower, N.L. 1999. Medial prefrontal cortices are unified by common connections with superior temporal cortices and distinguished by input from memory-related areas in the rhesus monkey. *J Comp Neurol* 410:343–67.

Barbas, H., and Pandya, D.N. 1989. Architecture and intrinsic connections of the prefrontal cortex in the rhesus monkey. *J Comp Neurol* 286:353–75.

Battig, K., Rosvold, H.E., and Mishkin, M. 1960. Comparison of the effect of frontal and caudate lesions on delayed response and alternation in monkeys. *J Comp Physiol Psychol* 53:400–4.

Bayer, S.A. 1985. Neurogenesis in the olfactory tubercle and islands of calleja in the rat. *Int J Dev Neurosci* 3:135–47.

Beach, T.G., Tago, H., and McGeer, E.G., 1987. Light microscopic evidence for a substance P-containing innervation of the human nucleus basalis of Meynert. *Brain Res* 408:251–57.

Belin, D., and Everitt, B.J. 2008. Cocaine seeking habits depend upon dopamine-dependent serial connectivity linking the ventral with the dorsal striatum. *Neuron* 57:432–41.

Belin, D., Mar, A.C., Dalley, J.W., Robbins, T.W., and Everitt, B.J. 2008. High impulsivity predicts the switch to compulsive cocaine-taking. *Science* 320:1352–55.

Bevan, M.D., Clarke, N.P., and Bolam, J.P. 1997. Synaptic integration of functionally diverse pallidal information in the entopeduncular nucleus and subthalamic nucleus in the rat. *J Neurosci* 17:308–24.

Brinley-Reed, M., and McDonald, A.J. 1999. Evidence that dopaminergic axons provide a dense innervation of specific neuronal subpopulations in the rat basolateral amygdala. *Brain Res* 850:127–35.

Calzavara, R., and Haber, S.N. 2006. Relationship between the ventralis anterior/lateralis and medialis dorsalis thalamocortical projections and parvalbumin-positive fibers and cells in areas 9, 46, and 6 of primate cortex. Abstract appearing at the Society for Neuroscience 2006 Annual Meeting.

Calzavara, R., Mailly, P., and Haber, S.N. 2007. Relationship between the corticostriatal terminals from areas 9 and 46, and those from area 8A, dorsal and rostral premotor cortex and area 24c: An anatomical substrate for cognition to action. *Eur J Neurosci* 26:2005–24.

Carlezon, W.A., and Wise, R.A. 1996. Rewarding actions of phencyclidine and related drugs in nucleus accumbens shell and frontal cortex. *J Neurosci* 16:3112–22.

Carmichael, S.T., and Price, J.L. 1994. Architectonic subdivision of the orbital and medial prefrontal cortex in the macaque monkey. *J Comp Neurol* 346:366–402.

———. 1995a. Limbic connections of the orbital and medial prefrontal cortex in macaque monkeys. *J Comp Neurol* 363:615–41.

———. 1995b. Sensory and premotor connections of the orbital and medial prefrontal cortex of macaque monkeys. *J Comp Neurol* 363:642–40.

———.1996. Connectional networks within the orbital and medial prefrontal cortex of Macaque monkeys. *J Comp Neurol* 371:179–207.

Catsman-Berrevoets, C.E., and Kuypers, H.G. 1978. Differential laminar distribution of corticothalamic neurons projecting to the VL and the center median. An HRP study in the cynomolgus monkey. *Brain Res* 154:359–65.

Chang, H.T., Penny, G.R., and Kitai, S.T. 1987. Enkephalinergic-cholinergic interaction in the rat globus pallidus: A pre-embedding double-labeling immunocytochemistry study. *Brain Res* 426:197–203.

Chikama, M., McFarland, N., Amaral, D.G., and Haber, S.N. 1997. Insular cortical projections to functional regions of the striatum correlate with cortical cytoarchitectonic organization in the primate. *J Neurosci* 1724:9686–9705.

Chronister, R.B., Sikes, R.W., Trow, T.W., and DeFrance, J.F. 1981. The organization of the nucleus accumbens. In *The Neurobiology of the Nucleus Accumbens*, eds. R.B. Chronister and J.F. DeFrance, pp. 97–146. Brunswick, ME: Haer Institute.

Comoli, E., Coizet, V., Boyes, J., Bolam, J.P., Canteras, N.S., Quirk, R.H., Overton, P.G., and Redgrave, P. 2003. A direct projection from superior colliculus to substantia nigra for detecting salient visual events. *Nat Neurosci* 6:974–80.

Cools, R., Clark, L., and Robbins, T.W. 2004. Differential responses in human striatum and prefrontal cortex to changes in object and rule relevance. *J Neurosci* 24:1129–35.

Corlett, P.R., Aitken, M.R., Dickinson, A., Shanks, D.R., Honey, G.D., Honey, R.A., Robbins, T.W., Bullmore, E.T., and Fletcher, P.C. 2004. Prediction error during retrospective revaluation of causal associations in humans: fMRI evidence in favor of an associative model of learning. *Neuron* 44:877–88.

Cromwell, H.C., and Schultz, W. 2003. Effects of expectations for different reward magnitudes on neuronal activity in primate striatum. *J Neurophysiol* 89:2823–38.

Darian-Smith, C., Tan, A., and Edwards, S. 1999. Comparing thalamocortical and corticothalamic microstructure and spatial reciprocity in the macaque ventral posterolateral nucleus (VPLc) and medial pulvinar. J Comp *Neurol* 410:211–34.

Delgado, M.R., Locke, H.M., Stenger, V.A., and Fiez, J.A. 2003. Dorsal striatum responses to reward and punishment: Effects of valence and magnitude manipulations. *Cogn Affect Behav Neurosci* 3:27–38.

Delgado, M.R., Miller, M.M., Inati, S., and Phelps, E.A. 2005. An fMRI study of reward-related probability learning. *Neuroimage* 24:862–73.

Deschenes, M., Veinante, P., and Zhang, Z.W. 1998. The organization of corticothalamic projections: Reciprocity versus parity. *Brain Res Brain Res Rev* 28:286–308.

DiFiglia, M., Aronin, N., and Martin, J.B. 1982. Light and electron microscopic localization of immunoreactive leu-enkephalin in the monkey basal ganglia. *J Neurosci* 2(3):303–20.

Dommett, E., Coizet, V., Blaha, C.D., Martindale, J., Lefebvre, V., Walton, N., Mayhew, J.E., Overton, P.G., and Redgrave, P. 2005. How visual stimuli activate dopaminergic neurons at short latency. *Science* 307:1476–79.

Draganski, B., Kherif, F., Kloppel, S., Cook, P.A., Alexander, D.C., Parker, G.J., Deichmann, R., Ashburner, J., and Frackowiak, R.S. 2008. Evidence for segregated and integrative connectivity patterns in the human Basal Ganglia. *J Neurosci* 28:7143–52.

Elliott, R., Newman, J.L., Longe, O.A., and Deakin, J.F. 2003. Differential response patterns in the striatum and orbitofrontal cortex to financial reward in humans: A parametric functional magnetic resonance imaging study. *J Neurosci* 23:303–7.

Erickson, S.L., and Lewis, D.A. 2004. Cortical connections of the lateral mediodorsal thalamus in cynomolgus monkeys. *J Comp Neurol* 473:107–27.

Everitt, B.J., and Robbins, T.W. 2005. Neural systems of reinforcement for drug addiction: From actions to habits to compulsion. *Nat Neurosci* 8:1481–89.

Ferry, A.T., Ongur, D., An, X., and Price, J.L. 2000. Prefrontal cortical projections to the striatum in macaque monkeys: Evidence for an organization related to prefrontal networks. *J Comp Neurol* 425:447–70.

Flaherty, A.W., and Graybiel, A.M. 1994. Input-output organization of the sensorimotor striatum in the squirrel monkey. *J Neurosci* 14:599–610.

Fox, C.H., Andrade, H.N., Du Qui, I.J., and Rafols, J.A. 1974. The primate globus pallidus. A Golgi and electron microscope study. *J.R. Hirnforschung* 15:75–93.

Frankle, W.G., Laruelle, M., and Haber, S.N. 2006. Prefrontal cortical projections to the midbrain in primates: Evidence for a sparse connection. *Neuropsychopharmacology* 31:1627–36.

Friedman, D.P., Aggleton, J.P., and Saunders, R.C. 2002. Comparison of hippocampal, amygdala, and perirhinal projections to the nucleus accumbens: Combined anterograde and retrograde tracing study in the Macaque brain. *J Comp Neurol* 450:345–65.

Friedman, D.P., Murray, E.A., O'Neill, J.B., and Mishkin, M. 1986. Cortical connections of the somatosensory fields on the lateral sulcus of macaques: Evidence for a corticolimbic pathway for touch. *J Comp Neurol* 252:323–47.

Fudge, J.L., and Haber, S.N. 2000. The central nucleus of the amygdala projection to dopamine subpopulations in primates. *Neuroscience* 97:479–94.

———. 2001. Bed nucleus of the stria terminalis and extended amygdala inputs to dopamine subpopulations in primates. *Neuroscience* 104:807–27.

Fudge, J.L., Kunishio, K., Walsh, C., Richard, D., and Haber, S.N. 2002. Amygdaloid projections to ventrome-dial striatal subterritories in the primate. *Neuroscience* 110:257–75.

Fuster, J.M. 2001. The prefrontal cortex--an update: Time is of the essence. *Neuron* 30:319–33.

Gaspar, P., Stepneiwska, I., and Kaas, J.H. 1992. Topography and collateralization of the dopaminergic projections to motor and lateral prefrontal cortex in owl monkeys. *J Comp Neurol* 325:1–21.

Ghashghaei, H.T., and Barbas, H. 2002. Pathways for emotion: Interactions of prefrontal and anterior temporal pathways in the amygdala of the rhesus monkey. *Neuroscience* 115:1261–79.

Giménez-Amaya, J.M., McFarland, N.R., de las Heras, S., and Haber, S.N. 1995. Organization of thalamic projections to the ventral striatum in the primate. *J Comp Neurol* 354:127–49.

Goldman-Rakic, P.S., Bergson, C., Krimer, L.S., Lidow, M.S., Williams, S.M., and Williams, G.V. 1999. The primate mesocortical dopamine system. In *Handbook of Chemical Neuroanatomy, Vol. 15: The Primate Nervous System, Part III*, eds. F.E. Bloom, A. Bjorklund, and T. Hokfelt, pp. 403–28. Amsterdam: Elsevier Science.

Gruber, A.J., Dayan, P., Gutkin, B.S., and Solla, S.A. 2006. Dopamine modulation in the basal ganglia locks the gate to working memory. *J Comput Neurosci* 20:153–66.

Haber, S.N. 1987. Anatomical relationship between the basal ganglia and the basal nucleus of Maynert in human and monkey forebrain. *Proc Natl Acad Sci USA* 84:1408–12.

Haber, S.N. 2009. Chapter 1: Anatomy and connectivity of the reward circuit. In *Handbook of Reward and Decision Making*, eds. J.C. Dreher and L. Tremblay, pp. 3–27. Amsterdam: Elsevier Inc. Pages 3–27.

Haber, S.N., Fudge, J.L., and McFarland, N.R. 2000. Striatonigrostriatal pathways in primates form an ascending spiral from the shell to the dorsolateral striatum. *J Neurosci* 20:2369–82.

Haber, S.N., Groenewegen, H.J., Grove, E.A., and Nauta, W.J.H. 1985. Efferent connections of the ventral pallidum. Evidence of a dual striatopallidofugal pathway. *J Comp Neurol* 235:322–35.

Haber, S.N., Kim, K.S., Mailly, P., and Calzavara, R. 2006. Reward-related cortical inputs define a large striatal region in primates that interface with associative cortical inputs, providing a substrate for incentive-based learning. *J Neurosci* 26:8368–76.

Haber, S.N. and Knutson, B. 2010. The reward circuit: Linking primate anatomy and human imaging", *Neuropsychopharmacology* 35:4–26.

Haber, S.N., Kunishio, K., Mizobuchi, M., and Lynd-Balta, E. 1995a. The orbital and medial prefrontal circuit through the primate basal ganglia. *J Neurosci* 15:4851–67.

Haber, S.N., Lynd, E., Klein, C., and Groenewegen, H.J. 1990. Topographic organization of the ventral striatal efferent projections in the rhesus monkey: An anterograde tracing study. *J Comp Neurol* 293:282–98.

Haber, S.N., Lynd-Balta, E., and Mitchell, S.J. 1993. The organization of the descending ventral pallidal projections in the monkey. *J Comp Neurol* 3291:111–29.

Haber, S.N., and McFarland, N.R. 1999. The concept of the ventral striatum in nonhuman primates. *Ann N Y Acad Sci* 877:33–48.

Haber, S.N., and Nauta, W.J.H. 1983. Ramifications of the globus pallidus in the rat as indicated by patterns of immunohistochemistry. *Neuroscience* 9:245–60.

Haber, S.N., Ryoo, H., Cox, C., and Lu, W. 1995b. Subsets of midbrain dopaminergic neurons in monkeys are distinguished by different levels of mRNA for the dopamine transporter: Comparison with the mRNA for the D2 receptor, tyrosine hydroxylase and calbindin immunoreactivity. *J Comp Neurol* 362:400–10.

Haber, S.N., and Watson, S.J. 1985. The comparative distribution of enkephalin, dynorphin and substance P in the human globus pallidus and basal forebrain. *Neuroscience* 14:1011–24.

Haber, S.N., Wolfe, D.P., and Groenewegen, H.J. 1990b. The relationship between ventral striatal efferent fibers and the distribution of peptide-positive woolly fibers in the forebrain of the rhesus monkey. *Neuroscience* 39:323–38.

Hassani, O.K., Cromwell, H.C., and Schultz, W. 2001. Influence of expectation of different rewards on behavior-related neuronal activity in the striatum. *J Neurophysiol* 85:2477–89.

Hedreen, J.C., and DeLong, M.R. 1991. Organization of striatopallidal, striatonigral, and nigrostriatal projections in the Macaque. *J Comp Neurol* 304:569–95.

Heimer, L. 1978. The olfactory cortex and the ventral striatum. In *Limbic Mechanisms*, eds. K.E. Livingston and O. Hornykiewicz, pp. 95–187. New York: Plenum Press.

Heimer, L., De Olmos, J.S., Alheid, G.F., Person, J., Sakamoto, N., Shinoda, K., Marksteiner, J., and Switzer, R.C. 1999. The human basal forebrain. Part II. In *Handbook of Chemical Neuroanatomy*, eds. F.E. Bloom, A. Bjorkland, and T. Hokfelt, pp. 57–226. Amsterdam: Elsevier.

Hikosaka, O., Miyashita, K., Miyachi, S., Sakai, K., and Lu, X. 1998. Differential roles of the frontal cortex, basal ganglia, and cerebellum in visuomotor sequence learning. *Neurobiol Learn Mem* 70:137–49.

Hoogland, P.V., Welker, E., and Van der Loos, H. 1987. Organization of the projections from barrel cortex to thalamus in mice studied with Phaseolus vulgaris-leucoagglutinin and HRP. *Exp Brain Res* 68:73–87.

Ilinsky, I.A., Jouandet, M.L., and Goldman-Rakic, P.S. 1985. Organization of the nigrothalamocortical system in the rhesus monkey. *J Comp Neurol* 236:315–30.

Ilinsky, I.A., Yi, H., and Kultas-Ilinsky, K. 1997. Mode of termination of pallidal afferents to the thalamus: A light and electron microscopic study with anterograde tracers and immunocytochemistry in Macaca mulatta. *J Comp Neurol* 386:601–12.

Ito, R., Robbins, T.W., and Everitt, B.J. 2004. Differential control over cocaine-seeking behavior by nucleus accumbens core and shell. *Nat Neurosci* 7:389–97.

Iwai, E., and Yukie, M. 1987. Amygdalofugal and amygdalopetal connections with modality-specific visual cortical areas in macaques (Macaca fuscata, M. mulatta, and M. fascicularis). *J Comp Neurol* 261:362–87.

Jolkkonen, E., and Pitkanen, A. 1998. Intrinsic connections of the rat amygdaloid complex: Projections originating in the central nucleus. *J Comp Neurol* 395:53–72.

Jones, E.G., 1998. The thalamus of primates. In *The Primate Nervous System*, Part II, eds. F.E. Bloom, A. Björklund, and T. Hökfelt, pp. 1–298. Amsterdam: Elsevier Science.

Kalivas, P.W., Volkow, N., and Seamans, J. 2005. Unmanageable motivation in addiction: A pathology in prefrontal-accumbens glutamate transmission. *Neuron* 45:647–50.

Kasanetz, F., Riquelme, L.A., Della-Maggiore, V., O'Donnell, P., and Murer, M.G. 2008. Functional integration across a gradient of corticostriatal channels controls UP state transitions in the dorsal striatum. *Proc Natl Acad Sci USA* 105:8124–29.

Kimura, M. 1986. The role of primate putamen neurons in the association of sensory stimulus with movement. *Neurosci Res* 3:436–43.

Klitenick, M.A., Deutch, A.Y., Churchill, L., and Kalivas, P.W. 1992. Topography and functional role of dopaminergic projections from the ventral mesencephalic tegmentum to the ventral pallidum. *Neuroscience* 502:371–86.

Knutson, B., Adams, C.M., Fong, G.W., and Hommer, D. 2001. Anticipation of increasing monetary reward selectively recruits nucleus accumbens. *J Neurosci* 21:RC159.

Kolomiets, B.P., Deniau, J.M., Mailly, P., Menetrey, A., Glowinski, J., and Thierry, A.M. 2001. Segregation and convergence of information flow through the cortico-subthalamic pathways. *J Neurosci* 21:5764–72.

Kuhnen, C.M., and Knutson, B. 2005. The neural basis of financial risk taking. *Neuron* 47:763–70.

Kuo, J., and Carpenter, M.B. 1973. Organization of pallidothalamic projections in the rhesus monkey. *J Comp Neurol* 151:201–36.

Lavoie, B., and Parent, A. 1991. Dopaminergic neurons expressing calbindin in normal and parkinsonian monkeys. *Neuroreport* 2 (10): 601–4.

Lecourtier, L., and Kelly, P.H. 2007. A conductor hidden in the orchestra? Role of the habenular complex in monoamine transmission and cognition. *Neurosci Biobehav Rev* 31:658–72.

Lehericy, S., Benali, H., Van de Moortele, P.F., Pelegrini-Issac, M., Waechter, T., Ugurbil, K., and Doyon, J. 2005. Distinct basal ganglia territories are engaged in early and advanced motor sequence learning. *Proc Natl Acad Sci USA* 102:12566–71.

Levy, R., Friedman, H.R., Davachi, L., and Goldman-Rakic, P.S. 1997. Differential activation of the caudate nucleus in primates performing spatial and nonspatial working memory tasks. *J Neurosci* 17:3870–82.

Lewis, D.A. 1992. The catecholaminergic innervation of primate prefrontal cortex. *J Neural Transm Suppl* 36:179–200.

Lidow, M.S., Goldman-Rakic, P.S., Gallager, D.W., and Rakic, P. 1991. Distribution of dopaminergic receptors in the primate cerebral cortex: Quantitative autoradiographic analysis using [3H] raclopride, [3H] spiperone and [3H]sch23390. *Neuroscience* 40 (3): 657–71.

Lynd-Balta, E., and Haber, S.N. 1994a. Primate striatonigral projections: A comparison of the sensorimotor-related striatum and the ventral striatum. *J Comp Neurol* 345:562–78.

———. 1994b. The organization of midbrain projections to the striatum in the primate: Sensorimotor-related striatum versus ventral striatum. *Neuroscience* 59:625–40.

———. 1994c. The organization of midbrain projections to the ventral striatum in the primate. *Neuroscience* 59:609–23.

Mai, J.K., Stephens, P.H., Hopf, A., and Cuello, A.C. 1986. Substance P in the human brain. *Neuroscience* 17 (3): 709–39.

Martinez-Murillo, R., Blasco, I., Alvarez, F.J., Villalba, R., Solano, M.L., Montero-Caballero, M.I., and Rodrigo, J. 1988. Distribution of enkephalin-immunoreactive nerve fibers and terminals in the region of the nucleus basalis magnocellularis of the rat: A light and electron microscopic study. *J Neurocytol* 17:361–76.

Matsumoto, M., and Hikosaka, O. 2007. Lateral habenula as a source of negative reward signals in dopamine neurons. *Nature* 447:1111–15.

May, P.J., McHaffie, J.G., Stanford, T.R., Jiang, H., Costello, M.G., Coizet, V., Hayes, L.M., Haber, S.N., and Redgrave, P. 2009. Tectonigral projections in the primate: A pathway for pre-attentive sensory input to midbrain dopaminergic neurons. *Eur J Neurosci* 29:575–87.

Mayberg, H.S. 2003. Positron emission tomography imaging in depression: A neural systems perspective. *Neuroimaging Clin N Am* 13:805–15.

McFarland, N.R., and Haber, S.N. 2000. Convergent inputs from thalamic motor nuclei and frontal cortical areas to the dorsal striatum in the primate. *J Neurosci* 20:3798–3813.

———. 2001. Organization of thalamostriatal terminals from the ventral motor nuclei in the macaque. *J Comp Neurol* 429:321–36.

———. 2002a. Thalamic connections with cortex from the basal ganglia relay nuclei provide a mechanism for integration across multiple cortical areas. *J Neurosci* 22:8117–32.

———. 2002b. Thalamic relay nuclei of the basal ganglia form both reciprocal and nonreciprocal cortical connections, linking multiple frontal cortical areas. The *J Neurosci* 22:8117–32.

McHaffie, J.G., Jiang, H., May, P.J., Coizet, V., Overton, P.G., Stein, B.E., and Redgrave, P. 2006. A direct projection from superior colliculus to substantia nigra pars compacta in the cat. *Neuroscience* 138:221–34.

Mena-Segovia, J., Ross, H.M., Magill, P.J., and Bolam, J.P. 2005. *The Pedunculopontine Nucleus: Towards a Functional Integration with the Basal Ganglia.* New York: Springer Science and Business Media.

Mesulam, M.-M., and Mufson, E.J. 1993. The insula of reil in man and monkey. In *Cerebral Cortex*, eds. A. Peters and E.G. Jones, pp. 179–225. New York: Plenum Press.

Meyer, G., Gonzalez-Hernandez, T., Carrillo-Padilla, F., and Ferres-Torres, R. 1989. Aggregations of granule cells in the basal forebrain (islands of Calleja): Golgi and cytoarchitectonic study in different mammals, including man. *J Comp Neurol* 284:405–28.

Middleton, F.A., and Strick, P.L. 2002. Basal-ganglia 'projections' to the prefrontal cortex of the primate. *Cereb Cortex* 12:926–35.

Milad, M.R., Wright, C.I., Orr, S.P., Pitman, R.K., Quirk, G.J., and Rauch, S.L. 2007. Recall of fear extinction in humans activates the ventromedial prefrontal cortex and hippocampus in concert. *Biol Psychiatry* 62:446–54.

Mink, J.W. 1996. The basal ganglia: Focused selection and inhibition of competing motor programs. *Prog Neurobiol* 50:381–425.

Morecraft, R.J., Geula, C., and Mesulam, M.-M. 1992. Cytoarchitecture and neural afferents of orbitofronal cortex in the brain of the monkey. *J Comp Neurol* 323:341–58.

Muhammad, R., Wallis, J.D., and Miller, E.K. 2006. A comparison of abstract rules in the prefrontal cortex, premotor cortex, inferior temporal cortex, and striatum. *J Cogn Neurosci* 18:974–89.

Nishijo, H., Ono, T., and Nishino, H. 1988a. Single neuron responses in amygdala of alert monkey during complex sensory stimulation with affective significance. *J Neurosci* 8:3570–83.

———. 1988b. Topographic distribution of modality-specific amygdalar neurons in alert monkey. *J Neurosci* 8:3556–69.

O'Doherty, J., Dayan, P., Schultz, J., Deichmann, R., Friston, K., and Dolan, R.J. 2004. Dissociable roles of ventral and dorsal striatum in instrumental conditioning. *Science* 304:452–54.

Olds, J., and Milner, P. 1954. Positive reinforcement produced by electrical stimulation of septal area and other regions of rat brain. *J Comp Physiol Psychol* 47:419–27.

Ono, T., Tamura, R., Nishijo, H., Nakamura, K., and Tabuchi, E. 1989. Contribution of amygdalar and lateral hypothalamic neurons to visual information processing of food and nonfood in monkey. *Physiol Behav* 45:411–21.

Padoa-Schioppa, C., and Assad, J.A. 2006. Neurons in the orbitofrontal cortex encode economic value. *Nature* 441:223–26.

Pagnoni, G., Zink, C.F., Montague, P.R., and Berns, G.S. 2002. Activity in human ventral striatum locked to errors of reward prediction. *Nat Neurosci* 5:97–98.

Parent, A., and De Bellefeuille, L. 1982. Organization of efferent projections from the internal segment of the globus pallidus in the primate as revealed by fluorescence retrograde labeling method. *Brain Res* 245:201–13.

Pasupathy, A., and Miller, E.K. 2005. Different time courses of learning-related activity in the prefrontal cortex and striatum. *Nature* 433:873–76.

Paus, T. 2001. Primate anterior cingulate cortex: Where motor control, drive and cognition interface. *Nat Rev Neurosci* 2:417–24.

Penfield, W., and Faulk, M.E. 1955. The insula: Further observations on its function. *Brain* 78:445–70.

Percheron, G., and Filion, M. 1991. Parallel processing in the basal ganglia: Up to a point. *Trends Neurosci* 14:55–59.

Pitkanen, A., and Amaral, D.G. 1998. Organization of the intrinsic connections of the monkey amygdaloid complex: Projections originating in the lateral nucleus. *J Comp Neurol* 398:431–58.

Porrino, L.J., Lyons, D., Smith, H.R., Daunais, J.B., and Nader, M.A. 2004. Cocaine self-administration produces a progressive involvement of limbic, association, and sensorimotor striatal domains. *J Neurosci* 24:3554–62.

Porrino, L.J., Smith, H.R., Nader, M.A., and Beveridge, T.J. 2007. The effects of cocaine: A shifting target over the course of addiction. Prog Neuropsychopharmacol *Biol Psychiatry* 31:1593–1600.

Preuss, T.M., and Goldman-Rakic, P.S. 1989. Connections of the ventral granular frontal cortex of macaques with perisylvian and somatosensory areas: Anatomical evidence for somatic representation in primate frontal association cortex. *J Comp Neurol* 282:293–316.

Price, J.L., Carmichael, S.T., and Drevets, W.C. 1996. Networks related to the orbital and medial prefrontal cortex; a substrate for emotional behavior? *Prog Brain Res* 107:523–36.

Ray, J.P., and Price, J.L. 1993. The organization of projections from the mediodorsal nucleus of the thalamus to orbital and medial prefrontal cortex in Macaque monkeys. *J Comp Neurol* 337:1–31.

Ricardo, J.A., and Koh, E.T. 1978. Anatomical evidence of direct projections from the nucleus of the solitary tract to the hypothalamus, amygdala, and other forebrain structures in the rat. *Brain Res* 153:1–26.

Roesch, M.R., and Olson, C.R. 2004. Neuronal activity related to reward value and motivation in primate frontal cortex. *Science* 304:307–10.

Russchen, F.T., Amaral, D.G., and Price, J.L. 1987. The afferent input to the magnocellular division of the mediodorsal thalamic nucleus in the monkey, Macaca fascicularis. *J Comp Neurol* 256:175–210.

Russchen, F.T., Bakst, I., Amaral, D.G., and Price, J.L. 1985. The amygdalostriatal projections in the monkey. An anterograde tracing study. *Brain Res* 329:241–57.

Saper, C.B., and Loewy, A.D. 1980. Efferent connections of the parabrachial nucleus in the rat. *Brain Res* 197:291–317.

Saunders, R.C., Rosene, D.L., and Van Hoesen, G.W. 1988. Comparison of the efferents of the amygdala and the hippocampal formation in the rhesus monkey: II. Reciprocal and non-reciprocal connections. *J Comp Neurol* 271:185–207.

Schneider, R.J., Friedman, D.P., and Mishkin, M. 1993. A modality-specific somatosensory area within the insula of the rhesus monkey. *Brain Res* 621:116–20.

Schultz, W. 2000. Multiple reward signals in the brain. *Nat Rev Neurosci* 1:199–207.

———. 2002. Getting formal with dopamine and reward. *Neuron* 36:241–63.

Schultz, W., Tremblay, L., and Hollerman, J.R. 2000. Reward processing in primate orbitofrontal cortex and basal ganglia. *Cereb Cortex* 10:272–84.

Selemon, L.D., and Goldman-Rakic, P.S. 1985. Longitudinal topography and interdigitation of corticostriatal projections in the rhesus monkey. *J Neurosci* 5:776–94.

———. 1990. Topographic intermingling of striatonigral and striatopallidal neurons in the rhesus monkey. *J Comp Neurol* 297:359–76.

Sherman, S.M., and Guillery, R.W. 1996. Functional organization of thalamocortical relays. *J Neurophysiol* 76:1367–95.

Showers, M.J.C., and Lauer, E.W. 1961. Somatovisceral motor patterns in the insula. *J Comp Neurol* 117:107–16.

Smith, K.S., and Berridge, K.C. 2007. Opioid limbic circuit for reward: Interaction between hedonic hotspots of nucleus accumbens and ventral pallidum. *J Neurosci* 27:1594–605.

Smith-Swintosky, V.L., Plata-Salaman, C.R., and Scott, T.R. 1991. Gustatory neural coding in the monkey cortex: Stimulus quality. *J Neurophysiol* 66:1156–65.

Spooren, W.P.J.M., Lynd-Balta, E., Mitchell, S., and Haber, S.N. 1996. Ventral pallidostriatal pathway in the monkey: Evidence for modulation of basal ganglia circuits. *J Comp Neurol* 3703:295–312.

Strick, P.L. 1976. Anatomical analysis of ventrolateral thalamic input to primate motor cortex. *J Neurophysiol* 39:1020–31.

Szabo, J. 1979. Strionigral and nigrostriatal connections. Anatomical studies. *Appl Neurophysiol* 42:9–12.

Taha, S.A., and Fields, H.L. 2006. Inhibitions of nucleus accumbens neurons encode a gating signal for reward-directed behavior. *J Neurosci* 26:217–22.

Takikawa, Y., Kawagoe, R., and Hikosaka, O. 2002. Reward-dependent spatial selectivity of anticipatory activity in monkey caudate neurons. *J Neurophysiol* 87:508–15.

Tanaka, S.C., Doya, K., Okada, G., Ueda, K., Okamoto, Y., and Yamawaki, S. 2004. Prediction of immediate and future rewards differentially recruits cortico-basal ganglia loops. *Nat Neurosci* 7:887–93.

Tindell, A.J., Smith, K.S., Pecina, S., Berridge, K.C., and Aldridge, J.W. 2006. Ventral pallidum firing codes hedonic reward: When a bad taste turns good. *J Neurophysiol* 96:2399–409.

Tremblay, L., and Schultz, W. 2000. Reward-related neuronal activity during go-nogo task performance in primate orbitofrontal cortex. *J Neurophysiol* 83:1864–76.

Turner, B.H., Mishkin, M., and Knapp, M. 1980. Organization of the amygdalopetal projections from modality-specific cortical association areas in the monkey. *J Comp Neurol* 191:515–43.

Turner, M.S., Lavin, A., Grace, A.A., and Napier, T.C. 2001. Regulation of limbic information outflow by the subthalamic nucleus: Excitatory amino acid projections to the ventral pallidum. *J Neurosci* 21:2820–32.

Ullsperger, M., and von Cramon, D.Y. 2003. Error monitoring using external feedback: Specific roles of the habenular complex, the reward system, and the cingulate motor area revealed by functional magnetic resonance imaging. *J Neurosci* 23:4308–14.

Vogt, B.A., Vogt, L., Farber, N.B., and Bush, G. 2005. Architecture and neurocytology of monkey cingulate gyrus. *J Comp Neurol* 485:218–39.

Volkow, N.D., Wang, G.J., Ma, Y., Fowler, J.S., Wong, C., Ding, Y.S., Hitzemann, R., Swanson, J.M., and Kalivas, P. 2005. Activation of orbital and medial prefrontal cortex by methylphenidate in cocaine-addicted subjects but not in controls: Relevance to addiction. *J Neurosci* 25:3932–39.

Volkow, N.D., Wang, G.J., Telang, F., Fowler, J.S., Logan, J., Childress, A.R., Jayne, M., Ma, Y., and Wong, C. 2006. Cocaine cues and dopamine in dorsal striatum: Mechanism of craving in cocaine addiction. *J Neurosci* 26:6583–88.

Wallis, J.D., and Miller, E.K. 2003. Neuronal activity in primate dorsolateral and orbital prefrontal cortex during performance of a reward preference task. *Eur J Neurosci* 18:2069–81.

Walton, M.E., Bannerman, D.M., Alterescu, K., and Rushworth, M.F. 2003. Functional specialization within medial frontal cortex of the anterior cingulate for evaluating effort-related decisions. *J Neurosci* 23:6475–79.

Watanabe, K., Lauwereyns, J., and Hikosaka, O. 2003. Neural correlates of rewarded and unrewarded eye movements in the primate caudate nucleus. *J Neurosci* 23:10052–57.

Westby, G.W., Keay, K.A., Redgrave, P., Dean, P., and Bannister, M. 1990. Output pathways from the rat superior colliculus mediating approach and avoidance have different sensory properties. *Exp Brain Res* 81:626–38.

Wilson, C.J. 2004. The basal ganglia. In *Synaptic Organization of the Brain*, ed. G.M. Shepherd, pp. 361–413. New York: Oxford University Press.

Wise, R.A. 2002. Brain reward circuitry: Insights from unsensed incentives. *Neuron* 36:229–40.

Wise, S.P., Murray, E.A., and Gerfen, C.R. 1996. The frontal cortex-basal ganglia system in primates. *Crit Rev Neurobiol* 10:317–56.

Wurtz, R.H., and Albano, J.E. 1980. Visual-motor function of the primate superior colliculus. *Annu Rev Neurosci* 3:189–226.

Yaxley, S., Rolls, E.T., and Sienkiewicz, Z.J. 1990. Gustatory responses of single neurons i the insula of the macaque monkey. *J Neurophysiol* 63:689–700.

Zaborszky, L., and Cullinan, W.E. 1992. Projections from the nucleus accumbens to cholinergic neurons of the ventral pallidum: A correlated light and electron microscopic double-immunolabeling study in rat. *Brain Res* 570:92–101.

Zahm, D.S. 1989. The ventral striatopallidal parts of the basal ganglia in the rat. II. Compartmentation of ventral pallidal efferents. *Neuroscience* 30:33–50.

Zheng, T., and Wilson, C.J. 2002. Corticostriatal combinatorics: The implications of corticostriatal axonal arborizations. *J Neurophysiol* 87:1007–17.

12 Multiple Reward Layers in Food Reinforcement

Ivan E. de Araujo

CONTENTS

12.1 INTRODUCTION

Why do animals eat? Virtually any organism must continuously procure from the environment the energy required to maintain vital biochemical processes. For most organisms—insects and mammals alike—the fuel needed to maintain cellular growth and function is obtained from exogenous sources, that is, "food." Therefore, ingesting fuels that can be either readily oxidized or stored in the body as energy reserves is the ultimate "reward" all living creatures are willing to work for.

In 1947, E. Adolph experimentally manipulated the amount of calories available to rats and noticed that, when next presented with free access to food, these animals adjusted caloric intake by immediately increasing the amount of food ingested (Adolph 1947). Upon considering such evidence, Adolph readily concluded that "rats eat for calories." These early observations suggested that animals regulate their intake as a response to the body's metabolic needs. In other words, the postingestive effects produced by metabolically efficient foods exert by themselves reinforcing effects on behavior—a hypothesis that, we shall see, has since then gained strong experimental support.

Adolph (1947) did also notice, however, that the regulation of food intake must be under many influences. Among these—besides caloric content itself—are the multiple sensory properties of foods, such as their taste, odor, texture, and appearance. To this list we should add those initially neutral environmental signals that incrementally acquire predictive functions via associative learning. These diverse sensory aspects of food stimuli can by themselves motivate behavior, by functioning as proximal or distal indices of nutritive value. The many detectable sensory features associated with the actual metabolic value of a food form "layers" of cues that gain motivational value by themselves, and are critical for the ability of an organism to locate fuels in a fast, efficient manner.

In other words, the behavioral processes controlling food intake must be understood as resulting from a summation of relatively independent "layers of reward" that act to sustain positive energy

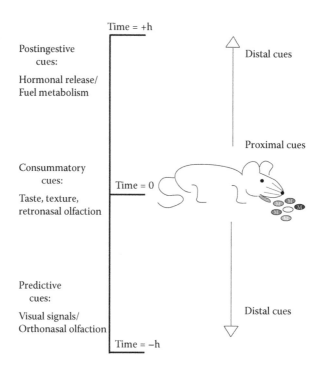

FIGURE 12.1   Proximal and distal cues associated with food reinforcement. Proximal cues can be defined as those sensory cues that are associated with the consummatory act of eating itself; it might include inputs such as the taste and the texture of a food, or its odor as detected via the nasopharynx route (retronasal olfaction). Distal cues, on the other hand, can be understood as those sensory cues that either precede the consummatory act ("preingestive" distal cues, such as visual or olfactory signals predicting food availability) or follow the consummatory act ("postingestive" distal cues, such as metabolic changes resulting from eating). Distal cues might gain reward value due to associations formed with ensuing proximal cues. Proximal cues themselves might also become reward predictors if consistently associated with ensuing positive postingestive effects. In this figure, the consummatory act takes place at time T = 0, and distal cues occur at variables times that might either precede (–h) or follow (+h) the consummatory act.

balance. For example, sugars can be considered as prototypical "bilayered" rewards: energy-carrying molecules that also are highly palatable* (i.e., capable of inducing intake via stimulation of sweet receptors). To these layers are associated reward signals that are either "proximal" to the consummatory act of eating itself (such as the taste of foods) or more "distal" (see Figure 12.1 for a scheme), such as anticipatory visual and olfactory cues predicting food delivery (Gottfried, O'Doherty, and Dolan 2003), or the rewarding postconsumptive effects produced by nutrients. Potentially, "preingestive" distal cues (i.e., those occurring prior to the act of ingestion) have different physiological functions from those associated with postingestive distal cues, particularly with regard to the encoding of predictive value. The implication is that different brain regions are involved in the representation of pre- and postingestive distal cues.

The question now arises, which of these more proximal and distal cues constitute the critical rewarding event in food preferences. It has been long thought that the main drivers of food intake are those sensations associated with the food's taste or palatability. In fact, the perceptual quality of sweet taste is so compelling and familiar to us that we might well be led to assume that the sensation

---

* In this chapter the terms "palatable" and "palatability" will refer to the orosensory rewarding properties of a stimulus that are sufficient to elicit intake independently of learning or previous experience. Although some authors favor using this term as relating to the overall set of factors exerting positive controls on intake, we would like in the present chapter to exclude from the definition of "palatable" those initially unpleasant compounds, such as mild irritants or bitter tastants, which might gain incentive value through association with postingestive effects.

of "sweetness" constitutes the essence of what we call "sugar reward" (idea that sweetness does not necessarily a reward make, see Chapter 6). Part of this impression certainly arises from our innate and almost universal attraction to non-caloric "artificial" sweeteners: their potent palatability seemingly indicates that calorie-independent sweetness is the central factor or "reward layer" governing our strong preferences for carbohydrate-rich compounds.

However, the formation of long-term food preferences is a complex process and it shouldn't be automatically assumed that animals will form preferences for palatable, sensory cues when those come unaccompanied by postingestive effects. In fact, as we will argue in this chapter, the critical reward event occurring during food intake seems to be the triggering of postingestive processes that follow food intake. More precisely, current evidence would suggest that postingestive effects are both necessary and sufficient for the formation of preferences towards calorie-rich nutrients. On the other hand, gustatory cues do not seem to have the ability to sustain long-term preferences if unaccompanied by postingestive factors. In other words, the experimental data available so far indicate that a hierarchy exists in which distal, delayed postingestive effects function as the principal factor regulating nutrient choice, at least over extended periods of time.

How can the brain regulate food intake, given that preferences seem to be under the simultaneous control of different distal and proximal events? One possibility is that a brain circuit exists that has the ability to respond to all behaviorally relevant distal and proximal cues. We will argue that the mammalian midbrain dopamine system is one such candidate circuit,[*] given its prominent role in food reward processing and its ability to detect and respond not only to proximal cues such as palatable tastes but also to distal cues that acquire predictive power through learning. In addition, and consistent with the preponderant role played by postingestive factors in food intake, we will also review recent evidence showing that the detection of nutrient processing from its absorption in the gut is sufficient to activate the mesolimbic dopamine system in the absence of gustatory or flavor stimulation. We will explore some of the implications of these findings and the many related questions that remain unanswered.

12.2 GUSTATORY REWARDS AND FOOD PREFERENCES

12.2.1 THE LABELED-LINE MODEL OF GUSTATORY CODING

Direct stimulation of some taste receptor cells can, by itself, exert reinforcing control on food intake. The psychophysical and neural bases of gustation have been reviewed elsewhere in this volume (see Chapter 6). The purpose of this section is to provide the reader with the basic terminology employed in the following sections.

An ongoing debate as to how taste qualities are coded has been the focus of attention of taste researchers for many years now. In one view, named the "across-fiber pattern" model, perceived taste qualities correspond to patterns of activity across afferent nerve populations (Erickson 2001). An alternative, opposing view proposes instead a "labeled-line" model, in which different taste qualities are encoded by separate peripheral circuits defined by subsets of taste cells expressing specific chemosensors in connection with the afferent neurons upon which they synapse. If, on one hand, the existence of broadly tuned gustatory neurons has been argued to give support to the across-fiber model (Boudreau et al. 1985; Frank, Bieber, and Smith 1988; Caicedo, Kim, and Roper 2002), evidence is mounting in favor of the existence of dedicated, labeled lines transducing taste signals to the central nervous system (for a different perspective see Lemon and Katz 2007).

---

[*] It must be noted that the involvement of dopamine and other amines in reward and punishment signaling is not restricted to mammalian species. In particular, *Drosophila* larvae are capable of forming aversive and appetitive associations between odorant and gustatory cues (Gerber and Stocker 2007), an ability that seems to depend on the two biogenic amines dopamine and octopamine respectively (Schwaerzel et al. 2003). However, because it remains unclear whether nutrients can directly stimulate amine neurons in insects, in this chapter we will restrict the discussion to data obtained from higher, mammalian animals. See Section 12.4 for further discussion.

The first line of evidence arises from the receptor distribution patterns on the tongue. Three types of taste sensations are known to be mediated by G-protein-coupled receptors, in their turn divided into two broad groups, namely, T1Rs and T2Rs, whose function is at least in some cells supported by the alpha component of the taste-specific G-protein gustducin (α-gustucin, McLaughlin, McKinnon, and Margolskee 1992). On one hand, the T1Rs mediate behavioral attraction to nutritive tastants via the sweet-detecting heterodimer T1R2/T1R3 (and possibly the homodimer T1R3/T1R3) and the L-amino acid-detecting T1R1/T1R3 heterodimers (Zhao et al. 2003). On the other hand, repulsive bitter taste sensations are mediated by the T2R family of receptors (Chrandrashekar et al. 2000). Importantly, it has been established that none of those receptors co-express in taste cells (Chrandrashekar et al. 2000). Therefore, the anatomical substrate is in place to support network-dedicated, labeled lines of taste transduction for sweet, L-amino acids, and bitter (Damak, Mosinger, and Margolskee 2008).

Another line of evidence supporting the existence of these dedicated lines relates to the concept of restricting the expression of some essential components of the intracellular taste signaling cascade to a limited, receptor-specific group of cells. For example, expressing the phospholipase Cβ2 only in taste cells that express the bitter receptor T2R5 (which selectively binds to cyclohexamide, a toxic bitter compound) results in normal detection of bitter but not sweet or L-amino acid tastes (Zhang et al. 2003). In addition, restricting the transgenic expression of a modified opioid receptor to cells that contain T1R2 receptors (i.e., in cells specifically tuned to sweet tastes) results in preferences for a tasteless synthetic opiate similar to those found for sucrose (Zhao et al. 2003). Conversely, when the same opiate receptor was selectively expressed in "bitter" cells (i.e., T2R-expressing taste cells), the animals displayed a rejection to the synthetic opiate in a manner that emulates rejection to bitter compounds. Finally, it is of note that in mice produced to express bitter T2R receptors in "sweet" cells (i.e., T1R2-expressing cells), a sugar-like attraction for bitter tastants is observed (Mueller et al. 2005). Such genetic manipulations therefore strongly suggest that behavioral responses produced by tastants are mediated by a highly specific subset of taste cells whose activation is necessary and sufficient to elicit stereotyped behavioral responses. In particular, these findings imply that different stereotyped behaviors are mediated by non-overlapping groups of taste cells, a fact that considerably weakens the possibility that taste transduction depends on distributed patterns of activity in the periphery.*

12.2.2   The Dopamine Mesolimbic Pathway and Palatability

How does the brain, more specifically its reward-processing centers, respond to the stimulation of T1R taste cells? Judging from the stereotypical, appetitive responses elicited by mere stimulation of "sweet" cells even when expressing exogenous receptors, one would expect that brain circuits involved in processing reward information must be sensitive to sensory stimulation independently of the physiological consequences of ingesting the ligands. The role of brain dopamine systems in mediating food reward, and in encoding stimulus palatability, has been well established. Dopamine antagonists attenuate the hedonic value of sweet-tasting nutrients, in that animals pretreated with either D1- or D2-type dopamine receptor antagonists behave toward high concentrations of sucrose solutions as if they were weaker than usual (Xenakis and Sclafani 1981; Geary and Smith 1985; Bailey, Hsiao, and King 1986; Wise 2006). Conversely, tasting palatable foods elevates dopamine levels in the nucleus accumbens (NAcc) of the ventral striatum (Hernandez and Hoebel 1988), a brain region largely implicated in food reinforcement (Kelley, Schiltz, and Landry 2005). In humans, striatal dopamine release directly correlates with the perceived hedonic value of food

---

* It must be noted that the existence of gustatory labeled lines in the periphery does not allow one to conclude that distributed encoding is not involved in representing taste information in the central nervous system. In fact, it seems that the intricate circuitry of the brain allows for taste information to be encoded by entire populations of non-sensory specific neurons (Katz, Simon, and Nicolelis 2001), and such representations might even involve precise temporal patterns of firing activity (Di Lorenzo 2003).

stimuli (Small, Jones-Gotman, and Dagher 2003). But is dopamine release induced by sweet palatability per se independent of carbohydrate metabolism? In fact, taste-elicited stimulation of the central dopamine systems seems to take place even in the absence of intestinal nutrient absorption. In "sham-feeding" studies where a catheter is implanted in the stomach to prevent nutrients from reaching the intestinal tract, accumbens dopamine levels increase in proportion to the concentration of the sucrose solution used to stimulate the intraoral cavity (Hajnal, Smith, and Norgren 2004).

It is therefore plausible to assume that the events leading to the stimulation of brain reward circuits via dopamine release are initiated within the oral cavity, upon the activation of taste receptors. This implies that the dopamine release effect in accumbens related to sweet taste stimulation must depend on the integrity of central taste relays conveying gustatory information to downstream brain circuits. In fact, the parabrachial nucleus seems to be required for accumbens dopamine levels to increase upon gustatory stimulation. In rodents, axonal fibers originating in the gustatory aspect of the nucleus of the solitary tract ascend ipsilaterally to the parabrachial nucleus, establishing this pontine structure surrounding the conjunctivum brachium as the second-order gustatory relay (see Chapter 6 in this volume; and Norgren and Leonard 1971, 1973; Norgren and Pfaffmann 1975). The parabrachial nucleus is typically divided into two main portions, medial and lateral, relative to the conjunctivum brachium (Reilly 1999). Concerning this division, both anatomical and electrophysiological evidence suggest that parabrachial nucleus taste neurons are located primarily in medial subnuclei (Perrotto and Scott 1976; Fulwiler and Saper 1984; Ogawa, Hayama, and Ito 1987).

Dopamine release upon palatable taste stimulation seems to be mediated by projections to limbic circuits originating in the parabrachial nucleus in a way that is independent of thalamic relays.* In fact, whereas one group of projections from PBN reach the insular cortex via the taste thalamic relay (Norgren and Wolf 1975), a second, separate pathway reaches the amygdala, lateral hypothalamus and the bed nucleus of the stria terminalis (Norgren 1976; Li, Cho, and Smith 2005). Thus, it has been shown that lesions to the PBN limbic, but not to the PBN thalamocortical, pathway blunt the dopaminergic response during intake of palatable tastants (Norgren and Hajnal 2005; Norgren, Hajnal, and Mungarndee 2006). Such effects were further confirmed by experiments using c-fos measurements (Mungarndee et al. 2004).

12.2.3 PALATABILITY AND THE FORMATION OF LONG-TERM FOOD PREFERENCES

Consistent with what is now known about the labeled-line molecular logic of taste transduction, it would be straightforward to assume that the sensation of sweetness exerts a potent attractive effect on brain reward systems and, therefore, on behavior. Indeed, both deprived and non-deprived animals will not only avidly consume sweet solutions, but also run through intricate mazes or incessantly press levers to obtain sweet rewards (Kare 1971). In general, the expression of those motivated behaviors will increase with the concentration of the sweet tastant in the solution.

The hedonic sensation of sweetness is innate in humans. This is demonstrated by the reactions observed in children upon their first exposure to sugary solutions: the newborns will immediately suck the solutions and produce characteristic facial expressions (Ganchrow, Steiner, and Daher 1983). In rats, pups as young as six days old are strongly attracted to sweetness, given their robust intake responses to sweet compounds such as sucrose, lactose, and saccharin (Hall and Bryan 1981). It is also well established that other species show similar attractions to sweet compounds at early ages (Houpt, Houpt, and Pond 1977).

The strength of the reinforcing properties of sweet compounds is illustrated by their ability to form associations with and elicit increases in intake of arbitrary flavors, an effect denominated "flavor-taste conditioning." In this case, a distinct and moderately pleasant flavor functions

---

* In primates, the parabrachial nucleus does not seem to integrate the central taste pathways, with solitary tract fibers projecting directly onto the taste portion of the thalamus (Scott and Small 2009). Thus the relationship between taste-induced dopamine release and the parabrachial nucleus remains to be determined in primate models.

as a conditioned stimulus at the same time as an innately pleasant stimulus such as sweet taste functions as the unconditioned stimulus. In this design, a flavor cue may gain motivational value through flavor-taste conditioning if its presentation is paired with another taste or flavor that is hedonically positive to the animal. For example, combining a given arbitrary flavor with a sweet tastant in a solution will increase intake levels of this flavor when experienced alone, presumably due to the learned previous associations between the flavor and the palatable sweet taste (for a review Myers and Sclafani 2006). In other words, the consummatory responses to an arbitrary flavor can be made to increase via conditioning when associated with a palatable flavor, such as those associated sweet compounds.

However, are the reinforcing properties of sweet tastes sufficiently strong to influence intake in the long term? Although the sensory hedonic properties linked to taste and flavor are associated with high levels of intake in the short term (i.e., from minutes to a few hours), there is currently no definite evidence that such sensory properties regulate or even enhance longer-term increases in caloric intake.

In fact, in a series of experiments performed during the 1980s, Naim, Kare and colleagues investigated the role of sensory input in long-term experiments in rats where the nutritional composition of the food was controlled (Naim et al. 1985, 1986; Naim, Brand, and Kare 1987). The first, overall goal of these experiments was to verify whether supplementing the diets with highly preferred flavors and textures was sufficient to enhance long-term caloric intake of foods low in sugar or fat content. The second aim was to verify whether an effect would be observed in flavor-supplemented diets containing higher levels of sugar or fat. This last experimental manipulation was intended to mimic the composition of high-caloric diets whose consumption is associated with the development of obesity.

The results of these experiments are, to say the least, intriguing. Overall, they strongly suggest that palatable flavors and textures incorporated into the diets did not induce overconsumption of calories at any time up to 23 days (Naim and Kare 1991). Most interestingly, the added flavors increased intake during the first five days, but not thereafter. Such results were not replicated for the case of high-fat diets (Naim et al. 1985). Taken together, these results suggest that adding flavors and textures to non-caloric diets did not affect intake in rats except for a short and transient effect on intake over the first five days of the experiment.

The associative strength of palatable non-caloric compounds can be directly contrasted to those of caloric but less palatable ones. Consistent with the above, Fedorchak and Bolles (1987) have shown that pairing flavor-conditioned stimuli with caloric compounds produces a stronger effect on flavor preferences compared to palatable ones. Exposures to fruit flavors were paired with either caloric ethanol (in doses that do not seem to produce major aversive/irritant effects), non-caloric sweet saccharin (sweet but non-caloric), or water. During the post-training two-bottle choice tests, it was concluded that flavors associated with ethanol were preferred over saccharin-paired and water-paired flavors by sated rats, and that food deprivation during the choice test enhanced this preference. In addition, flavors associated with 8% sucrose (caloric and sweet) were preferred over water-paired flavors during ad libitum testing, an effect that was enhanced by hunger. Overall, calorie-mediated preferences were stronger than taste-mediated preferences. Fedorchak and Bolles (1987) then concluded that hunger enhances the expression of calorie- but not taste-mediated conditioned flavor preferences.

Similar conclusions could be inferred concerning the more specific role of sweet taste in carbohydrate intake. One would say, as noted in the *Introduction*, that sweet taste would underlie the excessive intake of calories associated with sugars. However, adding sucrose to solid diets results in no gain, or even loss, of body weight (Cohen and Teitelbaum 1964). On the other hand, although sucrose solutions do enhance caloric intake and may lead to obesity, the same effect is observed with non-sweet carbohydrate solutions such as Polycose (a polysaccharide) solutions (Sclafani and Xenakis 1984). Even more revealing is the fact that rats, when offered solutions containing an unpalatable mixture of Polycose and SOA (a bitter tastant), will consume amounts of that solution comparable to that of an alternative solution containing a saccharin-Polycose mixture (Sclafani and Vigorito 1987). Therefore evidence is lacking for a preponderant role of sweet taste in excessive carbohydrate intake.

A more recent demonstration that palatable tastes, when unaccompanied by metabolic effects, lack the ability of inducing longer-term preferences was given in a study using mice, where animals were required to express their preferences for sipper positions previously associated with the delivery of either caloric sucrose or sucralose, a non-caloric but strongly sweet sucrose-derived compound (de Araujo et al. 2008). In this behavioral model, animals are allowed to form an association between a particular sipper in a behavioral test chamber and the postingestive effects produced by drinking from that sipper. This is accomplished in sweet-naïve animals by first determining the initial side-preferences using a series of preliminary two-bottle tests where both sippers contained water. Hungry and thirsty mice are then exposed to a conditioning protocol where alternating access to either water or caloric sucrose was given for six consecutive days. Conditioning sessions consisted of daily 30 minutes free access to either water (assigned to the same side of initial bias) or caloric sucrose (assigned to the opposite side) while access to the other sipper was blocked (see Figure 12.2a–c for a scheme of the behavioral task). As expected, wild-type, taste-enabled mice consumed significantly more sucrose than water during the conditioning, one-bottle sessions (Figure 12.2d). The conditioning sessions were then followed by two-bottle, water-water tests identical to those run to determine initial side biases. During test sessions, the mice reversed their initial side-preference biases by drinking significantly more water from the sipper that during conditioning sessions had been associated with caloric sucrose, with preference ratios of up to 80% (Figure 12.2f).

When the same conditioning experiments were performed on an additional group of naïve animals, but this time using a non-caloric sucralose solution, a similar result was observed during conditioning sessions in that the animals consumed significantly greater amounts of sucralose compared to water (in fact, intake levels were even higher than those observed for sucrose). However, and rather importantly, during the postconditioning two-bottle test sessions these animals failed to display preference for the sippers associated with the delivery of sucralose, with preference ratios around 50%. Such inability of palatable non-caloric, unlike caloric, compounds to induce the formation of side preferences in behavioral tests indicates that palatability per se is not sufficient to produce the behavioral modifications necessary for animals to develop long-term consummatory behaviors.

Taken together, the studies described above strongly suggest that orosensory factors play a relatively minor role in the long-term increase in energy intake and associated adiposity in rats and mice. Rather, it seems that postingestional factors, related to the caloric content and/or metabolic effects of nutrients, are a major regulator of food intake over extended periods of time. The behavioral and physiological aspects of postingestive factors are explored in the following sections.

12.3 POSTINGESTIVE REINFORCEMENT

12.3.1 POSTINGESTIVE REGULATION OF FOOD INTAKE

Postingestive factors refer to physiological events subsequent to the consummatory act of eating. Postingestive factors include both pre- and postabsorptive events. The former include events such as gastric distention and stimulation of nerve terminal sensors throughout the gut epithelium, whereas the latter relate to physiological responses that follow nutrient absorption by the gut such as fuel oxidation or deposition, along with increases in plasma hormonal levels. The question of whether the critical reinforcing postingestive event is pre- or postabsorptive remains pretty much open for debate (although see the considerations in this chapter). In any event, the importance of postingestive factors as main regulators of food intake is now generally accepted, and new methodologies allowing for the experimental analysis of postingestive effects are being actively pursued.

Inquiries on the possibility that postingestive effects could act as regulatory factors in the absence of orosensation were taken on as early as the 1950s. In an intriguing study, Miller and Kessen (1952)

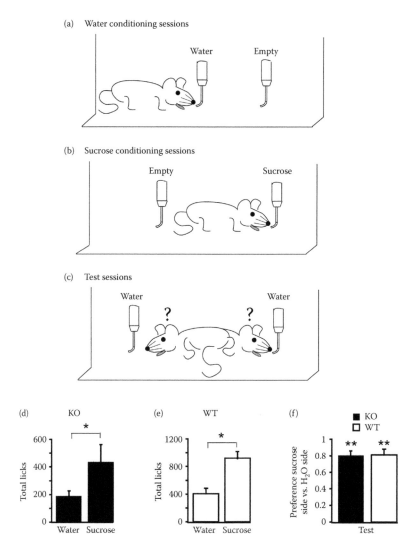

FIGURE 12.2 A conditioning protocol to study the reward value of postingestive effects independently of taste signaling. Behavioral protocols can be used for studying the formation of food preferences based on taste-independent, postingestive effects. Hungry and thirsty adult mice (including wild-type and sweet-insensitive *Trpm5* knockout mice, see text for details) can be exposed to a conditioning protocol where alternated access to either water or 0.8 M sucrose is given for 6 consecutive days. Conditioning sessions consist of daily 30-min free access to either water (available from a sipper located on the particular side of the cage where animals usually prefer to drink water, see panel a) or 0.8 M sucrose (offered in a sipper located on the opposite side, see panel b), while access to the second sipper is blocked. Water and sucrose sessions are alternated across the 6 consecutive days of conditioning. These conditioning sessions are then followed by two-bottle water choice tests (see panel c). During test sessions, both wild-type and knockout animals reverse their initial side-preference biases by drinking significantly more water from the sipper that during conditioning sessions had been associated with 0.8 M sucrose. The overall results of this experiment are as follows: During the single-bottle conditioning sessions, both sweet-insensitive knockout (d) and wild-type (e) animals consumed significantly more sucrose than water (*unpaired two-sample t-test, $p < .05$). During 10-minute-long two-bottle postconditioning test sessions where water was accessible from both sippers, a significant reversal of initial bias was observed in both genotypes as revealed by the measured preference ratios (f) (**independent t-test against 0.5, $p < .02$). Reversal of bias in KOs is indicative that animals successfully associated sipper-side with its postingestive effects. (Reprinted from *Neuron*, 57, de Araujo, I.E., et al., 930–41, Copyright 2008, with permission from **Elsevier**.)

infused either milk or saline intragastrically in rats depending on which arm of a T-maze they entered. The found that under a rather extended period of time (approximately 40 days) the rats developed the habit of entering the maze associated with milk infusion. Although the authors also noted that rats that were allowed to drink the solutions were able to learn the maze task much faster, this experiment constituted an early demonstration that postingestive effects are sufficient to sustain learned preferences for nutrients, although learning can be greatly facilitated by orosensory inputs.

Another series of pioneering experiments was performed by Epstein, Teitelbaum and colleagues (Epstein and Teitelbaum 1962; Epstein 1967). The researchers prevented rats from eating foods while allowing them to obtain intragastric infusions of nutrients by pressing a lever. They showed that the number of lever presses produced by the rats were essentially a function of the amount of nutrient infused via the gastric cannulae, with lower amounts of nutrients producing higher number of responses (presumably this corresponds to a compensatory response pattern). More precisely, in the cases where infused amounts were minimal, rats achieved stable levels of nutrient intake by increasing lever press frequency in such a way that body weight was eventually restored to normal levels. Similar experiments by Miller and Kessen (1952) further corroborated the principle that postingestive factors are sufficient to control caloric intake. It is also interesting to note that humans will also work to infuse diets intragastrically in order to maintain their necessary levels of nutrition (Jordan 1969).

Although it had become clear by the late 1960s that postingestive effects act as main regulators of food intake, some researchers became interested in the more specific question of whether these effects, specifically in the form of intragastric infusions, can influence or even increase preferences for certain foods. One elegant way to answer such a question would require experimental paradigms where the orosensory properties of food are paired with intragastric infusions. Holman (1968) introduced such a protocol, in which exposure of rats to a flavored drink for five minutes was immediately followed by an intragastric infusion of either a nutrient or water. The training sessions were six in total, with three of them allowing the animals to form an association between one flavor and the postingestive effects produced by infused nutrient, and the other three between another flavor and infused water. During subsequent two-bottle tests, Holman (1968) observed that the vast majority of rats preferred the flavor that had previously been paired with nutrient infusion. This seems to have been the first demonstration of what had come to be known as "flavor-nutrient conditioning," that is, that preferences for distinct flavor can develop as a function of the postingestive effects produced by the flavored food.

How likely is it, however, that paradigms where flavored drinks are paired with intragastric infusions of nutrients reproduce the physiological processes by which caloric food acquires motivational value through experience? Booth seems to have been the first one to attempt to produce conditioned flavor preferences using nutritive flavored foods (Booth 1972). In these experiments rats were fed different flavored foods that however varied with respect to their starch content. When the rats were given a choice between the different foods whose particular flavors had been previously associated with different starch contents, Booth (1972) observed that the animals preferred the flavored food associated with the highest caloric content (during the test, flavored foods were all presented at equally moderate caloric levels). Further variations of this experimental design include the interesting case where the differently flavored foods each associated with a specific caloric value are presented simultaneously to animals (Bolles, Hayward, and Crandall 1981), demonstrating that several days of exposure will eventually lead to a long-lasting preference for the flavored foods paired with higher caloric content.

It could therefore be concluded that pairing of distinct flavored drinks with intragastric infusion does realistically model the physiological processes that enhance flavor preferences following long-term exposure. This method has been successfully adopted by other researchers (Mather, Nicolaidis, and Booth 1978; Tordoff and Friedman 1988; Tordoff 1991; Sclafani 2001). One advantage of this method is that it can be employed virtually ad infinitum if one considers the number of all possible combinations involving distinct flavors, nutrients, and their concentrations. In fact, Sclafani and colleagues have been employing the flavor-nutrient conditioning method repeatedly, providing useful information concerning the relative reinforcing strength of different nutrients (reviewed in Sclafani 2001).

Despite its elegance and efficiency, the flavor-nutrient method based on intragastric infusions does not shed light on a crucial issue, namely whether ingestive behaviors can be actively sustained in the *absence* of oropharyngeal flavor sensation. In fact, the flavor-nutrient conditioning protocol necessarily pairs intragastric infusions of nutrients with flavor stimulation, leaving open the question of whether sensory stimulation by distinct flavors is required to sustain normal ingestive behaviors. Obviously, it must be noted that in the atypical case that flavor cues are absent, other sensory features associated with the food object would have to be present to allow for associations with ensuing postingestive effects to be formed; such non-orosensory cues include, for example, the spatial location of the food. In general, whatever the nature of the sensory cues involved, these internal representations or memories can then be rapidly accessed and retrieved through incentive learning processes when the animal next encounters the food object (as discussed in Chapter 13).

The question above can be more clearly addressed if the ability to detect the sensory properties of distinct flavors is prevented from occurring during the experiments. One way to achieve this refers to using genetically engineered animals lacking taste sensation. As explained above, a conditioning protocol can be used where wild-type mice will develop a preference for drinking, during postconditioning two-bottle water vs. water tests, from the sipper location previously associated with nutritive sucrose but not with the location associated with the non-nutritive sweetener sucralose (de Araujo et al. 2008). The same study also employed mice lacking a functional transient receptor potential channel M5 (Zhang et al. 2003). The TRPM5 ion channel is expressed in taste receptor cells (Perez et al. 2002) and is required for sweet, bitter, and amino acid taste signaling (Zhang et al. 2003). It was hypothesized in this study that sweet-blind trpm5 knockout mice would develop a preference for spouts associated with the presentation of sucrose solutions when allowed to detect the solutions' rewarding postingestive effects.

Once the insensitivity of KO mice to the orosensory reward value of sucrose was established, de Araujo et al. (2008) tested whether a preference for sippers associated with caloric sucrose solutions could develop in water- and food-deprived trpm5 knockout mice when they are allowed to form an association between a particular sipper in the test chamber and the postingestive effects produced by drinking from that sipper. As explained above, this was accomplished in sweet taste-naïve animals by first determining the initial side-preferences using a series of preliminary two-bottle tests where both sippers contained water and by exposing animals to conditioning sessions consisting of daily 30 minutes free access to either water (assigned to the same side of initial bias) or sucrose (assigned to the opposite side) while access to the other sipper was blocked (Figure 12.2a–c).

Analysis of the behavioral data across both wild-type and knockout mice revealed no significant genotype × stimulus interaction since during conditioning sessions both wild-type and knockout animals consumed significantly more sucrose than water (Figure 12.2d, e). In addition, during the postconditioning two-bottle tests, it was observed that both wild-type and knockout animals reversed their initial side-preference biases by drinking significantly more water from the sipper that during conditioning sessions had been associated with nutritive sucrose (Figure 12.2f). Now, when the same experiments were instead run using sucralose (palatable but lacking nutritive benefits), unlike for the sucrose case, a significant genotype × stimulus interaction was found because only the wild-type animals consumed significantly more sucralose than water during the conditioning sessions. Furthermore, during the two-bottle test sessions, conducted after conditioning to sucralose, knockout mice, like their wild-type counterparts, showed no preferences for sippers associated with the delivery of sucralose. Overall, these results provide evidence in favor of the hypothesis that postingestive effects can exert positive controls on ingestive (licking/swallowing) behaviors even in the absence of taste signaling or detection of distinct flavors.

12.3.2 THE POSTINGESTIVE REWARD SIGNAL

While there is little disagreement that postingestive factors produced by nutrients regulate food intake, much more controversial is the nature of the signal that acts on the brain as a postingestive reinforcer. Broadly speaking, the candidate signals can be classified into two groups, related to pre- and postabsorptive events. The former group concerns those sensing mechanisms that occur before nutrient absorption but simultaneous with the arrival of nutrients to the gut. The latter group refers to those events that occur following absorption, and non-exclusively includes a variety of signals such as fuel oxidation metabolites and changes in plasma hormonal levels. In this section, we review some of the mechanisms proposed so far to function as the postingestive reward signal.

12.3.2.1 Preabsorptive Mechanisms

12.3.2.1.1 Orosensory signals

As argued above, orosensory signals including taste or other flavor additives constitute weak reinforcers when employed as unconditioned stimuli during flavor conditioning paradigms. Overall, calorie-mediated preferences are stronger than taste-mediated preferences (Fedorchak and Bolles 1987; Naim and Kare 1991). In addition, mice lacking the cellular machinery required for taste transduction do develop robust preferences for sources of sucrose delivery (de Araujo et al. 2008). These facts strongly suggest that delayed oropharyngeal sensations do not account for the development of food preferences based on postingestive effects.

12.3.2.1.2 Gastric Signals

We could in principle start by ruling out as reward signal candidates the visceromechanical signals from the gut. These include pressure-related signals reaching the brain upon nutrient arrival in the stomach. In fact, because only nutrients, but not water, saline, or artificial sweeteners can produce postingestive reinforcement in flavor-nutrient conditioning paradigms, simple mechanical pressure on gut nerve terminals could not be the primary source of a central reinforcement signal.

It could be claimed, however, that different compounds might induce different transit/gastric emptying time-courses (Friedman, Ramirez, and Tordoff 1996), and that these differential patterns in gastric activity could act centrally to communicate information to the brain regarding the presence and processing of nutrients in the gut. This is consistent with the concept that liquid nutrient gastric emptying represents an interaction between gastric volume and nutrient-induced duodenal feedback (Moran et al. 1999). However, this hypothesis becomes less plausible if one considers the early studies performed by Tsang (1938) on gastrectomized hungry rats. Tsang removed over 90% of the stomach from rats before their behaviors in food-baited mazes and cages were studied. Interestingly, after 24 hours of fasting, gastrectomized rats were nearly as well motivated as normals in the first trial of the maze. These studies showed in general that gastrectomy will cause an increase in overall motivation, certainly because the animals sought to compensate for the decreased caloric intake resulting from having a drastically reduced stomach. Although Tsang (1938) noted that an empty stomach is likely to be a necessary condition for motivation for food, contractions of the stomach per se cannot underlie the source of the motivation.

Consistent with the above, it was later found that gastric vagotomy does not interfere with habitual feeding patterns in rats (Snowdon and Epstein 1970), and more recently it was shown that abdominal vagotomy does not interfere with flavor preferences conditioned by Polycose (Sclafani and Lucas 1996). Taken together, all of these lesion-based studies weaken the idea that gastric signals function as central reinforcement signals, suggesting that other mechanisms downstream to gastric emptying are involved.

12.3.2.1.3 Preabsorptive Intestinal Signals and Gut Taste Receptors

What evidence is there in favor of a preabsorptive postingestive reinforcement signal originating from the gut epithelium, more precisely in the intestinal tract? One simple way to test whether such mechanisms exist in the first place is to prevent nutrient absorption during flavor-nutrient conditioning tests and observe the ensuing behavioral responses. To test this hypothesis Elizalde and Sclafani (1988, 1990) exposed rats to two flavored drinks, only one of which contained the non-sweet polysaccharide Polycose. However, for one group of rats, the flavored Polycose drink also contained acarbose, an inhibitor of starch hydrolysis, effectively preventing Polycose-derived glucose from being absorbed in the duodenum. While the animals that were exposed to the Polycose-only drink developed robust preferences for the associated flavor, those exposed to the Polycose+acarbose drink did not. This finding challenges the idea that preabsorptive mechanisms constitute a major reinforcement signal in postingestive reward.

The above would indicate additionally that a fundamental role might be played by intestinal glucose transporters in postingestive reinforcement, at least for the case of carbohydrates. The intestine has the ability to modulate its glucose-absorptive capacity by regulating the expression levels of the intestinal sodium/glucose co-transporter 1 ("SGLT1," Dyer et al. 2003). Unfortunately, the precise contribution of SGLT1 (or related proteins such as SGLT3) expression in the gut to carbohydrate postingestive reinforcement has not yet been investigated. However, recent developments have demonstrated that gastrointestinal epithelial cells express many of the known (taste) receptors and downstream factors located in taste cells of lingual epithelium. This finding is important for the understanding of postingestive reinforcement because it potentially provides a molecular basis for gut chemosensation, whose signals could possibly be conveyed to the brain, informing of the presence and composition of nutrients in the lumen.

With regard to the intestinal epithelium, the alpha component of the taste-specific G-protein gustducin (α-gustducin, McLaughlin, Mckinnon, and Margolskee 1992) was shown to be present in brush cells of the rat proximal intestine (Hofer, Puschel, and Drenckhahn 1996), as well as in mouse intestinal endocrine cells and enteroendocrine cell lines (Wu et al. 2002). More recently it was shown that T1R receptors are also present in the rodent gut epithelium as well as in enteroendocrine cell lines (Dyer et al. 2005). Importantly, α-gustducin, T1R2, and T1R3, as well as other taste signaling proteins including the taste ion channel TRPM5, are co-expressed in some mouse and human enteroendocrine cells (Bezençon, le Coutre, and Damak 2007; Margolskee et al. 2007). These co-expression data are extremely relevant since the concomitant presence of at least a subset of these proteins in lingual taste cells is required for normal taste transduction. In general, these results strongly suggest that the intestinal epithelium can "taste" dietary composition and possibly provide a signal to the brain on postingestive events taking place after eating.

What evidence is currently available regarding the physiological functions of these gut-expressed taste signals? Although research is still in progress, convincing evidence exists that both α-gustducin and T1R receptors regulate the expression levels of SGLT1 in the intestine. More precisely, Margolskee and colleagues (2007) have used mouse knockouts lacking α-gustducin or T1R3 to show that the absence of either of these signals blocked the ability of dietary sugars, as well as of artificial sweeteners, to up-regulate the expression of SGLT1. Interestingly, SGLT1 expression levels in both types of knockout mice were similar to those of wild-type mice fed a low-carbohydrate diet. This indicates the existence, in addition to basal SGLT1 expression levels, of a signaling cascade pathway that is initiated by T1R3/α-gustducin activity to increase expression levels of SGLT1 in response to luminal sugars (Jang et al. 2007; Margolskee et al. 2007; Egan and Margolskee 2008).

However, do these data provide evidence that gut-expressed taste-related proteins play a role in postingestive reinforcement? The behavioral data describing postingestive regulation of sugar intake in TRPM5 knockout mice (de Araujo et al. 2008) contributes to the ongoing debate on whether nutrient-sensing by the gastrointestinal system makes use of taste-like transduction pathways to detect luminal contents and regulate nutrient absorption. Indeed, the fact that TRPM5 knockout animals developed a preference for the sipper locations associated with sucrose availability, whereas wild-type animals did not condition to sipper locations associated with sucralose (a non-caloric substance that activates the same taste transduction pathways as sucrose), indicates that the presence of the taste TRP channel M5 in the gastrointestinal tract (Bezençon, le Coutre, and Damak 2007) is neither necessary nor sufficient for sweet nutrients to act centrally as reinforcers. This behavioral observation was further strengthened by the fact that during sucrose intake knockout animals displayed changes in blood glucose levels comparable to those observed in wild-types. Furthermore, a similar conclusion was recently reached based on the long-term consummatory patterns produced by T1R3 knockout mice (Zukerman et al. 2008). It was observed in this study that T1R3 knockout mice did develop a preference for sucrose solutions over a few days of exposure. The authors attributed the experience-induced sucrose preference to a post-oral conditioned preference for non-sweet orosensory features of the sugar solutions (odor, texture, etc.). Independently of which cues were used by these animals, the results clearly indicate that T1R3, and likewise TRPM5, expression in the gut epithelium is not required for mice to detect the postingestive effects of sucrose.

Taken together, the results above demonstrate that taste proteins essential for sweet taste transduction in lingual epithelium are not required for sucrose to act centrally as a postingestive reinforcer, despite the putative role of these proteins in modulating SGLT1 expression in the gut epithelium. Judging from the available data, it seems more plausible to think that gut taste proteins are more specifically related to the controlled release of endocrine factors by the gut.

12.3.2.2 Postabsorptive Mechanisms

A number of postabsorptive mechanisms have been suggested to act as the reward signal controlling conditioned responses to food via postingestive reinforcement. We review some of the current evidence in their favor below.

12.3.2.2.1 Hepatic Mechanisms

The first investigator to propose a "hepatostatic" theory of food control was M. Russek (1970, 1981). Russek's theory seems to have evolved from his early observations that dogs displayed a reduced appetite for food when they were given hepatic-portal infusions of glucose. Even more interesting was the finding that conditioning the hepatic glucose infusions to the presentation of a light caused animals to reduce food intake while being exposed to the light even in the absence of infusions (Russek 1970). Although Russek considered these postingestive and associative effects to be of a preabsorptive nature, i.e., mediated by sensory activation of autonomic fibers, it established the then original concept that the central organ for glucose homeostasis might play a central role in feeding.

The study of the role of the liver in conditioned food intake was done in greater depth by M. Friedman, M. Tordoff and colleagues (Tordoff and Friedman 1988, 1994; Tordoff, Rawson, and Friedman 1991; Friedman et al. 1999; Friedman 2007). In one important experiment using rats, a flavor-nutrient design was employed where the unconditioned stimulus consisted of hepatic-portal infusions of either glucose or saline (Tordoff and Friedman 1986). During test sessions, the rats preferred the flavor that had been associated with hepatic (but not jugular) glucose infusions, demonstrating that glucose detection in the portal-hepatic system, bypassing the gastrointestinal tract, is sufficient to provide a brain with a postingestive reinforcement signal.

Different carbohydrates will differentially tax the liver with singular metabolic demands. Therefore, another way to assess the role of the liver in postingestive reinforcement relates to comparing the differential effects produced by different carbohydrates when employed as unconditioned stimuli in flavored food-nutrient conditioning paradigms. In fact, whereas glucose is absorbed and utilized as fuel by muscle, brain, and other tissues besides the liver, fructose greatly differs from glucose in the sense that its greater utilization occurs in the liver, with fructose being unable to cross the blood-brain barrier (Tordoff 1991). Consistent with a hepatostatic theory of conditioned food preferences, rats developed a strong preference for flavors paired with fructose drinks over those paired with glucose drinks. These experiments seem to indicate that the liver is a crucial site for the generation of reinforcement signals in conditioned food intake.

Intriguingly, later experiments comparing the effects of fructose and glucose in flavored drink-nutrient conditioning paradigms suggested that intragastric fructose infusions provide a weaker postingestive signal compared to glucose infusions (Sclafani, Fanizza, and Azzara 1999). The reason for such a discrepancy is presently not clear and might signify that liver-derived reinforcement signal is superseded by other signals generated during intragastric infusions of sugars. In any case this latter point, if anything, stresses the need for further research on the pathways conveying information of hepatic processes to the brain.

12.3.2.2.2   Metabolic Signals

Friedman has proposed that food intake patterns are ultimately controlled by signals generated during the oxidation of metabolic fuels (for reviews see e.g., Friedman 1989, 1991). This hypothesis is directly related to the assumption discussed above that the liver is a critical site for the generation of reinforcing signals given the prominent role of this organ in fuel metabolism. From a biological point of view at least, this is probably the most sensible hypothesis one could come up with, since any consummatory behaviors that do not eventually result in fuel utilization should ultimately have no reinforcement value.

This concept is rooted in the experimental observation that inhibition of both glucose utilization and fatty acid oxidation interfere with food intake (Friedman and Ramirez 1985; Friedman et al. 1999). For example, systemic administration of 2-deoxy-D-glucose (2-DG) elicits feeding in rats (Friedman and Tordoff 1986), presumably by inhibiting glycolysis in peripheral tissues and brain and depriving animals of usable energy. In fact, 2-DG is a glucose analogue that competes with glucose for transport through membrane channels and for phosphorylation by hexokinase (Sokoloff 1989), while inhibiting glycolysis at the phosphohexoisomerase step (Wick et al. 1957).

However, by assuming that a feeding-inducing stimulus is generated during intramitochondrial oxidative phosphorylation and ATP production (Friedman 1991), the theory must also sustain that such a stimulus is independent of the type of fuel used during oxidation. Therefore, one would predict an enhanced effect on food intake when metabolism of more than one type of fuel is inhibited (since by blocking only one type of fuel metabolism the others are left to operate as alternative fuel reserves). In fact, Friedman and Tordoff (1986) have shown that 2-DG and methyl palmoxirate, an inhibitor of fatty acid oxidation at the membrane transport level, act synergistically to alter food intake in rats.

The above raises the question of how changes in fuel oxidation can be detected by the nervous system in order to ultimately alter food intake. Friedman et al. proposed that such fuel-oxidative information could be conveyed to the brain via afferent hepatic nerves, including the hepatic branch of the vagus (Friedman 1991). In fact, liver functions are not limited to simultaneously handling metabolic processes related to several different fuels (fatty acid oxidation and synthesis, metabolizing of fructose and glycerol, glycogen synthesis/breakdown, etc.; Langhans, Egli, and Scharrer 1985b). In addition, manipulations of hepatic metabolism seem to affect food intake patterns (Friedman et al. 1999) and interfere with normal feeding responses elicited by fructose intake (Langhans, Egli, and Scharrer 1985a). Furthermore, severing hepatic nerves alters normal food intake patterns (Friedman and Sawchenko 1984), suggesting that these nerves constitute a neural route through which information about fuel oxidation levels is relayed to brain circuits.

However, as noted by Friedman (1991) himself, there is no evidence that these lesions are selective to afferent fibers. Therefore, it could be assumed that at least part of this effect might result from an inability of the brain (as the ultimate sensor of energy status via hormonal factors) to control fuel metabolism in the liver. In fact, severing of hepatic nerves does not seem to abolish some compensatory intake behaviors that follow deprivation in rats (Egli, Langhans, and Scharrer 1986). One possibility is that fuel utilization, in this case glucose utilization, controls food intake via direct sensing by neurons, since brain cells require relatively high levels of glucose to preserve cellular function. In summary, although the idea that hepatic fuel oxidation relates to a peripheral signal that is conveyed to the brain to ultimately control feeding is an exciting hypothesis, research is needed in order to further our understanding on how information on fuel oxidation occurring at the mitochondrial level ultimately reaches brain centers involved in postingestive-based control of feeding.

12.3.2.2.3   Brain Sensing of Plasma Circulating Factors

Plasma levels of factors released in the bloodstream in response to nutrient intake could provide the brain with a reinforcement signal via activation of nerve afferents, such as the abdominal branches of the vagus in the case of cholecystochinin (CCK), or by binding to their respective specific receptors expressed in the brain, as in the case of adiposity factors such as insulin and leptin.

The intestinal satiety factor CCK has been shown to produce conditioned flavor preferences in rats (Perez and Sclafani 1991), suggesting a role for this factor in conditioned food preferences. However, the same group has also shown that high doses of the CCK-A receptor antagonist devazepide fail to inhibit flavor preferences conditioned by intraduodenal Polycose infusions (Perez, Lucas, and Sclafani 1998). Therefore, it appears that CCK-related mechanisms might be more directly related to the satiating effects of intraintestinal infusions of carbohydrates rather than to their postingestive reinforcing effects.

The pancreatic factor insulin is released in the bloodstream in response to nutrient intake. Peripherally circulating insulin crosses the blood-brain barrier in proportion to serum insulin levels via a saturable transport mechanism (Margolis and Altszuler 1967; Woods and Porte 1977). In 1979, Woods and colleagues showed that insulin infusions into the ventricular system of the brain result in reduced food intake and body weight (Woods et al. 1979). In fact, insulin receptors were soon after found to be richly expressed in the arcuate nucleus of the hypothalamus (Van Houten et al. 1979). In this nucleus, which is of central importance to food intake control, insulin receptors co-express with the anorexigenic neuropeptides proopiomelanocortin, the precursor of melanocyte-stimulating hormone, and cocaine- and amphetamine-regulated transcripts, as well as with the orexigenic neural factors neuropeptide Y and agouti-related peptide (Plum, Belgardt, and Bruning 2006).

One would thus expect that insulin could provide the brain with a signal conveying information on the processing of nutrients in the gut. In fact, a role for insulin in postingestive reinforcement has been suggested by the findings of Oetting, Vanderweele and colleagues. Vanderweele et al. (1985) showed that sham-feeding rats have the ability to acquire preferences for flavored milks that have been paired with insulin injection over flavored milks paired with saline injections. This however has been put into question by an experiment performed by Ackroff, Sclafani, and Axen (1997), who showed that rats treated with streptozotocin (a drug model of insulin-deficient diabetes) display the same behaviors as normal rats by preferring a flavor that had been mixed with a glucose solution over a flavor that had been mixed with a fructose solution. In addition, both diabetic and non-diabetic rats acquired a preference for the flavor paired with intragastric infusions of glucose over flavors paired with fructose infusions. These results would indicate that normal insulin responses to glucose are not necessary for glucose-conditioned flavor preference.

12.3.2.2.4 Brain Dopamine Signaling

Conditioned preferences for caloric foods might also depend on brain-derived signals that are known to be involved in associations between unconditioned and conditioned reward stimuli. In fact, a role for dopamine signaling in flavor-nutrient conditioning is suggested by experiments employing administration of dopamine receptor antagonists in the nucleus accumbens. Rats treated with local infusions in nucleus accumbens of the D1-receptor antagonist displayed a dose-dependent reduction in intake of a flavor paired with intragastric infusions of glucose, compared with controls infused with saline (Touzani, Bodnar, and Sclafani 2008). Interestingly, the effect of dopamine signaling antagonism on postconditioning preference tests was less compelling. In any event, these results demonstrate that D1-like receptors in the nucleus accumbens are required for the acquisition, and possibly also for the expression, of glucose-conditioned flavor preferences.

One function that might be performed by dopamine signaling in accumbens during flavor-nutrient pairing is the strengthening of the associative link between conditioned flavors and unconditioned postingestive rewarding effects. It has in fact been proposed that the adaptive properties of increased accumbal dopamine release during taste stimulation are consistent with a role in associative learning (Di Chiara and Bassareo 2007). These authors have therefore proposed, consistent with the above, that release of dopamine in the nucleus accumbens, particularly in its more medial aspects (shell), following food intake might function to associate the gustatory properties of foods with their postingestive effects (Di Chiara and Bassareo 2007).

However, more recent findings suggest that the presence of taste or flavor stimulation is not required for the postingestive effects of foods to induce dopamine release in the nucleus accumbens. In fact, it has been found that in sweet-blind trpm5 knockout mice, caloric intake per se, independently of taste, was sufficient to increase extracellular dopamine levels in the nucleus accumbens (de Araujo et al. 2008). More precisely, it was first found in this study that the non-caloric sweetener sucralose intake produced significantly higher increases in dopamine levels in wild-type compared to knockout animals. These results are consistent with a role for dopamine signaling in accumbens derived from taste stimulation alone (Hajnal, Smith, and Norgren 2004). However, when the same comparison was performed with respect to sucrose, no differences were found between the dopamine release levels in wild-type and knockout mice. In other words, while sweet taste stimulation without caloric content produced, as expected, significant increases in accumbal dopamine levels in wild-type, taste-enabled animals but not in sweet-blind mice, caloric sucrose evoked the same levels of dopamine increase in both wild-type and knockout mice. These results therefore strongly suggest that even in the absence of taste transduction and/or palatability, nutrient intake has the ability to induce measurable tonic increases in accumbens dopamine. Thus palatability and postingestive factors both seem to increase dopamine levels independently in brain-reward circuits, though the nutrient-induced increases do not require the concomitant presence of flavor inputs, as had been suggested previously (Di Chiara and Bassareo 2007). It should be therefore inferred that events initiating in the gastrointestinal tract, most likely postabsorptive ones, regulate dopamine release in nucleus accumbens.

It remains unclear whether the same dopaminergic neurons are stimulated by these two independent pathways. If so, dopamine neurons would function within a brain circuit supporting the convergence of sensory (taste) and postingestive factors, possibly allowing for the formation of associations between them. Immunohistochemical and tracing techniques might be combined in future studies to determine whether dopamine neurons activated by metabolic cues are also targeted by taste projections from gustatory relays in the brainstem.

Whatever the mechanism inducing the dopaminergic rise in nucleus accumbens might be (i.e., flavor independent or not), neither of these two conjectures explains how postingestive factors gain access to midbrain dopamine cells in the first place. Some clues in this direction have been provided by the work by Figlewicz and colleagues, who have shown that the functional forms of the insulin and leptin receptors (Figlewicz et al. 2003), as well as of some of their substrates (Pardini et al. 2006), are richly expressed in dopaminergic neurons of the substantia nigra compacta and ventral tegmental area regions of the midbrain.

In fact, it has been demonstrated that leptin receptors expressed in dopaminergic neurons of the midbrain are functional and influence dopamine release (Fulton et al. 2006; Hommel et al. 2006). However, it is currently unknown whether brain leptin receptors play a role in postingestive reinforcement. In addition, the functional implications of insulin receptor expression in dopamine neurons have been little explored. Although it has been suggested that insulin infusions in the midbrain dopamine areas "decrease" the reward value of sucrose, since mice were found to reduce overall intake of sucrose solutions upon infusion (Figlewicz 2003; Figlewicz et al. 2006), this might simply imply that insulin receptor activation in midbrain dopamine areas provides the brain with a robust signal of caloric intake. In fact, if insulin were to decrease the reward value of foods, then we would expect that strongly insulin-releasing factors such as sugars or fats will become non-preferred over longer periods of time, which is obviously not the case. One interesting experiment testing the reinforcing power of insulin would consist of pairing non-caloric flavors with midbrain infusions of either insulin or vehicle and assessing their effects in postconditioning preference tests. An alternative explanation to the intriguing idea that brain insulin signaling decreases the reward value of sweet compounds consists in predicting that animals will develop a preference for flavors paired with insulin infusions. Although this would conflict with the results mentioned above stating that diabetic rats do display normal preferences for flavors associated with nutrients (Ackroff, Sclafani, and Axen 1997), it cannot be ruled out that diabetic rats might have developed compensatory mechanisms to low insulin levels, such as increases in plasma leptin. Therefore, the issue of whether brain insulin receptor signaling plays a role in postingestive reinforcement, and whether it interacts with leptin receptor signaling, remains open for future investigations.

Finally, another mechanism through which dopamine neurons could sense changes in physiological state refers to the possibility that dopamine neurons functions as glucosensors, i.e., that midbrain dopamine neurons can change their membrane potentials as a function of extracellular concentrations of glucose. It is currently unknown whether dopamine neurons express the cellular elements that seem to act as glucosensors in brain regions regulating energy homeostasis, such as the hypothalamus. For example, it would be interesting to verify whether glucokinase (hexokinase IV), which is the rate-limiting kinase in glucosensing (Levin 2006), is expressed in midbrain dopamine neurons, or whether dopamine neurons receive direct inputs from glucosensing neurons of the hypothalamus or brainstem.

In summary, it seems plausible that postingestive mechanisms operating as reinforcers in conditioned food preferences are of a postabsorptive nature, and most likely depend on signals that derive from fuel utilization (Figure 12.3). As we will see below, new research endeavors are needed to unveil the metabolic and central pathways involving the action of a postingestive reinforcement signal.

12.4 CONCLUSIONS AND FUTURE DIRECTIONS

In this chapter, we have argued that food reward consists of a multi-layered behavioral process, in which positive controls on intake are exerted by both proximal cues (with respect to eating, e.g., food flavor, the consummatory act) and distal cues (e.g., preingestive food-predictive signals, and postingestive physiological and metabolic effects). In short-term experiments performed on naïve animals, proximal cues seem to exert a stronger influence on intake. However, the bulk of evidence from past and current studies favors the view that the critical event regulating preferences for certain nutrients over others across extended periods of time consist primarily of postingestive effects. Furthermore, the postingestive effects relevant for the formation of conditioned food preferences seem to be of a postabsorptive nature, i.e., they seem to depend on fuel utilization/oxidation and/ or hormonal release. Finally, these behaviorally critical postabsorptive effects have been shown to activate brain reward pathways, including the mesolimbic brain dopamine system, an effect that does not depend on the concomitant presence of taste or flavor stimulation. In other words, although multiple layers of rewarding influences affect food intake, a hierarchy exists in which palatable compounds unable to provide sources for fuel oxidation eventually become less preferred in comparison to more energy-rich sources. From a biological point of view, a postingestive reinforcement signal that depends on postabsorptive events obviously provides great advantages over hypothetical preabsorptive mechanisms.

However, several crucial questions remain to be addressed concerning the identity and nature of the postingestive reinforcement signal. If on one hand the early observations by Adolph (1947) suggesting that animals "eat for calories" remain essentially correct, it remains to be demonstrated that different fuels providing the same amount of calories per gram will produce the same postingestive reinforcement effect. In fact, depending on the identity of the body part that turns out to be the crucial generator of postingestive reward signals, certain macronutrients might be more prone to act as postingestive reinforcers than others due to the intrinsic differences in how organs metabolize fuels.

For example, whereas several peripheral organs can make efficient usage of non-glucose fuels such as fatty acids, brain cells rely almost entirely on glucose or its immediate derivatives including lactate (Bouzier-Sore et al. 2006; Oltmanns et al. 2008; Pellerin 2008). This is consistent with the significant increases in feeding observed following inhibition of glucose utilization in the brain (Miselis and Epstein 1975). In addition, it is revealing that in one study, infusions of fatty acid oxidation inhibitors induced increases in food intake only upon co-administration with a glucose utilization inhibitor (Friedman and Tordoff 1986). These results favor the hypothesis that a crucial aspect of food reinforcement relates to glucose utilization and oxidation. Therefore, due to the critical reliance of the brain on glucose, it is plausible to hypothesize that glucose metabolism in brain

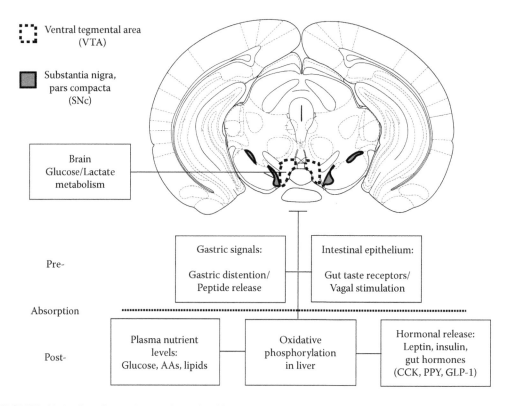

FIGURE 12.3 Putative pathways through which postingestive effects produced by eating nutrients might modulate dopamine activity and reward. Peripheral physiological signals generated by the ingestion of nutrients might gain access to the brain and modulate neural activity in midbrain dopamine areas, thereby acting as central reward signals. Postingestive peripheral signals might be generated as soon as nutrients make contact with the gastrointestinal tract (preabsorptive signals), and possibly involving the release of peptides from the epithelium of the stomach and/or intestine, as well as activation of nutrient-sensing membrane receptors expressed on vagal terminals or epithelial cells (such as gut taste receptors). Current evidence however favors the existence of postabsorptive, metabolic signals as the critical factors associated with postingestive reinforcement. Such factors might include postabsorptive hormonal release (including insulin and leptin secretion) or currently undetermined signals generated by oxidative phosphorylation of fuels in liver (see text for details). Finally, a putative but currently unexplored mechanism relates to the possibility that dopamine neurons act as true metabolic sensors, directly assessing the metabolic consequences of nutrient ingestion via intracellular catabolic processes. In any event, it must be noticed that these different putative postingestive reward signals are not mutually incompatible, and might act in concert to create a robust signal that reinforces food selections independently of taste signaling in the oral cavity. The figure displays some of the pre- and postabsorptive signals that might influence brain dopamine activity, thereby acting as true reinforcers. The figure displays on the top a coronal slice of the mouse brain that includes ventral midbrain areas. Regions of the mouse ventral midbrain containing dopamine-producing neurons correspond to the gray regions in the figure and include the ventral tegmental area (VTA, light gray delimited by black dashed borders) and the pars compacta of the substantia nigra (SNc, dark gray delimited by thick continuous borders). Obviously, it must be noted that these pathways also converge into other brain regions involved in the control of food intake, including the hypothalamus. AAs: amino acids.

cells is required for the generation of behaviorally relevant postingestive reward signals. Of particular interest is whether the blockage of glucose utilization in midbrain dopamine regions is sufficient to disrupt the formation of food preferences based on postingestive effects. Future research must determine the extent to which dopamine neurons of the brain reward pathways operate as true metabolic sensors, and whether metabolic sensing can influence dopamine-associated functions such as temporal difference learning and reward prediction (Hollerman and Schultz 1998; also see Chapters 14 and 15 in this volume).

It must be observed that none the of physiological mechanisms that might plausibly function as postingestive reward signals can offer an explanation for how the brain forms associations between sensory information that arises from flavor inputs and ensuing metabolic events taking place several minutes or even hours later. Although research on this topic remains to be undertaken, experiments involving aversive postingestive effects including conditioned taste aversion paradigms suggest that gustatory cortical circuits might play a fundamental role in actively sustaining sensory information in order to allow the formation of neural associations with ensuing postingestive effects (Berman and Dudai 2001). The precise role of the gustatory cortex in the regulation of postingestive reinforcement is yet another important topic for future investigation.

As a final comment, while the research presented here has overwhelmingly focused on vertebrate species, this should not be taken to imply that food reward research is limited to these organisms. On the contrary, the involvement of dopamine and other amines in reward and punishment signaling is equally important for invertebrate species. In insects such as *Apis mellifera*, *Gryllus bimaculatus*, and *Drosophila melanogaster*, the two biogenic amines dopamine and octopamine seem to be involved in punishment and reward learning, respectively (Schwaerzel et al. 2003; Unoki, Matsumoto, and Mizunami 2005; Selcho et al. 2009). More recently, it has been suggested that dopamine signaling via the dD1 receptor is essential for both reward and aversive processing in *Drosophila* (Kim, Lee, and Han 2007). Altogether, the current evidence suggests conserved dopaminergic mechanisms for reward processing across different species. In addition, the *Drosophila* analogue of the mammalian tyrosine hydroxylase enzyme catalyzes the rate-limiting step of dopamine synthesis and is expressed in all dopaminergic neurons (Friggi-Grelin et al. 2003), a finding that further indicates conserved mechanisms in reward learning. Future research must determine whether, in addition to olfactory cues conditioned to primary rewards, gustatory inputs and/or changes in physiological state have the ability to regulate dopamine release in *Drosophila*. In any event, the presence of dopamine-dependent reward processing in *Drosophila* reveals the opportunity for thorough investigations of the molecular bases of dopaminergic system sensitivity to metabolic cues.

REFERENCES

Ackroff, K., Sclafani, A., and Axen, K.V. 1997. Diabetic rats prefer glucose-paired flavors over fructose-paired flavors. *Appetite* 28:73–83.

Adolph, E.F. 1947. Urges to eat and drink in rats. *Am J Physiol* 151:110–25.

Bailey, C.S., Hsiao, S., and King, J.E. 1986. Hedonic reactivity to sucrose in rats: Modification by pimozide. *Physiol Behav* 38:447–52.

Berman, D.E., and Dudai, Y. 2001. Memory extinction, learning anew, and learning the new: Dissociations in the molecular machinery of learning in cortex. *Science* 291:2417–19.

Bezençon, C., le Coutre, J., and Damak, S. 2007. Taste-signaling proteins are coexpressed in solitary intestinal epithelial cells. *Chem Senses* 32:41–49.

Bolles, R., Hayward, L., and Crandall, C. 1981. Conditioned taste preferences based on caloric density. *J Exp Psychol Anim Behav Process* 7:59–69.

Booth, D.A. 1972. Satiety and behavioral caloric compensation following intragastric glucose loads in the rat. *J Comp Physiol Psychol* 72:412–32.

Boudreau, J.C., Sivakumar, L., Do, L.T., White, T.D., Oravec, J., and Hoang, N.K. 1985. Neurophysiology of geniculate ganglion (facial nerve) taste systems: Species comparisons. *Chem Sens* 10:89–127.

Bouzier-Sore, A.K., Voisin, P., Bouchaud, V., Bezancon, E., Franconi, J.M., and Pellerin, L. 2006. Competition between glucose and lactate as oxidative energy substrates in both neurons and astrocytes: A comparative NMR study. *Eur J Neurosci* 24:1687–94.

Caicedo, A., Kim, K.N., and Roper, S.D. 2002. Individual mouse taste cells respond to multiple chemical stimuli. *J Physiol* 544:501–9.

Chrandrashekar, J., Mueller, K.L., Hoon, M.A., Adler, E., Feng, L., Guo, W., Zucker, C.S., and Ryba, N.J.P. 2000. T2Rs function as bitter taste receptors. *Cell* 100:703–11.

Cohen, A.M., and Teitelbaum, A. 1964. Effect of dietary sucrose and starch on oral glucose tolerance and insulin-like activity. *Am J Physiol* 206:105–8.

Damak, S., Mosinger, B., and Margolskee, R.F. 2008. Transsynaptic transport of wheat germ agglutinin expressed in a subset of type II taste cells of transgenic mice. *BMC Neurosci* 9:96.

de Araujo, I.E., Oliveira-Maia, A.J., Sotnikova, T.D., Gainetdinov, R.R., Caron, M.G., Nicolelis, M.A., and Simon, S.A. 2008. Food reward in the absence of taste receptor signaling. *Neuron* 57:930–41.

Di Chiara, G., and Bassareo, V. 2007. Reward system and addiction: What dopamine does and doesn't do. *Curr Opin Pharm* 7:69–76.

Di Lorenzo, P.M. 2003. The neural code for taste in the brainstem: Response profiles. *Physiol Behav* 69:87–96.

Dyer, J., Salmon, K.S., Zibrik, L., and Shirazi-Beechey, S.P. 2005. Expression of sweet taste receptors of the T1R family in the intestinal tract and enteroendocrine cells. *Biochem Soc Trans* 33:302–5.

Dyer, J., Vayro, S., King, T.P., and Shirazi-Beechey, S.P. 2003. Glucose sensing in the intestinal epithelium. *Eur J Biochem* 270:3377–88.

Egan, J.M., and Margolskee, R.F. 2008. Taste cells of the gut and gastrointestinal chemosensation. *Mol Interv* 8:78–81.

Egli, G., Langhans, W., and Scharrer, E. 1986. Selective hepatic vagotomy does not prevent compensatory feeding in response to body weight changes. *J Auton Nerv Syst* 15:45–53.

Elizalde, G., and Sclafani, A. 1988. Starch-based conditioned flavor preferences in rats: Influence of taste, calories and CS-US delay. *Appetite* 11:179–200.

———. 1990. Flavor preferences conditioned by intragastric polycose infusions: A detailed analysis using an electronic esophagus preparation. *Physiol Behav* 47:63–77.

Epstein, A. 1967. Feeding without oropharyngeal sensations. In *The Chemical Senses and Nutrition*, eds. M.R. Kare, and O. Maller, pp. 263–280. Baltimore: The Johns Hopkins Press.

Epstein, A., and Teitelbaum, P. 1962. Regulation of food intake in the absence of taste, smell and other oropharyngeal sensations. *J Comp Physiol Psychol* 55:753–59.

Erickson, R.P. 2001. The evolution and implications of population and modular neural coding ideas. In *Advances in Neural Population Coding*, ed. M.A.L. Nicolelis, pp. 9–29. Amsterdam: Elsevier.

Fedorchak, P.M., and Bolles, R.C. 1987. Hunger enhances the expression of calorie – but not taste-mediated conditioned flavor preferences. *J Exp Psychol Anim Behav Process* 13:73–79.

Figlewicz, D.P. 2003. Insulin, food intake, and reward. *Semin Clin Neuropsychiatry* 8:82–93.

Figlewicz, D.P., Bennett, J.L., Naleid, A.M., Davis, C., and Grimm, J.W. 2006. Intraventricular insulin and leptin decrease sucrose self-administration in rats. *Physiol Behav* 89:611–16.

Figlewicz, D.P., Evans, S.B., Murphy, J., Hoen, M., and Baskin, D.G. 2003. Expression of receptors for insulin and leptin in the ventral tegmental area/substantia nigra (VTA/SN) of the rat. *Brain Res* 964:107–15.

Frank, M.F., Bieber, S.L., and Smith, D.V. 1988. The organization of taste sensibilities in hamster chorda tympani nerve fibers. *J Gen Physiol* 91:861–96.

Friedman, M.I. 1989. Metabolic control of food intake. *Bol Asoc Med PR* 81:111–13.

———. 1991. Metabolic control of caloric intake. In *Chemical Senses, Vol. 4: Appetite and Nutrition*, eds. M.I. Friedman, M.G. Tordoff, and M.R. Kare, 19–38. New York: Marcel Dekker.

———. 2007. Obesity and the hepatic control of feeding behavior. *Drug News Perspect* 20:573–78.

Friedman, M.I., Harris, R.B., Ji, H., Ramirez, I., and Tordoff, M.G. 1999. Fatty acid oxidation affects food intake by altering hepatic energy status. *Am J Physiol* 276:R1046–53.

Friedman, M.I., and Ramirez, I. 1985. Relationship of fat metabolism to food intake. *Am J Clin Nutr* 42:1093–98.

Friedman, M.I., Ramirez, I., and Tordoff, M.G. 1996. Gastric emptying of ingested fat emulsion in rats: Implications for studies of fat-induced satiety. *Am J Physiol* 270:R688–92.

Friedman, M.I., and Sawchenko, P.E. 1984. Evidence for hepatic involvement in control of ad libitum food intake in rats. *Am J Physiol* 247:R106–113.

Friedman, M.I., and Tordoff, M.G. 1986. Fatty acid oxidation and glucose utilization interact to control food intake in rats. *Am J Physiol* 251:R840–45.

Friggi-Grelin, F., Coulom, H., Meller, M., Gomez, D., Hirsh, J., and Birman, S. 2003. Targeted gene expression in Drosophila dopaminergic cells using regulatory sequences from tyrosine hydroxylase. *J Neurobiol* 54:618–27.

Fulton, S., Pissios, P., Manchon, R.P., Stiles, L., Frank, L., Pothos, E.N., Maratos-Flier, E., and Flier, J.S. 2006. Leptin regulation of the mesoaccumbens dopamine pathway. *Neuron* 51:811–22.

Fulwiler, E.E., and Saper, C.B. 1984. Subnuclear organization of the efferent connections of the parabrachial nucleus in the rat. *Brain Res Rev* 7:229–59.

Ganchrow, J.R., Steiner, J.E., and Daher, M. 1983. Neonatal facial expressions in response to different qualities and intensities of gustatory stimuli. *Infant Behav Dev* 6:473–84.

Geary, N., and Smith, G.P. 1985. Pimozide decreases the positive reinforcing effect of sham fed sucrose in the rat. *Pharmacol Biochem Behav* 22:787–90.

Gerber, B., and Stocker, R.F. 2007. The Drosophila larva as a model for studying chemosensation and chemosensory learning: A review. *Chem Sens* 32:65–89.

Gottfried, J.A., O'Doherty, J., and Dolan, R.J. 2003. Encoding predictive reward value in human amygdala and orbitofrontal cortex. *Science* 301:1104–7.

Hajnal, A., Smith, G.P., and Norgren, R. 2004. Oral sucrose stimulation increases accumbens dopamine in the rat. *Am J Physiol Regul Integr Comp Physiol* 286:R31–R37.

Hall, W.G., and Bryan, T.E. 1981. The ontogeny of feeding in rats: IV. Taste development as measured by intake and behavioral responses to oral infusions of sucrose and quinine. *J Comp Physiol Psychol* 95:240–51.

Hernandez, L., and Hoebel, B.G. 1988. Food reward and cocaine increase extracellular dopamine in the nucleus accumbens as measured by microdialysis. *Life Sci* 42:1705–12.

Hofer, D., Puschel, B., and Drenckhahn, D. 1996. Taste receptor-like cells in the rat gut identified by expression of alpha-gustducin. *Proc Natl Acad Sci USA* 93:6631–34.

Hollerman, J.R., and Schultz, W. 1998. Dopamine neurons report an error in the temporal prediction of reward during learning. *Nature Neurosci* 1:304–9.

Holman, G.L. 1968. Intragastric reinforcement effect. *J Comp Physiol Psychol* 69:432–41.

Hommel, J.D., Trinko, R., Sears, R.M., Georgescu, D., Liu, Z.W., Gao, X.B., Thurmon, J.J., Marinelli, M., and DiLeone, R.J. 2006. Leptin receptor signaling in midbrain dopamine neurons regulates feeding. *Neuron* 51:801–10.

Houpt, K.A., Houpt, T.R., and Pond, W.G. 1977. Food intake controls in the suckling pig: Glucoprivation and gastrointestinal factors. *Am J Physiol* 232:E510–14.

Jang, H.J., Kokrashvili, Z., Theodorakis, M.J., Carlson, O.D., Kim, B.J., Zhou, J., Kim, H.H., et al. 2007. Gut-expressed gustducin and taste receptors regulate secretion of glucagon-like peptide-1. *Proc Natl Acad Sci USA* 104:15069–74.

Jordan, H.A. 1969. Voluntary intra-gastric feeding: Oral and gastric contributions to food intake and hunger in humans. *J Comp Physiol Psychol* 68:498–506.

Kare, M.R. 1971. Comparative study of taste. In *Handbook of Sensory Physiology*, ed. L.M. Beidler, pp. 278–292. Berlin: Springer-Verlag.

Katz, D.B., Simon, S.A., and Nicolelis, M.A. 2001. Dynamic and multimodal responses of gustatory cortical neurons in awake rats. *J Neurosci* 21:4478–89.

Kelley, A.E., Schiltz, C.A., and Landry, C.F. 2005. Neural systems recruited by drug- and food-related cues: Studies of gene activation in corticolimbic regions. *Physiol Behav* 86:11–14.

Kim, Y.C., Lee, H.G., and Han, K.A. 2007. D1 dopamine receptor dDA1 is required in the mushroom body neurons for aversive and appetitive learning in Drosophila. *J Neurosci* 27:7640–47.

Langhans, W., Egli, G., and Scharrer, E. 1985a. Selective hepatic vagotomy eliminates the hypophagic effect of different metabolites. *J Auton Nerv Syst* 13:255–62.

———. 1985b. Regulation of food intake by hepatic oxidative metabolism. *Brain Res Bull* 15:425–28.

Lemon, C.H., and Katz, D.B. 2007. The neural processing of taste. *BMC Neurosci* 8 Suppl 3:S5.

Levin, B.E. 2006. Metabolic sensing neurons and the control of energy homeostasis. *Physiol Behav* 89:486–89.

Li, C.S., Cho, Y.K., and Smith, D.V. 2005. Modulation of parabrachial taste neurons by electrical and chemical stimulation of the lateral hypothalamus and amygdala. *J Neurophys* 93:1183–96.

Margolis, R., and Altszuler, N. 1967. Insulin in the cerebrospinal fluid. *Nature* 215:1375–76.

Margolskee, R.F., Dyer, J., Kokrashvili, Z., Salmon, K.S., Ilegems, E., Daly, K., Maillet, E.L., Ninomiya, Y., Mosinger, B., and Shirazi-Beechey, S.P. 2007. T1R3 and gustducin in gut sense sugars to regulate expression of Na+-glucose cotransporter 1. *Proc Natl Acad Sci USA* 104:15075–80.

Mather, P., Nicolaidis, S., and Booth, D.A. 1978. Compensatory and conditioned feeding responses to scheduled glucose infusions in the rat. *Nature* 273:461–63.

McLaughlin, S.K., McKinnon, P.J., and Margolskee, R.F. 1992. Gustducin is a taste-cell specific G protein closely related to transducins. *Nature* 357:563–69.

Miller, N.E., and Kessen, M.L. 1952. Reward effects of food via stomach fistula compared with those of food via mouth. *J Comp Physiol Psychol* 45:555–64.

Miselis, R.R., and Epstein, A. 1975. Feeding induced by intracerebroventricular 2-deoxy-D-glucose in the rat. *Am J Physiol* 29:1438–47.

Moran, T.H., Wirth, J.B., Schwartz, G.J., and McHugh, P.R. 1999. Interactions between gastric volume and duodenal nutrients in the control of liquid gastric emptying. *Am J Physiol* 276:R997–R1002.

Mueller, K.L., Hoon, M.A., Erlenbach, I., Chandrashekar, J., Zuker, C.S., and Ryba, N.J.P. 2005. The receptors and coding logic for bitter taste. *Nature* 434:225–29.

Mungarndee, S.S., Lundy, R.F., Caloiero, V.G., and Norgren, R. 2004. Forebrain c-fos expression following sham exposure to sucrose after central gustatory lesions: A quantitative study. Abstract from the Society of Neuroscience 2004 Annual Meeting.

Myers, K.P., and Sclafani, A. 2006. Development of learned flavor preferences. *Dev Psychobiol* 48:380–88.

Naim, M., Brand, J.G., Christensen, C.M., Kare, M.R., and Van Buren, S. 1986. Preference of rats for food flavors and texture in nutritionally controlled semi-purified diets. *Physiol Behav* 37:15–21.

Naim, M., Brand, J.G., and Kare, M.R. 1987. The preference-aversion behavior of rats for nutritionally-controlled diets containing oil or fat. *Physiol Behav* 39:285–90.

Naim, M., Brand, J.G., Kare, M.R., and Carpenter, R.G. 1985. Energy intake, weight gain and fat deposition in rats fed flavored, nutritionally controlled diets in a multichoice ("cafeteria") design. *J Nutr* 115:1447–58.

Naim, M., and Kare, M.R. 1991. Sensory and postingestional components of palatability in dietary obesity: An overview. In *Chemical Senses*, eds. M.I. Friedman, M.G. Tordoff, and M.R. Kare, pp. 109–26. New York: Marcel Dekker.

Norgren, R. 1976. Taste pathways to hypothalamus and amygdala. *J Comp Neurol* 166:17–30.

Norgren, R., and Hajnal, A. 2005. Taste pathways that mediate accumbens dopamine release by sapid sucrose. *Physiol Behav* 84:363–69.

Norgren, R., Hajnal, A., and Mungarndee, S.S. 2006. Gustatory reward and the nucleus accumbens. *Physiol Behav* 89:531–35.

Norgren, R., and Leonard, C.M. 1971. Taste pathways in rat brainstem. *Science* 173:1136–39.

———. 1973. Ascending central gustatory pathways. *J Comp Neurol* 150:217–38.

Norgren, R., and Pfaffmann, C. 1975. The pontine taste area in the rat. *Brain Res* 91:99–117.

Norgren, R., and Wolf, G. 1975. Projections of thalamic gustatory and lingual areas in the rat. *Brain Res* 92:123–29.

Ogawa, H., Hayama, T., and Ito, S. 1987. Response properties of the parabrachio-thalamic taste and mechano-receptive neurons in rats. *Brain Res* 68:449–57.

Oltmanns, K.M., Melchert, U.H., Scholand-Engler, H.G., Howitz, M.C., Schultes, B., Schweiger, U., Hohagen, F., Born, J., Peters, A., and Pellerin, L. 2008. Differential energetic response of brain vs. skeletal muscle upon glycemic variations in healthy humans. *Am J Physiol Regul Integr Comp Physiol* 294:R12–16.

Pardini, A.W., Nguyen, H.T., Figlewicz, D.P., Baskin, D.G., Williams, D.L., Kim, F., and Schwartz, M.W. 2006. Distribution of insulin receptor substrate-2 in brain areas involved in energy homeostasis. *Brain Res* 1112:169–78.

Pellerin, L. 2008. Brain energetics (thought needs food). *Curr Opin Clin Nutr Metab Care* 11:701–5.

Perez, C.A., Huang, L., Rong, M., Kozak, J.A., Preuss, A.K., Zhang, H., Max, M., and Margolskee, R.F. 2002. A transient receptor potential channel expressed in taste receptor cells. *Nat Neurosci* 5:1169–76.

Perez, C., Lucas, F., and Sclafani, A. 1998. Devazepide, a CCK(A) antagonist, attenuates the satiating but not the preference conditioning effects of intestinal carbohydrate infusions in rats. *Pharmacol Biochem Behav* 59:451–57.

Perez, C., and Sclafani, A. 1991. Cholecystokinin conditions flavor preferences in rats. *Am J Physiol* 260:R179–85.

Perrotto, R.S., and Scott, T.R. 1976. Gustatory neural coding in the pons. *Brain Res* 110:283–300.

Plum, L., Belgardt, B.F., and Bruning, J.C. 2006. Central insulin action in energy and glucose homeostasis. *J Clin Invest* 116:1761–66.

Reilly, C.A. 1999. The parabrachial nucleus and conditioned taste aversion. *Brain Res Bull* 48:239–54.

Russek, M. 1970. Demonstration of the influence of an hepatic glucosensitive mechanism on food intake. *Physiol Behav* 5:1207–9.

———. 1981. Current status of the hepatostatic theory of food intake control. *Appetite* 2:137–43.

Schwaerzel, M., Monastirioti, M., Scholz, H., Friggi-Grelin, F., Birman, S., and Heisenberg, M. 2003. Dopamine and octopamine differentiate between aversive and appetitive olfactory memories in Drosophila. *J Neurosci* 23:10495–10502.

Sclafani, A. 2001. Post-ingestive positive controls of ingestive behavior. *Appetite* 36:79–83.

Sclafani, A., Fanizza, L.J., and Azzara, A.V. 1999. Conditioned flavor avoidance, preference, and indifference produced by intragastric infusions of galactose, glucose, and fructose in rats. *Physiol Behav* 67:227–34.

Sclafani, A., and Lucas, F. 1996. Abdominal vagotomy does not block carbohydrate-conditioned flavor preferences in rats. *Physiol Behav* 60:447–53.

Sclafani, A., and Vigorito, M. 1987. Effects of SOA and saccharin adulteration on polycose preference in rats. *Neurosci Biobehav Rev* 11:163–68.

Sclafani, A., and Xenakis, S. 1984. Sucrose and polysaccharide induced obesity in the rat. *Physiol Behav* 32:169–74.

Scott, T.R., and Small, D.M. 2009. The role of the parabrachial nucleus in taste processing and feeding. *Ann N Y Acad Sci* 1170:372–77.

Selcho, M., Pauls, D., Han, K.A., Stocker, R.F., and Thum, A.S. 2009. The role of dopamine in Drosophila larval classical olfactory conditioning. *PloS One* 4:e5897.

Small, D.M., Jones-Gotman, M., and Dagher, A. 2003. Feeding-induced dopamine release in dorsal striatum correlates with meal pleasantness ratings in healthy human volunteers. *NeuroImage* 19:1709–15.

Snowdon, C.T., and Epstein, A. 1970. Oral and intragastric feeding in vagotomized rats. *J Comp Physiol Psychol* 71:59–67.

Sokoloff, L. 1989. Circulation and energy metabolism of the brain. In *Basic Neurochemistry*, eds. G. Siegel, R.W. Albers, and P. Molinoff, pp. 565–590. New York: Raven Press.

Tordoff, M.G. 1991. Metabolic basis of learned food preferences. In *Chemical Senses, Vol. 4: Appetite and Nutrition*, eds. M.I. Friedman, M.G. Tordoff, and M.R. Kare, 239–60. New York: Marcel Dekker.

Tordoff, M.G., and Friedman, M.I. 1986. Hepatic portal glucose infusions decrease food intake and increase food preference. *Am J Physiol* 251:R192–96.

———. 1988. Hepatic control of feeding: Effect of glucose, fructose, and mannitol infusion. *Am J Physiol* 254:R969–76.

———. 1994. Altered hepatic metabolic response to carbohydrate loads in rats with hepatic branch vagotomy or cholinergic blockade. *J Auton Nerv Syst* 47:255–61.

Tordoff, M.G., Rawson, N., and Friedman, M.I. 1991. 2,5-anhydro-D-mannitol acts in liver to initiate feeding. *Am J Physiol* 261:R283–88.

Touzani, K., Bodnar, R., and Sclafani, A. 2008. Activation of dopamine D1-like receptors in nucleus accumbens is critical for the acquisition, but not the expression, of nutrient-conditioned flavor preferences in rats. *Eur J Neurosci* 27:1525–33.

Tsang, Y.C. 1938. Hunger motivation in gastrectomized rats. *J Comp Physiol Psychol* 26:1–17.

Unoki, S., Matsumoto, Y., and Mizunami, M. 2005. Participation of octopaminergic reward system and dopaminergic punishment system in insect olfactory learning revealed by pharmacological study. *Eur J Neurosci* 22:1409–16.

Van Houten, M., Posner, B.I., Kopriwa, B.M., and Brawer, J.R. 1979. Insulin-binding sites in the rat brain: In vivo localization to the circumventricular organs by quantitative radioautography. *Endocrinology* 105:666–73.

Vanderweele, D.A., Oetting, R.L., Jones, R.E., and Deems, D.A. 1985. Sham feeding, flavor associations and diet self-selection as indicators of feeding satiety or aversive effects of peptide hormones. *Brain Res Bull* 14:529–35.

Wick, A.N., Drury, D.R., Nakada, H.I., and Wolfe, J.B. 1957. Localization of the primary metabolic block produced by 2-deoxyglucose. *J Biol Chem* 224:963–69.

Wise, R.A. 2006. Role of brain dopamine in food reward and reinforcement. *Philos Trans R Soc Lond B Biol Sci* 361:1149–58.

Woods, S., Lotter, E., McKay, L., and Porte, D. 1979. Chronic intracerebroventricular infusion of insulin reduces food intake and body weight of baboons. *Nature* 282:503–5.

Woods, S., and Porte, D. 1977. Relationship between plasma and cerebrospinal fluid insulin levels of dogs. *Am J Physiol* 233:E331–34.

Wu, S.V., Rozengurt, N., Yang, M., Young, S.H., Sinnett-Smith, J., and Rozengurt, E. 2002. Expression of bitter taste receptors of the T2R family in the gastrointestinal tract and enteroendocrine STC-1 cells. *ProcNatl Acad USA* 99:2392–97.

Xenakis, S., and Sclafani, A. 1981. The effects of pimozide on the consumption of a palatable saccharin-glucose solution in the rat. *Pharmacol Biochem Behav* 15:435–42.

Zhang, Y., Hoon, M.A., Chandrashekar, J., Mueller, K.L., Cook, B.W.D., Zucker, C.S., and Ryba, N.J. 2003. Coding of sweet, bitter, and umami tastes: Different receptor cells sharing similar signaling pathways. *Cell* 112:293–301.

Zhao, G.Q., Zhang, Y., Hoon, M.A., Chandrashekar, J., Erienbach, I., Ryba, N.J.P., and Zuker, C.S. 2003. The receptors for mammalian sweet and umami taste. *Cell* 115:255–66.

Zukerman, S., Glendinning, J.I., Margolskee, R.F., and Sclafani, A. 2008. T1R3 taste receptor is critical for sucrose but not Polycose taste. *Am J Physiol Regul Integr Comp Physiol* 296:R866–76.

13 Sensation, Incentive Learning, and the Motivational Control of Goal-Directed Action

Bernard W. Balleine

CONTENTS

13.1 INTRODUCTION

Recent analyses of goal-directed action have not only pointed to the importance of the learning processes through which actions and their consequences are encoded, but have also emphasized the performance factors that influence choice between actions and action initiation more generally (Dickinson and Balleine 2002; Hasselmo 2005; Balleine and Ostlund 2007). It is important to understand why this is the case. Although the value of acting—or not acting—can often appear to be obvious enough from an adaptive perspective, in fact the information that can be derived from the action-outcome association is not sufficient to determine a course of action; knowing that an action results in a particular outcome does not entail whether that action should be performed or not. Although it might appear adaptive to perform food-related actions more frequently when food deprived and less frequently when replete, this need not necessarily occur and often doesn't, something well documented in cases of eating disorders (Davis et al. 2004; Morrison and Berthoud 2007). In fact, what determines whether a specific action will be both selected *and* subsequently initiated is not just the identity but also the evaluation of the outcome associated with the selected action.

In order to decide on a course of action, therefore, both the consequences and the value of the consequences of various alternative actions need to be specified. Therefore, establishing the determinants of a course of action, whether in psychological or neural terms, requires an account of (i) the learning processes through which action-outcome associations are encoded and (ii) the motivational and emotional processes that establish the value of the consequences or outcomes of actions. In this chapter both of these processes will be described further although, because there have been a number of recent reviews of the learning processes underlying goal-directed action (Balleine, Delgado, and Hikosaka 2007; Balleine, Liljeholm, and Ostlund 2009; Balleine and Ostlund 2007; Yin, Ostlund, and Balleine 2008; Balleine and O'Doherty 2010), I will mainly focus on evidence relating to the behavioral and neural bases of incentive learning, i.e., the process by which we and other animals assign value to the consequences or goals of goal-directed actions. To provide the basis for the presentation of current research on this issue I will focus first on evaluative conditioning, which constitutes the motivational basis of affective processes generally, and then turn to animal models of goal-directed action to describe how this basic evaluative process is elaborated into an incentive learning process in the service of this capacity. I will then take up the issue of the neural bases of incentive learning and describe some recent research on this issue in the final section.

13.2   EVALUATIVE CONDITIONING

Since Pavlov (1927), students of learning have commonly divided sensory/perceptual events into those that are conditioned and those that are unconditioned. However, Pavlov made this distinction solely on the basis of the *behavioral* response that stimuli evoke on first presentation, whereas, in line with the increasingly cognitive emphasis of contemporary analyses, this distinction has largely given way to one based on the activation of event "representations." It is, therefore, an implication of current analyses that unconditioned responses (URs) are a function of the activation of the "representation" of the unconditioned stimulus (US). Although this shift in emphasis might appear quite subtle, the addition of this "representational" assumption has some unexpected consequences. Any UR worthy of the name should be evoked on the first presentation of the US *without the need for learning*, something that fits well enough with Pavlov's distinction between conditioning and URs. However, if, given the representational assumption, URs are determined by activation of the US representation, then these "representations" *must necessarily predate* the production of the UR and, therefore, also be innate.

Pavlov himself proposed that the US "representation," whatever form one might suppose it takes, follows rather than precedes the production of the UR. Although salivation to food placed in the mouth is immediate, Pavlov (1927, 23) suggested that "the effect of the sight and smell of food is not due to an inborn reflex, but to a reflex which has been acquired in the course of the animal's own individual existence." The cognitive perspective conflates, therefore, what Pavlov regarded as unconditioned about USs, the UR that they elicit on first contact, with what he argued is not, the association between their sensory-perceptual features (i.e., their taste, smell, visual, auditory, or textural features) and the physiologically based motivational systems activated by the detection of such things as calories, fluids, pains, and so on that supports their representation as independent events. For Pavlov, therefore, the motivating effect of the perceptual features of events capable of evoking URs was only established once the animal had experienced these features together with motivational activity, a process he referred to as "signalization" (ibid.).

At the time, there was little evidence to support this claim: Pavlov cited an unpublished experiment in which young dogs were found only to salivate to the sight of bread or meat after they had first been fed on these commodities. More recently, however, a more substantial literature on this subject has emerged referring not to signalization but to a process of "evaluative" conditioning, i.e., that learning process through which the motivational and affective properties of the US (and, by association, of the conditioned stimuli [CS]) are established. References to evaluative conditioning have, occasionally, surfaced in the past, although cloaked in quite different terms, for example, in the work of Young (1949) and in analyses of research in the 1940s and 1950s on what came to be called externalized or acquired drive (Bolles 1967). Additionally, Moll (1964) reports evidence consistent with Pavlov's "signalization" process in young rats. On their first experience with food deprivation Moll's rats ate substantially less than was required to make up their deficit or even to maintain them, although they rapidly learned to increase consumption over time and over presentations of the food. Similarly, Changizi, McGehee, and Hall (2002) observed that rat pups did not exhibit food-seeking behavior when food deprived unless they had previous experience with food deprivation and eating. Perhaps more surprisingly, Hall and colleagues also found that the same is true of water for thirsty rats (Hall, Arnold, and Myers 2000; Changizi, McGehee, and Hall 2002). Thus, for example, in pre-weanling rats or rats weaned onto a fluid diet, the induction of a strong, extra-cellular thirst was observed to have no immediate effect on water consumption relative to rats not made thirsty. After experience with water in the thirsty state had been allowed, however, subsequent induction of thirst produced an immediate increase in water consumption. The representation of specific foods and fluids as biologically significant for hungry or thirsty rats appears, therefore, to be acquired.

The procedures used to assess evaluative conditioning have obvious similarities to those used to generate conditioned taste preferences and aversions. With regard to the former, rats are generally first deprived of some essential commodity or other (e.g., nutrients, fluids, or, more specifically, sodium or calcium, etc.) after which a stimulus (usually a taste) is paired with the delivery of the deprived commodity, presented either in solution with the taste or via intragastric, intraduodenal, hepatic portal, or intravenous routes. Evidence for evaluative conditioning would be established if, relative to rats given the taste and the infusion of the deprived commodity unpaired, the paired group significantly increases their willingness to contact and consume the taste (Sclafani 1999). It has also been reported that treatments such as these increase the tendency of rats to show ingestive, orofacial fixed action patterns (FAPs) when the paired taste is contacted (Forestell and Lolordo 2003). Deprivation of one or other commodity appears to be necessary to generate conditioned taste preferences of this kind (Harris et al. 2000), suggesting that evaluative conditioning is modulated by visceral and humoral signals originating in regulatory processes such as those that control feeding, drinking, and so on (Sudakov 1990). Indeed, studies that have specifically manipulated deprivation states report, for example, that the acquisition of taste preferences by nutrient loads is strongly controlled by the degree of food deprivation (Harris et al. 2000). Studies of orofacial FAPs confirm that these reactions to taste stimuli are also modulated by motivational state. The taste reactivity patterns elicited by sugar solutions are augmented by hunger (Berridge 1991b), whereas those elicited by saline are enhanced by sodium appetite (Berridge et al. 1984).

Garcia (Garcia 1989; Garcia, Brett, and Rusiniak 1989) has argued that conditioned taste aversions too are best viewed as an example of evaluative conditioning (what he called "Darwinian conditioning"). This procedure involves the pairing of a, usually sweet, taste with the injection of an emetic agent, such as lithium chloride (LiCl). Subsequently, both orofacial FAPs shift from acceptance to those associated with rejection (Berridge 2000b) and the consumption of substances that contain that taste is strongly and enduringly altered by this pairing. Garcia, Brett, and Rusiniak (1989) argue that this effect reflects the formation of an association between taste afferents and brain stem autonomic centers that subsequent work has identified as the parabrachial nucleus for conditioned aversions (Reilly 1999) as well as, interestingly enough, for conditioned preferences (Sclafani et al. 2001). The site of integration appears to differ for evaluative conditioning involving olfactory, visual, auditory, and somatosensory features, and likely involves the amygdala (Holland, Petrovich, and Gallagher 2002) along with its afferents in the sensory cortex, brain stem, and hypothalamic nuclei.

This analysis is consistent with the view, illustrated in Figure 13.1, that evaluative conditioning reflects an association of the sensory features of an event (Se) with basic motivational processes (M), such as those that detect nutrients, salts, fluids, etc., to produce a primary incentive (or π), e.g., a food or drink of a particular kind. Interestingly, this form of conditioning appears to alter the evaluative significance of specific sensory events in a manner that appears not to

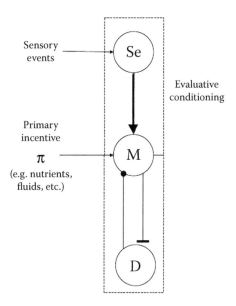

FIGURE 13.1   Schematic illustration of the associative processes proposed to mediate evaluative conditioning. Primary incentives (π) activate intrinsic motivational processes (M), themselves modulated by deprivation (or drive, D) conditions. Sensory changes (Se) contiguous with motivational activation associated with a specific sensory event (e.g., a particular kind of food, threat, etc.) strengthen connections between these systems, altering the subsequent evaluation of that sensory event.

depend on simple predictive learning.* As illustrated in Figure 13.1, and outlined previously (Balleine 2004)—but beyond the current analysis—the effect of this connection is itself directly affected by factors that modulate M (among which are primary drive states, D, such as hunger, thirst, etc.).

It is important to note that this analysis points to an additional, independent, and more fundamental learning process that precedes Pavlovian conditioning to establish the representation of the US on which that predictive learning is based. Within contemporary theorizing, it is not uncommon to read accounts of Pavlovian processes framed in informational terms; i.e., the CS "predicts" or "signals" the US or engenders an "expectancy" of the US (Rescorla 1973). But, given Pavlov's own analysis, it is clear that much of this talk of informational variables in Pavlovian conditioning must be predicated upon this more fundamental, evaluative learning process through which the representation of the specific US about which the cue provides information is acquired. However, if evaluative processes are critical to the acquisition and expression of predictive relations involving the US representation in Pavlovian conditioning, they are even more critical in determining what is learned in instrumental conditioning.

---

* In addition to these sources of evidence for evaluative conditioning, a substantial and controversial literature has emerged in humans when a specific sensory cue is paired with an event that produces a potent change in motivational and affective state (De Houwer, Thomas, and Baeyens 2001; De Houwer, Baeyens, and Field 2005). This literature is controversial particularly with respect to whether changes in responding to the cue reflect a change in evaluation or merely the encoding of new predictive information; i.e., whether the evaluation of the cue is changed because of what it predicts rather than because it has itself changed value (Pleyers et al. 2007). Generally, it is the latter finding, i.e., evidence of evaluative changes in situations where subjects are unable to retrieve predictive information (Baeyens et al. 1990; Dickinson and Brown 2007), that fits best with the current analysis and with animal experiments; i.e., with the view that evaluative conditioning reflects the relationship between an explicit sensory event and an active motivational/affective process that is not explicitly represented.

13.3 INSTRUMENTAL CONDITIONING

Although the ability to anticipate forthcoming events is clearly a highly adaptive capacity, predictive learning is not sufficient for successful adaptation in a changing environment. Although predictive learning can provide animals with the capacity to elicit anticipatory responses as a result of learning about associations between events (such as those arranged in Pavlovian conditioning), as argued above, the adaptive form of these responses is clearly determined by evolutionary processes rather than individual learning. As a consequence, an animal that can engage only in predictive learning is at the mercy of the stability of the causal consequences of these responses. In order for responses to remain adaptive in an unstable environment, an animal must be capable of modifying its behavioral repertoire in the face of changing environmental contingencies, i.e., it must be capable of learning to control responses *instrumental* to gaining access to sources of benefit and avoiding events that can maim or kill. Thus, this form of learning is often referred to as instrumental conditioning.

As has been described in detail in previous reviews (e.g., Dickinson and Balleine 2002; Balleine and Ostlund 2007), early conceptualizations of the instrumental learning process regarded it as a form of acquired reflex and, as such, emphasized the stimulus conditions within which specific adaptive responses were performed. (See Chapter 2 in this volume for a historical overview of this area.) Those responses resulting in appropriate feedback—"satisfaction" for Thorndike (1911) or drive reduction for Hull (1943)—were thought to become more strongly associated with those stimulus conditions with the feedback serving to reinforce that stimulus-response (S-R) association. According to this view, therefore, the outcome of an instrumental action does not enter into the associative structure controlling performance of the action and acts merely as a catalyst to help cement the association between stimulus and response. Although this was an influential position for much of the twentieth century, over the last 20 years considerable evidence has accumulated against this S-R conceptualization of what animals learn in instrumental conditioning, coming mainly from outcome devaluation studies. The logic behind these studies is straightforward (see Figure 13.2). Animals are first trained to perform an instrumental action for a particular outcome after which the value of the outcome is changed in some way and then the propensity to perform the action is assessed. If initial training merely establishes an S-R association reinforced by the outcome, subsequently changing the animals' evaluation of that outcome should have no impact on subsequent performance of that action because the relation between the action and the outcome is not represented within the S-R structure. (See Chapter 14 in this volume for further discussion of S-R associations and habit learning.) If, however, animals do encode the consequences of their actions during instrumental training, any subsequent change in the animals' evaluation of the outcome should be directly manifested in their performance. In fact, in recent years a large number of successful demonstrations of instrumental outcome-revaluation have been reported (Colwill and Rescorla 1986; Dickinson and Balleine 1994, 2002).

The first demonstrations of the outcome devaluation effect used a conditioned taste aversion procedure to change the animals' evaluation of the outcome. For example, Colwill and Rescorla (1985), following Adams and Dickinson (1981), trained hungry rats to perform two instrumental actions, lever pressing and chain pulling, with one action earning access to food pellets and the other earning access to a sucrose solution (see Figure 13.2). The rats were then given several trials in which they were allowed to consume one of these outcomes with the levers and chains withdrawn, and were then made ill by an injection of the emetic agent, LiCl. Over trials this treatment strongly suppresses consumption of the outcome paired with illness, an example of a conditioned taste aversion. All animals were then given a choice extinction test on the levers and chains conducted in extinction, i.e., in the absence of either of the outcomes. Although S-R accounts should predict no effect of this treatment, Colwill and Rescorla found that animals performed fewer of the action whose training outcome was subsequently paired with LiCl than the other action, indicating that the rats had indeed encoded the consequences of their actions. Subsequently, Rescorla (1990) demonstrated the potency of the outcome devaluation procedure showing that even motivationally incidental features

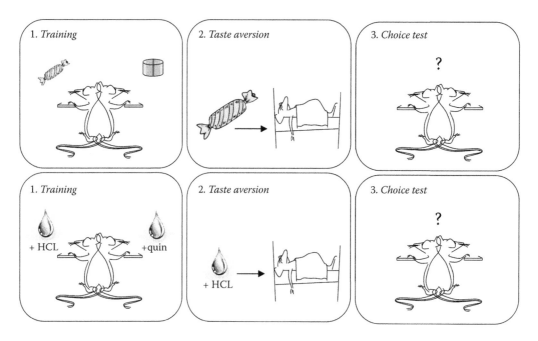

FIGURE 13.2 Outcome devaluation. In outcome devaluation treatments, changes in the value of the goal on the performance of goal-directed actions are assessed. Typically (top panels) rats are trained on two actions (here two levers), each earning a different outcome (left panel). After this phase, one or other outcome is devalued by taste aversion learning (center panel). When the outcome is no longer consumed, the tendency of the rats to perform the two actions is tested in the training situation in extinction. Typically, rats reduce performance on the action that, in training, delivered the now devalued outcome. The bottom panels illustrate the procedure of Rescorla (1990) demonstrating that this effect depends on taste processing (see also Balleine and Dickinson 1998a). Here the two outcomes were water with either hydrochloric acid or quinine added. Hence, changes in the tendency to perform one or other action must reflect the ability of the rats to integrate sensory information about outcome identity in training with the current incentive value established during devaluation.

of the instrumental outcome are encoded (see Figure 13.2, bottom panels). Thirsty rats were trained to lever press and to chain pull with both actions earning water. For one action, however, the water was made sour using a small quantity of hydrochloric acid whereas the other action earned water made bitter using a small quantity of quinine. An aversion was then conditioned to either the bitter or sour water by pairing it with LiCl after which an extinction test was conducted on the levers and chains. Despite the fact that the critical motivational feature, i.e., the fluidic property, was the same for both of the instrumental outcomes, Rescorla found that animals performed fewer of the action that, in training, had delivered the poisoned outcome, indicating that they had encoded the incidental sour and bitter taste features of the water outcomes as consequences of their instrumental actions (Figure 13.2).

Rescorla's (1990) demonstration is important. It shows that the evaluation of an outcome can be mediated by an arbitrary feature motivationally: its taste. If this devaluation treatment modified the degree of thirst or the animal's encoding of the motivationally relevant properties of fluid outcomes, then the performance of both of the actions should have been reduced on test and to a similar degree. As such, this finding provides evidence for a highly sensory-specific encoding of the instrumental outcome; one that allows for the modification of the value of a taste feature while leaving the value of features common to the other fluid outcome (e.g., temperature, texture, visual features, etc.) relatively unaffected. The importance of these demonstrations of the outcome revaluation effect lies in the fact that, together, they provide strong evidence that, in instrumental conditioning, *animals encode the specific features of the consequences or outcome of their instrumental actions.*

Furthermore, these studies helped confirm that instrumental performance is not only determined by the encoding of the action-outcome relation but also by the animals' current evaluation of the outcome.

Nevertheless, although this is a valuable conclusion from these studies, perhaps the most important question is left entirely unanswered by this analysis; i.e., how exactly does taste aversion learning, a treatment that causes quite general malaise, produce such sensory-specific changes in outcome value and, hence, such specific effects on the animals' choice between actions?

13.4   INCENTIVE LEARNING

Perhaps the simplest account of the way taste aversion learning works to devalue the instrumental outcome can be derived from general accounts of aversive conditioning. According to this, pairing the instrumental outcome with illness changes the evaluation of the outcome through the formation of a predictive association between the food or fluid and the aversive state induced by illness, such that the animal learns that the outcome now signals that aversive consequence. Garcia (1989) proposed a different mechanism. He suggested that the change in the evaluation of the outcome induced by taste aversion learning is not due to changing what the outcome predicts but *how it tastes*. Garcia related the change in taste to negative feedback from a system sensitive to illness that he identified as inducing a disgust or distaste reaction. *Hence, on this account it is the association between the taste and the motivational processes activated by illness that alters the subsequent response to those features.* It is important to see that this view implies that taste aversion learning involves two processes: (1) an effective pairing of the outcome with illness initially enables a connection between the sensory properties of the outcome and processes sensitive to illness; (2) the activation of this association when the outcome is subsequently contacted to generate a distaste reaction, allowing the animal to associate the outcome representation with that negative emotional state.

This account predicts, therefore, that, to induce outcome devaluation, it is not sufficient merely to pair the outcome with an injection of LiCl. Rather, a change in value is not induced until the second process is engaged when the outcome is again contacted. Unfortunately, the procedures used to induce outcome devaluation do not differentiate between these two accounts of outcome devaluation because taste aversion is usually induced using multiple pairings of the outcome with illness, allowing the animal to learn both the signaling relationship and the emotional effects of contact with an at least partially devalued outcome. If, however, a substantial aversion to the outcome could be conditioned with a single pairing of the outcome with illness, then these accounts of outcome devaluation make divergent predictions.

We have assessed this prediction in a number of experiments and have found consistent evidence in favor of Garcia's account. For example, in one study (Balleine and Dickinson 1991) we trained thirsty rats to lever press for access to a sugar solution. Illness was then induced by an injection of LiCl immediately after this session for one group of rats (IMM). A second, control group (DEL), was also made ill but after a delay sufficient to prevent any aversion being conditioned to the sugar solution (this is effectively an unpaired control). For reasons not of immediate relevance but that will become apparent, both groups, which we shall refer to as Group IMM-H2O and Group DEL-H2O, were then allowed to drink water in the test chamber on the next day in the absence of the lever before being tested for their willingness to press the lever on the third day in extinction, i.e., in the absence of the sugar. If animals have only to associate sugar with illness to show a devaluation effect, then any effective pairing should be sufficient to generate a change in lever press performance. If, however, they have to discover that the sugar solution is no longer valuable after the poisoning then giving a single sucrose-illness pairing *without allowing re-contact with the sucrose before the test* should have no effect on performance. Indeed, this latter effect is exactly what we observed: Group IMM-H2O, made ill immediately after earning the sugar solution and being re-exposed to water, pressed just as frequently as Group DEL-H2O, which had experienced the delayed illness. We demonstrated that the sucrose-illness pairing was effective. We found that the rats that received immediate illness had a strong aversion to the sugar solution in a subsequent punishment session in which lever pressing once again delivered the sugar solution. In this session, as soon as these animals started receiving and thus making contact with the devalued sugar, they stopped pressing, unlike the control rats, which showed sustained performance throughout the reacquisition session. This result suggests that, during re-acquisition, the animals experienced the nausea or disgust elicited by contact with the sugar and that this experience reduced incentive value of the outcome and, hence, performance of lever pressing.

If contact with the outcome after the devaluation treatment is required to reduce the incentive value of the outcome, giving this experience prior to the extinction test should induce a devaluation effect and reduce performance in the devalued group relative to the control groups. To test this possibility, a second pair of groups, Group IMM-SUC and Group DEL-SUC, were trained in exactly the same manner as the previous two groups except that they were allowed to contact the sugar solution, rather than water, on the day prior to the extinction test. This experience should have allowed immediately poisoned rats to discover their aversion and, therefore, to refrain from pressing the lever during the test the next day. This is just what happened; Group IMM-SUC pressed significantly less than either of the delay groups, Groups DEL-SUC and DEL-H2O, as well as Group IMM-H2O. The significance of this finding is that it allows us to conclude that outcome devaluation induced by taste aversion learning is not a form of predictive learning but rather is produced by feedback during consummatory contact with the poisoned outcome. If a change in incentive value were mediated solely by a signaling process, we should expect that pairing the sweet solution with illness would be sufficient to devalue it as a goal of instrumental performance. Indeed, allowing re-exposure to it in the absence of illness should serve to weaken, not strengthen, any signaling relation between the outcome and illness acquired in training. The fact that re-exposure increased the devaluation effect suggests that the representation of sucrose was modified as the goal of the rats' instrumental performance when they were allowed to consume it after the pairing with illness.

Together with other similar findings (Balleine and Dickinson 1992; Balleine, Paredes-Olay, and Dickinson 2005), the results of this experiment suggest that outcome devaluation depends upon the interaction of two learning processes. The first process involves the conditioning of an association between the outcome and processes that are activated by the induction of illness by LiCl. The failure of this learning process to directly impact on instrumental performance suggests that it is not, alone, sufficient to induce outcome devaluation. *Rather, it appears to be necessary for feedback from this first learning process to become explicitly associated with the specific sensory features of the outcome itself for devaluation to occur.* It would seem plausible to suppose that the first process is something akin to that derived above from Pavlov's analysis of the way animals acquire the US representation in Pavlovian conditioning. As such, this learning process is referred to here as *evaluative conditioning.* The second learning process appears to be critical for animals to establish the current rewarding properties of the instrumental outcome on the basis of evaluative processing of this kind. In the past, this second learning process has been identified as *incentive learning* (Dickinson and Balleine 1993, 1994, 1995).

The kind of structure that the above discussion suggests underlies the way that taste aversion learning acts to devalue the instrumental outcome is illustrated in Figure 13.3. Pairing the instrumental outcome with illness produces an association between a representation of the sensory properties of the outcome (O_{Se}) and a motivational structure sensitive to the effects of illness which, in line with previous analyses (Rozin and Fallon 1987; Dickinson and Balleine 1994), is identified as that reflecting or mediating disgust (M_{disg}). Thus, it is this association between O_{Se} and M_{disg} that is referred to above as underlying the initial conditioning connection following the taste-illness pairing and that, as described above and in previous analyses (e.g., Balleine 2001, 2004), is referred to as *evaluative conditioning* (refer Figure 13.1). It is proposed that establishing this association opens a feedback loop that provides the basis for a second learning process proposed to underlie outcome devaluation, i.e., that of *incentive learning*, that is engaged when the outcome is subsequently contacted. This contact activates the representation of the outcome that, through prior evaluative

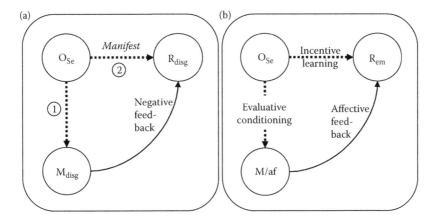

FIGURE 13.3 Schematic diagrams of incentive learning. (a) Taste aversion-induced devaluation has been found to engage two processes: (1) a latent process connecting the sensory properties of the outcome, most notably taste features, with a system sensitive to illness and productive of disgust (M_{disg}). This provides negative feedback in the form of a disgust response (R_{disg}) that provides the basis for (2) the second learning process engaged during re-exposure to the outcome when this disgust response is made manifest, referred to as *incentive learning*. (b) A general structure for outcome revaluation engaged by changes in the motivational significance of the instrumental outcome, or O_{Se}. Connections between sensory features of the outcome and specific motivational/affective structures, during evaluative conditioning, provide the basis for affective feedback in the form of an emotional response (R_{em}) when the outcome is contacted during incentive learning, allowing the encoding of the experienced value of the outcome.

conditioning, then activates the disgust system and this latter activity produces negative feedback, the disgust or distaste response (R_{disg}), which is contingent upon contact with the outcome. It is this feedback that is then associated with the representation of the food or fluid itself and that acts to change the incentive value assigned to the outcome.

One reason for proposing that pairing the instrumental outcome with illness conditions an association between the outcome representation and disgust is provided by evidence that antiemetics attenuate both the conditioning and expression of taste aversions (Balleine, Davies, and Dickinson 1995; Experiment 1). This result led us to predict that an antiemetic should also attenuate the effect of incentive learning in instrumental outcome devaluation. To test this prediction, thirsty rats were trained in a single session to perform two actions, lever pressing and chain pulling, with one action delivering the sucrose solution and the other delivering the saline solution on a concurrent schedule. Immediately after this training session all the rats were given an injection of LiCl. Over the next 2 days, the rats were re-exposed to both the sucrose and the saline solutions. Prior to one re-exposure session, rats were injected with an antiemetic (ondansetron) whereas prior to the other session they were injected with vehicle. The next day the rats were given a choice extinction test on the lever and chain. If re-exposure devalues the instrumental outcome via the ability of the outcome representation to access the disgust system, blocking the activity of that system with an antiemetic should be predicted to attenuate the effects of re-exposure such that, on test, the action that, in training, delivered the outcome subsequently re-exposed under the antiemetic should be performed more than the other action. In fact, this is exactly what was found (cf. Balleine, Davies, and Dickinson 1995; Experiment 2).

13.5 THE UBIQUITY OF INCENTIVE LEARNING: THE MOTIVATIONAL CONTROL OF GOAL-DIRECTED ACTION

The reason for emphasizing the role of incentive learning in instrumental outcome-devaluation effects is the fact that it also appears to be the process by which motivational states other than disgust, such as hunger and thirst, encode the value of goals such as foods and fluids. It is well established that the motivational state of rats is a major determinant of their instrumental performance; not surprisingly, hungry animals work more vigorously for a food reward than sated ones. But what current evidence suggests is that this is because a food-deprived state induces an animal to assign a higher incentive value to nutritive outcomes when they are contacted in that state and that this high rating of the incentive value of the outcome is then reflected in a more vigorous rate of performance. Although this suggestion stands contrary to general drive theories of motivation—that suppose that increments in motivation elicit their effects on performance by increases in general activation (Hull 1943)—there are good empirical grounds for arguing that motivational states often do not directly control performance (Dickinson and Balleine 1994, 2002; Balleine 2001). For example, Balleine (1992) trained groups of undeprived rats to lever press for a food reward (in different experiments this was either food pellets or a maltodextrin solution). After training, half of the rats were shifted to a food deprivation schedule whereas the remainder were maintained undeprived before both groups were given an extinction test on the levers. We found that performance of the groups on test did not differ even though the shift in motivational state was clearly effective; in a subsequent test where the animals could again earn the food pellets, the food-deprived rats pressed at a substantially higher rate than the undeprived rats. Although motivational state clearly did not exert any direct control over extinction performance in the absence of the food reward, we found, as in taste aversion learning, that motivational state could control performance if the rats were given the opportunity for incentive learning. This was achieved by allowing the rats consummatory contact with the instrumental outcome in the test motivational state prior to that test. We trained two further groups of rats to lever press when undeprived, before both groups were given the opportunity to consume the instrumental outcome when food deprived, prior to a test in which one group was tested undeprived and the other food deprived. Now a clear difference in performance emerged; rats tested when food deprived and previously allowed to consume the instrumental outcome when food deprived pressed at a higher rate than groups either allowed consummatory contact hungry but tested undeprived, or that were not allowed this consummatory contact at all (cf. Balleine 1992).

We were also able to confirm that this incentive learning effect depended upon the instrumental contingency. We trained undeprived rats to perform two actions, lever pressing and chain pulling, with one action earning access to food pellets and the other to a maltodextrin solution. All rats were then given a choice extinction test on the levers and chains. Prior to the test, however, the animals were given six sessions in which they were allowed to consume one instrumental outcome when food deprived and, on alternate days, the other outcome in the training, i.e., undeprived, state. On test, we found that animals performed more of the action that, in training, had delivered the outcome re-exposed in the food-deprived state prior to the test than the other action (Balleine 1992).

A natural interpretation of this incentive learning effect can be drawn from the analysis applied to taste aversion learning above and presented in Figure 13.3b. When undeprived, any nutrient-based motivational state should be relatively inactive and so any connection between the outcome and that state would be productive of relatively little motivational/affective feedback when the outcome is consumed. Hence, one should suppose that the value of the outcome based on that feedback would be relatively low. When food deprived, however, this activation should be increased and any motivational/affective feedback produced by the outcome-nutritive state connection should now be increased, resulting in an increased emotional response, an enhanced value being assigned to the outcome, and an increase in performance. In subsequent experiments we were able to test this account of deprivation-induced changes in the incentive value of food outcomes using the endogenous satiety peptide cholecystokinin (CCK) and an antagonist of the CCKA receptor, at which CCK binds to induce its satiety action, MK329 (devazepide). In one experiment (Balleine, Davies, and Dickinson 1995), for example, rats were trained undeprived to lever press and chain pull for the pellet and maltodextrin outcomes and then re-exposed to both outcomes when food deprived. During the re-exposure, however, one of the two outcomes was consumed after an injection of CCK whereas the other was not. If the state of nutritive motivation controls the feedback on which incentive learning is based, then CCK's satiety effects should be predicted to reduce the feedback induced by re-exposure to the outcome when food deprived. In fact, this is exactly what we found: when given a choice test between the two actions when food deprived, the rats reduced their performance of the action trained with the outcome re-exposed under CCK relative to the other action.

It should be noted that this role for incentive learning in instrumental performance following a shift in motivational state is not confined to post-training increases in food deprivation. The same general pattern of results was also found for the opposite shift, i.e., where rats were trained to lever press for food pellets when food deprived and then tested when undeprived. In this case, rats only reduced their performance when food deprivation was reduced if they were allowed to consume the instrumental outcome when undeprived prior to the test, an effect that also emerged in the choice situation in which the outcome of one action was exposed undeprived prior to the test whereas the other had only been contacted when food deprived (Balleine 1992; Balleine and Dickinson 1994). This effect was also found to depend on the state of nutritive motivation during re-exposure; we were able to block the reduction in value using the CCK antagonist devazepide administered during the re-exposure phase. As found with both taste aversion learning and increases in food deprivation, therefore, the influence of a reduction in food deprivation on incentive value appears to be mediated by the emotional response induced by nutritive/affective feedback, itself induced by initial evaluative conditioning, thereby allowing the formation of a connection between the outcome representation and this emotional response to establish outcome value (Figure 13.3b).

Finally, it is important to note just how general this role for incentive learning is in encoding the value of instrumental outcomes. Based on what is now a very large number of studies, the generality of the role of incentive learning in instrumental performance has been confirmed for a number of different motivational systems and in a number of revaluation paradigms. For example, in addition to taste aversion learning and increases and decreases in food deprivation, incentive learning has been found to mediate: (i) specific satiety-induced outcome devaluation effects (Balleine and Dickinson 1998a); (ii) shifts from water deprivation to satiety (Lopez, Balleine, and Dickinson 1992); (iii) changes in outcome value mediated by drug states (Balleine, Ball, and Dickinson 1994; Balleine, Davies, and Dickinson 1995; Hutcheson et al. 2001; Hellemans, Dickinson, and Everitt 2006); (iv) changes in the value of thermoregulatory rewards (Hendersen and Graham 1979); and (v) sexual rewards (Everitt and Stacey 1987; Woodson and Balleine 2002; see Balleine 2001; Dickinson and Balleine 1994, 2002 for reviews). In all these cases it is clear that animals have to learn about changes in the incentive value of an instrumental outcome through consummatory contact with that outcome before this change will affect performance.

13.6 ENCODING AND RETRIEVING INCENTIVE VALUE DURING DECISION MAKING

A final important issue to consider with regard to the function and representation of reward is the question of how reward value transfers from encoding to retrieval to inform decision making. In the experiments described above, the re-exposure and test phases are often conducted several days apart and there are no explicit internal or external cues that the rat can use to determine choice performance. Instead, the rats must rely on their *memory* of specific action-outcome associations and the current relative value of the instrumental outcomes. But how is value retrieved during this test?

One theory proposes that value is retrieved through the operation of the same processes through which it is encoded. This view is perhaps best exemplified by Damasio's (1996) *somatic marker hypothesis* according to which decisions based on the value of specific goals are determined by re-experiencing the emotional effects associated with contact with that goal. An alternative theory proposes that values, once determined through incentive learning, are encoded as abstract values (e.g., "X is good" or "Y is bad") and so are not dependent on re-experiencing the original emotional effects associated with contact with the goal and on encoding incentive value, for their retrieval (see Balleine and Dickinson [1998b] for further discussion).

We have conducted several distinct series of experiments to test these two hypotheses and in all of these the data suggest that, after incentive learning, incentive values are encoded abstractly and do not involve the original emotional processes that established those values for their retrieval (Balleine, Ball, and Dickinson 1994; Balleine and Dickinson 1994; Balleine, Davies, and Dickinson 1995; Balleine, Garner, and Dickinson 1995). Recall that, in previous studies assessing the role of basic motivation and affective processes in encoding changes in incentive value, we assessed the effect of the antiemetic ondansetron in blocking outcome devaluation induced by LiCl-induced taste aversion during re-exposure to that outcome after the outcome-illness pairing. In that study we found that the antiemetic could block the change in value induced during re-exposure; when, having trained to perform two actions for different outcomes, both outcomes were paired with illness with one re-exposed under the antiemetic and other not, rats preferred the action whose outcome was re-exposed under the antiemetic to the other action in a choice extinction test. The question we then asked was whether the antiemetic would influence the retrieval of incentive value in a similar manner to the way it affected the encoding of that value, as should be predicted by the somatic marker hypothesis.

Testing this prediction was a simple matter; we merely gave the rats the choice test after an injection of the antiemetic. If the antiemetic affects the motivational/affective processes that influence the *encoding* of incentive values during incentive learning, then blocking these processes on test should also block the *retrieval* of those values if the same motivational/affective processes are necessary for the retrieval of those values. Relative to a group not injected with the antiemetic, the effects of re-exposure should be diminished and both actions should be more similarly preferred on test. In contrast to this prediction, however, the antiemetic had no effect whatever during the test: the action whose outcome was re-exposed under the antiemetic was preferred to the same degree whether the rats were tested under the antiemetic or not.

We have also assessed this prediction by injecting rats with either CCK or devazepide, as appropriate, on test after increases or decreases in incentive value induced by post-training increases or decreases in food deprivation. Again, the question was whether the same effects would emerge when tested under CCK or devazepide or whether, as predicted by the somatic marker hypothesis, the retrieval of those values, and hence choice performance, would be influenced by the CCK or devazepide on test. Again, however, the results could not have been clearer; in both cases, whether we gave injections of devazepide or CCK on test, these injections had no effect on choice performance. Although these drugs were very effective in blocking the encoding of incentive value during actual contact with the outcome during incentive learning, they had no detectable effect whatsoever on the retrieval of those values during choice.

Hence, the motivational/affective processes that are engaged during incentive learning—and that contribute to the emotional feedback through which the incentive value of the outcome is encoded—are not involved during its retrieval. Rather, it appears that, once the appropriate learning experience has been provided, the incentive value of rewarding outcomes is encoded in a manner that does not require those processes to retrieve them. As presented in Figure 13.3b, the current view of the necessary conditions for encoding a change in the incentive value of an outcome is the contiguous presentation of the outcome and an emotional response. Once this is experienced, however, it appears that animals extract that value and store it in a manner that allows it to be retrieved independently of the current state of the motivational and affective processes that determined that value in the first place. The nature of this retrieval is at present not well understood. However, some hints can be gleaned from studies that have assessed the neural bases of incentive learning and that we describe in the next section.

13.7   THE NEURAL BASES OF INCENTIVE LEARNING

13.7.1   ENCODING INCENTIVE VALUE

As described above, considerable evidence suggests that the reward value of food is mediated by changes in taste processing. For example, specific satiety treatments have been found to be extremely effective in producing selective changes in the incentive value of instrumental outcomes and in the performance of actions that gain access to those outcomes over and above the effects of satiety on motivation for nutrients generally or even for specific macronutrients (Balleine and Dickinson 1998a, 1998c). In one study (Balleine and Dickinson 1998a) hungry rats were trained to press a lever and pull a chain with one action earning sour starch and the other salty starch before they were sated on either the salty or sour starch and given an extinction test on the lever and chain. Although both actions earned equivalent nutrients of a similar macronutrient structure, the rats still altered their choice performance to favor the action that, in training, delivered the outcome on which they were not sated; i.e., they were able to modify their choice based on changes in taste (Balleine and Dickinson 1998a).

Given these findings, one might expect neural structures involved in taste processing to be involved in incentive learning. In another series of experiments, therefore, we assessed the effects of cell-body lesions in the gustatory region of the insular cortex, for some time known to be involved in taste processing, although not taste detection (Braun, Lasiter, and Kiefer 1982), on specific satiety-induced devaluation and on incentive learning conducted after instrumental training when hungry after a shift to a sated state (Balleine and Dickinson 2000). Although these lesions had no effect on the ability of rats to detect changes in value when they actually contacted a specific outcome on which they were sated, they were deeply amnesic when forced to choose between two actions based on their memory of satiety-induced changes in value. These results suggest that the gustatory cortex operates as one component of an incentive system, acting to encode the taste features of the instrumental outcome as an aspect of the representation of the value of that outcome in memory. From this perspective, the gustatory insular cortex is not involved in detecting changes in incentive value per se; that would appear to require the integration of taste memory, involving the gustatory insular cortex, with an affective signal, apparently mediated by a different component of the incentive system (Balleine 2001). Thus, consistent with the analysis described above, changes in the value of the taste features of nutritive outcomes appear to be a function of emotional feedback, i.e., of the emotional response experienced contiguously with detection of the taste. If the emotional response is "pleasant," the value of the outcome should be correspondingly increased whereas if it is "unpleasant" it should be reduced. Hence, treatments that produce changes in palatability in rats, usually assessed by taste reactivity responses, are also those that most potently modify the value of instrumental outcomes (Berridge, Grill, and Norgren 1981; Berridge 2000a). For example, Rolls, Rolls, and Rowe (1983) provided a clear demonstration that, in humans, eating one particular food to satiety strongly reduces the pleasantness rating of that food but not other similar foods. Likewise, in rats, Berridge (1991a) demonstrated that, when sated on milk, ingestive taste reactivity responses were reduced and aversive taste reactivity responses were increased when milk, but not when sugar, was subsequently contacted. These kinds of data suggest, therefore, that satiety-induced changes in incentive value are not a product of general shifts in motivation but reflect variations in the association of taste features with specific emotional responses.

If incentive learning is determined by the association of sensory and emotional processes then, again, it can be supposed that neural structures implicated in the formation of associations of this kind are critically involved in this form of learning. *Indeed, the broader implication is that sensory brain regions themselves must be co-conspirators with "learning"-related brain regions, in order to effect changes in the incentive value of instrumental outcomes.* Herein lies a fundamental interface between sensation and reward, in keeping with the broader themes of this volume.

For example, the gustatory region of insular cortex maintains reciprocal connections with the nucleus accumbens (Brog et al. 1993) and basolateral amygdala (BLA) (Sripanidkulchai, Sripanidkulchai, and Wyss 1984; Yamamoto, Azuma, and Kawamura 1984) and, unsurprisingly, it is these structures that have been heavily implicated in reward. Many of these findings have relied on indirect means of assessing changes in incentive value derived from changes in discriminated approach in context preference studies (e.g., Cunningham and Noble 1992; Robledo and Koob 1993; Skoubis and Maidment 2003) and taste reactivity responses that, when independent of ingestion, have been argued to reflect the hedonic responses to reward (Pecina and Berridge 2005; Smith and Berridge 2007). More direct evidence has emerged from studies showing that lesions of the BLA rendered the instrumental performance of rats insensitive to devaluation by sensory-specific satiety, appearing no longer able to associate the sensory features of the instrumental outcome with its incentive value. The BLA has itself been heavily implicated in a variety of learning paradigms that have an evaluative component; for example this structure has long been thought to be critical for fear conditioning and has recently been reported to be involved in a variety of feeding-related effects including sensory-specific satiety (Malkova, Gaffan, and Murray 1997), the control of food-related actions (see below), and in food consumption elicited by stimuli associated with food delivery (Holland, Petrovich, and Gallagher 2002; Petrovich et al. 2002). Indeed, in two recent series of experiments we have found clear evidence of the involvement of the BLA in incentive learning.

In one series we found that lesions of the BLA rendered the instrumental performance of rats insensitive to outcome devaluation, apparently because they were no longer able to associate the sensory features of the instrumental outcome with its incentive value (Balleine, Killcross, and Dickinson 2003). More recently, we have confirmed this suggestion using post-training infusions of the protein-synthesis inhibitor anisomycin (Wang et al. 2005). It has now been well documented that both the consolidation of the stimulus-affect association that underlies fear conditioning and its reconsolidation after retrieval depends on the synthesis of new proteins in the BLA (Nader, Schafe, and LeDoux, 2000; Schafe et al. 2001). In a recent experiment, we first trained hungry rats to press two levers with one earning food pellets and the other a sucrose solution. After this training the rats were sated and given the opportunity for incentive learning, i.e., they were allowed to consume either the food pellets or the sucrose solution in the sated state. Immediately after this consumption phase, half of the rats were given an infusion of anisomycin, whereas the remainder were given an infusion of vehicle. In a subsequent choice extinction test, conducted on the two levers when sated, rats in the vehicle group performed fewer responses on the lever that, in training, delivered the outcome to which they were re-exposed (sated) prior to the test: i.e., the standard incentive learning effect (Balleine 1992). By contrast, the infusion of anisomycin completely blocked this shift in preference. To assess whether incentive learning is subject to reconsolidation involving the BLA, we gave all of the rats a second re-exposure episode to either the pellets or sucrose when sated such that, if they had been first given vehicle infusion then they were now given an anisomycin infusion, whereas if they were first given an anisomycin infusion they were now given a vehicle infusion. Although, again, vehicle-infused rats showed reliable incentive learning, those given the anisomycin infusion performed indifferently on the two levers despite the fact that these same rats had previously shown perfectly clear evidence of incentive learning after the first episode of re-exposure (Wang et al. 2005).

13.7.2 Opioid Processes in Reward versus Reactivity

Attention has also turned of late to the role of endogenous opioid peptides, which have long been postulated as mediators of an endogenous hedonic tone (Kosterlitz and Hughes 1975), and as potential conveyers of the affective properties of reward processes during incentive learning. To the extent that another single neurotransmitter/neuromodulatory system might be expected to fulfill the general role of "hedonic mediator," the endogenous opioids are obvious candidates (Koob and Le Moal 1997). Not only is exogenous administration of μ-opioid receptor agonists reinforcing but, conversely, administration of the general opioid antagonists naloxone and naltrexone is aversive in rodents (Grevert and Goldstein 1977a; Mucha and Iversen 1984; Mucha and Walker 1987) and produces dysphoria in humans (Grevert and Goldstein 1977b), suggesting the presence of an endogenous opioid tone maintaining a basal "hedonic state." In line with these suggestions, Berridge and colleagues (Pecina and Berridge 2000; Smith and Berridge 2007) have claimed, in a manner that contrasts with our claims above regarding the role of the BLA in incentive learning, that opiate processes in a network involving the ventral pallidum and nucleus accumbens shell play a central role in the hedonic processes through which the reward value of specific events and commodities is encoded. Using taste reactivity as an objective measure of hedonic processing in rodents, Berridge and colleagues found that morphine infused into either ventral pallidum or the accumbens shell enhanced taste reactivity reactions to a sucrose solution, an effect that was abolished when these structures were disconnected (Smith and Berridge 2005; Pecina, Smith, and Berridge 2006). Nevertheless, opioid peptide-containing neurons and receptors are present in multiple basal forebrain regions (Ding et al. 1996; Poulin et al. 2006) implicated not only in taste reactivity, but also in reward processing, most notably the opioid-rich BLA, with which the ventral pallidum and accumbens shell are interconnected (Johnson et al. 1994).

In a recently published series of experiments using rats as subjects, we found, consistent with Berridge's analysis, that the broad spectrum opioid antagonist naloxone infused into the ventral pallidum or accumbens shell blocked food deprivation-induced increases in licking-related, sucrose palatability responses (Wassum et al. 2009). However, naloxone infused into these structures had *no effect on incentive learning* conducted after an increase in food deprivation or on subsequent changes in instrumental performance. Conversely, we found that intra-BLA naloxone blocked the effect of incentive learning on instrumental performance without affecting sucrose palatability responses. Furthermore, in line with the general finding, previously described, that the processes engaged in the encoding and retrieval of incentive value differ, we found that this effect on intra-BLA naloxone was specific to the encoding of incentive value; intra-BLA naloxone did not affect the retrieval of previously updated incentive value information.

This anatomical double-dissociation of the opioid-mediated determinants of reward detection and of reward encoding provides a number of important insights into incentive learning. Firstly, it confirms the role of the BLA in incentive learning and extends our understanding of the neural mechanisms by which incentive learning functions by focusing on opioid receptor-related processes in that structure. Secondly, and perhaps more importantly from a theoretical perspective, it suggests that the basis for updating reward value is not the *palatability responses* induced by contact with the outcome during incentive learning. Although, as has been proposed in the past (James 1890), it is possible that the emotional processing of events is based on physical reactions to those events (we feel pleasure because we smile, sad because we cry, and so on), it appears that rats do not use the fact that they lick (or do not lick) a sugar solution as the basis for the value that they place on it, at least with respect to their decision to perform actions to gain access to that sugar.

With respect to the encoding of incentive value, previous effects of amygdala manipulations on feeding have been found to involve connections between the amygdala and the hypothalamus (Petrovich et al. 2002) and, indeed, it has been well reported that neuronal activity in the hypothalamus is primarily modulated by chemical signals associated with food deprivation and food ingestion, including various macronutrients (Seeley et al. 1996; Levin 1999; Woods et al. 2000; Wang et al. 2004). Conversely, through its connections with visceral brain stem, midline thalamic nuclei, and associated cortical areas, the hypothalamus is itself in a position to modulate motivational and nascent affective inputs into the amygdala. Together with the findings described above, these hypothalamic inputs, when combined with the amygdala's sensory afferents, provide the basis for the kind of associative process required to alter incentive value. As illustrated in Figure 13.3b, this neural structure is consistent with a simple feedback circuit within which the goal or reward value of a specific event is set and, indeed, can be re-set when subsequently contacted on the basis of the animal's current internal state.

13.7.3 RETRIEVING INCENTIVE VALUE: THE ROLE OF THE INSULAR CORTEX RECONSIDERED

Although the role of the gustatory region of the insular cortex is consistent with its involvement in the encoding of incentive value, it is also consistent with the *retrieval* of incentive value. The fact that rats with lesions of this region are able to detect changes in incentive value but are unable to recall those changes on test could be due to a failure to encode that change. However, it is also consistent with a deficit in retrieving incentive value once encoded. In the above analysis the focus is on role of the efferents of the gustatory insular cortex to the BLA in establishing the association between the sensory features of the outcome (in particular its taste features) with the emotional feedback induced during consummatory contact with the outcome to encode incentive value. However, it is possible that the well-documented efferents from amygdala to insular cortex (Sripanidkulchai, Sripanidkulchai, and Wyss 1984), previously proposed to form a part of a circuit mediating the retrieval of incentive value (Yamamoto 2006), are equally important to this effect and that, rather than blocking the detection of this relationship, the bilateral lesions of insular cortex served rather to affect the rats' ability to retrieve this relationship from memory.

To assess the role of BLA-gustatory insular connections in incentive learning, we have recently attempted to disconnect these structures using asymmetrical lesions (see Figure 13.4). Prior to instrumental training all rats were given a unilateral lesion of the BLA in one or the other hemisphere.

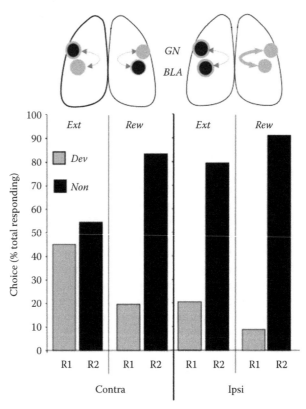

FIGURE 13.4 Effect of disconnection of gustatory insular cortex from the basolateral amygdala on choice. Rats were trained on two lever press actions for different outcomes before one outcome was devalued. Tests were conducted in extinction (Ext) and with the devalued and non-devalued rewards delivered contingent on lever pressing (Rew). Devaluation had a comparable effect on choice in both extinction and reward test in rats given ipilateral lesions (Ipsi) of gustatory insular cortex and basolateral amygdala (main effect of devaluation, $p < .05$; devaluation×test interaction; $F < 1$). In rats in which the lesions were in different hemispheres (Contra), effectively disconnecting these structures, significant devaluation was observed only in the rewarded test ($p < .05$) and not in extinction ($p > .05$); devaluation×test interaction, $p < .05$.

Half of the rats were also given a control, unilateral, lesion of the gustatory insular region ipsilateral to the BLA lesion (IPSI), whereas the other, disconnection, group was given the gustatory lesion in the hemisphere contralateral to the BLA lesion (CONTRA). The rats were then trained to perform two actions, with one action earning grain pellets and the other a polycose solution. After training, one or the other outcome was devalued by a specific satiety treatment, and then a choice test between the two actions was conducted in extinction. If retrieving the incentive value of the outcome depends on its encoding through an interaction of the gustatory insular cortex and the BLA, then the disconnection of these structures should replicate the deficit observed with gustatory insular lesions alone; i.e., the CONTRA group should show a deficit during the extinction test relative to the IPSI group. However, given that the gustatory insular and the BLA were only damaged unilaterally in both the IPSI and CONTRA groups, we should not (i) expect any effect of retrieval if reward was not based on the connection of these structures; and (ii) anticipate any effect of the lesion on a rewarded test when the devalued and non-devalued outcomes were presented and their value detected directly through consummatory contact and performance adjusted during the course of the test. In fact, as illustrated in Figure 13.4, this is exactly what we found. Although a normal devaluation effect was observed in the IPSI group in both the extinction test and the rewarded test, a devaluation effect only emerged in the CONTRA group in the rewarded test; no effect of devaluation emerged when this group was asked to choose between the two actions in extinction. As such, it appears that the retrieval of the value of the instrumental outcome, but not the detection of that change in value, depends on the interaction of the BLA and gustatory insular cortex, consistent with the argument that the gustatory insular region is critical for retrieving the incentive value of instrumental outcomes from memory.

This conclusion is consistent with other recent research suggesting that gustatory insular cortex is important for retrieving the value of specific food rewards. For example, in rats, Kerfoot et al. (2007) paired a tone with sucrose reward and found that, after the sucrose was devalued, the tone evoked significant activation of the gustatory insular region, as assessed by FOS-related immunoreactivity, consistent with tone-induced retrieval of the reduced value of the sucrose. Likewise, the gustatory insular region in humans is heavily activated by both the consumption of appetizing foods and pictures of those foods (Simmons, Martin, and Barsalou 2005). Interestingly, increased activation in gustatory cortex has been reported in obese female human subjects when either consuming or anticipating the consumption of chocolate milk shakes (Stice et al. 2008). Conversely, both reduced activity, measured using fMRI, and μ-opioid receptor binding, measured using PET, were observed in gustatory cortex in subjects suffering from bulimia nervosa, with the degree of binding correlating negatively with subsequent fasting behavior (Bencherif et al. 2005). Again, these findings are consistent with the role of the gustatory cortex in the retrieval of incentive value.

13.8   ACTION SELECTION AND INITIATION: SOME CONCLUDING COMMENTS

Deciding which of multiple alternative actions to perform requires animals to compare the consequences of those actions based on their incentive values. In order to make this comparison these incentive values must first be encoded through a process of incentive learning. The argument presented in this paper proposes that the only necessary feature of this learning process is the pairing of the sensory properties of some consequent event with an emotional response. However, this emotional response is not entirely arbitrary—evidence described above suggests that it is the product of an initial evaluative conditioning process through which these sensory properties become connected with basic motivational and affective processes and together produce the emotional feedback (cf. Figure 13.3). Thus, although sensory processes are in themselves affectively neutral and do not imply any particular course of action, through the functions of the evaluative and incentive learning processes, specific sensory events can elicit both emotional feedback (based on the former; e.g., Baeyens et al. 2000) and, independently, specific value-based decisions (cf. Balleine and Dickinson 1998b; Dickinson and Balleine 2009).

Although not particularly parsimonious, this model of instrumental reward does have the advantage of providing a basis for distinguishing between the functions of rewarding events and the behavioral effects of those events, in evaluative, Pavlovian, and instrumental conditioning. As has previously been demonstrated, although incentive learning is required before shifts in motivational conditions will affect instrumental actions, this is not true of Pavlovian conditioned responses, nor is it true of responses in evaluative conditioning procedures, like taste aversion and taste preference learning, although in these cases the consummatory tests that are often used can also provide the basis for incentive learning and so can sometimes confound these two. Nevertheless, the argument for distinct processes is bolstered by evidence that distinct neural systems are involved. The finding, described above, that opioid processes in ventral pallidum and the accumbens shell influence the palatability of rewarding foods but not instrumental incentive learning, whereas opioid processes in the BLA influence incentive learning but not palatability, is one such example. Likewise, the now expanding evidence that the influence of Pavlovian cues and incentive learning on choice is doubly dissociable both behaviorally (Corbit and Balleine 2003; Holland 2004) and neurally (Corbit, Muir, and Balleine, 2001; Ostlund and Balleine 2007b) demonstrates that the "representation" of the US and of the instrumental outcome that supports Pavlovian and instrumental conditioning, respectively, is not identical or supported by the same neural circuitry. In fact, these distinctions fall naturally from the model because, as has been described previously (Balleine 2001, 2004, 2005; Balleine and Ostlund 2007), the associative relations capable of generating the behavioral and functional differences mediated by the sensory-motivational association, the Pavlovian CS-US association (between a novel sensory event and the sensory-motivational dyad that underlies the US "representation"), and the incentive learning process are interrelated but not interdependent processes.

However, as argued above, to provide a full account of any decision not only requires the specification of both the learning and the incentive processes that contribute to instrumental performance but also requires an account of how these processes interact. Take the example of outcome devaluation described above. Recall that, after training on two actions for distinct outcomes, reducing the value of one outcome causes animals to alter their subsequent choice performance to favor the action that, in training, delivered the still valued outcome. It is clear that, to alter their performance in this manner without direct feedback, animals need: (i) to have learned what actions lead to what outcomes; and (ii) to have encoded the current relative values of the two outcomes. But to actually choose a particular course of action, they must also be able to integrate these two processes: the relative values of the outcomes have to inform them about which of the two actions to choose. Nevertheless, although establishing the nature of this integrative process is critical to understanding choice, we have, at present, only rudimentary information as to how this is achieved.

Recent research has started to suggest the kinds of processes likely to be involved. For example, evidence suggests that, in selecting and initiating actions, animals can reason both forwards (i.e., from actions to their likely consequences and so to an evaluation of those consequences and thence to action initiation or inhibition) and backwards (i.e., from a particularly desired outcome to the immediate action most likely to bring it about). At one level these appear to be distinct strategies. Thus, for example, discriminative stimuli appear to influence choice through action selection rather than by retrieving outcome values directly; outcome devaluation does not appear to influence discrimination performance per se (Colwill and Rescorla 1990). By contrast, the influence of outcomes themselves lies in their ability to control performance via their incentive value; changes in outcome value appear to influence action initiation and, indeed, the reinstatement of performance directly through the action-outcome association (Balleine and Ostlund 2007; Ostlund and Balleine 2007a).

However, at another level it seems likely that these strategies are mediated through a common architecture. The fact that the two processes that contribute to the integration of action-outcome and incentive value information in decision making (i.e., an action-outcome process and an action selection process based on stimulus-action associations) converge on a common, if massively distributed, action representation may provide the basis for amalgamating these processes into a single network. Although detailed discussion is beyond the scope of the current chapter, in other places we have provided a more detailed account of one means by which this could be achieved and the interested reader is referred there for more information (Dickinson and Balleine 1993, 1994, 2002; Balleine and Ostlund 2007).

The relationship between action selection and initiation is, of course, merely the tip of a much larger issue relating to the integration of cognitive and emotional processes more generally. There has long been something of a chasm separating theory and research into the learning and motivational systems that respond to emotional events and those involved purely in sensation, perception, discrimination, categorization, concept formation, language, attention, problem solving, and so on. Nevertheless, the fact that many fundamental capacities, including decision-making, require the smooth integration of all of these processes suggests it is a gap that can be bridged, and recognition of the central role that incentive learning plays in goal-directed action may provide at least one avenue for future research with this express aim. This is particularly true given the targets that current research has revealed, implicating the relationship between the fronto-striatal network that encodes action-outcome associations and the cortico-limbic network that unites sensory, motivational, and affective processes, the integration of which ultimately comprises the key determinants of incentive learning.

ACKNOWLEDGMENT

The preparation of this chapter was supported by a grant from the National Institute of Mental Health (#MH56446) and a Laureat Fellowship from the Australian Research Council.

REFERENCES

Adams, C.D., and Dickinson, A. 1981. Instrumental responding following reinforcer devaluation. *Q J Exp Psychol* 33B:109–21.

Baeyens, F., Eelen, P., Van den Bergh, O., and Crombez, G. 1990. Flavor-flavor and color-flavor conditioning in humans. *Learn Motiv* 21:434–55.

Balleine, B., Delgado, M., and Hikosaka, O. 2007. The role of the dorsal striatum in reward and decision-making. *J Neurosci* 27:8161–65.

Balleine, B., Liljeholm, M., and Ostlund, S. 2009. The integrative function of the basal ganglia in instrumental conditioning. *Behav Brain Res* 199:43–52.

Balleine, B., and O'Doherty, J. 2010. Human and rodent homologies in action control: Corticostriatal determinants of goal-directed and habitual action. *Neuropsychopharmacology* 35:48–69.

Balleine, B.W. 1992. Instrumental performance following a shift in primary motivation depends on incentive learning. *J Exp Psychol Anim Behav Process* 18:236–50.

———. Incentive processes in instrumental conditioning. In *Handbook of Contemporary Learning Theories*, ed. R.M.S. Klein, 307–66. Hillsdale, NJ: LEA.

———. 2004. Incentive behavior. In *The Behavior of the Laboratory Rat: A Handbook With Tests*, eds. I.Q. Whishaw and B. Kolb, 436–46. Oxford: Oxford University Press.

———. 2005. Neural bases of food seeking: affect, arousal and reward in corticostriatolimbic circuits. *Physiol Behav* 86:717–30.

Balleine, B.W., Ball, J., and Dickinson, A. 1994. Benzodiazepine-induced outcome revaluation and the motivational control of instrumental action in rats. *Behav Neurosci* 108:573–89.

Balleine, B.W., Davies, A., and Dickinson, A. 1995. Cholecystokinin attenuates incentive learning in rats. *Behav Neurosci* 109:312–19.

Balleine, B.W., and Dickinson, A. 1991. Instrumental performance following reinforcer devaluation depends upon incentive learning. *Q J Exp Psychol* 43B:279–96.

———. 1992. Signalling and incentive processes in instrumental reinforcer devaluation. *Q J Exp Psychol B* 45:285–301.

———. 1994. Role of cholecystokinin in the motivational control of instrumental action in rats. *Behav Neurosci* 108:590–605.

———. 1998a. The role of incentive learning in instrumental outcome revaluation by specific satiety. *Anim Learn Behav* 26:46–59.

———. 1998b. Consciousness: The interface between affect and cognition. In *Consciousness and Human Identity*, ed. J. Cornwell, 57–85. Oxford: Oxford University Press.

———. 1998c. Goal-directed instrumental action: contingency and incentive learning and their cortical substrates. *Neuropharmacology* 37:407–19.

———. 2000. The effect of lesions of the insular cortex on instrumental conditioning: Evidence for a role in incentive memory. *J Neurosci* 20:8954–64.

Balleine, B.W., Garner, C., and Dickinson, A. 1995. Instrumental outcome devaluation is attenuated by the anti-emetic ondansetron. *Q J Exp Psychol B* 48:235–51.

Balleine, B.W., Killcross, A.S., and Dickinson, A. 2003. The effect of lesions of the basolateral amygdala on instrumental conditioning. *J Neurosci* 23:666–75.

Balleine, B.W., and Ostlund, S.B. 2007. Still at the choice point: Action selection and initiation in instrumental conditioning. *Ann NY Acad Sci* 1104:147–71.

Balleine, B.W., Paredes-Olay, C., and Dickinson, A. 2005. The effect of outcome devaluation on the performance of a heterogeneous instrumental chain. *Int J Comp Psychol* 18:257–72.

Bencherif, B., Guarda, A., Colantuoni, C., Ravert, H., Dannals, R., and Frost, J. 2005. Regional mu-opioid receptor binding in insular cortex is decreased in bulimia nervosa and correlates inversely with fasting behavior. *J Nucl Med* 46:1349–51.

Berridge, K.C. 1991a. Modulation of taste affect by hunger, caloric satiety, and sensory-specific satiety in the rat. *Appetite* 16:103–20.

———. 1991b. Modulation of taste affect by hunger, caloric satiety, and sensory-specific satiety in the rat. *Appetite* 16:103–20.

———. 2000a. Measuring hedonic impact in animals and infants: Microstructure of affective taste reactivity patterns. *Neurosci Biobehav Rev* 24:173–98.

———. 2000b. Reward learning: Reinforcement, incentives, and expectations. *Psychol Learn Motiv* 40:223–78.

Berridge, K.C., Flynn, F.W., Schulkin, J., and Grill, H.J. 1984. Sodium depletion enhances salt palatability in rats. *Behav Neurosci* 98:652–60.

Berridge, K.C., Grill, H.J., and Norgren, R. 1981. Relation of consummatory responses and preabsorptive insulin release to palatability and learned taste aversions. *J Comp Physiol Psychol* 95:363–82.

Bolles, R.C. 1967. *Theory of Motivation*. New York: Harper & Row.

Braun, J.J., Lasiter, P.S., and Kiefer, S.W. 1982. The gustatory neocortex of the rat. *Physiol Psychol* 10:13–45.

Brog, J.S., Salyapongse, A., Deutch, A.Y., and Zahm, D.S. 1993. The patterns of afferent innvervation of the core and shell in the "accumbens" part of the rat ventral striatum:immunohistochemical detection of retrogradely transported fluoro-gold. *J Comp Neurol* 338:255–78.

Changizi, M.A., McGehee, R.M.F., and Hall, W.G. 2002. Evidence that appetitive responses for dehydration and food deprivation are learned. *Physiol Behav* 75:295–304.

Colwill, R.C., and Rescorla, R.A. 1986. Associative structures in instrumental learning. *Psychol Learn Motiv* 20:55–104.

Colwill, R.M., and Rescorla, R.A. 1985. Postconditioning devaluation of a reinforcer affects instrumental responding. *J Exp Psychol Anim Behav Proc* 11:120–32.

———. 1990. Effect of reinforcer devaluation on discriminative control of instrumental behavior. *J Exp Psychol Anim Behav Process* 16:40–47.

Corbit, L.H., and Balleine, B.W. 2003. Instrumental and Pavlovian incentive processes have dissociable effects on components of a heterogeneous instrumental chain. *J Exp Psychol Anim Behav Process* 29:99–106.

Corbit, L.H., Muir, J.L., and Balleine, B.W. 2001. The role of the nucleus accumbens in instrumental conditioning: Evidence of a functional dissociation between accumbens core and shell. *J Neurosci* 21:3251–60.

Cunningham, C., and Noble, D. 1992. Conditioned activation induced by ethanol: Role in sensitization and conditioned place preference. *Pharmacol Biochem Behav* 43:307–13.

Damasio, A.R. 1996. The somatic marker hypothesis and the possible functions of the prefrontal cortex. *Philos Trans R Soc Lond B Biol Sci* 351:1413–20.

Davis, C., Levitan, R.D., Muglia, P., Bewell, C., and Kennedy, J.L. 2004. Decision-making deficits and overeating: a risk model for obesity. *Obes Res* 12:929–35.

De Houwer, J., Baeyens, F., and Field, A.P. 2005. Associative learning of likes and dislikes: Some current controversies and possible ways forward. *Cogn Emot* 19:161–74.

De Houwer, J., Thomas, S., and Baeyens, F. 2001. Associative learning of likes and dislikes: A review of 25 years of research on human evaluative conditioning. *Psychol Bull* 127:853–69.

Dickinson, A., and Balleine, B.W. 1993. Actions and responses: The dual psychology of behaviour. In *Spatial Representation*, eds. N. Eilan, R. McCarthy, and M.W. Brewer, 277–93. Oxford: Basil Blackwell.

———. 1994. Motivational control of goal-directed action. *Anim Learn Behav* 22:1–18.

———. 2002. The role of learning in the operation of motivational systems. In *Learning, Motivation & Emotion*. Vol. 3 of *Steven's Handbook of Experimental Psychology*. 3rd ed. Ed. C.R. Gallistel, 497–533. New York: John Wiley.

———. 2009. Hedonics: The cognitive-motivational interface. In *Pleasures of the Brain*, eds. M. Kringelbach, and K. Berridge, 74–84. Oxford: Oxford University Press.

Dickinson, A., and Brown, K.J. 2007. Flavor-evaluative conditioning is unaffected by contingency knowledge during training with color-flavor compounds. *Learn Behav* 35:36–42.

Ding, Y.Q., Kaneko, T., Nomura, S., and Mizuno, N. 1996. Immunohistochemical localization of mu-opioid receptors in the central nervous system of the rat. *J Comp Neurol* 367:375–402.

Everitt, B.J., and Stacey, P. 1987. Studies of instrumental behavior with sexual reinforcement in male rats (Rattus norvegicus): II. Effects of Preoptic area lesions, castration and testosterone. *J Comp Psychol* 101:407–19.

Forestell, C.A., and Lolordo, V.M. 2003. Palatability shifts in taste and flavour preference conditioning. *Q J Exp Psychol* 56B:140–60.

Garcia, J. 1989. Food for Tolman: Cognition and cathexis in concert. In *Aversion, Avoidance and Anxiety*, eds. T. Archer, and L.-G. Nilsson, 45–85. Hillsdale, NJ: Lawrence Erlbaum.

Garcia, J., Brett, L., and Rusiniak, K.W. 1989. Limits of Darwinian conditioning. In *Contemporary Learning Theories: Instrumental Conditioning Theory and the Impact of Biological Constraints on Learning*, eds. S.B. Klein, and R.R. Mowrer, 181–203. Hillsdale, NJ: Lawrence Erlbaum.

Grevert, P., and Goldstein, A. 1977a. Some effects of naloxone on behavior in the mouse. *Psychopharmacology (Berl)* 53:111–13.

———. 1977b. Effects of naloxone on experimentally induced ischemic pain and on mood in human subjects. *Proc Natl Acad Sci USA* 74:1291–94.

Hall, W., Arnold, H., and Myers, K. 2000. The acquisition of an appetite. *Psychol Sci* 11:101–5.

Harris, J.A., Gorissen, M.C., Bailey, G.K., and Westbrook, R.F. 2000. Motivational state regulates the content of learned flavor preferences. *J Exp Psychol Anim Behav Process* 26:15–30.

Hasselmo, M. 2005. A model of prefrontal cortical mechanisms for goal-directed behavior. *J Cogn Neurosci* 17:1115–29.

Hellemans, K.G., Dickinson, A., and Everitt, B.J. 2006. Motivational control of heroin seeking by conditioned stimuli associated with withdrawal and heroin taking by rats. *Behav Neurosci* 120:103–14.

Hendersen, R.W., and Graham, J. 1979. Avoidance of heat by rats: Effects of thermal context on the rapidity of extinction. *LearnMotiv* 10:351–63.

Holland, P.C. 2004. Relations between Pavlovian-instrumental transfer and reinforcer devaluation. *J Exp Psychol Anim Behav Process* 30:104–17.

Holland, P.C., Petrovich, G.D., and Gallagher, M. 2002. The effects of amygdala lesions on conditioned stimulus-potentiated eating in rats. *Physiol Behav* 76:117–29.

Hull, C.L. 1943. *Principles of Behavior*. New York: Appleton.

Hutcheson, D.M., Everitt, B.J., Robbins, T.W., and Dickinson, A. 2001. The role of withdrawal in heroin addiction: Enhances reward or promotes avoidance? *Nat Neurosci* 4:943–47.

James, W. 1890. *The Principles of Psychology*. Boston: Holt.

Johnson, L.R., Aylward, R.L., Hussain, Z., and Totterdell, S. 1994. Input from the amygdala to the rat nucleus accumbens: Its relationship with tyrosine hydroxylase immunoreactivity and identified neurons. *Neuroscience* 61:851–65.

Kerfoot, E., Agarwal, I., Lee, H., and Holland, P. 2007. Control of appetitive and aversive taste-reactivity responses by an auditory conditioned stimulus in a devaluation task: A FOS and behavioral analysis. *Learn Mem* 14:581–89.

Koob, G., and Le Moal, M.D. 1997. Drug abuse: hedonic homeostatic dysregulation. *Science* 278:52–58.

Kosterlitz, H., and Hughes, J. 1975. Some thoughts on the significance of enkephalin, the endogenous ligand. *Life Sci* 17:91–96.

Levin, B.E. 1999. Arcuate NPY neurons and energy homeostasis in diet-induced obese and resistant rats. *Am J Physiol* 276:R382–87.

Lopez, M., Balleine, B.W., and Dickinson, A. 1992. Incentive learning and the motivational control of instrumental performance by thirst. *Anim Learn Behav* 20:322–28.

Malkova, L., Gaffan, D., and Murray, E. 1997. Excitotoxic lesions of the amygdala fail to produce impairment in visual learning for auditory secondary reinforcement but interfere with reinforcer devaluation effects in rheus monkeys. *J Neurosci* 17:6011–20.

Moll, R.P. 1964. Drive and maturation effects in the development of consummatory behavior. *Psychol Rep* 15:295–302.

Morrison, C.D., and Berthoud, H.R. 2007. Neurobiology of nutrition and obesity. *Nutr Rev* 65:517–34.

Mucha, R., and Iversen, S. 1984. Reinforcing properties of morphine and naloxone revealed by conditioned place preferences: A procedural examination. *Psychopharmacology (Berl)* 82:241–47.

Mucha, R., and Walker, M. 1987. Aversive property of opioid receptor blockade in drug-naive mice. *Psychopharmacology (Berl)* 93:483–88.

Nader, K., Schafe, G.E., and LeDoux, J.E. 2000. The labile nature of consolidation theory. *Nat Rev Neurosci* 1:216–19.

Ostlund, S.B., and Balleine, B.W. 2007a. Instrumental reinstatement depends on sensory- and motivationally-specific features of the instrumental outcome. *Learn Behav* 35:43–52.

———. 2007b. Orbitofrontal cortex mediates outcome encoding in Pavlovian but not instrumental conditioning. *J Neurosci* 27:4819–25.

Pavlov, I.P. 1927. Conditioned reflexes: an investigation of the physiological activity of the cerebral cortex. xv:430.

Pecina, S., and Berridge, K.C. 2000. Opioid site in nucleus accumbens shell mediates eating and hedonic 'liking' for food: map based on microinjection Fos plumes. *Brain Res* 863:71–86.

Pecina, S., and Berridge, K. 2005. Hedonic hot spot in nucleus accumbens shell: Where do mu-opioids cause increased hedonic impact of sweetness? *J Neurosci* 25: 11777–86.

Pecina, S., Smith, K.S., and Berridge, K.C. 2006. Hedonic hot spots in the brain. *Neuroscientist* 12:500–11.

Petrovich, G.D., Setlow, B., Holland, P.C., and Gallagher, M. 2002. Amygdalo-hypothalamic circuit allows learned cues to override satiety and promote eating. *J Neurosci* 22:8748–53.

Pleyers, G., Corneille, O., Luminet, O., and Yzerbyt, V. 2007. Aware and (dis)liking: Item-based analyses reveal that valence acquisition via evaluative conditioning emerges only when there is contingency awareness. *J Exp Psychol Learn Mem Cogn* 33:130–44.

Poulin, J.F., Chevalier, B., Laforest, S., and Drolet, G. 2006. Enkephalinergic afferents of the centromedial amygdala in the rat. *J Comp Neurol* 496:859–76.

Reilly, S. 1999. The parabrachial nucleus and conditioned taste aversion. *Brain Res Bull* 48:239–54.

Rescorla, R.A. 1973. Informational variables in Pavlovian conditioning. *Psychol Learn Motiv* 10:1–46.

———. 1990. Instrumental responses become associated with reinforcers that differ in one feature. *Anim Learn Behav* 18:206–11.

Robledo, P., and Koob, G. 1993. Two discrete nucleus accumbens projection areas differentially mediate cocaine self-administration in the rat. *Behav Brain Res* 55:159–66.

Rolls, E.T., Rolls, B.J., and Rowe, E.A. 1983. Sensory-specific and motivation-specific satiety for the sight and taste of food and water in man. *Physiol Behav* 30:185–92.

Rozin, P., and Fallon, A.E. 1987. A perspective on disgust. *Psych Rev* 94:23–41.

Schafe, G.E., Nader, K., Blair, H.T., and LeDoux, J.E. 2001. Memory consolidation of Pavlovian fear conditioning: A cellular and molecular perspective. *Trends Neurosci* 24:540–46.

Sclafani, A. 1999. Macronutrient-conditioned flavor preferences. In *Neural Control of Macronutrient Selection*, eds. H.-R. Bertoud, and R.J. Seeley, 93–106. Boca Raton, FL: CRC Press.

Sclafani, A., Azzara, A.V., Touzani, K., Grigson, P.S., and Norgren, R. 2001. Parabrachial nucleus lesions block taste and attenuate flavor preference and aversion conditioning in rats. *Behav Neurosci* 115:920–33.

Seeley, R.J., Matson, C.A., Chavez, M., Woods, S.C., Dallman, M.F., and Schwartz, M.W. 1996. Behavioral, endocrine, and hypothalamic responses to involuntary overfeeding. *Am J Physiol* 271:R819–23.

Simmons, W., Martin, A., and Barsalou, L. 2005. Pictures of appetizing foods activate gustatory cortices for taste and reward. *Cereb Cortex* 15:1602–8.

Skoubis, P., and Maidment, N. 2003. Blockade of ventral pallidal opioid receptors induces a conditioned place aversion and attenuates acquisition of cocaine place preference in the rat. *Neuroscience* 119:241–49.

Smith, K.S., and Berridge, K.C. 2005. The ventral pallidum and hedonic reward: Neurochemical maps of sucrose "liking" and food intake. *J Neurosci* 25:8637–49.

———. 2007. Opioid limbic circuit for reward: interaction between hedonic hotspots of nucleus accumbens and ventral pallidum. *J Neurosci* 27:1594–1605.

Sripanidkulchai, K., Sripanidkulchai, B., and Wyss, J.M. 1984. The cortical projection of the basolateral amygdaloid nucleus in the rat: A retrograde fluorescent dye study. *J Comp Neurol* 229:419–31.

Stice, E., Spoor, S., Bohon, C., Veldhuizen, M., and Small, D. 2008. Relation of reward from food intake and anticipated food intake to obesity: A functional magnetic resonance imaging study. *J Abnorm Psychol* 117:924–35.

Sudakov, K.V. 1990. Oligopeptides in the organization of feeding motivation: A systemic approach. *Biomed Sci* 1:354–58.

Thorndike, E.L. 1911. *Animal Intelligence*. New York: Macmillan.

Wang, R., Liu, X., Hentges, S.T., Dunn-Meynell, A.A., Levin, B.E., Wang, W., and Routh, V.H. 2004. The regulation of glucose-excited neurons in the hypothalamic arcuate nucleus by glucose and feeding-relevant peptides. *Diabetes* 53:1959–65.

Wassum, K., Ostlund, S., Maidment, N., and Balleine, B. 2009. Distinct opioid circuits determine the palatability and desirability of rewarding events. *PNAS* 106:12512–17.

Wang, S.H., Ostlund, S.B., Nader, K., and Balleine, B.W. 2005. Consolidation and reconsolidation of incentive learning in the amygdala. *J Neurosci* 25:830–35.

Woods, S.C., Schwartz, M.W., Baskin, D.G., and Seeley, R.J. 2000. Food intake and the regulation of body weight. *Annu Rev Psychol* 51:255–77.

Woodson, J.C., and Balleine, B.W. 2002. An assessment of factors contributing to instrumental performance for sexual reward in the rat. *Q J Exp Psychol B* 55:75–88.

Yamamoto, T. 2006. Neural substrates for the processing of cognitive and affective aspects of taste in the brain. *Arch Histol Cytol* 69:243–55.

Yamamoto, T., Azuma, S., and Kawamura, Y. 1984. Functional relations between the cortical gustatory area and the amygdala: Electrophysiological and behavioral studies in rats. *Exp Brain Res* 56:23–31.

Yin, H., Ostlund, S., and Balleine, B. 2008. Reward-guided learning beyond dopamine in the nucleus accumbens: The integrative functions of cortico-basal ganglia networks. *Eur J Neurosci* 28:1437–48.

Young, P.T. 1949. Food-seeking drive, affective process, and learning. *Psychol Rev* 56:98–121.

14 Reward Predictions and Computations

John P. O'Doherty

CONTENTS

14.1 INTRODUCTION

The ability to predict when and where a reward will occur enables humans and other animals to initiate behavioral responses prospectively in order to maximize the probability of obtaining that reward. Reward predictions can take a number of distinct forms depending on the nature of the associative relationship underpinning them (Balleine et al. 2008; and see Chapters 13 and 15 in this volume). The simplest form of reward prediction is one based on an associative "Pavlovian" relationship between arbitrary stimuli and rewards, acquired following experience of repeated contingent pairing of the stimulus with the reward. Subsequent presentation of the stimulus elicits a predictive representation of the reward, by virtue of the learned stimulus-reward association. This form of prediction is purely passive: it signals when a reward might be expected to occur and elicits Pavlovian conditioned reflexes, but does not inform about the specific behavioral actions that should be initiated in order to obtain it. By contrast, other forms of reward prediction are grounded in learned instrumental associations between stimuli, responses, and rewards, thereby informing about the specific behavioral responses that, when performed by the animal, lead to a greater probability of obtaining that reward. Instrumental reward predictions can be either goal directed (based on response-outcome associations and therefore sensitive to the incentive value of the outcome), or habitual (based on stimulus-response associations and hence insensitive to changes in outcome value) (Balleine and Dickinson 1998). In this chapter, we will review evidence for the presence of multiple types of predictive reward signal in the brain. We will also outline some of the candidate computational mechanisms that might be responsible for the acquisition of these different forms of reward predictions and evaluate evidence for the presence of such mechanisms in the brain.

14.2   STIMULUS-BASED PREDICTIONS

Studies of the neural basis of stimulus-based Pavlovian reward-prediction signals have focused on the amygdala in the medial temporal lobes, orbitofrontal cortex (OFC) on the ventral surface of the frontal lobes, and the ventral striatum in the basal ganglia.

Single-unit recording studies in both rodents and non-human primates have implicated neurons in these areas in encoding stimulus-reward associations. Schoenbaum, Chiba, and Gallagher (1998) recorded from single neurons in rat amygdala and OFC, while on each trial animals were presented with one of two different odor cues indicating whether or not a subsequent nose poke by the rat in a food well would result in the delivery of an appetitive sucrose solution or an aversive quinine solution. Neurons in both amygdala and OFC were found to discriminate between cues associated with the positive and negative outcomes, and some were also found to show an anticipatory response related to the expected outcome. Paton et al. (2006) found evidence for separate neuronal populations in monkeys responding to one of three Pavlovian cues predictive of the subsequent delivery of either a pleasant juice reward, no outcome, or an aversive air puff to the eye. Furthermore, activity in these neurons was found to change as a function of reversal of the associations—some neurons stopped responding to a specific cue following reversal, while other neurons reversed their cue selectivity. Cue-related and anticipatory responses have also been found in monkey OFC related to a monkey's behavioral preference for the associated outcomes, such that the responses of the neurons to the cue paired with a particular outcome depend on the relative preference of the monkey for that outcome compared to another outcome presented in the same block of trials (Tremblay and Schultz 1999).

Another brain region implicated in Pavlovian reward prediction is ventral striatum. Neurons in ventral striatum have also been found to reflect expected reward in relation to the onset of a stimulus presentation, and activity of neurons in this area increases as a function of the degree of progression through a task sequence ultimately leading to reward (Shidara, Aigner, and Richmond 1998; Cromwell and Schultz 2003; Day et al. 2006). Similar findings have emerged from functional neuroimaging studies in humans. Gottfried, O'Doherty, and Dolan (2002) reported activity in both amygdala and orbitofrontal cortex following presentation of visual stimuli predictive of the subsequent delivery of both a pleasant and an unpleasant odor. Gottfried, O'Doherty, and Dolan (2003) subsequently showed that activity in these regions inresponse to Pavlovian cues track the current incentive value of an associated unconditioned stimulus (UCS). Subjects underwent conditioning in which food odors (the UCSs) were paired with visual conditioned stimuli (CSs), and then subsequently the value of a food odor associated with one of the stimuli was decreased by feeding subjects to satiety on a food corresponding to that specific odor. Neural responses to presentation of the CS paired with the devalued odor were found to correspondingly decrease in OFC, amygdala, and ventral striatum from before to after the satiation procedure, whereas no such decrease was evident for the CS paired with the non-devalued odor. The above findings therefore implicate a network of brain regions involving the amygdala, ventral striatum, and OFC in the learning and expression of stimulus-based reward predictions.

14.3   ACQUISITION OF STIMULUS-BASED REWARD PREDICTIONS

14.3.1   Theoretical Models of Reward Prediction Learning

The finding of stimulus-related predictive reward signals in the brain raises the question of how such signals are acquired in the first place. Modern theories of reward learning suggest that such learning proceeds by means of a signal called a prediction error, which encodes the difference between the reward that is predicted and the actual reward that is delivered. This notion was originally instantiated in the Rescorla-Wagner (RW) model of classical conditioning (Rescorla and Wagner 1972) and more recently espoused in a class of models collectively known as reinforcement learning (Sutton and Barto 1998). According to the RW model, the process by which a CS comes to produce a conditioned response is represented by two variables: V_t, which is the strength of the conditioned response elicited on trial t, or more abstractly the value of the CS, and U, which is the mean value of the UCS. For conditioning to occur, through repeated contingent presentations of the CS and US, the variable V_t (which may be initially zero at $t = 1$) should, as trials progress, converge toward the value of U. On any given trial if the reward is presented we can set $U_t = 1$, and if the reward is not presented $U_t = 0$. At the core of the RW model is the aforementioned prediction error signal δ_t, which represents the difference between the current value of U_t and V_t ($\delta_t = U_t - V_t$), on each conditioning trial. Under these circumstances δ_t will be positive because $V_t < U_t$. The value of V_t is then updated in proportion to δ_t. Assuming that the reward is always delivered when the CS is presented, then over the course of learning V_t eventually converges to U, δ_t will tend to zero, and once this happens learning is complete. However, if on a particular trial the reward is suddenly omitted after the CS is presented, on this occasion δt will be less than zero, because $V_t > U_t$. This would subsequently result in a reduction of the value of V_t. The idea that prediction error signals can take on both positive and negative response characteristics is the central feature of this form of learning model, and as we shall see is at the core of modern views on how such learning might be implemented in the brain.

An important extension of the RW model is the Temporal Difference (TD) Learning Model (Sutton 1988). This model overcomes some of the initial limitations of the RW model such as an inability to learn sequential stimulus-based predictions (e.g., when one stimulus predicts another stimulus which in turn predicts reward), and a lack of sensitivity to the timing between stimulus presentation and reward delivery, which is known to be a critical factor in modulating the efficacy of conditioning (Sutton 1988; Dayan and Abbott 2001). The key difference between the TD model and the earlier RW model is that whereas the RW model is trial based and only concerned with estimating predicted reward pertaining to a particular stimulus *across* trials, the TD model is also concerned with estimating the future-predicted reward from discrete time-points i within a trial until the end of the trial (Schultz, Dayan, and Montague 1997). As a consequence, the TD prediction error signal has a much richer temporal profile than its RW cousin. In particular, the TD error would first generate a strong positive signal at the time of presentation of the UCS before learning is established, but this positive prediction error signal would then shift back in time within a trial over the course of learning (Figure 14.1a). By the time learning is complete, the TD error would be positive only at the time of presentation of the earliest predictive cue stimulus. Furthermore, on any occasion in the trial where greater reward is delivered than expected (or if the reward is delivered sooner or later than expected), then a positive prediction error would be elicited. Similarly, if less reward is delivered than expected at the specific time that it has previously been found to occur, then a negative prediction error will be elicited. For more details of this model and its properties see Montague, Dayan, and Sejnowski (1996) and Dayan and Abbott (2001).

14.3.2 Prediction Error Signals in the Brain

Evidence for reward prediction error signals was found in the activity patterns of dopamine neurons recorded from awake, behaving, non-human primates undergoing simple instrumental or classical conditioning tasks (Schultz, Dayan, and Montague 1997; Schultz 1998). The response profile of these neurons does not correspond to a simple RW rule but rather has more in common with that predicted by TD learning, showing each of the temporal response properties within a trial described above. Just like the TD error signal, these neurons increase their firing when a reward is presented unexpectedly but decrease their firing from baseline when a reward is unexpectedly omitted, respond initially at the time of the UCS before learning is established, but shift back in time within a trial to respond instead at the time of presentation of the cue once learning has taken place. Further evidence in support of this hypothesis has been garnered from recent studies using fast cyclical voltammetry assays of dopamine release in ventral striatum during Pavlovian reward conditioning, whereby dopamine released into ventral striatum exhibited a shifting profile, occurring initially at the time of reward but subsequently shifting back to occur at the time of presentation of the reward-predicting cue (Day and Carelli 2007). To test for evidence of a temporal difference prediction error signal in the human

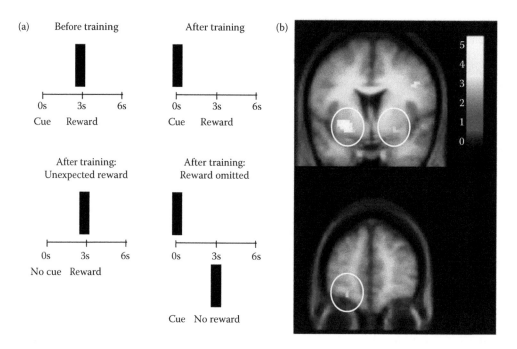

FIGURE 14.1 Temporal difference prediction error signals during Pavlovian reward conditioning in humans. (a) Properties of the temporal difference prediction error signal. This signal responds positively at the time of reward presentation before training, shifts to responding at the time of presentation of the predictive cue after training, and shows a decrease from baseline (negative signal) if an expected reward is omitted. (b) Results from an fMRI study testing for brain regions correlating with a temporal difference prediction error signal during classical conditioning in humans. Significant correlations with the TD prediction error signal were found in ventral striatum bilaterally (activated areas shown circled in top panel) and in orbitofrontal cortex (activation shown circled in bottom panel). (Reprinted from *Neuron*, 38, O'Doherty, J., Dayan, P., Friston, K., Critchley, H., Dolan, R.J., Temporal difference models and reward-related learning in the human brain, 329–37, Copyright 2003, with permission from **Elsevier**.)

brain, O'Doherty et al. (2003) scanned human subjects while they underwent a classical conditioning paradigm in which associations were learned between arbitrary visual fractal stimuli and a pleasant sweet taste reward. This study found significant correlations between a TD model and activity in a number of brain regions, most notably ventral striatum (ventral putamen bilaterally) (Figure 14.1b) and OFC, both prominent target regions of dopamine neurons. These results suggest that prediction error signals are present in the human brain during reward learning and that these signals conform to a response profile consistent with a specific computational model: TD learning. Another study by McClure, Berns, and Montague (2003) also revealed activity in ventral striatum consistent with a reward prediction error signal using an event-related trial-based analysis.

14.4 PREDICTIVE-REWARD SIGNALS OF INSTRUMENTAL ASSOCIATIONS

So far, we have considered reward predictions tied to the presentation of specific stimuli, which can often occur in situations where rewards are not in any way contingent on the performance of behavioral actions, but what about predictions related to the implementation of specific behavioral actions upon which rewards *are* contingent? A number of fMRI studies in humans have found evidence for expected reward signals in a specific brain region, the ventromedial prefrontal cortex, while subjects performed instrumental actions in order to obtain reward. Kim, Shimojo, and O'Doherty (2006) used a learning algorithm based on the RW rule to generate trial-by-trial predictions of expected reward as subjects made decisions between which of two possible actions to choose in order to obtain monetary reward in one condition or to avoid losing money in a different condition. Different actions were associated with distinct probabilities of either winning or losing money, such that in the reward condition one action was associated with a 60% probability of winning money and the other action with only a 30% probability of winning. To maximize their cumulative reward, subjects should learn to choose the action associated with the higher reward probability. In the avoidance condition, subjects were presented with a choice between the same probabilities, except in this context 60% of the time after choosing one action they avoided losing money, whereas this only occurred 30% of the time after choosing the alternate action. To minimize their losses, subjects should learn to choose the action associated with the 60% probability of loss avoidance. Model-generated expected value signals for the action chosen were found to be correlated on a trial-by-trial basis with BOLD responses in medial OFC and adjacent medial prefrontal cortex, which collectively can be described as ventromedial prefrontal cortex (vmPFC), in both the reward and avoidance conditions. In other words, activity in these areas was increased under situations where greater reward was expected for the action chosen (according to the learning algorithm) and decreased under conditions where less reward was expected for a given action (Figure 14.2). Similar

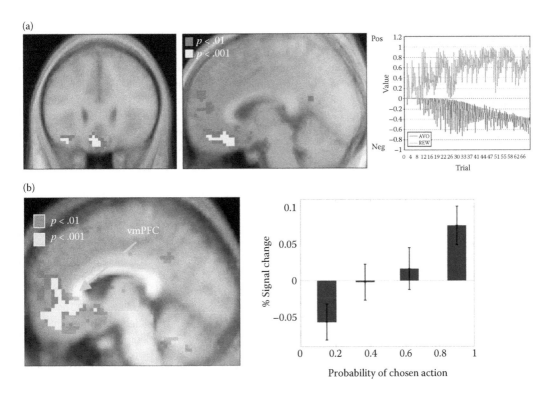

FIGURE 14.2 Reward prediction signals in ventromedial prefrontal cortex during reward-based action selection in humans. (a) Regions of ventromedial prefrontal cortex (medial and central orbitofrontal cortex extending into medial prefrontal cortex) correlating with expected value signals generated by a variant of the actor/critic model during an fMRI study of instrumental choice of reward and avoidance (left and middle panels). The model-predicted expected value signals are shown for one subject in the right panel for both the reward (top line) and avoidance (bottom line) conditions. (Data from Kim, H., Shimojo, S., and O'Doherty, J.P., *PLoS Biol* 4, e233, 2006. With permission.) (b) Similar regions of ventromedial prefrontal cortex correlating with model-predicted expected value signals during performance of a four-armed bandit task with non-stationary reward distributions (left panel). BOLD signal changes in this region are shown plotted against model predictions (right panel), revealing an approximately linear relationship between the expected value and BOLD signal changes in this region. (Data from Daw, N.D., O'Doherty, J.P., Dayan, P., Seymour, B., and Dolan, R.J., *Nature*, 441, 876, 2006. With permission.)

results were obtained by Daw et al. (2006), who used a four-armed bandit task in which "points" (that would later be converted into money) were paid out on each bandit. Again, activity in vmPFC correlated with the trial-by-trial estimate of expected reward attributable to the action (in this case bandit), chosen on that trial.

14.4.1 Goal-Directed Value Signals

However, although the above studies provide evidence of expected reward signals during reward-based action selection in vmPFC, they do not delineate the underlying associations upon which such signals depend. Such signals could be grounded in action-outcome associations, corresponding to the goal-directed component of instrumental conditioning, or alternatively could pertain to the encoding of habitual stimulus-response associations, which owing to the process of being incrementally strengthened as a result of prior reinforcement, would be stronger under situations where a given action yields greater reward than under situations where an action yields less reward. To address this question and determine whether human vmPFC is involved in encoding response-outcome (goal directed) or stimulus-response (habitual) associations, Valentin, Dickinson, and O'Doherty (2007) used an experimental design inspired by the animal learning literature (Balleine and Dickinson 1998). A key behavioral manipulation allowing one to determine whether behavior is under goal-directed or habitual control is to selectively devalue the outcome associated with that action, by for example feeding the animal to satiety on that outcome, and then testing the degree to which the animal persists in choosing that action following the devaluation process. If action selection is goal directed (and therefore depending on action-outcome associations), then the animal should immediately stop responding on the action associated with the devalued outcome. If, on the other hand, behavioral control is habitual and dependent on stimulus-response associations that do not link directly to the current value of the outcome, then the animal will persist in responding, as no representation of the outcome will be elicited during performance. In the Valentin, Dickinson, and O'Doherty study, subjects were scanned while they learned to choose instrumental actions associated with the subsequent delivery of different food rewards. Following training, one of these foods was devalued by feeding the subject to satiety on that food (similar to the approach used in Gottfried, O'Doherty, and Dolan 2003). The subjects were then scanned again, while being re-exposed to the instrumental choice procedure (in extinction). By testing for regions of the brain showing a change in activity during selection of the devalued action compared to that elicited during selection of the valued action from pre- to post-satiety, it is possible to identify regions showing sensitivity to the learned action-outcome associations, thereby revealing candidate areas responsible for goal-directed instrumental learning. The regions found to show such a response profile included vmPFC as well as an additional region of central OFC (Figure 14.3). These findings suggest that learned representations in vmPFC are sensitive to the current incentive value of the reward, ruling out a contribution of this region to habitual stimulus-response processing, which by definition would not show such sensitivity.

The aforementioned results would therefore appear to implicate vmPFC in encoding action-outcome and not stimulus-response associations. However, another possibility that cannot be ruled out on the basis of the Valentin, Dickinson, and O'Doherty study alone is that vmPFC may instead contain a representation of the discriminative stimulus (in this case the fractals), used to signal the different available actions, and that the learned associations in this region may be formed between these stimuli and the outcomes obtained (i.e., a Pavlovian rather than an instrumental association).

When comparing the results of the instrumental devaluation study by Valentin, Dickinson, and O'Doherty study with that of the purely Pavlovian devaluation study by Gottfried, O'Doherty, and Dolan (2003) reviewed in Section 14.2, there is some suggestion that the brain regions involved in instrumental associations may be at least partly dissociable. In the Gottfried, O'Doherty, and Dolan study, modulatory effects of reinforcer devaluation were found in central OFC but not vmPFC, whereas in the Valentin, Dickinson, and O'Doherty study evidence was found of instrumental devaluation effects in both central and medial OFC. This raises the possibility that medial OFC (part of vmPFC)

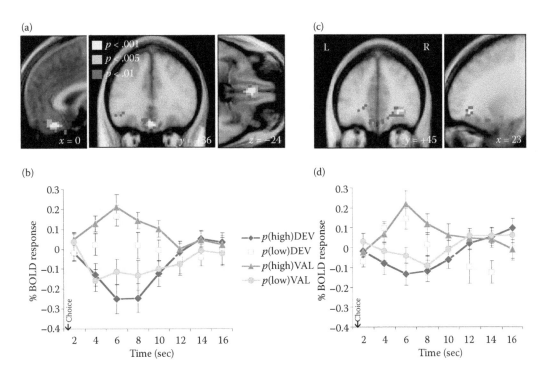

FIGURE 14.3 **(See Color Insert)** Regions of vmPFC and OFC showing response properties consistent with action-outcome learning. Neural activity during action selection for reward shows a change in response properties as a function of the value of the outcome with each action. Choice of an action leading to a high probability of obtaining an outcome that had been devalued (p(high)DEV) led to a decrease in activity in these areas whereas choice of an action leading to a high probability of obtaining an outcome that was still valued led to an increase in activity in the same areas. Devaluation was accomplished by means of feeding the subject to satiety on that outcome prior to the test period. (a) A region of medial OFC showing a significant modulation in its activity during instrumental action selection as a function of the value of the associated outcome. (b) Time-course plots derived from the peak voxel (from each individual subject) in the mOFC during trials in which subjects chose each one of the four different actions (choice of the high- vs. low-probability action in either the valued or devalued conditions). (c) A region of the right central OFC also showed significant modulation during instrumental conditioning. (d) Accompanying time-course plots from central OFC are shown. (Data from Valentin, V.V., Dickinson, A., and O'Doherty, J.P., *J Neurosci*, 27, 4019, 2007.)

may be more involved in the goal-directed component of instrumental conditioning whereas central OFC may contribute more to Pavlovian stimulus-outcome learning (as this area was found in both the Valentin, Dickinson, and O'Doherty study that did have a Pavlovian component and in the previous purely Pavlovian devaluation study). This proposal is compatible with the known anatomical connectivity of these areas in which central areas of OFC (Brodmann areas 11 and 13) receive input primarily from sensory areas, consistent with a role for these areas in stimulus-stimulus learning, whereas the medial OFC (areas 14 and 25) receives input primarily from structures on the adjacent medial wall of prefrontal cortex such as cingulate cortex, an area often implicated in response selection and/or reward-based action choice (Carmichael and Price 1995, 1996).

More compelling evidence of a role for vmPFC specifically in encoding action-related value signals has come from an fMRI study by Glascher et al. (2009). In this study, subjects participated in two distinct tasks: in the "action-based" task, subjects had to choose between performing one of two different physical motor responses (rolling a trackball vs. pressing a button) in the absence of explicit discriminative stimuli signaling those actions. Monetary rewards or losses were delivered on a probabilistic basis according to their choice of the different physical actions, and the rewards available on the different actions changed over time. Trial-by-trial model-predictive expected reward signals were generated for each action choice made by the subjects. Similar to the results found in studies where both discriminative stimulus information and action-selection components are present, in this task activity in vmPFC was found to track the expected reward corresponding to the chosen action (Figure 14.4). These results suggest that activity in vmPFC does not necessarily depend on the presence of discriminative stimuli, indicating that this region contributes to encoding of action-related value signals. In another, "stimulus-based," task, subjects performed action selection where decision options are denoted by the presence of specific discriminative stimuli; however, the two physical actions denoting the different choice options were randomly assigned (depending on random spatial position of the two discriminative stimuli). In common with the action-based reversal task, expected reward signals were also observed in vmPFC while subjects performed the stimulus-based task, consistent with a number of previous reports (Daw et al. 2006; Hampton, Bossaerts, and O'Doherty 2006; Kim, Shimojo, and O'Doherty 2006; Valentin, Dickinson, and O'Doherty 2007). Furthermore, in a conjunction analysis to test for regions commonly activated in both the action-based and stimulus-based choice conditions, robust activity was found in vmPFC. Overall, these findings could be taken to indicate that vmPFC contributes to both stimulus-based and action-based processes. An alternative possibility is that activity in vmPFC during the stimulus-based condition is, in common with that in the action-based condition, also being driven by goal-directed

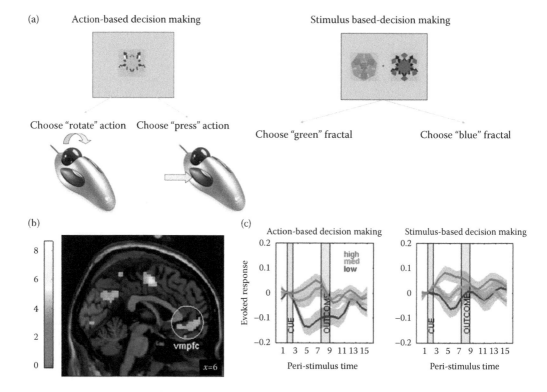

FIGURE 14.4 (See Color Insert) Expected reward representations in vmPFC during action-based and stimulus-based decision making. (a) Illustration of action-based decision-making task in which subjects must choose between one of two different physical actions that yield monetary rewards with differing probabilities, and a stimulus-based decision-making task in which subjects must choose between one of two different fractal stimuli randomly assigned to either the left or right of the screen, which yield rewards with differing probabilities. (b) A region of vmPFC correlating with expected reward during *both* action- and stimulus-based decision making. (c) Time courses from vmPFC during the action- and stimulus-based decision-making tasks plotted as a function of model-predicted expected reward from low (blue) to high (red). (From Glascher, J., Hampton, A.N., and O'Doherty, J.P., *Cereb Cortex*, 19, 483, 2009. With permission.)

action-outcome associations. Although in the stimulus-based task the particular physical motor response required to implement a specific decision varied on a trial-by-trial basis (depending on where the stimuli are presented), it is possible for associations to be learned between a combination of visual stimuli locations, responses, and outcomes. Thus, the common involvement of vmPFC in both the action- and stimulus-based reversal could be attributable to the possibility that this region is generally involved in encoding values of chosen actions but that those action-outcome relationships are encoded in a more abstract and flexible manner than concretely mapping specific physical motor responses to outcomes. The more flexible encoding of "actions" that this framework would entail may have parallels with computational theories of goal-directed learning in which action selection is proposed to occur via a flexible forward model system, which explicitly encodes the states of the world, the transition probabilities between those states, and the outcomes obtained in those states (Daw, Niv, and Dayan 2005). Overall, therefore, these findings suggest that vmPFC may play a role in encoding the value of chosen actions irrespective of whether those actions denote physical motor responses or more abstract decision options.

Additional evidence in support of a contribution of vmPFC to goal-directed learning and in encoding action-based value signals has come from a study by Tanaka, Balleine, and O'Doherty (2008). Apart from devaluing the outcome, another way to distinguish goal-directed from habitual behavior in the animal learning literature is to degrade the contingency. Contingency is the term used by behavioral psychologists to describe the relationship between an action and its consequences or outcome, which is defined in terms of the difference between two probabilities: the probability of the outcome given the action is performed and the probability of the outcome given the action is not performed (Hammond 1980; Beckers, De Houwer, and Matute 2007). Animals whose behavior is under control of the goal-directed system and who learn to perform an action for a reward under a situation of a highly contingent relationship between actions and outcomes will, following degradation of that contingency (by making the reward available even if the action is not performed), reduce their rate of responding on that action (Adams 1981; Dickinson and Balleine 1993). However, if animals are habitized, they will persist in responding on the action following contingency degradation, indicative of an insensitivity to action-outcome contingencies. Thus, another means apart from outcome devaluation to identify brain systems involved in goal-directed action-outcome learning and to discriminate those regions from those involved in habitual control is to assess neural activity tracking the degree of contingency between actions and outcomes during instrumental responding. To study this process in humans, Tanaka, Balleine, and O'Doherty abandoned the traditional trial-based approach, typically used in experiments using humans and non-human primates, in which subjects are cued to respond at particular times in a trial, for the unsignaled, self-paced approach more often used in studies of associative learning in rodents in which subjects themselves choose when to respond. Subjects were scanned with fMRI while in different sessions they responded on four different free operant reinforcement schedules that varied in the degree of contingency between responses made and rewards obtained. Consistent with the findings from the outcome devaluation study of Valentin, Dickinson, and O'Doherty, activity in two sub-regions of vmPFC (medial OFC and medial prefrontal cortex), as well as in dorsomedial striatum, was higher on average across a session when subjects were performing on a high-contingency schedule than in a session when they were performing on a low-contingency schedule (Figure 14.5). Moreover, in the sub-region of vmPFC identified on the medial wall, activity was found to vary not only with the degree of contingency overall on average across a schedule, but also with a locally computed estimate of the contingency between actions and outcome that tracks rapid changes in contingency over time within a session, implicating this specific sub-region of medial prefrontal cortex in the on-line computation of contingency between actions and outcomes.

When taken together, all of the evidence described above implicates vmPFC in encoding reward predictions based on action-outcome associations, which suggests that predictive reward representations in these medial parts of prefrontal cortex may be distinct from those in more central and lateral parts of OFC, amygdala, and ventral striatum, which, as reviewed in Section 14.2, may be

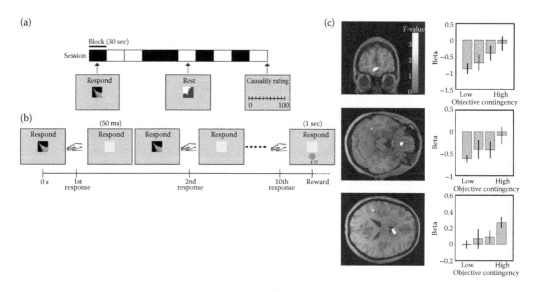

FIGURE 14.5 Brain regions tracking objective action-outcome contingency in humans. (a) Schematic of task design used in the human fMRI study by Tanaka, Balleine, and O'Doherty, (2008). Each experiment consisted of four sessions lasting 5 min each. A single session included five "RESPOND" blocks, in which the subject made button presses, and five "REST" blocks. At the end of each session, subjects rated how causal their button presses were in earning money on a scale from 0 to 100. (b) Example of the event schedule of the RESPOND block. When the subject pressed the button, the stimulus on the screen turned yellow for 50 ms. Rewards were then delivered according to the specific schedule being employed in that session. (c) Brain regions tracking global objective contingency were medial PFC (top), medial OFC (middle), and dorsomedial striatum (bottom). Bar plots show parameter estimates for each session plotted as a function of objective contingency (from lowest to highest). (From Tanaka, S.C., Balleine, B.W., and O'Doherty, J.P., *J Neurosci*, 28, 6750, 2008.)

more involved in encoding Pavlovian stimulus-outcome as opposed to instrumental action-outcome associations.

14.4.2 Habit Value Signals

The finding that predictive reward representations in vmPFC are based on action-outcome associations leaves open the question of where and how reward predictions based on habitual S-R associations are encoded. A reasonable hypothesis based on the rodent literature is that such signals could be present in a part of the dorsolateral striatum (Yin, Knowlton, and Balleine 2004). However, perhaps surprisingly, very little is known about where or how habitual value-signals are encoded in the human or primate brain more generally. Although a number of human fMRI studies have employed procedural learning paradigms, which are often assumed to invoke habit-like learning processes, in fact these studies have never attempted to determine whether procedurally learned behavior is in fact under goal-directed or habitual control by using either the outcome devaluation or contingency degradation probes described above. As a consequence, these studies have never been able to determine whether responses acquired in a procedural learning paradigm are, in fact, habitual, goal-directed, or a combination of both. However, in those few paradigms that have used the appropriate manipulations in humans, such as the studies by Valentin, Dickinson, and O'Doherty and Tanaka, Balleine, and O'Doherty, it is possible to test for the presence of habit-like value signals. In the Valentin study, such signals would take the form of being a predictive reward-signal that would discriminate between rewarding and non-rewarding actions during acquisition, but show no difference in activity between the devalued and valued actions following the devaluation procedure, whereas in the Tanaka, Balleine, and O'Doherty study, candidate habit signals would have been expected to emerge, particularly in situations where subjects were responding on a schedule with a low contingency between actions and outcomes. Notably in both of these studies, no clear evidence was found for predictive-reward signals consistent with habitual associative processes. The failure to find evidence for habit-like signals in these studies does not of course imply that those signals are not present. A number of factors could contribute to the lack of evidence for such signals in these studies. First of all, habitual processes by their very nature might be expected to be much less metabolically demanding such that the degree of blood oxygenation required to sustain them might be considerably less than would be the case for goal-directed representations. Therefore, perhaps the neural correlates of habit signals are much weaker and therefore more difficult to detect using standard BOLD imaging protocols than is the goal-directed component. Alternatively, perhaps the BOLD correlates of these different signals are only present in situations when behavior is being controlled by that system. In the Valentin study, it is known that behavior is under the control of the goal-directed system, because subjects showed significant reductions in their responses to the action associated with the devalued outcome compared to the action associated with the still valued outcome. In animal studies, one of the key behavioral manipulations required to demonstrate habitual control in rodents is to overtrain animals on a particular instrumental action (Balleine and Dickinson 1998). Actions that are exposed to only little or moderate training appear to be predominantly under goal-directed control. Thus, it is possible that in the Valentin, Dickinson, and O'Doherty study, the failure to observe clear evidence of habitual value-signals may be because subjects were exposed to only moderate training on the instrumental actions. Similarly, in Tanaka, Balleine, and O'Doherty, subjects were exposed to only brief 5 min sessions for each contingency, and therefore it is likely that behavior was also under goal-directed and not habitual control. A final possibility is that because humans probably possess enhanced cognitive control mechanisms compared to rodents, perhaps even after extensive training, human behavior might remain under goal-directed control. In order to address these questions, Tricomi, Balleine, and O'Doherty (2009) scanned subjects with fMRI while they performed on a variable interval schedule for food rewards. Subjects performed multiple training sessions each of 8 min duration during which they performed one of two different actions in particular trial blocks, which led to the delivery of one of two specific food outcomes (stored and consumed after the session). One group of subjects was overtrained by receiving four training sessions per day on three separate days, while another group received only moderate training by being exposed to only two sessions. Following the training sessions subjects were fed to satiety on one of the foods, thereby selectively devaluing that outcome. When tested in extinction, the overtrained group showed a tendency to maintain responding on the action associated with the now devalued outcome, remarkably analogous to behavior shown in rodents under similar circumstances, whereas the undertrained group quickly reduced their responding on the action associated with the devalued outcome. Moreover, activity in a posterior region of lateral striatum was found to increase over the course of training and was maximal on the final day of training in the overtrained group. These findings therefore indicate that humans do show behavioral evidence of habitual control following overtraining, just as in rodents, and that a region of posterior lateral striatum may be involved in such a process. This supports the idea that BOLD correlates of habit learning are increasingly detectable over the course of learning, perhaps reflecting the greater involvement of this system in controlling behavior as a function of training. It is also notable that the involvement of human posterolateral striatum in the expression of habitual behavior resonates with findings implicating a similar part of the striatum in this process in rodents (Yin, Knowlton, and Balleine 2004).

14.5 COMPUTATIONAL SIGNALS UNDERLYING LEARNING OF INSTRUMENTAL ASSOCIATIONS

We now consider computational theories about how instrumental associations, be they goal directed or habitual, might be acquired. In Section 14.3 we reviewed the RW model and its real-time extensions and showed how these models appear to provide a good characterization of neural signals underlying learning of stimulus-reward associations. For instrumental conditioning, a related class of models can be invoked collectively known as reinforcement learning (Sutton and Barto 1998). The core feature of reinforcement learning models is that in order to choose optimally between different actions, an agent needs to maintain internal representations of the expected reward available on each action, and then subsequently choose the action with the highest expected value. Also central to these algorithms is the notion of a prediction error signal that is used to learn and update expected values for each action through experience, just as in the RW learning model for Pavlovian conditioning described earlier. In one such model—the actor/critic—action selection is conceived as involving two distinct components: a critic, which learns to predict future reward associated with particular states in the environment, and an actor, which chooses specific actions in order to move the agent from state to state according to a learned policy (Barto 1992, 1995). The critic encodes the value of particular states in the world and as such has the characteristics of a Pavlovian reward prediction signal described above. The actor stores a set of probabilities for each action in each state of the world and chooses actions according to those probabilities. The goal of the model is to modify the policy stored in the actor such that over time those actions associated with the highest predicted reward are selected more often. This is accomplished by means of the aforementioned prediction error signal that computes the difference in predicted reward as the agent moves from state to state. This signal is then used to update value predictions stored in the critic for each state, but also to update action probabilities stored in the actor such that if the agent moves to a state associated with greater reward (and thus generates a positive prediction error), then the probability of choosing that action in future is increased. Conversely, if the agent moves to a state associated with less reward, this generates a negative prediction error and the probability of choosing that action again is decreased.

In order to distinguish regions of the brain involved in mediating the actor from the critic, O'Doherty et al. (2004) scanned hungry human subjects with fMRI while they performed a simple instrumental conditioning task in which they were required to choose one of two actions leading to juice reward with either a high or low probability (Figure 14.6a). Neural responses corresponding to the generation of prediction error signals during performance of the instrumental task were compared to that elicited during a control Pavlovian task in which subjects experienced the same stimulus-reward contingencies but did not actively choose which action to select. This comparison was designed to isolate the actor (which was hypothesized to be engaged only in the instrumental task) from the critic (which was hypothesized to be engaged in both the instrumental and Pavlovian control tasks). While dorsal striatum was correlated with prediction errors in the instrumental task only, by contrast ventral striatum was correlated with prediction errors in both the instrumental and Pavlovian tasks (Figure 14.6b). These findings thereby provided evidence of a possible ventral/dorsal trend within the striatum such that ventral striatum is more concerned with implementing the critic, while dorsal striatum is more involved in implementing the actor, confirming a initial proposal along these lines by Montague, Dayan, and Sejnowski (1996). However, while the above study provides evidence in support of the presence of an actor/critic type mechanism in the brain, it does not establish the causal role of such signals in learning these associations and in subsequently controlling behavior.

To address this issue, Schonberg et al. (2007) made use of an instrumental reward-conditioning task, which has the property that there is a high degree of variance across the population in the degree to which human subjects can successfully learn the task. Approximately 50% fail to converge in their choices toward the two options (out of the four available options) that yield the greatest probability of reward over 150 trials, whereas the other 50% tend to converge quite rapidly on the optimal choices. This property of the task provides a useful means of testing the degree to which reward-prediction error signals in dorsal striatum are engaged and can differentiate those subjects who successfully learn the instrumental associations from those who do not. To address this, both "learner" and "non-learner" groups were scanned while performing this task. Consistent with the possibility that reward-prediction errors in dorsal striatum are causally related to acquisition of instrumental reward associations, activity in dorsal striatum was significantly better correlated with reward prediction error signals

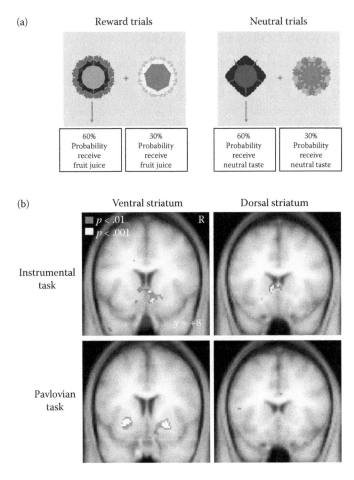

FIGURE 14.6 Prediction error signals underlying action selection for reward. (a) Schematic of instrumental choice task used by O'Doherty et al. (2004). On each trial of the reward condition, the subject chooses between two possible actions, one associated with a high probability of obtaining juice reward (60%), the other a low probability (30%). In a neutral condition, subjects also choose between actions with similar probabilities, but in this case they receive an affectively neutral outcome (tasteless solution). Prediction error responses during the reward condition of the instrumental choice task were compared to prediction error signals during a yoked Pavlovian control task. (b) Significant correlations with the reward prediction error signal generated by an actor/critic model were found in ventral striatum (ventral putamen extending into nucleus accumbens proper) in both the Pavlovian and instrumental tasks, suggesting that this region is involved in stimulus-outcome learning. By contrast, a region of dorsal striatum (anteromedial caudate nucleus) was found to be correlated with prediction error signals only during the instrumental task, suggesting that this area is involved in stimulus-response or stimulus-response-outcome learning. (Data from O'Doherty, J., Dayan, P., Schultz, J., Deichmann, R., Friston, K., and Dolan, R.J., *Science*, 304, 452, 2004. With permission.)

in learners than in non-learners. On the other hand, consistent with the actor/critic proposal, while reward-prediction error activity in ventral striatum was also weaker in the non-learners, this ventral striatum prediction error activity did not significantly differ between groups.

These results suggest a dorsal/ventral distinction within the striatum whereby ventral striatum is more concerned with Pavlovian or stimulus-outcome learning, while dorsal striatum is more engaged during learning of stimulus-response or stimulus-response-outcome associations. The suggestion that human dorsal striatum is specifically involved in situations when subjects need to select actions in order to obtain reward has received support from a number of other fMRI studies, both model based and trial based (Haruno et al. 2004; Tricomi, Delgado, and Fiez 2004).

14.6   COMPUTATIONAL MODELS OF INSTRUMENTAL LEARNING AND GOALS VERSUS HABITS

Another outstanding issue is to what extent does the distinction between goal-directed and habitual reward predictions described earlier map onto theories of computational reinforcement learning such as the actor/critic? Daw, Niv, and Dayan (2005) proposed that a reinforcement learning model such as the actor/critic is concerned purely with learning of habitual S-R value signals (see also Balleine, Daw, and O'Doherty 2008). According to this interpretation, action-value signals learned by an actor/critic would not be immediately updated following a change in the value of the reward outcome (such as by devaluation). Instead such an update would occur only after the model re-experiences the reward in its now devalued state and generates prediction errors that would incrementally modulate action-values. Alternatively, action-values learned by the actor-critic could reflect the strength of an association not only between stimuli and responses, but also between actions and outcomes. In that event, such a representation would show devaluation effects and hence meet the criteria of being goal directed. There is currently insufficient empirical data to distinguish between these possibilities. One way potentially to address this is to determine whether the reward-prediction error signal generated by dopamine neurons (which should be at the core of any reinforcement learning-like process) is sensitive to devaluation effects. If not, this would support the idea that RL models and the associated error signal is involved in learning habitual S-R associations; otherwise the idea that dopamine neurons may contribute to the goal-directed component of learning will have to be entertained.

Daw et al. (2005) also proposed an alternative model to the actor/critic to account for the goal-directed component of instrumental learning: a forward model. Unlike reinforcement learning, which develops approximate or "cached" values for particular actions based on prior experience with those actions, in the "forward model," values for different actions are worked out on-line by taking into account knowledge about the rewards available in each state and the transition probabilities between each state, and iteratively working out the value of each available option, analogous to how a real chess player or computer chess algorithm might work out which chess move to make next by explicitly thinking through the consequences of all possible moves. One property of this model is that value representations should be modulated not only incrementally, but instantaneously following detection of a change in the underlying state-space or structure of the decision problem.

Evidence that predictive reward representations in the brain are sensitive to changes in state has emerged from a study by Hampton, Bossaerts, and O'Doherty (2006). In this study subjects were asked to participate in a decision problem called probabilistic reversal learning in which two different actions yield reward with distinct probabilities. The key feature of the task is that the rewards available on the two actions are anti-correlated, that is, when one action has a high value the other has a low value, in terms of the amount of rewards that are obtained probabilistically on the two actions, and that after an unpredictable series of trials the reward probabilities on the two actions reverse. Hampton et al. compared two computational models in terms of how well they could account for human choice behavior on the task and for the pattern of neural activity in vmPFC during task performance. One of these algorithms incorporated the rules or structure of the decision problem as would be expected for a state-based inference mechanism, such that following a reversal the value of the two actions was instantly updated to reflect knowledge of the abstract structure. The other model was a simple reinforcement learning algorithm that did not incorporate the structure and thus would only learn to slowly and incrementally update values following successive reinforcements. Consistent with a role for vmPFC in state-based inference, predicted reward signals in this region were found to reflect the structure of the decision problem, such that activity was updated instantly following a reversal, rather than being updated incrementally as might be expected for a simple non-state-based reinforcement learning mechanism.

Although the forward model approach is appealing as a possible computational explanation for goal-directed learning, evidence in support of such a model is still rather preliminary. For now the main conclusion available on the basis of the current evidence is that prediction error signals certainly seem to play a role in the acquisition of instrumental associations, particularly through input into dorsal striatum, and that changes in state representations can result in the direct modulation of value signals in vmPFC. Further studies both at the level of human imaging and single-unit neurophysiology in other animals will be needed to establish the extent to which such signals contribute to learning of goal-directed value signals, habitual values, or both.

14.7 CONCLUSIONS

In this chapter, we have reviewed evidence for the existence of multiple types of predictive reward-signals in the human brain and for a number of different possible learning mechanisms that might underlie their acquisition. Pavlovian or stimulus-bound reward predictions appear to be present in three principle brain regions: OFC (particularly central and lateral areas), amygdala, and ventral striatum. Such signals may be learned via a stimulus-based reward prediction error signal originating in the phasic activity of dopamine neurons and projecting to ventral striatum and elsewhere. In addition, reward predictions based on instrumental action-outcome associations appear to be encoded in vmPFC, which incorporates medial OFC and adjacent medial prefrontal cortex, as well as the anterior medial striatum. Furthermore, a region of more posterior lateral striatum appears to be engaged once behavior is under habitual and not goal-directed control when following repetitive performance of a particular action: individuals no longer take into account the incentive value of outcomes while performing actions previously associated with those outcomes. There is also now considerable evidence to suggest that prediction error signals into dorsal striatum may play a direct role in the acquisition of instrumental reward associations.

An important future direction will be to establish clearly the extent to which stimulus-bound and action-bound predictions are neurally dissociable. Another pressing issue, not touched on in the present chapter, is that once the neural systems responsible for each different type of reward prediction have been delineated, it will also be important to begin to understand how these different types of reward-prediction systems interact together in order to ultimately control behavior. Although this question has already been extensively studied in the animal literature, it has, as yet, only received preliminary treatment in humans (Bray et al. 2008; Talmi et al. 2008).

REFERENCES

Adams, C.D. 1981. Variations in the sensitivity of instrumental responding to reinforcer devaluation. *Q J Exp Psychol* 34B:77–98.

Balleine, B.W., Daw, N.D., and O'Doherty J.P. 2008. Multiple forms of value learning and the function of dopamine. In *Neuroeconomics: Decision Making and the Brain*, eds. P.W. Glimcher, C.F. Camerer, R.A. Poldrack, and E. Fehr, 367–87. New York: Academic Press.

Balleine, B.W., and Dickinson, A. 1998. Goal-directed instrumental action: Contingency and incentive learning and their cortical substrates. *Neuropharmacology* 37:407–19.

Barto, A.G. 1992. Reinforcement learning and adaptive critic methods. In *Handbook of Intelligent Control: Neural, Fuzzy, and Adaptive Approaches*, eds. D.A. White and D.A. Sofge, 469–91. New York: Van Norstrand Reinhold.

———. 1995. Adaptive critics and the basal ganglia. In *Models of Information Processing in the Basal Ganglia*, eds. J.C. Houk, J.L. Davis, and B.G. Beiser, 215–32. Cambridge, MA: MIT Press.

Beckers, T., De Houwer, J., and Matute, H., eds. 2007. *Human Contingency Learning: Recent Trends in Research and Theory*. London: Psychology Press.

Bray, S., Rangel, A., Shimojo, S., Balleine, B., and O'Doherty, J.P. 2008. The neural mechanisms underlying the influence of pavlovian cues on human decision making. *J Neurosci* 28:5861–66.

Carmichael, S.T., and Price, J.L. 1995. Sensory and premotor connections of the orbital and medial prefrontal cortex of macaque monkeys. *J Compar Neurol* 363:642–64.

———. 1996. Connectional networks within the orbital and medial prefrontal cortex of macaque monkeys. *J Compar Neurol* 371:179–207.

Cromwell, H.C., and Schultz, W. 2003. Effects of expectations for different reward magnitudes on neuronal activity in primate striatum. *J Neurophysiol* 89:2823–38.

Daw, N.D., Niv, Y., and Dayan, P. 2005. Uncertainty-based competition between prefrontal and dorsolateral striatal systems for behavioral control. *Nat Neurosci* 8:1704–11.

Daw, N.D., O'Doherty, J.P., Dayan, P., Seymour, B., and Dolan, R.J. 2006. Cortical substrates for exploratory decisions in humans. *Nature* 441:876–79.

Day, J.J., Wheeler, R.A., Roitman, M.F., and Carelli, R.M. 2006. Nucleus accumbens neurons encode Pavlovian approach behaviors: Evidence from an autoshaping paradigm. *Eur J Neurosci* 23:1341–51.

Day, J.J., and Carelli, R.M. 2007. The nucleus accumbens and Pavlovian reward learning. *Neuroscientist* 13:148–59.

Dayan, P., and Abbott, L. 2001. *Theoretical Neuroscience*. Boston: MIT Press.

Dickinson, A., and Balleine, B.W. 1993. Actions and responses: The dual psychology of behaviour. In *Spatial Representation*, eds. N. Eilan, R. McCarthy, and M.W. Brewer, 277–93. Oxford: Basil Blackwell.

Glascher, J., Hampton, A.N., and O'Doherty, J.P. 2009. Determining a role for ventromedial prefrontal cortex in encoding action-based value signals during reward-related decision making. *Cereb Cortex* 19:483–95.

Gottfried, J.A., O'Doherty, J., and Dolan, R.J. 2002. Appetitive and aversive olfactory learning in humans studied using event-related functional magnetic resonance imaging. *J Neurosci* 22:10829–37.

———. 2003. Encoding predictive reward value in human amygdala and orbitofrontal cortex. *Science* 301:1104–7.

Hammond, L.J. 1980. The effect of contingency upon appetitive conditioning of free operant behavior. *J Exp Anal Behav* 34:297–304.

Hampton, A.N., Bossaerts, P., and O'Doherty, J.P. 2006. The role of the ventromedial prefrontal cortex in abstract state-based inference during decision making in humans. *J Neurosci* 26:8360–67.

Haruno, M., Kuroda, T., Doya, K., Toyama, K., Kimura, M., Samejima, K., Imamizu, H., and Kawato, M. 2004. A neural correlate of reward-based behavioral learning in caudate nucleus: A functional magnetic resonance imaging study of a stochastic decision task. *J Neurosci* 24:1660–65.

Kim, H., Shimojo, S., and O'Doherty, J.P. 2006. Is avoiding an aversive outcome rewarding? Neural substrates of avoidance learning in the human brain. *PLoS Biol* 4:e233.

McClure, S.M., Berns, G.S., and Montague, P.R. 2003. Temporal prediction errors in a passive learning task activate human striatum. *Neuron* 38:339–46.

Montague, P.R., Dayan, P., and Sejnowski, T.J. 1996. A framework for mesencephalic dopamine systems based on predictive Hebbian learning. *J Neurosci* 16:1936–47.

O'Doherty, J., Dayan, P., Friston, K., Critchley, H., and Dolan, R.J. 2003. Temporal difference models and reward-related learning in the human brain. *Neuron* 38:329–37.

O'Doherty, J., Dayan, P., Schultz, J., Deichmann, R., Friston, K., and Dolan, R.J. 2004. Dissociable roles of ventral and dorsal striatum in instrumental conditioning. *Science* 304:452–54.

Paton, J.J., Belova, M.A., Morrison, S.E., and Salzman, C.D. 2006. The primate amygdala represents the positive and negative value of visual stimuli during learning. *Nature* 439:865–70.

Rescorla, R.A., and Wagner, A.R. 1972. A theory of Pavlovian conditioning: Variations in the effectiveness of reinforcement and nonreinforcement. In *Classical Conditioning II: Current Research and Theory*, eds. A.H. Black and W.F. Prokasy, 64–99. New York: Appleton Crofts.

Schoenbaum, G., Chiba, A.A., and Gallagher, M. 1998. Orbitofrontal cortex and basolateral amygdala encode expected outcomes during learning. *Nat Neurosci* 1:155–59.

Schonberg, T., Daw, N.D., Joel, D., and O'Doherty, J.P. 2007. Reinforcement learning signals in the human striatum distinguish learners from nonlearners during reward-based decision making. *J Neurosci* 27:12860–67.

Schultz, W. 1998. Predictive reward signal of dopamine neurons. *J Neurophysiol* 80:1–27.

Schultz, W., Dayan, P., and Montague, P.R. 1997. A neural substrate of prediction and reward. *Science* 275:1593–99.

Shidara, M., Aigner, T.G., and Richmond, B.J. 1998. Neuronal signals in the monkey ventral striatum related to progress through a predictable series of trials. *J Neurosci* 18:2613–25.

Sutton, R.S. 1988. Learning to predict by the methods of temporal differences. *Mach Learn* 3:9–44.

Sutton, R.S., and Barto, A.G. 1998. *Reinforcement Learning*. Cambridge, MA: MIT Press.

Talmi, D., Seymour, B., Dayan, P., and Dolan, R.J. 2008. Human pavlovian-instrumental transfer. *J Neurosci* 28:360–68.

Tanaka, S.C., Balleine, B.W., and O'Doherty, J.P. 2008. Calculating consequences: Brain systems that encode the causal effects of actions. *J Neurosci* 28:6750–55.

Tremblay, L., and Schultz, W. 1999. Relative reward preference in primate orbitofrontal cortex. *Nature* 398:704–8.

Tricomi, E., Balleine, B.W., and O'Doherty, J.P. 2009. A specific role for posterior dorsolateral striatum in human habit learning. *Eur J Neurosci* 29:2225–32.

Tricomi, E.M., Delgado, M.R., and Fiez, J.A. 2004. Modulation of caudate activity by action contingency. *Neuron* 41:281–92.

Valentin, V.V., Dickinson, A., and O'Doherty, J.P. 2007. Determining the neural substrates of goal-directed learning in the human brain. *J Neurosci* 27:4019–26.

Yin, H.H., Knowlton, B.J., and Balleine, B.W. 2004. Lesions of dorsolateral striatum preserve outcome expectancy but disrupt habit formation in instrumental learning. *Eur J Neurosci* 19:181–89.

15 Orbitofrontal Cortex and Outcome Expectancies: Optimizing Behavior and Sensory Perception

Geoffrey Schoenbaum, Matthew R. Roesch,
Tom A. Stalnaker, and Yuji K. Takahashi

CONTENTS

15.1 INTRODUCTION

Orbitofrontal cortex has long been associated with adaptive, flexible behavior. Indeed the argument has been made that the ability of humans to adapt so rapidly to changing circumstances is, in part, linked to the expansion of this and other prefrontal regions. The association between the orbitofrontal cortex and adaptive behavior is apparent in accounts by Dr. John Harlow in 1868 (Harlow 1868) of the erratic, inflexible, stimulus-bound behavior of Phineas Gage, who reportedly suffered extensive damage to the orbital prefrontal regions (Damasio et al. 1994). Since then, increasingly refined experimental work has demonstrated repeatedly that damage to the orbitofrontal region, a set of loosely defined areas in the prefrontal regions overlying the orbits (Price 2007), impairs the ability of animals and humans to rapidly change their behavior in the face of changing contingencies and unexpected outcomes (see Chapter 16 in this volume for further considerations of this topic).

Two dominant hypotheses have been advanced to explain the role of orbitofrontal cortex in adaptive behavior. The first was the idea that orbitofrontal cortex is fundamentally critical to the ability to inhibit inappropriate or incorrect responses. The second was the suggestion that orbitofrontal cortex supports rapid changes in behavior because it serves as a rapidly flexible encoder of associative information. By this account, orbitofrontal cortex is faster at learning new information than other brain areas and thereby drives selection of the correct response or, perhaps, inhibits selection of the incorrect response.

More recently the orbitofrontal cortex has been shown to be critical to signaling of outcome expectancies—signals concerning the characteristics, features, and specific value of particular outcomes that are predicted by cues (and perhaps responses; though see Ostlund and Balleine 2007b) in the environment (Schoenbaum and Roesch 2005). Here we will argue that this function provides a better explanation for the role of the orbitofrontal cortex in adaptive behavior than either of the established hypotheses. First, we will review data from reversal learning tasks, showing that orbitofrontal cortex is critical for changing behavior in the face of unexpected outcomes. Then, we will provide a brief overview of the two dominant hypotheses, followed by data that directly contradict both accounts. Thereafter, we will review more recent evidence that orbitofrontal cortex is critical to signaling information about expected outcomes. As we will show, these signals are prominent in the neural activity and BOLD response in orbitofrontal cortex, and their role in guiding behavior is evident in deficits caused by orbitofrontal damage in a variety of behavioral settings in which outcomes must be used to guide normal behavior, even when contingencies are not changing. We will suggest that these same signals are also necessary for the detection of prediction errors when contingencies are changing, thereby facilitating changes in associative representations in other brain areas and, ultimately, behavior. Finally, we will also suggest that expectancy signals in orbitofrontal cortex might also impact early sensory regions, optimizing behavior through context-dependent firing and recall of environmental cues predicting future reward.

15.2   ORBITOFRONTAL CORTEX IS CRITICAL FOR ADAPTIVE BEHAVIOR

The role of orbitofrontal cortex in adaptive behavior is typically assessed experimentally using tasks that incorporate reversal learning. In these tasks, a subject is first trained that responding under one circumstance leads to reward and that responding under another circumstance will lead to non-reward or punishment. After stable responding is established, the contingencies or associations are reversed, and the subject must switch or reverse responding. This can be done in rats using a simple odor discrimination task. Rats are trained to sample odor cues at a centrally located port and then visit a nearby fluid well. Rats learn that one odor predicts sucrose, while a different odor predicts quinine. Rats like sucrose but want to avoid quinine, and so they learn to discriminate between the two odor cues, making "go" responses after sampling the positive, sucrose-predicting cue and withholding that response, called a "no-go," after sampling the negative, quinine-predicting cue. After they have learned this discrimination to a 90% performance criterion and are responding stably at that level, we can assess their ability to rapidly reverse their responses by switching the odor-outcome associations.

As illustrated in Figure 15.1, lesions of the orbitofrontal cortex cause a specific impairment in the ability of rats to rapidly reverse responding in this setting (Schoenbaum et al. 2003a). Lesions encompassed the dorsal bank of the rhinal sulcus, including the laterally located orbital and agranular insular regions. These areas have a pattern of connectivity with sensory regions, mediodorsal thalamus, basolateral amygdala, and ventral striatum that approximates that of orbital regions in primate species (Schoenbaum and Setlow 2001). Lesions did not include medial orbital areas, which have a different pattern of connectivity. Importantly, these lesions had no effect on the ability of the rats to acquire the odor problems (Figure 15.1a). Thus orbitofrontal-lesioned rats were able to detect and discriminate between the odor cues, were similarly motivated to respond for sucrose and to avoid quinine, and were able to modify responding when first introduced to a negative odor cue. However, these rats were markedly slower to modify responding when the odor-outcome associations of the final odor pair were reversed (Figure 15.1b); they required approximately twice as many trials as controls to relearn this discrimination after the first reversal, and this deficit reemerged when the problem was re-reversed. This is the classic reversal learning deficit that is associated with orbitofrontal damage. This same deficit has been shown repeatedly by a host of different labs over the past 50 years across species, designs, and learning materials (Teitelbaum 1964; Butter 1969; Jones and Mishkin 1972; Rolls et al. 1994; Bechara et al. 1997; Meunier, Bachevalier, and Mishkin

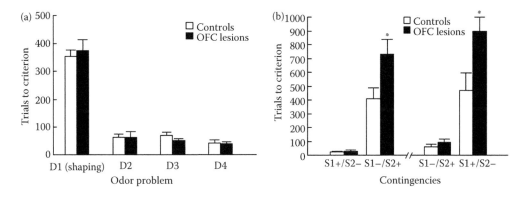

FIGURE 15.1 Effect of orbitofrontal lesions on rapid reversal of trained responses. Rats were trained to sample odors at a central port and then respond at a nearby fluid well. In each odor problem, one odor predicted sucrose and a second quinine. Rats had to learn to respond for sucrose and to inhibit responding to avoid quinine. Shown are trials required by controls and orbitofrontal-lesioned rats to meet a 90% performance criterion on a series of 4 of these odor problems (a) followed by two serial reversals of the final problem (b). (Adapted from Schoenbaum, G. et al., *Learning and Memory*, 10, 129, 2003. With permission.)

1997; Schoenbaum et al. 2002; Chudasama and Robbins 2003; Fellows and Farah 2003; McAlonan and Brown 2003; Hornak et al. 2004; Izquierdo, Suda, and Murray 2004; Pais-Vieira, Lima, and Galhardo 2007; Bissonette et al. 2008; Reekie et al. 2008). Indeed this may be one of the more reliable brain-behavior relationships of which we are aware, rivaling even that of hippocampus and declarative memory in its reproducibility and robustness.

15.3 WHY IS ORBITOFRONTAL CORTEX CRITICAL FOR ADAPTIVE BEHAVIOR?

Such a clear association between a particular brain area and a particular cognitive function begs an explanation. The search for an explanation has been dominated by two proposals. In this section, we will review these proposals and then discuss evidence that directly contradicts them.

15.3.1 ORBITOFRONTAL CORTEX AS INHIBITOR OF RESPONDING

The first and perhaps still most prevalent proposal is that orbitofrontal cortex is critical for adaptive behavior generally—and reversal learning in particular—because it plays a fundamental role in inhibiting inappropriate responses. This idea has a long history, going back at least 135 years to writings by David Ferrier (1876), who proposed, based on experimental work in dogs and monkeys, that the frontal lobes might be critical to suppressing motor acts in favor of attention and planning. Though also associated generally with prefrontal cortex (Mishkin 1964), the function of response inhibition has become more closely associated with orbitofrontal function in the past several decades, because it provides a very attractive description of the general and even specific behavioral effects of orbitofrontal damage (Fuster 1997). For example, this explanation fits well with the "orbitofrontal syndrome," which is typically described as a constellation of symptoms including impulsive, disinhibited, and perseverative responding (Damasio 1994), and of course it also provides an excellent description of reversal learning deficits (Teitelbaum 1964; Butter 1969; Jones and Mishkin 1972; Rolls et al. 1994; Bechara et al. 1997; Meunier, Bachevalier, and Mishkin 1997; Schoenbaum et al. 2002; Chudasama and Robbins 2003; Fellows and Farah 2003; McAlonan and Brown 2003; Hornak et al. 2004; Izquierdo, Suda, and Murray 2004; Pais-Vieira, Lima, and Galhardo 2007; Bissonette et al. 2008; Reekie et al. 2008) as well as deficits in detour-reaching and stop-signal tasks (Wallis et al. 2001; Dillon and Pizzagalli 2007; Eagle et al. 2008; Torregrossa, Quinn, and Taylor 2008).

However, this idea does not have very good predictive power outside of these settings. In fact, it is quite easy to find behaviors that have a high requirement for response inhibition that are not affected by orbitofrontal damage. For example, in many of the reversal studies described earlier, orbitofrontal-lesioned subjects are able to successfully inhibit the same responses during initial learning that they have such difficulty inhibiting after reversal (Rolls et al. 1994; Bechara et al. 1997; Meunier Bachevalier, and Mishkin 1997; Schoenbaum et al. 2002; Pais-Vieira, Lima, and Galhardo 2007). This includes our setting, in which rats have to inhibit a strongly pre-trained response at the fluid well during initial discrimination learning (Figure 15.1a) (Schoenbaum et al. 2003a). Before they are ever trained to withhold responding, rats in our studies typically receive 500–1000 shaping trials in which they sample an odorized air stream and then respond for reward. Yet when a new odor cue that predicts a negative outcome is introduced, they learn to inhibit this highly trained response at the same rate as controls and they also show the same improvement over several different odor discrimination problems. These data indicate that orbitofrontal cortex is not necessary, generally, for inhibiting pre-trained responses.

Similarly, orbitofrontal cortex is not required for inhibiting "pre-potent" or innate response tendencies. For example, in reinforcer devaluation tasks, animals with orbitofrontal lesions are readily able to withhold the selection or consumption of food that has been paired with illness or fed to satiety (Gallagher, McMahan, and Schoenbaum 1999; Baxter et al. 2000; Pickens et al. 2003; Izquierdo, Suda, and Murray 2004; Pickens et al. 2005; Burke et al. 2008). This ability can also be demonstrated in a reversal setting, as in a study by Murray et al. (Chudasama, Kralik, and Murray 2007). In this study, monkeys were allowed to choose between different size peanut rewards. To receive the larger amount, they had to select the smaller one. Thus the monkeys had to switch or reverse their innate bias toward selecting the reward that they wanted. Monkeys with orbitofrontal damage learned to do so just as well as controls. These data also indicate that orbitofrontal cortex is not necessary for inhibiting innate responses. Thus, while the inability to inhibit inappropriate responses is an effect or symptom of orbitofrontal damage that is evident in a number of settings, including reversals, this description does not provide an adequate all-inclusive definition of the underlying function that orbitofrontal cortex contributes to behavior.

15.3.2 Orbitofrontal Cortex as Flexible Encoder of Associative Information

More recently it has been proposed that the orbitofrontal cortex is critical to adaptive behavior because it is a rapidly flexible encoder of associative information, particular for associations between cues and appetitive and aversive outcomes (Rolls 1996). According to this hypothesis, the orbitofrontal cortex is better than other brain areas at rapidly encoding the new, correct associations and contributes to adaptive behavior by signaling this information to other areas, thereby driving selection of the correct response or, perhaps, inhibiting selection of the incorrect one.

This idea has its roots in single-unit recording studies reporting that cue-evoked neural activity in orbitofrontal cortex rapidly reflects associative information, under normal circumstances and also during reversal learning. This was first reported by Rolls and colleagues (Thorpe, Rolls, and Maddison 1983), who noted that single-units in monkey orbitofrontal cortex would often change their firing to objects, such as syringes, when their association with reward was changed, say by filling them with aversive saline rather than rewarding juices. The change in firing occurred rapidly after the monkeys experienced the unexpected outcome. Rolls has also reported similar reversal of associative encoding in single-units recorded in a visual discrimination task (Rolls et al. 1996). We've seen similar correlates in orbitofrontal neurons recorded in rats learning to reverse odor discriminations (Schoenbaum et al. 1999, 2003b; Stalnaker et al. 2006).

The prominent reversal of associative correlates in these orbitofrontal neurons during reversal learning and the poor reversal performance caused by orbitofrontal damage are clearly consistent with the idea that this area supports adaptive behavior because it is particularly efficient at learning new associative information. According to this proposal, orbitofrontal cortex acts as a rapidly flexible associative look-up table, deciphering the outcome associated with a particular cue after reversal more rapidly and with greater accuracy than any other brain region. This proposal has enormous explanatory power for these data; yet like the response inhibition proposal above, it is not consistent with more recent evidence that directly tests its predictions.

For example, if orbitofrontal cortex is an associative look-up table, one might expect reversal of encoding to be a dominant feature across an ensemble of orbitofrontal neurons. Yet a broader consideration of the neural correlates in orbitofrontal cortex reveals that this is not true (Stalnaker et al. 2006). This is illustrated in Figure 15.2, which shows the average response of all cue-selective neurons during a reversal. As a group, these neurons do not reverse or recode the associations across

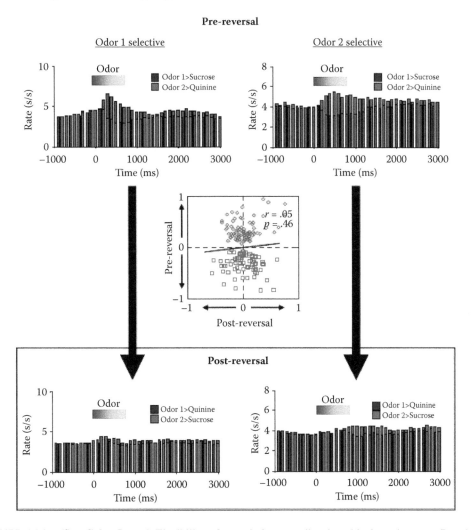

FIGURE 15.2 **(See Color Insert)** Flexibility of associative encoding in orbitofrontal cortex. Population response of neurons in orbitofrontal cortex identified as cue-selective during learning. Average activity per neuron is shown, synchronized to odor onset, during and after reversal. Inset scatter plot (middle part of figure) compares the cue-selectivity indices before (y axis) and after (x axis) reversal for all the cue-selective neurons used to construct the population histograms. Blue and red symbols show data for "Odor 1 Selective" neurons and "Odor 2 Selective" neurons, respectively. Unlike the single-unit example in Figure 15.1, the population responses failed to reverse and the cue-selectivity indices showed no correlation. (Adapted from Stalnaker, T.A. et al.: Abnormal associative encoding in orbitofrontal neurons in cocaine-experienced rats during decision-making. *Eur J Neurosci* 2006, 24, 2643, Copyright Wiley-VCH Verlag GmbH & Co. KGaA. With permission.)

reversal. The populations fail to reverse even though about 25% of the neurons in these populations do reverse encoding. The reason for this is apparent in the inset in Figure 15.2, which plots an index of cue-selectivity for each neuron before and after reversal. This plot shows that there is no relationship between cue-selectivity before and after reversal in orbitofrontal neurons. Thus reversal of encoding in orbitofrontal neurons can only be demonstrated by cherry-picking neurons that reverse from the overall population. This result is clearly inconsistent with the view of this region as a particularly flexible associative learning area, as suggested by reports focusing on reversal correlates in single-units.

This view is also contradicted by the relationship between reversal of encoding in orbitofrontal neurons and reversal performance. If orbitofrontal cortex were driving performance due to an ability to rapidly encode the new associations, then one should expect reversal of encoding to be associated with rapid reversal learning, whereas a failure to reverse encoding should be associated with slower learning. However we find exactly the opposite relationship (Stalnaker et al. 2006). Rats acquire reversals significantly more slowly when we observe reversal of encoding in orbitofrontal neurons.

And finally we now know that orbitofrontal cortex is far from unique in the reversal of associative encoding that it shows during reversal learning; cue-selective neurons in many other brain regions also reverse firing during reversal learning, and they do so much more rapidly and in greater proportions. This is particularly evident in the basolateral amygdala, where the majority of the cue-selective neurons—55%–60%—reverse firing (Schoenbaum et al. 1999; Saddoris, Gallagher, and Schoenbaum 2005; Stalnaker et al. 2007b). This is illustrated by the population responses in Figure 15.3, which switch cue-selectivity, and by the inset, which shows a significant inverse correlation between cue-selectivity before and after reversal in basolateral amygdala neurons. Similar results have also been reported in a Pavlovian reversal task in primates (Patton et al. 2006).

15.4 ORBITOFRONTAL CORTEX AND SIGNALING OF OUTCOME EXPECTANCIES

If orbitofrontal cortex is not critical for adaptive behavior either due to a special role in response inhibition or because this area is faster than other regions at encoding new associative information, then why is orbitofrontal cortex necessary for adaptive behavior? What underlying function does orbitofrontal cortex provide that facilitates changes in behavior when outcomes are not as expected?

In the next two sections, we will suggest that this underlying function involves signaling what have been called outcome expectancies—literally predictions about the characteristics, features, and specific value of outcomes that the animal expects to receive, given particular circumstances and cues in the environment (Schoenbaum and Roesch 2005). In this section, we will show that these signals are prominent in neural activity and BOLD responses in orbitofrontal cortex and argue that the importance of these signals is evident in deficits caused by orbitofrontal damage in a variety of behavioral settings in which outcomes must be used to guide normal behavior, even when contingencies are not changing. In the next section, we will suggest that these same signals are also necessary for the detection of prediction errors when contingencies are changing, thereby facilitating changes in associative representations in other brain areas and, ultimately, behavior.

15.4.1 NEURAL CORRELATES

Signals reflecting outcome expectancies are prominent in neural activity in orbitofrontal cortex. In addition to the cue-selective activity described above, orbitofrontal neurons also tend to fire in advance of meaningful events in trial-based tasks. Indeed this may be the most striking feature, overall, of neural activity in orbitofrontal cortex. Orbitofrontal neurons exhibit firing correlated with every event in the trial, and typically this activity is not triggered by these events but

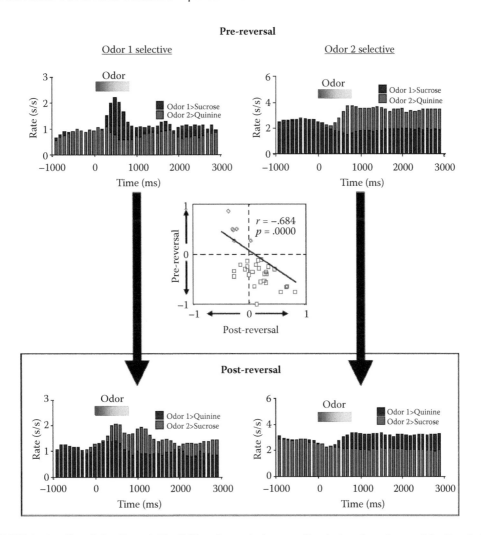

FIGURE 15.3 (See Color Insert) Flexibility of associative encoding in basolateral amygdala. Population response of neurons in basolateral amygdala identified as cue-selective during learning. Average activity per neuron is shown, synchronized to odor onset, during and after reversal. Inset scatter plot compares the cue-selectivity indices before (*y* axis) and after (*x* axis) reversal for all the cue-selective neurons used to construct the population histograms. Blue and red symbols show data for "Odor 1 Selective" neurons and "Odor 2 Selective" neurons, respectively. In contrast to similar data from orbitofrontal cortex in Figure 15.2, the population responses reversed, and the cue-selectivity indices showed a strongly significant inverse correlation. (Adapted from Stalnaker, T.A. et al., *Nat Neurosci*, 10, 949, 2007. With permission.)

rather increases in anticipation of them (Schoenbaum and Eichenbaum 1995; Lipton, Alvarez, and Eichenbaum 1999; Ramus and Eichenbaum 2000). In other words, event-related activity in orbitofrontal neurons anticipates these predictable and presumably value-laden events.

Such anticipatory activity is particularly strong prior to delivery of primary rewarding or aversive outcomes. We see this during discrimination learning, when distinct populations of neurons in orbitofrontal cortex develop selective firing prior to delivery of sucrose or quinine (Schoenbaum, Chiba, and Gallagher 1998). Often these neurons initially fire to one or the other outcome and then come to fire in anticipation of that outcome and, later, to cues that predict the outcome. Unlike reward-responsive dopamine neurons, which also transfer activity to predictive events (Montague, Dayan, and Sejnowski 1996; Hollerman and Schultz 1998; Waelti, Dickinson, and Schultz 2001; Bayer and Glimcher 2005; Pan et al. 2005; Roesch, Calu, and Schoenbaum 2007), these neurons do not stop firing to the rewards (Schoenbaum et al. 2003b; Stalnaker et al. 2006). As a result, their activity is not well described as a prediction error signal (see also later discussion below and Chapter 14); rather, activation of these neurons during progressively earlier periods in the trial is better explained as a representation of the actual outcome. Interestingly, this activity develops independently of choice performance, which is not orbitofrontal dependent, but instead develops in concert with changes in response latencies that seem to reflect active anticipation of particular outcomes.

Similar correlates have also been reported by a number of other investigators in rats, monkeys, and humans, in both single-unit activity and in BOLD response (Schoenbaum, Chiba, and Gallagher 1998; Tremblay and Schultz 1999; Schultz, Tremblay, and Hollerman 2000; Roesch, Taylor, and Schoenbaum 2006; van Duuren et al. 2007). For example, Schultz and colleagues have shown that orbitofrontal neurons exhibit outcome-expectant activity during a visual delayed response task (Tremblay and Schultz 1999). Monkeys were trained to respond to visual cues to obtain different food rewards. As illustrated by the single-unit example in Figure 15.4, many orbitofrontal neurons exhibited differential activity after responding, in anticipation of presentation of a particular food item. The monkey expected that food and the neuron's activity reflected that expectation. This

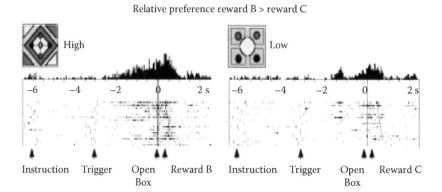

FIGURE 15.4 Outcome-expectant neural activity in monkey orbitofrontal cortex. Monkeys were trained to respond after presentation of visual cues to obtain different food rewards. The visual items each predicted a particular food, for which the monkeys had different preferences. Two items were presented in each block of trials. Shown is activity in a single unit recorded across two of these trial blocks. Firing increased after responding to reward and also in anticipation of reward. This activity is higher prior to the preferred reward in both blocks; critically this means that firing increased in anticipation of reward B, in the lower panel, when B became the preferred reward (compare to upper panel where B is the non-preferred reward). (Adapted from Tremblay, L. and Schultz, W., *Nature*, 398, 704, 1999. With permission.)

anticipatory activity did not simply reflect the physical or sensory attributes of the expected food reward, but rather reflected something about the value the monkey placed on this item. This was revealed by recording from the same neurons across blocks in which the relative preference for the different food rewards shifted. Outcome-expectant activity often changed to reflect changes in the relative value of a particular food across blocks. Although transitive encoding of the value of multiple rewards in orbitofrontal cortex has recently been shown in a randomized design (Padoa-Schioppa and Assad 2008), Schultz's original report is important because by reliably altering the relative values of available rewards, it provides an excellent demonstration that these anticipatory signals reflect the animal's judgment regarding the value of the expected outcome.

Notably, although anticipatory firing has also been observed in other areas (Watanabe 1996; Schoenbaum, Chiba, and Gallagher 1998; Hikosaka and Watanabe 2000; Tremblay and Schultz 2000; Shidara and Richmond 2002; Wallis and Miller 2003; Hikosaka and Watanabe 2004; Sugase-Miyamoto and Richmond 2005; Patton et al. 2006), studies that have compared such activity between different brain areas have found that it emerges first in orbitofrontal cortex (Schoenbaum, Chiba, and Gallagher 1998), and outcome-expectant activity in other areas—basolateral amygdala specifically—is reduced to near chance levels by even unilateral lesions of orbitofrontal cortex (Saddoris, Gallagher, and Schoenbaum 2005). This suggests that orbitofrontal neurons are, to some extent, constructing these representations based on afferent input, rather than simply receiving this information from other areas. Indeed, we would suggest that firing in anticipation of outcomes during delays or uncued periods of a task is just a special example of a more general function that orbitofrontal cortex fulfills, even during other periods in the task.

15.4.2 Behavioral Correlates

Of course, the neural firing patterns described above are only correlates of behavior. Alone they cannot provide conclusive evidence that orbitofrontal cortex is critical for signaling information about expected outcomes. For that one must rely on behavioral studies involving lesions or other manipulations. If orbitofrontal cortex is in fact critical for signaling outcome expectancies, then manipulations that disrupt or alter output from this brain area should preferentially disrupt behaviors that depend on information about outcomes.

This turns out to be an excellent description of many, if not most, orbitofrontal-dependent behaviors. A full accounting of the entire list is beyond the scope of this review; however, to illustrate we will describe a few examples in detail in addition to providing a list of others. The first comes from the odor discrimination task described above. At the same time rats are learning to withhold responding to an odor that predicts a negative outcome, they also show changes in the speed or latency at which they respond to the fluid well; they begin to respond faster after sampling a positive-predictive odor, as if expecting sucrose, and slower after sampling a negative-predictive odor, as if expecting quinine (Schoenbaum et al. 2003a). The difference in their response latencies becomes significant before accurate choice performance emerges and, as noted earlier, develops at the same time that neurons in the orbitofrontal cortex begin to show differential firing in anticipation of sucrose or quinine (Schoenbaum, Chiba, and Gallagher 1998). Furthermore, response latencies have been shown to be particularly sensitive to information about outcomes (Sage and Knowlton 2000). Manipulations affecting orbitofrontal cortex selectively abolish differential latency changes in our task and affect latency changes in other settings (Bohn, Giertler, and Hauber 2003a, 2003b; Schoenbaum et al. 2003a).

A second and perhaps more conclusive example showing a critical role for orbitofrontal cortex in the use of information about expected outcomes comes from studies using Pavlovian reinforcer devaluation (Holland and Rescorla 1975; Holland and Straub 1979). In these studies, an animal is trained that a cue predicts a particular reward. Subsequently, the value of the reward is reduced by pairing it with illness or selective satiation, and then the animal's ability to access and use that new value to guide the learned responding is assessed by presenting the cue by itself. Normal animals show reduced responding to the predictive cue, reflecting their ability to access and use cue-evoked representations of the reward and its current value. Monkeys and rats with orbitofrontal lesions fail to show this normal effect of devaluation (Gallagher, McMahan, and Schoenbaum 1999; Izquierdo, Suda, and Murray 2004; Machado and Bachevalier 2007); although these animals stop consuming the devalued or satiated food, they continue to respond to the cue that predicts that food to the same extent as non-devalued controls. Critically, the deficit caused by orbitofrontal damage is evident even when orbitofrontal cortex is present for the initial training and for devaluation (Pickens et al. 2003, 2005); thus the deficit seems to reflect a critical role for orbitofrontal cortex in mobilizing and using the learned information about the value of the expected outcome to guide or influence responding. This observation is consistent with data in monkeys and humans that neural activity in orbitofrontal cortex changes in real time as a result of satiation (Critchley and Rolls 1996; O'Doherty et al. 2000; Gottfried, O'Doherty, and Dolan 2003). This distinguishes the role of orbitofrontal cortex in this setting from that of other regions, like amygdala or mediodorsal thalamus, which seem to be necessary during the earlier phases (Hatfield et al. 1996; Malkova, Gaffan, and Murray 1997; Pickens et al. 2003; Wellmann, Gale, and Malkova 2005; Mitchell, Browning, and Baxter 2007).

A third example comes from work by Balleine and colleagues using Pavlovian-to-instrumental transfer. Transfer occurs when animals are independently trained to associate a cue with reward (Pavlovian stage) and to associate an instrumental response such as a lever press with reward (instrumental phase), after which they show an increased rate of instrumental responding in the presence of the Pavlovian cue (Estes 1948; Holland 2004). There are two forms of this transfer: a general form, observed when the rewards predicted by the cue and response are different, and a specific form, observed when the reward is the same in both cases. The specific form is thought to depend on the ability of the cue to evoke a representation of the particular reward linked to the instrumental response. Consistent with the proposal that orbitofrontal cortex is critical for such outcome signaling, Balleine and colleagues have reported that specific transfer is particularly sensitive to orbitofrontal lesions (Ostlund and Balleine 2007a). Again the orbitofrontal cortex, unlike areas such as basolateral amygdala (Corbit and Balleine 2005), appears to be required specifically at the time the information must be used to guide responding, since only post-training lesions were effective.

These examples indicate that one fundamental role of neural activity in orbitofrontal cortex is to signal information about expected outcomes to other brain areas. Other examples of orbitofrontal-dependent behaviors further corroborate this account. These would include delayed discounting (Mobini et al. 2002; Kheramin et al. 2003; Winstanley et al. 2004), conditioned reinforcement, and other second-order behaviors (Cousens and Otto 2003; Hutcheson and Everitt 2003; Pears et al. 2003; Burke et al. 2008), Pavlovian approach behaviors (Chudasama and Robbins 2003), the enhancement of discriminative responding by different outcomes (McDannald et al. 2005), and even cognitive and affective processes currently under investigation in human models, such as regret and counterfactual reasoning (Camille et al. 2004). In each case, normal performance would require the ability to signal, in real time, information about the features, characteristics, and values of outcomes predicted by cues and circumstances in the environment.

15.5   OUTCOME EXPECTANCIES AND ADAPTIVE BEHAVIOR

Of course, the orbitofrontal-dependent behaviors described above differ from adaptive behavior as assessed by reversal learning or related tasks, since they generally do not involve changes in established associations or response contingencies. For example, orbitofrontal inactivation impairs changes in cue-evoked responding after devaluation, even though nothing about the underlying associations has changed; only the value of the outcome has been altered. Similarly, damage to orbitofrontal cortex disrupts transfer, even when lesions are made after the rats have learned the cue-reward and response-reward associations underlying this phenomenon. These data indicate that signals regarding expected outcomes from orbitofrontal cortex influence judgment and decision making.

So how does signaling of expected outcomes help explain why orbitofrontal cortex is so important for modifying behavior when contingencies *are* changing? One possibility is that these signals may also be critical for driving updating of associative representations in other brain regions in the face of unexpected outcomes, particularly in subcortical areas, such as striatum and amygdala, which are strongly implicated in associative learning processes. According to classical learning theory (Rescorla and Wagner 1972), and its more modern cousin reinforcement learning (Sutton and Barto 1998), associative learning is driven by prediction errors (for further details see Chapter 14). A prediction error (δ) is calculated from the difference between the value of the outcome that is predicted by actions and cues in the environment (V), and the value of the outcome that is actually received (λ), according to the equation, $\delta = c(\lambda - V)$, where c reflects processes like surprise or attention, which can influence the rate of learning. There is now strong evidence that these prediction errors are signaled by phasic activity in midbrain dopamine neurons, as well as afferent regions such as habenula; such phasic firing is proposed to act as a teaching signal to stamp in associative representations in areas like striatum and amygdala. If signaling of expected outcomes by orbitofrontal cortex were also contributing to calculation of these prediction errors, essentially providing information necessary to compute V, then that would explain why orbitofrontal damage disrupts changes in behavior when contingencies are altered.

This proposal makes a number of testable predictions. For starters, it predicts that changes in associative representations in downstream regions, such as amygdala, should be dependent on orbitofrontal cortex. Consistent with this prediction, we have reported that associative encoding in basolateral amygdala in our reversal task is markedly less flexible in orbitofrontal-lesioned rats (Figure 15.5) than that in controls (Figure 15.3) (Saddoris, Gallagher, and Schoenbaum 2005). Neurons were less likely to become selective for predictive cues with training, and those that *were* present failed to reverse their cue-selectivity. Moreover, this downstream inflexibility appears to be the proximal cause of the orbitofrontal-dependent reversal deficit, since lesions or inactivation of basolateral amygdala abolishes the reversal deficit caused by orbitofrontal lesions (Stalnaker et al. 2007a). Since recoding of associations in orbitofrontal cortex lags recoding in basolateral amygdala (Schoenbaum et al. 1999), these results cannot be easily explained as reflecting rapid flexibility in orbitofrontal cortex. However they are fully consistent with the proposal that signaling of the old associations by orbitofrontal neurons facilitates encoding of the new associations by downstream areas and thereby changes behavior. Indeed, this is consistent with our report that better reversal performance is observed when cue-selectivity in orbitofrontal cortex fails to reverse (Stalnaker et al. 2006). Under our proposal, the explanation for this finding would be that when orbitofrontal neurons continue to encode pre-reversal associations, negative prediction error signals are facilitated, leading to faster learning.

Other data consistent with this proposal comes from a Pavlovian over-expectation task (Rescorla 1970). In this task, rats are first trained that several Pavlovian cues are each independent predictors of reward. Subsequently, two of the previously trained cues are presented together in compound, followed by the same reward. When the effect of this compound training on responding for the individual cues is assessed later in a probe test, a spontaneous reduction in responding is found. This reduced responding is thought to result from the violation of summed expectations for reward during compound training. That is, animals expect to get double the reward after the compound cues, whereas they actually only get the usual reward. Notably, unlike reversal learning, nothing about the outcome is changed. Instead, prediction errors are induced by directly manipulating the animals' expectations for reward. Furthermore, the learning induced by these manipulations can be dissociated from the use of the newly acquired information, since the former occurs during compound training and the latter in the probe test.

Using this task, we have found that reversible inactivation of orbitofrontal cortex during compound training prevents the later reduction in responding to the individual cues (Figure 15.6a, b) (Takahashi et al. 2009). This result cannot be explained as a simple deficit in using associative information encoded by orbitofrontal neurons to guide behavior, because orbitofrontal cortex is fully functional

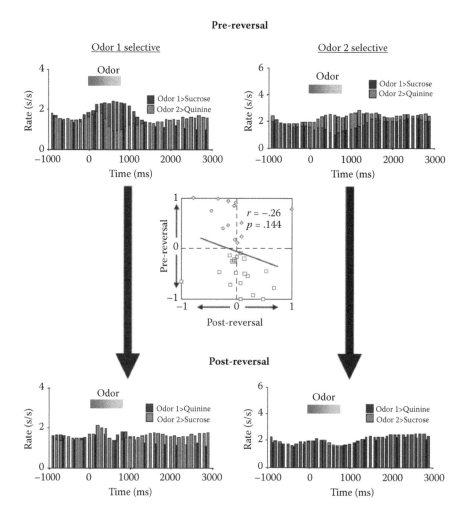

FIGURE 15.5 Flexibility of associative encoding in basolateral amygdala depends on input from orbito-frontal cortex. Population response of cue-selective neurons in basolateral amygdala in rats with ipsilateral lesions of orbitofrontal cortex. Average activity per neuron is shown, synchronized to odor onset, during and after reversal. Inset scatter plot compares the cue-selectivity indices before (*y* axis) and after (*x* axis) reversal for all the cue-selective neurons used to construct the population histograms. Blue and red symbols show data for "Odor 1 Selective" neurons and "Odor 2 Selective" neurons, respectively. Unlike the populations recorded in intact rats, illustrated in Figure 15.3, the population response recorded in orbitofrontal-lesioned rats did not reverse cue-selectivity, and the cue-selectivity indices showed no correlation. (Adapted from *Neuron*, 46, Saddoris, M.P., Gallagher, and M., Schoenbaum, G. Rapid associative encoding in basolateral amygdala depends on connections with orbitofrontal cortex. 321–31. Copyright 2005. With permission from **Elsevier**.)

at the time the behavior is assessed. Instead, it indicates that signals from orbitofrontal cortex are required for learning, which presumably occurs in other regions. Consistent with this idea, blockade of orbitofrontal NMDA receptors in a separate group of rats during compound training had no effect on over-expectation. The most coherent interpretation of these data is that orbitofrontal cortex contributes to teaching signals to update representations in other areas.

Interestingly, it has been suggested that orbitofrontal neurons might directly signal prediction errors (Schultz and Dickinson 2000; Rolls and Grabenhorst 2008). This proposal is supported by a number of brain imaging studies that have reported that neural activity in OFC, as reflected in BOLD signal, is correlated with errors in reward prediction (Nobre et al. 1999; Berns et al. 2001;

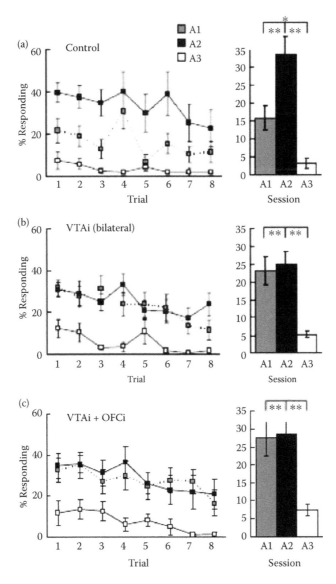

FIGURE 15.6 Effect of bilateral inactivation of ventral tegmental area, or contralateral inactivation of orbitofrontal cortex and ventral tegmental area, on changes in behavior after over-expectation. Rats in all groups conditioned normally and maintained responding during compound training (though only controls showed summation to the compound cue). Shown is food cup responding to the auditory cues during the critical probe test. A1 = compound-conditioned cue; A2 = control-conditioned cue; A3 = non-conditioned cue (CS-). Controls (a) exhibited weaker responding in this probe test to the cue that had been compounded. This decline in responding was not observed if ventral tegmental area had been inactivated bilaterally (b) during prior compound training or if ventral tegmental area and orbitofrontal cortex were disconnected via contralateral inactivation (c). (Adapted from *Neuron*, 62, Takahashi, Y. et al., The orbitofrontal cortex and ventral tegmental area are necessary for learning from unexpected outcomes, 269–80, Copyright 2009. With permission from **Elsevier**.)

O'Doherty et al. 2003; Dreher, Kohn, and Berman 2006; Tobler et al. 2006). For example, BOLD signal in regions within OFC increases abruptly when expectations for reward are not met (Nobre et al. 1999) and this signal conforms with formal learning theory predictions in a blocking paradigm (Tobler et al. 2006). Thus, OFC might contribute to learning during over-expectation if it were directly signaling reward prediction errors.

However, data from single-unit recording studies are not consistent with this proposal. Aside from anecdotal reports (Thorpe, Rolls, and Maddison 1983; Ramus and Eichenbaum 2000; Feierstein et al. 2006), there is very little evidence for appreciable error encoding by orbitofrontal neurons. Moreover, a direct comparison of prediction error correlates in dopamine neurons, which have been clearly demonstrated to signal reward prediction errors, against those in orbitofrontal neurons, shows unambiguously that the two areas do not signal this information similarly (Takahashi et al. 2009). As expected, dopamine neurons increased firing in response to unexpected reward and suppressed firing on reward omission (Figure 15.7a). These changes were inversely correlated, and the loss of activity to reward was also inversely correlated with the development of activity to predictive cues. However none of these features were evident in the activity of reward-responsive (or any other) orbitofrontal neurons. Instead these neurons actually tended to fire more to an expected reward. This is because the normal reward response is essentially shifted backward in time (Figure 15.7b), so that the orbitofrontal response develops *in anticipation* of the expected reward, while the dopamine neuron response develops *after* an unexpected reward. Activity shows a similar time course

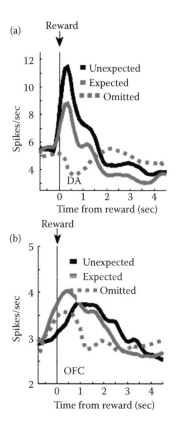

FIGURE 15.7 Activity in ventral tegmental area dopamine neurons and orbitofrontal neurons in response to unexpected and expected reward delivery and omission of an expected reward. As expected, dopamine neurons (a) exhibited greater firing in response to unexpected reward than to a reward rats were learning to expect. Subsequently the same neurons suppressed firing on omission of this reward. (Adapted from Roesch, M.R., et al., *Nat Neurosci,* 10, 1615, 2007. With permission.) Neurons in orbitofrontal cortex (b) fired to reward but did not show elevated activity when that same reward was delivered unexpectedly, nor did they suppress firing on omission. Instead they exhibited increased firing in anticipation of reward delivery. (Adapted from *Neuron,* 62, Takahashi, Y. et al., The orbitofrontal cortex and ventral tegmental area are necessary for learning from unexpected outcomes, 269–80, Copyright 2009, with permission from **Elsevier**.)

prior to reward on omission trials. Thus activity in orbitofrontal neurons signals reward predictions and not reward prediction errors.

The relationship between signaling of reward predictions by orbitofrontal neurons and reward prediction errors by dopamine neurons is consistent with the proposal that the two areas interact in calculating errors; when activity before reward in orbitofrontal cortex is low, consistent with low reward expectations, activity after reward in the dopamine neurons is high (and vice versa). Accordingly, inactivation of ventral tegmental area during compound training in the over-expectation task abolishes the normal decline in responding, as does unilateral inactivation of orbitofrontal cortex in one hemisphere and ventral tegmental area in the contralateral hemisphere (Figure 15.6c). These results, together, argue strongly that information regarding expected outcomes signaled by orbitofrontal neurons contributes to calculation of reward prediction errors signaled by dopamine neurons.

15.6 OUTCOME EXPECTANCIES AND ASSOCIATIVE ENCODING IN PIRIFORM CORTEX

Above we have argued that information regarding expected outcomes in orbitofrontal neurons influences prediction errors and associative learning in downstream areas more closely tied to decision making. However orbitofrontal cortex might also work upstream, modulating activity in more sensory-related structures such as piriform cortex. Piriform cortex, the largest of the olfactory cortical areas, is part of a bidirectional system involved in processing olfactory information (Haberly 2001). On one hand, piriform cortex, in particular anterior regions, has strong reciprocal connections with olfactory bulb and has traditionally been thought of as primary olfactory cortex. On the other hand, piriform cortex also receives substantial descending input from downstream brain areas, including orbitofrontal cortex and amygdala (Johnson et al. 2000; Majak et al. 2004). This relationship suggests that piriform cortex might not function solely as a primary sensory region, but also as an association cortex, integrating incoming olfactory information with descending input from higher-order association areas such as OFC (also see Chapter 5 in this volume for further details).

Consistent with this notion we have shown that activity to odor cues in anterior and posterior piriform cortex during performance of our go/no-go discrimination task (Roesch, Stalnaker, and Schoenbaum 2006, 2007) was indeed much more associative in nature than would be expected from a purely sensory region. For example, although odor-selective activity was present in many neurons in anterior piriform cortex, this activity was typically modulated by learning or reversal of simple odor discriminations. Interestingly, the prevalence of such correlates increased as we moved posteriorly in piriform cortex. This is consistent with the connectivity of anterior versus posterior areas with limbic structures and likely reflects strong input from amygdala (Datiche and Cattarelli 1996; Johnson et al. 2000; Majak et al. 2004; Illig 2005).

So what functions do interactions between orbitofrontal cortex, amygdala, and piriform cortex play in associative learning? Clearly, during learning, piriform cortex serves as a way station to orbitofrontal cortex and other areas, signaling sensory features of odor cues. Notably, pure sensory encoding in piriform cortex is unique among the areas we have recorded from, including orbitofrontal cortex, amygdala, and ventral striatum. However, this interaction is bidirectional; after learning, orbitofrontal cortex might actively modulate afferent input to piriform cortex (cf. Cohen et al. 2008). This might prepare neurons to recognize and respond to expected odors more rapidly or even promote certain neurons to fire in certain contexts (e.g., after learning or after reversal) but not others. Orbitofrontal input might even initiate activity in piriform cortex in the absence of any odor, allowing for recall of odors and odor-related associations (Haberly 2001). In either case, the strong reciprocal connections between these two areas (Johnson et al. 2000; Haberly 2001; Illig 2005) are likely to support appropriate behavioral responding to olfactory cues in the service of optimizing behavior. Of course, future work disrupting feedback from orbitofrontal cortex will be necessary to test whether this input is influencing encoding in piriform cortex as we propose.

15.7 FUTURE ISSUES

We have discussed evidence for the involvement of the orbitofrontal cortex in adapting behavior in the face of unexpected outcomes. Current evidence contradicts long-held ideas that this role reflects response inhibition or rapid flexibility of associative encoding. Instead we have suggested that it reflects contributions of orbitofrontal signaling to changing associative representations in other brain regions, mediated indirectly through support of prediction-error signaling by systems such as the midbrain dopamine neurons. This proposal is consistent with neurophysiological and behavioral studies published in the past decade revealing that orbitofrontal cortex is fundamentally critical to signaling outcome expectancies. These signals facilitate judgment about expected outcomes to guide behavior, even when contingencies are unchanged. Our proposal simply extends this idea to suggest that these same signals also facilitate learning when contingencies are changing.

This proposal bears striking similarities to ideas regarding the role of ventral striatum in so-called actor-critic models of reinforcement learning (Barto, Sutton, and Anderson 1983; Sutton and Barto 1998; O'Doherty et al. 2004). In these proposals, it is ventral striatum that is hypothesized to provide the information necessary to compute the "state value," or V, which is required for calculating reward prediction errors. We would suggest that orbitofrontal cortex plays a similar role, not instead of but in addition to contributions from ventral striatum and likely other regions. Indeed, much as there are multiple memory systems, there are likely to be multiple, parallel critics, each providing a particular type of information relevant to computing V. With regard to orbitofrontal cortex and ventral striatum, one possibility is that output from orbitofrontal cortex provides information regarding the value of the specific outcome that is predicted by Pavlovian cues in the environment. This is consistent with its proposed role in associative learning from devaluation and other tasks and also with recent data from our lab showing that orbitofrontal lesions prevent unblocking when one equally valued outcome is switched for another (Burke et al. 2008). At the same time, ventral striatum might provide information regarding the general affect or emotion that has been associated with the same predictive cue, again consistent with the clear involvement of this area in behavioral tasks, such as Pavlovian-to-instrumental transfer, that require such general affective properties (so-called motivational habits). One might also postulate the existence of an instrumental critic, perhaps in medial prefrontal regions, providing information about values predicted by actions (Valentin, Dickinson, and O'Doherty 2007).

Inputs from these areas might converge on midbrain areas, either directly or indirectly. For example, ventral striatum receives input from orbitofrontal cortex; thus ventral striatum could function as a super-critic, integrating information regarding the general affective and outcome-specific properties of Pavlovian cues and sending it off to be used in error signaling. Alternatively orbitofrontal cortex also projects to other areas, including directly to midbrain and could thereby bypass ventral striatum. Indeed both pathways may be utilized to influence learning in subtly different ways.

Importantly, all of these ideas are imminently testable. fMRI and single-unit studies combined with lesions and inactivation, and behavioral tasks that manipulate expectancies based on the specific and general values of actions and Pavlovian cues can directly address—either confirming or invalidating—hypotheses regarding these circuits. We believe there is already substantial experimental support for the simple ideas we have laid out here. However further work holds the potential to sketch out this interesting circuit and its functions in much more detail. This work will surely provide evidence contrary to our proposal, but it holds the promise of creating a truly useful framework for how brain circuits implement the simple associative learning processes that help us navigate our ever-changing world.

REFERENCES

Barto, A.G., Sutton, R.S., and Anderson, C.W. 1983. Neuron-like adaptive elements that can solve difficult learning control problems. *IEEE Transactions on Systems, Man, and Cybernetics* 13:834–46.

Baxter, M.G., Parker, A., Lindner, C.C.C., Izquierdo, A.D., and Murray, E.A. 2000. Control of response selection by reinforcer value requires interaction of amygdala and orbitofrontal cortex. *Journal of Neuroscience* 20:4311–19.

Bayer, H.M., and Glimcher, P.W. 2005. Midbrain dopamine neurons encode a quantitative reward prediction error signal. *Neuron* 47:129–41.

Bechara, A., Damasio, H., Tranel, D., and Damasio, A.R. 1997. Deciding advantageously before knowing the advantageous strategy. *Science* 275:1293–94.

Berns, G.S., McClure, S.M., Pagnoni, G., and Montague, P.R. 2001. Predictability modulates human brain response to reward. *Journal of Neuroscience* 21:2793–98.

Bissonette, G.B., Martins, G.J., Franz, T.M., Harper, E.S., Schoenbaum, G., and Powell, E.M. 2008. Double dissociation of the effects of medial and orbital prefrontal cortical lesions on attentional and affective shifts in mice. *Journal of Neuroscience* 28:11124–30.

Bohn, I., Giertler, C., and Hauber, W. 2003a. Orbital prefrontal cortex and guidance of instrumental behavior in rats under reversal conditions. *Behavioral Brain Research* 143:49–56.

———. 2003b. NMDA receptors in the rat orbital prefrontal cortex are involved in guidance of instrumental behavior under reversal conditions. *Cerebral Cortex* 13:968–76.

Burke, K.A., Franz, T.M., Miller, D.N., and Schoenbaum, G. 2008. The role of orbitofrontal cortex in the pursuit of happiness and more specific rewards. *Nature* 454:340–44.

Butter, C.M. 1969. Perseveration and extinction in discrimination reversal tasks following selective frontal ablations in Macaca mulatta. *Physiology and Behavior* 4:163–71.

Camille, N., Coricelli, G., Sallet, J., Pradat-Diehl, P., Duhamel, J.-R., and Sirigu, A. 2004. The involvement of the orbitofrontal cortex in the experience of regret. *Science* 304:1168–70.

Chudasama, Y., Kralik, J.D., and Murray, E.A. 2007. Rhesus monkeys with orbital prefrontal cortex lesions can learn to inhibit prepotent responses in the reversed reward contingency task. *Cerebral Cortex* 17:1154–59.

Chudasama, Y., and Robbins, T.W. 2003. Dissociable contributions of the orbitofrontal and infralimbic cortex to pavlovian autoshaping and discrimination reversal learning: Further evidence for the functional heterogeneity of the rodent frontal cortex. *Journal of Neuroscience* 23:8771–80.

Cohen, Y., Reuveni, I., Barkai, E., Maroun, M. 2008. Olfactory learning-induced long-lasting enhancement of descending and ascending synaptic transmission to the piriform cortex. *Journal of Neuroscience* 28:6664–69.

Corbit, L.H., and Balleine, B.W. 2005. Double dissociation of basolateral and central amygdala lesions on the general and outcome-specific forms of pavlovian-instrumental transfer. *Journal of Neuroscience* 25:962–70.

Cousens, G.A., and Otto, T. 2003. Neural substrates of olfactory discrimination learning with auditory secondary reinforcement. I. Contributions of the basolateral amygdaloid complex and orbitofrontal cortex. *Integrative Physiological and Behavioral Science* 38:272–94.

Critchley, H.D., and Rolls, E.T. 1996. Hunger and satiety modify the responses of olfactory and visual neurons in the primate orbitofrontal cortex. *Journal of Neurophysiology* 75:1673–86.

Damasio, A.R. 1994. *Descartes Error*. New York: Putnam.

Damasio, A.R., Grabowski, T., Frank, R., Galaburda, A.M., and Damasio, A.R. 1994. The return of Phineas Gage: Clues about the brain from the skull of a famous patient. *Science* 264:1102–5.

Datiche, F., and Cattarelli, M. 1996. Reciprocal and topographic connections between the piriform and prefrontal cortices in the rat: A tracing study using the B subunit of the cholera toxin. *Brain Research Bulletin* 41:391–98.

Dillon, D.G., and Pizzagalli, D.A. 2007. Inhibition of action, thought, and emotion: A selective neurobiological review. *Applied Previews of Psychology* 12:99–114.

Dreher, J.C., Kohn, P., and Berman, K.F. 2006. Neural coding of distinct statistical properties of reward information in humans. *Cerebral Cortex* 16:561–73.

Eagle, D.M., Baunez, C., Hutcheson, D.M., Lehmann, O., Shah, A.P., and Robbins, T.W. 2008. Stop-signal reaction-time task performance: Role of prefrontal cortex and subthalamic nucleus. *Cerebral Cortex* 18:178–88.

Estes, W.K. 1948. Discriminative conditioning 2. Effects of a Pavlovian conditioned stimulus upon a subsequently established operant response. *Journal of Experimental Psychology* 38:173–77.

Feierstein, C.E., Quirk, M.C., Uchida, N., Sosulski, D.L., and Mainen, Z.F. 2006. Representation of spatial goals in rat orbitofrontal cortex. *Neuron* 51:495–507.

Fellows, L.K., and Farah, M.J. 2003. Ventromedial frontal cortex mediates affective shifting in humans: Evidence from a reversal learning paradigm. *Brain* 126:1830–37.

Ferrier, D. 1876. *The Functions of the Brain*. New York: GP Putnam's Sons.

Fuster, J.M. 1997. *The Prefrontal Cortex*. 3rd edn. New York: Lippin-Ravencott.

Gallagher, M., McMahan, R.W., and Schoenbaum, G. 1999. Orbitofrontal cortex and representation of incentive value in associative learning. *Journal of Neuroscience* 19:6610–14.

Gottfried, J.A., O'Doherty, J., and Dolan, R.J. 2003. Encoding predictive reward value in human amygdala and orbitofrontal cortex. *Science* 301:1104–7.

Haberly, L.B. 2001. Parallel-distributed processing in olfactory cortex: New insights from morphological and physiological analysis of neuronal circuitry. *Chemical Senses* 26:551–76.

Harlow, J.M. 1868. Recovery after passage of an iron bar through the head. *Publications of the Massachusetts Medical Society* 2:329–46.

Hatfield, T., Han, J.S., Conley, M., Gallagher, M., and Holland, P. 1996. Neurotoxic lesions of basolateral, but not central, amygdala interfere with Pavlovian second-order conditioning and reinforcer devaluation effects. *Journal of Neuroscience* 16:5256–65.

Hikosaka, K., and Watanabe, M. 2000. Delay activity of orbital and lateral prefrontal neurons of the monkey varying with different rewards. *Cerebral Cortex* 10:263–71.

———. 2004. Long- and short-range reward expectancy in the primate orbitofrontal cortex. *European Journal of Neuroscience* 19:1046–54.

Holland, P.C. 2004. Relations between Pavlovian-Instrumental transfer and reinforcer devaluation. *Journal of Experimental Psychology: Animal Behavior Processes* 30:104–17.

Holland, P.C., and Rescorla, R.A. 1975. The effects of two ways of devaluing the unconditioned stimulus after first and second-order appetitive conditioning. *Journal of Experimental Psychology: Animal Behavior Processes* 1:355–63.

Holland, P.C., and Straub, J.J. 1979. Differential effects of two ways of devaluing the unconditioned stimulus after Pavlovian appetitive conditioning. *Journal of Experimental Psychology: Animal Behavior Processes* 5:65–78.

Hollerman, J.R., and Schultz, W. 1998. Dopamine neurons report an error in the temporal prediction of reward during learning. *Nature Neuroscience* 1:304–9.

Hornak, J., O'Doherty, J., Bramham, J., Rolls, E.T., Morris, R.G., Bullock, P.R., and Polkey, C.E. 2004. Reward-related reversal learning after surgical excisions in orbito-frontal or dorsolateral prefrontal cortex in humans. *Journal of Cognitive Neuroscience* 16:463–78.

Hutcheson, D.M., and Everitt, B.J. 2003. The effects of selective orbitofrontal cortex lesions on the acquisition and performance of cue-controlled cocaine seeking in rats. *Annals of the New York Academy of Science* 1003:410–11.

Illig, K.R. 2005. Projections from orbitofrontal cortex to anterior piriform cortex in the rat suggest a role in olfactory information processing. *Journal of Comparative Neurology* 488:224–31.

Izquierdo, A.D., Suda, R.K., and Murray, E.A. 2004. Bilateral orbital prefrontal cortex lesions in rhesus monkeys disrupt choices guided by both reward value and reward contingency. *Journal of Neuroscience* 24:7540–48.

Johnson, D.M., Illig, K.R., Behan, M., and Haberly, L.B. 2000. New features of connectivity in piriform cortex visualized by intracellular injection of pyramidal cells suggest that "primary" olfactory cortex functions like "association" cortex in other sensory systems. *Journal of Neuroscience* 20:6974–82.

Jones, B., and Mishkin, M. 1972. Limbic lesions and the problem of stimulus-reinforcement associations. *Experimental Neurology* 36:362–77.

Kheramin, S., Brody, S., Ho, M.-Y., Velazquez-Martinez, D.N., Bradshaw, C.M., Szabadi, E., Deakin, J.F.W., and Anderson, I.M. 2003. Role of the orbital prefrontal cortex in choice between delayed and uncertain reinforcers: A quantitative analysis. *Behavioral Processes* 64:239–50.

Lipton, P.A., Alvarez, P., and Eichenbaum, H. 1999. Crossmodal associative memory representations in rodent orbitofrontal cortex. *Neuron* 22:349–59.

Machado, C.J., and Bachevalier, J. 2007. The effects of selective amygdala, orbital frontal cortex or hippocampal formation lesions on reward assessment in nonhuman primates. *European Journal of Neuroscience* 25:2885–2904.

Majak, K., Ronkko, S., Kemppainen, S., and Pitkanen, A. 2004. Projections from the amygdaloid complex to the piriform cortex: A PHA-L study in the rat. *Journal of Comparative Neurology* 476:414–28.

Malkova, L., Gaffan, D., and Murray, E.A. 1997. Excitotoxic lesions of the amygdala fail to produce impairment in visual learning for auditory secondary reinforcement but interfere with reinforcer devaluation effects in rhesus monkeys. *Journal of Neuroscience* 17:6011–20.

McAlonan, K., and Brown, V.J. 2003. Orbital prefrontal cortex mediates reversal learning and not attentional set shifting in the rat. *Behavioral Brain Research* 146:97–30.

McDannald, M.A., Saddoris, M.P., Gallagher, M., and Holland, P.C. 2005. Lesions of orbitofrontal cortex impair rats' differential outcome expectancy learning but not conditioned stimulus-potentiated feeding. *Journal of Neuroscience* 25:4626–32.

Meunier, M., Bachevalier, J., and Mishkin, M. 1997. Effects of orbital frontal and anterior cingulate lesions on object and spatial memory in rhesus monkeys. *Neuropsychologia* 35:999–1015.

Mishkin, M. 1964. Perseveration of central sets after frontal lesions in monkeys. In *The Frontal Granular Cortex and Behavior*, eds. J.M. Warren, and K. Akert, pp. 219–41. New York: McGraw-Hill.

Mitchell, A.S., Browning, P.G., and Baxter, M.G. 2007. Neurotoxic lesions of the medial mediodorsal nucleus of the thalamus disrupt reinforcer devaluation effects in rhesus monkeys. *Journal of Neuroscience* 27:11289–95.

Mobini, S., Body, S., Ho, M.-Y., Bradshaw, C.M., Szabadi, E., Deakin, J.F.W., and Anderson, I.M. 2002. Effects of lesions of the orbitofrontal cortex on sensitivity to delayed and probabilistic reinforcement. *Psychopharmacology* 160:290–98.

Montague, P.R., Dayan, P., and Sejnowski, T.J. 1996. A framework for mesencephalic dopamine systems based on predictive hebbian learning. *Journal of Neuroscience* 16:1936–47.

Nobre, A.C., Coull, J.T., Frith, C.D., and Mesulam, M.M. 1999. Orbitofrontal cortex is activated during breaches of expectation in tasks of visual attention. *Nature Neuroscience* 2:11–12.

O'Doherty, J.P., Dayan, P., Friston, K., Critchley, H., and Dolan, R.J. 2003. Temporal difference models and reward-related learning in the human brain. *Neuron* 38:329–37.

O'Doherty, J., Dayan, P., Schultz, J., Deichmann, R., Friston, K.J., and Dolan, R.J. 2004. Dissociable roles of ventral and dorsal striatum in instrumental conditioning. *Science* 304:452–54.

O'Doherty, J., Rolls, E.T., Francis, S., Bowtell, R., McGlone, F., Kobal, G., Renner, B., and Ahne, G. 2000. Sensory-specific satiety-related olfactory activation of the human orbitofrontal cortex. *Neuroreport* 11:893–97.

Ostlund, S.B., and Balleine, B.W. 2007a. Orbitofrontal cortex mediates outcome encoding in Pavlovian but not instrumental learning. *Journal of Neuroscience* 27:4819–25.

———. 2007b. The contribution of orbitofrontal cortex to action selection. *Annals of the New York Academy of Science* 1121:174–92.

Padoa-Schioppa, C., and Assad, J.A. 2008. The representation of economic value in the orbitofrontal cortex is invariant for changes in menu. *Nature Neuroscience* 11:95–102.

Pais-Vieira, M., Lima, D., and Galhardo, V. 2007. Orbitofrontal cortex lesions disrupt risk assessment in a novel serial decision-making task for rats. *Neuroscience* 145:225–31.

Pan, W.-X., Schmidt, R., Wickens, J.R., and Hyland, B.I. 2005. Dopamine cells respond to predicted events during classical conditioning: Evidence for eligibility traces in the reward-learning network. *Journal of Neuroscience* 25:6235–42.

Patton, J.J., Belova, M.A., Morrison, S.E., and Salzman, C.D. 2006. The primate amygdala represents the positive and negative value of visual stimuli during learning. *Nature* 439:865–70.

Pears, A., Parkinson, J.A., Hopewell, L., Everitt, B.J., and Roberts, A.C. 2003. Lesions of the orbitofrontal but not medial prefrontal cortex disrupt conditioned reinforcement in primates. *Journal of Neuroscience* 23:11189–11201.

Pickens, C.L., Saddoris, M.P., Gallagher, M., and Holland, P.C. 2005. Orbitofrontal lesions impair use of cue-outcome associations in a devaluation task. *Behavioral Neuroscience* 119:317–22.

Pickens, C.L., Setlow, B., Saddoris, M.P., Gallagher, M., Holland, P.C., and Schoenbaum, G. 2003. Different roles for orbitofrontal cortex and basolateral amygdala in a reinforcer devaluation task. *Journal of Neuroscience* 23:11078–84.

Price, J.L. 2007. Definition of the orbital cortex in relation to specific connections with limbic and visceral structures and other cortical regions. *Annals of the New York Academy of Science* 1121:54–71.

Ramus, S.J., and Eichenbaum, H. 2000. Neural correlates of olfactory recognition memory in the rat orbitofrontal cortex. *Journal of Neuroscience* 20:8199–8208.

Reekie, Y.L., Braesicke, K., Man, M.S., and Roberts, A.C. 2008. Uncoupling of behavioral and autonomic responses after lesions of the primate orbitofrontal cortex. *Proceedings of the National Academy of Sciences* 105:9787–92.

Rescorla, R.A. 1970. Reduction in the effectiveness of reinforcement after prior excitatory conditioning. *Learning and Motivation* 1:372–81.

Rescorla, R.A., and Wagner, A.R. 1972. A theory of Pavlovian conditioning: Variations in the effectiveness of reinforcement and nonreinforcement. In *Classical Conditioning II: Current Research and Theory*, eds. A.H. Black, and W.F. Prokasy, pp. 64–99. New York: Appleton-Century-Crofts.

Roesch, M.R., Calu, D.J., and Schoenbaum, G. 2007. Dopamine neurons encode the better option in rats deciding between differently delayed or sized rewards. *Nature Neuroscience* 10:1615–24.

Roesch, M.R., Stalnaker, T.A., and Schoenbaum, G. 2006. Associative encoding in anterior piriform cortex versus orbitofrontal cortex during odor discrimination and reversal learning. *Cerebral Cortex* 17:643–52.

———. 2007. Associative encoding in anterior piriform cortex versus orbitofrontal cortex during odor discrimination and reversal learning. *Cerebral Cortex* 17:643–52.

Roesch, M.R., Taylor, A.R., and Schoenbaum, G. 2006. Encoding of time-discounted rewards in orbitofrontal cortex is independent of value representation. *Neuron* 51:509–20.

Rolls, E.T. 1996. The orbitofrontal cortex. *Philosophical Transactions of the Royal Society of London B* 351:1433–43.

Rolls, E.T., Critchley, H.D., Mason, R., and Wakeman, E.A. 1996. Orbitofrontal cortex neurons: Role in olfactory and visual association learning. *Journal of Neurophysiology* 75:1970–81.

Rolls, E.T., and Grabenhorst, F. 2008. The orbitofrontal cortex and beyond: From affect to decision-making. *Progress in Neurobiology* 86:216–44.

Rolls, E.T., Hornak, J., Wade, D., and McGrath, J. 1994. Emotion-related learning in patients with social and emotional changes associated with frontal lobe damage. *Journal of Neurology, Neurosurgery, and Psychiatry* 57:1518–24.

Saddoris, M.P., Gallagher, M., and Schoenbaum, G. 2005. Rapid associative encoding in basolateral amygdala depends on connections with orbitofrontal cortex. *Neuron* 46:321–31.

Sage, J.R., and Knowlton, B.J. 2000. Effects of US devaluation on win-stay and win-shift radial maze performance in rats. *Behavioral Neuroscience* 114:295–306.

Schoenbaum, G., Chiba, A.A., and Gallagher, M. 1998. Orbitofrontal cortex and basolateral amygdala encode expected outcomes during learning. *Nature Neuroscience* 1:155–59.

———. 1999. Neural encoding in orbitofrontal cortex and basolateral amygdala during olfactory discrimination learning. *Journal of Neuroscience* 19:1876–84.

Schoenbaum, G., and Eichenbaum, H. 1995. Information coding in the rodent prefrontal cortex. I. Single-neuron activity in orbitofrontal cortex compared with that in pyriform cortex. *Journal of Neurophysiology* 74:733–50.

Schoenbaum, G., Nugent, S., Saddoris, M.P., and Setlow, B. 2002. Orbitofrontal lesions in rats impair reversal but not acquisition of go, no-go odor discriminations. *Neuroreport* 13:885–90.

Schoenbaum, G., and Roesch, M.R. 2005. Orbitofrontal cortex, associative learning, and expectancies. *Neuron* 47:633–36.

Schoenbaum, G., and Setlow, B. 2001. Integrating orbitofrontal cortex into prefrontal theory: Common processing themes across species and subdivision. *Learning and Memory* 8:134–47.

Schoenbaum, G., Setlow, B., Nugent, S.L., Saddoris, M.P., and Gallagher, M. 2003a. Lesions of orbitofrontal cortex and basolateral amygdala complex disrupt acquisition of odor-guided discriminations and reversals. *Learning and Memory* 10:129–40.

Schoenbaum, G., Setlow, B., Saddoris, M.P., and Gallagher, M. 2003b. Encoding predicted outcome and acquired value in orbitofrontal cortex during cue sampling depends upon input from basolateral amygdala. *Neuron* 39:855–67.

Schultz, W., and Dickinson, A. 2000. Neuronal coding of prediction errors. *Annual Review of Neuroscience* 23:473–500.

Schultz, W., Tremblay, L., and Hollerman, J.R. 2000. Reward processing in primate orbitofrontal cortex and basal ganglia. *Cerebral Cortex* 10:272–83.

Shidara, M., and Richmond, B.J. 2002. Anterior cingulate: Single neuronal signals related to degree of reward expectancy. *Science* 296:1709–11.

Stalnaker, T.A., Franz, T.M., Singh, T., and Schoenbaum, G. 2007a. Basolateral amygdala lesions abolish orbitofrontal-dependent reversal impairments. *Neuron* 54:51–58.

Stalnaker, T.A., Roesch, M.R., Franz, T.M., Burke, K.A., and Schoenbaum, G. 2006. Abnormal associative encoding in orbitofrontal neurons in cocaine-experienced rats during decision-making. *European Journal of Neuroscience* 24:2643–53.

Stalnaker, T.A., Roesch, M.R., Franz, T.M., Calu, D.J., Singh, T., and Schoenbaum, G. 2007b. Cocaine-induced decision-making deficits are mediated by miscoding in basolateral amygdala. *Nature Neuroscience* 10:949–51.

Sugase-Miyamoto, Y., and Richmond, B.J. 2005. Neuronal signals in the monkey basolateral amygdala during reward schedules. *Journal of Neuroscience* 25:11071–83.

Sutton, R.S., and Barto, A.G. 1998. *Reinforcement Learning: An Introduction*. Cambridge MA: MIT Press.

Takahashi, Y., Roesch, M.R., Stalnaker, T.A., Haney, R.Z., Calu, D.J., Taylor, A.R., Burke, K.A., and Schoenbaum, G. 2009. The orbitofrontal cortex and ventral tegmental area are necessary for learning from unexpected outcomes. *Neuron* 62:269–80.

Teitelbaum, H. 1964. A comparison of effects of orbitofrontal and hippocampal lesions upon discrimination learning and reversal in the cat. *Experimental Neurology* 9:452–62.

Thorpe, S.J., Rolls, E.T., and Maddison, S. 1983. The orbitofrontal cortex: Neuronal activity in the behaving monkey. *Experimental Brain Research* 49:93–115.

Tobler, P.N., O'Doherty, J., Dolan, R.J., and Schultz, W. 2006. Human neural learning depends on reward prediction errors in the blocking paradigm. *Journal of Neurophysiology* 95:301–10.

Torregrossa, M.M., Quinn, J.J., and Taylor, J.R. 2008. Impulsivity, compulsivity, and habit: The role of the orbitofrontal cortex revisited. *Biological Psychiatry* 63:253–55.

Tremblay, L., and Schultz, W. 1999. Relative reward preference in primate orbitofrontal cortex. *Nature* 398:704–8.

———. 2000. Modifications of reward expectation-related neuronal activity during learning in primate orbitofrontal cortex. *Journal of Neurophysiology* 83:1877–85.

Valentin, V.V., Dickinson, A., and O'Doherty, J.P. 2007. Determining the neural substrates of goal-directed learning in the human brain. *Journal of Neuroscience* 27:4019–26.

van Duuren, E., Nieto-Escamez, F.A., Joosten, R.N.J.M.A., Visser, R., Mulder, A.B., and Pennartz, C.M.A. 2007. Neural coding of reward magnitude in the orbitofrontal cortex during a five-odor discrimination task. *Learning and Memory* 14:446–56.

Waelti, P., Dickinson, A., and Schultz, W. 2001. Dopamine responses comply with basic assumptions of formal learning theory. *Nature* 412:43–48.

Wallis, J.D., Dias, R., Robbins, T.W., and Roberts, A.C. 2001. Dissociable contributions of the orbitofrontal and lateral prefrontal cortex of the marmoset to performance on a detour reaching task. *European Journal of Neuroscience* 13:1797–1808.

Wallis, J.D., and Miller, E.K. 2003. Neuronal activity in primate dorsolateral and orbital prefrontal cortex during performance of a reward preference task. *European Journal of Neuroscience* 18:2069–81.

Watanabe, M. 1996. Reward expectancy in primate prefrontal neurons. *Nature* 382:629–32.

Wellmann, L.L., Gale, K., and Malkova, L. 2005. GABAA-mediated inhibition of basolateral amygdala blocks reward devaluation in macaques. *Journal of Neuroscience* 25:4577–86.

Winstanley, C.A., Theobald, D.E.H., Cardinal, R.N., and Robbins, T.W. 2004. Contrasting roles of basolateral amygdala and orbitofrontal cortex in impulsive choice. *Journal of Neuroscience* 24:4718–22.

16 The Neurology of Value

Lesley K. Fellows

CONTENTS

16.1 INTRODUCTION

Many disorders of the central nervous system affect judgment and decision making. Along with other impairments of "executive function," these symptoms have been challenging to understand within a classical neurological-localizationist framework, beyond a general link to the frontal lobes. Notwithstanding these difficulties, the fact that decision making deficits can emerge in neurological diseases provides a starting point for determining the neural circuits that underlie them. This method of enquiry has several advantages: from a clinical perspective, a better understanding of brain-behavior relationships can help in the diagnosis, prognosis, and treatment of disordered decision making. From a basic science perspective, this approach provides the capacity to test two main questions. The first concerns the behaviors themselves and the second the brain substrates of those behaviors.

Franz-Joseph Gall and his phrenologist colleagues were prepared to localize behaviors as complex as "benevolence" and "wit" to specific regions of the brain. Current views instead suggest that complex behaviors can be understood as depending on simpler component processes, which can in turn be localized to specific neural circuits, which interact to produce the complex behavior (Stuss and Alexander 2007). There are, in principle, many ways to dissect the complexity of decision making. Knowing how to carve up this complexity is perhaps the central challenge in these still-early days of the neuroscience of human decision making. If the aim is to provide a model that will be useful from a neurobiological perspective, then candidate component processes must relate in a meaningful way to the brain.

Studies of neurological patients can be particularly useful in testing whether putative component processes are, in fact, distinct (by showing, for example, that one process is impaired, and another spared, after brain damage). Furthermore, when such work is carried out in patients with defined brain injuries, inferences can be drawn about the brain substrates of the process in question. Thus, a neurological approach provides a useful window on decision making. This chapter will review work on the brain substrates of value-based decision making from this perspective, concentrating particularly on component process approaches. As we shall see, at least some of this work has addressed how "value" relates to choice objects in the world, and as such can be framed as a linkage of sensation and reward. Experimental studies of patients with focal brain injury aimed at delineating the neuroanatomical substrates of decision making will be the primary focus, but I will also touch on work examining the neurochemical modulation of choice. As mentioned, this basic science research also offers an interesting perspective on the mechanisms underlying the everyday difficulties of patients with dysfunction of systems important to decision and reward processing. I will return to this at the end of the chapter.

16.2 A SHORT HISTORY OF THE NEUROPSYCHOLOGY OF DECISION MAKING

There is now a substantial corpus of experimental work on decision making in patients with focal frontal lobe damage, and an emerging literature on the effects of dopaminergic and basal ganglia pathologies. However, perhaps the largest contribution of neurology to decision neuroscience has been to provide colorful anecdotes as introductory material for papers and talks. Indeed, it is remarkable to have come this far into the chapter without having mentioned Phineas Gage. This unfortunate nineteenth-century New Englander went from reliable construction foreman to existence proof of the importance of the frontal lobes in regulating behavior, after an industrial accident which sent a six-foot iron bar through his brain (Harlow 1868). Anecdotal reports of poor judgment after frontal injury continued in the ensuing century, culminating in an influential, detailed case study of the patient E.V.R. This patient was notable both for making poor choices and for having extreme difficulty in arriving at any decision at all, as sequelae of injury to orbitofrontal and ventromedial prefrontal cortex (PFC) (Eslinger and Damasio 1985). These adjacent sub-regions of PFC are commonly injured together, and will here be referred to together as the ventromedial frontal lobes (VMF; see Figure 16.1). Understanding the functions of this region has been a major focus of the experimental neuropsychological studies of decision making to date.

A clinical anecdote is weak evidence, particularly when the symptoms relate to complex behaviors, but a strong stimulus for creative hypothesis generation. Framing the altered behavior of E.V.R., and of other patients with VMF damage, as a deficit in decision making has contributed to the development of novel experimental measures of learning and decision making (Bechara, Damasio, and Damasio 2000; Bechara et al. 1997). In turn, this has led to links being made between these clinical observations and animal research on learning and reinforcement, and (separately) with decision theory developed in behavioral economics and psychology. Together, these approaches have begun

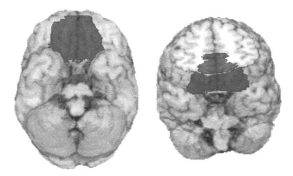

FIGURE 16.1 Sub-region of prefrontal cortex here designated as ventromedial frontal lobe (VMF), shown in darker gray on ventral (left panel) and partial coronal (right panel) sections on a three-dimensional view of the human brain.

to shed light on the brain substrates of decision and value, and are providing a fresh perspective on the constellation of clinical symptoms that follow frontal lobe injury (Fellows 2007a; Murray, O'Doherty, and Schoenbaum 2007).

Phrenology notwithstanding, decision making does not take place in only one region of the brain. Indeed, there is every reason to believe that decisions draw upon an extensive network of subcortical and cortical structures, and also to expect that fundamental signals of reward, punishment, and expectancy have a wide influence on neural processes beyond what might be considered as strictly decision making. This chapter will begin with a review of what has been learned about the role of VMF in decision making, primarily drawing from patient lesion studies. This will also provide the opportunity to discuss a range of candidate component processes likely to be important in decision making. I will then turn to other neuroanatomical substrates of decision making, including amygdala, insula, and striatum, and other regions within PFC, and review what is known about their contributions to these processes in humans. Finally, I will briefly discuss what has been learned about the neurochemical substrates of decision making in human subjects, and then consider the clinical implications of this basic science literature.

16.3   VENTROMEDIAL FRONTAL LOBE DAMAGE AFFECTS DECISION MAKING

Patients with VMF damage may have changes in behavior, judgment, and insight, often unaccompanied by demonstrable deficits on standard neuropsychological measures. This disconnect relates less to the mysterious functions of this region and more to the relatively narrow focus of standard neuropsychological tests of frontal lobe function. In an effort to capture the decision making deficits of these patients, Bechara and colleagues at the University of Iowa developed a new experimental measure that has come to be called the Iowa gambling task (IGT). This task was an empirical effort to capture the complexity of decisions about reward, punishment, and risk. The task "worked" in the sense that patients with VMF damage performed differently from healthy participants, but its complexity has resulted in controversy about the basis of this performance deficit and, indeed, the basis of the abnormal performance of many populations tested since, ranging from drug addicts to individuals with personality disorders (Bechara et al. 1994, 1997, 2005; Dunn, Dalgleish, and Lawrence 2006; Fellows and Farah 2005a; Maia and McClelland 2004; Tomb et al. 2002).

The IGT involves choosing amongst four decks of cards, for a total of 100 trials. After each choice, the player receives feedback in the form of a (typically hypothetical) monetary win, and, on some trials, a monetary loss as well. The feedback contingencies are such that two of the decks provide larger wins on every trial, but occasionally even larger losses. The other two decks provide smaller wins, but even smaller losses, making these decks more advantageous overall. The order of the cards in each deck is fixed. The majority of healthy participants begins with a preference for the high win decks, but gradually shift to choosing more often from the lower win (but also lower loss) decks. By contrast, patients with VMF damage tend to persist in choosing more often from the disadvantageous decks.

This impairment alone is interesting, but the initial work with this task took the question further by measuring autonomic responses, and by questioning participants at intervals about their explicit knowledge of the reinforcement contingencies. Healthy subjects seemed to be choosing from the better decks before they were able to explain why they were doing so, and also tended to have larger skin conductance responses prior to choosing from the disadvantageous decks as the task proceeded. By contrast, patients with VMF damage did not develop these discriminatory skin conductance responses prior to choosing and made disadvantageous choices even after they were able to report the task contingencies (Bechara et al. 1997). These data were interpreted as support for the somatic marker hypothesis of decision making, which proposes that risky decisions are taken based, in part, on embodied representations of that risk (colloquially, "gut feelings"), indexed by (or perhaps directly encoded by) autonomic changes that may not necessarily relate to conscious knowledge (Bechara, Damasio, and Damasio 2000; Damasio 1994; Dunn, Dalgleish, and Lawrence 2006).

Subsequent work has called into question many aspects of the interpretation of this experiment (reviewed in Dunn, Dalgleish, and Lawrence 2006). However, it is important not to throw the baby out with the bathwater: Patients with VMF damage are clearly impaired on the IGT and determining the mechanism underlying that impairment is likely to be informative. Furthermore, understanding the role of implicit learning from feedback (Pessiglione et al. 2008), and of autonomic signals (Critchley et al. 2003; Heims et al. 2004; Reekie et al. 2008), would seem important in developing a complete description of decision making, and of mapping that description onto the brain. Although the IGT was developed to test "decision making," it is perhaps more useful to frame it as a learning task. I will first discuss a series of experiments that shed light on the mechanism that may underlie the learning deficits of VMF patients on the IGT, and then review the literature on the role of this area in simpler forms of learning from feedback, before returning to consider whether VMF plays a role in decision making beyond the roles it seems to play in learning.

16.4   VENTROMEDIAL FRONTAL LOBE IS CRITICAL FOR FLEXIBLY LEARNING FROM FEEDBACK

As far back as the 1960s, studies in animals had established that lesions to OFC disrupted specific forms of reinforcement learning, including reversal learning and extinction. Importantly, these tasks involve learning the reward (or punishment) value associated with *stimuli*—objects in the world. Macaques with OFC resections were typically able to learn to choose between two objects based on whether the object was paired with a (food) reward or non-reward, but made errors when the stimulus-reinforcement contingencies were changed (so-called reversal learning), or persisted in choosing previously rewarded objects after they were no longer followed by reward (i.e., under conditions of extinction) (Butter 1969; Izquierdo, Suda, and Murray 2004; Jones and Mishkin 1972). The same impairment in rapidly updating stimulus-reinforcement contingencies, captured by simple reversal learning tasks (with feedback in the form of points or play money), has also been demonstrated in humans with VMF damage, with the critical region likely to be medial OFC (Fellows and Farah 2003; Hornak et al. 2004; Rolls et al. 1994).

Interestingly, the IGT contains a reversal learning requirement. Because the card order is fixed, players are initially lured into a preference for the disadvantageous decks; for the first several trials, these decks are, in fact, better choices. Patients with VMF damage develop this initial preference just like healthy subjects, but then seem to have difficulty overcoming this initial preference despite mounting losses. We hypothesized that this reversal learning feature of the IGT was critical for the impaired performance of VMF patients. We first established that VMF damage led to impaired reversal learning, tested with a very simple two-deck card game. We then administered a "shuffled" version of the IGT which eliminated the reversal learning requirement of that task, by making it clear in the initial trials that the high-win decks also hold large losses. Remarkably, patients with VMF damage performed this shuffled version as well as did healthy control subjects (Fellows and Farah 2005a). Furthermore, the extent to which their performance improved from the standard to the shuffled IGT was correlated with the extent of their reversal learning impairments as measured by the simple reversal learning task (Fellows and Farah 2003). These studies did not measure skin conductance responses, leaving the relationship between these findings and the somatic marker hypothesis an open question.

This experiment argues that VMF damage disrupts reversal learning, which in turn leads to poor IGT performance. However, it does not mean that the IGT is *only* a jumped up reversal learning task, nor does it follow that all impaired performance on the IGT is due to a fundamental reversal learning deficit. Clearly the IGT taps other processes. For example, as can be seen in Figure 16.2, at least some patients with PFC damage sparing VMF are also impaired on the IGT (Clark et al. 2003; Fellows and Farah 2005a; but see Bechara et al. 1998), even though such patients have no difficulty with simple reversal learning (Fellows and Farah 2003). Amygdala damage is also associated with poor IGT performance (Bechara et al. 1999), with more recent evidence suggesting that this phenomenon may be due to a role for amygdala in influencing value information in VMF (Hampton

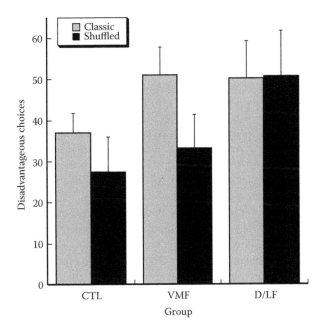

FIGURE 16.2 Mean number of choices from the disadvantageous decks over 100 trials in the classic version of the IGT (light bars) and the shuffled version (dark bars; see text for details) for patients with VMF damage, dorsal and/or lateral frontal lobe damage (D/LF), and healthy, demographically matched controls (CTL). Error bars show the 95% confidence intervals. The VMF group made significantly more disadvantageous choices compared to controls in the classic task, but did not differ from controls in the shuffled version. Interestingly, patients with D/LF damage were impaired in both conditions, compared to controls. (Data from Fellows, L.K. and Farah, M.J., *Cereb Cortex*, 15, 58, 2005.)

et al. 2007). At the least, the effects of damage to other brain regions on this task emphasize that IGT performance cannot be interpreted as a specific index of VMF function.

The critical role of VMF in reversal learning provides a simpler, more focused starting point for investigating the contributions of this region to learning from feedback, and so to decision making that relies on such learning. Despite its relative simplicity, reversal learning itself involves several component processes (Wheeler and Fellows 2008; Fellows 2007b, Murray, O'Doherty, and Schoenbaum 2007; Schoenbaum, Saddoris, and Stalnaker 2007). The specific pattern of reversal learning impairment shown by the patients we studied suggested that the difficulty relates to adjusting performance in response to unexpected negative feedback, as occurs on the critical reversal trial. Patients with VMF damage characteristically failed to change tack in response to that feedback. We wondered if this related to a specific deficit in learning from negative feedback, and examined that possibility using a probabilistic stimulus-reinforcement task that allowed us to separately probe learning from negative and positive feedback.

Learning in a probabilistic (as opposed to a fully deterministic) environment is more difficult, and patients with frontal lobe damage were, in general, less likely to reach criterion. Of those who were able to learn enough to move to the test phase of the task, patients with VMF damage showed the predicted selective deficit in learning from negative feedback. These patients learned to associate stimuli with positive feedback as well as controls, but were unable to learn to avoid the stimuli more often associated with negative feedback. Patients with frontal damage sparing VMF learned about the same from both forms of feedback, as did the healthy controls (Wheeler and Fellows 2008).

This finding argues for a critical role for VMF in learning from negative feedback and provides evidence that, at least in some contexts, learning from positive and negative feedback is dissociable. It leaves open what the specific role of VMF in this process might be. Is the learning itself occurring within VMF, or does VMF play a permissive role in facilitating or allowing such learning to occur in other neural systems (Stalnaker et al. 2007)? I will return to these questions after reviewing what is known about the role of VMF in decision contexts that do not involve learning.

16.5 VENTROMEDIAL FRONTAL LOBE DAMAGE DISRUPTS DECISION MAKING UNDER CERTAINTY AND UNDER RISK

A different perspective on VMF function has arisen from a literature focusing not on learning, but on economic value. The value of a stimulus (i.e., a decision option) is not an intrinsic property, but a product of cognition. To the dismay of normative economics theorists, even the value of money is changeable: Most of us would much rather have $10 today than the same amount in a year (Ainslie 2001). Value is not only affected by delay: it is influenced by other contextual factors, such as what else is available, and by the internal state of the organism. For example, the value of a slice of chocolate cake will depend on whether the other menu options are lemon tart or soda crackers, and will depend on whether you are hungry, or just finishing an all-you-can-eat chocolate binge. In this sense, value can be likened to more strictly sensory-perceptual properties of stimuli, such as color or size. While these are quantifiable in absolute terms, their perception can be influenced by context, such as surrounding colors, and fine-grained distinctions between different hues (or heights, etc.) are typically easier to establish in relative rather than absolute terms. Can value be understood as a higher-order sensory feature of objects? If so, is this only a superficial analogy, does it reflect analogous neural mechanisms (Rushworth, Mars, and Summerfield 2009), or is it carried in the same channels? Although single-unit work has shown that learned reward value of stimuli can be reflected in neural activity in posterior cortical regions also engaged in higher-order sensory processing (Sugrue, Corrado, and Newsome 2004), I am not aware of any work addressing whether value-related functions are disrupted in human subjects with disorders of sensory processing (a topic briefly touched on in Chapter 10 in this volume).

More work on the neural representations of value has been done in OFC, with several lines of evidence in animal models suggesting that neurons in this area encode the reinforcement value of stimuli (Rolls 2000; Schoenbaum et al. 2003; Padoa-Schioppa and Assad 2006, 2008). There are also functional imaging data supporting the view that OFC activity relates to current stimulus value in healthy human subjects (Elliott et al. 2003; O'Doherty et al. 2001; O'Doherty 2004; Plassmann, O'Doherty, and Rangel 2007; Small et al. 2001).

If neurons within VMF encode the value of stimuli, the next question might be: to what end? How are these value representations engaged to influence behavior? An obvious possibility is that these representations directly guide decisions. How do you choose between two good options? You check in with VMF, find out which one has the greatest value, and select it. In this formulation, VMF is functioning as a "value look-up table" (Schoenbaum, Saddoris, and Stalnaker 2007). The prediction that follows is that VMF damage should disrupt even the simplest forms of decision making that require such information. Simple preference judgments are a common example of such decisions. Chocolate or vanilla? Red wine or white? Such decisions would seem to involve the simplest sort of value comparison. Further, depending on the specific choice, they need not draw on recent learning, or involve consideration of risk or probability. As anyone who has stood in line at an ice cream stand can attest, such preference judgments can still be very difficult. Does VMF play a necessary role in such choices?

We addressed this question with a simple experiment, asking patients with such damage to indicate their preference between pairs of stimuli, in different categories: foods, colors, and famous people. While there is no right or wrong answer in such choices, we reasoned that the hypothesized degraded representation of stimulus value would lead to inconsistencies in these preferences. Applying the principle of transitivity, we tested the consistency of preferences across this series of choices. If subjects preferred A to B, and B to C, they should prefer A to C. When they chose C over A, this was considered an error (or in formal economic terms, an "irrational" choice). As predicted, patients with VMF damage (and not patients with frontal damage elsewhere) were more inconsistent in their preferences, an experimental demonstration of the clinical reports of "capriciousness" after frontal lobe damage that go back to Phineas Gage (Fellows and Farah 2007).

A fundamental deficit in assigning relative value to stimuli may be the basis for the deficits in more complex forms of decision making after VMF damage reported in several other studies. One more complex form of choice under certainty is so-called multi-attribute decision making. Here, subjects must integrate information about options across multiple domains, such as the rent, location, and size of apartments, in order to make a choice. Even when all information is available, these can be very difficult decisions. In a process-focused study of such decision making, we found that VMF damage led to a very different strategy of information acquisition, and to different final choices, again compared to either non-VMF frontal damage, or to demographically matched control subjects. One explanation for the different strategy in those with VMF damage is that they were acquiring information in a way that minimized the need to make relative value comparisons (Fellows 2006).

If VMF damage disrupts choices under certainty, it is perhaps not surprising that it also disrupts choices made more difficult through considerations of risk. The somatic marker hypothesis suggests that there is something special about risky decision making, and the initial studies in this area argued that VMF was important in that "something special." However, introducing either risk (known uncertainty) or ambiguity (unknown uncertainty) into decision making might be expected simply to increase the level of difficulty in determining value, enhancing a fundamental deficit in "valuing" in patients with VMF damage. An early effort to deconstruct the IGT led to the development of a task that explicitly tested decision making under (known) risk. Now called the Cambridge gambling task, this task offers gambles with varying (but explicitly provided) probabilities, and subjects choose how much money to stake on each trial. Although initial results were somewhat inconsistent (Clark et al. 2003; Mavaddat et al. 2000; Rogers et al. 1999), what would seem to be the definitive study using this task has demonstrated that VMF damage does not affect the ability to choose the higher probability option, but that such patients are less risk-averse than healthy subjects, in that they systematically bet more at every level of risk. Interestingly, they remain sensitive to differences in risk: although their bets are high, they do scale with probability (Clark et al. 2008). Other work, using quite different tasks, has also demonstrated less risk-aversion after ventral frontal damage (Hsu et al. 2005; Shiv et al. 2005). Risk in such tasks, as in life, involves potential losses, raising the possibility that these observations reflect a generic underweighting of losses (whether experienced or anticipated) after VMF damage, which would be consistent with the results of the learning experiments discussed above.

Other decisions involve comparing not simply the anticipated value of current options (what will be), but also the value of what might have been. In choice paradigms under uncertainty (i.e., between gambles) where subjects will learn the outcome of both their choice, and of the other, unchosen option, they often choose so as to minimize regret. Regret is an interesting response, because it involves projecting relative value: anticipating the value of a given option in light of the possible outcomes of other options. Even objectively positive outcomes are downgraded in subjective value when the subject learns that the alternative gamble, had they chosen it, would have provided an even better outcome (Mellers 2000). One intriguing study suggests that VMF damage disrupts decision making that involves consideration of regret: unlike healthy subjects, patients with such damage did not choose so as to minimize regret, and did not report emotional responses consistent with the experience of regret, although they were more pleased when they won rather than lost on the choice they did make (Camille et al. 2004).

Regret can be seen as a higher-order value comparison, in which the subject considers not only the predicted value of the chosen option in and of itself, but how that value will be affected by the outcome of the non-chosen option, given that they know they will learn that other outcome. Other higher-order value comparisons include envy, when the value of an outcome is compared against the (higher) value of an outcome received by someone else, and schadenfreude (gloating), when the subjective value of an outcome is increased by the knowledge that another person had a worse outcome. The ability to recognize these emotional states in hypothetical scenarios has been shown to be affected by VMF damage (Shamay-Tsoory, Tibi-Elhanany, and Aharon-Peretz 2007). The authors cast this deficit in a "theory-of-mind" framework, but (as they also suggest) it may be reasonable to interpret these emotions as requiring particularly difficult higher-order relative value estimates. A similar explanation might be advanced for the effects of VMF damage on moral decision making (Koenigs et al. 2007).

16.6   VALUE, EXPECTATION, AND LEARNING

It remains unclear whether the effects of VMF damage on stimulus-value learning (and particularly reversal learning) and on the ability to track relative stimulus value relate to a common underlying process. The region damaged in such patients is large enough to suppose that multiple processes may be affected. Furthermore, heterogeneity of damage, practical limitations in how accurately that damage can be ascertained with anatomical neuroimaging, and small sample sizes all constrain the ability to provide a definitive answer to this question. In principle, identifying patients who show dissociable patterns of performance on these two tasks would directly address this question, even if the anatomical substrates could not be definitively established. In practice, this assumes that the tasks are comparable in their indexing of the process (or processes) of interest. In fact, both the preference judgment and (to a lesser degree) the reversal learning tasks used to date are uncomfortably close to ceiling, making it difficult to make strong statements about this interesting question.

The alternative—that the two abilities relate to a single function of VMF—is worth considering. One model that could accommodate both effects is as follows: current anticipated value of an option is represented in OFC. This information is used to compare the (anticipated) values of multiple options in order to choose between them. It is also used to compare the value of an outcome after a choice with the anticipated value, in order to determine whether the value was equal to, less than, or greater than expected. Occurrence of a less than expected value (in particular) then engages learning mechanisms, both to update value projections for subsequent decisions, and perhaps to make a change of behavior more likely (Schoenbaum, Saddoris, and Stalnaker 2007). Such a formulation also aligns with hypotheses outlined in Chapter 15 in this volume.

16.7   VALUE AND TIME

Time is an important variable in decision making and can influence choice in several ways. Perhaps most obviously, time is a variable in choices that require delay of gratification. The ability to withhold a response to an immediately available reward, in order to gain a larger reward after a delay, is a classic test of self-control (Ainslie 2001; Mischel, Shoda, and Rodriguez 1989) and failure to withhold such responses is a classic clinical sign of frontal lobe dysfunction. Real-life decisions sometimes take this form, but often future considerations are less well defined. For example, rewards that are delayed are also often less certain, confounding time and probability. In decisions that involve repeated choice, time can be inextricably linked with learning.

Understanding the interplay between time and value requires clear operationalization of the time process in question. Temporal discounting is one well-studied construct meeting this criterion. Temporal discounting refers to the characteristic loss of (subjective) value of a reward as a function of the delay to its delivery. This loss of value can be measured in various ways and can typically be fitted with a hyperbolic function. Over short timescales (seconds) and actually experienced reward, temporal discounting has been observed in pigeons and rats. Similar temporal discounting effects are seen in adult humans, although this has most commonly been measured over much longer timescales (days to years), and with monetary rewards, whether real or hypothetical (Ainslie 2001). Temporal discounting measured in this way seems to have ecological validity. For example, groups known for real-life impulsive behaviors, such as drug addicts, discount the value of reward more steeply than do healthy control groups (Kirby, Petry, and Bickel 1999; Petry, Bickel, and Arnett 1998). Functional imaging studies have investigated the neural substrates of discounting, implicating both frontal and sub-cortical regions (Kable and Glimcher 2007; McClure et al. 2004).

We tested the hypothesis that VMF contributes to the ability to maintain subjective value despite delay, administering a standard temporal discounting measure to a group of patients with such damage, compared to both healthy control subjects and to patients with damage affecting lateral frontal lobes. We also included a non-frontal brain-injured control group, because we were concerned that the experience of a significant illness might, in and of itself, affect the interplay between time and reward. Somewhat to our surprise, we were unable to detect any systematic effect of brain injury, whether involving VMF or elsewhere, on temporal discounting rate. The hypothetical nature of the task may be one explanation for the null finding, although the same patients do have difficulties with other hypothetical tasks, including the preference judgments described above. Perhaps more importantly, a general "value representation" role for VMF would not predict a temporal direction to any deficit following VMF damage, at least if immediate and delayed reward values are both being represented in this region. Instead, one might predict an increase in inconsistent choices (leading to a "noisier" fit to the temporal discounting function, rather than a change in the slope of that function), due to a general degradation of the fidelity of value representations, just as was observed in the preference judgment experiment. This was not a planned outcome measure in the experiment just described, but would be worth testing in future work.

By contrast, VMF damage did affect a second measure of future thinking that would seem to have a bearing on decision making. Using a measure first developed to study "future orientation" as a personality trait, we asked how far into the future participants spontaneously projected themselves when asked to do so in a particular context. Healthy older subjects thought ahead in time almost 15 years, on average, whereas those with VMF damage considered much shorter time windows of 5–6 years. This foreshortening of future time was not solely a non-specific effect of brain injury or illness, in that it was significantly more marked in the VMF group than in the other brain-injured (and comparably disabled) control groups (Fellows and Farah 2005b).

Recent work has begun to ask interesting questions about the relation of memory to future thinking, showing, for example, that amnesia due to temporal lobe damage is associated with an impaired ability to imagine the future (reviewed in Schacter, Addis, and Buckner 2008). This work again highlights the role of learning and memory in decision making: we generate expectations about the future based on past experience, and the memory of that past experience frames both the content and the temporal window of decision making. Furthermore, it appears that common brain networks are involved in both remembering and predicting.

16.8 VALUE AND EMOTION

There appear to be close links between value and emotion. Value assessment may incorporate emotional responses and in social situations assessment of the likely value of a choice may rely on predicting or detecting emotional responses in others. Intriguingly, damage to VMF disrupts various emotional processes. Patients with VMF damage have an altered experience of complex emotions such as embarrassment (Beer et al. 2003, 2006). There is less consensus as to whether the experience of basic emotions is affected by such damage (Berlin, Rolls, and Kischka 2004; Hornak et al. 2003; Roberts et al. 2004; Gillihan et al. 2010). VMF damage can also disrupt the ability to recognize emotions in others, whether from voice or facial expression (Heberlein et al. 2008; Hornak et al. 2003). It is tempting to speculate that these various deficits are somehow linked and to wonder whether they may in turn be linked to deficits in decision making (Rolls 1999). However, as with preference and reinforcement learning, it may be that these abilities are linked only by anatomical coincidence (i.e., rely on nearby but distinct regions within VMF) rather than because they share the same component processes. More work is needed to resolve these issues. Demonstration of dissociations within individual patients may be particularly helpful in this regard, although this enterprise is complicated by a lack of consensus on the component processes of each of these complex abilities, and on the appropriate measures of these processes.

16.9 BEYOND VMF: DORSOMEDIAL PFC MAY LINK VALUE TO ACTIONS

There is mounting evidence from electrophysiological studies in monkeys (Amiez, Joseph, and Procyk 2006; Shidara and Richmond 2002), and from ERP and fMRI studies in humans (Brown and Braver 2007; Gehring and Willoughby 2002; Rushworth and Behrens 2008) that activity in dorsomedial PFC (anterior cingulate cortex [ACC] and adjacent medial PFC) also reflects reward value. While value can be readily conceived of as attached to a stimulus (that cake looks *really* good), it can also, somewhat less intuitively, be seen as a feature of an action (e.g., pulling the lever on a slot machine) (Kennerley et al. 2006; Rushworth and Behrens 2008). One view is that dorsomedial PFC is involved in decision making at the level of action; that it biases responses based on the anticipated value of the action. The values of stimulus and response are often confounded in real-life decisions, but experimental work suggests that the neural substrates of these two value representations are different, at least in animal models (Rushworth et al. 2007). Lesion experiments in monkeys indicate that stimulus-value and action-value learning can be dissociated, with ACC playing a critical role only in the latter (Rudebeck et al. 2008). In a task in which two different actions were possible, but only one was consistently associated with reward, lesioned animals did not sustain the rewarded action over multiple trials, suggesting a deficit in developing an integrated sense of the value of each action (Kennerley et al. 2006). Whether this region plays a necessary role in human decision making, and if so, the nature of that role, has yet to be examined systematically.

However, this region has been intensively studied in relation to conflict and error monitoring in humans (Botvinick 2007; Carter and van Veen 2007; Ridderinkhof et al. 2004). There has been conflicting evidence as to whether intact medial PFC is necessary for conflict monitoring (di Pellegrino, Ciaramelli, and Ladavas 2007; Fellows and Farah 2005c; Stuss et al. 2001). We have recently found that damage to this region disrupts error prediction in several tasks (Modirrousta and Fellows 2008a, 2008b), consistent with converging fMRI, ERP, and computational evidence (Brown and Braver 2005, 2008; Burle et al. 2008; Hester et al. 2005). This finding is relevant to understanding the performance of the ACC-lesioned monkeys described above, who could adjust their immediate response after feedback, but failed to develop a predictive model of action value based on experience. In many contexts, including the conflict tasks (such as the Stroop task) that have been the main focus of performance monitoring studies, action value is all-or-none, right or wrong. In decision contexts, it can be more nuanced (Amiez, Joseph, and Procyk 2005; Oliveira, McDonald, and Goodman 2007; Rushworth and Behrens 2008). One model of dmPFC function that could account for these varied findings is that this region represents the predicted value of an ongoing action in general (Mansouri, Tanaka, and Buckley 2009; Rushworth and Behrens 2008). Further work will be needed to test this possibility in humans, however.

16.10 THE ROLE OF THE INSULA IN REWARD AND DECISION

Insular cortex has long been associated with central autonomic control and interoception in general (Craig 2003), and with the emotional/motivational aspects of pain processing more specifically (Price 2000). This region is implicated in decision making tasks on the basis of functional imaging findings (Kuhnen and Knutson 2005; Paulus et al. 2003). There is some early evidence that insula plays a necessary role in certain forms of decision making. Pure insula lesions are quite rare, with damage typically also disrupting white matter tracts connecting frontal and parietal cortex, and cortex to basal ganglia, including amygdala, as well as variable encroachment on ventrolateral prefrontal, motor, somatosensory, and lateral temporal lobe. With these provisos in mind, damage involving the insula has been shown to increase the amount of money bet in the Cambridge gambling task, to a degree comparable to what was seen after VMF damage (Clark et al. 2008). Unlike VMF patients, however, insula damage was associated with a loss of sensitivity to probability information; that is, the size of bets did not systematically scale with the likelihood of winning.

A second study relevant to understanding the potential role of insular cortex in decision and value processes examined the effects of acquired insula injury in addicted smokers on their addiction. While those with insula damage were no more likely to quit smoking than patients with damage elsewhere, insula damage was associated with a reduction in the subjective urge to smoke (i.e., craving) (Naqvi et al. 2007). More systematic study of the effects of insula damage is needed to clarify the mechanisms underlying the intriguing observations in both of these studies, and to understand the links between these findings and earlier work on the role of this area in mediating the emotional and behavioral relevance of physical pain (see Price 2000 for review).

16.11   STRIATAL MECHANISMS IN DECISION MAKING

There is little doubt that value-based learning involves striatal mechanisms, based on an abundance of evidence from animal studies (Smith et al. 2009) and functional neuroimaging in humans (Izuma, Saito, and Sadato 2008; Knutson and Cooper 2005; Pessiglione et al. 2008; Seymour et al. 2007), and at least one recent study showing that striatal damage disrupts feedback-driven learning, particularly reversal learning (Bellebaum et al. 2008).

Patients with Parkinson's disease (PD) are perhaps the most commonly studied proxy for striatal dysfunction in humans. Dopamine denervation in the striatum (particularly the putamen) is the hallmark of this degenerative disorder. However, pathological findings are heterogeneous across patients and over time other monoaminergic systems and the cortex itself are also variably affected. Differences between PD and healthy participants therefore may not reflect (only) dopamine-striatal dysfunction. Studies that use a within-subject design, comparing a single group of patients on and off dopamine replacement therapy, can be interpreted with more confidence: observed effects are likely due to dopamine, although may not be due to dopamine acting specifically in the striatum. One such study showed that dopamine modulates reinforcement learning. Patients with PD learned better from positive feedback when on their dopamine replacement therapy and better from negative feedback when off (Frank, Seeberger, and O'Reilly 2004). This pattern was predicted by a neurocomputational model of dopamine actions in the striatum (Frank 2005). Other work is also consistent with a role for dopamine in striatally mediated trial and error learning in patients with PD (see Shohamy et al. 2008 for review).

There has been interesting, albeit limited, work on the contributions of other neurochemicals to aspects of reinforcement learning and decision making. It has been shown, for example, that pharmacological manipulation of serotonin (but not noradrenaline) signaling in healthy subjects can influence learning from probabilistic feedback (Chamberlain et al. 2006). Other work has suggested that serotonin modulates learning from negative feedback specifically (Cools, Robinson, and Sahakian 2008).

16.12   AMYGDALA AND VALUE

There is evidence that the amygdala is critical for at least some forms of decision making. The amygdala has close, bi-directional connections with orbitofrontal cortex (Ghashghaei and Barbas 2002), and there is evidence that both flexible stimulus-reinforcement learning and classical conditioning rely critically on interactions between these two regions in animal models (Baxter et al. 2000; Roberts, Reekie, and Braesicke 2007; Schoenbaum and Roesch 2005; LeDoux 2003; Seymour and Dolan 2008). Interactions between amygdala and VMF have been demonstrated in human subjects during reversal learning (Hampton et al. 2007). Lesions to amygdala were reported to impair IGT performance in humans and this structure was proposed as important for signaling somatic states in the somatic marker hypothesis (Bechara, Damasio, and Damasio 2003,

Bechara et al. 1999). For the moment, the role of amygdala in human decision making has not been specified in much detail.

16.13   HEDONICS AND DECISION MAKING

The chapter so far has largely treated value as a concept, rather than as a subjective experience. While decisions can, in principle, be guided by dry economic analysis, everyday experience suggests that many aspects of value are "felt": pleasure and pain can be anticipated in vivid detail, risky choices can summon strong emotional responses, complete with autonomic changes, and near-misses can be played and re-played in the imagination. Further, one may be more or less explicitly aware of value information. Embodied and implicit aspects of value anticipation and experience have been important in neuroscience theories of human decision making, perhaps most prominently in the somatic marker hypothesis (Damasio 1994). Here, I will briefly discuss work related to this theme.

What are the neural substrates of choices made based on "hunches" or "gut feelings," and are these distinct from more explicit aspects of decision making? This has been a much-debated feature of the IGT (Bechara et al. 2005; Dunn, Dalgleish, and Lawrence 2006; Maia and McClelland 2004, 2005). One way to frame this issue is to ask: can implicit representations of value guide choice? Work on implicit learning suggests that the general answer is yes (Yin and Knowlton 2006). Indeed, neural circuits involved in reinforcement learning (striatum, amygdala) can encode value, and influence choice, in the absence of explicit awareness of either the value estimate or, remarkably, even of the stimulus (Pessiglione et al. 2008). A related question, whether autonomic signals guide choice, also remains unsettled (Bechara et al. 1997; Heims et al. 2004; North and O'Carroll 2001; Tomb et al. 2002), although it seems that appetitive autonomic responses may be altered by OFC lesions without necessarily affecting behavior in marmosets (Reekie et al. 2008), and dissociation between autonomic and cognitive performance has been reported after ACC damage in humans (Critchley et al. 2003).

A third issue is the relation between value-guided choice and the subjective experience of value. In principle, lesion studies might be able to shed light on this question. For example, do lesions to VMF sufficient to disrupt value-based decision making also disrupt emotional experience related to value? Unsurprisingly, the answer turns out to be complicated. Part of this complexity is a measurement problem. What aspects of emotion should be affected and how should these be captured experimentally? As reviewed above (and see Hornak et al. 2003), damage to VMF does appear to disrupt some aspects of emotional experience, including complex emotions like embarrassment or regret, but other aspects of emotional experience, such as transiently evoked happiness and sadness, remain intact (Gillihan et al. 2010). If these patients are not "emotionally numb," do they experience pleasure? Again, there is the problem of how to measure pleasure. Anecdotally, such patients do not typically report anhedonia in clinical interviews (in contrast to patients with major depression, for example, in whom anhedonia is often a prominent feature). Subjective report of hedonic experience, for example as measured by self-report questionnaire (e.g., Snaith-Hamilton Pleasure Scale), is typically within the normal range (Fellows, unpublished data). However, self-report is problematic in general because of demand effects, and a particular problem in such patients, who may lack insight into their subjective state.

16.14   VALUE AGNOSIA OR APRAXIA OF CHOICE?

The findings reviewed in this chapter suggest a network of brain regions that, in various combinations, contribute to decision making. Perhaps the strongest evidence is for a critical role for VMF in decision making and flexible stimulus-reinforcement learning. More speculatively, dorsomedial PFC may be involved in linking value to action, with the contributions of other regions such as insula, amygdala, and striatum likely, although as yet less well-specified in humans. This body of work, together with converging evidence reviewed elsewhere in this volume, make a strong case for the plausibility of a "neurology of value." Indeed, the mechanisms that track the value of decision options, whether stimuli or actions, make an important link between perception and action. The frontal lobes in general are classically considered as important for "goal-directed behavior." In the lab, goals are often assigned by the experimenter. In real life, we must judge for ourselves what is worth pursuing. This fundamental aspect of choice can rest on very basic "good" or "bad" determinations, on more subtle "better"/"best" judgments, or on very abstract "as good as it gets around here, and better than what the other guy got" or perhaps "better than I deserve" estimations.

From a clinical point of view, disruptions of these mechanisms may yield a complete inability to decide, but more commonly seem to lead to "wrong" choices. Given the inconsistencies, biases, and irrationality that mark our everyday decisions at the best of times (Kahneman 2003; Kahneman and Tversky 1979; Tversky 1969), a neurologist might be hard-pressed to distinguish clinical impairments in decision making. The tongue-in-cheek definition of "poor judgment" as when the patient's decision runs contrary to the doctor's recommendation is not much help. Nevertheless, patients with brain injury may show a variety of deficits that can be framed as impaired judgment, altered motivation, or changes in personality. The latter encompasses many symptoms, but may include decisions that are "out of character."

Can the cognitive neuroscience findings on the neural substrates of decision making reviewed in this chapter be applied to a better understanding of these clinical presentations? This basic science has the potential to offer both novel conceptual frameworks from which to identify component processes of motivation, evaluation, and decision, and the empirical tools to measure these processes. Conceptually, impairments of decision making have been considered under the umbrella of executive dysfunction. Presumably, this is partly because they occur after frontal lobe injury, and partly because decision making would seem to be closely related to other complex, higher-order cognitive abilities such as those involved in abstraction, planning, and problem solving. However, the analysis presented here builds from relatively simple concepts of reward and punishment influencing basic aspects of learning and choice in dynamic environments. This raises alternative perspectives on clinical impairments of decision making. For example, the value of a stimulus can be seen as a higher-order "sensory" property, combining information from primary sensory modalities with learned reward (or punishment) information, interpreted in the organism's current context. Considered in this way, disruption of evaluation can be conceived of as an impairment of higher-order sensory processing, i.e., an agnosia (see Chapter 10 in this volume). Such patients can perceive and identify an object, but fail to determine its (current, relative) worth, which will necessarily impair decision making about it.

A similar analysis can be applied, albeit more tentatively, to actions: actions must be linked to value to be adaptive. It is not enough to know "how" to carry out a higher-order behavior; there must also be an authentic "why" driving that behavior. Degraded links between action and value could lead to the kind of purposeless, environmentally triggered but contextually inappropriate behavior that is a hallmark of some forms of frontal damage. Again, existing clinical terminology might be adapted to capture this syndrome. The inability to perform a complex action despite intact basic motor function is termed apraxia. Such patients may have difficulty correctly manipulating tools, for example, or *difficulty showing how to* manipulate tools in the absence of direct access to those tools. Disorders in which actions are not guided by value can be considered as an "apraxia of choice," or "value apraxia." A patient so afflicted might manipulate the tool correctly, but employ it for some contextually inappropriate, i.e., valueless, purpose.

I am loathe to burden cognitive neurology with further variations of either apraxia or agnosia, both terms that can confuse as much as they clarify. However, I do think that the conceptualization of value as a higher-order sensory property, or as a factor influencing action selection, can be a useful heuristic in parsing the component processes of decision making. The experimental neuropsychology reviewed in this chapter, and the field of decision neuroscience more generally, is making inroads into identifying component processes of decision making and their neural substrates.

Importantly, in many cases these components can be traced to much simpler aspects of sensory processing and reinforcement learning, both relatively well-understood in animal models. This framework helps to "rescue" decision making from the murky domain of complex executive function, providing more traction for understanding this aspect of goal-directed human behavior. Still very much a nascent field, this area of study has the potential to provide novel insights into human behavior, and into disorders of human behavior common in both neurology and psychiatry.

ACKNOWLEDGMENTS

Martha Farah, Elizabeth Wheeler, and Michael Frank contributed to the work discussed here. This research would not be possible without the generous participation of patients, their families, and the clinicians who care for them. I acknowledge operating support from NIH (NIDA R21DA022630), CIHR (MOP-77583), and the Parkinson Society of Canada, and salary support from the CIHR's Clinician-Scientist program and the Killam Trust.

REFERENCES

Ainslie, G. 2001. *Breakdown of Will*. Cambridge: Cambridge University Press.
Amiez, C., Joseph, J.P., and Procyk, E. 2005. Anterior cingulate error-related activity is modulated by predicted reward. *Eur J Neurosci* 21:3447–52.
———. 2006. Reward encoding in the monkey anterior cingulate cortex. *Cereb Cortex* 16:1040–55.
Baxter, M.G., Parker, A., Lindner, C.C., Izquierdo, A.D., and Murray, E.A. 2000. Control of response selection by reinforcer value requires interaction of amygdala and orbital prefrontal cortex. *J Neurosci* 20:4311–19.
Bechara, A., Damasio, A.R., Damasio, H., and Anderson, S.W. 1994. Insensitivity to future consequences following damage to human prefrontal cortex. *Cognition* 50:7–15.
Bechara, A., Damasio, H., and Damasio, A.R. 2000. Emotion, decision making and the orbitofrontal cortex. *Cereb Cortex* 10:295–307.
———. 2003. Role of the amygdala in decision-making. *Ann N Y Acad Sci* 985:356–69.
Bechara, A., Damasio, H., Damasio, A.R., and Lee, G.P. 1999. Different contributions of the human amygdala and ventromedial prefrontal cortex to decision-making. *J Neurosci* 19:5473–81.
Bechara, A., Damasio, H., Tranel, D., and Anderson, S.W. 1998. Dissociation of working memory from decision making within the human prefrontal cortex. *J Neurosci* 18:428–37.
Bechara, A., Damasio, H., Tranel, D., and Damasio, A.R. 1997. Deciding advantageously before knowing the advantageous strategy. *Science* 275:1293–95.
———. 2005. The Iowa Gambling Task and the somatic marker hypothesis: Some questions and answers. *Trends Cogn Sci* 9:159–62; discussion 162–64.
Beer, J.S., Heerey, E.A., Keltner, D., Scabini, D., and Knight, R.T. 2003. The regulatory function of self-conscious emotion: Insights from patients with orbitofrontal damage. *J Pers Soc Psychol* 85:594–604.
Beer, J.S., John, O.P., Scabini, D., and Knight, R.T. 2006. Orbitofrontal cortex and social behavior: Integrating self-monitoring and emotion-cognition interactions. *J Cogn Neurosci* 18:871–79.
Bellebaum, C., Koch, B., Schwarz, M., and Daum, I. 2008. Focal basal ganglia lesions are associated with impairments in reward-based reversal learning. *Brain* 131:829–41.
Berlin, H.A., Rolls, E.T., and Kischka, U. 2004. Impulsivity, time perception, emotion and reinforcement sensitivity in patients with orbitofrontal cortex lesions. *Brain*.
Berridge, K.C. 2007. The debate over dopamine's role in reward: The case for incentive salience. *Psychopharmacology (Berl)* 191:391–431.
Botvinick, M.M. 2007. Conflict monitoring and decision making: Reconciling two perspectives on anterior cingulate function. *Cogn Affect Behav Neurosci* 7:356–66.
Brown, J.W., and Braver, T.S. 2005. Learned predictions of error likelihood in the anterior cingulate cortex. *Science* 307:1118–21.
———. 2007. Risk prediction and aversion by anterior cingulate cortex. *Cogn Affect Behav Neurosci* 7:266–77.
———. 2008. A computational model of risk, conflict, and individual difference effects in the anterior cingulate cortex. *Brain Res* 1202:99–108.
Burle, B., Roger, C., Allain, S., Vidal, F., and Hasbroucq, T. 2008. Error negativity does not reflect conflict: A reappraisal of conflict monitoring and anterior cingulate cortex activity. *J Cogn Neurosci* 20:1637–55.

Butter, C. 1969. Perseveration in extinction and in discrimination reversal tasks following selective frontal ablations in macaca mulatta. *Physiol Behav* 4:163–71.

Camille, N., Coricelli, G., Sallet, J., Pradat-Diehl, P., Duhamel, J.R., and Sirigu, A. 2004. The involvement of the orbitofrontal cortex in the experience of regret. *Science* 304:1167–70.

Carter, C.S., and van Veen, V. 2007. Anterior cingulate cortex and conflict detection: An update of theory and data. *Cogn Affect Behav Neurosci* 7:367–79.

Chamberlain, S.R., Muller, U., Blackwell, A.D., and Clark, L., Robbins, T.W., and Sahakian, B.J. 2006. Neurochemical modulation of response inhibition and probabilistic learning in humans. *Science* 311:861–63.

Clark, L., Bechara, A., Damasio, H., Aitken, M.R., Sahakian, B.J., and Robbins, T.W. 2008. Differential effects of insular and ventromedial prefrontal cortex lesions on risky decision-making. *Brain* 131:1311–22.

Clark, L., Manes, F., Antoun, N., Sahakian, B.J, and Robbins, T.W. 2003. The contributions of lesion laterality and lesion volume to decision-making impairment following frontal lobe damage. *Neuropsychologia* 41:1474–83.

Cools, R., Robinson, O.J., and Sahakian, B. 2008. Acute tryptophan depletion in healthy volunteers enhances punishment prediction but does not affect reward prediction. *Neuropsychopharmacology* 33:2291–99.

Craig, A.D. 2003. Interoception: The sense of the physiological condition of the body. *Curr Opin Neurobiol* 13:500–5.

Critchley, H.D., Mathias, C.J., Josephs, O., O'Doherty, J., Zanini, S., Dewar, B.K., et al. 2003. Human cingulate cortex and autonomic control: Converging neuroimaging and clinical evidence. *Brain* 126:2139–52.

Damasio, A.R. 1994. *Descartes'* Error: Emotion, Reason, and the Human Brain. New York: Avon Books.

di Pellegrino, G., Ciaramelli, E., and Ladavas, E. 2007. The regulation of cognitive control following rostral anterior cingulate cortex lesion in humans. *J Cogn Neurosci* 19:275–86.

Dunn, B.D., Dalgleish, T., and Lawrence, A.D. 2006. The somatic marker hypothesis: A criticial evaluation. *Neurosci Biobehav Rev* 30:239–71.

Elliott, R., Newman, J.L., Longe, O.A., and Deakin, J.F. 2003. Differential response patterns in the striatum and orbitofrontal cortex to financial reward in humans: A parametric functional magnetic resonance imaging study. *J Neurosci* 23:303–7.

Eslinger, P.J., and Damasio, A.R. 1985. Severe disturbance of higher cognition after bilateral frontal lobe ablation: Patient EVR. *Neurology* 35:1731–41.

Fellows, L.K. 2006. Deciding how to decide: Ventromedial frontal lobe damage affects information acquisition in multi-attribute decision making. *Brain* 129:944–52.

———. 2007a. Advances in understanding ventromedial prefrontal function: The accountant joins the executive. *Neurology* 68:991–95.

———. 2007b. The role of orbitofrontal cortex in decision making: A component process account. *Ann N Y Acad Sci* 1121:421–30.

Fellows, L.K., and Farah, M.J. 2003. Ventromedial frontal cortex mediates affective shifting in humans: Evidence from a reversal learning paradigm. *Brain* 126:1830–37.

———. 2005a. Different underlying impairments in decision-making following ventromedial and dorsolateral frontal lobe damage in humans. *Cereb Cortex* 15:58–63.

———. 2005b. Dissociable elements of human foresight: A role for the ventromedial frontal lobes in framing the future, but not in discounting future rewards. *Neuropsychologia* 43:1214–21.

———. 2005c. Is anterior cingulate cortex necessary for cognitive control? *Brain* 128:788–96.

———. 2007. The role of ventromedial prefrontal cortex in decision making: Judgment under uncertainty or judgment per se? *Cereb Cortex* 17:2669–74.

Fellows, L.K., Heberlein, A.S., Morales, D.A., Shivde, G., Waller, S., and Wu, D.H. 2005. Method matters: An empirical study of impact in cognitive neuroscience. *J Cogn Neurosci* 17:850–58.

Frank, M.J. 2005. Dynamic dopamine modulation in the basal ganglia: A neurocomputational account of cognitive deficits in medicated and nonmedicated Parkinsonism. *J Cogn Neurosci* 17:51–72.

Frank, M.J., Seeberger, L.C., and O'Reilly, R.C. 2004. By carrot or by stick: Cognitive reinforcement learning in parkinsonism. *Science* 306:1940–43.

Gehring, W.J., and Willoughby, A.R. 2002. The medial frontal cortex and the rapid processing of monetary gains and losses. *Science* 295:2279–82.

Ghashghaei, H.T., and Barbas, H. 2002. Pathways for emotion: Interactions of prefrontal and anterior temporal pathways in the amygdala of the rhesus monkey. *Neuroscience* 115:1261–79.

Gillihan, S.J., Xia, C., Padon, A.A., Heberlein, A.S., Farah, M.J., and Fellows, L.K. 2010. Contrasting roles for lateral and ventromedial prefrontal cortex in transient and dispositional affective experience. *Soc Cog Affect Neurosci*. ePub ahead of print 10 May 2010.

Hampton, A.N., Adolphs, R., Tyszka, M.J., and O'Doherty, J.P. 2007. Contributions of the amygdala to reward expectancy and choice signals in human prefrontal cortex. *Neuron* 55:545–55.

Harlow, J.M. 1868. Passage of an iron rod through the head. *Pub Mass Med Soc* 2:327.

Heberlein, A.S., Padon, A.A., Gillihan, S.J., Farah, M.J., and Fellows, L.K. 2008. Ventromedial frontal lobe plays a critical role in facial emotion recognition. *J Cogn Neurosci* 20:721–33.

Heims, H.C., Critchley, H.D., Dolan, R., Mathias, C.J., and Cipolotti, L. 2004. Social and motivational functioning is not critically dependent on feedback of autonomic responses: Neuropsychological evidence from patients with pure autonomic failure. *Neuropsychologia* 42:1979–88.

Hester, R., Foxe, J.J., Molholm, S., Shpaner, M., and Garavan, H. 2005. Neural mechanisms involved in error processing: A comparison of errors made with and without awareness. *Neuroimage* 27:602–8.

Hornak, J., Bramham, J., Rolls, E.T., Morris, R.G., O'Doherty, J., Bullock, P.R., et al. 2003. Changes in emotion after circumscribed surgical lesions of the orbitofrontal and cingulate cortices. *Brain* 126:1691–1712.

Hornak, J., O'Doherty, J., Bramham, J., Rolls, E.T., Morris, R.G., Bullock, P.R., et al. 2004. Reward-related reversal learning after surgical excisions in orbito-frontal or dorsolateral prefrontal cortex in humans. *J Cogn Neurosci* 16:463–78.

Hsu, M., Bhatt, M., Adolphs, R., Tranel, D., and Camerer, C.F. 2005. Neural systems responding to degrees of uncertainty in human decision-making. *Science* 310:1680–83.

Izquierdo, A., and Murray, E.A. 2007. Selective bilateral amygdala lesions in rhesus monkeys fail to disrupt object reversal learning. *J Neurosci* 27:1054–62.

Izquierdo, A., Suda, R.K., and Murray, E.A. 2004. Bilateral orbital prefrontal cortex lesions in rhesus monkeys disrupt choices guided by both reward value and reward contingency. *J Neurosci* 24:7540–48.

Izuma, K., Saito, D.N., and Sadato, N. 2008. Processing of social and monetary rewards in the human striatum. *Neuron* 58:284–94.

Jones, B., and Mishkin, M. 1972. Limbic lesions and the problem of stimulus–reinforcement associations. *Exp Neurol* 36:362–77.

Kable, J.W., and Glimcher, P.W. 2007. The neural correlates of subjective value during intertemporal choice. *Nat Neurosci* 10:1625–33.

Kahneman, D. 2003. A perspective on judgment and choice: Mapping bounded rationality. *Am Psychol* 58:697–720.

Kahneman, D., and Tversky, A. 1979. Prospect theory: An analysis of decisions under risk. *Econometrica* 47:263–91.

Kennerley, S.W., Walton, M.E., Behrens, T.E., Buckley, M.J., and Rushworth, M.F. 2006. Optimal decision making and the anterior cingulate cortex. *Nat Neurosci* 9:940–47.

Kirby, K.N., Petry, N.M., and Bickel, W.K. 1999. Heroin addicts have higher discount rates for delayed rewards than non-drug-using controls. *J Exp Psychol: Gen* 128:78–87.

Knutson, B., and Cooper, J.C. 2005. Functional magnetic resonance imaging of reward prediction. *Curr Opin Neurol* 18:411–17.

Koenigs, M., Young, L., Adolphs, R., Tranel, D., Cushman, F., Hauser, M., et al. 2007. Damage to the prefrontal cortex increases utilitarian moral judgements. *Nature* 446:908–11.

Koob, G.F., Ahmed, S.H., Boutrel, B., Chen, S.A., Kenny, P.J., Markou, A., et al. 2004. Neurobiological mechanisms in the transition from drug use to drug dependence. *Neurosci Biobehav Rev* 27:739–49.

Kuhnen, C.M., and Knutson, B. 2005. The neural basis of financial risk taking. *Neuron* 47:763–70.

LeDoux, J. 2003. The emotional brain, fear, and the amygdala. *Cell Mol Neurobiol* 23:727–38.

Maia, T.V., and McClelland, J.L. 2004. A reexamination of the evidence for the somatic marker hypothesis: What participants really know in the Iowa gambling task. *Proc Natl Acad Sci USA* 101:16075–80.

———. 2005. The somatic marker hypothesis: Still many questions but no answers: Response to Bechara et al. *Trends Cogn Sci* 9:162–64.

Mansouri, F.A., Tanaka, K., and Buckley, M.J. 2009. Conflict-induced behavioural adjustment: A clue to the executive functions of the prefrontal cortex. *Nat Rev Neurosci* 10:141–52.

Mavaddat, N., Kirkpatrick, P.J., Rogers, R.D., and Sahakian, B.J. 2000. Deficits in decision-making in patients with aneurysms of the anterior communicating artery. *Brain* 123 (10): 2109–17.

McClure, S.M., Laibson, D.I., Loewenstein, G., and Cohen, J.D. 2004. Separate neural systems value immediate and delayed monetary rewards. *Science* 306:503–7.

Mellers, B.A. 2000. Choice and the relative pleasure of consequences. *Psychol Bull* 126:910–24.

Mischel, W., Shoda, Y., and Rodriguez, M.I. 1989. Delay of gratification in children. *Science* 244:933–38.

Modirrousta, M., and Fellows, L.K. 2008a. Dorsal medial prefrontal cortex plays a necessary role in rapid error prediction in humans. *J Neurosci* 28:14000–5.

————. 2008b. Medial prefrontal cortex plays a critical and selective role in 'feeling of knowing' meta-memory judgments. *Neuropsychologia* 46:2958–65.

Murray, E.A., O'Doherty, J.P., and Schoenbaum, G. 2007. What we know and do not know about the functions of the orbitofrontal cortex after 20 years of cross-species studies. *J Neurosci* 27:8166–69.

Naqvi, N.H., Rudrauf, D., Damasio, H., and Bechara, A. 2007. Damage to the insula disrupts addiction to cigarette smoking. *Science* 315:531–34.

North, N.T., and O'Carroll, R.E. 2001. Decision making in patients with spinal cord damage: Afferent feedback and the somatic marker hypothesis. *Neuropsychologia* 39:521–24.

O'Doherty, J., Kringelbach, M.L., Rolls, E.T., Hornak, J., and Andrews, C. 2001. Abstract reward and punishment representations in the human orbitofrontal cortex. *Nat Neurosci* 4:95–102.

O'Doherty, J.P. 2004. Reward representations and reward-related learning in the human brain: Insights from neuroimaging. *Curr Opin Neurobiol* 14:769–76.

Oliveira, F.T., McDonald, J.J., and Goodman, D. 2007. Performance monitoring in the anterior cingulate is not all error related: Expectancy deviation and the representation of action-outcome associations. *J Cogn Neurosci* 19:1994–2004.

Padoa-Schioppa, C., and Assad, J.A. 2006. Neurons in the orbitofrontal cortex encode economic value. *Nature* 441:223–26.

————. 2008. The representation of economic value in the orbitofrontal cortex is invariant for changes of menu. *Nat Neurosci* 11:95–102.

Paulus, M.P., Rogalsky, C., Simmons, A., Feinstein, J.S., and Stein, M.B. 2003. Increased activation in the right insula during risk-taking decision making is related to harm avoidance and neuroticism. *Neuroimage* 19:1439–48.

Pessiglione, M., Petrovic, P., Daunizeau, J., Palminteri, S., Dolan, R.J., and Frith, C.D. 2008. Subliminal instrumental conditioning demonstrated in the human brain. *Neuron* 59:561–67.

Pessoa L. 2008. On the relationship between emotion and cognition. *Nat Rev Neurosci* 9:148–58.

Petry, N.M., Bickel, W.K., and Arnett, M. 1998. Shortened time horizons and insensitivity to future consequences in heroin addicts. *Addiction* 93:729–38.

Plassmann, H., O'Doherty, J., and Rangel, A. 2007. Orbitofrontal cortex encodes willingness to pay in everyday economic transactions. *J Neurosci* 27:9984–88.

Price, D.D. 2000. Psychological and neural mechanisms of the affective dimension of pain. *Science* 288:1769–72.

Reekie, Y.L., Braesicke, K., Man, M.S., and Roberts, A.C. 2008. Uncoupling of behavioral and autonomic responses after lesions of the primate orbitofrontal cortex. *Proc Natl Acad Sci USA* 105:9787–92.

Ridderinkhof, K.R., Ullsperger, M., Crone, E.A., and Nieuwenhuis, S. 2004. The role of the medial frontal cortex in cognitive control. *Science* 306:443–7.

Roberts, A.C., Reekie, Y., and Braesicke, K. 2007. Synergistic and regulatory effects of orbitofrontal cortex on amygdala-dependent appetitive behavior. *Ann N Y Acad Sci* 1121:297–319.

Roberts, N.A., Beer, J.S., Werner, K.H., Scabini, D., Levens, S.M., and Knight, R.T., et al. 2004. The impact of orbital prefrontal cortex damage on emotional activation to unanticipated and anticipated acoustic startle stimuli. *Cogn Affect Behav Neurosci* 4:307–16.

Rogers, R.D., Everitt, B.J., Baldacchino, A., Blackshaw, A.J., Swainson, R., and Wynne, K., et al. 1999. Dissociable deficits in the decision-making cognition of chronic amphetamine abusers, opiate abusers, patients with focal damage to prefrontal cortex, and tryptophan-depleted normal volunteers: Evidence for monoaminergic mechanisms. *Neuropsychopharmacology* 20:322–39.

Rolls, E.T. 1999. *The Brain and Emotion*. Oxford: Oxford University Press.

————. 2000. The orbitofrontal cortex and reward. *Cereb Cortex* 10:284–94.

Rolls, E.T., Hornak J., Wade D., and McGrath J. 1994. Emotion-related learning in patients with social and emotional changes associated with frontal lobe damage. *J Neurol Neurosurg Psychiatry* 57:1518–24.

Rorden, C., and Karnath, H.O. 2004. Using human brain lesions to infer function: A relic from a past era in the fMRI age? *Nat Rev Neurosci* 5:813–19.

Rudebeck, P.H., Behrens, T.E., Kennerley, S.W., Baxter, M.G., Buckley, M.J., Walton, M.E., et al. 2008. Frontal cortex subregions play distinct roles in choices between actions and stimuli. *J Neurosci* 28:13775–85.

Rudebeck, P.H., and Murray, E.A. 2008. Amygdala and orbitofrontal cortex lesions differentially influence choices during object reversal learning. *J Neurosci* 28:8338–43.

Rushworth, M.F., and Behrens, T.E. 2008. Choice, uncertainty and value in prefrontal and cingulate cortex. *Nat Neurosci* 11:389–97.

Rushworth, M.F., Behrens, T.E., Rudebeck, P.H., and Walton, M.E. 2007. Contrasting roles for cingulate and orbitofrontal cortex in decisions and social behaviour. *Trends Cogn Sci* 11:168–76.

Rushworth, M.F., Mars, R.B., and Summerfield, C. 2009. General mechanisms for making decisions? *Curr Opin Neurobiol* 19:75–83.

Schacter, D.L., Addis, D.R., and Buckner, R.L. 2008. Episodic simulation of future events: Concepts, data, and applications. *Ann N Y Acad Sci* 1124:39–60.

Schoenbaum, G., and Roesch, M. 2005. Orbitofrontal cortex, associative learning, and expectancies. *Neuron* 47:633–36.

Schoenbaum, G., Saddoris, M.P., and Stalnaker, T.A. 2007. Reconciling the roles of orbitofrontal cortex in reversal learning and the encoding of outcome expectancies. *Ann N Y Acad Sci* 1121:320–35.

Schoenbaum, G., Setlow, B., Saddoris, M.P., and Gallagher, M. 2003. Encoding predicted outcome and acquired value in orbitofrontal cortex during cue sampling depends upon input from basolateral amygdala. *Neuron* 39:855–67.

Seymour, B., Daw, N., Dayan, P., Singer, T., and Dolan, R. 2007. Differential encoding of losses and gains in the human striatum. *J Neurosci* 27:4826–31.

Seymour, B., and Dolan, R. 2008. Emotion, decision making, and the amygdala. *Neuron* 58:662–71.

Shamay-Tsoory, S.G., Tibi-Elhanany, Y., and Aharon-Peretz, J. 2007. The green-eyed monster and malicious joy: The neuroanatomical bases of envy and gloating (schadenfreude). *Brain* 130:1663–78.

Shidara, M., and Richmond, B.J. 2002. Anterior cingulate: Single neuronal signals related to degree of reward expectancy. *Science* 296:1709–11.

Shiv, B., Loewenstein, G., Bechara, A., Damasio, H., and Damasio, A.R. 2005. Investment behavior and the negative side of emotion. *Psychol Sci* 16:435–39.

Shohamy, D., Myers, C.E., Kalanithi, J., and Gluck, M.A. 2008. Basal ganglia and dopamine contributions to probabilistic category learning. *Neurosci Biobehav Rev* 32:219–36.

Small, D.M., Zatorre, R.J., Dagher, A., Evans, A.C., and Jones-Gotman, M. 2001. Changes in brain activity related to eating chocolate: From pleasure to aversion. *Brain* 124:1720–33.

Smith, K.S., Tindell, A.J., Aldridge, J.W., and Berridge, K.C. 2009. Ventral pallidum roles in reward and motivation. *Behav Brain Res* 196:155–67.

Stalnaker, T.A., Franz, T.M., Singh, T., and Schoenbaum, G. 2007. Basolateral amygdala lesions abolish orbitofrontal-dependent reversal impairments. *Neuron* 54:51–58.

Stuss, D.T., and Alexander, M.P. 2007. Is there a dysexecutive syndrome? *Philos Trans R Soc Lond B Biol Sci* 362:901–15.

Stuss, D.T., Floden, D., Alexander, M.P., Levine, B., and Katz, D. 2001. Stroop performance in focal lesion patients: Dissociation of processes and frontal lobe lesion location. *Neuropsychologia* 39:771–86.

Sugrue, L.P., Corrado, G.S., and Newsome, W.T. 2004. Matching behavior and the representation of value in the parietal cortex. *Science* 304:1782–87.

Tomb, I., Hauser, M., Deldin, P., and Caramazza, A. 2002. Do somatic markers mediate decisions on the gambling task? *Nat Neurosci* 5:1103–4; author reply 1104.

Tversky, A. 1996. Intransitivity of preferences. *Psych Rev* 76.

Wheeler, E.Z., and Fellows, L.K. 2008. The human ventromedial frontal lobe is critical for learning from negative feedback. *Brain* 131:1323–31.

Yin, H.H., and Knowlton, B.J. 2006. The role of the basal ganglia in habit formation. *Nat Rev Neurosci* 7:464–76.

Part IV

Civilized Sensory Rewards
(Distinctly Human Rewards)

17 Perfume

Rachel S. Herz

CONTENTS

17.1 PERFUME QUALITY AND ART

Contemporary perfumes contain from tens to hundreds of ingredients and are comprised of (1) essential oils derived from natural aromatic plant extracts and/or synthetic aromatic chemicals which are classified by structural group (e.g., alcohols, esters, aldehydes, and terpenes); (2) fixatives, natural or synthetic substances used to reduce the evaporation rate, increase perceived odor strength, and improve stability; and (3) solvents, the liquid in which the perfume oil is dissolved in. The typical solvent solution is 98% ethanol and 2% water.

The science of perfume is chemistry and the aromatic result is artistry. Indeed, especially in France, perfume creation is treated as high art. According to one French perfume executive: "... perfumers consider that what they create is great art and that because they are French the world should come on bended knee and consider itself lucky to be blessed with their creations... [they say] 'I launched this and that perfume, and my perfumes are wonderful, fabulous, they lost five million dollars, but who cares, they're objects of art that will live love forever and conform to my immortal, pure aesthetic'" (Burr 2007, 53–54). Perfume briefs, the perfume company's instruction to the perfumer of what the perfume should smell (*be*) like, are equally artistic and vague. For example, Parfums Dior posed this brief for *Pure Poison* (2004): "What is it like to have something soft and hard at the same time?" (Burr 2007, 55). More typical examples are: "Give us the scent of a warm cloud floating in a fresh spring sky over Sicily raining titanium raindrops on a woman with emerald eyes" (Burr 2007, 55).

The language of perfumery bears witness to its inherently aesthetic qualities. Fragrances are designated according to their *concentration* level, the scent *family* they belong to, and the *notes* in the scent. The concentration level of the perfume oil in a fine fragrance indicates its intensity and its predicted duration on the skin. The more concentrated the perfume, the stronger the scent and the longer it will last. Although there is variability within the definitions, there are four major perfume concentration classifications. *Parfum* contains between 15% and 30% aromatic compounds; *eau de parfum* contains 8–15% aromatic compounds; *eau de toilette* ranges from 4% to 8% aromatic compounds; and *eau de cologne* contains between 2% and 5% aromatic compounds. *Eau de cologne* was originally invented by Italian perfumers living in Köln (Cologne) Germany in the 1700s and was made from rosemary and citrus essences dissolved in wine. However, the term "cologne" has become a generic for a weakly concentrated perfume and/or a man's fine fragrance.

The second category of perfume grouping is by scent family and scent family subtype. Scent families are designated with traditional classification terms (originating from around 1900) and modern terms (since 1945). The main scent families are Floral, Chypre, Fougère, Marine/Ozonic, Oriental, Citrus, Green, and, most recently, Gourmand. Scent family subtypes include terms such as fresh, aldehyde, amber, fruity, spicy, woody, and animalic. For further information on scent family types, see Moran (2000).

Perfumery can be likened to the nose as music is to the ear. In keeping with this aesthetic connotation, the third classification of a perfume's olfactory quality is described in musical metaphors. The combination of ingredients in a perfume is called a "composition" and it has three "notes" that unfold over time. The first note is called the *top note*, or *head note*, and it produces the immediate impression of the perfume. Top notes consist of small, light molecules with high volatility that evaporate quickly. *Middle notes* (also called *heart notes*) emerge just before the top notes have dissipated. Scents from this note class appear anywhere from two minutes to one hour after the application of a perfume. *Base notes* (or *bottom* or *dry down*) appear while the middle notes are fading. Compounds of this class are often the fixatives used to hold and boost the strength of the lighter top and middle notes. Base notes are large, heavy molecules that evaporate slowly and are usually not perceived until 30 minutes after the application of the perfume. Some base notes can still be detectable 24 hours or more after application. The varying evaporation rates of different molecules in a perfume mean that a perfume will not smell the same when it is first put on as it does three hours later.

Perfume qualities are described in musical metaphors not solely because of the aesthetic relationship between perfume and music but because there are so few specific words dedicated to olfactory experience. Anthropologists have found that in all known languages, there are fewer words that refer explicitly to our experience of smells than there are for any other sensation (Classen, Howes, and Synnott 1994). In English, *aromatic*, *fragrant*, *pungent*, *redolent*, and *stinky* exhaust the list of adjectives that specifically describe olfactory stimuli and nothing else. More common terms used to describe odors, like *floral* or *fruity*, are references to the odor-producing objects (flowers and fruits), not the odors themselves. We also borrow terms from other senses; chocolate smells *sweet*, grass smells *green*, and so on (Herz 2005, 2008).

Various possibilities explain why our sense of smell and language are so disconnected. First, unlike other sensory systems, olfactory information does not need to be integrated in the thalamus prior to processing in the cortex, and it is argued that the thalamus has relevance for language. Second, a large body of evidence indicates that the majority of olfactory processing occurs in the right hemisphere of the brain, whereas language processing is known to be dominated by the left hemisphere (see Royet and Plailly 2004, for review). It has also been suggested that odors are hard to name because of competition between odor and language processing for cognitive resources that share the same neural substrates (Lorig 1999). This latter theory is supported by a magnetoencephalographic study which showed that the presence of an odor altered the semantic processing of words and degraded word encoding, but did not influence nonsemantic processing (Walla et al. 2003).

17.2 BRIEF HISTORY OF PERFUME

The word perfume comes from the Latin *"per fumum,"* meaning *through smoke*. It originated about 4000 years ago among the Mesopotamians in the form of incense. The original aromatic essences used in perfumery were herbs and spices like coriander and myrtle; flowers were not used until much later. Perfumery, the art of making perfume, then traveled to Egypt where it was initially only used in rituals for the gods or pharaohs. Indeed the ancient Egyptians had a God of Perfume, Nefertem (Figure 17.1a). Nerfertem didn't begin his mythical life as a God, but through legend and by his association with the highly aromatic, and possibly narcotic, blue water-lily flower, he rose to become the divine representation of perfume and luck. The personal wearing of scent was first recorded by the Egyptians who put flowers, herbs, and spices into wax cones that they wore on their heads; as the wax melted the aromatic mixture flowed out and perfumed them.

The Persian philosopher and physician Avicenna (ca. 980–1037) introduced the process of extracting oils from flowers by distillation: the method of boiling a liquid mash through which chemicals with different properties can be separated. This method is still used today.

The Etruscans revered perfume to the point that Etruscan women were *never* without it. The Etruscan spirit of adornment "Lassa" is a naked winged female carrying a perfume bottle. She is depicted on the engraved brass mirrors that dead Etruscan women were buried with to accompany them to the afterlife. The Romans were also great connoisseurs of perfumes and gladiators are said to have applied a different scented lotion to each area of their body before a contest. But as Christianity rose with its severe and simple attitudes towards adornment, perfume evaporated in the mist. Fortunately, this austere attitude towards self-scenting was not to be a permanent obstacle to the fragrance-seeking nose.

Perfumery came to Europe in the fourteenth century when, in 1307, at the behest of Queen Elizabeth of Hungary, the first modern perfume made of scented oils blended in an alcohol solution was produced. This perfume was thereafter referred to as "Hungary Water." As European cities were becoming more and more fetid, the prevalence of donning perfume became ever greater. In the Renaissance perfume fashion was truly *reborn* and France became the epicenter of perfume development and culture, a position it has retained ever since. By the sixteenth and seventeenth centuries, the craze of perfuming everything was so extensive that even pets and jewelry were daubed with their owner's favorite scents. And if scenting your pet sounds like an eccentricity of the past, welcome to the world of today's perfume and pet fanatics. A spate of canine fine fragrances are currently on the market from high-end fashion designers like Juicy Couture, who offer *Pawfum*, at $60.00 for a 1-oz bottle of eau de parfum spray, as well as boutique perfumers, such as Renee Ryan, who in 2006 launched *Sexy Beast*, described as "a unisex blend of bergamot and vanilla-infused musk combined with natural patchouli, mandarin and nutmeg oils" (the 1.7-oz bottle sells for $50). For a thrifty alternate you can purchase *Oh My Dog!* by Etienne de Swardt, a mere $30.00 per 3.7 oz bottle (Figure 17.1b). The list goes on and cat lovers are not excluded.

By the end of the eighteenth century perfume was enjoying the status of high fashion and the higher one's importance the better one's fragrance. In 1709 a French perfumer proposed that the different classes should be scented differently. He concocted a royal perfume for the aristocracy, a bourgeois perfume for the middle classes, but said the poor were only deemed worthy of disinfectant. The court of Louis XV, king of France from 1715 to 1774, was known as "*La Cour Parfumée*" (The Perfumed Court), and the aristocracy were expected to wear a different perfume for every day of the week. So cherished were these objects of the nose that Marie Antoinette's perfume bottle has been recently recreated, and in an extremely limited edition a 25 ml flask may be purchased for 8,000 Euros (approximately $10,290 US dollars (Figure 17.1c).

From the Renaissance into the nineteenth century, perfume wearing and perfume type were ungendered, and men and women adorned them equally. The deodorizing drive of the mid-nineteenth century, however, led to a demise of perfume and a new conservative outlook towards it. Due to the promotion of germ theory and the understanding that filth (which usually smells) carried illness, scents of all kinds began to be perceived as evil. Perfume receded to the background and took on a muted public image, and wearing fragrance became gender stereotyped. Sweet floral blends were deemed exclusively feminine, while sharper, woodsy, pine, and cedar notes were characterized as masculine. In the early to mid-twentieth century, men with any credible social position had stopped wearing fragrance altogether and were only expected to smell of clean skin and tobacco, while women of respectable social standing were expected to smell only faintly of floral notes (Classen, Howes, and Synnott 1994). Only prostitutes and the déclassé dared wear the once prestigious heavy and animalic scents of earlier generations.

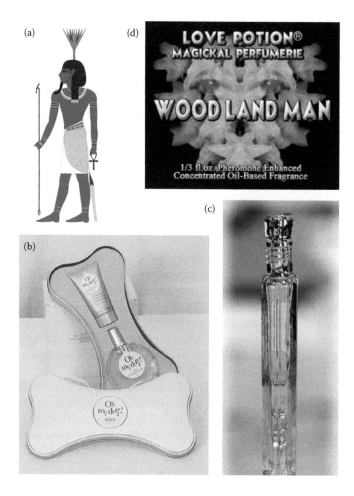

FIGURE 17.1 **(See Color Insert)** A limited pictorial survey of perfume and its vessels. (a) Nefertum, the Egyptian God of perfume and luck. (From file "Nefertem.svg" by Wikimedia Commons user Jeff Dahl. File at http://commons.wikimedia.org/wiki/File:Nefertum.svg.) (b) *Oh My Dog!* cologne for dogs by Etienne de Swardt. (From Elizabeth duPar at www.ohmydogstore.com. With permission.) (c) Marie Antoinette's perfume bottle *Eau de Sillage*. (Permission grated by original author. For copyright details of this image please contact chapter author Rachel Herz.) (d) *Love Potion* perfume explicitly marketed as a potent human sex pheromone. (From Love Potion Woodland Man; artwork by Ravon, Mara Fox at LovePotionPerfume.com. With permission.)

A break in American perfume repression came unexpectedly during the otherwise conservative era of the 1950s. *Chanel No. 5*, created in 1921, was the fifth fragrance in a line developed by Ernest Beaux for Gabrielle "Coco" Chanel. *Chanel No. 5* enjoyed popularity in France and Europe after its inception, but became a blockbuster when it was launched in the United States in the early 1950s, and Marilyn Monroe famously said that all she wore to bed was "two drops of *Chanel No. 5*." Since the mid-1950s, *Chanel No. 5* has been the most famous perfume in the world and it continues to outsell many of its modern rivals. *Chanel No. 5* was also the first fragrance to be created using synthetic chemicals. Before synthetics were used, scents faded quickly and perfumes had to be continuously reapplied.

Among the classic and trendy fine fragrances of today there are also many unusual options for the perfume esthete. Among the more atypical are those available from Demeter Fragrances, which boasts over 150 scents ranging from the playful to the shocking, including: *Holy Water, Dust, Playdoh, Funeral Home, Paperback,* and *Gin and Tonic*—the art savvy can even find *"This is* *not a Pipe."** Demeter has not restricted itself to the entirely iconoclastic demographic and wisely has capitalized on one of the top trends in fine fragrance today—food—with eau de *Birthday Cake* and *Sushi*. Demeter isn't the only company vying for the nose of the daring gourmand. The Stilton Cheese Makers Association, located in Surrey, England, recently launched *Eau de Stilton* as part of their 2006 campaign to encourage people to eat more of their cheese. The perfume, blended by a Manchester-based aromatics company, features a "symphony of natural base notes including yarrow, angelica seed, clary sage and valerian." According to Nigel White, a company spokesman, Eau de Stilton "recreates the earthy and fruity aroma" of the cheese "in an eminently wearable perfume." And if you want to mix the savory with sex, Burger King has the new cologne for you. *Flame* is, according to the company, "the scent of seduction, with a hint of flame-broiled meat." Red meat is a favorite food among men (Drewnowski1992; Weaver and Brittin 2001). However, if this cologne is aimed at a heterosexual male market then advertisers may have a problem. Women do not rank red meat as a highly preferred food (Drewniewski et al. 1992; Kubberod et al. 2002) and as such *Flame* is more likely to attract hungry men than lusty women.

On the subject of lust and women, *Chanel No. 5* was not only hurtled into fame by a sex goddess but is also tainted with the odors of sex itself. *Chanel No. 5* is in the scent category "floral aldehyde" and is composed of aldehydes, jasmine, rose, ylang ylang, iris, amber, and patchouli notes. But *Chanel No. 5* also contains secretions from the perineal glands of the civet cat—secretions with a strong musky, fecal odor. The anal secretions from the Himalayan civet cat, musk deer, and beaver (castoreum), and vomit from sperm whales (ambergris), have been historically used as perfume fixatives. Notably, with pressure from animal rights groups, the Chanel company claimed that as of 1998 natural civet was replaced with a synthetic substitute. Many of the most popular perfumes are laden with synthetic fecal notes, such as indole. *Eternity* (1988) by Calvin Klein is claimed to be one of the most indolic perfumes ever. The great French perfumer Jacques Guerlain once said that perfumes should smell of "the underside of my mistress." The scents he created, such as *Jicky* (1889) and *Shalimar* (1925), were tinged with vaginal and anal smells. Christian Dior's perfume called *Dioressence* (1969) was dubbed *le parfum barbare* (the barbaric perfume) and smelled of animalic and fecal notes. It is an interesting social observation that within the conservative palate of the 1950s and our modern obsession with ridding the body of its own body odor, the rise of perfumes that are redolent with fecal and bodily scents re-emerged. Yet, the presence of these funky, animalic notes was rarely advertised.

We are still obsessed with bathing and eliminating our natural body odor, but curiously a new wave of fragrances are now being marketed that are deliberately aimed at recreating the scents of body funk. *Secretions Magnifiques*, by L'Etat Libre d'Orange, is claimed to smell like a mix of blood, sweat, saliva, and semen, and even comes in a box emblazoned with a cartoon penis squirting semen. *Les Liaisons Dangereuses* by Kilian is "an orgy in a perfume bottle, a fragrance steeped in the scents of group sex…" "Bodies slick with sweat, hot with the odors of sexual favors," claims Kilian Hennessy, the perfumer who concocted it for By Kilian, his upscale, upstart Parisian perfume house. And Tom Ford, the fashion designer, reportedly told Estée Lauder executives that he wanted *Black Orchid*, his first fragrance, to smell like "a man's crotch." The irony of this new fetish for bodily and sexually scented perfumes is that our own natural body odor elicits serious responses and consequences for our sexual desire and reproductive biology, especially for women (see Section 17.4).

17.3 THE SOCIOLOGY OF PERFUME OR WHY WE WEAR PERFUME

The reasons for wearing perfume have been found to vary with psychological and demographic factors. In a study on women's psychology of fragrance that this author conducted for the Sense of Smell Institute in 2003 (http://www.senseofsmell.org/papers/R. Herz Survey Study Final Report w. tables.doc), it was found that the socio-cultural factors that affect wearing perfume in North America are very much like those that affect clothing fashion. Young women were found to be most

---

* Allusion to the surrealist artist Rene Magritte.

conformist in their perfume preferences, picking fragrances that are popular and/or that their peers are wearing. Women in their 40s were most individualistic and choose perfumes that they personally like without much consideration for outside opinion, while women aged 60 and above tended to choose perfumes that significant others had told them they enjoyed. A similar study conducted in the late 1980s for the Sense of Smell Institute showed comparable age-based trends but was a decade advanced. For example, women in their 30s were most oriented towards self-pleasing and women in their mid-40s and older were more oriented towards pleasing others (http://www.senseofsmell.org/papers/S._Schiffman_Ages_&_Stages_1991_1.doc). It is noteworthy that the differences in these data parallel the differences in the perceived role of women in society between the 1980s and the twenty-first century; 40 really is the new 30.

In Herz's 2003 fragrance study, mood was another factor shown to influence perfume choice. Women often reported that they chose a particular scent because it had a positive effect on their mood and/or was consistent with their mood at that time. Similarly, personality influenced perfume preferences depending upon how one saw oneself (e.g., "dramatic" = heavy, oriental fragrances, or "sporty" = light, fresh fragrances). The situation or aim of the occasion (e.g., a romantic encounter or a job interview) was also a factor in what types of fragrances women selected.

Recent research has not addressed the effects of fragrance wearing on person perception. However, two studies from the 1980s suggested that female use of fragrance can elicit a negative reaction from men in a professional context (Baron 1983, 1986). The studies assessed female confederate job applicants when they wore or did not wear a popular perfume (*Jontue*). In both studies male subjects (interviewers) devalued the candidates' job-related abilities when they wore fragrance, especially if it was combined with the candidate displaying other positive non-verbal cues. The women interviewers (subjects) did not show this negative bias. There are several possible reasons for this finding, which may not make it generalizable to the present. First, in the mid 1980s women had not achieved the current degree of equality and decrease in sexist valuations that they are currently (at least overtly) afforded. Secondly, although *Jontue* was selected as the most highly liked perfume from a pre-test sample, it may nevertheless have elicited a non-professional connotation among the male subjects, who were young (college students), presumably did not have prior experience in the real hiring process, and who may have only ever experienced this perfume in romantic settings. The context in which odors are presented in is a very powerful determinant of both their connotation and denotation (Herz and von Clef 2001; Zellner, Bartoli, and Eckard 1992). Thus, the meaning of *Jontue* to the male subjects may have led to a "dating" association and been incongruent with the "professional" setup. Incongruence between a fragrance and a context also results in negative evaluations (Fiore, Yah, and Yoh 2000).

As mentioned earlier, the gendering of perfume is a recent phenomenon and the socio-cultural factors that contribute to men and women's wearing of perfume are different. Since the Renaissance, women have been wearing fragrance. By contrast, the early twentieth century saw the rejection of fragrance by men. However, by the end of the twentieth century, fragrance for men had again become a prestigious fashion statement. In the United States, men's prestige fragrance (fragrances sold at department stores for at least $50.00) topped $900 million in 2006, with *Acqua di Gio* by Georgio Armani being the current number one seller (Herz 2007).

Men and women also differ in their psychological motivation for buying/wearing fragrance. Most heterosexual men principally wear fragrances that the women in their lives have given them rather than those they buy for themselves, and when men do buy fragrances they do so primarily because they believe it will attract women. By contrast, women predominantly acquire fragrance by purchasing it for a multitude of reasons, as outlined above (Herz 2007).

17.4   THE BIOLOGY OF BODY ODOR AND SEXUAL ATTRACTION

Traditional research on gender differences in human sexual behavior describes men as especially motivated by a woman's good looks and women as most motivated by a man's resource potential (see Buss and Schmidt 1993 for review). The explanations given for these differences are based on *parental investment theory* (Trivers 1972). The tenets of parental investment theory are based on the *selfish gene* motivation (Dawkins 1976) of striving to have the greatest representation of one's genes in future generations—the measure of one's reproductive success—in concert with the differential physical and behavioral costs and benefits that males and females incur in parenting.

In the cost-benefit analysis of human reproductive behavior, females invest far greater energy, resources, and time in reproduction than men, and women can only become pregnant from one man at a time (high cost). However, the benefit for women is that they are always sure of maternity; certain that the child they are investing in represents their genes and therefore facilitates their reproductive success. By contrast, the cost of reproduction for males is low but prior to the advent of genetic testing paternity was never certain. Thus, there is only a probabilistic relationship between male investment in a given child and the fact that he is investing in his own genetic material. Extrapolating from these principals, male mate selection strategies should favor being predisposed towards women who are most likely to be fertile, thus raising the probability that his genes will be passed onto future generations. By contrast, female mate selection strategies should ensure the selection of men who are most likely to secure offspring survival and thus increase the likelihood that her children (genes) will survive to be reproductively viable.

A number of cross-cultural studies have shown that males consider the female features of full lips, clear and smooth skin, clear eyes, high activity level, and a waist-hip ratio of 7:10 as attractive. These features are in fact signals of youthfulness, fertility, and potential child-bearing ability (Buss and Schmitt 1993; Hens 1995; Furnham, Tan, and McManus 1997). Thus, male mate-search strategies are predominately based on the evaluation of physical cues to fertility.

By contrast, the most adaptive strategy for females is to find males who signal that they can secure offspring survival. One way in which males can signal this ability is in the amount of material resources they can commit to a particular female and her children. A number of survey studies have indeed found that women show preferences for males with higher earning potential, ambition, and industriousness (Buss 1989; Buss and Schmitt 1993; Greenlees and McGrew 1994; Landolt, Lalumiere, and Quinsey 1995). However, far beyond the ability to take care of a woman and her children, the most important factor for a woman's reproductive success is the likelihood that her children will be healthy enough to survive and reproduce children themselves. Therefore women should be most concerned with signals indicating a man's physical health and most importantly signals that would indicate that the children they may conceive together will be maximally healthy.

Physical health is principally determined by one's immune system—specifically the genes of the major histocompatibilty complex (MHC) or human leukocyte antigens (by convention the term MHC will be used here). The MHC comprises over 50 alleles and is more polymorphic in extent than any known physiological system. No two individuals other than identical twins share the same set of MHC alleles. In an ideal situation, genetic compatibility between a specific mating couple would confer MHC allele combinations to offspring that maximize disease protection from invading micro-organisms and minimize deleterious recessive mutations. Thus, MHC alleles of parents should be dissimilar; in this way the positive genes of each are most likely to be expressed and the recessive mutations least likely to be replicated. Extrapolating to behavioral strategies, females should seek males whose MHC alleles are maximally dissimilar to her own. In other words, females should be sensitive to cues that are indicative of a male's immunological genotype.

Several studies have shown that MHC genotype influences mate choice in mice (Egid and Brown 1989; Potts, Manning, and Wakeland 1991; Yamazaki et al. 1976). Mice who are genetically identical except for minor variations in MHC loci will preferentially select mates who are dissimilar at these same loci. This discrimination has been shown to be based on the odor type of the mice (Boyse, Beauchamp, and Yamazaki 1987; Egid and Brown 1989; Yamazaki et al. 1979). Importantly, it is the female mouse which most actively makes these odor-based selections (Eklund, Egid, and Brown 1992).

Recent research among humans has also shown that MHC type plays a role in the selection of heterosexual mates. Studies on the North American Hutterite community have revealed that mate choice is influenced by an avoidance of spouses with a high degree of allele overlap with self (Ober 1999). Moreover, a number of studies by Ober and colleagues have shown that Hutterite couples with high rates of shared MHC antigens have lower fecundity and higher miscarriage rates than couples with low rates of shared MHC alleles (Ober et al. 1992, 1997, 1998). The negative consequences of MHC similar mating in the general population are further supported by fertility clinic data where higher rates of MHC similarities between couples are associated with a greater likelihood of infertility and recurrent spontaneous abortions (Ho et al. 1990; Thomas et al. 1985; Tiercy, Jeannet, and Mach 1990; Weckstein et al. 1991).

In the laboratory, Wedekind, Seebeck, Bettens, and Paepke (1995) found that females who were not on birth control pills preferred the smell of T-shirts worn by men within the test sample whose MHC genes most differed from their own. These women also reported that the scents of the T-shirts they preferred reminded them of their current and/or ex-mates. Thus, as with rodents, human mate selection appears to be influenced by female preferences for male body odors that correlate with MHC dissimilarity. As an interesting aside, one study has suggested that individuals with similar MHC types tend to share preferences for specific perfume ingredients (e.g., heliotrope) (Milinksi and Wedekind 2001). However, this finding has not been replicated or elaborated.

Recently, two survey studies found that women rank how a man smells as the most important physical trait in their choice of a sexual partner and that it is more important than all social and material status factors (Herz and Cahill 1997; Herz and Inzlicht 2002). Importantly, whether a man smelled especially attractive as a function of his natural body odor or the use of fragrance were both highly influential for female choice. That is, a man can seduce a woman on the basis of artificial fragrance and potentially mask a body odor that would indicate genetic incompatibility. This suggests that perfume can be a biological liability. Not only might an unsuitable biological match yield a less than optimally healthy child, the mismatch also increases the difficulty of conceiving (Ho et al. 1990; Thomas et al. 1985; Tiercy et al. 1990; Weckstein et al. 1991).

As professional women in the developed world postpone childbearing to their later 30s and early 40s, along with the increased infertility that comes with age, the male use of fragrance, and hence masking of incompatible genetics during courtship, may be a contributing factor to the current epidemic of infertility within this demographic. That is, women may be disproportionally selecting men who are genetically incompatible with them because they are unable to discern their "real" body odor during the initial phase of the relationship. After they discover what their man's real body odor is like it is usually too late because of the emotional attachment now formed to him. That is, because of the facility of the olfactory system to learn the emotional significance of odors through associative learning, the scent of a genetically incompatible mate will become attractive as a function of being experienced with other rewarding aspects of the suitor (a love bond, his charm, his earning power, etc.); see Section 17.6.

17.5 PERFUME, BODY ODOR, AND THE QUESTION OF HUMAN PHEROMONES

The discussion of body odor and attraction invariably leads to the question of human pheromones. Research on this topic is currently unresolved. We do not have the organ nor corresponding neural tissue to perceive pheromones as other mammals do, and data obtained for the most substantiated human pheromone, menstrual synchrony effects (e.g., McClintock 1971; Stern and McClintock 1998), has been criticized on statistical grounds (Wilson 1992; Weller and Weller 1997). Nevertheless, several new findings suggest that we may transmit and adorn aroma-chemicals that influence our sexual motivations.

The chemical androstadienone is a steroid derivative of the male sex hormone testosterone and its presence in body fluids (e.g., sweat) is higher in males than in females. For these reasons, androstadienone has been studied as a potential human sexual pheromone. In several studies androstadienone was shown to improve women's mood, *but* only when women were in the presence of men (Jacob, Hayreh, and McClintock 2001; Lundstrom and Olsson 2005). In the presence of a female experimenter, androstadienone had no impact on the participants. Another study with a male experimenter found that androstadienone increased women's self-rated sexual arousal and cortisol levels (Wyert et al. 2007). These results have led to the speculation that androstadienone is a "modulator" pheromone for women in certain social contexts—in the presence of men. However, the levels of androstadienone that women were exposed to in these studies were a million times higher than the amount a normal male actually emits. Thus, the ecological validity of androstadienone as a human pheromone is still questionable (Herz 2008).

Notably, male use of artificial fragrances can directly augment a woman's attraction towards him. We showed that women are more sexually responsive to a man's scent than any other physical attribute and that this effect holds both for his true body odor and artificial fragrance (Herz and Inzlicht 2002). Moreover, recent work has shown that women visually judge men as more attractive if the men are wearing fragrance (Roberts et al. 2009). In this study, women were shown videos of men pretending to introduce themselves to an attractive woman. Half of the men in the videos had been wearing scented deodorant for two days as well as during the video taping and the other half had not. Women rated the videotapes of men who had been wearing fragrance as significantly more attractive than the fragrance-free gents, even though the women judges could not perceive any fragrance. It was further found that the men wearing the scented deodorant felt more confident than the unscented men and that the more a man liked his deodorant scent, the more confident he felt. Herz (2003) also found that 90% of all women tested in the fragrance study reported feeling more confident when they wore fragrance than when they did not (http://www.senseofsmell.org/papers/R. Herz Survey Study Final Report w. tables.doc). Thus the feelings of self-confidence inspired by wearing fragrance can alter the wearer's behavior in a manner that increases their attractiveness to others, independent of whether those who judge them as attractive can also smell them.

With respect to chemicals emitted by women, it has been shown that the odors from breast-feeding women increased sexual desire among other women by 24% if they had a partner and 17% if they were single (Spencer et al. 2004). Thus, female body odors that vary with hormonal status may influence human sexual arousal.

Estrus is a physiological phase of the reproductive cycle of female mammals, including primates, during which there is increased female sexual receptivity, proceptivity, and attractiveness (Lange, Hartel, and Meyer 2002; Gangestad, Thornhill, and Garves-Apgar 2005). The traditional view for human reproductive biology holds that female estrus has become "lost" or "hidden" over evolutionary time (e.g., Burt 1992), presumably to promote continuous male interest and thus facilitate long-term pair-bonding and infant care-giving. However, recent studies have suggested that women during the most fertile phase of their menstrual cycle (ovulation) are most attractive to males. This increased attractiveness is manifested through superior facial attractiveness and body symmetry (Roberts et al. 2004; Manning et al. 1996), higher verbal creativity and fluency (Symonds et al. 2004), and more appealing body odor (Havlicek et al. 2006; Singh and Bronstad 2001).

Intriguingly, a recent field study on female attractiveness and hormonal status revealed that professional female lap dancers earned 80% more in tips from male patrons during ovulation than during the menses phase of their cycles (Miller, Tybur, and Jornda 2007). Thus, human females may indeed have an estrus phase and, like other mammals, are maximally attractive to men when a sexual encounter is most likely to lead to pregnancy. This finding should not be overinterpreted however, as there were no independent assessments of the dancers' performances and therefore it is not known whether they truly did perform without variation from day to day. Female libido is known to be higher during ovulation, as are moodiness and physical discomfort during menstruation. Therefore, the dancers may not have realized that they behaved more sensually during fertile days and less so while menstruating. Moreover, what the male patrons were responding to—the dancers' scent, looks, moves, or demeanor—is not known. At present the cause of this provocative finding is still a mystery.

17.6 THE UNIQUE CONNECTION BETWEEN OLFACTION AND EMOTION

Odors are uniquely capable of eliciting emotion. The first and most basic response we have to any scent is liking/disliking—a hedonic emotional response. Beyond simple liking, odors have the capacity to elicit full-blown emotional episodes that can be debilitating enough to require medical/psychiatric intervention, as in the recapitulation of past trauma, and/or to bring one to overwhelming joy. Why olfaction is so potently tied to emotion is due to the mechanism by which odors come to acquire meaning, associative learning, and the neurological organization of olfaction and emotion. (For further details see Chapter 5.)

Associative learning explains both how odors become liked and disliked, as well as how odors elicit deeply emotional memory associations and behavioral responses (Bartoshuk 1991; Engen 1991; Herz 2001). After being paired with an emotionally meaningful event, a previously neutral odor can reactivate the original event, such that when later encountered the odor itself elicits the emotions that were originally paired with it, along with the consequent cognitive, behavioral, and physiological sequelae of those emotions (Herz, Beland, and Hellerstein 2004; Herz, Schankler, and Beland 2004; Herz 2007, 2009a).

Evidence supporting the associative learning hypothesis for odor perception comes from a variety of sources. First, no stereotypical responses to odors have been found in newborns, unlike the innate hedonic responses that are shown to sweet and bitter tastes. It has, however, been repeatedly shown that when presented with odorants that they have probably never encountered before, infants and children often display very different preferences from those of adults (e.g., Soussignan et al. 1997). For instance, infants find the smells of sweat and feces pleasant (Engen 1982; Stein, Ottenberg, and Roulet 1958), and toddlers do not hedonically differentiate between odorants that adults find either very unpleasant (e.g., butyric acid, found in rancid foods) or pleasant (e.g., amyl acetate, which smells like banana). Only one published study has reported that young children (three-year-olds) have adult-like responses to certain odors (Schmidt and Beauchamp 1988). However, this experiment has been criticized on methodological grounds (Engen and Engen 1997).

The olfactory system is fully functional by the third month of gestation (Schaal, Marlier, and Soussignan 1995, 1998; Winberg and Porter 1998). Therefore odor learning begins before birth, as fetuses are exposed to odorant molecules from the volatile substances that their mothers consume. This fact also supports associative learning. Mennella and colleagues found that infants of mothers who consumed distinctive-smelling volatiles (e.g., garlic, alcohol, cigarette smoke) during pregnancy or lactation showed preferences for these smells compared to infants who had not been exposed to these scents (Mennella and Beauchamp 1991, 1993; Mennella, Johnson, and Beauchamph 1995). These early learned odor preferences can continue to influence food and flavor (primarily produced by smell) preferences in later childhood (Mennella and Garcia 2000) and adulthood (Haller et al. 1999).

Associative learning can even negate flavor preference trends. In a recent study of infant formula acceptance, it was shown that when neonates were exposed to an "offensive" formula flavor, they accepted it and showed preference for this formula in later childhood. However, infants not exposed to this flavor showed negative responses to it when tested; naive adults also evaluated the flavor as unpleasant (Mennella and Beauchamp 2002). Notably feeding, in addition to providing nutrition, is an opportunity for close physical contact and emotional bonding. In both rodents and humans, association through affectionate cuddling also induces preferences for specific (yet arbitrary) scents, such as cherry oil or mother's perfume (Balough and Porter 1986; Davis and Porter 1991; Lott, Sullivan, and McPherson 1989; Schleidt and Genzel 1990; Sullivan et al. 1991).

Cross-cultural data provides further support that associative learning, rather than hardwired responses, dictate olfactory preferences. No empirical data have shown cross-cultural consensus in odor evaluation for either common "everyday" odors (Ayabe-Kanamuura et al. 1998; Schleidt, Hold, and Attila 1981) or even "offensive" scents. Indeed, in a recent study undertaken by the U.S. military to create a "stink bomb" it was impossible to find an odor (including U.S. army issue latrine scent) that was unanimously considered unpleasant across various ethnic groups (Dilks, Dalton, and Beauchamp 1999). The following example illustrates how associated emotion is at the root of these effects.

In the mid-1960s, in Britain, Moncrieff (1966) asked adult respondents to provide hedonic ratings to a battery of common odors. A similar study was conducted in the United States in the late 1970s (Cain & Johnson, 1978). Included in both studies was the odorant methyl salicylate (wintergreen). Notably, in the British study, wintergreen was given one of the lowest pleasantness ratings, whereas in the U.S. study it was given the highest pleasantness rating. The reason for this difference can be explained by history. In Britain, the smell of wintergreen is associated with medicine and particularly for the participants in the 1966 study with analgesics that were popular during World War II, a time that these individuals would not remember fondly. Conversely, in the United States, the smell of wintergreen is exclusively a candy mint smell and one that has very positive connotations.

Empirical evidence supporting the emotional associative learning hypothesis for odor perception was shown in an experiment assessing autonomic responses to the odor of dental cement (eugenol). Patients who had previous negative experiences with dentist visits and who were fearful of dental procedures showed autonomic responses clearly indicative of fear when exposed to eugenol in the laboratory, while patients with no prior negative dental history did not (Robin et al. 1998). Most recently, laboratory studies directly aimed at testing the associative learning hypothesis for olfaction showed that a novel odorant could be made to be perceived as good or bad as a function of the emotional associations (good or bad) that were learned to it (Herz, Beland, and Hellerstein 2004).

The highly associable and emotionally evocative properties of odors are further substantiated on neuroanatomical grounds. Olfactory efferents have a uniquely direct connection with the neural substrates of emotional processing (Cahill et al. 1995; Turner, Mishkin, and Knapp 1980). Only two synapses separate the olfactory nerve from the amygdala, a structure critical for the expression and experience of emotion (Aggleton and Mishkin 1986) and human emotional memory (Cahill et al. 1995); and only three synapses separate the olfactory nerve from the hippocampus, which is critically involved in various declarative memory functions and associative learning (Eichenbaum 2001; Schwerdtfeger, Buhl, and Germroth 1990). Associative learning of specific cues (e.g., odors) to emotion is also mediated by the amygdala (LeDoux 1998) and the orbitofrontal cortex (OFC), to which olfactory processing is highly localized (see Rolls 1999). Moreover, the OFC is responsible for assigning affective valence—that is, the reward value of stimuli (Davidson, Putnam, and Larson 2000)—and fMRI experiments have found specific neural activations in the OFC for pleasant versus unpleasant odors (Gottfried et al. 2002; Anderson et al. 2003; Rolls, Kringelbach, and de Araujo 2003).

Due to the intimate neural circuitry between the olfactory cortex and the amygdala, emotional responses triggered by odors can occur instantly upon odor exposure and thus seem immediate and without conscious/cognitive mediation. Often after the emotion is experienced the associated event comes to mind, but the immediacy and priority of the emotional response makes the phenomenology of olfactory experiences different from other stimuli. In odor-evoked recall the emotional response comes first and then the formation of a cognitive association returns (though not always) later. By contrast, for all other triggers of mood and memory, the cognitive meaning precedes the assessment of emotional meaning (Herz 2007). Odors also elicit more emotional and evocative memories than any other sensory cues (see Herz 2009b for review). Indeed we have demonstrated this in an fMRI experiment concerning perfume (Herz et al. 2003).

For our study, we interviewed potential female (only) volunteers to determine whether a specific perfume could be identified that elicited a positive autobiographical memory. The criterion for participant selection was recalling a positive, personal memory in which both the smell and sight of a perfume figured. For example, one participant stated that her memory was: "A trip to Paris when I was in 4th grade and me sitting and watching my mother while she was getting ready to go out and the *Opium* perfume that she used which was on her vanity." Individuals who met these criteria were invited to participate in the main experiment and one to two months later were scheduled for fMRI testing.

Testing followed a block design in which three blocks of 16 trials were administered consisting of four stimuli and clean air. Each trial lasted 30 seconds and involved the participant smelling or seeing their personally meaningful perfume (experimental odor [EO], experimental visual [EV]) and smelling or seeing a generic unmarketed perfume (control odor [CO], control visual [CV]). Each stimulus was presented twice per block, alternating with air. The order of stimulus presentation was randomly determined across blocks and subjects. The first trial of every block and between each stimulus trial was an air-only trial. During the air-only trials participants were asked to clear their mind as much as possible. During the olfactory and visual trials, participants were asked to consider whether the stimulus evoked a memory, and if so to remain thinking about that memory while the stimulus was present. After scanning, participants were presented with their EO, EV, and the CO and CV, and asked to rate the emotionality of their experience that accompanied it during the scanning procedure.

fMRI analyses revealed significantly greater activation in the amygdala and hippocampal regions during recall to the odor of the personally significant perfume (EO) than to any other stimulus. This is particularly noteworthy because odors generally elicit activation in these limbic structures. Thus the present finding was due to the distinctive emotionality of the perfume-evoked memory that was elicited and was not an olfactory artifact. Furthermore, behavioral testing confirmed that participants experienced significantly more emotion when exposed to their personally meaningful fragrance than any other stimulus (see Figure 17.2). This result is the first neurobiological demonstration that the subjective experience of the emotional potency of perfume-evoked memory is specifically correlated with heightened activation in the amygdala-hippocampal region during recall.

Associative learning and the immediate neural and emotional responses that odors elicit explain how some odors have earned the reputation of having "aromatherapeutic" effects, such as

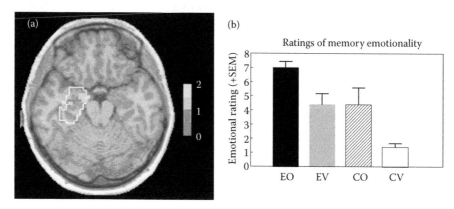

FIGURE 17.2 (a) Activation for the experimental odor (EO) in the amygdala. The positive activation difference for the comparison EO vs. EV+CO+CV is shown. EO = experimental odor; EV = experimental visual; CO = control odor; CV = control visual. The slice shown is at Z = −16 mm inferior to anterior commissure (AC). The maximum intensity difference of 1.65 (MR units) appeared at 14, 8, −16, relative to the AC, corresponding to left hemisphere Brodmann areas 28 and 34. (From *Neuropsychologia*, 42, Herz, R.S. et al. Neuroimaging evidence for the emotional potency of odor-evoked memory, 371–78, Copyright 2004, with permission from **Elsevier**.) (b) Mean emotion ratings given during memory elicitation to each stimulus.

increasing positive mood and heart rate. Note, however, that an odor which elicits such a response is probably doing so by triggering a learned emotional association, which produces a specific mood and physiological consequence. For example, a scent connected to feeling happy and energized can indeed elicit the feelings of energy and vigor as well as heightened heart-rate or blood pressure. It should be noted that no direct pharmacological properties for odors have yet been demonstrated (Herz 2009a).

17.7 IS PERFUME DISTINCTLY HUMAN?

Are humans the only animals to adorn themselves with odor for the purposes of self-pleasure and to attract conspecifics? Many animals are known to cover themselves in scents for the purposes of defense or predation. For example, wolves will roll their bodies in carcasses or feces, presumably because this masks their own scent, which thus enables them to invisibly ambush prey. Adorning with scent therefore has adaptive purposes for other animals besides humans, but do any other animals deliberately don artificial fragrance for the purpose of pleasure?

According to researchers at the Dallas Zoo, who were trying to find scents they could use to lure ocelots into mating areas as well as move them away from dangerous areas such as highways, it was serendipitously discovered that a bottle of *Obsession for Men* cologne was the solution. The felines had paid no attention to various natural wild scents, but when one researcher tested her boyfriend's bottle of *Obsession*, the ocelots immediately started rolling around in the scent and rubbing themselves with it. According to the zoo curator: "They tried to adorn themselves with it" (*New Scientist*, April 10, 1999). It was speculated that this was because of the musky indolic components in the perfume. Female ocelots in particular went wild for *Obsession for Men*. Male ocelots were also drawn to the scent, but for unknown reasons the male response is more variable. The use of artificial fragrance by zoo animals is not unique to ocelots, as captive gorillas have been known to rub perfume-sample cards from magazines on themselves. However, this behavior by our primate cousins may be due to the deprived context of captivity. Thus, whether we are alone or not in the personal application of fragrance for pleasure is currently unclear.

Nonetheless, the human drive towards perfume seems unparalleled. The drive is hedonic and sensual as well as psychological and manipulative. We use perfume to deliberately create pleasure and to attempt to manipulate the mood and behavior of others, especially the passion of a desired conspecific. Indeed, the holy grail of the perfume industry is to find a chemical(s) that when added to fragrance would increase the sexual interest of an intended other (Figure 17.1d).

17.8 IS PERFUME A SENSORY REWARD?

The interface between sensation and reward has been a central topic of this volume. The question therefore must be asked whether perfume is a sensory reward. On the surface the answers appears to be yes. Perfume is both intrinsically and extrinsically a sensory reward above all else. Intrinsically, perfume is created for and used for pure pleasure more than for any other function. Extrinsically, the use of perfume has rewarding properties in sexual benefits by attracting mates, and perfume may also facilitate social status and attractiveness. For example, the use of perfume can mask body-odor signals of poor health and hygiene, and because women find a man's scent such a powerful attractant, perfuming may particularly benefit the biological fitness of men who may not otherwise be attractive.

However, as mentioned earlier, there are biological perils to perfume. Perfuming oneself, especially as a man, is not particularly beneficial for the individual or species from the point of view of attracting female mates who will be optimal biological matches and hence maximize reproductive success and child health. The benefits of female perfume use are questionable too. If females are emitting chemical signals indicative of their fertility, as possibly suggested by the lap-dancer data (Miller, Tybur, and Jornda 2007), then potential mates would be biologically served to know it.

The second assessment of the reward value of perfume is whether it serves as a reinforcer for human behavior. There have been no empirical investigations regarding the reinforcing value of perfume. However, observation of human behavior suggests that perfume is an inherently reinforcing stimulus, to the extent that it is deliberately sought out and its application is repeatedly performed with the apparent outcome of pleasure. Thus, within a classical conditioning paradigm one might predict that a favored perfume could act like an unconditioned stimulus, such that another neutral stimulus would gain reinforcement value after being paired with the perfume. This is anecdotally supported by the experience we may have upon meeting a stranger who is wearing the same perfume as someone we have very fond feelings towards—such encounters can generate a positive bias in our attitudes towards the stranger. This supposition however needs to be put to rigorous test. In all likelihood, a favored perfume initially comes to be preferred as a consequence of appetitive classical conditioning, for example, the association of a particular (hedonically inert) scent with positive emotions or with a loved one. After the perfume becomes endowed with conditioned value, it can then function as a *second-order* conditioned reinforcer (Holland and Rescorla 1975), conferring positive value on other stimuli through conditioned chaining. Indeed, the use of emotionally valenced odors as conditioned reinforcers has been demonstrated in both psychological (Todrank et al. 1995; Hermann et al. 2000) and neuroimaging (Gottfried, O'Doherty, and Dolan 2002, 2003; Gottfried and Dolan 2004) studies of human olfactory associative learning.

17.9 CONCLUSIONS AND THE FUTURE

It is evident that the scented human body has biological significance for mate selection and reproductive success. The chemicals in our sweat, both our unique body-odor and possibly other chemosignals that may or may not be smelled per se, have a strong bearing on our fecundity. Additionally, adorning artificial scent increases mate attraction through various mechanisms such as increasing self-confidence or seducing a biologically incompatible mate.

What is not yet determined is whether perfume in itself is a true reward. From a neurobiological perspective, it has been shown that positive and negative odors elicit different patterns of activation (Gottfried et al. 2002; Anderson et al. 2003; Rolls, Kringelbach, and de Araujo 2003; Winston et al. 2005). Earlier work on hemispheric lateralization suggested that the left hemisphere may be dominant for positive emotional experience (Davidson 1984) and in this regard it is notable that we found a tendency for more robust activation to the personally meaningful perfume in our fMRI study in participants' left hemispheres (Herz et al. 2003). However, a left laterality effect for personally positive odors is not entirely consistent with other neurological findings addressing odor pleasantness (Zald and Pardo 1997; Royet et al. 2003). Methodological differences in the aims of the limited studies conducted may be responsible for the lack of consistency and here is undoubtedly an area in which more work needs to be done.

Another neurological avenue for assessing perfume's rewarding properties involves the role of the OFC. The OFC is involved in both reward and olfactory processing. However, the degree to which perfume experience specifically recruits the OFC and perhaps sub-regions thereof is not known and would be a very interesting topic to study. It is also known that the dopaminergic neurotransmitter system is highly involved in the experience of reward, and that dopaminergic pathways innervate the OFC and amygdala (Schienle et al. 2009). However, to date there are no data in humans assessing whether hedonically positive odors have a particularly activating effect on dopaminergic transmission. Future research comparing emotional and non-emotional olfactory stimuli and the neurotransmitter systems subserving them is a new topic awaiting pioneering pursuit.

From a cognitive-behavioral perspective, it has been shown that perfumes can elicit highly emotional associations and responses. However, the perfume in question must first have acquired emotional significance for the smeller through associative learning. Using behavioral paradigms in concert with neurobiological techniques, the responses elicited by a perfume before and after its association to an emotionally appetitive event could be examined. This methodology would elucidate the interplay between conscious and neurobiological mechanisms and the ways in which fragrance elicits pleasure and reward.

This chapter has reviewed various facets of the perfume experience and related topics of human chemosignal emissions in seduction, sexual advertising, and reproductive success. The present review has made it clear that although it is a tantalizing area where speculation and anecdote abound, there has been little empirical research. Most importantly, perfume has barely been studied from a neurobiological perspective or evaluated as a reinforcing stimulus. Future innovative research in perfume neuroscience and reward will be a great adornment to the field.

REFERENCES

Aggleton, J.P., and Mishkin, M. 1986. The amygdala: Sensory gateway to the emotions. In *Emotion: Theory, Research and Experience. Vol 3: Biological Foundations of Emotion*, eds. R. Plutchik and H. Kellerman, pp. 281–99. Orlando: Academic Press.

Anderson, A.K., Christoff, K., Stappen, I., Panitz, D., Ghahremani, D.G., Glover, G., Gabrieli, J.D., and Sobel, N. 2003. Dissociated neural representations of intensity and valence in human olfaction. *Nature Neuroscience* 6:196–202.

Ayabe-Kanamuura, S., Schicker, I., Laska, M., Hudon, R., Distel, H., Kobabyakaw, T., and Saito, S. 1998. Differences in perception of everyday odors: A Japanese-German cross-cultural study. *Chemical Senses*, 23:31–38.

Balough, R.D., and Porter, R.H. 1986. Olfactory preference resulting from mere exposure in human neonates. *Infant Behavior and Development* 9:395–401.

Baron, R.A. 1983. Sweet smell of success? The impact of pleasant artificial scents on evaluations of job applicants. *Journal of Applied Psychology* 68:709–13.

———. 1986. Self-presentation in job interviews: When there can be too much of a good thing. *Journal of Applied Social Psychology* 16:16–28.

Bartoshuk, L.M. 1991. Taste, smell and pleasure. In *The Hedonics of Taste*, ed. R.C. Bolles, pp. 15–28. Hillsdale, NJ: Lawrence Erlbaum.

Boyse, E.A., Beauchamp, G.K., and Yamazaki, K. 1987. The genetics of body scent. *Trends in Genetics* 3:97–102.

Burr, C. 2007. *The Perfect Scent*. New York: Henry Holt.

Burt, A. 1992. Concealed ovulation and sexual signals in primates. *Folia Primatologica* 58:1–6.

Buss, D.M. 1989. Sex differences in human mate preferences: Evolutionary hypotheses tested in 37 cultures. *Behavioral and Brain Sciences* 12:1–49.

Buss, D.M., and Schmitt, D.P. 1993. Sexual strategies theory: An evolutionary perspective on human mating. *Psychological Review* 100:204–32.

Cahill, L., Babinsky, R., Markowitsch, H.J., and McGaugh, J.L. 1995. The amygdala and emotional memory. *Nature* 377:295–96.

Cain, W.S., and Johnson, F. Jr. 1978. Lability of odor pleasantness: Influence of mere exposure. *Perception* 7:459–65.

Classen, C., Howes, D., and Synnott, A. 1994. *Aroma: The Cultural History of Smell*. New York: Routledge.

Davidson, R.J. 1984. Hemispheric asymmetry and emotion. In *Approaches to Emotion*, eds. K. Scherer and P. Ekman, pp. 39–57. Hillsdale, NJ: Erlbaum.

Davidson, R.J., Putnam, K.M., and Larson, C.L. 2000. Dysfunction in the neural circuitry of emotion regulation – a possible prelude to violence. *Science* 289:591–94.

Davis, L.B., and Porter, R.H. 1991. Persistent effects of early odor exposure on human neonates. *Chemical Senses* 16:169–74.

Dawkins, R. 1976. *The Selfish Gene*. New York: Oxford University Press.

Dilks, D.D., Dalton, P., and Beauchamp, G.K. 1999. Cross-cultural variation in responses to malodors. *Chemical Senses* 24:599.

Drewnowski, A., Kurth, C., Holden-Wiltse, J., and Saari, J. 1992. Food preferences in human obesity: Carbohydrate versus fats. *Appetite* 18:207–21.

Egid, K., and Brown, J.L. 1989. The major histocompatibility complex and female mating preferences in mice. *Animal Behaviour* 38:548.

Eichenbaum, H. 2001. The hippocampus and declarative memory: Cognitive mechanisms and neural codes. *Behavioural Brain Research* 127:199–207.

Eklund, A., Egid, K., and Brown, J.L. 1992. Sex differences in the use of the major histocompatibility complex for mate selection in congenic strains of mice. In *Chemical Signals in Vertebrates*, eds. R.L. Doty and D. Muller-Scwarze, pp. 213–17. New York: Plenum Press.

Engen, T. 1982. *The Perception of Odors*. Toronto: Academic Press.

———. 1991. *Odor Sensation and Memory*. New York: Praeger.

Engen, T., and Engen, E.A. 1997. Relationship between development of odor perception and language. *Enfance* 1:125–40.

Fiore, A.M., Yah, X., and Yoh, E. 2000. Effects of product display and environmental fragrancing on approach responses and pleasurable experiences. *Psychology & Marketing* 17:27–54.

Furnham, A., Tan, T., and McManus, C. 1997. Waist-to-hip ratio preferences of body shape: A replication and extension. *Personality and Individual Differences* 22:539–49.

Gangestad, S.W., Thornhill, R., and Garver-Apgar, C.E. 2005. Changes in women's sexual interests and their partners' mate retention tactics across the menstrual cycle: Evidence for shifting conflicts of interest. *Proceedings of the Royal Society of London B* 269:975–82.

Gottfried, J.A., Deichmann, R., Winston, J.S., and Dolan, R.J. 2002. Functional heterogeneity in human olfactory cortex: An event-related functional magnetic resonance imaging study. *Journal of Neuroscience* 22:10819–28.

Gottfried, J.A., and Dolan, R.J. 2004. Human orbitofrontal cortex mediates extinction learning while accessing conditioned representations of value. *Nature Neuroscience* 7:1144–52.

Gottfried, J.A., O'Doherty, J., and Dolan, R.J. 2002. Appetitive and aversive olfactory learning in humans studied using event-related functional magnetic resonance imaging. *Journal of Neuroscience* 22:10829–37.

Gottfried, J.A., O'Doherty, J., and Dolan, R.J. 2003. Encoding predictive reward value in human amygdala and orbitofrontal cortex. *Science* 301:1104–7.

Greenlees, I.A., and McGrew, W.C. 1994. Sex and age differences in preferences and tactics of mate attraction: Analysis of published advertisements. *Ethology and Sociobiology* 15:59–72.

Haller, R., Rummel, C., Henneberg, S., Pollmer, U., and Koster, E.P. 1999. The influence of early experience with vanillin on food preference in later life. *Chemical Senses* 24:465–567.

Havlicek, J., Dvorakova, R., Baros, L., and Flegr, J. 2006. Non-advertized does not mean concealed: Body odour changes across the human menstrual cycle. *Ethology* 112:81–90.

Hens, R. 1995. Waist-to-hip ratio and attractiveness: Replication and extension. *Personality and Individual Differences* 19:479–88.

Hermann, C., Ziegler, S., Birbaumer, N., and Flor, H. 2000. Pavlovian aversive and appetitive odor conditioning in humans: Subjective, peripheral, and electrocortical changes. *Experimental Brain Research* 132:203–15.

Herz, R.S. 2001. Ah, sweet skunk: Why we like or dislike what we smell. *Cerebrum* 34:31–47.

———. 2003. http://www.senseofsmell.org/papers/R. Herz Survey Study Final Report w. tables.doc.

———. 2005. The unique interaction between language and olfactory perception and cognition. In *Trends in Experimental Psychology Research*, pp. 91–109. New York: Nova Science.

———. 2007. *The Scent of Desire: Discovering Our Enigmatic Sense of Smell*. New York: William Morrow/ HarperCollins.

———. 2008. Olfaction. In *Sensation & Perception*. 2nd ed. pp. 330–58. Sunderland, MA: Sinauer Associates.

———. 2009a. Aromatherapy facts and fictions: A scientific analysis of olfactory effects on mood, physiology and behavior. *International Journal of Neuroscience* 119:263–90.

———. 2009b. Basic processes in human olfactory cognition: Current questions and future directions. *Annals of the New York Academy of Sciences* 1170:313–17.

Herz, R.S., Beland, S.L., and Hellerstein, M. 2004. Changing odor hedonic perception through emotional associations in humans. *International Journal of Comparative Psychology* 17:315–39.

Herz, R.S., and Cahill, E.D. 1997. Differential use of sensory information in sexual behavior as a function of gender. *Human Nature* 8:275–86.

Herz, R.S., Eliassen, J.C., Beland, S.L., and T. Souza. 2004. Neuroimaging evidence for the emotional potency of odor-evoked memory. *Neuropsychologia* 42:371–78.

Herz, R.S., and Inzlicht, M. 2002. Gender differences in response to physical and social signals involved in human mate selection: The importance of smell for women. *Evolution and Human Behavior* 23:359–64.

Herz, R.S., Schankler, C., and Beland, S. 2004. Olfaction, emotion and associative learning: Effects on motivated behavior. *Motivation and Emotion* 28:363–83.

Herz, R.S., and von Clef, J. 2001. The influence of verbal labeling on the perception of odors: Evidence for olfactory illusions? *Perception* 30:381–91.

Ho, H.N., Gill, T.J., Nsieh, R.P., Hsieh, H.J., and Lee, T.Y. 1990. Sharing of human leukocyte antigens in primary and secondary recurrent spontaneous abortions. *American Journal of Obstetrics and Gynecology* 163:178–88.

Holland, P.C., and Rescorla, R.A. 1975. Second-order conditioning with food unconditioned stimulus. *Journal of Comparative and Physiological Psychology* 88:459–67.

Jacob, S., Hayreh, D.J., and McClintock, M.K. 2001. Context-dependent effects of steroid chemosignals on human physiology and mood. *Physiology and Behavior* 74:15–27.

Kubberod, E., Ueland, O., Rodbotten, M., Westad, F., and Risvik, E. 2002. Gender specific preferences and attitudes towards meat. *Food Quality and Preference* 13:285–94.

Landolt, M.A., Lalumiere, M.L., and Quinsey, V.L. 1995. Sex differences in intrasex variations in human mating tactics: An evolutionary approach. *Ethology & Sociobiology* 16:3–23.

Lange, I.G., Hartel., A., and Meyer, H.H.D. 2002. Evoluation of oestrogen functions in vertebrates. *Journal of Steroid Biochmiestry and Molecular Biology* 83:219–26.

LeDoux, J. 1998. Fear and the brain: Where have we been, and where are we going? *Biological Psychiatry* 44:1229–38.

Lorig, T. 1999. On the similarity of odor and language perception. *Neuroscience and Biobehavioral Reviews* 23:391–98.

Lott, I.T., Sullivan, R.M., and McPherson, D. 1989. Associative olfactory learning occurs in the neonate. *Neurology* 39:110.

Lundstrom, J.N., and Olsson, M.J. 2005. Subthreshold amounts of social odorant affect mood, but not behavior, in heterosexual women when tested by a male, but not a female experimenter. *Biological Psychology* 70:197–204.

Manning, J.T., Scutt, D., Whiothosue, G.H., Kinnster, S.J., and Walton, J.M. 1996. Asymmetry and the menstrual cycle in women. *Ethology and Sociobiology* 17:129–43.

McClintock, M.K. 1971. Menstrual synchrony and suppression. *Nature* 229:244–45.

Mennella, J.A., and Beauchamp, G.K. 1991. The transfer of alcohol to human milk: Effects on flavor and the infant's behavior. *New England Journal of Medicine* 325:981–85.

———. 1993. The effects of repeated exposure to garlic-flavored milk on the nursling's behavior. *Pediatric Research* 34:805–8.

———. 2002. Flavor experiences during formula feeding are related to preferences during childhood. *Early Human Development* 68:71–82.

Mennella, J.A., and Garcia, P.L. 2000. Children's hedonic response to the smell of alcohol: Effects of parental drinking habits. *Alcoholism: Clinical and Experimental Research* 24:1167–71.

Mennella, J.A., Johnson, A., and Beauchamp, G.K. 1995. Garlic ingestion by pregnant women alters the odor of amniotic fluid. *Chemical Senses* 20:207–9.

Milinski, M and Wedekind, C. 2001. Evidence for MHC-correlated perfume preferences in humans. *Behavioral Ecology* 12:140–49.

Miller, G., Tybur, J.M., and Jornda, B.D. 2007. Ovulatory cycle effects on tip earnings by lap dancers: Economic evidence for human estrus? *Evolution and Human Behavior* 28:375–81.

Moncreiff, R.W. 1966. *Odour Preferences*. New York: Wiley.

Moran, J. 2000. *Fabulous Fragrances II*. La Quinta, CA: Crescent House Publishing.

Ober, C. 1999. Studies of HLA, fertility and mate choice in a human isolate. *Human Reproduction Update* 5:103–7.

Ober, C., Elias, S., Kostyu, D.D., and Hauck, W.W. 1992. Decreased fecundability in Hutterite couples sharing HLA-DR. *American Journal of Human Genetics* 50:6–14.

Ober, C., Hyslop, T., Elias, S., Weitkamp, L.R., and Hauck, W.W. 1998. Human leucocyte antigen matching and fetal loss: Results of a 10 year prospective study. *Human Reproduction* 13:33–38.

Ober, C., Weitkamp, L.R., Cox, N., Dytch, H., Kostyu, D., and Elias, S. 1997. HLA and mate choice in humans. *American Journal of Human Genetics* 61:497–504.

Potts, W.K., Manning, C.J., and Wakeland, E.K. 1991. Mating patterns in seminatural populations of mice influenced by MHC genotype. *Nature* 352:619–21.

Roberts, S.C., Havlicek, J., Flegr, J., Hrruskova, M., Little, A.C., Jones, B.C. et al. 2004. Female facial attractiveness increases during the fertile phase of the menstrual cycle. *Proceedings of the Royal Society of London Series B* 271S5:S270–72.

Roberts, S.C., Little, A.C., Lyndon, A., Roberts, J., Havlicek, J., and Wright, R.I. 2009. Manipulation of body odour alters men's self-confidence and judgments of their visual attractiveness by women. *International Journal of Cosmetic Science* 31:47–54.

Robin, O., Alaoui-Ismaili, O., Dittmar, A., and Vernet-Mauri, E. 1998. Emotional responses evoked by dental odors: An evaluation from autonomic parameters. *Journal of Dental Research* 77:1638–46.

Rolls, E.T. 1999. *The Brain and Emotion.* Oxford: Oxford University Press.

Rolls, E.T., Kringelbach, M.L., and de Araujo, I.E., 2003. Different representations of pleasant and unpleasant odours in the human brain. *European Journal of Neuroscience* 18:695–703.

Royet, J.-P., and Plailly, J. 2004. Lateralization of olfactory processes. *Chemical Senses* 29:731–45.

Royet, J.P., Plailly, J., Delon–Martin, C., Kareken, D.A., and Segebarth, C. 2003. fMRI of emotional responses to odors: Influence of hedonic valence and judgment, handedness, and gender. *Neuroimage* 20:713–28.

Schaal, B., Marlier, L., and Soussignan, R. 1995. Responsiveness to the odor of amniotic fluid in the human neonate. *Biology of the Neonate* 671:397–406.

———. 1998. Olfactory function in the human fetus: Evidence from selective neonatal responsiveness to the odor of amniotic fluid. *Behavioral Neuroscience* 112:1438–49.

Schienle, A., Schafer, A., Hermann, A., and Vaitl, D. 2009. Binge-eating disorder: Reward sensitivity and brain activation to images of food. *Biological Psychiatry* 65:654–61.

Schiffman, S.S. 1991. http://www.senseofsmell.org/papers/S._Schiffman_Ages_&_Stages_1991_1.doc.

Schleidt, M., and Genzel, C. 1990. The significance of mother's perfume for infants in the first weeks of their life. *Ethology and Sociobiology* 11:145–54.

Schleidt, M., Hold, B., and Attila, G. 1981. A cross-cultural study on the attitude towards personal odors. *Journal of Chemical Ecology* 7:19–31.

Schmidt, H.J., and Beauchamp, G.K. 1988. Adult-like odor preferences and aversions in three year old children. *Child Development* 59:1136–43.

Schwerdtfeger, W.K., Buhl, E.H., and Germroth, P. 1990. Disynaptic olfactory input to the hippocampus mediated by stellate cells in the entorhinal cortex. *Journal of Comparative Neurology* 292:163–77.

Singh, D., and Bronstad, P.M. 2001. Female body odour is a potential cue to ovulation. *Proceedings of the Royal Society of London Series B* 268:797–801.

Soussignan, R., Schaal, B., Marlier, L., and Jiang, T. 1997. Facial and autonomic responses to biological and artificial olfactory stimuli in human neonates: Re-examining early hedonic discrimination of odors. *Physiology & Behavior* 62:745–58.

Spencer, N.A., McClintock, M.K., Sellergren, S.A. Bullivant, S., Jacob, S., and Mennella, J.A. 2004. Social chemosignals from breastfeeding women increase sexual motivation. *Human Behavior* 46:362–70.

Stein, M., Ottenberg, M.D., and Roulet, N. 1958. A study of the development of olfactory preferences. *Archives of Neurological Psychiatry* 80:264–66.

Stern, K., and McClintock, M.K. 1998. Regulation of ovulation by human pheromones. *Nature* 392:177–79.

Sullivan, R.M., Taborsky-Barba, S., Mendoza, R., Itano, A., Leon, M., Cotman, C., Payne, T.F., and Lott, T. 1991. Olfactory classical conditioning in neonates. *Pediatrics* 87:511–18.

Symonds, C.S., Gallagher, P., Thompson, J.M., and Young, A.H. 2004. Effects of the menstrual cycle on mood, neurocognitive and neuroendocrine function in healthy premenopausal women. *Psychological Medicine* 34:93–102.

Thomas, M.L., Harger, J.H., Wagner, D.K., Rabin, B.S., and Gill III, T.J. 1985. HLA sharing and spontaneous abortion in humans. *American Journal of Obstetrics and Gynecology* 151:1053–58.

Tiercy, J.M., Jeannet, M., and Mach, B. 1990. A new approach for the analysis of HLA class II polymorphism: HLA oligotyping. *Blood Review* 4:9–15.

Todrank, J., Byrnes, D., Wrzesniewski, A., and Rozin, P. 1995. Odors can change preferences for people in photographs: A cross-modal evaluative conditioning study with olfactory USs and visual CSs. *Learning and Motivation* 26:116–40.

Trivers, R. 1972. Parental investment and sexual selection. In *Sexual Selection and the Descent of Man*, ed. B. Campbell. Chicago: Aldine-Atherton.

Turner, B.H., Mishkin, M., and Knapp, M. 1980. Organization of the amygdalopetal projections from modality-specific cortical association areas in the monkey. *Journal of Comparative Neurology* 191:515–43.

Walla, P., Hufnagl, B., Lehern, J., Mayer, D., Lindinger, G., Imhof, H., Deeke, L., and Lang, W. 2003. Olfaction and depth of word processing: A magnetoencephalographic study. *Neuroimage* 18:104–16.

Weckstein, L.N., Patrizio, P., Balmaceda, J.P., Asch, R.H., and Branch, D.W. 1991. Human leukocyte antigen compatibility and failure to achieve a viable pregnancy with assisted reproductive technology. *Acta European Fertility* 22:103–7.

Wedekind, C., Seebeck, T., Bettens, F., and Paepke, A.J. 1995. MHC-dependent mate preference in humans. *Proceedings of the Royal Society of London B* 260:245–49.

Weaver, M.R., and Brittin, H.C. 2001. Food preferences of men and women by sensory evaluation versus questionnaire. *Family and Consumer Sciences Research Journal* 29:288–301.

Weller, L., and Weller, A. 1997. Menstrual variability and the measurement of menstrual synchrony. *Psychoneuroendocrinology* 22:115–28.

Wilson, H.C. 1992. A critical review of menstrual synchrony research. *Psychoneuroendocrinology* 17:565–91.

Winberg, J., and Porter, R.H. 1998. Olfaction and human neonatal behaviour: Clinical implications. *Acta Paediatrica* 87:6–10.

Winston, J.S., Gottfried, J.A., Kilner, J.M., and Dolan, R.J. 2005. Integrated neural representations of odor intensity and affective valence in human amygdala. *Journal of Neuroscience* 25:8903–7.

Wyert, C., Webster, W.W., Chen, J.H., Wilson, S.R. McClary, A., Khan, R.M., and Sobel, N. 2007. Smelling a single component of male sweat alters levels of cortisol in women. *Journal of Neuroscience* 27:1261–65.

Yamazaki, K., Boyse, E.A., Mike, V., Thaler, H.T., Mathieson, B.J., Abbot, J., Boyse, J., Zayas, Z.A., and Thomas, L. 1976. Control of mating preferences in mice by genes in the major histocompatibility complex. *Journal of Experimental Medicine* 144:1324–35.

Yamazaki, K., Yamaguchi, M., Baranoski, L., Bard, J., Boyse, E.A., and Thomas, L. 1979. Recognition among mice. Evidence from the use of a Y-maze differentially scented by congenic mice of different major histocompatibility types. *Journal of Experimental Medicine* 150:755–76.

Zald, D.H., and Pardo, J.V. 1997. Emotion, olfaction, and the human amygdala: Amygdala activation during aversive olfactory stimulation. *Proceedings of the National Academy of Sciences USA* 948:4119–24.

Zellner, D.A., Bartoli, A.M., and Eckard, R. 1991. Influence of color on odor identification and liking ratings. *American Journal of Psychology* 104:547–61.

18 Visual Art

Anjan Chatterjee

CONTENTS

18.1 INTRODUCTION

A person stands transfixed before a Mark Rothko abstract painting oblivious to everything else. Embedded in this scene are questions that strike at the very heart of this book. What is the sensation being experienced? If the color or form of the painting were altered slightly, would the experience be the same? Why do some visitors to the museum glance at the same painting and shrug their shoulders before being absorbed by a Cezanne landscape or a Rembrandt portrait? Why do visitors bother to gaze at paintings at all? What is the nature of the reward that compels them to travel distances, pay entrance fees, and negotiate crowds to stare at pieces of canvas? What, if anything, distinguishes this rewarding experience from the pleasure of gazing at an attractive person or from the anticipation of a good meal?

In this chapter, I explore these issues in aesthetics through the lens of cognitive neuroscience. The term aesthetics is used broadly here, to encompass the perception, production, and response to art, but also to include the responses to objects and scenes that evoke a response that could be considered aesthetic. The nature of this response is something to which I shall return later in this chapter. I start by reviewing recent comments on the relationship of art and the brain made by visual neuroscientists. I then describe a framework that might guide research in neuroaesthetics. Following that, I review empirical work conducted thus far. Finally, I suggest how progress could be made in this nascent field.

At the outset, I should be clear about limits to defining art (Carroll 2000). Some philosophers have claimed that defining art with necessary and sufficient conditions is not possible (Weitz 1956). In response to such claims, recent theoreticians have defined art by its social and institutional (Dickie 1969) or its historical context (Danto 1964). Cognitive neuroscientists are unlikely to address sociological or historical conceptions of art. They are likely to sidestep definitional issues and focus on accepted examples of artwork or properties of these works as probes for experiments.

18.2   OBSERVATIONS ON THE RELATIONSHIP OF ART AND THE BRAIN

With rare exceptions, it is not clear that neuroscientists consider aesthetics worthy of inquiry. And some aestheticians probably consider neuroscientific inquiry into aesthetics an abomination. Only recently has neuroscience joined a tradition of empirical aesthetics that dates back to Fechner in the nineteenth century (Fechner 1876; and see Chapter 2 in this volume for more information about Fechner). The first wave of writings on aesthetics by neuroscientists points to parallels between art and organizational principles of the brain.

Zeki (1999) argued forcefully that no theory of aesthetics is complete without an understanding of its neural underpinnings. He suggested that the goals of the nervous system and of artists are similar. Both are driven to understand essential attributes of the world. The nervous system decomposes visual information into different components, such as color, luminance, and motion. Similarly, many artists, particularly within the last century, isolated different visual attributes. For example, Matisse emphasized color and Calder emphasized motion. Zeki suggests that artists, like visual neuroscientists, endeavor to uncover important distinctions in the visual world. In doing so, they discover modules of the visual brain. Recently, Cavanagh (2005) has similarly claimed that the artists' goals resemble those of the nervous system. He observes that paintings often violate the physics of shadows, reflections, colors, and contours. Rather than adhering to physical properties of the world, these paintings reflect perceptual shortcuts used by the brain. Artists, in experimenting with forms of depiction, discovered what psychologists and neuroscientists are now identifying as principles of perception.

Livingstone (2002) and Conway (Conway and Livingstone 2007) focused on how artists make use of complex interactions between different components of vision in creating their paintings. The dorsal (where) and ventral (what) processing distinction is a central tenet in visual neuroscience (Ungerleider and Mishkin 1982) (though see Chapter 8 in this volume). The dorsal stream is sensitive to contrast differences, motion, and spatial location. The ventral stream is sensitive to simple form and color. Livingstone suggests that the shimmering quality of water or the sun's glow on the horizon seen in some impressionist paintings (e.g., the sun and surrounding clouds in Monet's *Impression Sunrise*) is produced by isoluminant objects distinguishable only by color. The dorsal stream is insensitive to these elements of the image. Since the dorsal stream identifies motion (or the lack thereof) and spatial location, Livingstone argues that isoluminant forms are not fixed with respect to motion or spatial location and are experienced as unstable or shimmering. Conversely, since shape can be derived from luminance differences, she argues that artists use contrast to produce shapes, leaving color for expressive rather than descriptive purposes (as in Derain's portrait of Matisse). Livingstone highlights the combinatorial properties of visual attributes. Artists use these combinatorial properties to produce specific aesthetic effects.

Ramachandran and Hirstein (1999) proposed a set of perceptual principles underlying aesthetic experiences. For example, they emphasize the "peak shift" phenomena as offering insight into the aesthetics of abstract art by relying on Tinbergen's (1954) work. Tinbergen demonstrated that seagull chicks beg for food from their mothers by pecking on a red spot near the tip of the mother's beak. It turns out that a disembodied long thin stick with three red stripes near the end evokes an exaggerated response from these chicks. Ramachandran and Hirstein propose that neural structures that evolved to respond to specific visual stimuli respond more vigorously (a shift in their peak response) to underlying primitives of that form even when the viewer is not aware of the primitive. Their insight is that abstract art may be tapping into such visual primitives.

These examples of recent comments reflect an emerging recognition by neuroscientists that visual aesthetics are an important part of human experience, which ought to conform to principles of neural organization. Intriguing parallels exist between the concerns and techniques of artists and the organization of the visual brain. Some components of visual aesthetics should be amenable to empirical methods from cognitive neuroscience. The challenge ahead is to transform these comments into testable hypotheses and experimental paradigms.

18.3 A FRAMEWORK FOR NEUROAESTHETICS RESEARCH

A cognitive neuroscience research program in visual aesthetics rests on two principles (Chatterjee 2002). First, visual aesthetics, like vision in general, has multiple components. Second, an aesthetic experience emerges from a combination of responses to different components of a visual object. The process by which humans visually recognize objects offers a framework from which to consider these components. Investigations can be focused on these components and on their combinatorial properties.

The nervous system processes visual information both hierarchically and in parallel (Van Essen et al. 1990; Zeki 1993; Farah 2000). The levels of this processing can be classified as early, intermediate, and late vision (Marr 1982). Early vision extracts simple elements from the visual environment, such as color, luminance, shape, motion, and location (Livingstone and Hubel 1987; Livingstone 1988). These elements are processed in different parts of the brain. Intermediate vision segregates some elements and groups others together to form coherent regions in what would otherwise be a chaotic and overwhelming sensory array (Biederman and Cooper 1991; Grossberg, Mingolla, and Ros 1997; Vecera and Behrmann 1997; Ricci, Vaishnavi, and Chatterjee 1999). Late vision selects which of these coherent regions to scrutinize and evokes memories from which objects are recognized and meanings attached (Chatterjee 2003a; Farah 2000).

This sequence of visual processing is likely to be reflected in aesthetics (Chatterjee 2003b) (for a related model, see Helmut et al. 2004). Any work of art can be decomposed into its early, intermediate, and late vision components, and individual works of visual art (paintings) can be identified that exemplify each of these different componential stages (Figure 18.1). Aesthetic writings commonly distinguish between form and content (e.g., Russell and George 1990; Woods 1991). Similarly, scientists observe that early and intermediate vision process form and later vision processes content. Figure 18.2 shows a model of how the neuroscience of visual aesthetics might be mapped. The early features of an art object might be its color and its spatial location. These elements would be grouped together to form larger units in intermediate vision. Such grouping occurs automatically. The neural basis of grouping is not well understood, but it likely involves extrastriate cortex (Biederman and Cooper 1991; Grossberg, Mingolla, and Ros 1997). Grouping creates "unity in diversity," a central notion of compositional balance.

If compositional form is apprehended automatically by intermediate vision, then sensitivity to such form should also be automatic. Subjects are sensitive to compositional form "at a glance," with exposures as short as 50 msec (Locher and Nagy 1996). Intriguingly, preference for form predominates when images are shown over short exposure times, while preference for detail predominates when images are shown for slightly longer times (Ognjenovic 1991). Combinations of early and intermediate visual properties (e.g., color, shape, composition) engage attentional circuits mediated by frontal-parietal neural networks. Attentional modulation of early vision (Motter 1993, 1994; Shulman et al. 1997; Watanabe et al. 1998) is likely to contribute to a more vivid experience of the stimulus.

Beyond perception in visual aesthetics, two other aspects of aesthetics are important. The first is the emotional response to an aesthetic image; the second is how aesthetic judgments are made. The anterior medial temporal lobe, medial and orbitofrontal cortices, and subcortical structures mediate emotions in general, and reward systems in particular (Schultz, Dayans, and Mortague 1997; O'Doherty et al. 2001; Elliott, Friston, and Dolan 2000; Delgado et al. 2000; Breiter et al. 2001; Berridge and Kringelbach 2008). Aesthetic judgments about stimuli, as measured by preference ratings or appraisals, are likely to engage widely distributed circuits, most importantly the dorsolateral frontal and medial frontal cortices.

18.4 EMPIRICAL EVIDENCE: LESION STUDIES

Investigations of patients with brain damage have contributed greatly to our understanding of cognitive and affective systems. This approach also has substantial promise in advancing neuroaesthetics. Diseases of the brain can impair our ability to speak or comprehend language, to

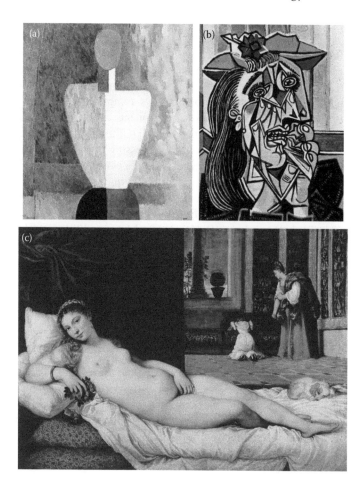

FIGURE 18.1 (See Color Insert) Visual aesthetics recapitulates visual processing: a hierarchical progression as documented through the brushstrokes of three renowned artists. Each painting depicts the female figure, through successive levels of representational complexity, from Malevich (early) and Picasso (intermediate) to Titian (late). (a) *Torso*, 1928–32 (oil on canvas) by Kazimir Severinovich Malevich (1878–1935). (Reproduced from *Torso*, 1928–32 (oil on canvas) by Kazimir Severinovich Malevich (1878–1935) State Russian Museum, St. Petersburg, Russia/Bridgeman Art Library. With permission.) (b) *Weeping Woman*, 1937 (oil on canvas) by Pablo Picasso (1881–1973). © 2010 Estate of Pablo Picasso/Artists Rights Society (ARS). (With permission: Reproduction, including downloading of Picasso works, is prohibited by copyright laws and international conventions without the express written permission of Artist Rights Society (ARS), New York.) (c) *Venus of Urbino*, 1538 (oil on canvas) by Titian (Tiziano Vecellio) (c. 1488–1576).

coordinate movements, to recognize objects, to apprehend emotions, and to make logical decisions. By contrast, while damage to the brain can certainly impair the ability to produce art, paradoxically, in some cases art abilities seem to improve. Brain damage can create a disposition to produce visual art, provide artists with a unique visual vocabulary, add to artists' descriptive accuracy, and enhance their expressive powers. These paradoxical improvements offer unique insights into the creative underpinnings of artistic output. They are reviewed elsewhere (Chatterjee 2004, 2006, 2009) and will not be discussed further in this chapter, in keeping with the present focus on sensation and reward.

Studies of people with brain damage also can advance our understanding of the perception and experience of art. Some people with brain damage probably do not perceive art in the same way that non-brain-damaged individuals do and their emotional responses to artwork may very well differ

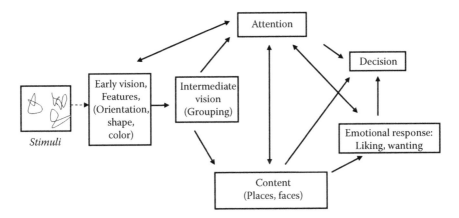

FIGURE 18.2 A general information-processing model to guide research in neuroaesthetics. See text for details.

from those of people without brain damage (cf. Chapter 16). However, neuropsychological investigations of aesthetic perception to date are non-existent. There is no adequate instrument to provide basic quantitative assessments of a person's apprehension artwork. We are currently developing such a tool, The Assessment of Art Attributes (Chatterjee et al. 2010). This assessment assumes that the perception of art can be organized along different perceptual and conceptual attributes (see Table 18.1). Using such an assessment, one could begin to investigate groups of patients and ascertain the relationship of brain damage to selective deficits or enhancements in art perception. Much remains to be learned if we can develop adequate methods and measurements for this line of inquiry.

18.5 EMPIRICAL EVIDENCE: IMAGING STUDIES OF BEAUTY

Beauty is integral to most people's concept of aesthetics (Jacobsen et al. 2004). Of course, not all art is beautiful and artists do not always intend to produce beautiful things. But beauty remains a central concept in discussions of art. Understanding the neural basis of the apprehension of and response to beauty might give us insight into the apprehension of and response to visual art. Facial beauty has received particular attention.

The response to facial beauty is likely to be deeply encoded in our biology. Cross-cultural judgments of facial beauty are quite consistent (Etcoff 1999; Perrett, May, and Yoshikawa 1994; Jones

TABLE 18.1
A Pervasive Concern in Empirical Aesthetics has been Distinguishing between Form and Content of Artwork

Form	Content
Hue	Depictive accuracy
Saturation	Animacy
Stroke/contour	Emotionality
Depth	Abstraction
Balance	Fantasy
Complexity	Symbolism

This list of attributes represents one version of how this distinction might be made in experimental studies (Chatterjee et al. 2008).

and Hill 1993). Adults and children within and across cultures agree in their judgments of facial attractiveness (Langlois et al. 2000), suggesting that universal principles of facial beauty exist. Similarly, infants look longer at attractive faces within a week of being born and the effects of facial attractiveness on infants' gaze generalize across race, gender, and age by 6 months (Langlois et al. 1991; Slater et al. 1998). Thus, the disposition to engage attractive faces is present in brains that have not been modified greatly by experience. Some components of beauty are shaped further by cultural factors (Cunningham et al. 2002), but the universal components are likely to have distinct neural underpinnings.

Several studies report that attractive faces activate neural circuitry involved in reward systems, including the orbitofrontal cortex, the nucleus accumbens, the ventral striatum (Kampe et al. 2001; Aharon et al. 2001; O'Doherty et al. 2003; Ishai 2007; Kranz and Ishai 2006), and the amygdala (Winston et al. 2007). These regional activations are interpreted as reflecting emotional valences attached to attractive faces (Senior 2003). The particular emotional valences are those involved in the expectation of rewards and the satisfaction of appetites. The idea that attractive faces are rewarding stimuli, at least for men, is evident behaviorally. Heterosexual men discount higher future rewards for smaller immediate rewards with attractive female faces (Wilson and Daly 2004). Presumably these patterns of neural activation reflect ways in which attractive faces influence mate selection (Ishai 2007).

Perceptual features of faces, such as averageness, symmetry, the structure of cheek bones, the relative size of the lower half of the face, and the width of the jaw, influence people's judgments of facial beauty (Grammer and Thornhill 1994; Enquist and Arak 1994; Penton-Voak et al. 2001). Winston et al. (2007) found left posterior occipito-temporal activity was enhanced by facial attractiveness. Similarly, Kranz and Ishai (2006) found greater activations in the lateral fusiform gyrus for attractive female faces than for unattractive female faces.

We conducted a study in which participants judged the attractiveness or matched the identity of pairs of faces. Attractiveness judgments evoked neural activity within a distributed network involving ventral visual association cortices and parts of dorsal posterior parietal and prefrontal cortices (Chatterjee et al. 2009). We interpreted the parietal, medial, and dorsolateral frontal activations as representing the neural correlates of the attention and decision-making components of this task. We also found positively correlated activity within the insula and negatively correlated activations within the anterior and posterior cingulate cortex. We inferred that these patterns represent the emotional responses to attractiveness. Importantly, when subjects matched the identity of faces, attractiveness continued to evoke neural responses in ventral visual areas, with a strength indistinguishable from that when participants considered beauty explicitly, suggesting that this ventral occipital region responds to beauty automatically.

Facial attractiveness is apprehended automatically (Palermo and Rhodes 2007; Olson and Marshuetz 2005) and has pervasive social effects beyond its specific role in mate selection. Attractive individuals are considered intelligent, honest, pleasant, and natural leaders (Kenealy, Frude, and Shaw 1988; Lerner et al. 1991; Ritts, Patterson, and Tubbs 1992), and are viewed as having socially desirable traits, such as strength and sensitivity (Dion, Berscheid, and Walster 1972). The cascade of neural events that bias social decisions is likely to be triggered by an early perceptual response to attractiveness. We proposed that neural activity within ventral visual cortices in response to facial attractiveness serves as the initial trigger for this cascade. The fact that this ventral occipital region of activation extended beyond cortical regions especially sensitive to faces per se raises the possibility that this area may be responsive to aesthetic objects more generally (Chatterjee et al. 2009).

18.6 EMPIRICAL EVIDENCE: IMAGING STUDIES OF ART

Very few studies have used art to examine the neural bases of aesthetics. While the goals in these studies were similar, their experimental approaches differed and the results at first glance appear varied. Kawabata and Zeki (2004) asked participants to rate abstract, still life, landscape, or portraitures paintings as beautiful, neutral, or ugly. They used a fixed-effects analysis in a 3×4 factor event-related design with event types segregated as beautiful, neutral, or ugly in one of the four painting categories. Not surprisingly, they found that the pattern of activity within the ventral visual cortex varied depending on whether subjects were looking at portraits, landscapes, or still-lives. In orbitofrontal cortex (BA 11) they found greater activity for beautiful than for ugly or neutral stimuli. In the anterior cingulate (BA 32) and left parietal cortex (BA 39), they found greater activity for beautiful than for neutral stimuli. Only activity within the orbitofrontal cortex increased with the beauty of all the painting types and the authors interpreted this activity as representing the neural underpinnings of the aesthetic emotional experience.

Vartanian and Goel (2004) used images of representational and abstract paintings in an fMRI study. They found that activity within the occipital gyri bilaterally and the left anterior cingulate increased with preference ratings. They also found that activity within the right caudate decreased as preference ratings decreased. Representational paintings evoked more activity within the occipital poles, the precuneus, and the posterior middle temporal gyrus than did abstract paintings.

Cela-Conde et al. (2004) used magnetoencephalography to record event potentials when participants viewed images of artworks and photographs. Participants judged whether or not the images were beautiful. Beautiful images evoked greater neural activity than not-beautiful images over the left dorsolateral prefrontal cortex with a latency of 400–1000 msec. The authors infer that this region is involved in making aesthetic judgments.

Jacobsen et al. (2005) used a different strategy to investigate the neural correlates of beauty in an fMRI study. Rather than use actual artworks as their stimuli, they used a set of geometric shapes designed in the laboratory. Participants judged whether the images were beautiful or whether the images were symmetric. Participants found symmetric patterns more beautiful than non-symmetric ones. Aesthetic judgments more than symmetry judgments activated the medial frontal cortex (BA 9/10), the precuneus, and the ventral prefrontal cortex (BA 44/47). The left intraparietal sulcus was conjointly active for symmetry and beauty judgments. Both beauty and complexity of the images evoked activity within orbitofrontal cortex. In a follow-up study using the same stimuli (Hofel and Jacobsen 2007), they found that beauty generated a lateral positive evoked potential in a temporal window between 360 and 1225 msec.

De Tommaso, Sardaro, and Livrea (2008) also used artworks that were rated as beautiful, neutral, and ugly. They found that gazing at beautiful paintings raised pain thresholds and at the same time inhibited P2 evoked potentials. The generators of the P2 potentials were localized to the anterior cingulate. These results suggest that aesthetic objects even in this laboratory setting are sufficiently engaging as to distract participants from unpleasant environmental stimuli.

One might be disheartened that these studies, all investigating aesthetics, report different patterns of activation. Nadal et al. (2008) suggest that the results of these studies are compatible within the general model (see Figure 18.2) that I proposed (Chatterjee 2003b). Engaging visual properties of paintings increase activity within ventral visual cortices (Vartanian and Goel 2004). Aesthetic judgments activate parts of dorsolateral prefrontal and medial prefrontal cortices (Cela-Conde et al. 2004; Jacobsen et al. 2005), and emotional responses to these stimuli activate the orbitofrontal (Kawabata and Zeki 2004; Jacobsen et al. 2005) as well as anterior cingulate (Kawabata and Zeki 2004; Vartanian and Goel 2004; de Tommaso, Sardaro, and Livrea 2008) cortices.

18.7   WHY ART?

Returning to the questions posed at the beginning of this chapter, why do people stand in hermetically sealed buildings that we call museums and stare at patches of paint on canvases? And why at some canvases, and not others? Any answer to these questions is necessarily speculative. I suggest that three factors are at play: the drive to beauty, the aesthetic attitude, and the institutional frameworks that promote and display art.

18.7.1 DRIVE FOR BEAUTY

Most people are drawn to beauty. As we have suggested in our own work, neural responses to attractive faces occur automatically, even when people are not explicitly judging beauty (Chatterjee et al. 2009). Three kinds of evolutionary arguments are made for the attraction to beauty. The first and most obvious is the way that beauty influences mate selection. When it comes to faces and bodies, the link between mate selection and beauty is clear (Rhodes et al. 2002). Attractive features represent phenotypic attributes that are desirable in selecting mates, such as genetic health and levels of immunocompetence (Etcoff 1999; Grammer et al. 2003; Penton-Voak et al. 2001; Perrett et al. 1998; Symons 1979; Thornhill and Gangestad 1999). (Also see Chapter 17 for more discussion of genetic health vis a vis olfactory phenotypes.) In this view, the nervous system has evolved to be attracted to specific configurations of facial features that signal "good genes," configurations that we have come to regard as beautiful. A variation of this view is the "costly signal" proposal (Zahavi and Zahavi 1997). Male birds attract female birds with extravagant plumage or elaborate songs that appear to be maladaptive. They interfere with movement and also attract predators. The costly signal proposal is that such displays advertise the unusual vigor of the displayer. The displayer can afford to indulge in these seemingly maladaptive behaviors because they are so fit to begin with. Art making requires considerable time and effort and, as a costly display, would similarly advertise one's fitness in the competition for mates.

Others have argued that conceptualizing art as derived from an attraction to beauty that is directly linked to mating behavior takes an unnecessarily narrow stance on its adaptive significance for humans. Dissanayake (2008) marshals considerable evidence for art's role in promoting social cohesion. She rejects relatively recent and specifically Western notions embedded in aesthetic discussions, such as the importance of novelty or individual creativity, and takes an ethnographic view of the adaptive significance of art to humanity. For her, the behavior of "making special" is critical to art. Ordinary objects, movements, patterns, and sounds are transformed into something extraordinary by exaggeration, repetition, embellishment, and so on (Brown and Dissanayake 2009). Beauty, virtuosity, costliness, and emotional investment are all ingredients in the process of making something special. By focusing on the ritualistic nature of art making and appreciation she emphasizes the adaptive importance of art in enhancing cooperation, and encouraging cohesion and continuity within local societies.

A different kind of argument for why we have come to regard things as beautiful has to do with the nature of mental processing involved when apprehending objects (Rentschler et al. 1999). On this account preferences arise as a by-product of a general information-processing mechanism. As mentioned before, a leading candidate for such a mechanism is the extraction of a prototype, or the central exemplar of a category. People prefer prototypes of different kinds of stimuli, such as color (Martindale and Moore 1988) and music (Smith and Melara 1990). Faces would presumably be another category of stimuli subject to this biased preference for prototypes (Halberstadt and Rhodes 2000). A variation on the information-processing account for preference is the idea that people prefer to look at things that are processed "fluently." Fluency, or the ease with which one processes objects, is rendered by specific physical features of objects as well their conceptual characteristics. Features of the object, such as symmetry and figure ground relationships, as well as the experiences of the viewer, influence fluent processing. The important point for our discussion is that processing fluency is associated with a positive affective response and aesthetic pleasure (Reber, Schwarz, and Winkielman 2004; Armstrong and Detweiler-Bedell 2008). People like what they process easily. Thus, familiarity as established by mere exposure and which contributes to processing fluency influences people's preferences for simple displays in laboratories (Moreland and Zajonc 1976). These influences also extend beyond the laboratory. Familiarity also influences which impressionist paintings are regarded highly (Cutting 2007).

Given that specific configurations of physical objects contribute to the experience of beauty, regardless of whether this contribution is driven by mating desires and rituals, promotes social cohesion, or facilitates processing, how do these experiences relate to the aesthetic experience? Herein lies a paradox, as I describe below.

18.7.2 THE AESTHETIC ATTITUDE

Evolutionary arguments for the importance of beauty emphasize its adaptive significance. Mate selection, social cohesion, and better information processing all have utility. The point of adaptation is to be useful in propagating the species. This utilitarian casting to aesthetics is at odds with an idea proposed in the eighteenth century (Kant 1790/1987) that the aesthetic attitude is one of "disinterested interest." While Kant's idea is by no means agreed on by all aestheticians, I base my speculations on its central role. Aesthetic pleasures are self-contained. They do not intrinsically encompass additional desires. That is not to say that an artwork cannot evoke utilitarian desires, such as the desire to own it or to display it to impress others. However, these rewards are not part of the aesthetic experience.

Can neuroscience contribute to an understanding of disinterested interest? Berridge and colleagues have drawn a distinction between "liking" and "wanting" (Wyvell and Berridge 2000; Berridge and Kringelbach 2008). Liking seems to be instantiated in the nucleus accumbens shell and the ventral pallidum mediated by opioid and GABAergic neurotransmitter systems. By contrast, the mesolimbic dopaminergic system, which includes the nucleus accumbens core, might mediate wanting, and cortical structures such as the cingulate and orbitofrontal cortex contribute to further conscious modulations of these liking and wanting experiences. This liking/wanting distinction is made in a rodent model with experiments using sweet and bitter tastes. Whether it generalizes to humans and in response to visual stimuli remains to be seen. However, the idea of a self-contained reward system could be the neural basis for aesthetic disinterested interest.

I suggest that neural circuitry for liking is an exaptation within our reward systems. That is, it is co-opted within our reward systems, cleaved off related reward systems with clear utilitarian designs of satisfying appetitive desires. Perhaps it developed to allow humans to maintain some distance from the objects of desire and has now come to serve the aesthetic attitude. However, at the extremes of aesthetic experiences lies a potential problem: it would be maladaptive. The most profound aesthetic experiences involve a refined liking, often described as awe or feeling the sublime, in which wanting has been tossed aside, but also in which individuals lose themselves in the experience. As mentioned earlier (de Tommaso, Sardaro, and Livrea 2008), they might not even feel pain! Imagine an early human in the savannah struck in rapture at a beautiful sunset, oblivious to the predator lying in wait in the tall grass. Or a modern human crossing the street suddenly immobilized by the soft evening light on a beautiful building, oblivious to cars whizzing by. The cost of the aesthetic experience is rendering the individual vulnerable. Entering an aesthetic attitude is dangerous.

18.7.3 THE INSTITUTIONAL CONTEXT FOR ART

The display and contemplation of objects that evoke aesthetic experiences are better done in safe havens. The havens could be protective caves at Lascaux or sanctuaries of medieval Europe. Until recently much art was displayed in religious institutions. The religious rapture such imagery can evoke is close to the aesthetic attitude. One is lifted beyond utilitarian desires and everyday concerns and in fact lifted beyond oneself. Religious meditations while gazing at visual icons serve as a vehicle to spirituality (Nouwen 1987).

In secular societies, I suggest that private collections and museums play the role of providing a sanctuary within which one can safely contemplate images. They also bracket aesthetic experiences and publicly recognize the sense in which artworks are "special," as described by Dissanyake (2008). Of course providing a sanctuary for and bracketing that which is special are not the only roles for museums. As discussed extensively by contemporary theoreticians, museums are embedded within a broader "art world" that encompasses a rich environment of comment, critique, and internal dialogues (Dickie 1969; Danto 1964), and is far removed from getting lost in a beautiful sunset. Thus, within internal dialogues of the art world, things regarded as ugly or incomprehensible by most might be highlighted and promoted. The drive to beauty that I suggest is the initiating trigger for artistic experiences might be regarded as naïve by some that establish institutional norms for art. But these movements are recent institutional embellishments on general human tendencies. Evolutionary biology in general and neuroscience in particular has little to contribute to this level of analysis of conceptions of art.

18.8   CONCLUDING COMMENTS

The neuroscience of visual aesthetics is in its infancy. With a field so wide open, progress in any direction would be an advance. Here, I suggest four areas worthy of focus. (1) A finer-grained exploration of the nature and contributions of sensations to aesthetic experiences. As already mentioned, visual art like any visual object can be decomposed into distinct attributes, such as color, line, texture, form, and so on. How do these attributes contribute to the aesthetic experience? How do we measure the contributions of these attributes? We have begun to develop a scale to do so for lesion studies (Chatterjee et al. 2010), but much work needs to be done.

How much of the aesthetic experience resides in a perceptual experience and how much resides in the emotional response to artwork? Paintings of landscapes are likely to activate the parahippocampus, still-lives the lateral occipital cortex, and portraits the fusiform gyrus. Does beauty modify these activations further? Perhaps these activations simply reflect category-specific activations evoked by perception itself, and the aesthetic work is done within reward systems. However, many feel that we perceive beautiful objects more vividly than non-beautiful objects. Some studies show neural responses to beauty within the ventral occipito-temporal cortex. Are these activations modulated by attention or is there an independent aesthetic factor that modulates neural activity? Moreover, within the ventral visual cortex, do general "visual beauty detectors" exist?

(2) Understanding the nature of aesthetic judgment. People vary in their aesthetic sensitivities. Aesthetic sensitivity has been referred to as a "T-factor" for taste (Eysenck 1941; Eysenck and Hawker 1994). People can also develop taste with training. Behavioral studies consistently show differences in the way that art-experienced individuals and art-naïve individuals look at artwork (Locher, Stappers, and Overbeeke 1999). Understanding the neural basis for taste and the ways aesthetic judgment might be modified with training would be of great interest.

The studies conducted thus far suggest that parts of the dorsolateral and medial prefrontal cortex are involved in making aesthetic judgments. These studies are unable to distinguish whether these brain activations are specific to aesthetic judgments or are part of neural systems that makes judgments regardless of the domain under consideration. Do aesthetic judgments engage neural circuits that are not engaged in other judgments?

(3) Characterizing the "reward" of aesthetic experiences. The imaging studies reviewed here implicate the orbitofrontal cortex, the anterior and posterior cingulate, the ventral striatum including the nucleus accumbens, the caudate, and the amygdala in one or another study as mediating the emotional response to beauty or to artwork. Presumably these structures differ in their functions (Berridge and Kringelbach 2008). Clearly a better sense of how these structures contribute to an overall emotional aesthetic experience is needed. Such an understanding would be necessary, for example, to address what is meant by "sublime," an emotional experience mentioned frequently in aesthetics (Kant 1790/1987), but not in affective neuroscience.

(4) Constraining evolutionary theories. Evolutionary biology and psychology provide principles that help organize and interpret complex behavior. The universal nature of producing ornamentations and being pleased by these ornamentations certainly encourages evolutionary analyses of art. The challenge for evolutionary aestheticians is how to constrain this theorizing and test hypotheses. Evidence is gathered from disparate sources to argue for example whether art is adaptive (and in what way) or whether it is a by-product of adaptation. Comparative neuroscience has not contributed to the evolutionary discussions about art, but may provide some constraints on the theorizing. Even if such contributions from neuroscience were to be made, these hypotheses remain inherently post

hoc (not unlike my speculations on "why art?"). The research enterprise lacks clear prospective tests of hypotheses and explicit conditions of falsifiability of experimental sciences.

In conclusion, aesthetics and a concern for beauty are central human experiences. They are integrally related to a specific set of sensations and rewards. While a small but dedicated group of investigators has been engaged in empirical aesthetics since the mid-nineteenth century, only recently has neuroscience wandered into this area. Some of the most basic questions about neuroaesthetics remain to be answered. The basic frameworks and methods from cognitive and affective neuroscience are in place for neuroaesthetics to mature as a science.

ACKNOWLEDGMENT

I would like to thank Lisa Santer for a critical reading of an earlier draft of this chapter and Jay Gottfried for pushing me to speculate beyond my natural inclinations.

REFERENCES

Aharon, I., Etcoff, N., Ariely, D., Chabris, C.F., O'Connor, E., and Breiter, H.C. 2001. Beautiful faces have variable reward value: fMRI and behavioral evidence. *Neuron* 32:537–51.

Armstrong, T., and Detweiler-Bedell, B. 2008. Beauty as an emotion: The exhilarating prospect of mastering a challenging world. *Review of General Psychology* 12 (4): 305–29.

Berridge, K., and Kringelbach, M. 2008. Affective neuroscience of pleasure: Reward in humans and animals. *Psychopharmacology* 199 (3):457–80.

Biederman, I., and Cooper, EE. 1991. Priming contour-deleted images: Evidence for intermediate representations in visual object recognition. *Cognitive Psychology* 23:393–419.

Breiter, HC., Aharon, I., Kahneman, D., Dale, A., and Shizgal, P. 2001. Functional imaging of neural response to expectancy and experience of monetary gains and losses. *Neuron* 30:619–39.

Brown, S., and Dissanayake, E. 2009. The arts are more than aesthetics: Neuroaesthetics as narrow aesthetics. In *Neuroaesthetics*, eds. M. Skov, and O. Vartanian, pp. 43–57. Amityville, NY: Baywood Publishing Company.

Carroll, N., ed. 2000. *Theories of Art Today*. Madison, WI: The University of Wisconsin Press.

Cavanagh, P. 2005. The artist as neuroscientist. *Nature* 434 (7031): 301–7.

Cela-Conde, C.J., Marty, G., Maestu, F., Ortiz, T., Munar, E., Fernandez, A., Roca, M., Rossello, J., and Quesney, F. 2004. Activation of the prefrontal cortex in the human visual aesthetic perception. *PNAS* 101 (16): 6321–25.

Chatterjee, A. 2002. Universal and relative aesthetics: A framework from cognitive neuroscience. Paper read at International Association of Empirical Aesthetics, at Takarazuka, Japan.

———. 2003a. Neglect. A disorder of spatial attention. In *Neurological Foundations of Cognitive Neuroscience*, ed. M.D'Esposito, pp. 1–26. Cambridge, MA: The MIT Press.

———. 2003b. Prospects for a cognitive neuroscience of visual aesthetics. *Bulletin of Psychology and the Arts* 4:55–59.

———. 2004. The neuropsychology of visual artists. *Neuropsychologia* 42:1568–83.

———. 2006. The neuropsychology of visual art: Conferring capacity. *International Review of Neurobiology* 74:39–49.

———. 2009. Prospects for a neuropsychology of art. In *Neuroaesthetics*, eds. M. Skov and O. Vartanian, pp. 131–43. Amityville, NY: Baywood Publishing Company.

Chatterjee, A., Thomas, A., Smith, S.E., and Aguirre, G.K. 2009. The neural response to facial attractiveness. *Neuropsychology* 23:135–143.

Chatterjee, A., Widick, P., Sternschein, R., and Smith II, W.B. 2010. The assessment of art attributes. *Empirical Studies of the Arts* 28 (2): 207–22.

Conway, B.R., and Livingstone, M.S. 2007. Perspectives on science and art. *Current Opinion in Neurobiology* 17 (4): 476–82.

Cunningham, M.R., Barbee, A.P., and Philhower, C.L. 2002. Dimensions of facial physical attractiveness: The intersection of biology and culture. In *Facial Attractiveness. Evolutionary, Cognitive, and Social Perspectives*, eds. G. Rhodes and L. Zebrowitz. Westport, CT: Ablex.

Cutting, J.E. Mere exposure, reproduction, and the impressionist canon. In *Partisan Canons*, ed. A. Brzyski, pp. 79–93. Durham: Duke University Press.

Danto, A.C. 1964. The artworld. *Journal of Philosophy* 61: 571–84.

de Tommaso, M., Sardaro, M. and Livrea, P. 2008. Aesthetic value of paintings affects pain thresholds. *Consciousness and Cognition* 17:1152–62.

Delgado, M.R., Nystrom, L.E., Fissell, K., Noll, D.C., and Fiez, J.A. 2000. Tracking the hemodynamic responses for reward and punishment. *Journal of Neurophysiology* 84:3072–77.

Dickie, G. 1969. Defining art. *American Philosophical Quarterly* 6:253–56.

Dion, K., Berscheid, E., and Walster, E. 1972. What is beautiful is good. *Journal of Personality and Social Psychology* 24:285–90.

Dissanayake, E. 2008. The arts after Darwin: Does art have an origin and adaptive function? In *World Art Studies: Exploring Concepts and Approaches*, eds. K. Zijlemans, and W. van Damme, pp. 241–63. Amsterdam: Valiz.

Elliott, R., Friston, K.J., and Dolan, R.J. 2000. Dissociable neural responses in human reward systems. *Journal of Neuroscience* 20:6159–65.

Enquist, M., and Arak, A. 1994. Symmetry, beauty and evolution. *Nature* 372 (6502): 169–72.

Etcoff, N. 1999. *Survival of the Prettiest*. New York: Anchor Books.

Eysenck, H.J. 1941. The empirical determination of an aesthetic formula. *Psychological Review* 48 (1): 83–92.

Eysenck, H.J., and Hawker, G.W. 1994. The taxonomy of visual aesthetic preferences: An empirical study. *Empirical Studies of the Arts* 12 (1): 95–101.

Farah, M.J. 2000. *The Cognitive Neuroscience of Vision*. Malden, MA: Blackwell.

Fechner, G. 1876. *Vorschule der Aesthetik*. Leipzig: Breitkopf & Hartel.

Grammer, K., Fink, B., Moller, A.P., and Thornhill, R. 2003. Darwinian aesthetics: Sexual selection and the biology of beauty. *Biological Review* 78:385–407.

Grammer, K., and Thornhill, R. 1994. Human (Homo sapiens) facial attractiveness and sexual selection: The role of symmetry and averageness. *Journal of Comparative Psychology* 108 (3): 233–42.

Grossberg, S., Mingolla, E. and Ros, W.D. 1997. Visual brain and visual perception: How does the cortex do perceptual grouping? *Trends in Neuroscience* 20:106–11.

Halberstadt, J., and Rhodes, G. 2000. The attractiveness of non-face averages: Implications for an evolutionary explanation of the attractiveness of average faces. *Psychological Science* 11:285–89.

Helmut, L., Benno, B., Andries, O., and Dorothee, A. 2004. A model of aesthetic appreciation and aesthetic judgments. *British Journal of Psychology* 95:489–508.

Hofel, L., and T. Jacobsen. 2007. Electrophysiological indices of processing aesthetics: Spontaneous or intentional processes? *International Journal of Psychophysiology* 65 (1): 20–31.

Ishai, A. 2007. Sex, beauty and the orbitofrontal cortex. *International Journal of Psychophysiology* 63 (2): 181–85.

Jacobsen, T., Buchta, K., Kohler, M., and Schroger, E. 2004. The primacy of beauty in judging the aesthetics of objects. *Psychological Reports* 94:1253–60.

Jacobsen, T., Schubotz, R.I., Hofel, L., and v Cramon, D.Y. 2005. Brain correlates of aesthetic judgments of beauty. *Neuroimage* 29:276–85.

Jones, D., and Hill, K. 1993. Criteria of facial attractiveness in five populations. *Human Nature* 4 (3): 271–96.

Kampe, K.K.W., Frith, C.D., Dolan, R.J., and Frith, U. 2001. Reward value of attractiveness and gaze. *Nature* 413:589.

Kant, I. 1790/1987. *Critique of Judgment*, trans. W.S. Pluhar. Indianapolis: Hackett.

Kawabata, H., and Zeki, S. 2004. Neural correlates of beauty. *Journal of Neurophysiology* 91 (4): 1699–1705.

Kenealy, P., Frude, N., and Shaw, W. 1988. Influence of children's physical attractiveness on teacher expectations. *The Journal of Social Psychology* 128:373–83.

Kranz, F., and Ishai, A. 2006. Face perception is modulated by sexual preference. *Current Biology* 16:63–68.

Langlois, J.H., Kalakanis, L.E., Rubenstein, A.J., Larson, A.D., Hallam, M.J., and Smoot., M.T. 2000. Maxims or myths of beauty: A meta-analytic and theoretical review. *Psychological Bulletin* 126:390–423.

Langlois, J.H., Ritter, J.M., Roggman, L.A, and Vaughn, L.S. 1991. Facial diversity and infant preferences for attractive faces. *Developmental Psychology* 27 (1): 79–84.

Lerner, R., Lerner, J., Hess, L., and Schwab, J. 1991. Physical attractiveness and psychsocial functioning among early adolescents. *Journal of Early Adolescence* 11:300–20.

Livingstone, M. 2002. *Vision and Art: The Biology of Seeing*. New York: Abrams.

Livingstone, M., and Hubel, D.H. 1987. Psychophysical evidence for separate channels for the perception of form, color, movement, and depth. *The Journal of Neuroscience* 7:3416–68.

Livingstone, M. and Hubel, D. 1988. Segregation of form, colour, movement, and depth: Anatomy, physiology, and perception. *Science* 240:740–49.

Locher, P.J., Stappers, P.J., and Overbeeke, K. 1999. An empirical evaluation of the visual rightness theory of pictorial composition. *Acta Psychologica* 103 (3): 261–80.

Locher, P., and Nagy, Y. 1996. Vision spontaneously establishes the percept of pictorial balance. *Empirical Studies of the Arts* 14 (1): 17–31.

Marr, D. 1982. *Vision. A Computational Investigation into the Human Representation and Processing of Visual Information*. New York: W.H. Freeman and Company.

Martindale, C., and Moore, K. 1988. Priming, prototypicality, and preference. *Journal of Experimental Psychology: Human Perception and Performance* 14:661–67.

Moreland, R.L., and Zajonc, R.B. 1976. A strong test of exposure effects. *Journal of Experimental Social Psychology* 12:170–79.

Motter, B.C. 1993. Focal attention produces spatially selective processing in visual cortical areas V1, V2, and V4 in the presence of competing stimuli. *Journal of Neurophysiology* 70:909–19.

———. 1994. Neural correlates of attentive selection for color or luminance in extrastriate area V4. *Journal of Neuroscience* 14:2178–89.

Nadal, M., Munar, E., Capo, M.A., Rosselo, J., and Cela-Conde, C.J. 2008. Towards a framework for the study of the neural correlates of aesthetic preference. *Spatial Vision* 21 (3): 379–96.

Nouwen, H.J.M. 1987. *Behold the Beauty of the Lord: Praying with Icons*. Notre Dame, IN: Ave Maria Press.

O'Doherty, J., Kringelbach, M.L., Rolls, E.T., Hornack, J., and Andrews, C. 2001. Abstract reward and punishment representations in the human orbitofrontal cortex. *Nature Neuroscience* 4:95–102.

O'Doherty, J., Winston, J., Critchley, H., Perret, D., Burt, D.M., and Dolan, R.J. 2003. Beauty in a smile: The role of orbitofrontal cortex in facial attractiveness. *Neuropsychologia* 41:147–55.

Ognjenovic, P. 1991. Processing of aesthetic information. *Empirical Studies of the Arts* 9 (1):1–9.

Olson, I.R., and Marshuetz, C. 2005. Facial attractiveness is appraised in a glance. *Emotion* 5:498–502.

Palermo, R., and Rhodes, G. 2007. Are you always on my mind? A review of how face perception and attention interact. *Neuropsychologia* 45:75–92.

Penton-Voak, I.S., Jones, B.C., Little, A.C., Baker, S., Tiddeman, B., Burt, D.M., and Perrett, D.I. 2001. Symmetry, sexual dimorphism in facial proportions and male facial attractiveness. *Proceedings of the Royal Society of London: Series B* 268:1617–-23.

Perrett, D.I., Lee, K.J., Penton-Voak, I., Rowland, D., Yoshikawa, S., Burt, D.M., Henzi, S.P., Castles, D.L., and Akamatsu, S. 1998. Effects of sexual dimorphism on facial attractiveness. *Nature* 394:884–87.

Perrett, D.I., May, K.A., and Yoshikawa, S. 1994. Facial shape and judgements of female attractiveness. *Nature* 368:239–42.

Ramachandran, V.S., and Hirstein, W. 1999. The science of art: A neurological theory of aesthetic experience. *Journal of Consciousness Studies* 6 (6–7):15–51.

Reber, R., Schwarz, N., and Winkielman, P. 2004. Processing fluency and aesthetic pleasure: Is beauty in the perceiver's processing experience? *Personality & Social Psychology Review (Lawrence Erlbaum Associates)* 8 (4): 364–82.

Rentschler, I., Jüttner, M., Unzicker, A., and Landis, T. 1999. Innate and learned components of human visual preference. *Current Biology* 9 (13): 665–71.

Rhodes, G., Harwood, K., Yoshikawa, S., Miwi, N., and McLean, I. 2002. The attractiveness of average faces: Cross-cultural evidence and possible biological basis. In *Facial Attractiveness. Evolutionary, Cognitive, and Social Perspectives*, eds. G. Rhodes and L. Zebrowitz. Westport, CT: Ablex.

Ricci, R., Vaishnavi, S., and Chatterjee, A. 1999. A deficit of preattentive vision: Experimental observations and theoretical implications. *Neurocase* 5 (1): 1–12.

Ritts, V., Patterson, M., and Tubbs, M. 1992. Expectations, impressions, and judgments of physically attractive students: A review. *Review of Educational Research* 62:413–26.

Russell, P.A., and George, D.A. 1990. Relationships between aesthetic response scales applied to paintings. *Empirical Studies of the Arts* 8 (1): 15–30.

Schultz, W., Dayans, P., and Montague, P.R. 1997. A neural substrate of prediction and reward. *Science* 275:1593–99.

Senior, C. 2003. Beauty in the brain of the beholder. *Neuron* 38:525–28.

Shulman, G.L., Corbetta, M., Buckner, R.L., Raichle, M.E., Fiez, J.A., Miezin, F.M., and Petersen, S.E. 1997. Top-down modulation of early sensory cortex. *Cerebral Cortex* 7:193–206.

Slater, A., Von Der Schulenburg, C., Brown, E., Badenoch, M., Butterworth, G., Parsons, S.and Samuels. C. 1998. Newborn infants prefer attractive faces. *Infant Behavior and Development* 21 (2): 345–54.

Smith, D.J., and Melara. R.J. 1990. Aesthetic preference and syntactic prototypicality in music: 'Tis the gift to be simple. *Cognition* 34:279–98.

Symons, D. 1979. *The Evolution of Human Sexuality*. Oxford: Oxford University Press.

Thornhill, R., and. Gangestad, S.W 1999. Facial attractiveness. *Trends in Cognitive Sciences* 3 (12): 452–60.

Tinbergen, N. 1954. *Curious Naturalist*. New York: Basic Books.

Ungerleider, L.G., and Mishkin, M. 1982. Two cortical visual systems. In *Analysis of Visual Behavior*, eds. D.J. Ingle, M.A. Goodale, and R.J.W. Mansfield. Cambridge: MIT Press.

Van Essen, D.C., Feleman, D.J., DeYoe, E.A., Ollavaria, J., and Knierman, J. 1990. Modular and hierarchical organization of extrastriate visual cortex in the macaque monkey. *Cold Springs Harbor Symp Quant Biol* 55:679–96.

Vartanian, O., and Goel, V. 2004. Neuroanatomical correlates of aesthetic preference for paintings. *NeuroReport* 15 (5): 893–97.

Vecera, S.P., and Behrmann, M. 1997. Spatial attention does not require preattentive grouping. *Neuropsychology* 11:30–43.

Watanabe, T., Sasaki, Y., Miyauchi, S., Putz, B. Fujimaki, N., Nielsen, M., Takino, R., and Miyakawa, S. 1998. Attention-regulated activity in human primary visual cortex. *Journal of Neurophysiology* 79:2218–21.

Weitz, M. 1956. The role of theory in aesthetics. *Journal of Aesthetics and Art Criticism* 15:27–35.

Wilson, M., and Daly, M. 2004. Do pretty women inspire men to discount the future? *Proceedings of the Royal Society of London* 271:177–79.

Winston, J.S., O'Doherty, J., Kilner, J.M., Perrett, D.I., and Dolan, R.J. 2007. Brain systems for assessing facial attractiveness. *Neuropsychologia* 45:195–206.

Woods, W.A. 1991. Parameters of aesthetic objects: Applied aesthetics. *Empirical Studies of the Arts* 9 (2):105–14.

Wyvell, C.L., and K.C. Berridge. 2000. Intra-accumbens amphetamine increases the conditioned incentive salience of sucrose reward: Enhancement of reward "wanting" without enhanced "liking" or response reinforcement. *Journal of Neuroscience* 20:8122–30.

Zeki, S. 1993. *A Vision of the Brain*. Oxford: Blackwell Scientific Publications.

———. 1999. Art and the brain. *Journal of Consciousness Studies* 6:76–96.

———. 1999. *Inner Vision: An Exploration of Art and the Brain*. New York: Oxford University Press.

19 Music

David H. Zald and Robert J. Zatorre

CONTENTS

19.1 INTRODUCTION

Why do humans take pleasure from sequences of tones? This question has perplexed many scientists and philosophers over the years. The great science fiction writer Arthur C. Clarke (1953) went so far as to suggest that an alien race would be puzzled and astonished by the amount of time we spend listening to sounds that have no apparent purpose or utility. Yet, humans consistently rank music among the top ten things that bring pleasure, usually above such things as money, food, or art (Dubé and LeBel 2003). So why do temporally organized pitch sequences as found in music bring such pleasure? In the present chapter, we outline some of the critical mechanisms that make music both rewarding and motivating, and attempt to link these factors to recent neuroscientific insights into the networks that mediate reward.

19.2 ORIGINS

Because of the seemingly superfluous nature of music to human survival, many theorists have speculated as to why humans evolved musical abilities at all. Indeed, Darwin (1871) declared, "As neither the enjoyment nor the capacity of producing musical notes are faculties of the least use to man in reference to his daily habits of life, they must be ranked amongst the most mysterious with which he is endowed." Proposed explanations have ranged from advantages in mate selection to the idea that music represented a prelinguistic form of communication (see Fitch 2006 for a discussion of the evidence for and against these hypotheses). At the other end of the spectrum lies the idea that music is a consequence of another set of adaptations, such as those required for language and other higher cognitive functions, rather than a process that was itself adaptive (Gould and Lewontin 1979). Perhaps more extreme, Pinker (1997) suggests that music serves little adaptive function, but provides an ability to "push our pleasure buttons." A more nuanced approach to this question has recently been provided by Patel (2007), who point out that it is not necessary to assume some specific advantage in natural selection in order to explain music's appearance, as it may well be the consequence of other adaptations, but that this would by no means indicate that music hence serves no purpose or is not itself valuable.

From an evolutionary perspective, it is notable that our closest primate relatives show little in the way of developed organized pitch and rhythmic sequencing. Some bird species have been reported to entrain to external sounds, such as musical rhythms, with body movements (Patel et al. 2009; Schachner et al. 2009), although there is as yet no evidence that they respond in terms of human-like metrical representations, that they do so to any self-produced sounds, or that this a natural behavior in the wild. Among primates, several ape species will beat their chest or other surfaces (Fitch 2006), but they do not entrain to rhythmic sounds. Probably the closest example of pitch sequencing occurs in gibbons, who make "duetting" calls together (Geissmann 2000). However, there is no evidence that gibbons or other nonhuman primates routinely learn melodies. Indeed, to the extent that it has been tested, nonhuman primates show little enjoyment or preference for music. When presented with a choice between musical stimuli and silence, tamarins and marmosets prefer silence (McDermott and Hauser 2007), suggesting the uniqueness of human's development of musical interests. So in spite of that harmonious adage, music may *not* tame the savage beast.

19.3   NEURAL CIRCUITRY INVOLVED IN MUSICAL PLEASURE

Before discussing theoretical proposals for the pleasure-inducing effects of music, we will review the empirical evidence that informs current hypotheses. At present, there is a small, but growing, literature that directly addresses the neural substrates for music-induced pleasure. Although this field is still in its early stages, the interpretation of these studies rests on a reasonably solid foundation of research in other domains. In particular, the large body of animal literature on pathways associated with auditory processing, reward, and motivation (as explored in Chapter 9) provides a critical basis for interpreting these studies. But since musical enjoyment appears to be a uniquely human trait, it is only in paradigms adapted to studying human responses that we can really address this issue. Data on this topic come from three primary sources: lesion studies, functional neuroimaging, and psychophysiology.

19.3.1   LESION STUDIES

We begin with an evaluation of the human lesion literature in relation to musical emotion. Whereas a relatively extensive experimental literature exists on the effects of brain lesions for music perception (for review, see Stewart et al. 2006), few of these studies have systematically explored musical emotions in general, or the emotions associated with pleasure in particular. There are reasonably clear reports that perceptual disorders can lead to a complete loss of pleasure derived from music; in cases where musical perception is so disordered that it sounds completely distorted, such a lack of pleasure is of course to be expected (e.g., Griffiths et al. 1997). In some instances, this phenomenon may also be part of a broader auditory agnosia where not only music, but all sounds, become incomprehensible to the patient (e.g., Habib et al. 1995; also see Chapter 10 in this volume). The lesion sites associated with such disorders are typically in the posterior temporal region, often bilaterally or sometimes on the right alone (Stewart et al. 2006). A clear example of this inability to derive emotional content because of perceptual loss was demonstrated by Peretz et al. (2001), who found that a patient unable to distinguish consonance from dissonance was also unable to indicate the degree of pleasantness or unpleasantness associated with dissonance. Interestingly, the lesions in this individual, which were primarily within auditory cortices bilaterally, did not overlap at all with the cortical and paralimbic regions associated with the evaluation of affective dissonance identified by Blood et al. (1999) in a neuroimaging study of healthy individuals (see below). This finding strongly suggests that the patient's inability to evaluate the affective value of dissonant music was a consequence of her perceptual deficit, since the neural regions associated with affect processing per se were intact, and since she had no generalized deficit in processing emotions in other domains.

In distinction to such cases, there are also reports of primary musical affect disorders subsequent to brain lesions. For example, Griffiths et al. (2004) report a case of a man who lost his affective response to music (in particular the experience of chills to some of his favorite pieces), following an infarct involving the left amygdala and insular regions. Unlike the patients described above, this individual had no disorder of musical perception that could account for the loss of pleasure he experienced. Another example is provided by a well-controlled experimental study of patients with amygdala damage (Gosselin et al. 2005) who rated music that usually elicits fear as expressing less fear than a control group. There was no accompanying perceptual disorder in this population either, leading to the conclusion that music perception and affective responses can be dissociated. In a parallel study, Gosselin et al. (2006) found that patients with damage to the parahippocampal region were relatively insensitive to the unpleasantness of dissonant music. The investigators reported that the degree of damage to this structure, but not to other structures (amygdala or hippocampus), correlated with affective judgments, indicating a likely dissociation between the roles of these structures. A similar finding regarding unpleasantness judgments was more recently reported by Khalfa et al. (2008), who additionally found that patients with left medial temporal damage were less likely to perceive the positive affect in music pieces that are usually rated as happy. It should be noted that these impairments may not be specific to music, but may instead reflect more general deficits in affective processing. For example, amygdala damage such as studied by Gosselin et al. (2006) also usually leads to abnormally low affective arousal in response to fearful faces, suggesting the presence of a general multimodal deficit in response to fear, rather than a deficit specific to music.

The picture that emerges from these studies is that musical perception and emotion can be dissociated in one direction but not the other. That is, perceptual deficits, if sufficiently severe, lead to a loss of musical emotional response, whereas an isolated loss of emotional response can occur in the absence of any perceptual disorder. What this likely means is that the affective evaluation depends on a distinct neural system from that required for perceptual analysis, but that this affective system depends upon input from the perceptual system. This would explain why damage to superior temporal or frontal cortices can wipe out both musical perception and emotion, but damage to other emotion-related structures only affects emotional and not perceptual musical abilities.

19.3.2 Neuroimaging

Neuroimaging studies of musical emotion extend the picture provided by clinical studies and also allow a relatively direct examination of the neural substrates that may be involved in healthy individuals. As with any technique, however, attention to experimental design is critical and converging evidence from other sources is also essential. One critical design issue which is especially important in neuroimaging is the need to validate the emotional experience of the person being studied. Not all imaging experiments have determined whether the listeners being scanned in fact experienced the intended emotion, which would seem an essential element for understanding the observed findings. But obtaining judgments of emotions from listeners is not easy, because evaluating one's affective response may itself change the nature of the emotion felt. Such judgments may also be easily biased, as there may be a demand characteristic in the experiment, leading to participant responses based on their expectations or normative conventions, rather than their actual experience of a given emotion. In this respect, the psychophysiological measures mentioned in the next section are especially useful as providing both an objective and reliable way to determine emotion without conscious intervention from the individual (it is difficult, for instance, to "fake" changes in skin conductance at will). But since psychophysiological measures are relatively non-specific, some type of subjective behavioral appraisal will almost certainly still be necessary in most circumstances.

One of the first studies to apply functional imaging to musical emotion was carried out by Blood et al. (1999), who examined changes in regional cerebral blood flow elicited by varying degrees of pleasant and unpleasant music (Figure 19.1). Pleasantness was specifically modulated by parametric manipulations of dissonance. The study uncovered a pattern of reciprocal brain activity in paralimbic and neocortical areas as a function of changes in dissonance, and hence pleasantness, as measured via behavioral ratings. Specifically, increased pleasantness was associated with recruitment of subcallosal and orbitofrontal cortex, whereas increasing dissonance, leading to unpleasant evaluations, resulted in parahippocampal cortical activity (Figure 19.1c). The latter structure has strong connections to, and strong functional connectivity with, the amygdala (Stein et al. 2007) and has been found to be active during unpleasant visual stimulation (Lane et al. 1997), although with less consistency than the amygdala. As noted above, the study by Gosselin et al. (2006) indicates that the parahippocampus is critical for dissonance-induced unpleasant emotion, as patients with lesions in this area fail to show the normal subjective response to dissonant music despite intact musical perception. Blood et al.'s observation that the ventromedial prefrontal regions are active in association with pleasant responses is also consistent with a large body of evidence from neuroimaging studies that indicate higher activation in ventromedial regions during processing of positive vs. negative stimuli or outcomes (Kringelbach and Rolls 2004). It may be noted that this pattern of positively valenced responses is not completely universal across stimulus modalities or ventromedial subregions (e.g., see Gottfried et al. 2006), but it is striking in its frequency. The reciprocal coupling between blood flow increases in ventromedial prefrontal cortex and decreases in the parahippocampal gyrus in the study by Blood et al. (1999) suggests the presence of a functional

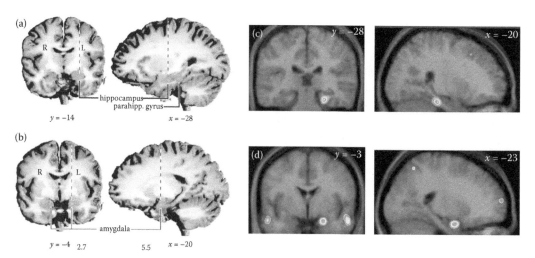

FIGURE 19.1 (See Color Insert) Changes in regional cerebral blood flow during unpleasant (dissonant) and pleasant music. (a and b) Blood oxygenation signal increases associated with listening to unpleasant dissonant music: activation was detected in the hippocampus, parahippocampal gyrus, and amygdala. (Modified from Koelsch, S. et al.: Investigating emotion with music: An fMRI study. *Human Brain Mapping*. 27, 239, 2006. Copyright Wiley-VCH Verlag GmbH & Co. KGaA. With permission.) (c) Increases in cerebral blood flow in the parahippocampal area associated with unpleasant dissonant music. (Modified by permission from Macmillan Publishers Ltd. [*Nature Neuroscience*] Blood, A.J. et al. 2, 382, copyright 1999.) (d) Decreases in cerebral blood flow in amygdala associated with highly pleasurable music. (Modified from Blood, A.J., Zatorre, R.J., *Proc Natl Acad Sci*, 98, 11818, 2001.) Note similarity across studies in recruitment of parahippocampal region (a and c) to negatively valenced music; and in activation or deactivation of amygdala to negatively or positively valenced music, respectively (b and d).

network, such that pleasant or unpleasant stimuli simultaneously influence brain activity in opposite directions in different regions.

The essential findings of Blood et al. (1999) were replicated in a subsequent fMRI neuroimaging study by Koelsch et al. (2006), who reported very similar patterns of reciprocal brain activity using a set of stimuli in which unpleasantness was also generated by dissonance (Figure 19.1a, b). A particularly interesting finding of this latter study is that neural responses associated with both pleasant and unpleasant music tended to increase over time, indicating that there is a certain time course to the affective response. Unlike Blood et al. however, Koelsch and colleagues also observed significant modulation of the amygdala, such that there was relative activation of this structure for unpleasant stimuli, coupled with relative deactivation for pleasant music. A subsequent paper from this group extended this result by demonstrating that the amygdala is activated not only by dissonance, but also by unexpected consonant chords when they are perceived as unpleasant due to the harmonic context (Koelsch 2008b, 2008c); these findings are consistent with this region's general sensitivity to negatively valenced stimuli (see Zald 2003 for review).

Although the above studies intended to examine both pleasure and displeasure, the nature of the stimuli used makes it more likely that they primarily probed the unpleasant side of the response pattern, since the consonant stimuli used to elicit pleasure were not selected specifically so as to maximize the individual listeners' pleasure. In these studies, listeners rated the consonant music as more pleasant than the dissonant music but this does not mean that they were experiencing a high state of pleasure as such. This brings us to a critical aspect of music which is sometimes overlooked: individual differences in preferences. It is sometimes difficult to elicit strong and consistent emotions if musical pieces are selected in advance for a study without taking into account different tastes amongst listeners (Carter, Mintun, and Cohen 1995; Thaut and Davis 1993). Among the individual factors that are important to consider in eliciting musical pleasure are age, musical acculturation, personality, and musical training (Grewe et al. 2007b), to say nothing of social factors that no doubt influence musical pleasure. Although listeners can broadly agree that certain musical excerpts exhibit certain emotions (Vieillard et al. 2008), unanimity in emotion elicited by music is not at all guaranteed (Gabrielsson 2001). What evokes musical bliss in one person may be anathema to another (readers of this chapter are invited to compare their musical recording collection to that of their parents or to that of their children, for an informal confirmation of this hypothesis).

One way to take individual tastes into account is to allow participants of a study to select their own favorite music that they know elicits strong emotional reactions. This approach is not without difficulties, since individual responses may be contaminated by associations and memories. Furthermore, when listening to familiar music it is more difficult to dissociate anticipatory reactions from more direct reactions. Nonetheless, it provides a way to probe intense pleasure, and this was the aim of a PET study by Blood and Zatorre (2001), who examined the neural substrates associated with musical chills. Blood and Zatorre had participants select music that they knew to elicit chills, excluding any with verbal content or explicit associations that may have contributed to their emotional impact. Taking advantage of the fact that one person's favorite music leaves another one cold, they paired their subjects so that the chills-inducing music used in one person became the control for the other, and vice versa. This manipulation ensured that across the entire sample none of the differences measured in the neural response could be driven by physical differences in the stimuli used (a recurring problem in many of the above-cited studies). A number of changes in psychophysiological variables (heart rate, respiration, muscle tension) were observed when listeners heard their own music, but not control music, thus validating their subjective reports of the occurrence of chills during several of the scan periods. The study revealed a widespread pattern of blood flow changes in the brain associated with the presence of chills. Activity increases were found in several neocortical sites (insula, supplementary motor area, anterior cingulate, orbitofrontal cortex), as well as in subcortical areas, including notably the ventral striatum (Figure 19.2a) and dorsomedial midbrain. Deactivations were also noted in the amygdala (Figure 19.1d) and adjacent hippocampus, as well as in ventromedial prefrontal cortex.

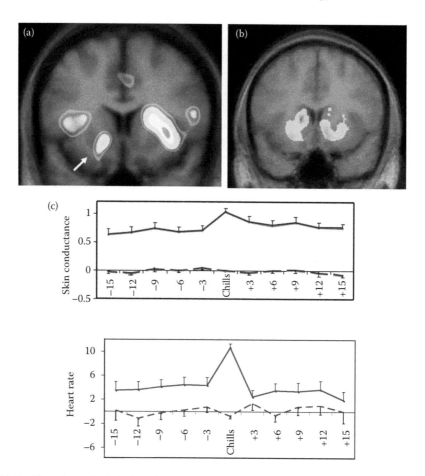

FIGURE 19.2 Neural and physiological changes as a consequence of chill-inducing music. (a) Coronal section showing increased cerebral blood flow response in left ventral striatum (arrow) during highly pleasurable music that elicits chills. (Modified from Blood, A.J., Zatorre, R.J., *Proc Natl Acad Sci*, 98, 11818, 2001.) (b) Coronal section showing increased dopamine release in dorsal and ventral striatum, as measured with raclopride PET, during highly pleasurable music that elicits chills. (The figure is derived from data reported in Salimpoor et al., in press). (c) Psychophysiological tracings indicating temporal profile of skin conductance and heart rate responses to chill-inducing music. Note peak in response coinciding with moment when listeners report chills. Dashed line represents responses to the same musical piece for listeners who did not find that music pleasurable. (Modified from Salimpoor, V. et al. *PloS One*, 4, 1–14, 2009. Creative Commons Attribution License.)

Given the importance of the nucleus accumbens and other striatal regions in reward and motivation, these findings suggest that music may resemble biologically rewarding stimuli in its ability to engage similar neural circuitry. Indeed, the pattern of brain activity associated with intensely pleasurable music is similar in many respects to that seen with other relevant stimuli, such as consumption of chocolate (Small et al. 2001) and even administration of cocaine (Breiter et al. 1997). The neocortical changes also parallel some of the responses seen with these other emotion-inducing paradigms, although it is difficult to know which brain responses are specific and which ones are general, to the extent that they are all simultaneously present. Indeed, this question remains an important issue for future studies to work out, particularly as arousal-related vs. pleasure-related responses are hard to dissociate with this type of experimental design.

Of particular interest are the decreases in amygdala and hippocampus that accompanied the chills, similar to those observed subsequently by Koelsch et al. (2006). This interaction may suggest that the euphoric feeling associated with pleasant music is mediated via gating of behaviorally antagonistic approach/withdrawal mechanisms, given the important role for the amygdala in negative emotional experience, and other aversive emotions. Thus the pleasure of music may be due both to positive engagement of brain areas related to reward and inhibition of areas mediating negative affective states. In considering this hypothesis, we note that many sensory stimuli that are experienced as intense, and even positively arousing, can also lead to activation of the amygdala, such that some investigators interpret amygdala activations in regards to salience rather than valence per se (see Zald 2003 for review). To date, positive music remains one of the only examples of a positively arousing stimulus capable of decreasing amygdala activity. This suggests that music may be unusual in its ability to downregulate the amygdala (although the exact meaning of deactivations in neuro-imaging studies remains open to interpretation).

Engagement of the ventral striatal area during music listening is unlikely to be linked exclusively to the experience of chills. Although simple consonant music such as used by Blood et al. (1999) is evidently not sufficient to recruit this system, more naturalistic music may do so even in the absence of chills, as shown by several studies (Koelsch et al. 2006; Brown, Martinez, and Parsons 2004; Menon and Levitin 2005; Mitterschiffthaler et al. 2007). These studies all find activity within the ventral striatum during presentation of music that is enjoyable, or happy, although it is difficult to be certain that it is the enjoyment, as opposed to some other feature of the music (attentional engagement, arousal, etc.) that is relevant if one only compares music to a "resting state" (Brown, Martinez, and Parsons 2004), or compares normal music to scrambled music (Menon and Levitin 2005), because such analyses are nonspecific. There is also the related problem of comparing conditions in which the stimuli differ in terms of physical parameters, so that it is difficult to dissect out whether a given effect is stimulus-driven, or related to the emotion; this could be why three of these studies (Brown, Martinez, and Parsons 2004; Koelsch et al. 2006; Mitterschiffthaler et al. 2007) report changes in auditory cortex in response to pleasurable music, even though it is unlikely that auditory cortex itself mediates the emotional responses of interest. Alternatively, this modulation of auditory cortex may reflect a top-down influence of emotion on early sensory cortical areas, as a similar modulation of visual cortex is seen when individuals view emotional pictures (Lang et al. 1998).

There is also some uncertainty about whether the measured brain response to music in these studies reflects experienced pleasure, a happy emotional experience congruent with the valence of the music (Mitterschiffthaler et al. 2007), or an absence of experienced displeasure when consonant music is contrasted with the unpleasantness of dissonant (Koelsch et al. 2006) or scrambled music (Menon and Levitin 2005). As discussed in more depth below, the distinction between experienced pleasure and the emotional valence expressed in the music is likely critical to interpreting the observed activations in these studies. Humans frequently take pleasure in music that is negatively valenced. This dissociation between pleasure and valence is particularly striking for sad music (Levinson 1990). Indeed sad musical pieces induce pleasurable chills even more often than happy music (Panksepp 1995), despite their negative emotional valence. While the precise interpretation of ventral striatal activations will require further investigation, it seems likely that the ventral striatal response is directly related to some aspect of positive emotion or pleasure elicited during music listening. Menon and Levitin (2005) further observe significant functional connectivity between the ventral striatum and the dopamine midbrain region, as well as hypothalamus, orbitofrontal cortex, and insula, which supports the idea that the striatal response is part of a network associated with emotion processing.

The findings of these studies all point to the mesolimbic reward system as important for music-induced pleasure, leading to the hypothesis that the dopamine system may be mediating the affective response. This hypothesis was recently tested and confirmed by Salimpoor et al. (in press). Using raclopride as a dopamine-specific PET radioligand, these investigators showed that there was dopamine release in both ventral and dorsal portions of the striatum while listeners experienced highly pleasurable pieces of music that induced chills (Figure 19.2b), as compared to control music, which was rated as neutral and that did not produce chills. These data provide direct support for the hypothesis that dopamine is released in the mesolimbic reward system while listening to pleasurable music.

In a complementary investigation (Salimpoor et al. in press) the same participants were tested with the same stimuli using fMRI. A functional dissociation in BOLD activations was observed between ventral and dorsal striatal regions. The dorsal striatal BOLD response was associated with time periods prior to the occurrence of chills, when listeners were anticipating maximum pleasure, while the ventral striatal BOLD response was found during the epochs in which chills were reported, and hence represent the highest pleasure moments. Finally, BOLD activity in dorsal and ventral regions also differentially predicted behavioral responses: number of chills experienced correlated with dorsal BOLD activity, whereas intensity of chills correlated with ventral striatal activity, in keeping with the idea that dorsal responses are related to expected events, while ventral responses are related to the experience of pleasure. Because the PET data do not provide temporal information, the precise relationship between dopamine release as measured with PET and these dissociable BOLD responses is uncertain, but to the extent that they are related (see Knutson and Gibbs 2007; Buckholtz 2010 for a discussion of this issue), the fMRI data raise the possibility of two phases of dopamine release, or topographically specific effects of dopamine release during highly rewarding music listening: one associated with musical expectancies and the other to resolution. We return to the importance of temporal features in understanding music reward below.

In summary, the extant data indicate that music activates key meso-corticolimbic areas that have previously been implicated as biological substrates of reward and emotion, and that this activity involves dopaminergic neural responses (cf. Chapters 4 and 11 in this volume). This knowledge is critical in understanding the neurobiological basis of musically mediated emotion, because it links the mechanisms involved in emotion associated with music to those associated with biologically relevant stimuli. However, the specific mechanisms through which these areas become activated remain largely unexplored. That is to say, the findings do not provide information on why organized combinations of sounds modulate the brain areas in question.

19.3.3 PSYCHOPHYSIOLOGY

There are a number of studies of the effects of music on psychophysiological indices, such as skin conductance, heart rate, and respiration. These studies generally agree that music can modulate these markers of physiological and mood states quite dramatically and consistently. Furthermore, there are strong relationships between subjective ratings of various affective states and accompanying physiological responses, which is important as a validation of behavioral report relative to an objective index of internal state. One of the first to study the psychophysiological responses to music conveying different emotions was Krumhansl (1997), who showed that music rated as happy, sad, or fearful gave rise to different patterns of electrodermal, cardiac, and temperature measures. Consistent with these findings, Rickard (2004) demonstrated that skin conductance in particular could be a sensitive indicator of the intensity of affective responses to musical excerpts. Skin conductance is among the most widely used psychophysical measures as it provides a robust index of arousal. Similarly, Steinbeis et al. (2006) and Koelsch et al. (2008) observed that electrodermal changes occurred when unexpected chords were presented in a harmonic sequence. Similar reactions were reported by Grewe et al. (2007a) in response to unfamiliar music, particularly at points where a new voice or section was introduced. These responses suggest a specific arousal or orientation response to music when it is unknown or deviates from expectancies.

One of the interesting features of music is that it generates expectancies based both on veridical knowledge of a piece of music (i.e., when one has heard the piece before) and implicit knowledge of the rules of the musical system that one has learned (Huron 2006; Meyer 1956). The latter may be likened to knowledge of syntax in language. The fact that psychophysiological changes are elicited by chord sequences that are not in themselves specifically emotion-inducing (e.g., dissonant) (Koelsch 2008), and also by unfamiliar music heard for the first time, suggests that physiological responses are generated in part by abstract knowledge of how musical antecedents lead to conclusions. These data provide an initial clue of the importance of expectancies in how we respond to music. We will return to the issue of expectancies later in this chapter, but note here that such expectancies play a prominent role in current theories of why music is rewarding.

Turning to the emotional states associated with pleasure, a number of studies have examined the psychophysiology associated with the experience of musical "chills," which is a useful marker of highly arousing and strongly positive emotion that appears to be qualitatively distinct from other emotional responses (Panksepp 1995). Several studies using electrophysiological and other techniques have specifically focused on the chills phenomenon. One of the first papers to probe this effect was carried out by Goldstein (1980), who tested the hypothesis that endogenous opioids might mediate musically induced chills by administering naloxone, an opiate blocker, prior to exposure to music. The results were inconclusive, however, as only three of ten listeners reported reduction of chills from this manipulation, and it is unclear whether this was a specific or more general effect. Several more recent studies (Craig 2005; Rickard 2004; Guhn, Hamm, and Zentner 2007; Grewe et al. 2007b; Salimpoor et al. 2009) have probed the psychophysiology associated with episodes when listeners report they are experiencing chills. The most consistent finding from these studies is that chills are accompanied by increases in skin conductance, along with heart rate and respiration, and that they are elicited at relatively consistent points in the music for any given listener, though not necessarily in a consistent manner across individual listeners (Figure 19.2c). In fact, Salimpoor et al. (2009a) showed that individual psychophysiological responses can be independent of psychoacoustical factors, and instead are strongly linked to listeners' experience of pleasant emotion. Such findings validate via an objective measurement that individual subjective reports of chills are indeed reflective of a distinct emotional response, although, as mentioned above, skin conductance changes are also elicited to surprising stimuli that do not necessarily induce pleasure. More importantly, these psychophysiological measurements implicate the autonomic nervous system in a physiological response which is no doubt mediated centrally, hence motivating studies on the neural basis of these reactions.

19.4   WHY IS MUSIC REWARDING?

A truly comprehensive discussion of the reasons music is rewarding is beyond the scope of this chapter as there are likely multiple complementary answers to this question and we cannot do each of these mechanisms justice. The question is closely related to the larger topic of whether music induces "true" emotions and what mechanisms allow this to happen. (See Juslin and Vastfjall 2008 and related commentaries for a larger discussion of these issues.) Nevertheless, a useful starting point is to consider what sort of positive feelings are induced by music. Zentner, Grandjean and Scherer (2008) recently reported the results of a confirmatory factor analysis of mood ratings induced by music. Seven primary positive factors were defined: (1) Wonder (including terms such as happy, amazed, and moved); (2) Transcendence (including terms such as inspired and feelings of spirituality); (3) Tenderness (with terms such as love, affectionate, and sensual); (4) Nostalgia (with terms such as sentimental and dreamy); (5) Peacefulness (with terms such as calm and relaxed); (6) Power (with terms such as energetic and triumphant); and (7) Joyful Activation (with terms such as stimulated, joyful, and animated). The first 5 primary factors loaded on a second-order factor, which the author labeled Sublimity, while Power and Joyful Activation loaded on a second order factor of Vitality. Based on these self-report findings, it is easy to speculate that there are several different mechanisms leading to pleasure in music, ranging from being impressed by the skill of the composer or musicians (Wonder), to reductions of tension (Peacefulness), and engagement of energy (Vitality).

19.4.1 AROUSAL

The induction of vitality and peacefulness by music is notable in that these states are closely tied to arousal. This sort of generalized up- or down-regulation of arousal is widely seen in how people use music to aid alertness, to help propel activities (such as during exercise), or to calm or soothe (e.g., lullabies). As others (Berlyne 1971; North and Hargreaves 1997) have previously speculated, gross changes in arousal may be explained through a brainstem (ascending reticular activating formation) mechanism. This arousal mechanism may be modulated by limbic structures, in particular the amygdala, which sends significant projections to the brainstem (Silvestri and Kapp 1998). Given the apparent up- and down-regulation of amygdalar activity by music, we speculate that the amygdala plays a critical role in this process of arousal manipulation.

The modulatory ability of music over ascending projection systems may also be related to its ability to act as an analgesic. It has been known for some time (Gardner, Licklider, and Weiss 1960) that music can be effective in reducing or controlling pain in people undergoing various types of medical treatments (e.g., Voss et al. 2004; Nilsson, Unosson, and Rawal 2005). A recent behavioral study has demonstrated that this effect is not due to generalized arousal, and is specific to pleasurable music, as no pain reduction is seen with music not judged to be pleasant (Roy, Peretz, and Rainville 2008). Moreover, in a recent study, Zhao and Chen (2009) found that pain reduction was similar for happy and sad music of similar perceived pleasantness, thus indicating that the pleasantness rather than the specific mood of the music is most significant in inducing analgesia (and further highlighting the idea that pleasure can be induced from either sad or happy music). Given these facts, it is tempting to conclude that the music-induced pleasure response, which we know to include both upregulation of dopaminergic reward circuitry and, perhaps especially relevant, downregulation of amygdala, may also cause a modulation of ascending nociceptive information. The details by which this interaction occurs will no doubt be a topic of intense future study.

The relationship of peacefulness and vitality to other mood states warrants special attention, because they provide a direct link to broad models of mood. Vitality is part of positive affect, which is marked by terms such as active and interested (Watson and Tellegen 1985), and comprises one of the two higher-order factors of mood. As such, the induction of vitality may be viewed as a direct modulation of positive affect. Peacefulness has a more complicated relationship to the general structure of mood. Tellegen originally suggested that peaceful calm states represented the low end of negative affect, as they reflect an absence of negative states such as anxiety or distress. Work carried out by Zald (unpublished) suggests that calm cannot be simply viewed as the low end of negative affect, as ratings of calm correlate more highly with positive affect terms than negative affect terms. Regardless of their specific position in affective space, it is clear that humans highly value feelings of vitality and feelings of peacefulness. Indeed, humans across cultures show a willingness to pay money for foods, herbs, medicines, illegal drugs, and sensory experiences that increase feelings of calm and vitality.

Yet, manipulation of arousal is almost certainly just one of several features that contributes to music's appeal. For instance, it is difficult to explain feelings of wonder, nostalgia, or transcendence through an arousal mechanism. Indeed, in reviewing these sorts of experiences, it is difficult to link such states to either arousal or the larger literature on affect and mood, in that most studies of mood do not include terms tapping these types of traits. There are only a few studies that have attempted to understand the neural substrates of these types of experiences, and as of yet, we would argue that the data are too limited to suggest a model of their neural substrates. Indeed, the literature on the neural substrates of awe is essentially nonexistent. Several methodological features will also make these domains difficult to capture using musical stimuli, leading us to suspect that it will be some time before we have a handle on each of the domains described by Zentner, Grandjean and Scherer (2008).

19.4.2 Emotional Communication and Contagion

Another avenue for exploring music's rewarding properties arises from the simple idea that music is a form of communication, which like nonverbal vocal cues can provide information to the listener about the emotional or motivational state of the person making the sounds. Although there has long been speculation about the parallels between nonverbal vocal and musical communication (see for instance Spencer 1857), it is only relatively recently that empirical research has addressed this issue. Much of this work is summarized by Juslin and Laukka (2003), who observe that music and vocalizations have similar decoding accuracy for individual emotions, use similar acoustic features to convey specific emotions, and show similar developmental trajectories. Based on these similarities, Juslin and Laukka argue that the development of music in humans is a specific outgrowth of adaptive advantages that were gained by being able to encode and decode emotions through vocalizations.

In treating music as a mode of emotional communication, it is useful to consider the concept of emotional contagion, through which the expression of an emotion leads to the experience of the same emotion in an observer. This concept has been primarily applied in relation to the effects of emotional facial expressions, where it has repeatedly been found that viewing emotional facial expressions induces similar emotional experiences in the viewer (Preston and de Waal 2002; Wild, Erb, and Bartels 2001). This process happens even in response to faces that are presented quite briefly (Lishner, Cooter, and Zald 2008), at time spans far shorter than a typical musical passage. Much less attention has been given to emotional contagion in the vocal sphere, but it is clear that vocalizations elicit subjective emotional responses in listeners. For instance, the sound of voiced laughter robustly induces positive affect in the listener (Bachorowski and Owren 2001).

To the extent that music uses similar acoustic cues as nonverbal vocalizations, it may tap into similar contagion-like features. This leads to a set of specific hypotheses, namely, that the experience of reward or other emotions from music will lead to activation of the same regions as when the person experiences that emotion in a natural setting, or when the emotion is invoked by an emotion contagion-like process from another sensory modality such as facial expressions. Thus, for instance, since exposure to happy facial expressions tends to engage the ventral striatum (particularly in those with high empathy; Chakrabarti, Bullmore, and Baron-Cohen 2006), we would predict similar engagement by happy music. As discussed earlier in the chapter, this hypothesis finds support in the small but increasing literature on emotion and music. For example, Mitterschiffthaler et al. (2007) report that happy music engages the ventral (and dorsal) striatum. This is an intriguing convergence, although we would caution against treating the striatal activations as necessarily reflecting the experience of happiness, per se. Interestingly, unlike some of the above-described studies which have emphasized pleasantness rather than mood (e.g., happy, sad) in relation to the engagement of the striatum, the data of Mitterschiffthaler and colleagues indicate that sad music did not produce a similar activation in the striatum. However, since no ratings of pleasantness were provided, it is unclear whether the pieces were perceived as highly pleasant.

If, as Juslin and Laukka (2003) argue, music is closely tied to nonverbal vocal communications, we would predict that the neural correlates of emotions induced by music would be particularly similar to those arising from nonverbal vocalizations that are not musical in nature. A few neuroimaging studies have examined neural responses to nonverbal emotions (Fecteau et al. 2007; Johnstone et al. 2006; Pourtois et al. 2005). Although multiple areas showed activation in response to happy or positive emotional vocalizations, there is a decided absence of striatal activations in these studies, making it unlikely that striatal activations during music can be directly related to pathways evolved to detect emotional features in vocalizations. A stronger argument may be made linking affective vocalization and music processing in portions of temporal cortex, as both modalities cause widespread temporal cortical activations, although studies utilizing both types of affective auditory information in the same subjects are needed to determine the degree of specific convergence that occurs. Some evidence also emerges for convergence in the amygdala, where Fecteau et al.

(2007) observed significant bilateral activations for both positive and negative vocalizations. We note, however, that this contrasts with the data presented above in which chills were associated with deactivation of the amygdala.

There is also an important caveat in attempting to extend the emotional contagion concept to music. Specifically, the emotional contagion paradigm has largely been explored in regards to discrete emotional expressions. For instance, a study may expose someone to a fearful face for a matter of a few seconds. In contrast, the conveyance of emotion in music is typically played out over a more extended period of time. This difference in temporal dynamics may limit the generalizability of the empirical literature on emotion contagion to music. Nevertheless, the idea that moods, and not just discrete emotions, are contagious seems reasonable, and thus emotional contagion remains a viable explanation for at least some of the rewarding properties of music conveying happiness.

Several theorists have argued that emotional contagion is at least partially mediated by sensory-motor interactions. This idea converges with recent work on the so-called mirror neuron system, which describes neural activity in premotor cortex as a mechanism for translating observed actions into motor responses. The role of this system in perception of emotion, empathy, and social cognition in general has been discussed by several investigators (for review see de Gelder 2006), the basic idea being that the interpretation of sensory signals is mediated at least in part by the mirror neuron system, as it provides a way of modeling the intended outcome of the motor actions of others. If music taps into a similar system, it stands to reason that modeling or mimicking of emotions expressed by music may be one way in which music may induce emotion (Juslin 2005; Jackendoff and Lerdahl 2006; Molnar-Szakacs and Overy 2006). For example the acoustical features of sad or subdued music (e.g., slow tempo, low intensity, sustained tones, smooth transitions between tones) are compatible with the physical movements associated with sadness or depression (slowed action, low intensity of movements and vocalizations, etc.). Conversely, music typical of happiness or excitement tends to be loud, fast, with abrupt changes, and is associated with rapid, high-energy movements. These motor features are explicitly expressed in dance, but can also impact the tempo and vigor of other physical activities such as walking and running. Sensory-motor interactions as mediated by auditory and premotor regions may provide the link between listening and moving, since premotor cortices are often recruited during music perception in the absence of overt movement, just as rhythmic motor actions can elicit auditory cortex activity (Zatorre, Chen, and Penhune 2007). In turn, this interaction may enhance or even create an affective response because of the close relationship between emotions and their motor manifestations. The psychophysiological changes associated with music listening reviewed above might also provide afferent feedback to this same system, thereby further augmenting the affective state.

19.4.3   The Problem of Unhappy Music

Regardless of the method through which music invokes emotion, it is easy to see why individuals would select music that induces a happy mood. However, people frequently report enjoying music that induces negative affective states. Zentner, Grandjean, and Scherer (2008) suggest the presence of two primary factors that capture negative affect in music: Tension and Sadness. Why would someone actively choose to listen to music that causes tension or sadness, when such mood states are generally viewed as unpleasant or even aversive? No empirical literature really addresses this issue. However, some attempts have been made to explain this on theoretical grounds. Scherer (2004) proposes a distinction between aesthetic and other "utilitarian" emotions, and suggests that aesthetic emotions are experienced in a manner that is detached from urgency or pragmatic, self-oriented concerns. Lacking this urgency, feelings induced by music may be experientially strong, but nevertheless lack the full physiological arousal that would arise when these emotions are induced by events that would directly impact the person in day-to-day life. This type of aesthetic emotion is not limited to music, but also occurs in various types of arts and literature. Such emotions may be experienced as rewarding because they augment our engagement with the art form, and allow us to focus our attention on the art form and away from other real-world distractions. Yet, there is still something unsatisfying about such an explanation in that we often listen to music without any specific aim of escapism. We suspect that if you asked people buying a compact disc or a ticket to a concert, few would suggest the purchase was based on a desire to escape or a desire to experience emotions. Something else must act as a reinforcer here.

19.4.4 EXPECTANCY AND PREDICTION CONFIRMATION

Among the many different mechanisms postulated to explain the reward of music, one of the most refined ideas involves the rewards associated with expectancy and prediction confirmation. This approach argues that the reward of music does not need to arise through the specific induction of mood states or discrete emotions, but rather music is rewarding because of properties intrinsic to how we process sequential events. In his classic book, *Emotion and Meaning in Music*, Leonard Meyer (1956) argued that music's ability to evoke emotion primarily derives from expectations. In the more recent book, *Sweet Anticipation*, David Huron (2006) expands on this idea, arguing for the presence of five components (labeled by the acronym ITPRA) that link expectations in music to reward. These include imagining responses, tension responses, prediction responses, reaction responses, and appraisal responses. Imagining responses refer to situations where we complete music in our imagination (ahead of the music's actual completion). Tension responses refer to the pre-outcome preparation (motoric and perceptual) that occurs in anticipation of music's next step or resolution. Prediction responses involve the results of a comparison between the person's prediction and the actual outcome of the music: when the music is accurately predicted it is experienced as rewarding (for the sake of clarity, we will refer to this as prediction confirmation, and use the term prediction more generically to refer to the act of forecasting future or subsequent events). Reaction responses reflect a quick response to the actual outcome, be it positive or negative. Finally, appraisal responses reflect the slower conscious determination of the meaning of the outcome.

The idea that prediction confirmation is rewarding is well supported. Indeed, psychological studies have strongly demonstrated that confirmation of predictions leads to positive affect (Mandler 1975). While such studies have rarely examined musical prediction, the structure of music lends itself well to predictions. Sections and phrases of music often have clear beginnings, middles, and ends, with the beginning and middle sections often providing clear clues as to how the section or phrase will end. Thus, music provides a series of micro- (note to note and phrase to phrase) and macro- (section to section) level outcomes. If you are asked to sing "Happy Birthday to __," most people will not stop at the "to," but include the outcome "you." In this way music is filled with successive streams of micro- and macro-level events, each of which provides an opportunity to predict where the music will go and subsequently provides an opportunity to verify whether those predictions are met or not.

Once one is familiar with a particular musical tradition, it is possible to make reasonable predictions about multiple features of music, even if one has never heard the specific piece of music before. One need not be consciously aware of these predictions. Rather, there are statistical (actuarial) features within music traditions that direct the expectations of a listener without the listener necessarily being aware of these statistical properties and lead to implicit knowledge of musical rules, or "syntax." For example, in Western music, one can predict frequent pitch changes of two semitones, since these are far more common than other pitch changes (for a review of the statistical properties of music see Huron 2006). The size of steps also can be predicted based on whether the melody is ascending or descending (descending uses smaller steps). The direction of the next pitch change can be predicted based on the size of the current pitch change, with small intervals typically followed by changes in the same direction and large intervals more frequently followed by movement in the opposite direction (melody regression). If you hear a first step in a melody descend, there is a 70% probability that the next step will also descend. Melodies frequently follow an arch-shaped pattern, rising in the middle and returning at their end. The total distribution of tones relative to the root can be predicted for the whole piece: even limiting notes to a given scale, certain tones, such as fifths, occur much more frequently than sixth and sevenths. Similarly, in many musical styles, chord progressions often have fairly set sequences, such as 12-bar blues, or resolutions to the tonic. While the typical listener is probably not consciously aware of most of these statistical properties, they nevertheless are likely to shape their expectations. In many regards, this is equivalent to the implicit pattern learning that develops during exposure to serially presented visual stimuli, which can be learned through repetition, even without the participants' explicit awareness of a pattern (Nissen and Bullemer 1987). In the visual realm, such implicit learning develops for both sequences of visual stimuli, as well as configurations of stimuli (Chun 2000) (also see Chapter 5 for further discussion of sensory configurational learning). While psychological research has focused more on the visual sphere, empirical studies make evident that implicit learning also occurs in the auditory domain (Dennis, Howard, and Howard 2006; Tillmann and McAdams 2004; van Zuijen et al. 2006). Because speech and music unfold over time, such implicit learning of sequences may be particularly critical, as the information must be perceived in sequence to have accurate meaning. Indeed, this sort of statistical learning of patterns and sequences lies at the heart of models of language acquisition, and appears to play a similarly prominent role in infants' acquisition of music (Saffran 2003).

As a consequence of the statistical properties of music, clear expectancies emerge. Such expectancies have motivating properties. We innately want to complete sequences. Singing "Happy birthday to _____," not only leads to a prediction of "you," but produces a desire to complete the phrase. Try to play an ascending major scale, but leave out the octave at the top of the scale. Perhaps even more dramatically, play the same scale in descending order but end on the second rather than the root. If you are like many people, there is an urge for completion. The resolution of this tension is experienced as pleasurable (this formulation is similar to drive theories [and even Freudian ideas about catharsis—see Chapter 2], in which need states are experienced as tension, and the removal of the need is experienced as rewarding). The need for completion is certainly weaker than our basic physiological needs, but it may nevertheless be experienced quite powerfully. Urges for sequence completion can be so great that they lead to clinically significant problems. This is most clearly seen in patients with obsessive-compulsive disorder, who become extremely distressed when unable to complete a behavior satisfactorily. The need for completion may reflect the nature of information storage. We often store information (or actions) in chunks, such that once started, the rest of the sequence is triggered automatically. This chunking often happens with no conscious effort (Bower and Winzenz 1969). Indeed, it may only be with conscious effort that the remainder of a well-learned sequence can be suppressed. Because of the sequential nature of music, such chunking is essential, since the memorization of a piece of music involves hundreds or even thousands of notes in sequence, which far outstrips the roughly seven pieces of information that we can typically hold on-line in short-term memory (Miller 1956).

To the extent that the desire for completion is motivating, it can be manipulated by a delay or obstruction of the predicted outcome. Composers frequently take advantage of this feature by inserting extra chords or notes before the resolution. Similarly, slowing the tempo can delay the expected resolution. Such manipulations can heighten the motivation for completion, and increase the pleasure of the prediction confirmation when the expected closure finally happens.

19.4.5 DOPAMINE EXPECTATION AND PREDICTION

Studies of brain systems involved in reward have long emphasized the critical role of the dopamine (DA) system. Reinforcing drugs of abuse such as cocaine and amphetamine engage this system, and animals will perform repetitive operant responses to the exclusion of meeting other basic needs in order to stimulate DA projections (Wise 1998). For many years animal studies emphasized DA release simply as a neural substrate for reward. Consistent with such a model, the rapid buildup of DA caused by many drugs of abuse induces euphoria. However, over the last two decades a number of critical insights into DA functioning have emerged that point to DA's critical role in processes that are linked to prediction and anticipation.

DA cells show two types of firing (Grace 1991). The first is a tonic pacemaker-like firing, which provides a general statewise level of DA in target areas. This statewise activity may produce prolonged effects, such as determining the amount of motivated responding an individual is willing to make to obtain a reward. The second type of DA cell firing involves phasic bursts (brief trains) of firing. This second type of firing appears exquisitely tuned to the prediction of rewards. These cells fire when a reward is unpredicted, or underpredicted (Mirenowicz and Schultz 1994; Schultz 1998). By contrast, the cells do not fire in response to the receipt of fully predicted rewards, but rather fire to the cues that predicted its occurrence (further details provided in Chapter 14). For instance if a bell sound always precedes a reward (the bell is a conditioned stimulus), the cells will fire when the bell occurs, rather than firing for the actual rewarding outcome (Schultz, Apicella, and Ljungberg 1993). Critically, the pattern of DA cell firing fits a temporal difference learning model, in which learning occurs when there is an error in the prediction of an outcome (Schultz and Dickinson 2000). A positive prediction error occurs whenever the outcome is better than expected, while a negative prediction error occurs when an outcome is worse than expected. The phasic firing of DA neurons corresponds closely to a positive prediction error that is transmitted to multiple target brain regions simultaneously. Functional MRI studies have demonstrated that activity in the midbrain (presumably in DA neurons) conforms to this type of positive prediction error (D'Ardenne et al. 2008). This DA release is ramified in the ventral striatum, which consequently shows responses that also track positive prediction errors (Yacubian et al. 2006).

If we accept the basic tenet that music is rewarding at least in part due to its opportunity to provoke and confirm predictions, then DA is likely to be released during the learning process, especially when the listener is in the early process of becoming familiar with the piece. A person who is learning or relearning a piece of music will be able to successfully predict more aspects of the music. As such, initial parts of a sequence will take on the positive reward value that was previously associated just with the sequence's completion. In other words, when initially hearing the music, the biggest DA surge would likely arise from the closure of a phrase or a section, whereas with repetition, the beginning of the phrase will provoke more of the DA release. To be clear here, we are not suggesting that knowledge of the piece leads to a lack of pleasure in the piece or motivation to hear the piece through. However, knowledge of the piece will dramatically alter the motivational and rewarding experience of the music and the temporal features of these experiences.

19.4.6 Dissociating Positive Prediction Errors and Prediction Confirmation

There is a major paradox that arises when trying to integrate a temporal difference learning model with the idea that prediction confirmation is rewarding. Specifically, if prediction confirmation drives reward, the more the person is able to predict the music, the more they should like it. By contrast, temporal difference models suggest that with too much learning, there will be little positive prediction error, and therefore DA release will decrease and perhaps be limited to just the first few notes of a section or a piece (just enough to identify the piece, since the rest would be highly predictable) (see Figure 19.3). There are at least three situations that provide a test of whether greater prediction confirmation leads to more or less reward. These include prior exposure, conventionality/prototypicality, and musical fluency.

Prior exposure to a piece of music makes it more predictable. Thus, if accurate prediction is rewarding, one would expect music that is familiar to be better liked or enjoyed than music that is not familiar. This contention is supported by mere exposure effects in which the prior exposure to a stimulus increases preferences for that stimulus (Zajonc 1968). This effect was originally described in the visual domain, but subsequent studies have confirmed its presence for melodies (Wilson 1979). That is, the mere repetition of a melody makes it more preferred and pleasant. Such mere exposure effects can work both at the level of a total piece (listening to the piece performed on

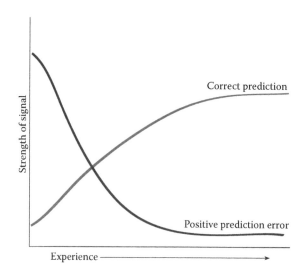

FIGURE 19.3   The contrasting effects of repeated exposures on the strength of positive prediction errors and prediction confirmation.

multiple occasions) or within a piece of music. Most music traditions include repetitions or repetitions with modest variations within the same piece of music. Thus, even if one has never heard a piece before, a motif, melody, or chord progression may be repeated multiple times, such that by the later parts of the piece, the key themes have become more familiar. When the variations are modest, the expectations may be fairly explicit and the listener may be consciously aware of the repetition. With more dramatic variations, the repetition may be less transparent, but nevertheless the aspects that have not changed (for instance the underlying chord progression, bass line, or rhythmic pattern) will still provide an implicit basis for prediction based on the previous exposures to the section.

On first listening, music that is more conventional/prototypical will also be easier to predict than music that strays from a musical tradition. Thus, pieces that conform to standard statistical features will be perceived positively. The flip side to this axiom is that music that grossly defies conventions will at least initially be viewed as unappealing because of its prevention of accurate predictions. The classic tale of the disastrous 1913 debut of Stravinsky's *Rite of Spring* is a case in point: the audience, unable to properly predict the music and unsettled by the unorthodox choreography, appeared to despise the piece. And yet, with repetition and exposure to its innovations, the piece has taken its place as one of the triumphs of its period.

Finally, fluency (i.e., the person's level of musical listening skill) should influence people's musical enjoyment, with individuals with greater musical fluency finding it easier to predict the music. The issue of fluency ties into a larger issue in general aesthetics. Reber, Schwarz, and Winkielman (2004) emphasize the idea that stimuli that are fluently processed are inherently pleasant, leading to increased positive affect. In music, this is partially supported. Musical novices tend to prefer simple, easier to predict, pieces over more challenging pieces (Smith and Melara 1990), which is consistent with a fluency view.

There are however limits to some of these prediction confirmation effects. While changes in preference ratings remain high after multiple exposures, they often asymptote after multiple repetitions (Bornstein 1989). Thus, at a certain point there is little further gain. Repetition of pieces close in time (for instance within a single extended session) leads to reductions in liking after as few as eight repetitions (Schellenberg, Peretz and Vieillard 2008), suggesting a cap on the benefits of repetition. There appears an inverted *U*-shaped curve related to familiarity, in which repetitions producing greater familiarity produce increases in liking, but this effect peaks after multiple exposures and can be antagonized by a countervailing process related to over-familiarity, if repetitions occur too frequently after initial learning.

Similarly, fluency can lead to reduced liking of pieces that are too easy. Individuals with greater musical fluency take less pleasure from easy musical pieces relative to more challenging (less easy to predict) pieces (Smith and Melara 1990). If prediction confirmation alone were driving reward and preference, the appreciation of the simpler pieces should remain high even in the musically fluent. These data suggest that there is not a singular relationship between prediction and musical pleasure. Rather, the experience of pleasure more likely represents a summation or weighted integration of correct prediction and positive prediction error, with the relative weighting varying based on the extent of prior exposure to the musical piece or genre.

An additional feature of DA coding may lead to its greater engagement early in learning. Specifically, the DA system has long been recognized to play a role in responses to novelty. Novel stimuli trigger SN/VTA firing (Ljungberg, Apicella, and Schultz 1992). Several recent fMRI studies observed blood oxygen level-dependent (BOLD) responses in the SN/VTA region when healthy humans anticipated or viewed novel pictures or associations (Bunzeck and Duzel 2006; Wittmann et al. 2007). Kakade and Dayan (2002) suggest that this type of novelty-induced phasic firing of DA neurons provides a motivating "exploration bonus" that encourages exploration of stimuli or environments. In the context of music, this may produce a signal boost towards pursuing new musical experiences, which like positive prediction errors partially counteracts mere exposure effects in determining musical preferences. In most individuals this signal boost is probably weaker than the effects of expectancy and prediction, and generally limits individuals willing to explore radically different musical genres than ones that they already know. However, personality factors such as novelty seeking, and breadth of musical exposure, may lead some individuals to gain a greater novelty bonus for musical exploration.

19.4.7 Music, Pleasure, and Wanting

In the above discussion, we have frequently emphasized the experience of pleasure that arises from listening to music. We have also emphasized DA-striatal reward circuitry as an important substrate for affective aspects of music. However, it is critical to distinguish between different aspects of reward processing, which can be divided into reward learning, reward wanting, and reward liking. In addition to its role as a phasic reward learning cue, current models of DA function emphasize its role as a substrate for motivating behavior to obtain rewards. These models argue that DA firing itself does not underlie the subjective experience of pleasure, but rather provides a basis for the subjective experience of wanting (Berridge and Robinson 1998). For instance, manipulations of the DA system (particularly those involving the ventral tegmental area's projections to the nucleus accumbens) will alter the amount an animal is willing to work for a reward, but not its apparent consummatory pleasure (liking) of the reward (Salamone, Cousins, and Snyder 1997). By contrast, Berridge and Robinson argue that the experience of pleasure (i.e., liking) is linked instead to the endogenous opioid system.[*] To the extent that rapid increases in nucleus accumbens DA leads to euphoria, it is argued to arise secondary to its ability to stimulate beta-endorphins (Roth-Deri, Green-Sadan, and Yadid 2008).

Wanting and liking reflect different temporal phases of reward processing, with wanting corresponding to an anticipatory, appetitive phase, and liking corresponding to a consummatory phase. This anticipatory-consummatory temporal relationship parallels the repeated prediction-confirmation phases of music. It also parallels the fMRI findings of the study by Salimpoor et al. (in press) in which separate dorsal and ventral striatal activations emerged respectively in the anticipation of chill-inducing sections of music, and during the maximally pleasurable chill period itself. Although PET imaging of DA release lacks the temporal resolution to distinguish these two phases, the observation of temporally discrete fMRI activations in the same locations as DA release suggests

---

[*] Although, as discussed in Chapter 13, opioid processes in the basolateral amygdala may also help mediate incentive learning and the encoding of incentive value, casting a slightly different light on the opiate "liking" hypothesis.

a potential biphasic dopaminergic response. The topography of these findings is intriguing in that recent addiction research has suggested greater dorsal striatal involvement in anticipation (craving) and greater ventral striatal (nucleus accumbens) engagement during intoxication (euphoria) (Koob and Volkow 2010).* We note in this regard music can have an almost compulsive-addictive feel to it, as the motivation to hear the music once started can be enormous. Indeed, for music fans, just imagining a few bars of a piece of music may produce a significant desire to experience the music.

19.4.8 Dopamine and Uncertainty

We have already noted that factors such as fluency, conventional/prototypicality, and repeated exposure will influence the degree of positive prediction errors and influence phasic dopamine firing. Music that is highly conventional, too easy, too repetitive, or that has been heard too many times will not produce positive prediction errors. There has to be a chance of being wrong.

Uncertainty of outcome brings us to another aspect of DA cell firing. Specifically, DA cells have been observed to show a distinct response related to unpredictability. In an electrophysiological study with monkeys (Fiorillo, Tobler, and Schultz 2003), it was observed that 29% of SN/VTA cells demonstrated firing which increased from the time a conditioned stimulus was presented to the time at which the outcome of the trial was given (i.e., the reward was or was not received). This sustained increase was maximal when probability was at 50% chance of reward or no reward, and was larger when the value of the potential reward was increased (or when the size of the difference between two potential rewards was increased). This extended period of firing may have two effects. First, given dopamine's motivational role, it may increase wanting or desire for the possible outcome. Such effects may contribute to numerous behaviors, such as gambling and watching sports, both of which produce substantial engagement of motivation when the outcome is unknown. Second, it may produce a state in which the subjective reward of the outcome is enhanced when it occurs (perhaps by priming the opioid system). Not all rewards are experienced equally; winning money produces greater reward than getting money that you knew was coming. Indeed, people will be more excited about winning a small amount of money than getting a larger amount of money that they already knew was coming. A win by one's favorite sports team over another team will be far more enjoyable when the outcome was not assured.

In the same manner, we suspect that uncertainty plays a major role in music appreciation, both increasing the motivation for the desired outcome and priming the subjective reward of prediction confirmation. Indeed, there may be ways of preventing 100% predictability even in a known piece of music. We noted previously that delaying a resolution may increase an urge for the resolution; the function of an appoggiatura in music is the perfect illustration of this phenomenon. Similarly, by inserting extra notes before a resolution, a composer may provide some uncertainty over whether the progression will end or go some place different; this phenomenon perhaps finds its extreme realization in the device common in nineteenth-century Western music known as a cadenza, which can extend a simple cadence sometimes for many minutes. Perhaps even more strikingly, the composer can insert notes that in the majority of other pieces lead in a different direction. In other words, the specific experience of the piece may run counter to a larger actuarial store of information about music, such that even if the piece is fully known, there is conflicting actuarial data on how the music should proceed. A common example of this effect is provided by the so-called deceptive cadence, in which a dominant chord, which would typically lead to a resolution back toward the tonal center or tonic chord, leads instead to a different chord (the submediant, in music-theoretic parlance), which is not expected, but is still perceived as an acceptable option. This situation sets up a tension between the implicit knowledge of musical syntax rules learned from many prior exposures to the typical chord progression, and the veridical long-term memory trace associated with the piece itself that one may have heard on multiple

---

* A different perspective is provided in Chapter 14, where it is argued that ventral and dorsal striatum are involved in Pavlovian and instrumental conditioning, respectively.

occasions (see Huron 2006 for additional discussion). In live performance situations, the music may become even more unpredictable as the performance is slightly different every time, and the performer has considerable leeway to emphasize some events over others, to introduce delays at particular moments, or in some cases to improvise entirely new musical passages.

The type of DA firing that occurs with uncertainty behaves in a temporally distinct manner from the positive prediction error signaling. Whereas the positive prediction error signal is brief and punctate, occurring when an outcome is better than expected (or a not fully expected cue occurs), the firing of DA neurons during unpredictability is maintained until the outcome is known. Thus its effects may be far more sustained and powerful. It also provides a focus not just on whether the outcome occurs, but when the outcome occurs. The longer the delay, the greater the motivation for reaching the resolution. Musically, such heightening can occur by extending the number of notes before a resolution, slowing the tempo, or even coming to a full rest before providing the expected outcome. In his 2006 treatise on music and prediction, David Huron never mentions issues of DA coding. However, his characterization of when individuals experience tension or desire for resolution appears highly consistent with the present formulation, as he describes that tension may be particularly large when statistical characteristics lead to a strong prediction of an outcome, but do not directly move to that outcome. He notes that it is in these situations where listeners often express the most clear motivation or desire when listening to music, using terms such as "yearning," which is certainly in the same subjective arena as the "wanting," "craving," or "compelled" experiences that are often described in relation to DA functioning.

Interestingly, tension and resolution lie at the heart of aesthetic theory. Dewey (1934) argued that aesthetics come from a conversion of resistance and tensions towards a resolution. There has to be some tension and movement towards resolution for a piece to be valued. A piece that is totally conventional in its movement to resolutions will lack impact. Rather, some degree of ambiguity or unpredictability is necessary to create tension. This may be at the essence of why expert fluent listeners may not particularly enjoy pieces that are too simple and easy. The rewards associated with correct prediction appear to be minimal if there was no risk of being wrong. Thus, there may be a threshold of uncertainty or difficulty that must be surpassed before prediction confirmation provides a reward. A similar process likely occurs with media other than music: books and TV shows aimed at young children are often unappealing to older children and adults (we suspect most readers who were ever forced as adults to watch an episode of the children's TV show *Teletubbies* are nodding their heads in agreement while reading this).

Armstrong and Detweiler-Bedell (2008) make a distinction between the types of rewards associated with fluency and those associated with works that provide greater challenges. They suggest that fluency alone may be associated with a piece of art being perceived as pretty, whereas true beauty in art emerges when greater levels of processing must be invested to understand (resolve) the piece. In other words, there has to be an ambiguity in order to reach a high reward. To the extent that music can provide both uncertainty and repeated opportunities for resolution and prediction confirmation, it is capable of moving beyond the simply pretty to the arena that Armstrong and Detweiler-Bedell refer to as beauty.

19.5   CONCLUSIONS

We have articulated several routes through which music may be experienced as rewarding, including the manipulation of arousal and emotional contagion, the engagement of "aesthetic emotions," and by tapping mechanisms involved in coding predictions and uncertainty. Thus, rather than proposing a singular mechanism through which music becomes rewarding, we suggest that there are multiple mechanisms that enter into the process. These different mechanisms rely on several different neural circuits, including those involved in regulating arousal, motivation, and reward prediction. We have particularly emphasized mesolimbic and paralimbic processes as central to the experience of reward in music. These areas are of course also present in mammals in general. Yet, music does not appear to act as a reward for any other species. We suspect that these species lack the cortical auditory processing architecture that would allow them to organize, learn, anticipate, and predict long sequences and combinations of tones and timbres in a manner that produces inherent reward. Hence, we would once again emphasize that musical pleasure arises from an interaction between biologically ancient reward mechanisms and much more recently evolved cortical systems which are highly modifiable by individual experience and culture.

An observant reader may note that in presenting a model of musical reward that focuses on anticipation and prediction, we have provided little data from studies that were designed to directly test these models. However, if, as we argue, features related to anticipation, prediction confirmation, and uncertainty in music engage dopaminergic, limbic, and paralimbic mechanisms, and their associated cortical loops, then it should be possible to test these ideas by manipulating these features with appropriate neuroimaging, electrophysiological, and neuropsychological paradigms. Such studies would move the field beyond a simple statement that music engages areas involved in emotion, to a more comprehensive assessment of the neural mechanisms that underlie our continued desire for, and pleasure from, sequences of tones.

REFERENCES

Armstrong, T., and Detweiler-Bedell, B. 2008. Beauty as an emotion: The exhilarating prospect of mastering a challenging world. *Review of General Psychology* 12:305–29.

Bachorowski, J.A., and Owren, M.J. 2001. Not all laughs are alike: Voiced but not unvoiced laughter readily elicits positive affect. *Psychological Science* 12:252–57.

Berlyne, D. 1971. *Aesthetics and Psychobiology.* New York: Appleton-Century Crofts.

Berridge, K.C., and Robinson, T.E. 1998. What is the role of dopamine in reward: Hedonic impact, reward learning, or incentive salience? *Brain Research. Brain Research Reviews* 28:309–69.

Blood, A.J., and Zatorre, R.J. 2001. Intensely pleasurable responses to music correlate with activity in brain regions implicated in reward and emotion. *Proceedings of the National Academy of Sciences* 98:11818–23.

Blood, A.J., Zatorre, R.J., Bermudez, P., and Evans, A.C. 1999. Emotional responses to pleasant and unpleasant music correlate with activity in paralimbic brain regions. *Nature Neuroscience* 2:382–87.

Bornstein, R.F. 1989. Exposure and affect: Overview and meta-analysis of research, 1968–1987. *Psychological Bulletin* 106:265–89.

Bower, G.H., and Winzenz, D. 1969. Group structure coding and memory for digit series. *Journal of Experimental Psychology Monographs* 80:1–17.

Breiter, H.C., Gollub, R.L., Weisskoff, R.M., Kennedy, D.N., Makris, N., Berke, J.D., Goodman, J.M., et al. 1997. Acute effects of cocaine on human brain activity and emotion. *Neuron* 19:591–611.

Brown, S., Martinez, M., and Parsons, L. 2004. Passive music listening spontaneously engages limbic and paralimbic systems. *NeuroReport* 15:2033–37.

Buckholtz, J.W., Treadway, M.T., Cowan, R.L., Woodward, N.D., Benning, S.D., Li, R., Ansari, M.S., et al. 2010. Mesolimbic dopamine reward system hypersensitivity in individuals with psychopathic traits. *Nature Neuroscience* 13:419–21.

Bunzeck, N., and Duzel, E. 2006. Absolute coding of stimulus novelty in the human substantia nigra/VTA. *Neuron* 51:369–79.

Carter, C.S., Mintun, M., and Cohen, J.D. 1995. Interference and facilitation effects during selective attention – An (H2O)-O-15 PET study of Stroop task performance. *Neuroimage* 2:264–72.

Chakrabarti, B., Bullmore, E., and Baron-Cohen, S. 2006. Empathizing with basic emotions: Common and discrete neural substrates. *Social Neuroscience* 1:364–84.

Chun, M.M. 2000. Contextual cueing of visual attention. *Trends in Cognitive Science* 4:170–78.

Craig, D.G. 2005. An exploratory study of physiological changes during chills induced by music. *Music Scientiae* 9:273.

D'Ardenne, K., McClure, S.M., Nystrom, L.E., and Cohen, J.D. 2008. BOLD responses reflecting dopaminergic signals in the human ventral tegmental area. *Science* 319:1264–67.

Darwin, C. 1871. *The Descent of Man and Selection in Relation to Sex.* John Murray: London.

de Gelder, B. 2006. Towards the neurobiology of emotional body language. *Nature Reviews Neuroscience* 7:242–49.

Dennis, N.A., Howard, J.H., and Howard, D.V. 2006. Implicit sequence learning without motor sequencing in young and old adults. *Experimental Brain Research* 175:153–64.

Dewey, J. 1934. *Art as Experience*. New York: Penguin Group.

Dubé, L., and LeBel, J.L. 2003. The content and structure of laypeople's concept of pleasure. *Cognition and Emotion* 17:263–295.

Fecteau, S., Belin, P., Joanette, Y., and Armony, J.L. 2007. Amygdala responses to nonlinguistic emotional vocalization. *Neuroimage* 36:480–87.

Fiorillo, C.D., Tobler, P.N., and Schultz, W. 2003. Discrete coding of reward probability and uncertainty by dopamine neurons. *Science* 299:1898–1902.

Fitch, W.T. 2006. The biology and evolution of music: A comparative perspective. *Cognition* 100:173–215.

Gabrielsson, A., ed. 2001. *Emotions in Strong Experiences with Music*. Oxford: Oxford University Press.

Gardner, W., Licklider, J., and Weiss, A. 1960. Suppression of pain by sound. *Science* 132:32–33.

Geissmann, T. 2000. Duet songs of the siamang, Hylobates syndactylus: I. Structure and organisation. *Primate Report* 56:33–60.

Goldstein, A. 1980. Thrills in response to music and other stimuli. *Physiological Psychology* 8:126–29.

Gosselin, N., Peretz, I., Johnsen, E., and Adolph, R. 2006. Amygdala damage impairs emotion recognition from music. *Neuropsychologia* 129:2585–92.

Gosselin, N., Peretz, I., Noulhiane, M., Hasboun, D., Beckett, C., and Baulac, M. 2005. Impaired recognition of scary music following unilateral temporal lobe excision. *Brain* 128:628.

Gottfried, J., Small, D., and Zald, D.H. 2006. The chemical senses. In *The Orbitofrontal Cortex*, eds. D.H. Zald and S.L. Rauch, pp. 125–72. Oxford University Press.

Gould, S.J., and Lewontin, R.C. 1979. The spandrels of San Marco and the panglossian paradigm: A critique of the adaptationist programme. *Proceedings of the Royal Society of London Series B* 205:581–98.

Grace, A.A. 1991. Phasic versus tonic dopamine release and the modulation of dopamine system responsivity – a hypothesis for the etiology of schizophrenia. *Neuroscience* 41:1–24.

Grewe, O., Nagel, F., Kopiez, R., and Altenmuller, E. 2007a. Emotions over time: Synchronicity and development of subjective, physiological, and facial affective reactions to music. *Emotion* 7:774–88.

———. 2007b. Listening to music as a re-creative process—Physiological, psychological and psychoacutical correlates of chills and strong emotions. *Music Perception* 24:297–314.

Griffiths, T., Warren, J., Dean, J., and Howard, D. 2004. 'When the feeling's gone': A selective loss of musical emotion. *Journal of Neurology, Neurosurgery & Psychiatry* 75:344–45.

Griffiths, T.D., Rees, A., Witton, C., Cross, P.M., Shakir, R.A., and Green, G.G. 1997. Spatial and temporal auditory processing deficits following right hemisphere infarction. A psychophysical study. *Brain* 120:785–94.

Guhn, M., Hamm, A., and Zentner, M.R. 2007. Physiological and muscio-acustic correlates of the chill response. *Music Perception* 24:473–83.

Habib, M., Daquin, G., Milandre, L., Royere, M., Rey, M., and Lanteri, A. 1995. Mutism and auditory agnosia due to bilateral insular damage – role of the insula in human communication. *Neuropsychologia* 33:327–39.

Huron, D. 2006. *Sweet Anticipation: Music and the Psychology of Expectation*. Cambridge, MA: MIT Press.

Jackendoff, R., and Lerdahl, F. 2006. The capacity for music: What is it and what's special about it? *Cognition* 100:33–72.

Johnstone, T., van Reekum, C.M., Oakes, T.R., and Davidson, R.J. 2006. The voice of emotion: An FMRI study of neural responses to angry and happy vocal expressions. *Social Cognitive and Affective Neuroscience* 1:242–49.

Juslin, P.N. 2005. From mimesis to chatarsis: Expression, perception, and induction of emotion in music. In *Musical Communication*, eds. D. Miell, R. MacDonald, and D. Hargreaves, pp. 85–115. New York: Oxford University Press.

Juslin, P.N., and Laukka, P. 2003. Communication of emotions in vocal expression and music performance: Different channels, same code? *Psychological Bulletin* 129:770–814.

Juslin, P.N., and Vastfjall, D. 2008. Emotional responses to music: The need to consider underlying mechanisms. *Behavioral and Brain Sciences* 31:559–75.

Kakade, S., and Dayan, P. 2002. Dopamine: Generalization and bonuses. *Neural Network* 15:549–59.

Khalfa, S., Roy, M., Rainville, P., Dalla Bella, S., and Peretz, I. 2008. Role of tempo entrainment in psychophysiological differentiation of happy and sad music? *International Journal of Psychophysiology* 68:17–26.

Knutson, B., and Gibbs, S.E. 2007. Linking nucleus accumbens dopamine and blood oxygenation. *Psychopharmacology (Berl)* 191:813–22.

Koelsch, S., Fritz, T., Cramon, D., Muller, K., and Friederici, A.D. 2006. Investigating emotion with music: An fMRI study. *Human Brain Mapping* 27:239–50.

Koelsch, S., Kilches, S., Steinbeis, N., and Schelinski, S. 2008. Effects of unexpected chords and of performer's expression on brain responses and electrodermal activity. *PLoS One* 3:e2631.

Koob, G.F., and Volkow, N.D. 2010. Neurocircuitry of addiction. *Neuropsychopharmacology* 35:217–38.

Kringelbach, M.L., and Rolls, E.T. 2004. The functional neuroanatomy of the human orbitofrontal cortex: Evidence from neuroimaging and neuropsychology. *Progress in Neurobiology* 72:341–72.

Krumhansl, C.L. 1997. An exploratory study of musical emotions and psychophysiology. *Canadian Journal of Experimental Psychology* 51:336–53.

Lane, R.D., Reiman, E.M., Bradley, M.M., Lang, P.J., Ahern, G.L., Davidson, R.J., and Schwatz, G. 1997. Neuroanatomical correlates of pleasant and unpleasant emotion. *Neuropsychologia* 35:1437–44.

Lang, P.J., Bradley, M.M., Fitzsimmons, J.R., Cuthbert, B.N., Scott, J.D., Moulder, B., and Nangia, V. 1998. Emotional arousal and activation of the visual cortex: An fMRI analysis. *Psychophysiology* 35:199–210.

Levinson, J. 1990. Music and negative emotions. In *Music, Art and Metaphysics: Essays in Philosophical Aesthetics*, ed. J. Levinson, pp. 306–35. Ithaca: Cornell University Press.

Lishner, D., Cooter, A.B., and Zald, D.H. 2008. Addressing measurement limitations in affective rating scales: Development of an empirical valence scale. *Cognition and Emotion* 22:180–92.

Ljungberg, T., Apicella, P., and Schultz, W. 1992. Responses of monkey dopamine neurons during learning of behavioral reactions. *Journal of Neurophysiology* 67:145–63.

Mandler, G. 1975. *Mind and Emotion*. New York: John Wiley.

McDermott, J., and Hauser, M.D. 2007. Nonhuman primates prefer slow tempos but dislike music overall. *Cognition* 104:654–68.

Menon, V., and Levitin, D.J. 2005. The rewards of music listening: Response and physiological connectivity of the mesolimbic system. *NeuroImage* 28:175–84.

Meyer, L.B. 1956. *Emotion and Meaning in Music*. Chicago: University of Chicago Press.

Miller, G.A. 1956. The magical number seven, plus or minus two: Some limits to our capacity for processing information. *Psychological Review* 63:81–97.

Mirenowicz, J., and Schultz, W. 1994. Importance of unpredictability for reward responses in primate dopamine neurons. *Journal of Neurophysiology* 72:1024–27.

Mitterschiffthaler, M.T., Fu, C.H., Dalton, J.A., Andrew, C.M., and Williams, S.C. 2007. A functional MRI study of happy and sad affective states induced by classical music. *Human Brain Mapping* 28:1150–62.

Molnar-Szakacs, I., and Overy, K. 2006. Music and mirror neurons: From motion to 'e'motion. *Social Cognitive & Affective Neuroscience* 1:235–41.

Nilsson, U., Unosson, M., and Rawal, N. 2005. Stress reduction and analgesia in patients exposed to calming music postoperatively: A randomized controlled trial. *European Journal of Anaesthesiology* 22:96–102.

Nissen, M.J., and Bullemer, P. 1987. Attentional requirements of learning: Evidence from performance measures. *Cognitive Psychology* 19:1–32.

North, A.C., and Hargreaves, D.J. 1997. Liking, arousal potential, and the emotions expressed by music. *Scandinavian Journal of Psychology* 38:45–53.

Panksepp, J. 1995. The emotional sources of "chills" induced by music. Music Perception. *Music Perception* 13:171–207.

Patel, A.D. 2007. *Music, Language and the Brain*. Oxford University Press: Oxford, U.K.

Patel, A.D., Iversen, J.R., Bregman, M.R., and Schuiz, I. 2009. Experimental evidence for synchronization to a musical beat in a nonhuman animal. *Current Biology* 19:827–30.

Peretz, I., Blood, A., Penhun, V., and Zatorre, R. 2001. Cortical deafness to dissonance. *Brain* 124:928–40.

Pinker, S. 1997. *How the Mind Works*. New York: Norton.

Pourtois, G., de Gelder, B., Bol, A., and Crommerlinck, M. 2005. Perception of facial expressions and voices and of their combination in the human brain. *Cortex* 41:49–59.

Preston, S.D., and de Waal, F.B. 2002. Empathy: Its ultimate and proximate bases. *Behavioral and Brain Sciences* 25:1–20.

Reber, R., Schwarz, N., and Winkielman, P. 2004. Processing fluency and aesthetic pleasure: Is beauty in the perceiver's processing experience? *Personality and Social Psychology Review* 8:364–82.

Rickard, N.S. 2004. Intense emotional responses to music: A test of the physiological arousal hypothesis. *Psychology of Music* 32:371–88.

Roth-Deri, I., Green-Sadan, T., and Yadid, G. 2008. beta-Endorphin and drug-induced reward and reinforcement. *Progress in Neurobiology* 86:1-21.

Roy, M., Peretz, I., and Rainville, P. 2008. Emotional valence contributes to music-induced analgesia. *Pain* 134:140–47.

Saffran, J.R. 2003. Statistical language learning mechanisms and constraints. *Current Directions in Psychological Science* 12:110–14.

Salamone, J.D., Cousins, M.S., and Snyder, B.J. 1997. Behavioral functions of nucleus accumbens dopamine: Empirical and conceptual problems with the anhedonia hypothesis. *Neuroscience & Biobehavioral Reviews* 21:341–59.

Salimpoor, V., Benvoy, M., Longo, G., Cooperstock, J., and Zatorre, R. 2009. The rewarding aspects of music listening are related to degree of emotional arousal. *PloS One* 4:1–14.

Salimpoor, V.N., Benovoy, M., Larcher, K., Dagher, A., and Zatorre, R.J. In press. Anatomically distinct dopamine release during anticipation and experience of peak emotion to music. *Nature Neuroscience*.

Schachner, A., Brady, T.F., Pepperberg, I.M., and Hauser, M.D. 2009. Spontaneous Motor Entrainment to Music in Multiple Vocal Mimicking Species. *Current Biology* 19:831–36.

Schellenberg, E.G., Peretz, I., and Vieillard, S. 2008. Liking for happy- and sad-sounding music: Effects of exposure. *Cognition and Emotion* 22:218–37.

Scherer, K.R. 2004. Which emotions can be induced by music? What are the underlying mechanisms? And how can we measure them? *Journal of New Music Research* 33:239–51.

Schultz, W. 1998. Predictive reward signal of dopamine neurons. *Journal of Neurophysiology* 80:1–27.

Schultz, W., Apicella, P., and Ljungberg, T. 1993. Responses of monkey dopamine neurons to reward and conditioned stimuli during successive steps of learning a delayed response task. *Journal of Neuroscience* 13:900–13.

Schultz, W., and Dickinson, A. 2000. Neuronal coding of prediction errors. *Annual Review of Neuroscience* 23:473–500.

Silvestri, A., and Kapp, B. 1998. Amygdaloid modulation of mesopontine peribrachial neuronal activity: Implications for arousal. *Behavioral Neuroscience* 112:571–88.

Small, D.M., Zatorre, R.J., Dagher, A., Evans, A.C., and Jones-Gotman, M. 2001. Changes in brain activity related to eating chocolate - From pleasure to aversion. *Brain* 124:1720–33.

Smith, J.D., and Melara, R.J. 1990. Aesthetic preference and syntactic prototypicality in music – Tis the gift to be simple. *Cognition* 34:279–98.

Spencer, H. 1857. The origin and function of music. *Fraser's Magazine* 56:396–408.

Stein, J.L., Wiedholz, L.M., Bassett, D.S., Weinberger, D.R., Zink, C.F., Mattay, V.S., and Meyer-Lindenberg, A. 2007. A validated network of effective amygdala connectivity. *NeuroImage* 36:736–45.

Steinbeis, N., Koelsch, S., and Sloboda, J. 2006. The role of harmonic expectancy violations in musical emotion. *Journal of Cognitive Neuroscience* 18:1380–93.

Stewart, L., von Kreigstein, K., Warren, J.D., and Griffiths, T.D. 2006. Music and the brain: Disorders of musical listening. *Brain* 128:2533–53.

Thaut, M.H., and Davis, W.B. 1993. The influence of subject-selected versus experiment-chosen music on affect, anxiety, and relaxation. *Journal of Music Therapy* 30:210–23.

Tillmann, B., and McAdams, S. 2004. Implicit learning of musical timbre sequences: Statistical regularities confronted with acoustical (dis)similarities. *Journal of Experimental Psychology-Learning Memory and Cognition* 30:1131–42.

van Zuijen, T.L., Simoens, V.L., Paavilainen, P., Naatanen, R., and Tervaniemi, M. 2006. Implicit, intuitive, and explicit knowledge of abstract regularities in a sound sequence: An event-related brain potential study. *Journal of Cognitive Neuroscience* 18:1292–1303.

Vieillard, S., Peretz, I., Gosselin, N., Khalfa, S., Gagnon, L., and Bouchard, B. 2008. Happy, sad, scary and peaceful musical excerpts for research on emotions. *Cognition and Emotion* 22:720–52.

Voss, J., Good, M., Yates, B., Baun, M., Thompson, A., and Hertzog, M. 2004. Sedative music reduces anxiety and pain during chair rest after open-heart surgery. *Pain* 112:197–203.

Watson, D., and Tellegen, A. 1985. Toward a consensual structure of mood. *Psychological Bulletin* 92:426–57.

Wild, B., Erb, M., and Bartels, M. 2001. Are emotions contagious? Evoked emotions while viewing emotionally expressive faces: Quality, quantity, time course and gender differences. *Psychiatry Research* 102:109–24.

Wilson, W.R. 1979. Feeling more than we can know: Exposure effects without learning. *Journal of Personality and Social Psychology* 37:811–21.

Wise, R.A. 1998. Drug-activation of brain reward pathways. *Drug and Alcohol Dependence* 51:13–22.

Wittmann, B.C., Bunzeck, N., Dolan, R.J., and Duzel, E. 2007. Anticipation of novelty recruits reward system and hippocampus while promoting recollection. *Neuroimage* 38:194–202.

Yacubian, J., Glascher, J., Schroeder, K., Sommer, T., Braus, D.F., and Buchel, C. 2006. Dissociable systems for gain- and loss-related value predictions and errors of prediction in the human brain. *Journal of Neuroscience* 26:9530–37.

Zajonc, R.B. 1968. Attitudinal effects of mere exposure. *Journal of Personality and Social Psychology* 9:1–27.

Zald, D.H. 2003. The human amygdala and the emotional evaluation of sensory stimuli. *Brain Research Reviews* 41:88–123.

Zatorre, R.J., Chen, J.L., and Penhune, V.B. 2007. When the brain plays music: Auditory-motor interactions in music perception and production. *Nature Reviews Neuroscience* 8:547–58.

Zentner, M., Grandjean, D., and Scherer, K. 2008. Emotions evoked by the sound of music: Characterization, classification, and measurement. *Emotion* 8:494–521.

Zhao, H., and Chen, A.C. 2009. Both happy and sad melodies modulate tonic heat pain. *Journal of Pain* 10:953–60.

Index

T - #0469 - 071024 - C8 - 254/178/21 - PB - 9780367382933 - Gloss Lamination